T0297228

Advances in Intelligent Systems and Computing

Volume 397

Series editor

Janusz Kacprzyk, Polish Academy of Sciences, Warsaw, Poland
e-mail: kacprzyk@ibspan.waw.pl

About this Series

The series "Advances in Intelligent Systems and Computing" contains publications on theory, applications, and design methods of Intelligent Systems and Intelligent Computing. Virtually all disciplines such as engineering, natural sciences, computer and information science, ICT, economics, business, e-commerce, environment, healthcare, life science are covered. The list of topics spans all the areas of modern intelligent systems and computing.

The publications within "Advances in Intelligent Systems and Computing" are primarily textbooks and proceedings of important conferences, symposia and congresses. They cover significant recent developments in the field, both of a foundational and applicable character. An important characteristic feature of the series is the short publication time and world-wide distribution. This permits a rapid and broad dissemination of research results.

More information about this series at http://www.springer.com/series/11156

L. Padma Suresh · Bijaya Ketan Panigrahi
Editors

Proceedings of the International Conference on Soft Computing Systems

ICSCS 2015, Volume 1

Editors
L. Padma Suresh
Noorul Islam Centre for Higher Education
Kumaracoil, Tamil Nadu
India

Bijaya Ketan Panigrahi
IIT Delhi
New Delhi
India

ISSN 2194-5357 ISSN 2194-5365 (electronic)
Advances in Intelligent Systems and Computing
ISBN 978-81-322-2669-7 ISBN 978-81-322-2671-0 (eBook)
DOI 10.1007/978-81-322-2671-0

Library of Congress Control Number: 2015953797

Springer New Delhi Heidelberg New York Dordrecht London

Printed on acid-free paper

Springer (India) Pvt. Ltd. is part of Springer Science+Business Media (www.springer.com)

Preface

The volumes contain the papers presented at the *International Conference On Soft Computing Systems* (ICSCS) held during 20 and 21 April 2015 at Noorul Islam Centre for Higher Education, Noorul Islam University, Kumaracoil. ICSCS 2015 received 504 paper submissions from various countries across the globe. After a rigorous peer-review process, 182 full length articles were accepted for oral presentation at the conference. This corresponds to an acceptance rate of 36 % and is intended for maintaining the high standards of the conference proceedings. The papers included in these AISC volumes cover a wide range of topics on Genetic Algorithms, Evolutionary Programming and Evolution Strategies such as AIS, DE, PSO, ACO, BFA, HS, SFLA, Artificial Bees and Fireflies Algorithm, Neural Network Theory and Models, Self-organization in Swarms, Swarm Robotics and Autonomous Robot, Estimation of Distribution Algorithms, Stochastic Diffusion Search, Adaptation in Evolutionary Systems, Parallel Computation, Membrane, Grid, Cloud, DNA, Quantum, Nano, Mobile Computing, Computer Networks and Security, Data Structures and Algorithms, Data Compression, Data Encryption, Data Mining, Digital Signal Processing, Digital Image Processing, Watermarking, Security and Cryptography, Bioinformatics and Scientific Computing, Machine Vision, AI methods in Telemedicine and eHealth, Document Classification and Information Retrieval, Optimization Techniques and their applications for solving problems in these areas.

In the conference separate sessions were arranged for delivering the keynote address by eminent members from various academic institutions and industries. Three keynote lectures were given in two different venues as parallel sessions on 20 and 21 April 2015. In the first session, Dr. Pradip K. Das, Associate Professor, Indian Institute of Technology, Guwahati gave a talk on “Trends in Speech Processing” in Venue 1 and Prof. D. Thukaram, Department of EEE, IISC, Bangalore, gave a lecture on “Recent Trends in Power System” and covered various evolutionary algorithms and its applications to Power Systems in Venue 2. In the second session, Dr. Willjuice Iruthayarajan, Professor, Department of EEE, National Engineering College, Kovilpatti, gave a talk on “Evolutionary

Computational Algorithm for Controllers" at Venue 1. All these lectures generated great interest among the participants of ICSCS 2015 in paying more attention to these important topics in their research work. Four sessions were arranged for electrical-related papers, five sessions for electronics-related papers, and nine sessions for computer-related papers.

We take this opportunity to thank the authors of all the submitted papers for their hard work, adherence to the deadlines and for suitably incorporating the changes suggested by the reviewers. The quality of a refereed volume depends mainly on the expertise and dedication of the reviewers. We are indebted to the Program Committee members for their guidance and coordination in organizing the review process.

We would also like to thank our sponsors for providing all the support and financial assistance. We are indebted to our Chairman, Vice Chancellor, Advisors, Pro Vice Chancellor, Registrar, faculty members and administrative personnel of Noorul Islam Centre for Higher Education, Noorul Islam University, Kumaracoil, for supporting our cause and encouraging us to organize the conference on a grand scale. We express our heartfelt thanks to Profs. B.K. Panigrahi and S.S. Dash for providing valuable guidelines and suggestions in the conduct of the various parallel sessions in the conference. We also thank all the participants for their interest and enthusiastic involvement. Finally, we thank all the volunteers whose tireless efforts in meeting the deadlines and arranging every detail meticulously made sure that the conference could run smoothly. We hope the readers of these proceedings find the papers useful, inspiring and enjoyable.

L. Padma Suresh
Bijaya Ketan Panigrahi

Organization Committee

Chief Patron

Dr. A.P. Majeed Khan, President, NICHE
Dr. A.E. Muthunayagam, Advisor, NICHE
Mr. M.S. Faizal Khan, Vice President, NICHE

Patron

Dr. R. Perumal Swamy, Vice Chancellor, NICHE
Dr. S. Manickam, Registrar, NICHE
Dr. N. Chandra Sekar, Pro-Vice Chancellor, NICHE
Dr. K.A. Janardhanan, Director (HRM), NICHE

General Chair

Dr. B.K. Panigrahi, IIT, Delhi

Programme Chair

Dr. L. Padma Suresh, Professor and Head, EEE

Programme Co-chair

Dr. I. Jacob Raglend, Professor, EEE
Dr. G. Glan Devadhas, Professor, EIE
Dr. R. Rajesh, Professor, Mech
Dr. V. Dharun, Professor, Biomedical

Coordinator

Prof. V.S. Bindhu, NICHE
Prof. K. Muthuvel, NICHE

Organizing Secretary

Dr. D.M. Mary Synthia Regis Prabha, NICHE
Prof. K. Bharathi Kannan, NICHE

Executive Members

Dr. S. Gopalakrishnan, Controller of Exams, NICHE
Dr. A. Shajin Nargunam, Director (Academic Affairs), NICHE
Dr. B. Krishnan, Director (Admissions), NICHE
Dr. Amar Pratap Singh, Director (Administration), NICHE
Dr. M.K. Jeyakumar, Additional Controller, NICHE
Prof. S.K. Pillai, NICHE
Dr. J. Jeya Kumari, NICHE
Prof. F. Shamila, NICHE
Prof. K. Subramanian, NICHE
Dr. P. Sujatha Therese, NICHE
Prof. J. Arul Linsely, NICHE
Dr. D.M. Mary Synthia Regis Prabha, NICHE
Prof. H. Vennila, NICHE
Prof. M.P. Flower Queen, NICHE
Prof. S. Ben John Stephen, NICHE
Prof. V.A. Tibbie Pon Symon, NICHE
Prof. Breesha, NICHE
Prof. R. Rajesh, NICHE

Contents

About the Editors

Dr. L. Padma Suresh obtained his doctorate from MS University and Dr. M.G.R. University, respectively. He is presently working as Professor and Head in the Department of Electrical and Electronics Engineering, Noorul Islam University, Kumaracoil, India. Dr. Suresh is well known for his contributions to the field in both research and education contributing over 75 research articles in journals and conferences. He is the editorial member of International Journal of Advanced Electrical and Computer Engineering and also served as reviewer for various reputed journals. He is a life member of the Indian Society for Technical Education. He also served in many committees as Convener, Chair and Advisory member for various external agencies. His research is currently focused on Artificial Intelligence, Power Electronics, Evolutionary Algorithms, Image Processing and Control Systems.

Dr. Bijaya Ketan Parnigrahi is an Associate Professor in the Electrical and Electronics Engineering Department in Indian Institute of Technology Delhi, India. He received his Ph.D. degree from Sambalpur University. He is serving as chief editor of the International Journal of Power and Energy Conversion. He has contributed more than 200 research papers in reputed indexed journals. His interests include Power Quality, FACTS Devices, Power System Protection and AI Application to Power System.

About the Editors

Dr. L. Padma Suresh obtained his doctorate from MS University and Dr. MGR University, respectively. He is presently working as Professor and Head in the Department of Electrical and Electronics Engineering, Noorul Islam University, Kumaracoil, India. Dr. Suresh is well known for his contributions to the field in both research and education contributing over 75 research articles in journals and conferences. He is the editorial member of International Journal of Advanced Electrical and Computer Engineering and also served as reviewer for various reputed journals. He is a life member of the Indian Society for Technical Education. He also served in many committees as Convener, Chair and Advisory member for various external agencies. His research is currently focused on Artificial Intelligence, Power Electronics, Evolutionary Algorithms, Image Processing and Control Systems.

Dr. Bijaya Ketan Panigrahi is an Associate Professor in the Electrical and Electronics Engineering Department in Indian Institute of Technology Delhi, India. He received his Ph.D. degree from Sambalpur University. He is serving as chief editor of the International Journal of Power and Energy Conversion. He has contributed more than 200 research papers in reputed indexed journals. His interests include Power Quality, FACTS Devices, Power System Protection, and AI Application to Power System.

Design and Implementation of an Energy Harvester for Low-Power Devices from Vibration of Automobile Engine

Ann Agnetta Chandru, S. Sakthivel Murugan and V. Keerthika

Abstract The need for electrical power supply has spurred different technologies to cater the requisite. Energy harvesting from ambient vibration is a promising method to satisfy the indispensible power requirements. Piezoelectric energy harvesting (PEH) systems are perfectly used to convert mechanical source into a usable electrical form. Hence it requires a system to be developed to harvest the electrical power. In automobile industry, battery is the vital device for power supply and is charged using the fuel. This in turn would have an impact on fuel consumption of the automobile. An alternate for this can be the unused vibration from the engine mount to charge the battery. A practical model of the system is developed to verify the experimental results with an innovation on device design with a MAX 756-based regulatory circuit. The harvester feeds the appliances within the vehicle.

Keywords Piezoelectric energy harvesting · Ambient vibration · Power management · Signal conditioning · Low power application

1 Introduction

To cater the electricity requirements in present day scenario to few amounts, various energy harvesting systems exist. An effort is taken to develop a hardware model to harvest the electrical energy. The unused vibration can be picked up at a particular frequency range and converted using different sensing elements like piezoelectric sensors, actuators and cantilevers and is supplied to appropriate

A.A. Chandru (✉) · S. Sakthivel Murugan · V. Keerthika
ECE Department, SSN College of Engineering, Chennai, India
e-mail: annagnetta21@gmail.com

S. Sakthivel Murugan
e-mail: sakthivels@ssn.edu.in

V. Keerthika
e-mail: vkeerthikasindhu@gmail.com

L.P. Suresh and B.K. Panigrahi (eds.), *Proceedings of the International Conference on Soft Computing Systems*, Advances in Intelligent Systems and Computing 397, DOI 10.1007/978-81-322-2671-0_1

appliances based on the power output level [1]. The piezoelectric concept-based energy harvesting system is developed using the unused vibration in a car engine can be used to power various low power application within the car that take up power from the existing battery. The objective is to design and develop a high efficient harvester using thin film piezoelectric cantilever based on the PZT effect with the aid of unused vibration from the automobile for low power applications in the range of 5 V and hence an eco friendly recycled energy harvesting system will be modelled.

The two key motivational points that led to the development of this harvesting system are as follows:

- To recycle the unused vibration of the car engine mount.
- To reduce the weak points of batteries used till date and reduce the overhead on the existing batteries.

Energy harvesting is a most fascinating concept among renewables as they are supply independent and would result in the development of self-sustained system. The renewable systems are supporting green energy production and its usage is being increased day to day. PEH is based on the basic piezoelectric concept for electrical generation usually done in two different modes 3–1 and 3–3 modes [2].

2 Energy Harvester Overview

Vibrational energy is considered as the source in the proposed system. Ambient vibration is the second largest source available next to solar power [3]. Vibration can be picked up from different mechanically moving equipments. The vibration is measured in terms of the frequency with which it vibrates and the amount of acceleration [4]. Most of the engines in present days are not an exemption to vibration, though most of them have reduced noise emission. Due to drastic development in the new types of vibration sensor, even the slightest mechanical changes are capable of producing large power outputs. The vibration energy in the range of up to 100 Hz is expected to be used to generate electrical energy.

The energy harvester proposed takes up vibration produced by the car engine and is converted into electrical energy in its usable form based on the piezoelectric effect. The effect explains that a physical change in size of the piezoelectric material would generate equal amount of output in terms of electrical energy, and power management has to be done to store the obtained energy or to supply immediately to the demanding device.

The block diagram would aid us to know about the overall system in a pictorial representation as shown in Fig. 1.

The energy harvester is the piezoelectric micro-thin film cantilever device. The power management module includes two important functions within it; they are the signal conditioning and power management module. The resultant output can be used to feed the applications ranging to low power scale that exist within the system

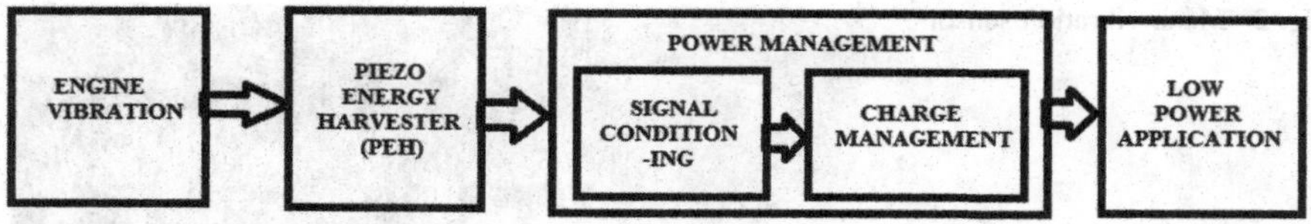

Fig. 1 Block diagram of the energy harvester

where we obtain the vibrational source. The following chapter would explain briefly about the working and the concepts of each module. They are projected on the state-of-the-art micro-thin film cantilever piezoelectric sensor and it would provide electrical energy and it would be supplied to the low power appliances. Before being supplied to the appliances, it is necessary to develop a proper power management and signal conditioning circuits.

3 Design Structure

The anticipated system would obtain its input from the unused vibration prevailing in the car engine and is the mechanical source. The vibration is taken as the input to the entire system and is obtained from the car engine. Car is being used by most of the people in every nook and corner of the world. It is no wonder that all the cars will have an engine. Most of the engines are prone to experience few levels of vibration. The vibrations that occur in a car engine can be used for energy harvesting using the concept of piezoelectricity. The proposed system can take up a vibration of 100 Hz and this can be obtained from the car during normal running condition of the car engine within few seconds after the vehicle has started to move and an acceleration of 1 g.

4 Piezoelectric Sensor

The vibration sensor that satisfies our design requirement is a micro-thin film cantilever type of device which obeys piezoelectric effect and is shown in Fig. 2. The cantilever shape is opted due to its efficiency and performance characteristics cited by Jiang bo Yaun et al. in [5]. The one chosen does not require any input power and it takes up vibration in the range of 70–00 Hz and would produce an output DC voltage of 0.7 V and this has a built-in rectifier circuit which is a bridge rectifier.

The dimension of the vibration sensor is shown in Fig. 3 sizing as 0.98 × 0.52 in. (25 × 13 mm) and can resist a temperature range of 0–70 °C. This sensor is compact, less heavy, easy to handle and cost-effective. The sensor has a cantilever shape which is similar to a springboard. The vibration applied can be compared with the person jumping on a springboard and used in 3–1 mode configuration.

Fig. 2 Mean vibration sensor

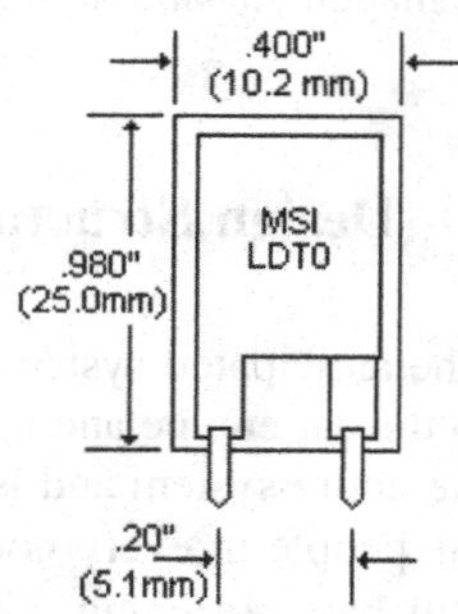

Fig. 3 Dimension of vibration sensor

5 Signal Conditioning

The output obtained from the vibration sensor is AC output and it cannot be directly fed to the low power applications and so rectification has to be done. The rectifier is in-built along with the sensor and is a bridge rectifier consisting of diodes. The diodes used were germanium diodes and a capacitor to perform smoothing to reduce ripple content which is carried out in the next part of the system. The obtained output is 0.7 V and cannot be directly stored in the battery and hence has to undergo signal conditioning to store it in the battery. The experimental setup for the rectifier is shown in Fig. 4 and the sensor can be optimally attached and mounted on a single board.

Germanium diodes has an advantage that it will conduct at a forward voltage of about 0.15 V, but a silicon diode will not start to conduct until a forward voltage of 0.6 V is reached.

To limit ripple to the expected value, the required capacitor size is proportional to the load current and must be inversely related to the source frequency and also the total number of output peaks of the rectifier circuit designed per input cycle.

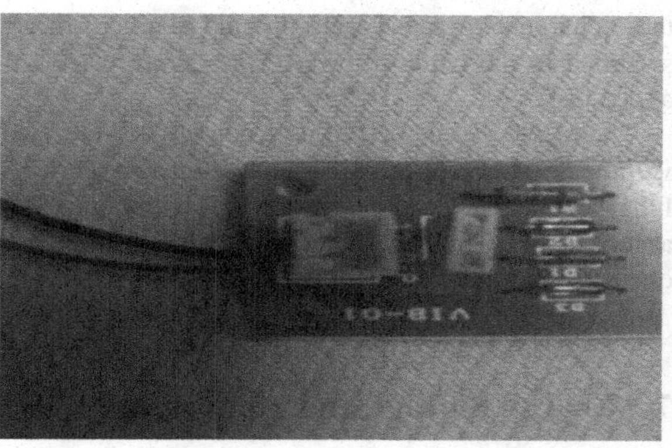

Fig. 4 Experimental setup of converter circuit

6 Charge Management

The rechargeable battery is charged using the PZT harvester and it is then boosted up to 5 V, which can be used as a supply power for the appliances that exist within the car. The experimental setup for the boost converter is the one which was developed to obtain the required power and was used to feed few applications such as a 12 V battery, stepper motor and it was also proven to be used for charging a mobile phone uninterruptedly.

The 0.7 V can be stored in a rechargeable battery and before that; it must be made up to 1.5 V. The MAX 756 IC can be used to boost up the 1.5 V into 3 or 5 V. The circuit is built using the circuit shown below. The DC–DC converter also called as booster circuit is designed to satisfy the design requirement. The experimental booster circuit configuration is shown in Fig. 5 and it is supplied with a rechargeable 1.2 V battery and it is recharged using the booster circuit output. Until

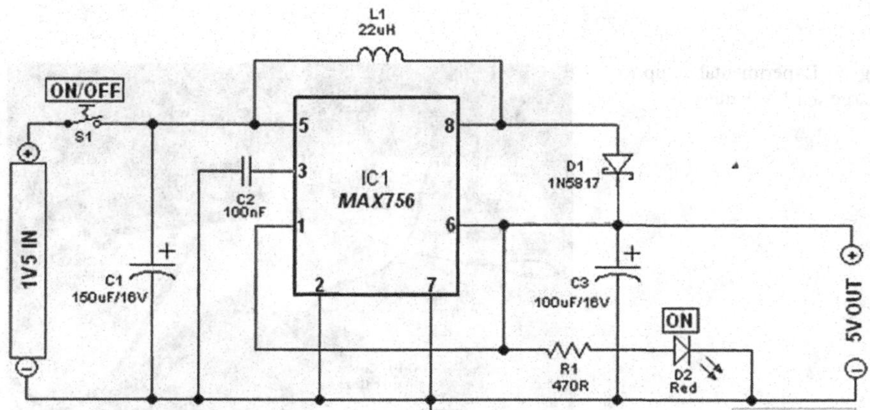

Fig. 5 DC–DC booster circuit

the rechargeable battery gets initially charged, the connected appliance might be switched to the battery power. This output can be directly fed to the low power appliances.

The MAX 756 IC is a CMOS step up DC–DC switching regulator for low input voltage or battery powered system. It operates down to 0.7 V input supply voltage. It has an efficiency of 87 % at 200 mA with 60 μA quiescent current. It can produce a current of 20 μA in shutdown mode. 500 kHz is its maximum switching frequency. It has a low-battery detector (LBI/LBO). Its a 8-Pin DIP and SO Packages.

7 Energy Storage and Application

The obtained boosted voltage can be fed to a 12 V 1.2 mAh lead–acid battery and then fed to appropriate appliances. This can be fed to the appliances that exist within the car and has a voltage range up to 12 V or even 5 V. The sensors and miniature devices that are inbuilt in the car close to the engine can be fed with the resultant power with minimal wiring. However, the initial start up of the devices is carried out from the car battery due to the delay experienced to charge the 1.2 V rechargeable batteries. The output can also be designed to simultaneously feed few low power appliances.

8 Experimental Prototype

The experimental setup of the proposed system is in Fig. 6. The vibration of 100 Hz was fed as the input to the vibration sensor and the output was rectified to 0.7 V DC and it was boosted to 5 V using a circuit design using MAX 756 IC and the output

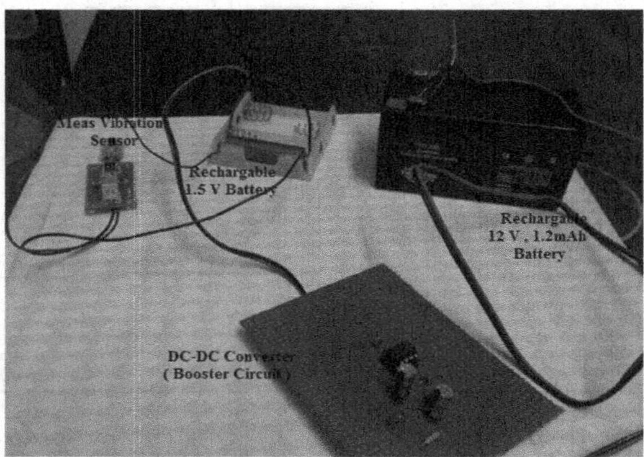

Fig. 6 Experimental setup to charge a 12 V battery

was stored in a 12 V battery. The boosted output can also be fed to the appliances that exist within the car and require a voltage level of 3.3 or 5 V.

The proposed system and the experimental setup can be used to run various low power applications that are viable in an automobile. The output was demonstrated to charge a mobile phone and supply a stepper motor.

The battery was charged without any additional charging circuitry as the output was stable and to recharge a 12 V battery, it requires a minimum of 3.3 V to recharge and is considered to be the minimum pickup voltage. The stepper motor is demonstrated as it is required in a car to lift the window shutters and few mechanically controlled rotor operation. The stepper motor used required an input of 3.3 V. The experimental setup is as shown in Fig. 7

Mobile charging was experimented as it requires 5 V to charge and was obtained from the designed harvester. This can be done uninterruptedly when the vibration at the input side is consistent when the car is running. The experimental setup by the proposed system is shown in Fig. 8.

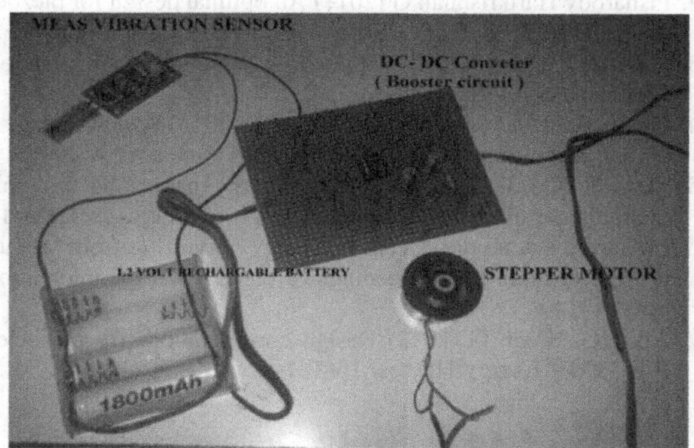

Fig. 7 Experimental setup to run a stepper motor

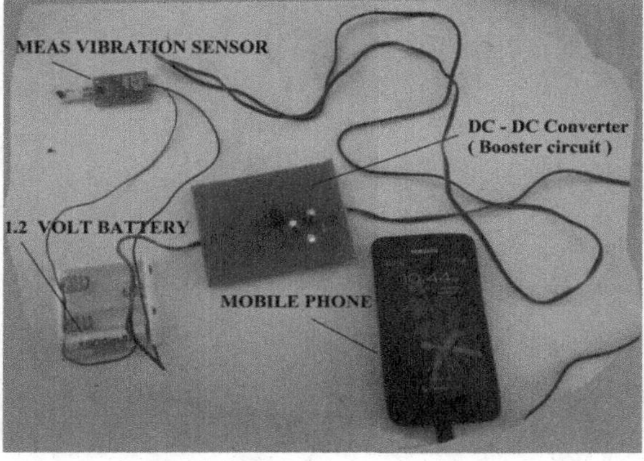

Fig. 8 Experimental setup to charge a mobile phone

9 Conclusion

The obtained output is aimed at low power application and the developed system is tested to recharge a 12 V lead–acid battery to run a stepper motor and charge a mobile phone. However, appliances has to depend on the car battery for few minutes due to the delay caused to charge the 1.2 V rechargeable battery and which in turn has to supply the DC–DC regulator designed to result in two different configurations. This system will be capable to supply the appliances as we travel and leads to a self-sustained energy harvester and would result in reduction of fuel consumption by the battery.

References

1. Ottman GK, Hofmann HF, Bhatt AC, Lesieutre GA (2002) Adaptive piezoelectric energy harvesting circuit for wireless remote power supply. IEEE Trans Power Electron 17(5):669–676
2. Pisharody Harikrishnan G (2011) An optimal design for piezoelectric energy harvesting system. In: IEEE PES Innovative Smart Grid Technologies, pp 6–11
3. Land Z, Tang X (2013) Large-scale vibration energy harvesting. J Intell Mater Syst Struct 24(11):1405–1430
4. Dhakar L, Huicong Liu FEH, Tay CL (2013) A new energy harvester design for high power output at low frequencies. Elsevier Sens Actuators A 199:344–352
5. Yuan J, Xie T, Chen W (2008) Energy harvesting with piezoelectric cantilever. In: IEEE international ultrasonic's symposium proceedings, pp 1397–1400
6. Mohareri, Arzanpour S (2011) Energy harvesting from vibration of a hydraulic engine mount using a turbine. In: Proceedings of the 2011 IEEE international conference on mechatronics, pp 134–140
7. Ali WG, Nagib G (2012) Design considerations for piezoelectric energy harvesting systems. IEEE Conf Energy Harvest 10(3):1–6

Modern Optimization-Based Controller Design for Speed Control in Flyback Converter-Driven DC Servomotor Drive

S. Subiramoniyan and S. Joseph Jawhar

Abstract Implementation of proportional-integral (PI) controller for speed control of direct current (DC) servomotor drive is an emerging trend in recent years. PI controller is a simple effective method and it needs tuning of the control parameters to improve the performance of the converters. The local treatment of the parameter tuning is no longer possible and it is thus essential to design a suitable topology for flyback converters that has to reduce the overshoot and settling time. Here an advanced PI control algorithm has been framed that optimize the PI controller parameters using ant colony optimization and cuckoo search algorithm. The PI control algorithm is implemented in FPGA to drive the DC servomotor drive. Framing the conventional system with optimized PI control algorithm shows significant reduction of overshoot and settling time and works effectively for DC servomotor drive.

Keywords ACO · Cuckoo search · ISE · Objective function · Overshoot time · Settling time

1 Introduction

Nowadays, DC servomotor is widely used in industries due to its wide range of speed control even if its maintenance cost is higher than the other motors like induction motor, etc. [1]. The speed control of DC motor is very interesting from research point of view and hence several methods are proposed in this field [2].

S. Subiramoniyan (✉)
Sathyabama University, Chennai, India
e-mail: sbynjec@gmail.com

S. Subiramoniyan · S. Joseph Jawhar
AP/EEE, Jayamatha Engineering College, Aralvaimozhi, India

S. Joseph Jawhar
Arunachala College of Engineering for Women, Vellichanthai, India

L.P. Suresh and B.K. Panigrahi (eds.), *Proceedings of the International Conference on Soft Computing Systems*, Advances in Intelligent Systems and Computing 397, DOI 10.1007/978-81-322-2671-0_2

Flyback converters [3] have been widely used in DC servomotor because of their relative simplicity and their excellent performance for multi-output applications [4]. They can save cost and volume compared with the other converters, especially in low power applications.

One of the most important factors in the power efficiency of flyback converters is the isolation transformer (coupled inductor) [5, 6]. Even though power losses in the main switch and diode are worth considering, that of the transformer is one of the dominant factors in terms of efficiency for flyback converters [7]. This can be controlled by an optimized controller that has to replace manual tuning of parameters. Adaptive control techniques have been proposed by researchers assuming linearized system models. These controllers have the ability to cope with small changes in system parameters such as valve flow coefficients, the fluid bulk modulus, and variable loading [8].

The proportional-integral (PI) controllers are very stable and have fast control due to the enhanced properties of the current-mode circuits used in the design [9, 10]. Regarding the simplicity and the nature of the PI controller, they are highly suitable for the immense use in a variety of different control systems [11].

Considering that the PI controller parameters have great impact on the responses of the complete system, tuning of the parameters is a significant task [12, 13].

Many random search methods, such as genetic algorithm (GA) have recently received much interest for achieving high efficiency and searching global optimal solution in space [14, 15]. Tuning of PID parameters is yet difficult because of variable system parameters. PSO is an efficient technique, drawn for optimization of PI parameters in recent years [16].

Syed and Ying [17] developed a fuzzy control approach with minimal rules to intelligently control engine power and speed behavior in a power option toward achieving these goals.

Abidin et al. [18] developed a decentralized PI tuning method using PSO algorithm. Utomo [19] proposed voltage tracking of a dc–dc flyback converter using neural network control. Bagis [20] proposed determination of the PID controller parameters by modified genetic algorithm.

Awouda and Mamat [21] carried a work on optimizing PID tuning parameters using gray prediction algorithm. Mahdi [22] proposed an optimization of PID controller parameters based on genetic algorithm for nonlinear electromechanical actuator.

The existing system designed with fuzzy technique requires large constraints and variables. The PSO optimization technique struck in selection of a single solution. The limitations of genetic algorithm are its slow convergence and lagging in rank-based fitness function. Also no such methods are available to effectively reduce the overshoot and settling time by optimizing the K_p and K_i values.

To overcome the above drawbacks here a new PI control algorithm has been introduced that makes use of the fine-tuned values of K_p and K_i optimized by ant colony optimization (ACO) and cuckoo search in maintaining low overshoot as well as settling time with improved performance.

1.1 Traditional PI Controller

Proportional-integral (PI) controllers have been used for industrial purpose due to their simplicity, easy designing method, low cost, and effectiveness. Due to presence of nonlinearity in the system, conventional PI controller is not very efficient.

For effective working of DC servomotor drive, the PI controller parameters namely K_p and K_i have to be tuned to achieve reduction in overshoot and settling time. The flyback converter has to be modulated with the PI controller that has to use an optimization algorithm for optimizing the values for K_p and K_i.

1.2 Modeling the Transformer Equivalent Circuit

The basic design of transformer equivalent circuit to be implemented for flyback converter is shown in Fig. 1.

1.3 Tuning of PI Controller Parameters

It is desired to select optimized values for K_p and K_i that reduce the overshoot and settling time. The objective function can be framed as minimization of ISE to eliminate small errors and the transfer function can be formulated as

$$F_1(x) = \int_0^{\infty} |e(t)|^2 \mathrm{d}t \tag{1}$$

with respect to the constraints $0 <= K_p <= 5, 0 <= K_i <= 5$ and the fitness function can be evaluated as

$$Ft_i = \frac{1}{Fi(x)} \tag{2}$$

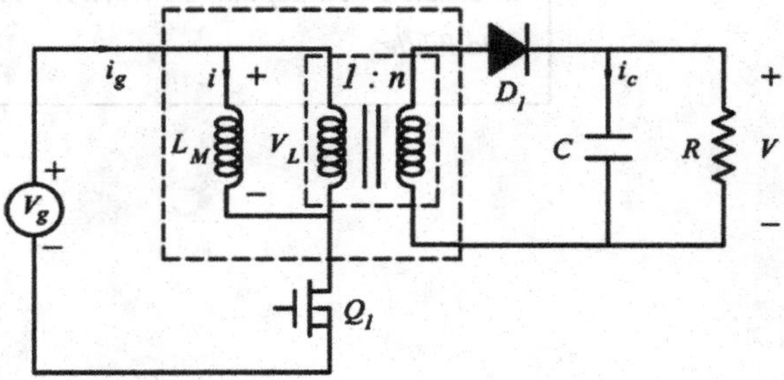

Fig. 1 Design of transformer equivalent circuit

2 Ant Colony Optimization

The algorithm is formulated for PI controller tuning by setting the objective functions as minimization of ISE. From the initial solution set, ACO searches for the optimized one in each step and stops once it grabs the optimized solution. This is clearly explained in the following Algorithm 1.

Algorithm 1 Ant colony optimization for PI parameters

♦Initialization:

a. Set initial parameters that are system: function

b. Set initial pheromone trails value CT.

c. Each ant is individually placed on initial state with empty memory.

♦While termination conditions not meet do

a. Construct Ant Solution ST(x):

Each ant constructs a path by successively applying the transition function the probability of moving from state to state depend on: as the attractiveness of the move, and the trail level of the move.

b. Apply Local Search

c. Best Tour check:

If there is an improvement, update it in ST(x).

d. Update CT:

- Evaporate bad solution from ST(x).

- For each ant perform the "ant-cycle" pheromone update.

- Reinforce the best tour with a set number of "elitist ants" performing the "ant-cycle"

d. Create a new population based on (2)

End While

3 Cuckoo Search

Algorithm 2 presents the working procedure of cuckoo search for PI controller.

Algorithm 2 Pseudo code of the cuckoo search (CS)

```
begin
Objective function f(x), x = (x1, ..., xd)T
Generate initial population of n host nests xi (i = 1, 2, ..., n)
while (t <MaxGeneration) or (stop criterion)
Get a cuckoo randomly by L'evy flights evaluate its
quality/fitness Fi
Choose a nest among n (say, j) randomly
if (Fi > Fj ),
replace j by the new solution;
end
A fraction (pa) of worse nests
        are abandoned and new ones are built;
Keep the best solutions
(or nests with quality solutions);
Rank the solutions and find the current best
end while
Postprocess results and visualization
end
```

4 Simulation Model

Figure 2 shows the simulation diagram of the system using MATLAB SIMULINK.

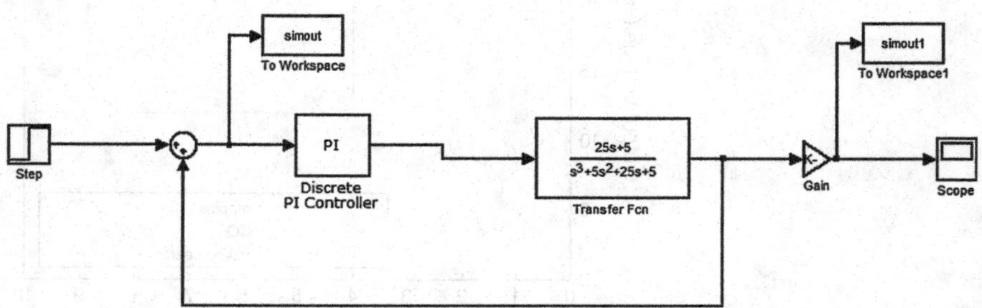

Fig. 2 Software implementation of the system

4.1 *Simulation Results*

4.1.1 Flyback Converter Subjected to an Input Voltage Variation

The optimized algorithm is implemented in MATLAB with $K_p = 1.29$ and $K_i = 9.63$. The input voltage is varied from 12 to 25 volts both in an increasing and decreasing manner. The set point of the output voltage is 12 V. The effectiveness of the controller with respect to overshoot and settling time is studied.

Figures 3 and 4 show the output voltage plotted against time. It is found that the controller acts very effectively and it maintains the constant output voltage irrespective of the input voltage variation.

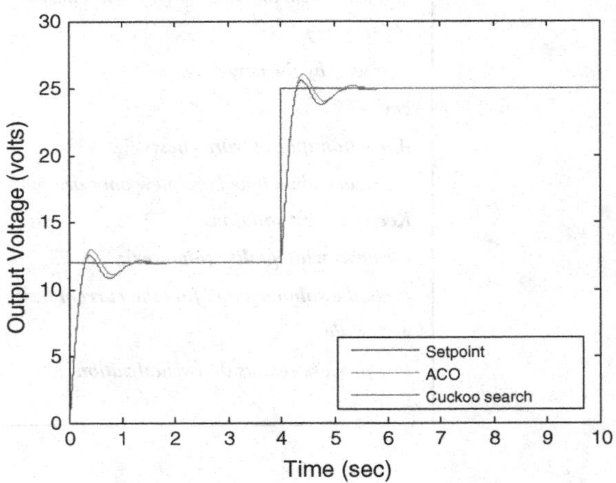

Fig. 3 Performance evaluation of ACO and cuckoo search in case of increased input voltage variation

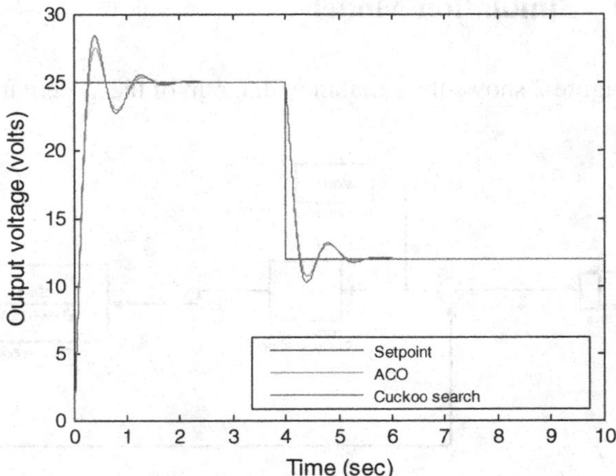

Fig. 4 Performance evaluation of ACO and cuckoo search in case of decreased input voltage variation

4.1.2 Flyback Converter Subjected to Load Variations

The flyback converter is subjected to a variation of load from 3 to 6 Ω both in an increasing and decreasing manner. The effectiveness of the controller with respect to the overshoot and settling time at the time of load variations is studied.

Figures 5 and 6 show the output voltage plotted against time. It is found that the controller acts very effectively and it maintains the constant output voltage irrespective of a variation of load from 3 to 6 Ω.

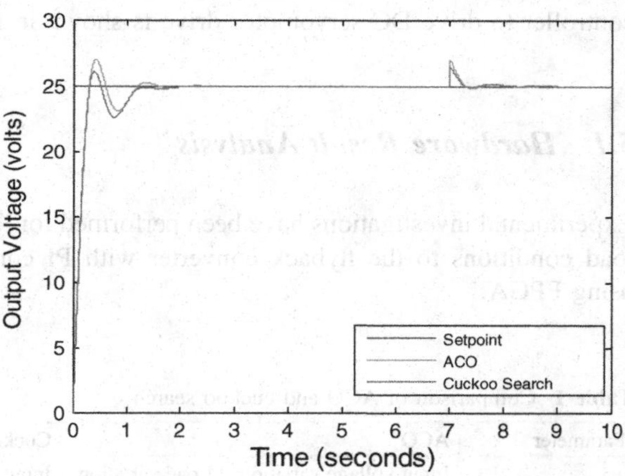

Fig. 5 Performance evaluation of ACO and cuckoo search in case of increased load variation

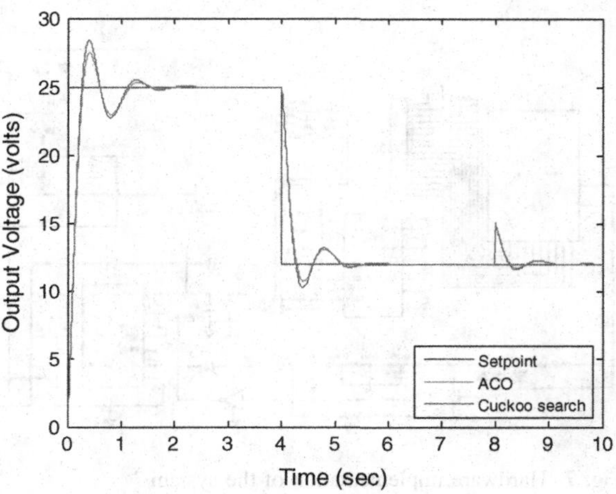

Fig. 6 Performance evaluation of ACO and cuckoo search in case of decreased load variation

4.1.3 Performance Comparison

Comparison has been made between the performance of ACO and cuckoo search and is presented in Table 1.

From Table 1, it can be inferred that the performance of cuckoo search is better than ACO in terms of overshoot and settling time.

5 Hardware Implementation

The components as well as the circuits used in the design of FPGA-based PI controller to drive DC servomotor drive is shown in Fig. 7.

5.1 Hardware Result Analysis

Experimental investigations have been performed for the various input voltages and load conditions to the flyback converter with PI control algorithm implemented using FPGA.

Table 1 Comparison of ACO and cuckoo search

Parameter	ACO		Cuckoo search	
	Input voltage variation	Load variation	Input voltage variation	Load variation
Overshoot (V)	1	0.9	0.5	0.4
Settling time (s)	1	3	0.8	2.8

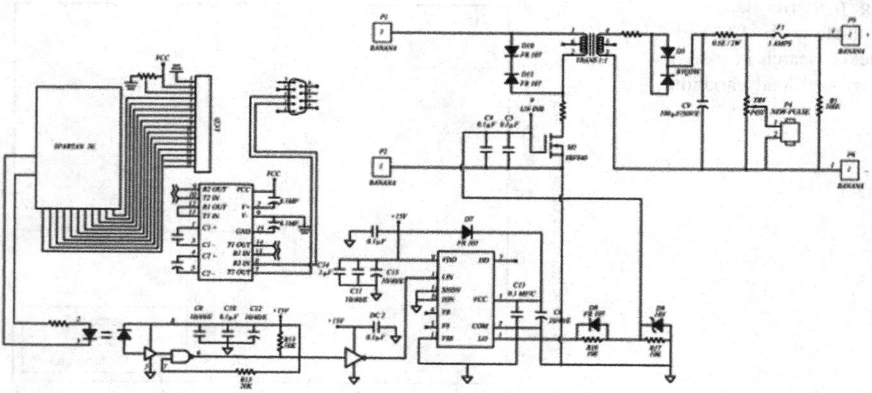

Fig. 7 Hardware implementation of the system

5.2 Flyback Converter Subjected to an Input Voltage Variation

The PI control algorithm is implemented in FPGA-based PI controller to drive the actual circuit of the flyback converter with the tuned values of K_p and K_i obtained by ACO and cuckoo search. The input voltage is varied from 1 to 4 V. The set point of the output voltage is 6 V. The effectiveness of the controller with respect to overshoot and settling time is studied.

Figure 8 shows the output voltage plotted with respect to time. It is found that the controller acts very effectively and maintains the constant output voltage of 6 V irrespective of the input voltage variation. The peak overshoot voltage at the time of input voltage variation is 60 % and the settling time is 90 ms.

Figure 9 shows the variation of output voltage versus time. It is found that the controller acts very effectively and it maintains the constant output voltage of 6 V, irrespective of a variation of input voltage from 2 to 4.2 V. The peak overshoot voltage at the time of input voltage variation is 10 % and the settling time is 75 ms.

5.3 Flyback Converter Subjected to Load Variations

The flyback converter is subjected to a variation of load from 3 to 6 Ω both in an increasing and decreasing manner. The effectiveness of the controller with respect to the overshoot and settling time at the time of load variations is studied.

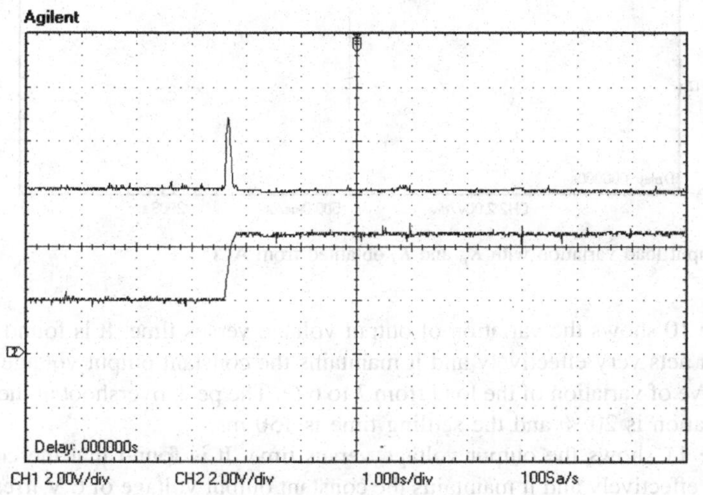

Fig. 8 Input voltage variation with K_p and K_i obtained from ACO

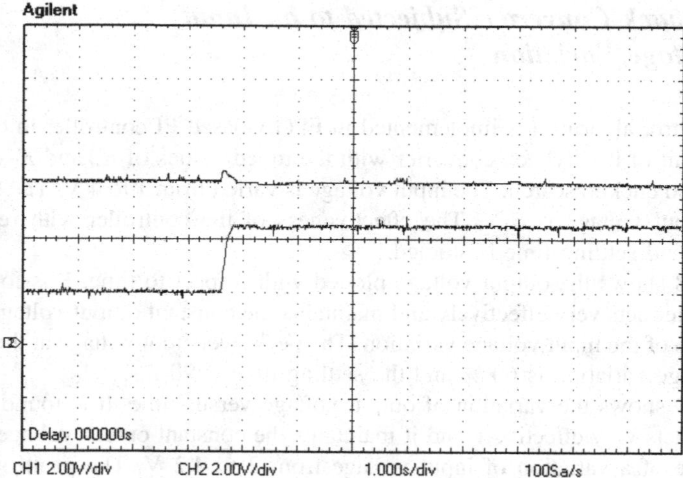

Fig. 9 Input voltage variation with K_p and K_i values obtained from cuckoo search

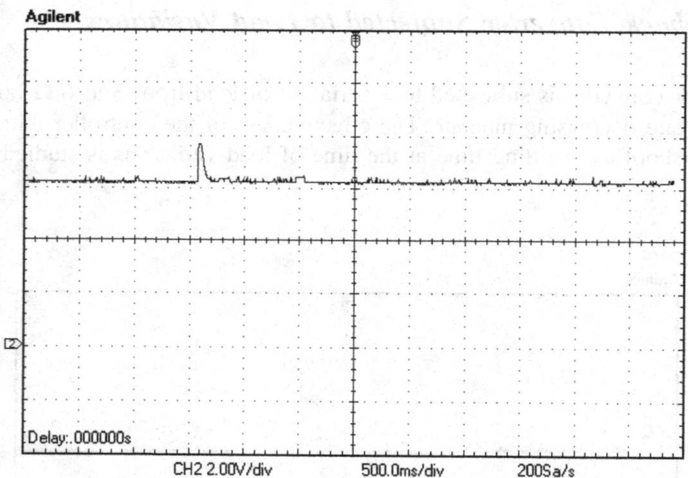

Fig. 10 Input load variation with K_p and K_i obtained from ACO

Figure 10 shows the variation of output voltage versus time. It is found that the controller acts very effectively and it maintains the constant output voltage of 6 V, irrespective of variation of the load from 3 to 6 Ω. The peak overshoot at the time of load variation is 20 % and the settling time is 150 ms.

Figure 11 shows the output voltage versus time. It is found that the controller acts very effectively and it maintains the constant output voltage of 6 V irrespective

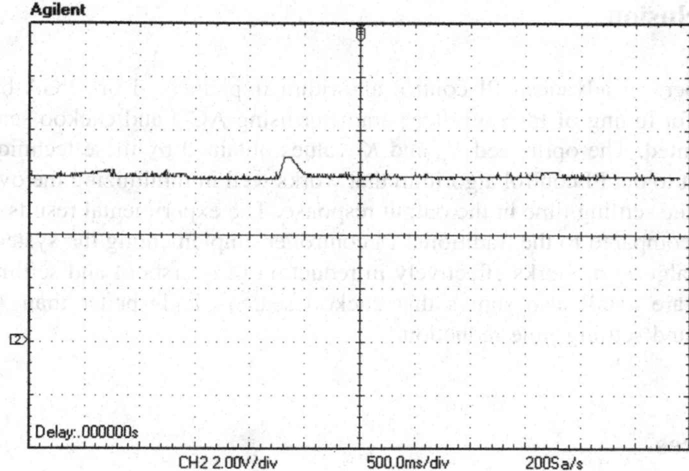

Fig. 11 Input load variation with K_p and K_i obtained from cuckoo search

of the variation of load from 3 to 6 Ω. The peak over shoot at the time of load variation is 40 % and the settling time is 100 ms.

5.4 Performance Evaluation

From the hardware result analysis, the performance of PI control algorithm implemented in FPGA for minimization of overshoot and settling time with different optimized PI controller parameter values obtained from ACO and cuckoo search are discussed in Table 2.

By implementing the above topology the performance of flyback converter increased and it works effectively in electrical equipments.

Table 2 Performance evaluation of PI control algorithm

Set point = 6 V				
Parameter	**ACO**		Cuckoo search	
	Input voltage variation	Input load variation	Input voltage variation	Input load variation
Overshoot (V)	60 %	20 %	10 %	40 %
Settling time (ms)	90	150	40	100

6 Conclusion

In this paper, an advanced PI control algorithm implemented in FPGA-based PI controller for tuning of PI controller parameter using ACO and cuckoo search has been presented. The optimized K_p and K_i values obtained by these techniques are fed as input to the PI control algorithm that works well in minimizing the overshoot as well as the settling time in the output response. The experimental results showed that when compared to the traditional PI controller, implementing the system using PI control algorithm works effectively in reduction of overshoot and settling time. The hardware result also shows that cuckoo search yields better than ACO in overshoot and settling time reduction.

References

1. Zhang F, Yan Y (2009) Novel forward-flyback hybrid bidirectional DC–DC converter. IEEE Trans Ind Electron 56:1578–1584
2. Singh V, Garg VK (2014) Tuning of PID controller for speed control of DC motor using soft computing techniques—a review. Int J Appl Eng Res 9:1141–1148
3. Hong S-S, Ji S-K, Jung Y-J, Roh C-W (2010) Analysis and design of a high voltage flyback converter with resonant elements. J Power Electron. 10(2):107–114
4. Tacca HE (1998) Single-switch two-output flyback-forward converter operation. IEEE Trans Power Electron 13:903–911
5. Rong P, Chen W, Lu Z (2009) A novel active clamped dual switch flyback converter. IEEE power electron motion control conference, pp 1277–1281
6. Lee J-H, Park J-H, Jeon JH (2011) Series-connected forward-flyback converter for high step-up power conversion. IEEE Trans Power Electron 26:3629–3641
7. Kim D-H, Park J-H (2013) High efficiency step-down flyback converter using coaxial cable coupled-inductor. J Power Electron 13:214–222
8. Ayman AA (2011) PID parameters optimization using genetic algorithm technique for electrohydraulic servo control system. Intell Control Autom 2:69–76
9. Erdal C, Toker A, Acar C (2001) A new proportional-integral-derivative (pid) controller realization by using current conveyors and calculating optimum parameter tolerances. J Electric Electron 1:267–273
10. YaPing L, ShengChun Y, Ke W, Dan Z (2011) Research on PI controller tuning for VSC-HVDC system. IEEE international conference on advanced power system automation and protection, pp 261–264
11. Bassi SJ, Mishra MK, Omizegba EE (2011) Automatic tuning of proportional–integral–derivative (pid) controller using particle swarm optimization (pso) algorithm. Int J Artif Intell Appl (IJAIA) 2:25–34
12. Wang L, Ertugrul N (2010) Selection of PI compensator parameters for VSCHVDC system using decoupled control strategy. IEEE universities power engineering conference (AUPEC), pp 1–7
13. Popadic B, Dumnic B, Milicevic D, Katic V, Corba Z (2013) Tuning methods for PI controller —comparison on a highly modular drive. IEEE
14. Rai P, Shekher V, Prakash O (2012) Determination of stabilizing parameter of fractional order PID controller using genetic algorithm. IJCEM Int J Comput Eng Manag 15:24–32
15. Ohri J, Kumar N, Chinda M An improved genetic algorithm for PID parameter tuning. Recent Adv Electr Comput Eng 191–198

16. Girirajkumar SM, Kumar AA, Anantharaman N (2010) Tuning of a PID controller for a real time industrial process using particle swarm optimization. IJCA special issue on evolutionary computation for optimization techniques (ECOT), pp 35–40
17. Syed FU, Yin H (2006) Rule-based fuzzy gain-scheduling P1 controller to improve engine speed and power behavior in a power-split hybrid electric vehicle. IEEE, pp 284–289
18. Abidin NZ, Sahlan S, Wahab NA (2013) Optimization tuning of pi controller of quadruple tank process. IEEE Australian control conference, pp 331–335
19. Utomo WM, Yi SS, Buswig YMY, Haron ZA, Bakar AA, Ahmad MZ (2012) Voltage tracking of a DC-DC flyback converter using neural network control. Int J Power Electron Drive Syst (IJPEDS) 2:35–42
20. Bagis A (2007) Determination of the PID controller parameters by modified genetic algorithm for improved performance. J Inf Sci Eng 23:1469–1480
21. Awouda AE, Mamat RB Optimizing PID tuning parameters using grey prediction algorithm. Int J Eng (IJE) 4:26–36
22. Shayma'a AM (2014) Optimization of PID controller parameters based on genetic algorithm for non-linear electromechanical actuator. Int J Comput Appl 94:11–20

EA-Based Optimization of Hybrid T-Slot Patch Antenna

Anindita Das, Mihir Narayan Mohanty, Laxmi Prasad Mishra and L. Padma Suresh

Abstract Evolutionary algorithm for optimization is the popular one and is referred to genetic algorithm. It is based on soft computing method. It has occupied a very good space in optimization techniques. In the same fashion it has been also applied for antenna parameters. Microstrip antenna is an essential component of wireless communication systems. The optimization leads for improving the performance of the system. As handsets improve in compactness with multifunctions in recent age, the antennas for these equipments have come under the spotlight. In this paper, the optimization has been applied for the design of hybrid slot antenna. The design is based on the T-slot combination with the L-slot microstrip patch. The shape will provide the broad bandwidth which is required for the operation of next generation wireless systems. Evolutionary algorithm-based optimization has been utilized in HFSS-ANSOFT to optimize the microstrip line feed antenna dimensions in order to obtain reliable return loss and high directivity. The dielectric constant and thickness of the antenna is 4.4 and 1.6 mm, respectively. The patch antenna is analyzed for different metrics for performance. A comparative analysis between nonoptimized patch design and optimized patch design has also been presented.

Keywords Microstrip antenna · T-slot · L-slot · Evolutionary algorithm · Multiband

1 Introduction

Biological concepts are used in the genetic algorithm due to which it has been named as evolutionary algorithm. The population, chromosomes, represent solutions to a problem. The chromosomes are represented as binary sequences. Genetic

A. Das · M.N. Mohanty (✉) · L.P. Mishra
Department of ECE, ITER, SOA University, Bhubaneswar, Odisha, India
e-mail: mihirmohanty@soauniversity.ac.in

L. Padma Suresh
NICE, Noorul Islam University, Kanyakumari, India

L.P. Suresh and B.K. Panigrahi (eds.), *Proceedings of the International Conference on Soft Computing Systems*, Advances in Intelligent Systems and Computing 397, DOI 10.1007/978-81-322-2671-0_3

algorithms (GA) are a method for solving optimization or search problems inspired by biological processes of inheritance, mutation, natural selection, and genetic crossover. Genetic algorithms (GA's) are adaptive methods, which can be used to solve search and optimization problems. The principles behind the power of GA's are based upon the genetic processes of biological organisms which over many generations, evolve according to the principles of natural selection and *survival of the fittest*.

Antennas are used for wireless communication systems, but also more and more for other applications such as navigation, security, surveillance, and various medical sensor systems. Recently wireless systems demand antennas where space value is quite limited. New technologies for wideband, multifunction and multielement antenna systems will be developed, and integration with electronics plays an important role. Antennas for wireless devices are required to have the compactness, multiband operation as well as broad bandwidth. Such antennas are highly essential for both commercial and military applications. Each antenna operates in a single or dual frequency bands, where different antenna is required for various applications [1–3]. To subsidize this problem, multiband antenna is where a single antenna can operate with many frequency bands. One of the techniques to construct a multiband antenna is by etching a T-slot in the antenna geometry [4]. Hence, antenna is one of the most important design issues in modern wireless communication systems. Since antennas are dependent on frequency, they have to be designed to operate for certain frequency bands. Though the antenna can satisfy the reduction in size, other factors like bandwidth, efficiency are taken care of. The idea of microstrip antenna arose from utilizing printed circuit technology and transmission lines [5, 6]. A microstrip antenna, in its simplest form, consists of a rectangular shape patch mounted on a substrate backed by a ground plane.

In many applications where size, volume, cost, performance, and ease of fabrication are very much required, patch antennas are preferred and made this field as the fastest growing segments in communication industry. These antennas have been used in various fields such as mobile communication, radar, GPS system, bluetooth, space technology, aircraft, missiles, satellite communication etc. [1, 2]. The patch is generally made of conducting material with any possible shape. The radiating patch and the feed lines are to be photoengraved on the dielectric substrate [7]. The reduction in size with gain and increase in bandwidth has become a major consideration in the microstrip patch antennas. Various techniques for bandwidth and gain enhancement have been suggested such as cutting slot in a patch. The application of multiband communication systems with combinations of frequency bands is increasing, whereby the international roaming is progressing globally, the communication capacity is expanding and new functions are being added including GPS (1.57 GHz) and Bluetooth (2.4 GHz) [8]. Therefore, it is to be expected, that the handsets should be compatible with multiband in future. In such multiband systems, a multiband antenna is one of the key devices as it is well suited for all the frequency bands without resort to multiple antennas [2, 9].

The paper is organized as follows: Sect. 2 presents a design approach for combined T-L slot rectangular patch antenna, Sect. 3 gives a brief about optimization using EA, Sect. 4 presents simulation results and finally in Sect. 5 conclusion is drawn.

2 Combined T-L Slot Patch Antenna Design

Microstrip patch antennas have been widely used in many applications because of their low profile and easy manufacturing process. However, most patch antennas provide a wide beamwidth and low radiation efficiency [10]. To overcome the disadvantages of the patch antenna, it is required to design optimized antenna for an effective performance. Many optimization algorithms are considered in the literature [10–12]. Genetic algorithm is one of the global optimization techniques and has been used in this work for optimizing the patch shape and size to have better performance of the antenna. It was exactly used to optimize the patch length, the slot dimensions, and the dimensions of the feed line [12]. The calculation of the dimensions of the patch antenna is based on the following relation [7]:

(a) Width of Patch

$$W = \frac{c}{2f_0}\sqrt{\frac{2}{\varepsilon_r + 1}}$$

(b) Effective Dielectric Constant

$$\varepsilon_{\text{reff}} = \frac{\varepsilon_r + 1}{2} + \frac{\varepsilon_r - 1}{2}\left[1 + 12\frac{h}{W}\right]^{-\frac{1}{2}}$$

(c) Due to fringing effects the change in dimension of length

$$\Delta L = 0.412\,h\frac{(\varepsilon_{\text{reff}} + 0.3)\left(\frac{W}{h} + 0.264\right)}{(\varepsilon_{\text{reff}} - 0.258)\left(\frac{W}{h} + 0.8\right)}$$

(d) Length of Patch

$$L = \frac{c}{2f_0\sqrt{\varepsilon_{\text{reff}}}} - 2\Delta L$$

where

f_0 = Operating Frequency
ε_r = Dielectric Constant
h = Substrate Thickness

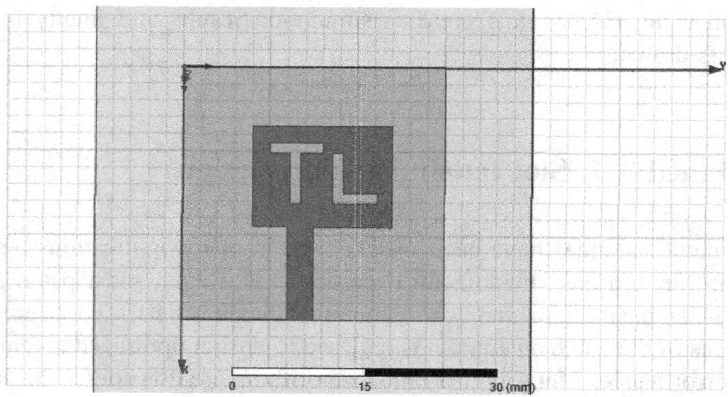

Fig. 1 Combined T-L slot microstrip patch antenna design

The configuration of the proposed antenna is shown in the Fig. 1.

The proposed antenna is designed on a layer substrate having a relative permittivity of 4.4 (FR4 Epoxy) and loss tangent of $\tan\delta = 0.02$. The substrate thickness is 1.6 mm. This antenna is excited by microstrip line feed. Generally, 50 Ω microstrip line is used to feed the radiating patch. In this work, two slots T and L have been etched on the surface of the antenna. The length and width of the vertical arms of both T and L slots are taken as 6 and 1.2 mm, respectively. The length and width of the horizontal arm of T slot are considered as 6 and 1.2 mm, respectively where as the length and width of the base arm of L slot are taken as 4 and 1.2 mm.

3 Optimization Method

The implementation of the GA method starts with the initial population as the first approximation of the solution. This generation genes are selected randomly. The fitness function is used to evaluate each member of the population. New members are produced in successive generations. It is usually consistent with the rules of roulette. Members with the larger segments of the roulette wheel have better chances to create a new generation. Members of the new generation modifies by crossovers and mutations. With an initially assumed crossover probability, both a pair of members and the crossover points are selected randomly. In a mutation process, genes may change their values [12–14]. As GA manipulates matrices, it is normally implemented using MATLAB. A number of binary digits are assigned to each variable so that the required accuracy of this variable is obtained in the final solution. The digital value of each variable is converted to analog value. Afterward the objective function (O) is evaluated. Then the relative fitness of each chromosome (C_i) is determined and is defined as

$$O = \sum_{i=1}^{n} \text{eval}_i[C_i]. \tag{1}$$

The illustration for the antenna parameter design based on the optimization technique is given as follows [15]:

Step 1 Enter the center frequency, dielectric constant, and substrate thickness in patch calculator programmed by MATLAB.

Step 2 Use the outputs (W, L) where W and L represent the width and length of the patch, respectively, for designing T-slot patch antenna in HFSS.

Step 3 Analyze the performance of the patch antenna designed in terms of return loss.

Step 4 If the return loss is better than −20 dB, then the proposed antenna is optimized; otherwise go to Step 3.

4 Results

The return losses of the nonoptimized antenna are found to be −20 dB at 6.7 GHz which is shown in Fig. 2 and its corresponding bandwidth is 250 MHz. The return losses of the optimized antenna are found to be −15 dB at 3.75 GHz and −23 dB at 7 GHz as shown in Fig. 3 where its corresponding bandwidths are found to be 150 and 400 MHz. There is an increase in the bandwidth of the antenna. Figure 4 shows the VSWR of the nonoptimized antenna which is found to be 1.02 at 6.7 GHz

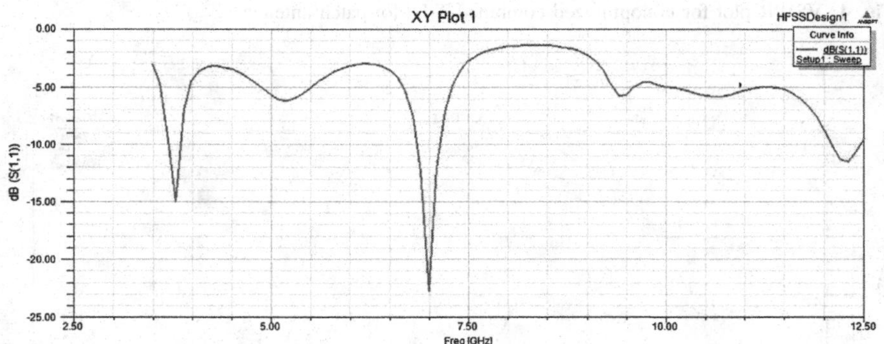

Fig. 2 Return loss plot for nonoptimized combined T-L slot patch antenna

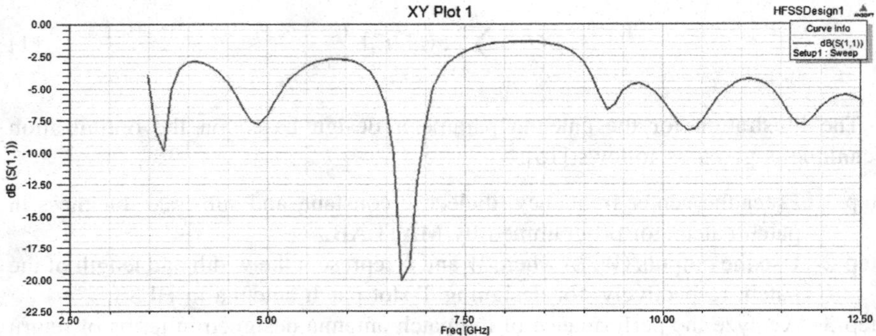

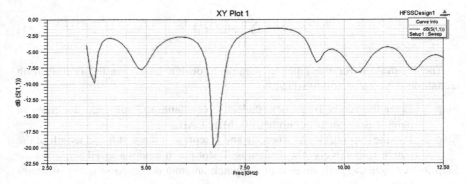

Fig. 3 Return loss plot for optimized combined T-L slot patch antenna

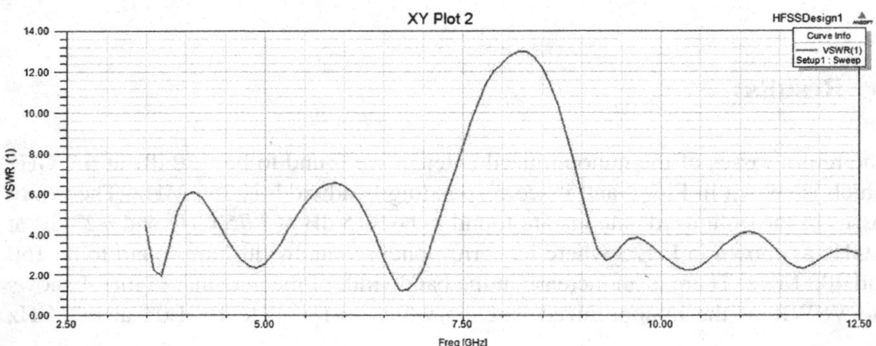

Fig. 4 VSWR plot for nonoptimized combined T-L slot patch antenna

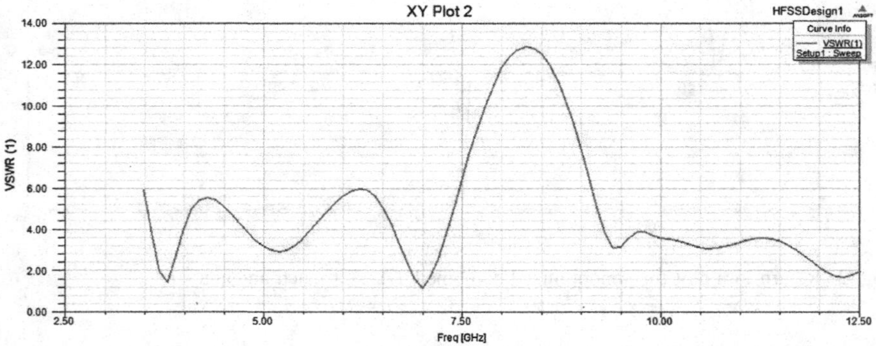

Fig. 5 VSWR plot for optimized combined T-L slot patch antenna

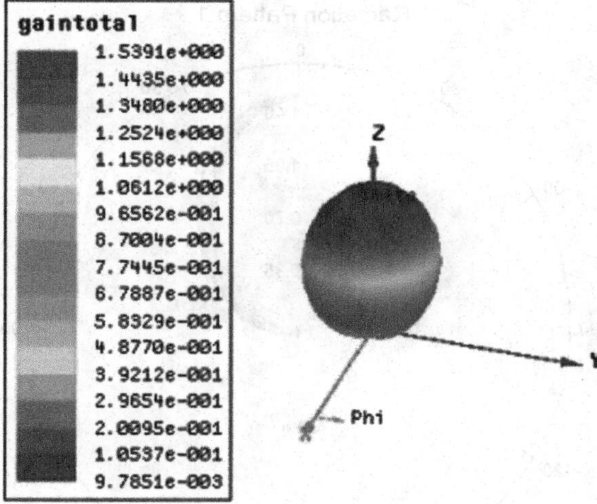

Fig. 6 3D polar plot of gain for optimized combined T-L slot patch antenna

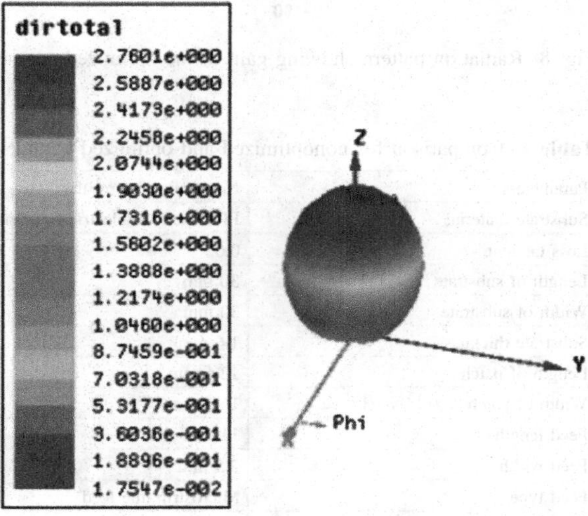

Fig. 7 3D polar plot of directivity for optimized combined T-L slot patch antenna

where as Fig. 5 shows the VSWR of the optimized antenna which are found to be 1.065 at 3.75 GHz and 1.01 at 7 GHz. Figure 6 shows the gain of the optimized antenna whose value is found to be 1.59 (2.01 dB) and Fig. 7 denotes the directivity of the optimized antenna and is found to be 2.76 (4.4 dB). Figure 8 shows the multiple radiation pattern of the optimized antenna and its gain. Table 1 is the comparison among nonoptimized and optimized parameters of the proposed antenna. For these results the conclusion can be drawn about the optimized antenna and presented in the following section.

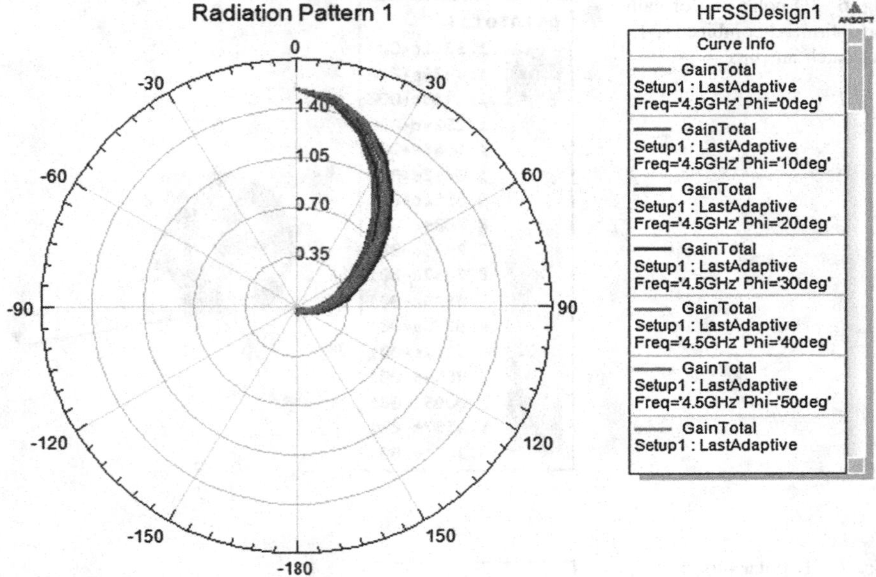

Fig. 8 Radiation pattern showing gain of the optimized antenna

Table 1 Comparison for nonoptimized and optimized parameters

Parameters	Non-optimized results	Optimized results
Substrate material	FR4 epoxy (dielectric constant = 4.4)	FR4 epoxy
Loss tangent	0.02	0.02
Length of substrate	30 mm	30 mm
Width of substrate	30 mm	30 mm
Substrate thickness	1.6 mm	1.6
Length of patch	12.6 mm	12 mm
Width of patch	16.6 mm	16 mm
Feed length	11 mm	10.4 mm
Feed width	2.5 mm	3 mm
Feed type	Microstrip line feed	Microstrip line feed
Feed location	(19, 12, 0)	(19.6, 12, 0)
'T' slot cut length and width	6, 1.2 mm	6, 1.2 mm
'L' slot cut vertical arm length and width	6, 1.2 mm	6, 1.2 mm
'L' slot cut base arm length and width	4, 1.2 mm	4, 1.2 mm
Return loss	−20 dB at 6.7 GHz	−15 dB at 3.75 GHz and −23 dB at 7 GHz
VSWR	1.02 at 6.7 GHz	1.065 at 3.75 GHz and 1.01 at 7 GHz
Bandwidths	250 MHz	150, 400 MHz

5 Conclusion

An antenna is designed to operate on multibands. The antennas are useful in the design of single antenna. A rectangular microstrip antenna with hybridization of T-slot and L-slot are described in this paper for multiband applications. This antenna is microstrip line fed and its structure is designed with the help of dielectric constant, substrate height, and resonant frequency. Antenna properties such as return loss, gain, VSWR, directivity and bandwidth are analyzed and discussed in this project. Design and analysis of this antenna are done by using high frequency structure simulator (HFSS). The GA is very precise and fast when compared to other techniques because it encodes the parameters and the optimization is done with the encoded parameters. Genetic algorithm is used to optimize the parameters. In the result section it has been exhibited. Patch antenna is suitable for mobile phone handsets moderately and accordingly has been developed. The antenna developed in this work has been confirmed to have better characteristics in terms of broad bandwidth and multiband operation.

References

1. Satpathy SK, Srinivasan V, Ray KP, Kumar G (1998) Compact microstrip antennas for personal mobile communication. In: IEEE region 10 international conference on global connectivity in energy, computer, communication and control, vol 2. pp 245–248
2. Wong BF, Lo YT (1984) Microstrip antenna for dual frequency operations. IEEE Trans Antenna Propag 32:938–943
3. Egashira S, Nishiyama E, Sakitani A (1990) Stacked microstrip antenna with wide band and high gain. In: Proceedings IEEE antennas and propagation society international symposium, vol 3. pp 1132–1135
4. Tiwari R, Singh SP, Yadav R, Yadav PK, Rao VK (2014) Dual band T slot rectangular microstrip patch antenna. Int J Emerg Technol Adv Eng 4(5):50–54
5. Sullivan PS, Schaubert DH (1986) Analysis of an aperture coupled microstrip antenna. IEEE Trans Antennas Propag 34:977–984
6. Himdi M, Daniel JP, Terret C (1989) Analysis of aperture-coupled microstrip antenna using the cavity method. Electron Lett 25:91–92
7. Balanis CA (1997) Antenna theory analysis and design. Wiley, New York
8. Rappaport TS (2009) Wireless communications: priciples and practice. Pearson Education
9. Garg R, Bharti P, Bahl I (2000) Microstrip antenna design handbook. Artech House, Boston
10. Milligan T (1985) Modern antenna design. McGraw Hill, New York
11. Herscovici N, Osorio MF, Peixeiro C (2002) Miniaturization of rectangular microstrip patches using genetic algorithms. IEEE Antennas Wirel Propag Lett 1:94–97
12. Vashist S, Soni MK, Singhal PK (2013) Genetic approach in patch antenna design. Int J Emerg Sci Eng 1(9)
13. Haupt RL, Warner D (2007) Genetic algorithms in electromagnetics. Wiley, New Jersey
14. Mohanty MN, Routray A, Kabisatpathy P (2010) Optimisation of features using evolutionary algorithm for EEG signal classification. Int J Comput Vision Robot 1(3):297–310
15. Errifi H, Baghdad A, Badri A (2014) Design and optimization of aperture coupled microstrip patch antenna using genetic algorithm. Int J Innovative Res Sci 3(5):12687–12694. ISSN:2319-8753

AADHAR License Management (ALM) System in India Using AADHAR IDs

Lalin L. Laudis, P.D. Saravanan, S. Anand and Amit Kumar Sinha

Abstract Number of accident count reported elevates with respect to the number of vehicles on road. Several reasons are being spotted as responsible for road accidents. One of the major causes is because of non-licensees on road (a driver without a valid license). Though, several measures are taken by the concerned departments, it is complicated to manage the non-licensees on road. Moreover, another meticulous cause for accidents on road is violation of driving rules (helmet, seatbelt, etc.). This research work concentrates on solving these two problems that would perhaps reduce the number of accidents to a considerable amount. The proposed AADHAR license management (ALM) system employs AADHAR ID which is implemented in INDIA as an identification card for every individual. In addition, this work proposes a new version of AADHAR ID which embeds the license within it. Furthermore, the management system employs iris scanning. Hence, the identity of the driver is ensured for perfection. Since, the AADHAR ID is embedded with a memory unit and unique identification number (UIN), it would store the iris pattern of the card holder and without its verification with the actual driver, the cardholder cannot drive the vehicle. This system is believed to reduce the non-licensees on road and would ensure the drivers to adopt the safety precautionary measure that has to be taken during driving.

Keywords License management · AADHAR ID · Iris processing · Biometrics · Pattern recognition · ALM

L.L. Laudis (✉) · P.D. Saravanan · A.K. Sinha
School of Electrical, Vel Tech University, Chennai, India
e-mail: lalin.vlsi13@veltechuniv.edu.in

P.D. Saravanan
e-mail: saravanan.est13@veltechuniv.edu.in

A.K. Sinha
e-mail: amitkumarsinha@veltechuniv.edu.in

S. Anand
Department of Electronics and Communication,
V.V. College of Engineering, Tirunelveli, India
e-mail: anands@vvcoe.org

L.P. Suresh and B.K. Panigrahi (eds.), *Proceedings of the International Conference on Soft Computing Systems*, Advances in Intelligent Systems and Computing 397, DOI 10.1007/978-81-322-2671-0_4

1 Introduction

The growth in transportation and the increased vehicles on road are a sort of threat on this era. Indeed, the number of road accidents is directly proportional to the number of vehicles on road. There may be a number of reasons behind every road accident. But, one of the causes for road accidents is the unlicensed drivers on road. Furthermore, another scary cause on road accidents may be proclaimed as the negligence of regulations. An United States study reveals that 20 %, i.e., about one in every five of all fatal crashes in U.S. involve drivers who should never have been on road due to the fact that they were unlicensed [1, 2]. In the same token, a 2002 study found that in the territory of Ontario, data from 1996–2003 determines that approximately 2000 fatal and injury crashes occur each year involving unlicensed drivers (Malefant et al. 2002) [2]. Moreover, a 2003 U.K. report submitted by the Department of Transport found that there were around a million unlicensed drivers on U.K roads [1, 2]. It has been reported in [3, 2], that 18 % of drivers in fatal crashes in the year 2007–2009 were unlicensed. Also in the year 1993–1997, 13.8 % of fatal crashes were by unlicensed drivers [2]. Thus the rate of accidents caused by unlicensed drivers on road is found to be increasing as days go by found that there is a very low awareness on penalty for unlicensed driving among those offenders who admitted driving.

A very similar issue to the unlicensed drivers is that the drivers who disobey the regulations of traffic. For two wheeler users, 90 countries which represent 77 % of the world population have comprehensive helmet law [3]. It has been reported that wearing a helmet would reduce death on accidents by 40 % and the serious injuries by 70 % [3]. With reference to the world health organization (WHO) report on safety 2013, the number of drivers who wear helmet range from 10 % in Ghana and Jamaica to almost 100 % in Netherlands and Switzerland [4].

As far as a four wheeler driving is concerned, wearing seatbelts reduces the risk of fatal injury by 40–50 % for drivers and front seat occupants, and between 25 and 75 % for rear seat occupants [2, 4]. Also, researchers are in progress with methodologies that may reduce the hazards for the rear seat occupants through implementation of comprehensive seat belt amendments. 111 countries that represent 69 % of the world's population now do have comprehensive seat belt laws covering all occupants [4]. Finally, another scaring issue is that drivers who are not eligible to hold a license drive the vehicle. This may include suspended or revoked licenses else expired or canceled or denied licenses. The verification of drivers on road individually is not an effortless process, rather on scanning the globalization process; it is quite a hard-hitting process. In observance with all the above-mentioned scenarios, it can be realized that there is an urge in need of a system which prohibits,

a. A driver without a valid license on road.
b. A driver from driving when the precautionary deeds are neglected.

For a perfect verification of the person and license, the so far adopted methodologies does not works effectively as they are not secure and forgery can be possibly done. Hence sophisticated methods viz, biometric scanning has to be followed up for verification. The biological verification may include fingerprint scanning, iris scanning, finger vein scanning etc., within which the most suited biometric verification techniques must be identified. Moreover, the traditional licenses, which do not have any biometric identification facility has to be replaced by using AADHAR Ids (for INDIA).

The requirements of the above-mentioned points are framed into a system. This paper work aims in design of a system that satisfies the constrains (a) and (b). The existing systems to ban unlicensed drivers taking the roads are discussed in Sect. 2, and their shortcomings are formulated (For two wheelers and for four wheelers). Section 3, briefs the proposed methodology for designing the system and how it overcomes the shortcomings of the existing system is reviewed. Section 4 identifies a sophisticated biometric methodology for verification. Section 5 proposes a new AADHAR ID which could be introduced in future to manage licenses in INDIA.

2 Existing Methodologies

The process of monitoring the non-licensees on road is very complicated, that too in a densely populated country as India, it is nearly impossible for perfect monitoring. However, certain methodologies are being adopted to reduce the ride of non-licensees on road. A traditional method that involves a random check of the vehicles on road. This completely relay on man power. Traffic policemen are employed for license verification and monitoring the vehicles to follow the traffic regulations. This is the most commonly used technique in almost several developing countries. The authorities randomly check the congested spots and verify the vehicles. However, this method has its own drawbacks:

a. A driver can find an alternative way from the verification spot.
b. The authorities may be lenient toward the known personalities.
c. An emergency situation of the drivers cannot be understood by the officials and hence it results delay in time for the sense of duty.
d. There is no assurance that the penalty collected is handed over to the government.

The (d) is a more sensitive national issue in developing countries like India.

Another approach of license management which was proposed in [5], employs fingerprint usage. This method is effective to an extent. The equipment is embedding with a smart card reader where the fingerprint of the driver is embedded. When a driver needs to ride the vehicle, the smart card must be inserted in the card reader, and the fingerprint of the driver is verified by a fingerprint scanner. When both the fingerprints match, the vehicle can be operated, else the ignition is closed. However, the proposed system in [5] have significant drawbacks.

(a) The license holder can ignite the vehicle and allow the nonlicensee to drive the vehicle.
(b) When a small injury in the thumb occurs, the vehicle cannot be driven.
(c) Chances are there that the fingerprint patter fades due to high mechanical work in hands.
(d) It also failed to ensure the driver to be a valid license holder.

A similar methodology, replacing fingerprint technology with finger vein technology was proposed by Divya et al. [2, 6]. However, the drawbacks continued as in [5].

3 Proposed Design

With consideration of the problems mentioned in Sect. 1 and the existing technologies in Sect. 2, there is an immediate need for a system that manages the license using a biometric approach. The proposed design of license management overcomes all the drawbacks of the previously proposed works. The system can be employed for two wheelers and four wheelers as well. The highlights of the proposed system are

(a) The traditional license is embedded in AADHAR Ids.
(b) Unauthorized usage of vehicles is stopped.
(c) Non-licensees on road could be eradicated.
(d) Safety precautions are ensured properly.

The system employs a hardware and software module which works simultaneously for perfection. The system completely works based on iris scanning of the driver and AADHAR ID. The entire operation of the system can be summarized as, initially the device checks for the presence of key in the vehicle. When it is, then the device scanner for the AADHAR ID switches itself ON. The used is directed to insert the AADHAR ID in which the iris pattern of the driver is embedded. Moreover, the iris pattern of the driver is obtained on the spot by the camera in the helmet (for two wheelers) and by the front camera (for four wheelers). When the system identifies the two iris pattern to be the same, the system ensures the validation of the identity card. Once all these confirmations are done, then the ignition could be initiated. Else, the system shuts down the ignition of the vehicle. This process is repeated for every 60 s, since after the ignition there may be a chance of handling the vehicle by a non-licensee. Hence to ensure the proper identity of the driver, the system continuously verifies the identification for every 60 s until the ignition if OFF.

Since the AADHAR ID is employed with unique identification number (UIN), the license holder could be easily traced by the department easily.

The block diagram of the proposed system is figured in Fig. 1.

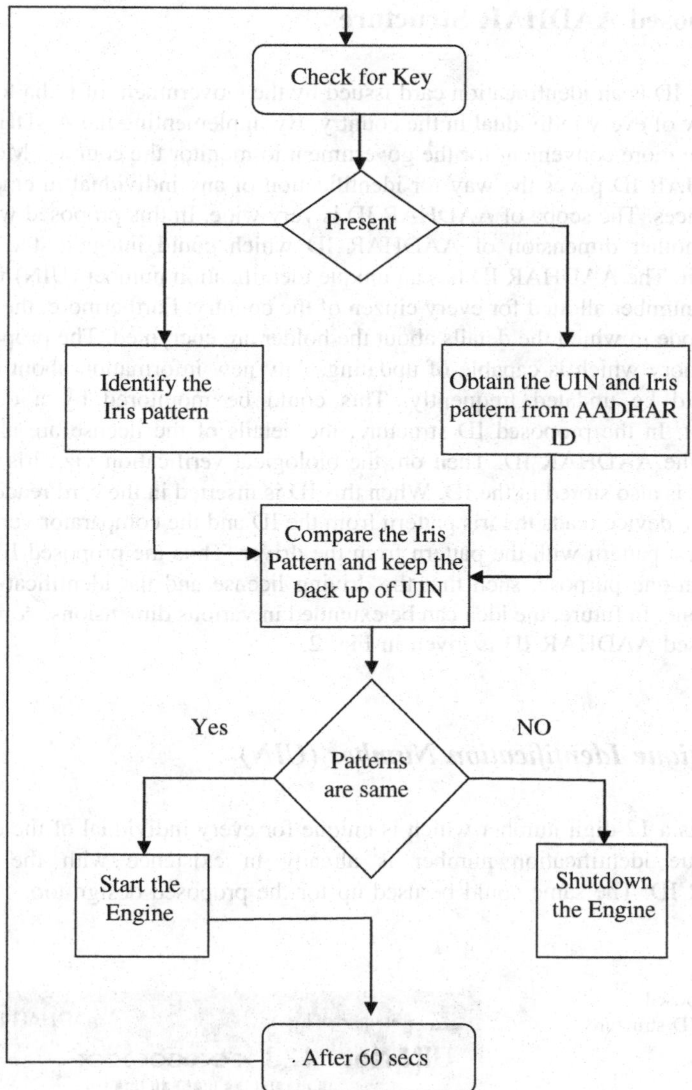

Fig. 1 Block diagram of proposed system

Another interesting concept of the proposed design is that the AADHAR ID has a memory element which could store the details of the driven vehicle. Thus, the proposed design employs iris pattern scanning and AADHAR ID's UIN and ensure the proper handling of the vehicle.

In addition, the card reader is equipped to identify whether the card is valid or not.

4 Proposed AADHAR Structure

AADHAR ID is an identification card issued by the Government of India to ensure the identity of every individual in the country. By implementing the AADHAR ID, it would be more convenient for the government to monitor the country. Moreover, the AADHAR ID paves the way for identification of any individual in emergency circumstances. The scope of AADHAR ID is very wide. In this proposed work, we suggest another dimension of AADHAR ID which could integrate the driving license in it. The AADHAR ID has an unique identification number (UIN) which is a 12-digit number allotted for every citizen of the country. Furthermore, the ID also has a barcode in which the details about the holder are encrypted. The proposed ID has a memory which is capable of updating. Any new information about the personal could be updated frequently. This could be monitored by a dedicated department. In the proposed ID structure, the details of the license are also integrated in the AADHAR ID. Then on, the biological verification viz., iris pattern, fingerprint is also stored in the ID. When this ID is inserted in the card reader in the vehicle, the device reads the iris pattern from the ID and the comparator verifies the obtained iris pattern with the pattern from the driver. Thus the proposed ID serves as a two-in-one purpose, such that the driving license and the identification card becomes one. In future, the idea can be extended in various dimensions. A model of the proposed AADHAR ID is given in Fig. 2.

4.1 Unique Identification Number (UIN)

The UIN is a 12-digit number which is unique for every individual of the country. The unique identification number is already in existence with the present AADHAR ID. The same could be used up for the proposed design too.

Fig. 2 Proposed AADHAR ID structure

4.2 Bar Code

A bar code is a unique way of identification. It may be a series of parallel lines with varying width and spacing and the barcode is an optical machine-readable representation of data relating to the object to which it is attached. Moreover, a bad core is a one-dimensional representation of data [7]. In the proposed AADHAR structure, a readable bar code is also employed to store data about the user.

4.3 Memory Unit

The memory unit of the proposed AADHAR ID is capable of storing considerable amount of data which include the biometric identification viz., fingerprint, iris pattern. Moreover, the memory unit is designed such that it is readable.

4.4 QR Code

It may also be termed as a two-dimensional bar code. A QR code can be expanded as Quick Response Code and is the trademark for a type of matrix barcode. A QR code uses four standardized encoding modes (numeric, alphanumeric, byte/binary, and kanji) to efficiently store data; extensions may also be used [8].

5 Biometric Iris Scanning

The biometric iris scanning-based identification is used in this proposed system. In the biometric method of identification and verification, human iris scanning is the foremost way of providing precise output. A human iris are identical in nature, it does not tie up with other iris even among identical twins in entire population.

In a self-directing technology, user recognition can be classified based on two classes, (I) identification, (II) verification. An actual process of verification means one-to-one juxtaposition. After scanning the individual/user iris, the information is yielded for authentication of the individual/user in the predetermined user record. And the comparison results ensures, whether the user is licensed or unlicensed. Identification is other way of authenticating the user. An individual/user eye image capture by the camera/sensor from some sort of distance and iris is focused to identify where area of light and dark fall. Eventually, the captured image is compared with predetermined user record. The flow diagram in the process of iris scanning is pictured in Fig. 3.

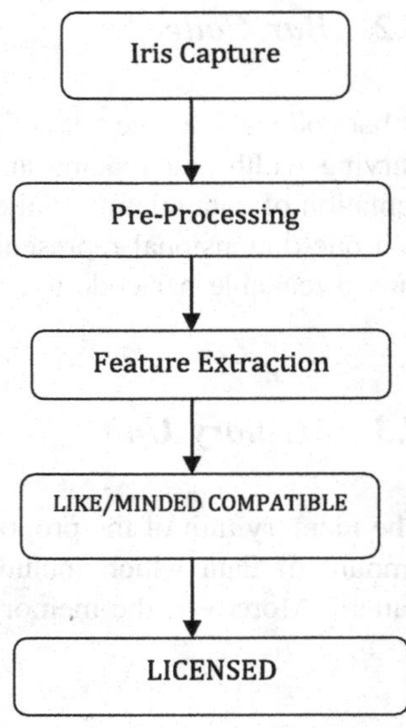

Fig. 3 Flow diagram for human iris scanning

5.1 *Iris Capturing*

A human iris biometric technology [9] deals with identifying a user iris pattern code extracted from the image of user eye. Human iris has unique pattern for the entire population. The iris characteristics are not affected even by using the eye lens. This obtained user iris pattern code is compared with the predetermined user record.

5.2 *Preprocessing*

In a preprocessing stage, the exact property of iris image is obtained with the peak and valley of the pixel. The image data will be improved and it suppresses unpropitious distortion for further processing. A pixel brightness transformations, geometric transformation, image restoration without changing the position of the original image.

5.3 *Feature Extraction*

An iris pattern code is build by demodulation as explained in [10]. The complex 2D-Gabor wavelets process is used to extract the original structure of the iris as sequence phasor.

6 Results and Discussions

The proposed model was simulated using verilog HDL code and implemented in Spartan 3E FPGA kit. The signals and the output for the model is tabulated as below:

From the Table 1, the signal K represent the presence of key in the equipment, SC represents the scanned output from the equipment (from helmet in case of two wheelers, and eyeglass in case of four wheelers). The signal AID is the signal from the AADHAR ID which holds the biometric iris pattern. CMP is the compared output by comparing the two signals SC and PR. The uniqueness of the system is that only when the three signals K, SC, and PR are logic 1, the system operates. Else, under any conditions if the system is in logic 0, i.e., the system shutsdown itself. Moreover, the above-mentioned flow diagram was coded using verilog HDL and simulated using Xilinx ISE simulator. The output of the simulator was verified based on the input and output combination mentioned in Table 1.

The output screen shots are shown in Figs. 4 and 5. The implementation of the proposed system in FPGA kit is pictured in Fig. 5, where it is implemented in Spartan 3E FPGA kit. The program was downloaded into the FPGA kit and the output logics were verified based on Table 1. It may be noted from Fig. 4 that the output from the iris comparator and smart card are virtually assumed and fed to the FPGA kit.

Table 1 Input and output combination

Si NO	K	SC	AID	CMP	Output
1.	1	1	1	1	1
2.	0	x	x	x	0
3.	1	x	x	0	0

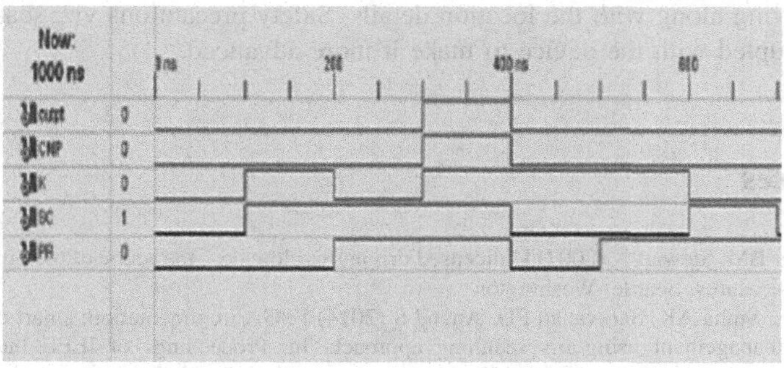

Fig. 4 Output window on simulation of the proposed model

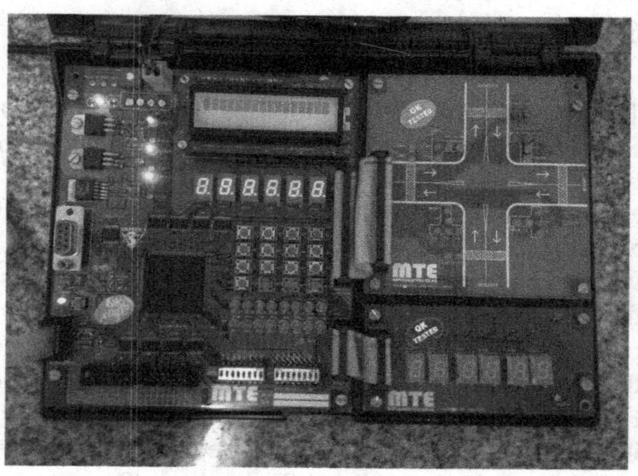

Fig. 5 FPGA implementation of the proposed model

7 Conclusion and Future Enhancements

The proposed system ensures that the vehicle can be driven only by authorized driver and that too with a valid driving license embedded in AADHAR ID. Hence on using the proposed system the rate of accidents can be considerably reduced and non-licensees on the road can also be eradicated. The proposed method can be extended by implementing theft control mechanisms, drowsiness detection of driver while driving, etc. Moreover, the AADHAR ID can also be extended by implementing tracking devices into it. Complex image processing algorithms can be used for processing the iris which could decrease the processing speed of the system. The proposed module can be interfaced with GPS/GSM module which could be of great use in near future. The combined module can be used to monitor from remote location about the vehicle. The data can be used to monitor about the person who is driving the vehicle, by this way theft can be minimized as it would help to find the person driving along with the location details. Safety precautions viz, seatbelt can also be coupled with the device to make it more advanced.

References

1. Sweedler BM, Stewart K (2007) Unlicensed driving worldwide—the scope of the problem and countermeasures. Seattle, Washington
2. Laudis L, Sinha AK, Saravanan PD, Anand S (2014) FPGA implmentation: smart card based license management using iris scanning approach. In: Proceedings of IEEE international conference on science, engineering & management research, pp 1–5
3. Griffin LI, DeLazerda S (2011) Unlicensed to to kill. AAA Foundation for Traffic Safety, pp 2–15
4. World Health Organization (2013) Global status report on road safety

5. Ashwin S, Loganathan S, Santosh Kumar S, Sivakumar P (2013) Prototype of a fingerprint based licensing system for driving. In: International conference on information communication and embedded systems (ICICES), pp 974–987
6. Divya D, Padmasarath S (2013) Finger vein based licensing and authentication scheme using GSM. IOSR J Comput Eng 15(3):30–35
7. http://en.wikipedia.org/wiki/Barcode
8. Denso-wave (2013) QR code features. Archived from the original on 20109-15. Accessed 3 Oct 2011
9. Iris Localization using Daughman's algorithm. http://www.bth.se/fou/cuppsats.nsf/
10. Daughmann J Mathematical explanation for iris recognation. http://www.cl.cam.ac.uk/~jgd1000/math.html

Selection of Anchor Nodes in Time of Arrival for Localization in Wireless Sensor Networks

K. Raghava Rao, T. Ravi Kumar and C. Venkatnaryana

Abstract Localization in wireless sensor network plays crucial role in emerging wireless communication applications. In this paper, the maximum lambda and minimum error (MaxL-MinE) localization technique is implemented with different configuration of nodes in MATLAB. It is found that the accuracy of position error is reduced as the number of anchor nodes increases. The outcome of this study would be useful for developing real-time wireless sensor network (WSN) localization models.

Keywords WSN · ToA · TDOA · RSS · DOA · MaxL-MinE

1 Introduction

A WSN is to observe environmental or physical conditions, such as light, sound, temperature, pressure, received signal strength, etc. and to helpfully transfer the data to a monitoring station through the network. Sensor position determination plays a major role in WSN and provides the location information of each senor. The sensor location can be estimated using either global positioning systems (GPS) or self-positioning WSN networks. Many location-based algorithms are developed by the authors [1]. The self-positioning methods based on the received signal strength (RSS), angle of arrival (AoA), time of arrival (ToA), and time difference arrival (TDOA) characteristics provide means for a low-cost and low power utilizing technique for position finding with good accuracy. The main challenge of WSN is

K. Raghava Rao (✉) · T. Ravi Kumar · C. Venkatnaryana
Department of ECM, KL University, Guntur, AP, India
e-mail: raghavarao@kluniversity.in

T. Ravi Kumar
e-mail: trkumar5@gmail.com

C. Venkatnaryana
e-mail: narayana7@gmail.com

L.P. Suresh and B.K. Panigrahi (eds.), *Proceedings of the International Conference on Soft Computing Systems*, Advances in Intelligent Systems and Computing 397, DOI 10.1007/978-81-322-2671-0_5

the position determination of unknown nodes with known node locations. The direction of arrival (DOA), received signal strength (RSS), time difference of arrival (TDOA), and time of arrival (ToA) techniques of the emitted signal are generally used to determine the source localization. The accuracy of ToA technique is high as compared to TDOA, RSS, and DOA methods [2]. In this paper, the ToA technique is implemented for unknown node selection using anchor nodes with different sensor placement scenarios. Good placement and selection of anchor nodes are needed for improving the positional accuracy. The estimation algorithms such as nonlinear least squares (NLS), maximum likelihood ratio (MLR), linear least squares (LLS), and weighted least squares (WLS) are tested under various scenario locations.

2 Source Localization Technique Based on ToA

ToA method is the one-way propagation time of the signal traversing between the anchor nodes and unknown nodes. This implies that synchronization of all sensors is essential to get the required information of ToA. The distance can be calculated by the product of ToA and light velocity [2]. Minimum three anchor nodes are required for position estimation. The typical anchor selection sensor configuration is given in Fig. 1.

Given the position of anchor nodes (x_i, y_i), the unknown node position (x, y), and the estimated distance d_i between the known or anchor nodes and the unknown node, the estimated position $(\hat{x}, \hat{y})$ is given by

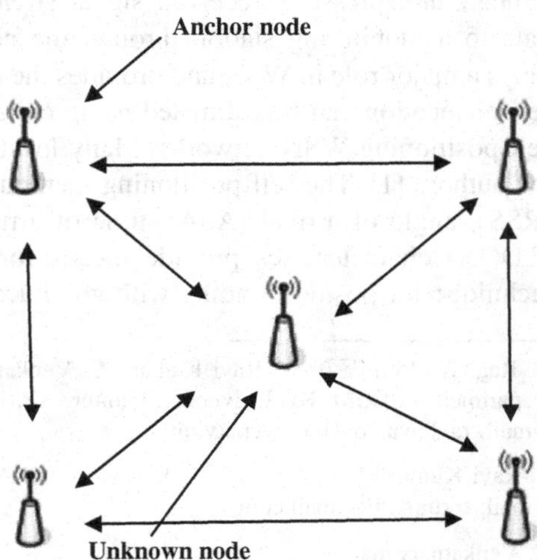

Fig. 1 Typical sensor configuration

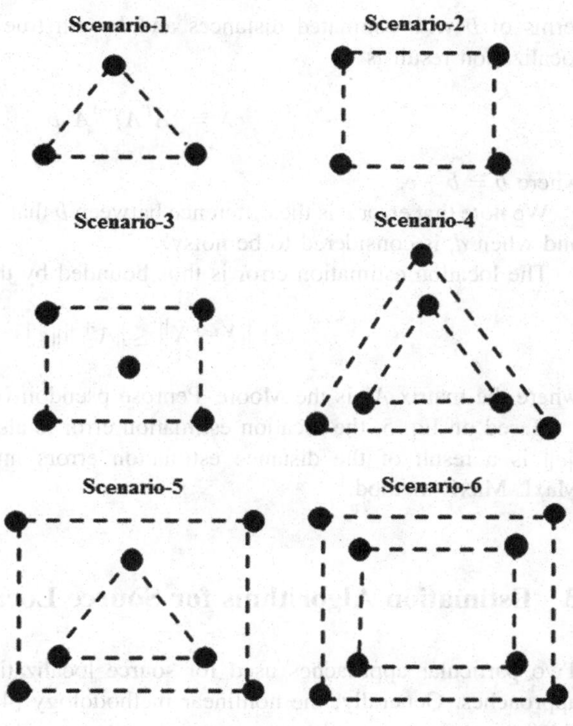

Fig. 2 Patterns for optimal anchor deployment

$$(\hat{x}, \hat{y}) = \underset{x,\, y}{\arg\min} \sum_{i=1}^{N} [\mathrm{sqrt}((x_i - x)^2 + (y_i - y)^2) - d_i]^2 \tag{1}$$

where N is the number of anchor nodes. The above nonlinear least squares (NLS) equation can be linearized by introducing a geometric constraint in the formulation of the equation. Starting with the $N \geq 2$ equations and subtracting the constraint (Fig. 2),

$$\frac{1}{N}\sum_{i=1}^{N}\left[(x_i - x)^2 + (y_i - y)^2\right] = \frac{1}{N}\sum_{i=1}^{N} d_i^2 \tag{2}$$

The objective is to minimize the location estimation error introduced by LLS. For an ideal case $x = [x, y]^T$ corresponds to the following.

$$X = (A^T A)^{-1} A^T b \tag{3}$$

However, the estimate of the distances could be influenced by noise, bias, and measurement error. Thereby expressing the resulting distance estimation error e in terms of $\tilde{b}$ with estimated distances and b with true distances as $\tilde{b} = b + e$, the localization result is

$$\tilde{X} = (A^T A)^{-1} A^T \tilde{b} \tag{4}$$

where $\tilde{b} = b + e$.

We note that error e is the difference between b that is obtained when d_i is perfect and when d_i is considered to be noisy.

The location estimation error is thus bounded by the following:

$$\|X - \tilde{X}\| \leq \|A^+\| \|e\| \tag{5}$$

where the matrix A^+ is the Moore–Penrose pseudoinverse of the matrix A.

Based on Eq. 5, the location estimation error is also effected by $\|e\|$. The term $\|e\|$ is a result of the distance estimation errors introduced by ranging in the MaxL-MaxE method.

3 Estimation Algorithms for Source Localization

Two particular approaches used for source localization are nonlinear and linear approaches. Generally, the nonlinear methodology [4] directly employs Equation $r = f(x) + n$, where r is the measurement vector, x is the unknown source position to be determined, $f(x)$ is a known nonlinear function in x, and n is an additive zero-mean noise vector, to solve for x by minimizing the least squares (LS) or the weighted least squares (WLS) cost function constructed from the following error function

$$E_{\text{nonlinear}} = r - f(\tilde{X}), \tag{6}$$

where $\tilde{X} = [\tilde{x}\tilde{y}]$ is the optimization variable for x. Meanwhile, the linear techniques convert the nonlinear into a set of linear equations in x

$$b = AX + q \tag{7}$$

where q is the transformed noise vector. From Eq. 7, we put up

$$e_{\text{linear}} = b - A\tilde{X} \tag{8}$$

LLS [5], WLLS [6], and subspace [7] estimators results can be obtained by applying the techniques LS or WLS on Eq. 8.

4 Results and Discussion

CRLB, ML, NLS, LLS and WLS, WLS2, and subspace localization estimation algorithms are developed in MATLAB simulations. Six configurations are considered for the analysis. Anchor node locations are considered in 2D coordinate systems. Distance between anchor nodes are calculated. ToA information is generated using random MATLAB function. Linear (WLS, WLS2, LLS, subspace algorithm) and nonlinear (ML and NLS) estimations are carried out. The mean square error is computed and evaluated for several configurations.

4.1 Scenario 1

Input anchors ([0,0], [10,0], and [5,10]) are given as input to MaxL-Min E algorithms. The target position is [2, 3]. The MSE values are calculated. Figure 3 shows the mean square position error with respect to SNR values for three anchor nodes. MSE error varies from 25 to −40 dB for the CRLB technique. It is evident that MSE decreases as SNR increases. NLS and ML estimation algorithms closely follow each other for 40 dB SNR values above. The lower SNR values of ML estimation algorithm give lesser MSE error. Figure 4 shows that the estimation algorithms closely follow each other for 40 dB SNR values above. The lower SNR values of the two-step WLS estimation algorithm give less MSE error.

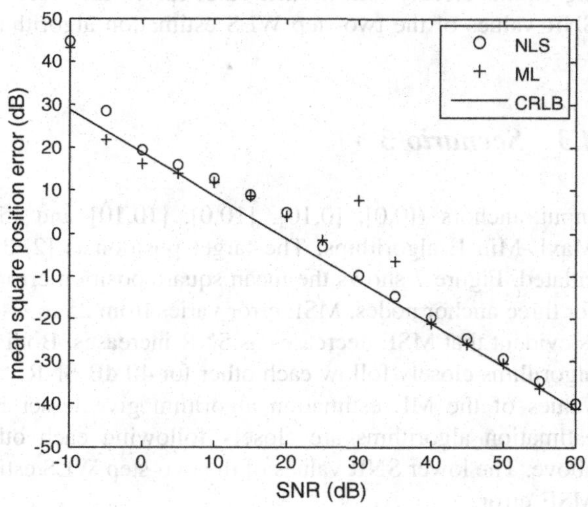

Fig. 3 MSE comparison of NLS, ML, and CRLB with respect to SNR for scenario 1

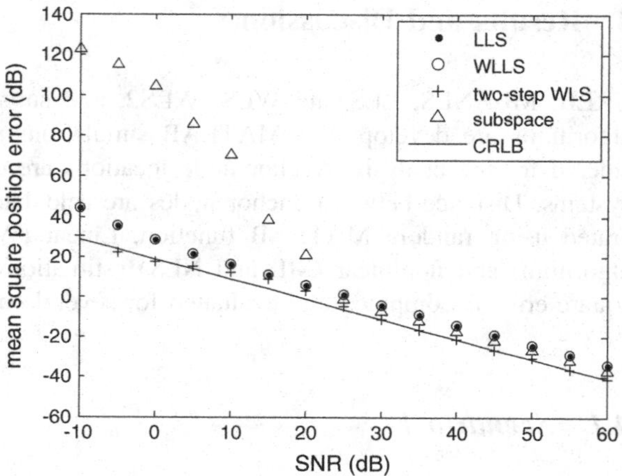

Fig. 4 MSE comparison of LLS, WLLS, WLS2, subspace, and CRLB with respect to SNR for scenario 1

4.2 *Scenario 2*

Input anchors ([0,0], [0,10], [10,0], and [10,10]) are given as input to MaxL-Min E algorithms. The target position is [2, 3]. The MSE values are calculated. Figure 5 shows the mean square position error with respect to SNR values for three anchor nodes. MSE error varies from 25 to −40 dB for CRLB technique. It is evident that MSE decreases as SNR increases. The NLS and ML estimation algorithms closely follow each other for 40 dB SNR values and above. The lower SNR values of the ML estimation algorithm give lesser MSE error. Figure 6 shows the estimation algorithms closely follow each other for 40 dB SNR values and above. The lower SNR values of the two-step WLS estimation algorithm give lesser MSE error.

4.3 *Scenario 3*

Input anchors ([0,0], [0,10], [10,0], [10,10] and [5,5]) are given as input to MaxL-Min E algorithms. The target position is [2, 3]. The MSE values are calculated. Figure 7 shows the mean square position error with respect to SNR values for three anchor nodes. MSE error varies from 25 to −40 dB for CRLB technique. It is evident that MSE decreases as SNR increases. Both the NLS and ML estimation algorithms closely follow each other for 40 dB SNR values above. The lower SNR values of the ML estimation algorithm give lesser MSE error. Figure 8 shows estimation algorithms are closely following each other for 40 dB SNR values above. The lower SNR values of the two-step WLS estimation algorithm give lesser MSE error.

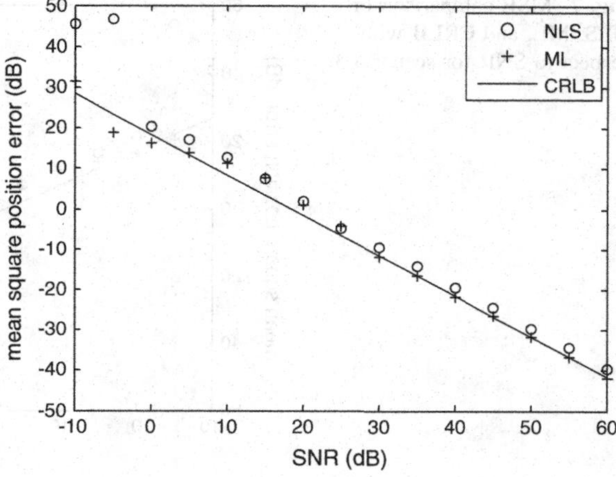

Fig. 5 MSE Comparison of NLS, ML, and CRLB with respect to SNR for scenario 2

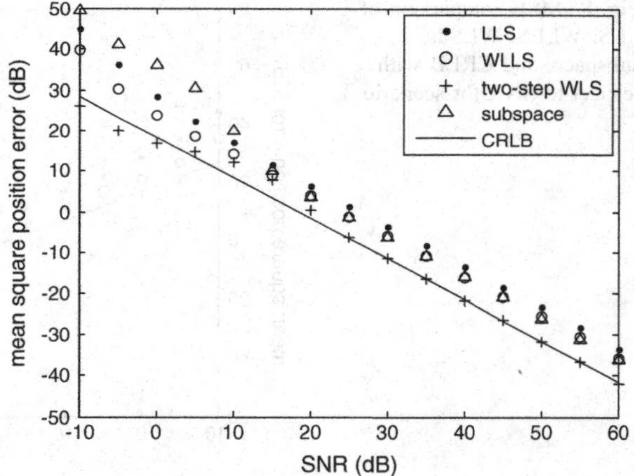

Fig. 6 MSE comparison of LLS, WLLS, WLS2, subspace, and CRLB with respect to SNR for scenario 2

4.4 Scenario 4

Input anchors ([0,0], [10,0], [5,10], [2.5,2.5], [7.5,2.5], and [5,7.5]) are given as input to MaxL-Min E algorithm. The target position is [2, 3]. The MSE values are calculated. Figure 9 shows the mean square position error with respect to SNR values for three anchor nodes. MSE error varies from 25 to −40 dB for the CRLB technique. It is evident that MSE decreases as SNR increases. NLS and ML estimation algorithms closely follow each other to CRLB for 40 dB SNR values above as well as for the lower SNR values. Figure 10 shows the estimation algorithms are closely following each other for 40 dB SNR values above. The lower SNR values of the two-step WLS estimation algorithm give lesser MSE error.

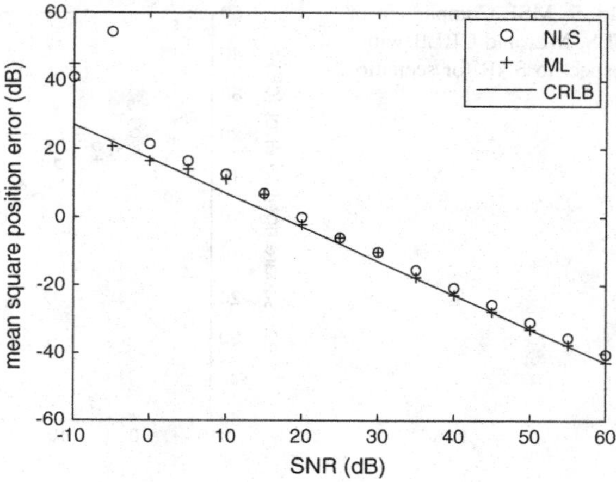

Fig. 7 MSE comparison of NLS, ML, and CRLB with respect to SNR for scenario 3

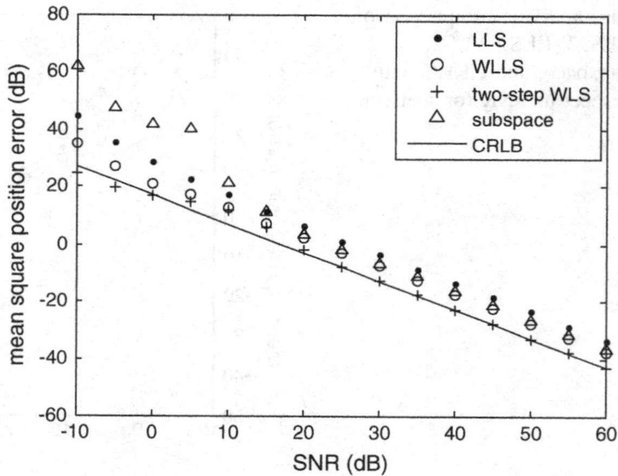

Fig. 8 MSE comparison of LLS, WLLS, WLS2, subspace, and CRLB with respect to SNR for scenario 3

4.5 *Scenario 5*

Input anchors ([0,0], [10,0], [0,10], [10,10], [2.5,2.5], [7.5,7.5] and [5,7.5]) are given as input to MaxL-Min E algorithm. The target position is [2, 3]. The MSE values are calculated. Figure 11 shows the mean square position error with respect to SNR values for three anchor nodes. The MSE error varies from 25 to −40 dB for the CRLB technique. It is evident that MSE decreases as SNR increases. NLS and ML estimation algorithms are closely following each other to CRLB for 40 dB SNR values above. The lower SNR values of the ML estimation algorithm follow the

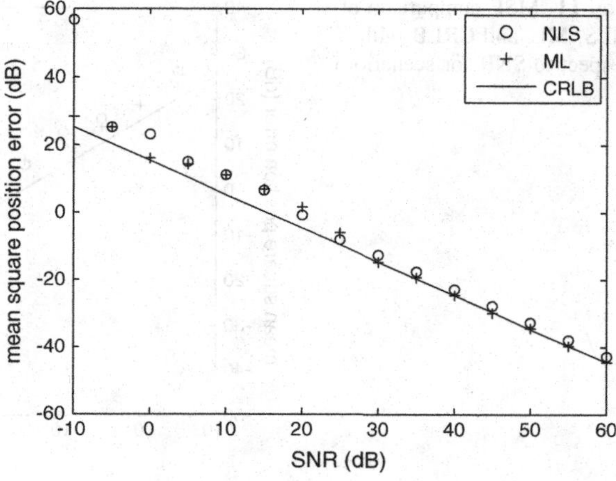

Fig. 9 MSE comparison of NLS, ML, and CRLB with respect to SNR for scenario 4

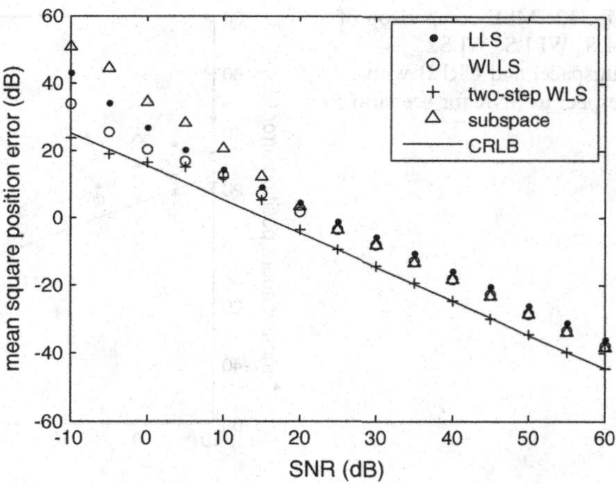

Fig. 10 MSE comparison of LLS, WLLS, WLS2, subspace, and CRLB with respect to SNR for scenario 4

CRLB. Figure 12 shows the estimation algorithms are closely following each other for 40 dB SNR values above. The lower SNR values of the two-step WLS estimation algorithm give lesser MSE error.

4.6 Scenario 6

Input anchors ([0,0], [10,0], [0,10], [10,10], [2.5,2.5], [7.5,2.5], [2.5,7.5] and [7.5,7.5]) are given as input to MaxL-Min E algorithm. The target position is [2, 3]. The MSE values are calculated. Figure 13 shows the mean square position error with respect to SNR values for three anchor nodes. MSE error varies from 25 to

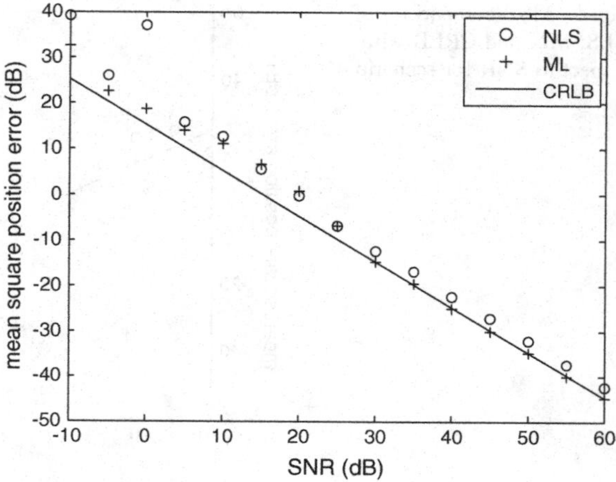

Fig. 11 MSE comparison of NLS, ML, and CRLB with respect to SNR for scenario 5

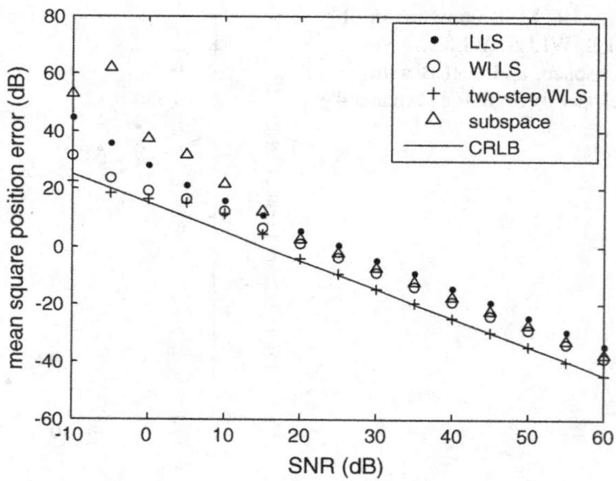

Fig. 12 MSE comparison of LLS, WLLS, WLS2, subspace, and CRLB with respect to SNR for scenario 5

−40 dB for the CRLB technique. It is evident that MSE decreases as SNR increases. NLS and ML estimation algorithms are closely following each other to CRLB for 40 dB SNR values above. Figure 14 shows that the estimation algorithms are closely following each other for 40 dB SNR values above. The lower SNR values of two-step WLS estimation algorithm give lesser MSE error. Figure 15 shows the comparison of MSE for NLS, ML, and CRLB algorithms with respect to the number of receivers. Minimum three anchor nodes are needed to find the location of an unknown node. Figure 16 shows the comparison of MSE for NLS, ML, and

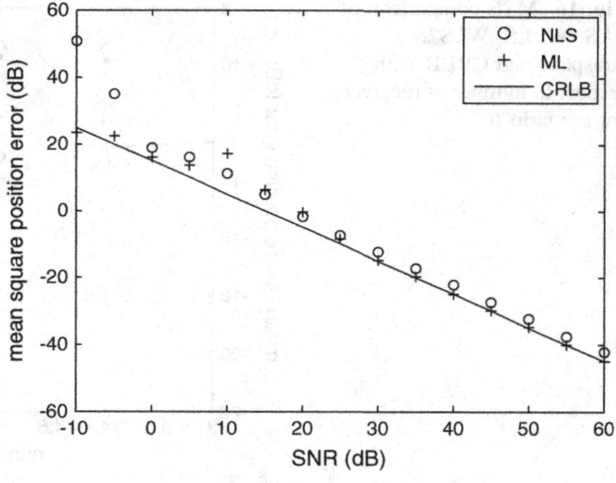

Fig. 13 MMSE comparison of NLS, ML, and CRLB with respect to SNR for scenario 6

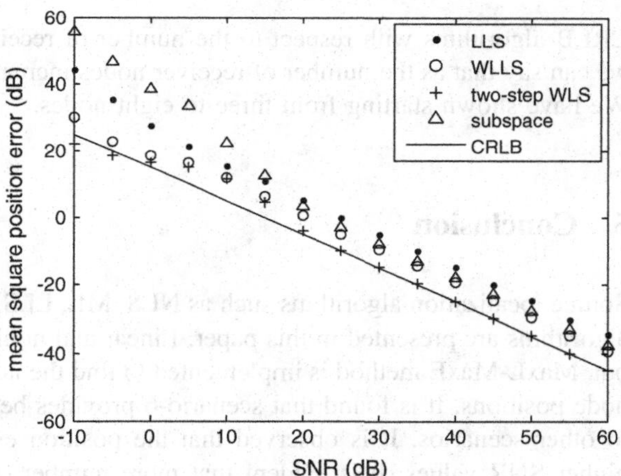

Fig. 14 MSE comparison of LLS, WLLS, WLS2, subspace, and CRLB with respect to SNR for scenario 6

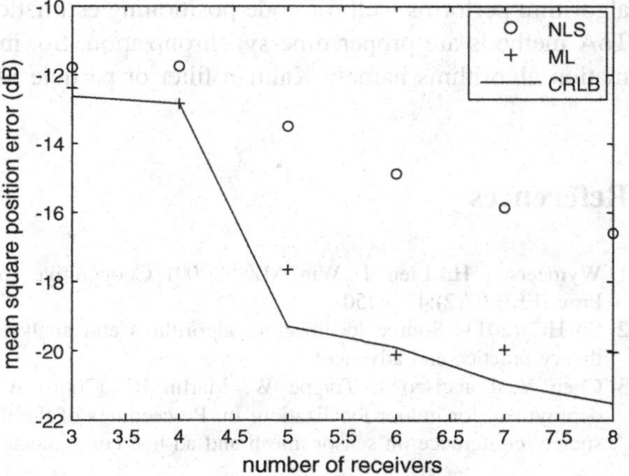

Fig. 15 MSE comparison of NLS, ML, and CRLB with respect to number of recievers for scenario 6

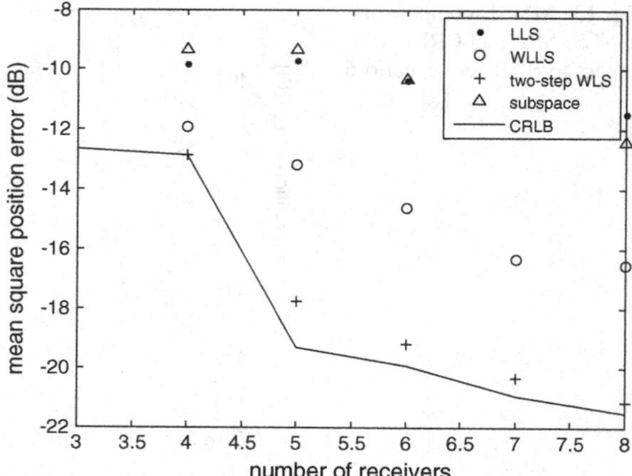

Fig. 16 MSE comparison of LLS, WLLS, WLS2, subspace and CRLB with respect to number of receivers for scenario 6

CRLB algorithms with respect to the number of receivers. From Figs. 15 and 16, we can say that as the number of receiver nodes increases the MSE error decreases. We have shown starting from three to eight nodes.

5 Conclusion

Source localization algorithms such as NLS, ML, LLS, WLS, WLS2, and subspace algorithms are presented in this paper. Linear and nonlinear estimations are carried out. MaxL-MaxE method is implemented to find the node position using the anchor node positions. It is found that scenario-6 provides better positioning as compared to other scenarios. It is observed that the position estimation error increases for higher SNR values. It is evident that more number of receivers provide accurate position estimations. Among all other estimation algorithms, the WLS2 estimation algorithm performs well for node positioning estimation. The main disadvantage of ToA methods are proper time synchronization. So, in future, recursive-based estimation algorithms namely Kalman filter or particle filter will be implemented.

References

1. Wymeersch H, Lien J, Win MZ (2009) Cooperative localization in wireless networks. Proc IEEE 97(2):427–450
2. So HC (2011) Source localization: algorithms and analysis. Handbook of position location theory practice and advances
3. Chen Y, Francisco J, Trappe W, Martin RP (2006) A practical approach to landmark deployment for indoor localization. In: Proceedings of the third annual IEEE communications society conference on sensor, mesh and ad hoc communications and networks (SECON)

4. Torrieri DJ (1984) Statistical theory of passive location systems. IEEE Trans Aerosp Electron Syst 20(2):183–198
5. Chen JC, Hudson RE, Yao K (2002) Maximum—likelihood source localization and unknown sensor location estimation for wideband signals in the near fi eld. IEEE Trans Signal Process 50 (8):1843–1854
6. Cheung KW, So HC, Ma W-K, Chan YT (2004) Least squares algorithms for time—of—arrival based mobile location. IEEE Trans Signal Process 52(4):1121–1128
7. So HC, Chan KW (2007) A generalized subspace approach for mobile positioning with time—of—arrival measurements. IEEE Trans Signal Process 55(10):5103–5107

Predicting the Risk of Diabetes Mellitus to Subpopulations Using Association Rule Mining

Murari Devakannan Kamalesh, K. Hema Prasanna, B. Bharathi, R. Dhanalakshmi and R. Aroul Canessane

Abstract Diabetes is one of the major concerns for majority of the population. Detecting the risk of diabetes earlier in patients is essential for taking appropriate treatment at the right time. Using data from electronic medical records, risk prediction can be done with the help of association rule mining. Association rule mining generates some sets of rules which will be useful for risk prediction in subpopulations. Based on the analysis the amount of risk is estimated, and hence appropriate treatment can be done. In order to summarize the rules, four summarization techniques were analyzed and a comparison was made between them. All the four methods showed appropriate results but the last method called bottom-up summarization (BUS) was most suitable and used for accurate results. BUS technique showed the subpopulations with high risk of diabetes. A detailed analysis of the bottom-up summarization produced results with accuracy.

Keywords Diabetes · Subpopulations · Association rule mining · Summarization techniques

---

M.D. Kamalesh (✉) · K.H. Prasanna · B. Bharathi · R. Aroul Canessane
Faculty of Computing, Sathyabama University, Chennai, India
e-mail: kamal2gd@gmail.com

K.H. Prasanna
e-mail: hema.prassu30@gmail.com

B. Bharathi
e-mail: bharathivaradhu@gmail.com

R. Aroul Canessane
e-mail: aroulcanessane@gmail.com

R. Dhanalakshmi
Department of Computer Science and Engineering, Jeppiaar Engineering College, Chennai, India
e-mail: dhanajovi@gmail.com

L.P. Suresh and B.K. Panigrahi (eds.), *Proceedings of the International Conference on Soft Computing Systems*, Advances in Intelligent Systems and Computing 397, DOI 10.1007/978-81-322-2671-0_6

1 Introduction

Diabetes is an ineradicable disease that is affecting millions of people in the world. It is a chronic disorder that affects the amount of sugar present in blood. If a body does not fabricate enough insulin or is not able to use its own insulin efficiently then it is said to be affected with diabetes. Insulin is a hormone liberated from the pancreas to control the amount of glucose in blood. About 8.3 % of the populations have diabetes and it is the seventh leading cause of death in United States. Diabetes leads to many complications, such as, high cholesterol and blood pressure that increases the risk of heart strokes. This high blood sugar produces the symptoms like increased thirst, frequent urination, and increased hunger. Diabetes includes a group of diseases namely hypertension, hyperlipidemia, etc. It is also death causing disease.

Due to the desperate need to recognize patients with elevated risk of diabetes well in advance, innumerable diabetes risk scores have been developed. Subpopulations with high risk scores are served more dynamically. These scores only furnish a calibration of the risk. They employ distinct factors without considering the collaborations between them. Certain risk factors are considered to assess the risk of diabetes in patients. These can be: age, medications, and comorbidities which are commonly available in electronic medical records. As huge set of risk factors are considered, large number of rules will be generated which complicates the risk prediction. To prevail over this, a number of summarization techniques were proposed (Table 1).

2 Related Works

Agarwal and Srikant [1] have discovered algorithms for mining association rules. He proposed two algorithms for considering the issue of ascertaining association rules between items in a huge database. The features of the two proposed algorithms (Apriori and AprioriTid) were combined into AprioriHybrid. AprioriHybrid has good maximizing possessions with regard to the transaction size and number of

Table 1 Factors considered for risk prediction [9]

Risk Factor	Description
Age	Age ≤ 10, 10 < Age ≤ 24, 24 < Age ≤ 45, Age > 45
Conditions	body mass index I ≥ 30, blood pressure ≥ 140 high-density lipoprotein < 40 for male and < 50 for female total cholesterol ≥ 200, triglyceride ≥ 150
Medication	Acearb, beta-blocker, calcium-channel blocker, diuretics, fibrates, statin, aspirin
Comorbities	Hypertension, current smoker, ischemic heart disease

items in the Database. Collins et al. [2] "Systematically reviewed and desperately predicted the management and coverage of methods used to flourish risk prediction models for anticipating the risk of having future risk of developing type 2 diabetes in adults."

Jin et al. [3] have proposed two new methods k-regression and tree regression to segregate the whole set of frequent item sets in order to reduce the restoration error. "The K-regression approach, employs a K-means type of clustering method, assures that the total restoration error achieves a local minimum. The tree-regression approach uses a decision-tree type of top-down partition process."

Xin et al. [4] have extracted redundancy-aware top-k patterns from a large set of frequent patterns. "Proposed a greedy algorithm which approximates an optimal solution with performance bound O(log k), where k is number of reported patterns." Srikant et al. [5], various fast algorithms for mining association rules have been developed. "They unified constraints that are Boolean expressions over the existence of items in the association algorithms. Combined three algorithms for mining association rules."

Wang et al. [6] have utilized probabilistic models for summarization. "Items are contemplated as random variables and Markov random field models are constructed on these variables based on frequent itemset patterns." Chandola et al. [7] have focused on summarization of a dataset of transactions with categorical results on optimization problems. "A probing algorithm was presented and different heuristics are investigated to generate a suitable summary for a set of transactions."

Kim et al. [8] have analyzed comorbities in patients with T2DM by using association rule mining (ARM). "The Dx Analyze Tool was proposed for cleaning data and data mart construction. The associations for comorbidities was revealed."

3 Methods and Procedures

3.1 Association Rule Mining

Association rule mining aims at finding interesting relations or associations between datasets in large databases. It is meant for identifying active rules disclosed in databases using distinct measures of attractiveness. ARM extracts frequent patterns among sets of data in databases. The most widely used example of ARM is market basket analysis. Let *X* be an indicator that shows whether a patient has a corresponding risk factor and denotes an item matrix. If *A* and *B* are two item sets, then the association rule is of the form *A* => *B* which implies that *B* is likely to apply to a patient given that *A* applies. Here, *A* is the antecedent and *B* is the consequent. The significance of the relation is indicated through support and confidence.

Support is the number of happenings of disease *A* and *B* from all the diseases.

$$\text{support}\,(AB) = \frac{\text{support}(AB)}{\text{no.of transactions in } D} \quad (1)$$

Confidence is the number of happenings of disease A from all the diseases.

$$\text{Confidence}\left({}^{A}\!/_{B}\right) = \frac{\text{Support}(AB)}{\text{Support}(A)} \quad (2)$$

In ARM, any item can appear in the antecedent of one rule and in the consequent of another.

3.2 Apriori Algorithm

Apriori is an algorithm used for finding frequent itemsets with candidate generation. It is a bottom-up approach which is performed level by level.

Candidate Generation: Let K be a set of frequent itemsets. The main idea behind apriori candidate generation is that if an itemset Y has minimum support, so do all the subsets of Y after all the candidates have been generated. Multiple scans of database is required to generate the candidate sets.

3.3 Rule Set and Database Summarization

The main aim of rule set summarization is to compress a ruleset 'L' to a smaller set 'A' such that L can be recovered from A with marginal loss of information. Dataset summarization aims to present the dataset 'D' into smaller set 'A' such that the original dataset D can be recovered from A with marginal loss.

There are number of techniques to solve these problems among which one method is sequential coverage.

Sequential Coverage algorithm:

Initially, a set ε of itemsets is procreated which forms the collation of itemsets from which the set of rules A are tabbed. μ is empty and is composed iteratively. In each individual iteration the rule E in ε which minimizes L is collected and added to μ. To avert selecting the same rule frequently, its influence is removed.

Algorithm:
Input: Set I of itemsets, number k of summary rules.
Output:Set μ of itemsets , s.t. μ minimizes the criterion L.
Generate an extended set ε of itemsets based on I
$\mu = \Phi$
While | μ | < k do
$A = \arg\ \min_{E \in \varepsilon} L(E)$
Add A to μ
Remove the effect of A
end while

The primary goal of summarization based on greedy set coverage is to find summarized set 'A' by greedily selecting the rulesets with minimum loss of information. A ruleset 'A' covers a rule 'I' if I is subset of A.

Thus, the similarity function among two sets A and I expressed as the relative patient coverage (RPC).

$$\mathrm{RPC}(A, I) = \frac{|D_A \cap D_I|}{|D_A \cup D_I|} \tag{3}$$

where, D_A represents the patients in D covered by rule A [9].

3.4 Summarization Techniques

The large set of rules generated by ARM can be summarized with the help of summarization techniques. In this section, we will discuss about four summarization techniques: APRX-Collection, RPGlobal, Top-K, and BUS [9].

a. *APRX-Collection*

This technique forms a superset of itemsets by adding one or two itemsets which are not present in the set. Then an itemset is selected from the superset which covers most of the itemsets from the actual set having a false positive rate less than a user defined parameter 'α'. The loss criterion for an itemset present in superset is:

$$\begin{cases} C_{\mathrm{aprx}}(E) = -|S_E|, & \text{if false positiverate} < \alpha \\ 0, & \text{otherwise} \end{cases} \tag{4}$$

where S_E represents set of items in actual itemsets covered by selected itemset [9].

The drawback of APRX-collection algorithm is that it does not consider redundancy in item set and there will be dilution of risk.

b. *RP-Global*

RP-Global overcomes the drawback of APRX-collection as it does not form a superset. It is similar to APRX-collection which addresses the two drawbacks by taking patient coverage into consideration. It forms the summarized itemsets from the actual itemset. Itemsets that differ from the itemset in patient coverage by not more than 1-δ. Where δ is a user specified parameter.

The loss criterion can be calculated as:

$$\begin{cases} C_{\mathrm{rpglobal}}(E) = -|S_E|, & \text{if } \forall I \in S_E,\ \mathrm{RPC}(E, I) > 1 - \delta \\ 0, & \text{otherwise} \end{cases} \tag{5}$$

where, I is item set and RPC is the RPC [9].

c. *Top-K*

This technique includes the concept of significance and redundancy. Significance corresponds to the risk and when multiple item sets cover the same item set then redundancy arises.

The redundancy of an item set *A* with respect to the other item set *I* is given by:

$$\text{redundancy}(A, I) = \text{RPC}(A, I)\min(y(A), y(I)) \tag{6}$$

where, min(*y*(*A*), *y*(*I*)) can be considered as part of risk possessed by the item set already selected [9].

When no item set is covered by both *A* and *I*, redundancy is '0' and if both are identical then *y*(*A*) = *y*(*I*).

The selection criterion for redundancy-aware top-*K* is:

$$C_{\text{topk}}(E) = \text{redundancy}(A, E). \tag{7}$$

Hence, (*A*, *E*) is maximum as *E* is already present in itemset [9].

d. *BUS*

Bottom-up summarization designates significance to the expression of the itemsets to the patient coverage information. If any of the transaction is not covered by the itemset then the transaction can be added to the summary. BUS gradually selects the itemsets *E* from the actual itemsets which increases the support and data coverage.

The selection [9] criterion:

$$C_{\text{bus}}(E) = -|D_E| - \text{DC}(E) \tag{8}$$

where *D* is the transaction that is added to the summary due to adequacy of the coverage and DC is the data coverage of the itemset '*E*' [9]. Of all the techniques BUS was optimal and showed suitable summaries. The selection criterion can be tested with the hybrid algorithm for optimal testing [10].

4 Conclusion

Diabetes is a chronic syndrome that is becoming crucial day-to-day. Association rule mining accompanied to a summarization technique renders a decisive tool for clinical environment. Using electronic data from EMR uncovers new knowledge. Four summarization techniques were analyzed and found that bottom-up summarization (BUS) produced optimal results. Untimely detection of diabetes mellitus makes human life buoyant. Further, risk prediction can be enhanced by using ARM and an approach called support vector machine (SVM). SVM mainly deals with classification of different patterns and helps to automatically guide new patients for the risk of diabetes through trained electronic medical records. With the help of SVM suitable medicines for new patients can be recommended based on the classified patient's data with high precision.

References

1. Agrawal R, Srikant R (1994) Fast algorithms for mining association rules. In: VLDB conference
2. Collins GS, Mallett S, Omar O, Yu L-M (2011) Developing risk prediction models for type 2 diabetes: a systematic review of methodology and reporting. BMC Medicine
3. Jin R, Abu-Ata M, Xiang Y, Ruan N (2008) Effective and efficient itemset pattern summarization: Regression- based approach. In: ACM international conference on knowledge discovery and data mining
4. Xin D, Cheng H, Yan X, Han J (2006) Extracting redundancy-aware top-k patterns. In: ACM international conference on knowledge discovery and data mining
5. Srikant R, Vu Q, Agrawal R (1997) Mining association rules with item constraints. In: American association for artificial intelligence
6. Wang C, Parthasarathy S, Summarizing itemset patterns using probabilistic models. In: ACM international conference on knowledge discovery and data mining
7. Chandola V, Kumar V (2006) Summarization-compressing data into an informative representation. Knowl Inf Syst
8. Kim HS, Shin AM, Kim MK, Kim YN (2012) Comorbidity study on type 2 diabetes mellitus using data mining. Korean J Intern Med 27
9. Simon GJ, Member, IEEE, Caraballo PJ, Therneau TM, Cha SS, Castro MR, Li PW (2014) Extending association rule summarization techniques to assess risk of diabetes mellitus. IEEE Trans Knowl Data Eng
10. Albert MJ, Ravi T (2015) Structural software testing: hybrid algorithm for optimal test sequence selection during regression testing. Int J Eng Technol 7(1)

Enhancement of Endoscopic Image Using TV-Image Decomposition

B. Jamlee Ludes and Suresh R. Norman

Abstract Endoscope is an instrument used to examine the interior of a hollow organ of a body. The endoscopes are inserted directly into the organ. The input images from the endoscopy camera are digital and are received at the computer USB port. This paper introduces a new approach for histogram equalization which involves the application of the Total Variation (TV) model to extract cartoon-texture component from the image. The undesired artifacts such as intensity saturation and over-enhancement are overcome by this approach. The experimental results show that the proposed method using a combination of TV Texture-Enhanced Histogram equalization (TE-HE) with Dark Channel Prior (DCP) strategy is an effective approach for contrast enhancement. Thus this paper shows the enhanced and original image at the same time so that doctors can clearly see the results and correctly predict diseases easily.

Keywords Contrast enhancement · TV image decomposition · Histogram · DCP

1 Introduction

The objective of image enhancement is to process a given image so that the result should contain more information than the original image. The purpose of image enhancement is to sharpen the image features. Image enhancement techniques help in improving the visibility of any portion or feature. The enhancement does not increase the inherent information content of the image data, but it can increase the dynamic range of the features so that they can be detected easily. Image enhancement means improving the contrast of the image and reducing noise. Image

B. Jamlee Ludes (✉) · S.R. Norman
ECE Department, SSN College of Engineering, Kalavakkam, Chennai, India
e-mail: jamleeludes@gmail.com

S.R. Norman
e-mail: sureshrnorman@ssn.edu.in

L.P. Suresh and B.K. Panigrahi (eds.), *Proceedings of the International Conference on Soft Computing Systems*, Advances in Intelligent Systems and Computing 397, DOI 10.1007/978-81-322-2671-0_7

enhancement methods can be based on either spatial or frequency domain techniques. Spatial domain techniques are performed to the image plane and are based on direct manipulation of pixels in an image. The frequency domain enhancement method can enhance an image by convoluting the image with a linear, position invariant operator. This convolution is performed in frequency domain with Discrete Fourier Transform (DFT) on every picture element. The Adaptive Histogram Equalization (AHE) is used to improve the contrast in images [1]. The disadvantages of this approach are it will amplify the noise and it fails to preserve the brightness with respect to the input image. Automatic contrast enhancement (ACE) by global histogram modification is a technique of image enhancement for visual inspection [2]. The disadvantage of this method is that it will over-enhance the image. The above disadvantages are overcome by the proposed method.

2 Overview of the Proposed Contrast Enhancement Algorithm

The proposed contrast enhancement approach involves a multi-step algorithm that applies the TV [3–7] model to obtain the cartoon-texture decomposition of the input image. An overview of the proposed contrast enhancement algorithm is depicted in Fig. 1. The input image captured from endoscopic camera is digital. In the preprocessing step the input image is passed through notch filter to remove 50 Hz noise from the image. A notch filter is a band-stop filter with a narrow stopband. They are more often used for noise reduction or elimination of undesirable artifacts in a signal.

Figure 1 shows that after the extraction of the cartoon-texture image components from the image, contrast enhancement is achieved by applying a nonlinear histogram warping process. Figure 2 shows the overall setup and components used to capture the image.

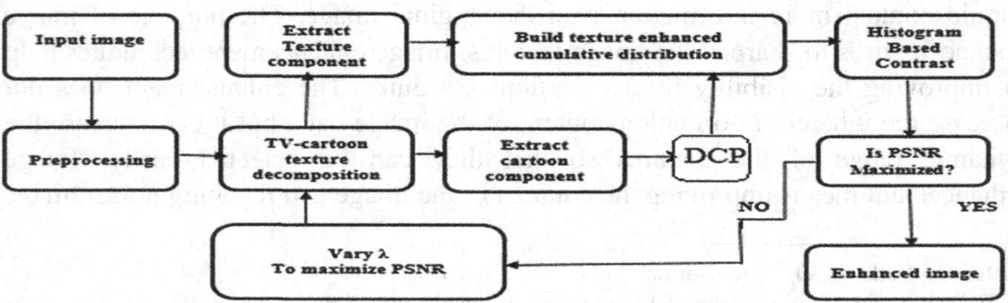

Fig. 1 Overview of the proposed contrast enhancement algorithm

Fig. 2 Endoscopic camera connected to the laptop via USB port

2.1 Total Variation (TV) Cartoon-Texture Decomposition

The main objective of this process is to extract the texture component, as this contains meaningful information in the input image and help us to reject the undesirable intensity transitions. Total variation (TV) method has been widely used for data denoising, face recognition, texture enhancement, and blind deconvolution. Using this variational approach, the input image can be decomposed as shown in Eqs. (1) and (2)

$$O = c + t \tag{1}$$

$$t = O - c \tag{2}$$

where O denotes the original image, c denotes the cartoon, and t denotes texture components respectively. The cartoon component c, which contains the de-textured components from input image I, can be determined using the following expression (3)

$$\min \int |\nabla c| + \lambda \|I - c\| d\Omega \tag{3}$$

where Ω denotes the image domain and λ denotes the lagrange multiplier which is inversely proportional to the strength of the data smoothing process.

2.2 Procedure to Extract Cartoon Component

The procedure described below helps us extract the cartoon component from the test image. The bilateral filter [4] is applied first to input image which is a nonlinear technique that can blur an image while respecting strong edges. It depends only on two parameters that indicate the size and contrast of the features to preserve. Once cartoon component is extracted, the texture component from the image can be obtained by subtracting original test image from the cartoon component for that image. Cartoon image is applied to DCP strategy [8]. After that histogram transformation is performed and the PSNR value is calculated. The dark channel is prior

[8] used for visibility restoration. The degraded image is represented below in Eq. (4)

$$I(x) = J(x)t(x) + A(1 - t(x)) \tag{4}$$

where $I(x)$ denotes degraded image, $J(x)$ $t(x)$ represents direct radiation due to scene radiance and its decay and $A(1 - t(x))$ represents airlight which results from previously scattered light. The original image can be obtained using the following expressions (5) and (6)

$$J(x) = \frac{I(x) - A(1 - t(x)}{t(x)} \tag{5}$$

$$t(x) = 1 - w^{\min}_{x \in \Omega(m,n)} I(x)/A \tag{6}$$

where A is the maximum intensity value. The procedure to extract the cartoon component is described below in Fig. 7.

Apply bilateral filter to the input image

↓

Verify that the input image exists and valid

↓

Convert the sRGB to CIELab colour space

↓

Determine gradient magnitude (E) of luminance E=√(G_x^2+G _y^2) where⟦ G⟧_(x)= ∂G/∂x and G_(y)= ∂G/∂y

↓

Create a simple edge map using the gradient magnitudes, E=G

↓

Determine if (E<minimum edge strength) = 0 where minimum edge strength=0.3

↓

Apply soft luminance quantization

↓

Transform back to sRGB color space

Flowchart 1 Describes the procedure to extract the cartoon component from the image

2.3 Texture-Enhanced Histogram Equalization

After the extraction of cartoon-texture component shown in Fig. 3, the next step of the proposed algorithm involves the calculation of the histogram transformation that is applied for contrast enhancement. The aim is to avoid the intensity saturation and over-enhancement effects that are introduced by the other histogram equalization process.

The idea behind this method is to implement a local intensity mapping process that modifies the shape of the histogram. The first step is to identify the pixels that are associated with strong textures after the application of the cartoon texture decomposition. The binary texture map is obtained by applying Eq. (7).

$$t_{ij}^{b} = \begin{cases} 1, & \left|I_{ij} - c_{ij}\right| > p \\ 0, & \text{otherwise} \end{cases}, \; p \in R^{+} \tag{7}$$

where t^{b} is the binary texture map and $p = 1.0$. After the application of binary mapping with the input image, the next step is the evaluation of the neighborhood around each texture pixel.

The next step involves the identification of extreme intensity values within the neighborhood that will be used to alter the intensity distribution H_c that is calculated from the cartoon component. The histogram calculation H_c is shown in Eqs. (8)–(10).

$$H_c = \bigcup_{p \in M} h_c(p), \tag{8}$$

$$h_c(p) = \int_{(i,j) \in \Omega} \delta\left(c_{ij}, p\right) \mathrm{d}\Omega \tag{9}$$

where

$$\delta\left(c_{ij}, p\right) = \begin{cases} 1, c_{ij} = p \\ 0, c_{ij} \neq p \end{cases} \tag{10}$$

The procedure for texture–enhanced histogram transformation is described in the following steps:

Step 1: For every pixel(i, j) belongs to image domain if texture pixel is equal to one.

Step 2: Then construct 3×3 neighborhood around each texture pixel.

Step 3: Calculate lowest and highest intensity values.

Step 4: If those intensity values lie in p then increase the intensity distribution by one.

Step 5: Repeat the above steps for all pixels in the test.

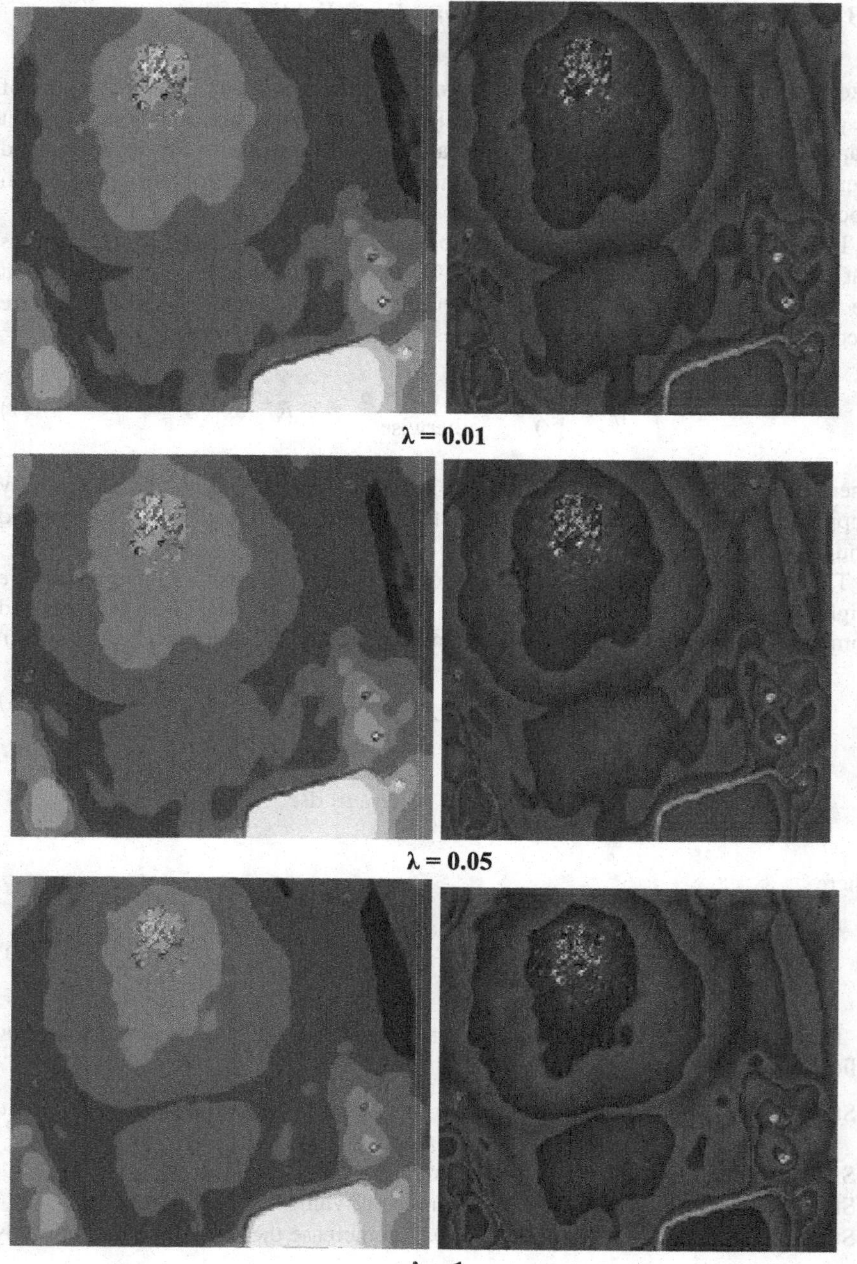

Fig. 3 Cartoon-texture component for different values of the parameter λ. *Left* Cartoon images. *Right* Texture images

The TV-based image decomposition is controlled by the parameter λ which is inversely proportional to the level of smoothing in the cartoon component of the image.

3 Experimental Results

Figure 4 shows the original and enhanced image for mouth image captured using the endoscopic camera. Figure 5 shows the histogram plot for the original and enhanced image. Figure 6 shows another test image captured using that camera. Table 1 shows the PSNR values for test images using various techniques.

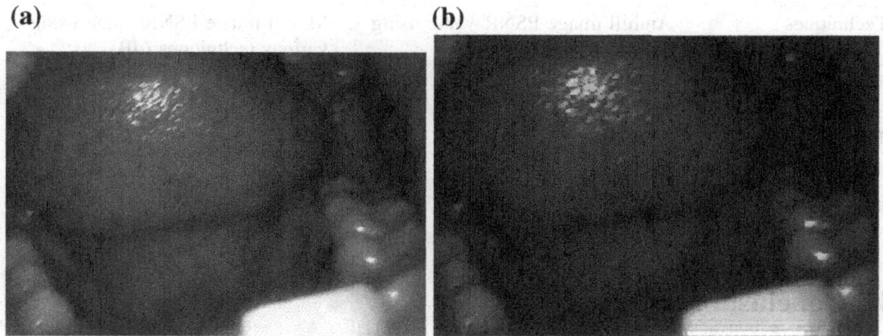

Fig. 4 **a** Input image and **b** Enhanced image

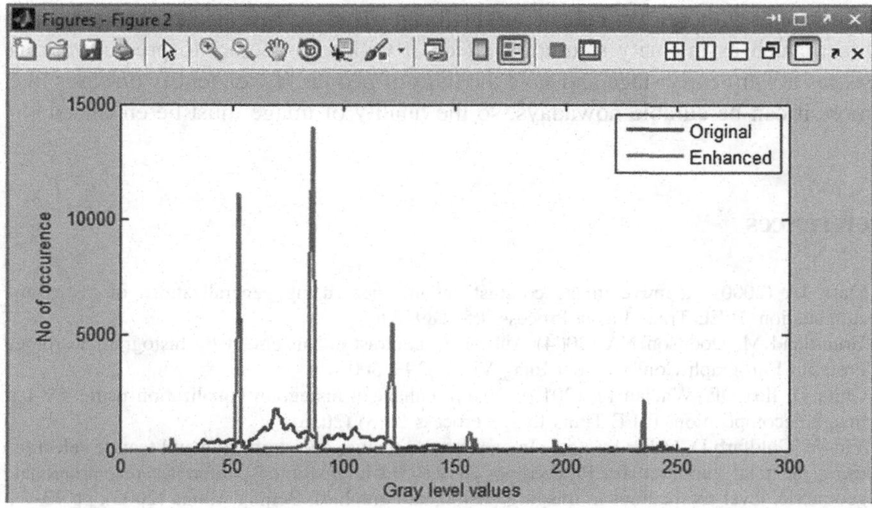

Fig. 5 Histogram plot for original image (*Blue line*) and enhanced image (*Green line*). *x*-axis represents gray level values and *y*-axis represents number of occurrence of those values

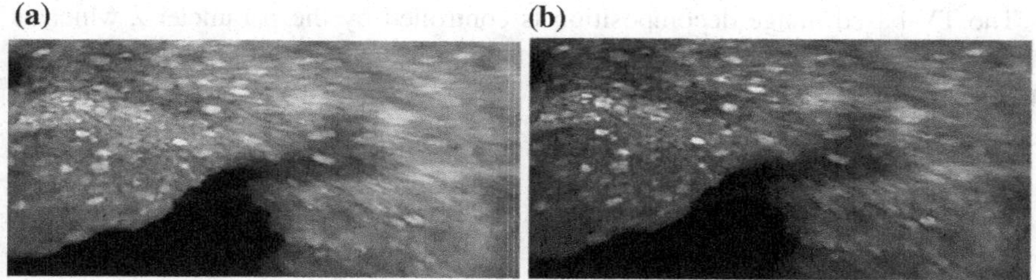

Fig. 6 **a** Input Anthill image and **b** enhanced Anthill image

Table 1 PSNR value for test images

Techniques	Anthill image-PSNR value using various techniques (dB)	Mouth image-PSNR value using various techniques (dB)
TV with DCP (proposed method)	39.9197	38.9790
TV without DCP	34.8094	35.8145
Adaptive HE	32.2610	30.4581
HE	30.0382	29.0353

4 Conclusion

The major aim of this paper was to introduce the new approach which involves the use TV model to obtain cartoon-texture decomposition approach for contrast enhancement without over enhancing the image. This method also avoids undesired artifacts such as intensity saturation effect. It will help doctors to predict various diseases at very early stage and save the lives of people. If we identify diseases like tumors it can be curable nowadays, so the quality of image must be enhanced.

References

1. Stark JA (2000) Adaptive image contrast enhancement using generalizations of histogram equalization. IEEE Trans Image Process 9(5):889–896
2. Grundland M, Dodgson NA (2004) Automatic contrast enhancement by histogram warping. Proc 5th Eurograph Conf Comput Imag Vis 32:293–300
3. Ghita O, Ilea DE, Whelan PF (2013) Texture enhanced histogram equalization using TV-L1 image decomposition. IEEE Trans Image Process 22(8) (2013)
4. Yin W, Goldfarb D, Osher S (2005) Image cartoon-texture decomposition and feature selection using the total variation. In: Proceedings of the 3rd international conference on variational, geometric, level set methods in imaging, vision, and graphics, Beijing, China (2005), pp 73–84
5. Yin W, Goldfarb D, Osher S (2007) The total variation regularized L1 model for multiscale decomposition. Multiscale Model Simul 6(1):190–211

6. Duval V, Aujol FJ, Gousseau Y (2009) The TV-L1 model: a geometric point of view. Multiscale Model Simul 8(1):154–189
7. Breuß M, Brox T, Bürgel A, Sonar T, Weickert J (2006) Numerical aspects of TV flow. Numer Algorithms 41(1):79–101
8. He K, Sun J, Tang X (2009) Single image haze removal using dark channel prior. CVPR, pp 1956–1963

Design of FIS-Based Model for Emotional Speech Recognition

R. Ram, H.K. Palo, Mihir N. Mohanty and L. Padma Suresh

Abstract Human beings have emotions associated with their acts and speeches. The emotional expressions vary with moods and situations. Speech is an important medium through which people express their feelings. Prosodic, spectral, and other parameters of speech vary with the emotions. The ability to represent the emotional speech varies with the type of features chosen. In an attempt to recognize such an emotional content of speech, one of the spectral features (linear prediction coefficients), have been first tested by the fuzzy interference system. Next to it hybridization of LPC features with different prosodic features were compared with LPC features for recognition accuracy. Results show that the hybridization of features can classify emotions better with the FIS system.

Keywords Speech synthesis · Emotional speech · Features · Fuzzy · FIS

1 Introduction

Human speech is mostly associated with some form of emotions. Neutral speech claims to have no emotional contents by researchers, but it is not accurately proved [1]. Hence emotion recognition is a major thrust area, where researchers' of speech community tends to focus in recent years. Speech emotion recognition is an important aspect in the signal processing area. It has many practical applications such as voice dialing, call conferencing, medical analysis, military, air traffic controllers, and many more. Speech emotion recognition is important to know

R. Ram · H.K. Palo · M.N. Mohanty (✉)
Department of ECE, ITER, Siksha 'O' Anusandhan University,
Bhubaneswar, Odisha, India
e-mail: mihirmohanty@soauniversity.ac.in

L. Padma Suresh
Department of EEE, NICE, Noorul Islam University, Nagerkoil, Kanyakumari, TN, India

L.P. Suresh and B.K. Panigrahi (eds.), *Proceedings of the International Conference on Soft Computing Systems*, Advances in Intelligent Systems and Computing 397, DOI 10.1007/978-81-322-2671-0_8

about the human mood. There have been many recognition techniques developed by different researchers [2–6]. But the selection of an accurate classifier and the extraction of robust features are still challenging. Fuzzy theory can be a vital technique in emotional speech processing for recognition purpose [7–11]. Also the emotion-based work has been proposed by authors in [12], using different feature extraction techniques.

Section 1 describes the introduction part, Sect. 2 gives the idea about feature extraction and in Sect. 3 the proposed work is described. Results and performance have been compared in Sect. 4. Section 5 concludes the work.

2 Feature Extraction

Features are the parametric representation of a signal. Hence, it is necessary and important to extract the efficient features for processing. In this work, three different types of features are extracted as listed below:

2.1 *Linear Prediction Analysis*

Linear prediction analysis is one of the most important speech analysis techniques [3, 5]. It has been used as a feature to reduce the computational complexity. So that it can minimize the prediction error. It is represented as

$$\tilde{x}(n) = \sum_{m=1}^{N} b[m]x[n-m] \tag{1}$$

2.2 *Zero Crossing Rate*

Zero crossing rate (ZCR) is another basic acoustic feature of an emotional speech signal [12]. It indicates the high frequency content of an emotional signal such as angriness. If a signal crosses the zero line more times, it implies a rapidly varying signal. It is represented as

$$Z(n) = \frac{1}{2N}\sum_{n=0}^{N-1} s(m) \cdot w(n-m) \tag{2}$$

2.3 Short Time Energy

Short time energy (STE) signifies the energy associated with a segmented signal [12]. Emotional speech energy is thus time varying. In this experimental study, the short time energy of different emotional speech is calculated. It is represented as

$$E_{\mathrm{T}} = \sum_{N=-\infty}^{\infty} s^2(n) \tag{3}$$

3 Proposed Work

3.1 Fuzzy Inference System

A fuzzy inference system (FIS) maps the inputs to the outputs based on the fuzzy set theory. It is a rule-based fuzzy system. Linguistic fuzzy rules and fuzzy logic are the concepts on which FIS depends. The system can simulate nonlinear behaviors with the help of them. Mamdani, Sugeno (TSK), and Tsukamoto are popular fuzzy inference techniques used in fuzzy systems [7, 8].

3.2 Design of FIS Model for Recognition

The intelligent model which is used for synthesis is based on the fuzzy logic. Fuzzy logic uses terms of natural language to approximate human decision making. It discards the quantitative terms of language for its implementation [8, 11, 13]. Thus, mapping of an input space into an output shape can be done conveniently. Fuzzy logic generally has blurred boundary fuzzy sets (Fig. 1).

The STE and ZCR were found for a particular emotional speech and their mean value was calculated. The linear prediction coefficients (LPCs) were extracted for that particular emotional speech and the mean of STE and ZCR was concatenated with linear prediction coefficient matrix. This process of extracting standard and modified linear coefficients was continued for happy, sad, and angry emotions. These features obtained using the above described methods were passed to the given model. Training and testing procedures were carried out for all emotional speech classes and the respective errors are calculated. A rule-based FIS from MATLAB has been used to optimize the accuracy. To take advantage of computational efficiency, Sugeno FIS is selected. As Sugeno FIS works well with adaptive and optimization techniques, it suits our purpose.

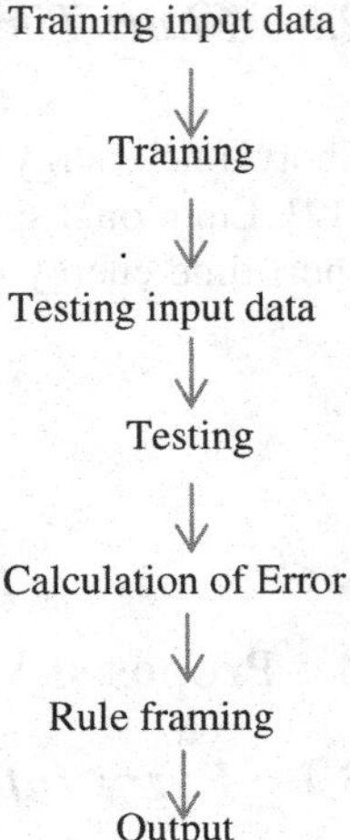

Fig. 1 A fuzzy synthesis model

3.3 *Training and Testing*

All the features are extracted using linear prediction analysis techniques. These features have been used to train and test the fuzzy model. The system is trained for all classes of emotional speech samples to provide a multi-environment training platform. It will naturally improve the classification accuracy. All the data are sent to the FIS as input parameters and the rules are framed to get the output.

4 Results and Discussion

The databases used in this experiment are recorded by different students and faculty members of ITER College. The speech samples are of different emotions containing sadness, angriness, and happiness. The database contains a total of 100 utterances. Fuzzy logic is used in this work for synthesizing the model. Fuzzy logic is also known as the rule-based fuzzy. The outputs of the rules can be viewed in this work using the default LPC coefficients. In order to observe the crisp value of the emotion, three rules were set. The average of linear prediction coefficients, short time energy, and zero crossing rate for "sadness," "happiness," and "anger" emotions were calculated and tabulated in Table 1. The training and testing data of FIS are shown in Figs. 2 and 3.

Table 1 Prosodic feature comparison of different emotions

Mean features	Angry emotion	Happy emotion	Sad emotion
LP coefficient	0.0154	0.0243	0.0123
Short time energy	11.1392	9.4624	7.3005
Zero crossing rate	0.0720	0.0615	0.0512

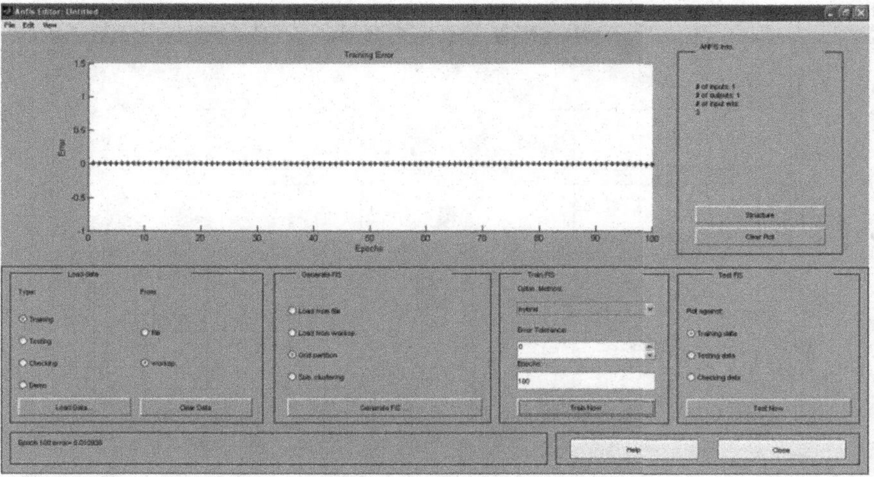

Fig. 2 Training data

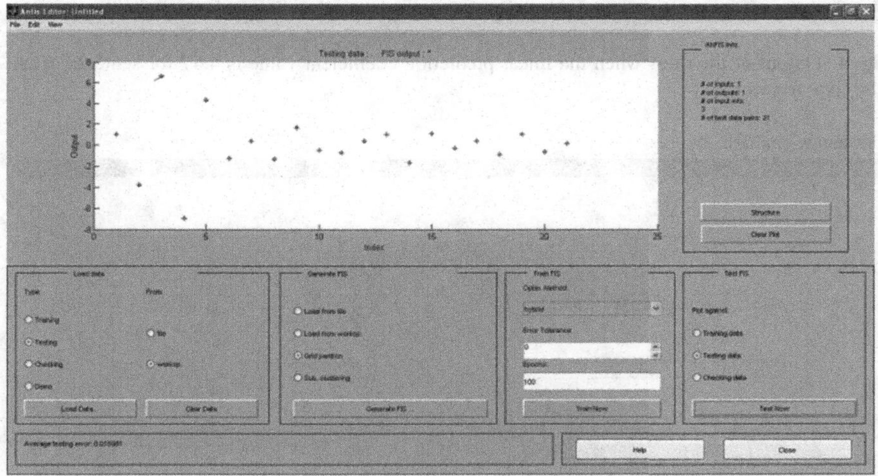

Fig. 3 Testing data

4.1 Fuzzy with Prosodic Features

Plots of fuzzy output for different emotions (i.e., "happiness," "sad," and "anger") have been shown in Figs. 4, 5 and 6. Figure 4 provides the fuzzy output when the linear prediction coefficient value is 4.47 for sentence *Have you gone insane*. Fuzzy outputs when the linear prediction coefficient value is 4.82 for the sentence *I am so*

For angry emotion:

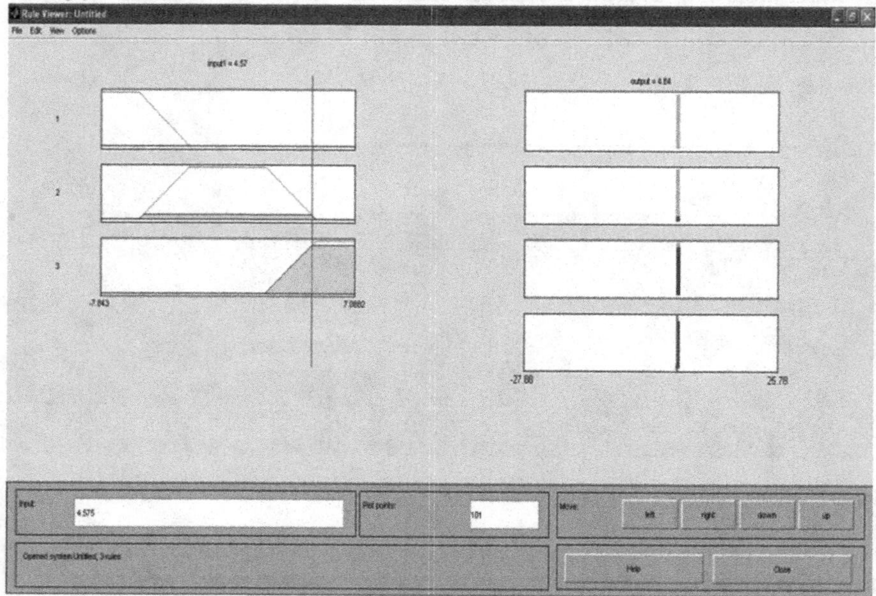

Fig. 4 Output of the rules when the linear prediction coefficient value is 4.47 for sentence *Have you gone insane*

For happy emotion:

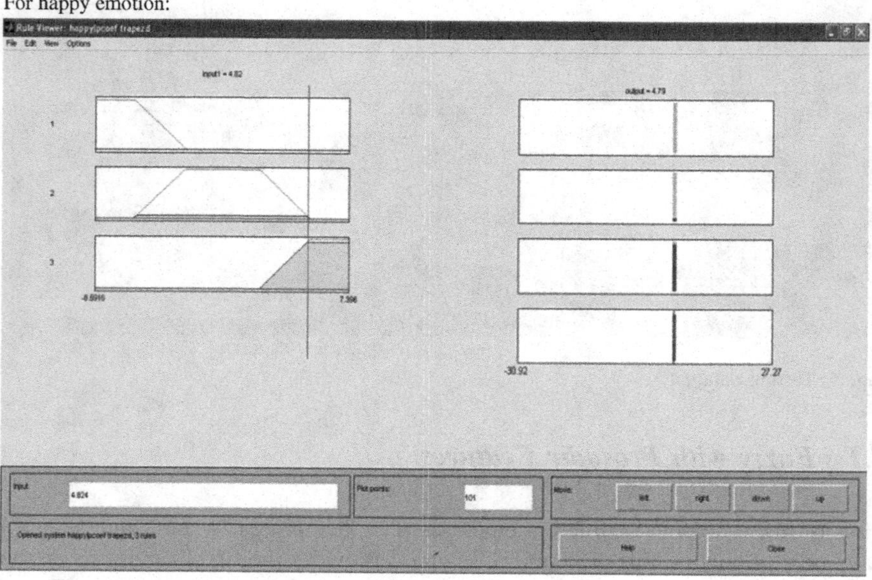

Fig. 5 Output of the rules when the linear prediction coefficient value is 4.82 for the sentence *I am so happy*

For sad emotion

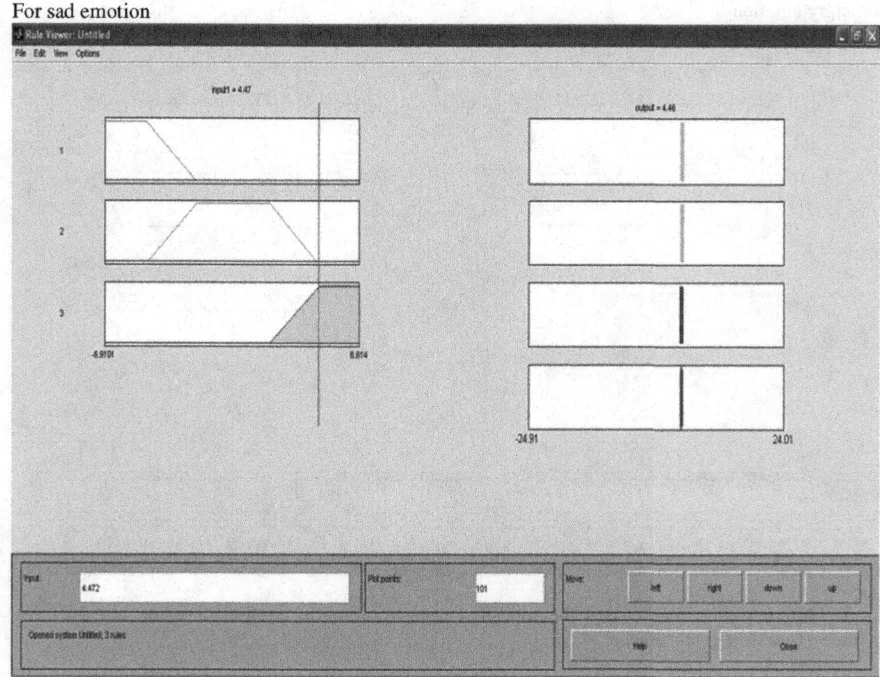

Fig. 6 Output of the rules when the linear prediction coefficient value is 4.47 for the sentence *I am sad*

happy are plotted in Fig. 5. Fuzzy output plot when the linear prediction coefficient value is 4.47 for the sentence *I am sad* is in Fig. 6. From the output it is observed that, the coefficient value of linear prediction for the input matches with the output of fuzzy system.

Table 2 Fuzzy performance on different emotions

Emotions	Average accuracy in Standard LPCs (%)	Average accuracy of modified LPCs with ZCR (%)	Average accuracy of modified LPCs with STE (%)
Anger	98.126	99.218	99.85
Happiness	98.435	99.327	99.88
Sadness	95.355	99.352	99.90

For angry emotion

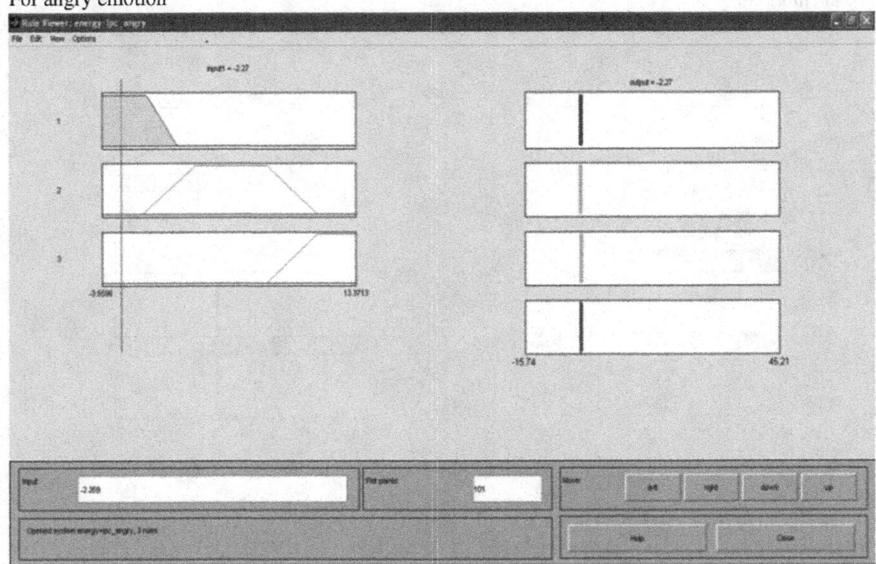

Fig. 7 Output of the rules when the modified linear prediction coefficient value is −2.27 for sentence *Have you gone insane*

For happy emotion

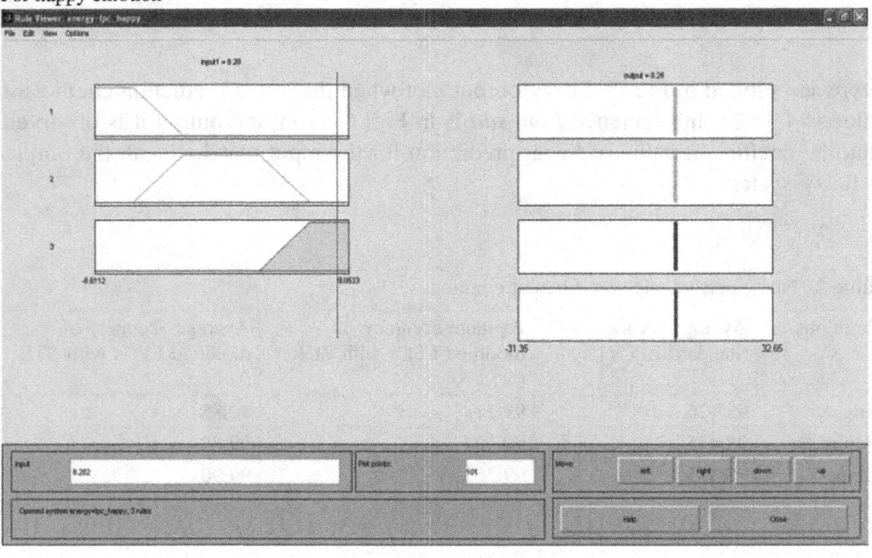

Fig. 8 Output of the rules when the linear prediction coefficient value is 8.28 for the sentence *I am so happy*

For sad emotion

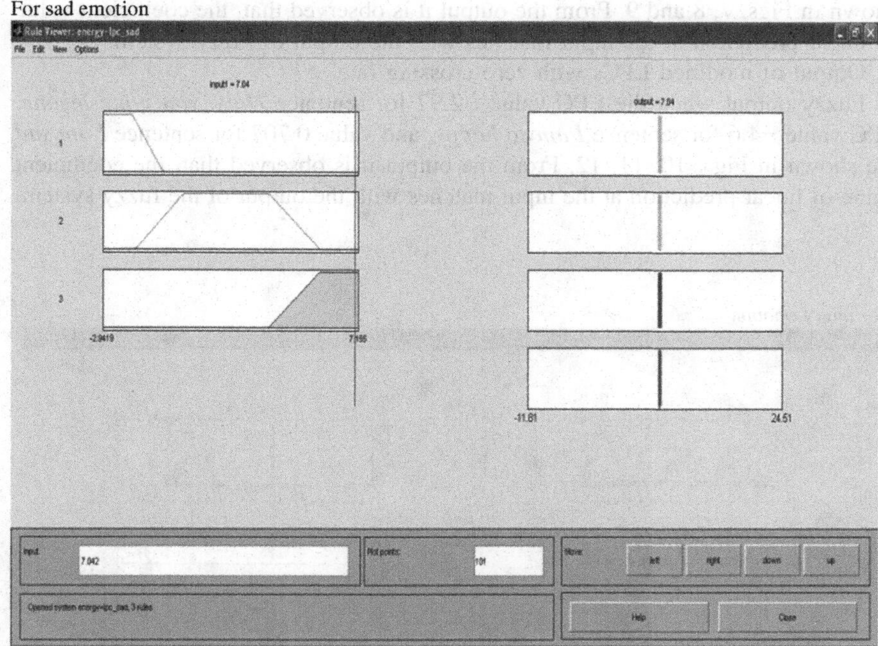

Fig. 9 Output of the rules when the linear prediction coefficient value is 7.04 for the sentence *I am sad*

4.2 *Fuzzy with Hybrid Features*

The modified values of LPC coefficients have been extracted by hybridization of LP coefficients with ZCR and STE separately as shown in Table 2. These values have been used for training and testing the FIS as earlier. Fuzzy outputs for different emotions (i.e., "Happiness," "sad," and "anger") with these hybrid features have been plotted.

Output of Modified LPCs with Short time energy: Plot of fuzzy output with the LPC value −2.27 for sentence *Have you gone insane*, LPC value is 8.28 for sentence *I am so happy* and the value is 7.04 for sentence *I am sad,* have been shown in Figs. 7, 8 and 9. From the output it is observed that, the coefficient value of linear prediction at the input matches with the output of fuzzy system.

Output of modified LPCs with zero crossing rate

Fuzzy output when the LPC value −2.97 for sentence *Have you gone insane,* LPC value −4.6 for sentence *I am so happy*, and value 0.707 for sentence *I am sad* are shown in Figs. 10, 11, 12. From the output it is observed that, the coefficient value of linear prediction at the input matches with the output of the fuzzy system.

For angry emotion

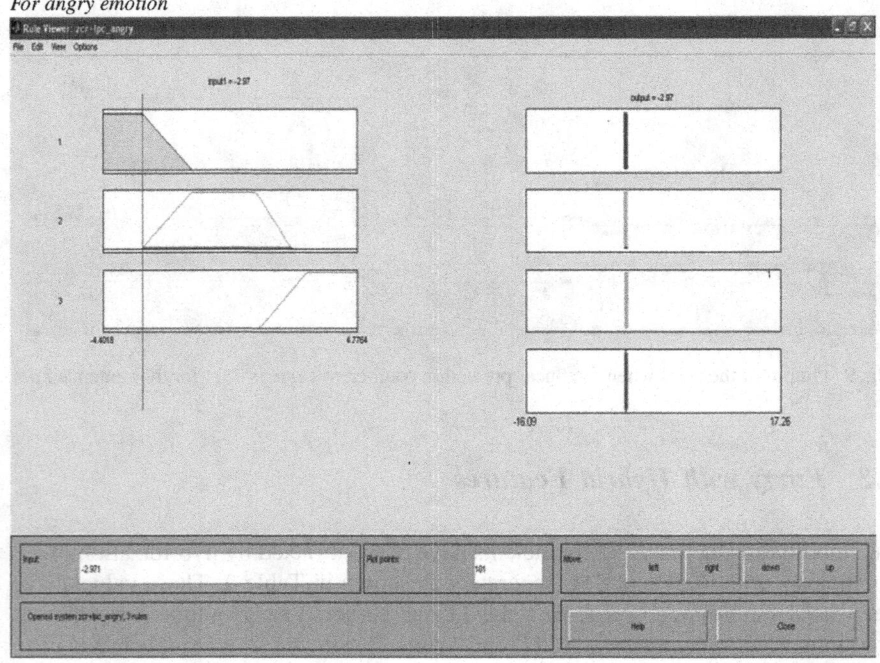

Fig. 10 Output of the rules when the linear prediction coefficient value is −2.97 for sentence *Have you gone insane*

For happy emotion

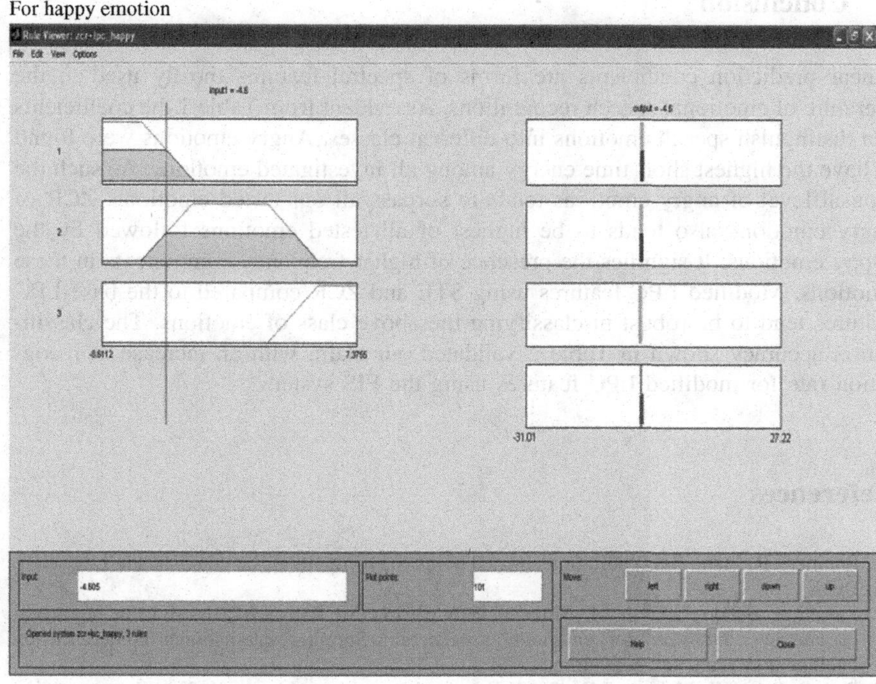

Fig. 11 Output of the rules when the linear prediction coefficient value is −4.6 for sentence *I am so happy*

For sad emotion

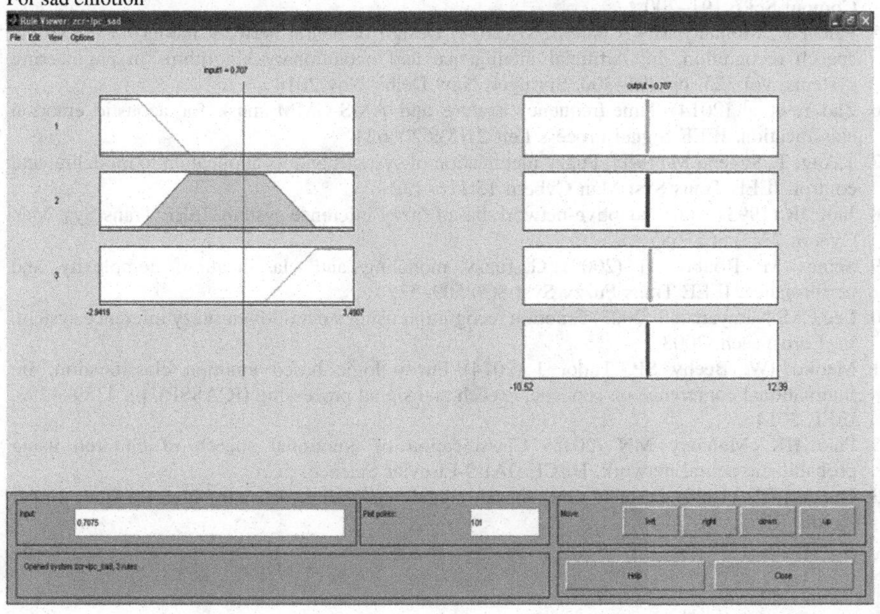

Fig. 12 Output of the rules when the linear prediction coefficient value is −0.707 for sentence *I am sad*

5 Conclusion

Linear prediction coefficients are forms of spectral features mostly used in the literature of emotional speech recognitions. As evident from Table 1 the coefficients can distinguish speech emotions into different classes. Angry emotions were found to have the highest short time energy among all investigated emotions. As such the arousal level of angry emotions tends to surpass all our tested emotions. ZCR of angry emotions also tends to be highest of all tested emotions followed by the happy emotions. It signifies the presence of higher frequency components in these emotions. Modified LPC features using STE and ZCR compared to the base LPC features tend to be robust in classifying the above class of emotions. The classification accuracy shown in Table 2 validated our claim with an increase in recognition rate for modified LPC features using the FIS system.

References

1. Rabiner LR, Juang BH (2003) fundamental of speech recognition, 1st edn. Pearson Education, Delhi
2. Smruti S, Sahoo J, Dash M, Mohanty MN (2014) An approach to design an intelligent parametric synthesizer for emotional speech. In: Springer international conference on intelligent systems and computing, 2014
3. Ram R, Palo HK, Mohanty MN (2013) Emotion recognition with speech for call centres using LPC and spectral analysis. Int J Adv Comput Res 3(11):189–194
4. Javidi MM, Roshan EF (2013) Speech emotion recognition by using combinations of C5.0, neural network (NN), and support vector machines (SVM) classification methods. J Math Comput Sci 6:191–200
5. Palo HK, Mohanty MN, Chandra M (2014) Design of neural network model for emotional speech recognition. In: Artificial intelligence and evolutionary algorithms in engineering systems, vol 325, pp 291–300, Springer, New Delhi, Nov 2014
6. Zao L et al (2014) Time-frequency feature and AMS-GMM mask for acoustic emotion classification. IEEE Signal Process Lett 21(5):620–624
7. Takagi T, Sugeno M (1985) Fuzzy idenfication of systems and its application to modeling and control. IEEE Trans Syst Man Cybern 15:116–132
8. Jang JR (1993) Anfis: adaptive-network-based fuzzy inference system. IEEE Trans Syst Man Cybern 23(3):665–685
9. Setnes M, Roubos H (2000) Ga-fuzzy modeling and classification: complexity and performance. IEEE Trans Fuzzy Syst 8(5):509–522
10. Lee CM, Narayanan S (2003) Emotion recognition using a data-driven fuzzy inference system. In: Eurospeech, 2003
11. Matiko JW, Beeby SP, Tudor J (2014) Fuzzy logic based emotion classification. In: International conference on acoustic, speech and signal processing (ICASSP), pp 4389–4393, IEEE 2014
12. Palo HK, Mohanty MN (2015) Classification of emotional speech of children using probabilistic neural network. IJECE (IAES-Elsevier Science) 5(2)
13. Quatieri TF (1996) Discrete-time speech signal processing, 3rd edn. Prentice-Hall, Upper Saddle River

Discrimination and Detection of Face and Non-face Using Multilayer Feedforward Perceptron

K.S. Gautam and T. Senthil Kumar

Abstract The paper proposes a face detection system that locates and extracts faces from the background using the multilayer feedforward perceptron. Facial features are extracted from the local image using filters. In this approach, feature vector from Gabor filter acts as an input for the multilayer feedforward perceptron. The points holding high information on face image are used for extraction of feature vectors. Since Gabor filter extracts features from varying scales and orientations, the feature points are extracted with high accuracy. Experimental results show the multilayer feedforward perceptron discriminates and detects faces from non-face patterns irrespective of the illumination changes.

Keywords Face detection · Multilayer feedforward perceptron · Pattern recognition · Computer vision · Gabor filter

1 Introduction

Human–computer interaction is a sector under ongoing researches which hold many applications including information management, search, and computer vision. It prepares us to address our needs with respect to the technology by building usable interfaces and by expressing needful system functionalities. Face detection is one among the technologies that interface the interaction between human and computers. Detection of faces is a simple task for human beings, but a challenging one for machines. The multilayer perceptron is trained for the task of discrimination. Here a multilayer feedforward perceptron is used for detecting faces in the grays-

K.S. Gautam (✉) · T. Senthil Kumar
Department of Computer Science and Engineering, Amrita Vishwa Vidyapeetham, Coimbatore 641042, Tamil Nadu, India
e-mail: lancer1589@gmail.com

T. Senthil Kumar
e-mail: t_senthilkumar@cb.amrita.edu

L.P. Suresh and B.K. Panigrahi (eds.), *Proceedings of the International Conference on Soft Computing Systems*, Advances in Intelligent Systems and Computing 397, DOI 10.1007/978-81-322-2671-0_9

cale image. The discrimination ability is increased by the cause of the downsampled Gabor Wavelet faces. The exact spot in which the feature points are spotted provides high information regarding face. The perceptron takes feature vectors from the Gabor filter as input and acts as a discriminator.

2 Related Works

2.1 Support Vector Machines: Osuna et al. [1]

Edgar Osuna, Robert Freund, and Federico Girosi [1] worked with the support vector machines (SVMs) and their applications. The proposed system can detect human faces (front view) in grayscale images. Prior knowledge of faces is not needed for this system. Decision surfaces are detected using a training equivalent that is capable of solving a linear constrained quadratic programming problem. The limitation of this approach is that when the data points are squared, the need for the memory also rises. For separating face and non-faces, a decision boundary is created. The decision boundaries are created in such a way that they have a large margin. As a result, there will be less number of misclassifications.

- Resolution of the training images: 19×19
- Expected output for face: 1
- Expected output for non-face: -1

The misclassified non-face images are stored as negative examples. The image recycling process is done and the image is cut into 19×19 windows; thus each and every window is preprocessed as mentioned before and then classified by SVM. The identified faces are being bordered with a rectangle. This gives a detection rate of 97.1 %.

2.2 Decision Trees: Huang et al. [2]

Jeffery Huang, Srinivas Gutta, and Harry Wechsler [2] put forth an algorithm for face detection using decision trees. The aim of this approach is to separate face and non-face images. They used 2,340 face images from the FERET database for discriminating face and non-face images. Set of rules is framed for object discrimination given the training set. The rules used form a decision tree that can detect the facial features. The images undergo preprocessing histogram equalization and detection of edges before they are given into the classifier and so the process speedups. Each 8×8 identified window is cropped and labeled for discriminating face or non-face using the decision tree induced.

- Training set images: 12
- Resolution of the image: 256 × 384
- Windows tagged correctly: 2759
- Windows tagged incorrectly: 15673.

Finally, postprocessing is done to find the presence of face by aggregating the face labels. The presence or absence of face is identified using a threshold measure. The numbers of face labels are counted in deciding the presence of face. Then a threshold value is set to check the presence of face. If the obtained value is greater than the threshold, then the presence of the face is confirmed. The accuracy of the built face detection system is quoted 96 %.

3 Creation and Training of Multilayer Feed Forward Perceptron

Multilayer perceptron (MLP) is a feedforward neural network with input layer, hidden layer, and output layer as shown in Table 1. In our approach, we have created a single hidden layer and their data flow is unidirectional as shown below in Fig. 1 [3]. Learning algorithm used for the network is backpropagation.

Table 1 Name of the layer and number of neurons present

Name of layer	Number of neurons
Input layer	Vector made of by $n \times n$ pixel input images
Output layer	Single neuron (1 when face image is present)
Hidden layer	n neurons

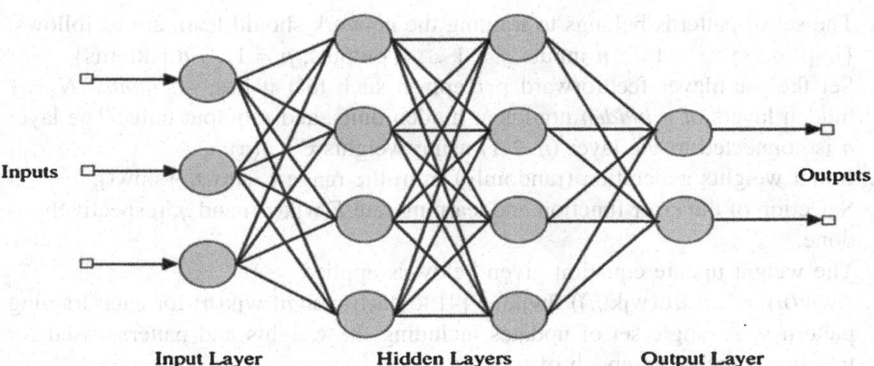

Fig. 1 Architecture of multilayer feedforward perceptron

The functioning of a neuron j in the hidden layer is as follows:

$$y_i = f\left(\sum_{j=1}^{m} w_{ij}x + b_i\right) \quad (1)$$

W_{1i} denotes weight of ith neuron; b_i is the threshold; x_i is an input of the neuron. Similarly, the output layer activity is

$$Sj = \sum_{\text{input}} w_{ji}x_i \quad (2)$$

In our system, 27 × 18 pixels represent the resolution of the images containing human face and non-face, hence the input layer constitutes 2160 neurons. The training function used in our approach is scaled conjugate gradient since the training function takes a less number of iterations to converge on comparison with other conjugate gradient algorithms. In our approach, the scaled conjugate gradient training function took 18 iterations and results in detection rate with accuracy compared with other training functions.

3.1 Training of Multilayer Feedforward Perceptron

The non-face feature vectors are assigned with −0.9 and face feature vectors assigned to 0.9. The data are being trained using the multilayer feedforward perceptron for 18 iterations. The number of iterations varies with the number of face and non-face images used for training. In our approach, we have used 80 face images and 70 non-face images of size 27*18 for training the multilayer feedforward perceptron.

Training Multilayer Feedforward Perceptron:

1. The set of patterns belongs to training the network should learn are as follows: {inip, outjp : i = 1 … n inputs, j = 1 … n outputs, p = 1 … n patterns}.
2. Set the multilayer feedforward perceptron such that it has: –n *inputs*, $N-1$ hidden layers of n *hidden* nonlinear hidden units, and n output units. The layer n is connected to the layer (n − 1) using weights € *wij*(n).
3. Initial weights generation (randomly) from the range [−smwt, +smwt].
4. Selection of the error function and learning rate E(wjk(n)) and η, respectively, is done.
5. The weight update equation given below is applied Δwjk(n) = − η ∂E(wjk(n)) ∂wjk(n) [4] to each weight wjk(n) for each training pattern p. A single set of updates including the weights and patterns used for training is called an epoch of training.
6. Repeat 5 till the network error function is minimized.

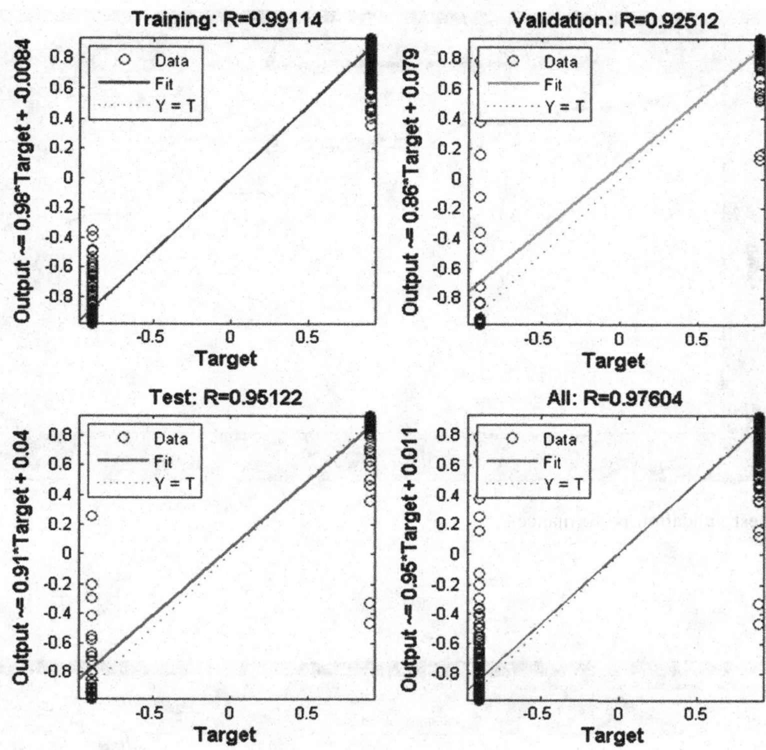

Fig. 2 Regression plot (training session)

The training, validation, and testing data are shown Fig. 2. The dashed line is the target value representation. The dark line best fits linear regression representations. R stands for the relationship between target and output. When R turns unity, it represents the exact relationship in a linear manner between the output and target so the data indicate a good fit. From the process of validation and test outcomes, we find R is greater than 0.9

Best validation performance is on 12th iteration as shown in Fig. 3 where the number of epoch is plotted against the mean square error. The blue and green bars represent training and validation data, respectively, and the red bar represents testing data. Outliers are the data points whose fit is comparatively bad than the majority of data as shown in Fig. 4.

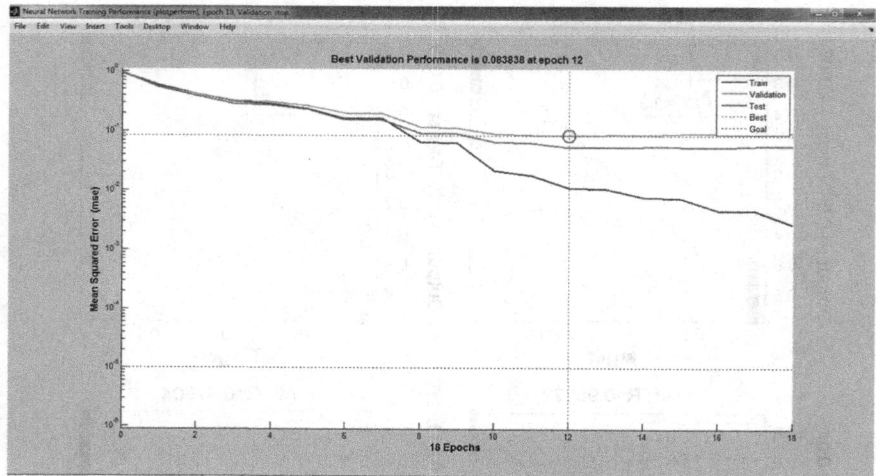

Fig. 3 Best validation performance

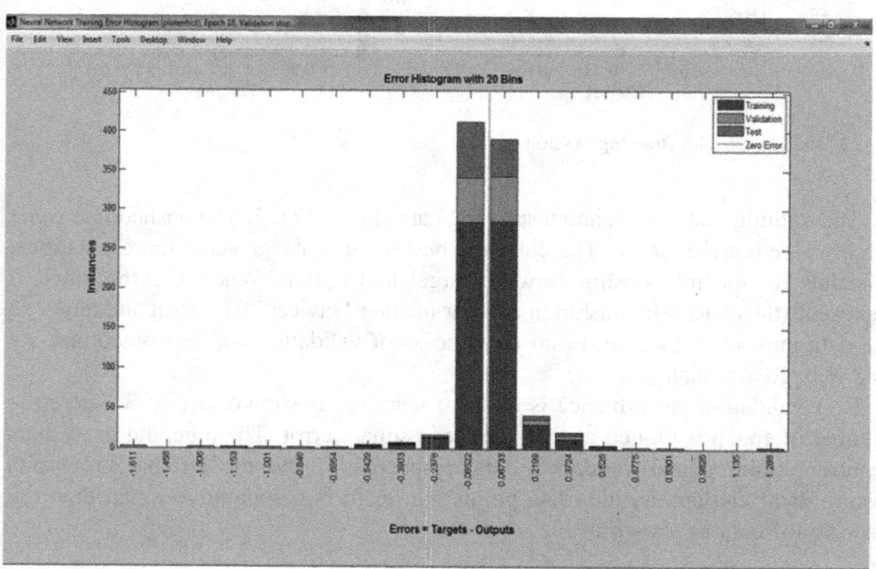

Fig. 4 Error histogram (training session)

4 Feature Extraction and Working of Multilayer Feedforward Perceptron

4.1 Feature Extraction

Feature extraction for this approach has two steps: they are localizing the feature points and computing the feature vectors. Feature vectors are extracted from the face in particular points such that it is capable of providing high-information content. The locations and number of feature points are not fixed. Gabor filter is used for feature extraction in our approach. The filter acts similar to mammalian visual cortex cells and extracts features from different orientations and scales. Gabor features are extracted at five scales and eight orientations as shown in Fig. 5. Each filter should be convolved with the image to get 40 (8*5 = 40) different representations called response matrices for the same image where each image gives a feature vector (Fig. 6).

Then the response matrix is converted to feature vector.

With the face image on the filter, peaks are identified. This process is done by searching the position in a window W_0 (size $W \times W$) using the procedure below: the location of feature point is at (x_0, y_0), if

$$Rj(x_0, y_0) = \max(x, y \in wo)(Rj(x, y)) \tag{3}$$

$$Rj(x_0, y_0) > 1/N_1N_2 \sum_{x=1}^{N_1} \sum_{y=1}^{N_2} Rj(x, y) \tag{4}$$

$J = 1, 2, 3\ldots40$

Rj represents the response of the image (face) to the jth Gabor filter. N_1 and N_2 are the sizes of face peaks of the responses [5]. The window size used to search feature points in Gabor filter responses is 9×9. By using the above process to each filter, the feature map is built for face. As a composition of Gabor wavelet transform

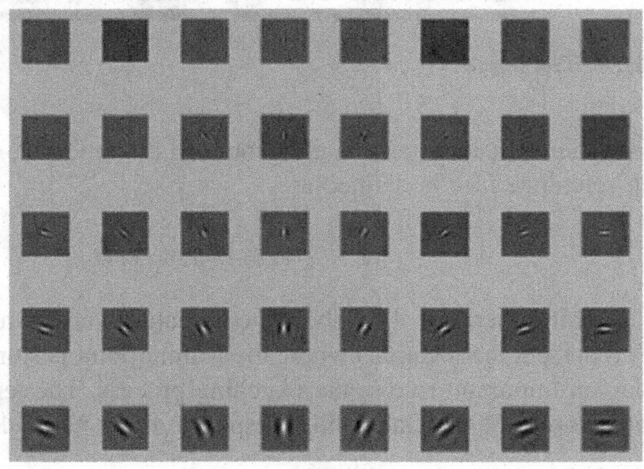

Fig. 5 Gabor filter with 5 scales and 8 orientations

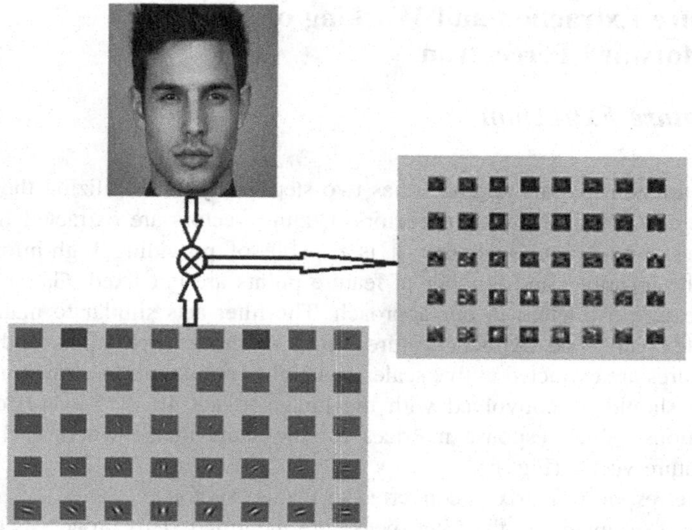

Fig. 6 Convolving image and filter

Fig. 7 Test image

coefficients, feature vectors are generated at the feature points. *n*th feature vector of *i*th reference face is defined as

$$v_{\text{in}} = \{x_n, y_n, R_{ij}(x_n, y_n)j = 1, \ldots .40\} \tag{5}$$

Totally, there are 40 Gabor filters; feature vectors hold 42 components. X_n and Y_n represent the locations where the feature point is stored and both the coordinates play an important role in the matching process. The remaining 40 components are the samples of the Gabor filter response at the particular point.

4.2 Working of Multilayer Feedforward Perceptron

In this section, the multilayer perceptron will check all faces in the window and around them. The result will be the output of the multilayer feedforward perceptron. The experimental analysis is done using the CMU face detection dataset [6]; the system is checked for its performance using non-dataset images which are manually clicked. One of the manually clicked image is used as test image shown in Fig. 7 above and the process flow is given below for reference. The process is described step by step here with results obtained while the system is running.

The yellow points represent the possible region where face is found as shown in Fig. 8 and the feature points get dilated which is the possibility of image after face matching as shown in Fig. 9. The matched features on the test image are shown in Fig. 10. Finally, the face images are detected with a rectangular bounding box as shown in Figs. 11 and 12.

4.3 Steps in Matching the Face Feature

Step 1 In Fig. 8, select the pixel that holds colored point (yellow) and then set it to normal.

Step 2 With the pixel at the center in the test image, cut the image with size 27 × 18 from it.

Step 3 From the image obtained after cutting with the size 27 × 18, feature extraction is done and the multilayer feedforward perceptron is made to give data that is trained and feature vector of the test image as input.

Step 4 (i) If the output is lesser than −0.97, the center pixel encircling 7 × 7 pixels set normal and update the image of Fig. 8 and go to step 1 else follow as below. (ii) If the output is lesser than −0.5 go to step 1; else continue.

Fig. 8 Possible region were face is found

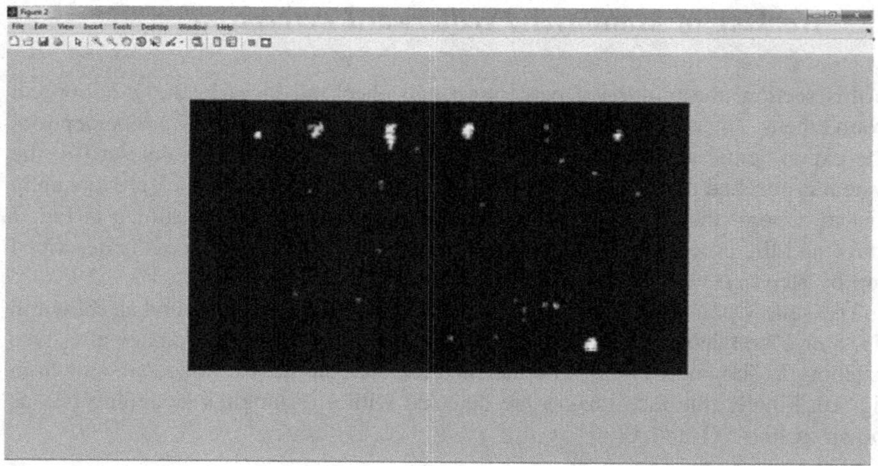

Fig. 9 Possibility of image after face matching

(iii) If the output is greater than 0.97, then update the 27 × 18 surrounding pixel values to normal and go to Step 5. (iv) If the output is less than 0.5 or greater than 0.5, then go to the next step.

Step 5 The encircling three pixels are checked and the image of 27 × 18 is cropped considering its center and switch to the third step. If the output is greater than 0.97, all 27 × 18 pixels are made normal.

Step 6 If the output is greater than 0.5, then in the image, Fig. 9 set their corresponding pixel to one. Switch on the first step to the final step till all the colored pixels (yellow) are not be normal.

5 Experimental Analyzes

Test images are being randomly taken from the CMU face detection dataset [6] and the results are tabulated with performance metrics such as detection rate which is the precision, recall, true positive, false positive, and false negative. This system is built using Matlab R2013a. Results are tabulated for analysis of the efficiency of the system. The performance measured in our approach is in terms of average face detection rate which is 81.78 %, shows that the proposed system is consistently good in detecting faces. Ten images are taken randomly from the CMU face detection dataset [6] and their prediction analysis is discussed in the tabulations given below.

Fig. 10 Matched features in the test image

Fig. 11 Detected face in the test image 1

5.1 Face Detection on Varying Illuminations

Experimental analysis is done with higher (Fig. 13), extreme (Fig. 14), and low illuminations (Fig. 15), and their corresponding detection rates are 83.33, 66.66, and 83.33 % (with five false positives), respectively.

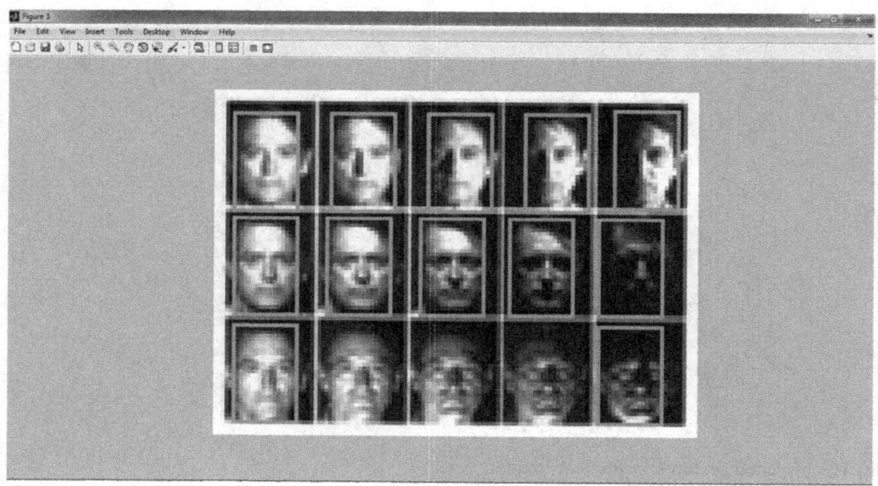

Fig. 12 Detected face from test image 2 from CMU dataset

Fig. 13 Higher illumination

True positive is the total number faces detected correctly, false positive denotes the non-detected faces in the image, and false negative denotes the non-faces detected as faces as shown in Table 2. The total faces present in the image and the

Fig. 14 Extreme illumination

Fig. 15 Lower illumination

number of faces detected are tabulated and analyzed. The ratio between the faces detected from the image and the total number of faces in the image gives the detection rate as shown in Table 3.

Table 2 Images and their corresponding predictions

#image	True positive	False positive	False negative
1	60	0	15
2	6	2	0
3	12	3	0
4	7	0	0
5	1	1	0
6	2	1	0
7	22	13	6
8	10	0	5
9	5	1	2
10	3	0	2

Table 3 Images and their detection rate

#image	Face detected	Total faces in the image	Detection rate/precision
1	60	60	100
2	6	8	75
3	12	15	80
4	7	7	100
5	1	2	50
6	2	3	66.66
7	22	35	62.857
8	10	10	100
9	5	6	83.33
10	3	3	100

$$\text{Face detection rate} = (\text{face detected}/\text{total number of faces in the image}) * 100$$
$$\text{(Or)}$$
$$\text{Precision} = (\text{true positive}/\text{true positive } + \text{false positive}) * 100$$
$$(\text{Here } N = 10)$$
$$\text{Average Detection rate} = \text{Total detection rate of } N\text{images}/\text{N}$$
$$\text{Average Recall rate} = \text{Total Recall of} N\text{images}/N$$

In our approach, the average detection rate of ten images taken from the CMU face detection dataset [6] is 81.78 %.

6 Conclusion

Our proposed approach uses Gabor filters for feature extraction which extracts Gabor features at five scales and eight orientations. The representations of frequency and orientation of Gabor filter are similar to the visual cortex system in mammalian brains. This makes Gabor filters an appropriate tool for texture representation. The feature vector from the Gabor filter is given as input to the trained multilayer feedforward perceptron that discriminates faces and non-face images. Experimental analysis is made with images from CMU dataset [6]. A manually clicked image is also presented as for detection of faces as shown in Fig. 7. The system works consistently good for all the images irrespective of the backgrounds and illumination changes with an average detection rate of 81.78 %.

References

1. Osuna E, Freund R, Girosi F (1997) Training support vector machines: an application to face detection. Computer vision and pattern recognition, 1997. In: Proceedings, 1997 IEEE computer society conference on. IEEE
2. Huang J, Gutta S, Wechsler H (1996) Detection of human faces using decision trees. In: Proceedings of the second international conference on automatic face and gesture recognition. IEEE
3. http://www.doc.ic.ac.uk/~nd/surprise_96/journal/vol4/cs11/report.html
4. http://www.cs.bham.ac.uk/~jxb/INC/l7.pdf
5. Wang X-Y et al (2013) Robust color image retrieval using visual interest point feature of significant bit-planes. Digital Signal Process 23(4):1136–1153
6. http://vasc.ri.cmu.edu/idb/html/face/frontal_image

A Conjectural Study on Machine Learning Algorithms

Abijith Sankar, P. Divya Bharathi, M. Midhun, K. Vijay and T. Senthil Kumar

Abstract Artificial Intelligence, a field which deals with the study and design of systems, which has the capability of observing its environment and does functionalities which aims at maximizing the probability of its success in solving problems. AI turned out to be a field which captured wide interest and attention from the scientific world, so that it gained extraordinary growth. This in turn resulted in the increased focus on a field—which deals with developing the underlying conjectures of learning aspects and learning machines—machine learning. The methodologies and objectives of machine learning played a vital role in the considerable progress gained by AI. Machine learning aims at improving the learning capabilities of intelligent systems. This survey is aimed at providing a theoretical insight into the major algorithms that are used in machine learning and the basic methodology followed in them.

Keywords Machine learning algorithms · Supervised learning · Unsupervised learning · Bagging · Boosting · KNN · Random forests · Logistic regression · Decision trees · Naïve bayes · *k*-Means clustering · Partitional clustering · Divisive clustering · Hierarchical clustering · Agglomerative clustering

1 Introduction

Machine Learning is one of the key areas associated to artificial intelligence and is the most sought out academic area in all the elite universities. The academic as well as industrial significance of machine learning is so high that world is now an edible entity for machine learning. Breaking down petabytes of data is a real complex task

A. Sankar (✉) · P. Divya Bharathi · M. Midhun · K. Vijay · T. Senthil Kumar
Department of Computer Science and Engineering, Amrita School of Engineering, Amrita Vishwa Vidyapeetham, Coimbatore, India
e-mail: abijith.1skn@gmail.com

T. Senthil Kumar
e-mail: t_senthilkumar@cb.amrita.edu

L.P. Suresh and B.K. Panigrahi (eds.), *Proceedings of the International Conference on Soft Computing Systems*, Advances in Intelligent Systems and Computing 397, DOI 10.1007/978-81-322-2671-0_10

and machine learning, being an excellent artificial intelligence tool, helps us in doing so and lets us find out meaningful senses out of this complex world. Problems once deemed to be unsolvable are being easily solved nowadays using machine learning techniques. It is inevitable to have a clear understanding of the basic algorithms and methods used in machine learning applications before delving deep into developing systems and solving problems using this technique. This paper first deals with the categorization of machine learning into three learning aspects: supervised, unsupervised, and bagging and boosting. This is followed by a brief introduction to the three learning methods. Effort has been put into providing the conjectural aspects of various algorithms that fall under different learning methods and are discussed separately.

2 Materials and Methods

Literature review The paper itself is aimed at providing familiarization to machine learning algorithms and this study is not based on any references. There are no many papers available or presented with the kind of comparison that we have drawn. We have given citation to the papers that we have referred based on the algorithms and methods.
Context and background of the work carried out This study is very highly focused and it is of high importance because of increasing usage of machine learning techniques in the market. Machine learning is used in a wide variety of areas as a major problem solving methodology. This serves as a major motivation for us to do a complete review in this sector and to ready a research review article. This is also going to serve as major motivation for researchers to carry out their research in this area.
Machine learning Machine learning deals with the development, analysis, and study of algorithms that can learn from data [1]. Rather than following stepwise instructions that comes in the form of programs, machine learning take inputs and does its functionality by creating models out of it [2]. Using these models, useful and meaningful predictions and decisions can be made. Machine learning has a wide variety of applications in the fields of search engine, bioinformatics, robot locomotion, computational finance etc. Figure 1 illustrates a broad categorization of some machine learning techniques.

2.1 *Supervised Learning*

Supervised learning is one of the main branches of machine learning which deals with finding out functionalities from labeled data [3]. Data which are properly classified are considered as labeled data. A set of training examples are used to train the dataset. The training examples comprises of a known set of input data and the associated

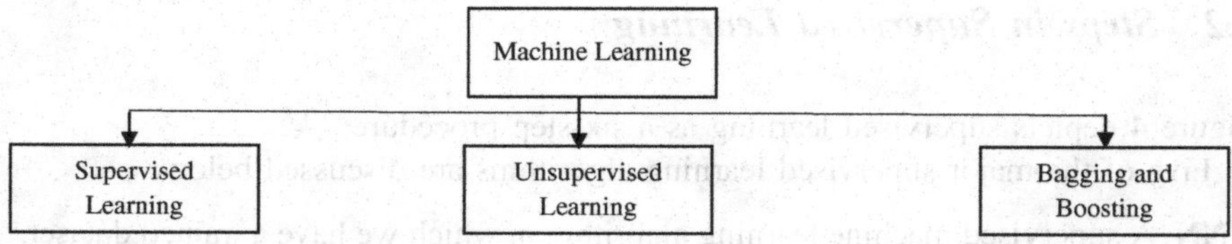

Fig. 1 Topics discussed under machine learning

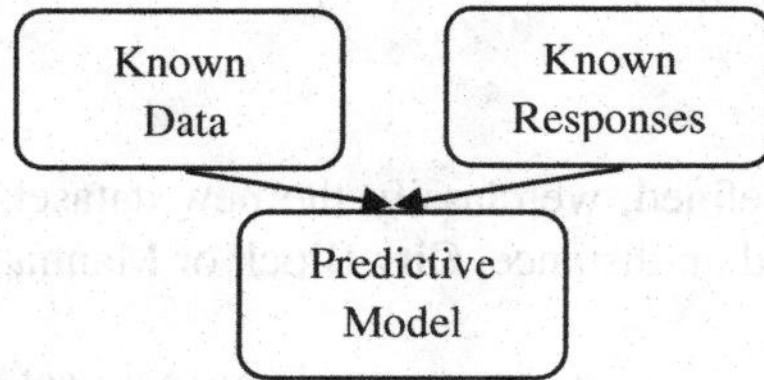

Fig. 2 Supervised learning: step 1

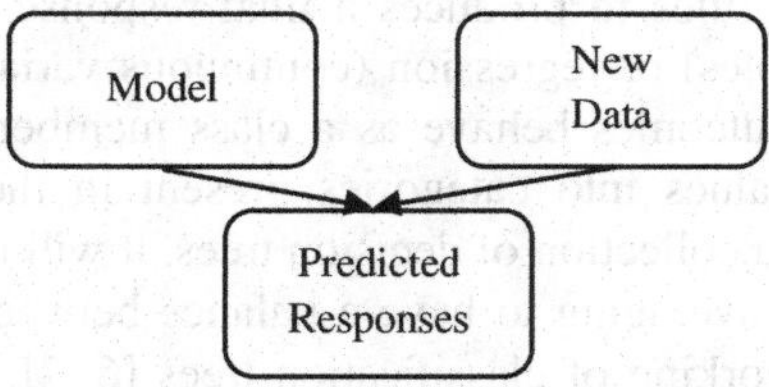

Fig. 3 Supervised learning: step 2

responses to that data so as to build a model that makes reasonable predictions, when provided with the response of a new data (as shown in Figs. 2 and 3).

A very simple example of such a prediction can be sought out as whether a cricket team will win their next match or not. This is done on the basis of the data obtained from the team's previous matches and considering its performances in several aspects such as bowling, batting, fielding etc.

Supervised learning can be broadly classified into two:

- Classification: This is done in cases when the responses are usually nominal and are limited to having a very few known values such as true or false
- Regression: When we have the responses in the form of real numbers, this method can be used.

2.2 Steps in Supervised Learning

Figure 4 depicts supervised learning as a six-step procedure:

Five of the major supervised learning algorithms are discussed below:

KNN A supervised machine learning algorithm in which we have a trained dataset. A new piece of data is compared with the existing data and classified to where it belongs. We have a k value (an integer value mostly less than 20); based on the majority of classes, we would place the new piece of data [4].

Steps involved:

1. A trained dataset.
2. Based on Kth value defined, we classify the new dataset.
3. We make use of Euclidian distance, City Block or Manhattan Distance formulas for classification.
4. Based on the majority of class, we group them (new set of clusters).
5. Repeat step 2, until we get constant set of clusters.

Random forests Random forests are a collection of simple decision trees, which uses independent binary rules to produces a final response. It is used in classification (categorical variables) or regression (continuous variables) applications [5]. In classification, these outcomes behave as a class membership or classify these independent predictor values into categories present in the dependent variable. Since random forest has a collection of decision trees, it will not have high variance as it uses the method of averaging to have a balance between two extremes.

Figure 5 shows the working of classification trees [6, 7].

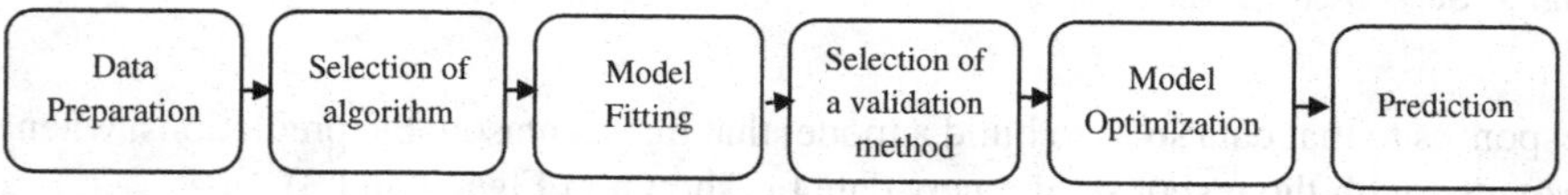

Fig. 4 Six-step methodology of supervised learning

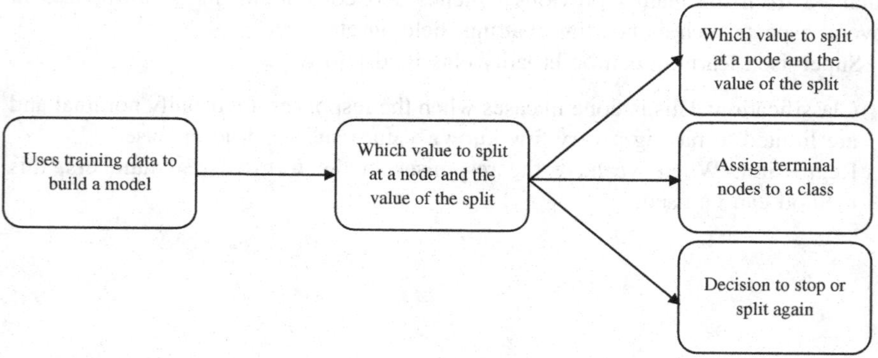

Fig. 5 Working of classification tree

Decision trees Decision trees are one of the most important supervised learning methods. It has a tree structure that has close resemblances to a flowchart. Rectangles are used to denote the internal nodes and ovals represent leaf nodes. The root node is labeled by an attribute. An internal node is comprised of two or more child nodes. The internal nodes do a conditional checking which tests the value of a particular attribute expression and decides which child node to select for the tree traversal. The edges between nodes are labeled by attribute values. The class labels are attached to every leaf node. There are two major phases associated with decision trees [8]:

- Growth phase.
- Pruning phase.

Growth phase In this, a top down approach is used for tree construction. Figure 6 illustrates the methodology.

Pruning phase In this, a bottom up approach is used for tree construction. Figure 7 shows the methodology.

The decision tree algorithms which are used most frequently are CART (Classification and Regression Trees), C4.5, and ID3 (Iterative Dichotomiser).

CART This algorithm, which was introduced by Breiman, uses a statistical dispersion measure called Gini Index as an attribute selection mechanism to build the decision tree. CART produces binary trees [9]. Accuracy is improved in this by using a cost complexity bottom up approach to remove the missing values and hence the unreliable branches.

C4.5 This algorithm makes use of a threshold value to split the attribute values into two. Those above the threshold fell into one branch and those below, to the other. A measure called gain ratio, which is the ratio of information gain to the inherent information within a dataset, is used to build the tree. Accuracy is improved by using a method called pessimistic pruning for the removal of branches that are unnecessary.

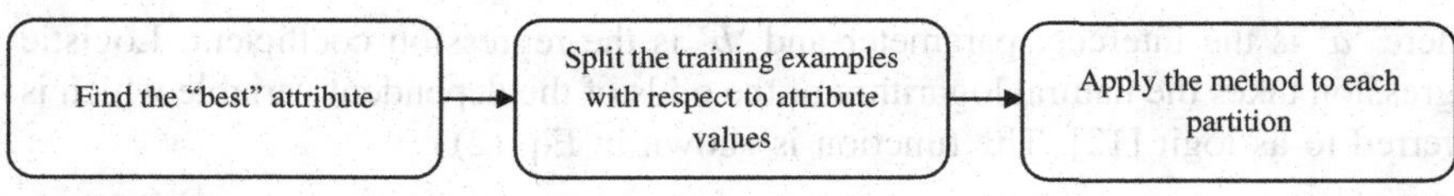

Fig. 6 Growth phase of a decision tree

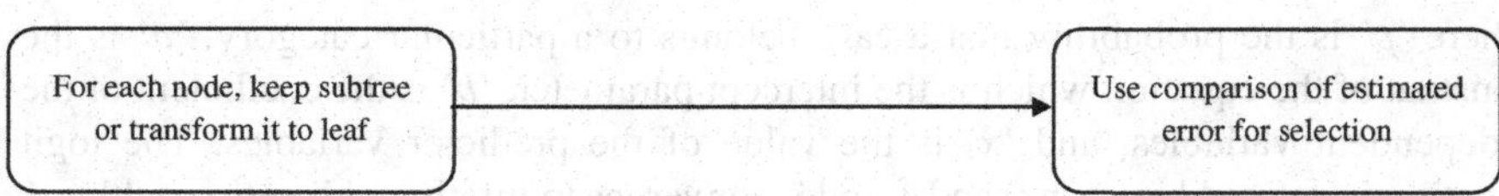

Fig. 7 Pruning phase of a decision tree

ID3 ID3 uses information gain from attribute values for partitioning [10]. The algorithm does not work when there is noisy data. Hence, data preprocessing has to be done so that noisy data is removed. Pruning is not supported by this algorithm.

Logistic regression Logistic regression involves more a probabilistic view of classification. It is an approach to predicting a discrete outcome. The relationship between the dependent variable and independent variable can be represented by using a logistic regression model if the dependent variable is categorical. Prediction of the value of dependent variable is done using the probability scores. It uses a maximum likelihood estimation and the values of the parameters are adjusted in an iterative manner until the maximum likelihood value is obtained. The main uses of logistic regression are prediction of group membership and it gives us knowledge of the strengths and relationships among the variables. The two types of logistic regression are as follows:

- Binary logistic regression
- Multinomial logistic regression

Binary logistic regression It is used in modeling the relationship between the dependent variable which is dichotomous and the independent variable which is either continuous or categorical. The outcome is mostly coded as "0" or "1" and there is attention required on the modeling binary response.

Multinomial logistic regression It is used in modeling the relationship between the dependent variable which is not dichotomous but comprised of more than two cases. It is a simple extension of binary logistic regression and the outcome can have three or more possible types [11].

Odds and odds ratio Odds is defined as the ratio of probability of success of an event to the probability of its failure. The odds ratio is defined as the ratio of two odds relative to different events. If the probability of occurrence of the event is p, the probability of nonoccurrence of the event is $1 - p$, then odd (event) = $p/(1 - p)$. Logistic function is given by the generalized formula given in Eq. (1)

$$P = \frac{e^{a+bX}}{1 + e^{a+bX}} \tag{1}$$

where 'a' is the intercept parameter and 'b' is the regression coefficient. Logistic regression takes the natural logarithm of the odds of the dependent variable which is referred to as logit [12]. The function is shown in Eq. (2)

$$\text{logit}(p) = \log\left[{}^{p}/_{(1-p)}\right] = \ln\left[{}^{p}/_{(1-p)}\right] = a + b_2x_2 + b_3x_3 + \cdots \tag{2}$$

where 'p' is the probability that a case belongs to a particular category, 'a' is the constant of the equation which is the intercept parameter, 'b' is the coefficient of the independent variables, and 'x' is the value of the predictor variables. The logit function is defined between 0 and 1 and it converges to infinity while approaching 1 and converges to negative infinity while approaching 0. It is smooth and symmetric and is the inverse of the logistic function. Regression coefficients represent the change in the logit for each unit change in the predictor. There are various techniques by which these coefficients are tested for inclusion or elimination from the model. It includes Wald test, Likelihood-Ratio test, and Case-control sampling.

Training a logistic regression model The "*b*" need to be optimized so as to give the best reproduction of training set labels. It is usually done by numerical approximation of maximum likelihood. Stochastic gradient descent may be used on very large datasets. Logistic regression gives good accuracy for simple data sets and it is easily extended to multiple classes. It is quick to train, very fast at classifying unknown records. It gives a natural probabilistic view of class predictors. The effect due to directly minimizing squared-error is called as overfitting. It arises when data is very high dimensional and training data is sparse. Regularization is one way to reduce overfitting.

Naive Bayes Naive Bayes is a very efficient, effective, and simple algorithm for machine learning and data mining [13]. It belongs to the family of probabilistic classifiers based on the application Bayes' theorem with independent assumptions. It assumes that the presence or absence of a feature of a class is unrelated to the presence or absence of any other feature [14]. For example a vegetable to be considered to be a potato if it is brown, spherical, and about 1″–3″ in diameter. These features may exist upon other features or depend on each other; naive Bayes classifier considers these properties to independently contribute to the probability that this vegetable is a potato [15]. Naive Bayes classifiers can be trained very efficiently in a supervised learning depending on the precise nature of the probability model [16]. Bayes theorem provides a way of calculating the posterior probability, $P(a|y)$ from $P(a), P(y)$ and $P(y|a)$. The probability of a class given predictor is given in Eq. (3).

$$P(a|y) = \frac{P(y|a)P(a)}{P(y)} \tag{3}$$

where $P(y|a)$: Probability of predictor given class;$P(a|y)$: Probability of class given predictor; $P(a)$: Probability of class; $P(y)$: Probability of predictor

Multinomial Naive Bayes It is the implementation of Naive Bayes for multinomially distributed data and it is used in text classification (data represented as tf-idf vectors). For each class y, the distribution is parametrized by vectors $\theta_y = (\theta_{y1}, \ldots, \theta_{yn}) n$ being the number of features and θ_{yi} is the probability $P(x_i|y)$ of feature i in a sample belonging to class y. Estimation of parameters θ_y is done by a smoothed version of maximum likelihood using Eq. (4).

$$\theta' = \frac{N_{yi} + \alpha}{N_y + \alpha n} \tag{4}$$

where $N_{yi} = \sum_{x \in T} x_i$ is the number of times feature i appears in a sample of class y and $N_y = \sum_{i=1}^{|T|} N_{yi}$ is the total count of all features for class y [17]. Setting $\alpha = 1$ is called Laplace smoothing, and $\alpha < 1$ is called Lidstone smoothing.

2.3 Unsupervised Learning

Unsupervised Learning includes many techniques to summarize and explain certain key features of data such as density estimation in statistics. Most of the methods in unsupervised learning make use of data mining concepts for preprocessing. The difficulty in unsupervised learning is to find the structure of unlabeled data as the learner is provided only with unlabeled data. Here, a framework can be developed based on the need of machine's goal to represent the input which is then used for decision making. The main examples for unsupervised learning are clustering and dimensionality reduction [18].

Clustering Clustering is a technique that is used to find the common groups which consists of similar data. The cluster is formed by placing the inputs that are similar in a group and placing the inputs that are different in different separate clusters. This method can also be called as an unsupervised learning method because no initial labeling of data is done, which differentiates this from supervised learning. The effectiveness and the result of a clustering technique relies on the algorithm used, the function for calculating the distance and the way in which it is applied.

Two types of algorithms are used for clustering:

- Partitional clustering
- Hierarchical clustering

Partitional clustering Partitional clustering can be done by implementing k-means algorithm [19]. The set of inputs S be $\{z_1, z_2, \ldots, z_n\}$, where $z_i = (z_{i1}, z_{i2}, \ldots, z_{ix})$ is a vector in a real-valued space $Z <= R^x$ where x is given by the number of attributes in the given set of data. Now, this algorithm will help to classify the given set of data into k number of clusters. The clusters will have cluster center, called centroid, respectively. The value of k can be specified manually by the user. Figure 8 shows the working of k-means algorithm for a given k.

The convergence criterion is as follows:

- The data points to the separate clusters should be the same and should not be rearranged.

The centroid should remain constant.

Hierarchical clustering In hierarchical clustering, nested tree-like structures are formed which is named as a dendrogram [20]. Figure 9 shows this structure.

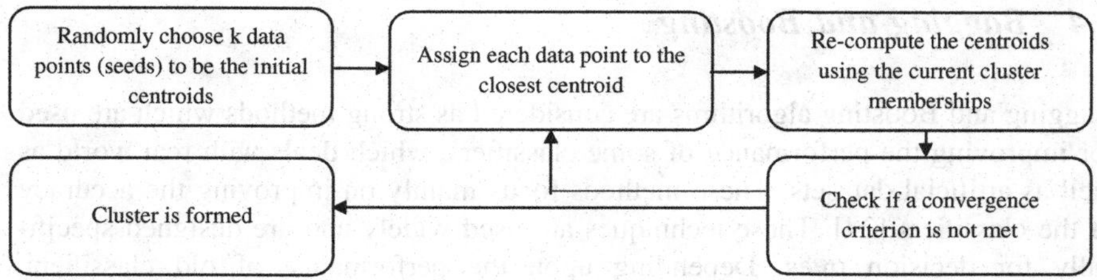

Fig. 8 Steps in *k*-means algorithm

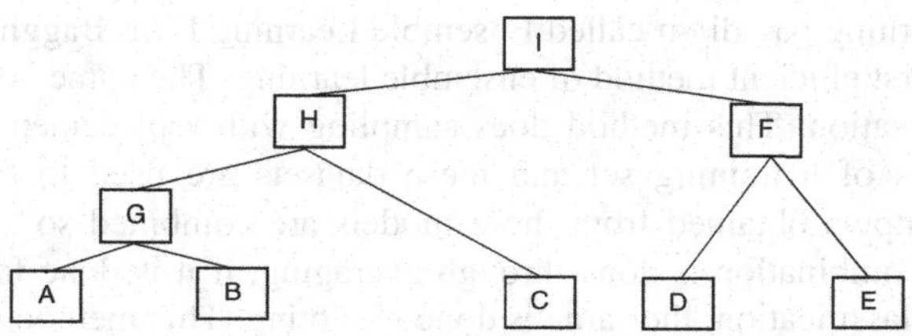

Fig. 9 Structure of hierarchical clustering

There are two types of hierarchical clustering [21]:

- Divisive clustering
- Agglomerative clustering

Divisive clustering This uses bottom up approach. It starts with all data points in one cluster, the root. This splits the root into a set of child clusters. Each child cluster is recursively divided further in each recursive step. Finally, this stops only when singleton clusters of individual data points remain. But, this method is not widely used. So we go for agglomerative clustering [22].

Agglomerative clustering This is a top down approach. In this method, the dendrogram is formed from the bottom level and groups the most similar which are the nearest pair of clusters. When all the data points are fitted into a single cluster, this method comes to an end [23]. Figure 10 shows the steps involved.

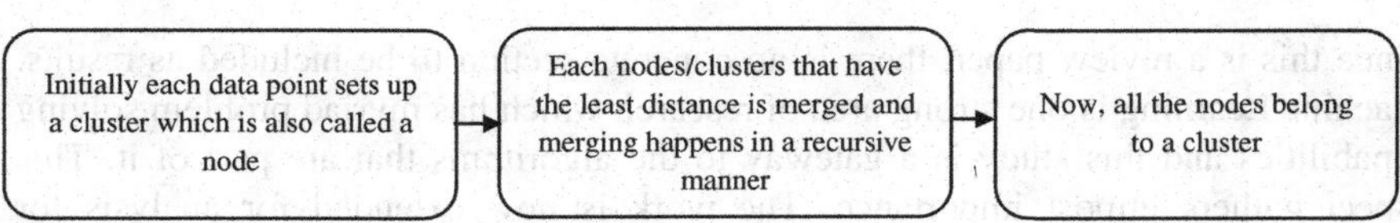

Fig. 10 Agglomerative clustering: steps

2.4 Bagging and Boosting

Bagging and Boosting algorithms are considered as strong methods which are used for improving the performance of some classifiers, which deals with real world as well as artificial datasets. These methods focus mainly on improving the accuracy of the classifiers [24]. These techniques are used widely and are designed specifically for decision trees. Depending upon the performance of old classifiers, boosting methods have the ability to adaptively alter the distribution of training data set, whereas Bagging methods are devoid of this property. [25]

Bagging When multiple learners are trained to solve the same problem, it becomes the machine learning paradigm called Ensemble Learning [26]. Bagging turned out to be the very first efficient method of ensemble learning. The name is derived from bootstrap aggregation. This method does sampling with replacement, i.e., it uses various versions of a training set and these datasets are used to train different models. The outputs obtained from these models are combined so as to obtain a single output. Combination is done through averaging, if it is done for regression, else if it is for classification, the same is done by voting. This method proves out to be really useful in the case of linear models that are unstable.

Boosting Construction of a good single classification rule becomes tedious when the training dataset has a small size when compared to highly dimensional data. The classifiers that are constructed upon such training sets will be biased and will have large variance due to poor estimation of classifier parameters [27]. Boosting is another ensemble meta algorithm which is a part of machine learning, which mainly focuses on reducing the variance and bias that occurs in supervised learning. The weak classifiers are repeatedly learned depending upon a distribution and these learners are added to a final strong classifier. With respect to the accuracy of the weak learner, they are added to the strong classifier by associating typical weights. After the addition of the weak learner, data reweighting is done and in this process, the misclassified samples gain weight and those which are correctly classified lose weight. Hence the misclassifications of the old weak learners form the point of interest of future weak learners.

3 Results and Discussion

Since this is a review paper, there is no concrete section to be included as results. Machine Learning is one strong area of research which has myriad problem solving capabilities and this study is a gateway to the algorithms that are part of it. This aspect gathers utmost importance. The work is now extended for analysis for Amrita School of Engineering students in Coimbatore and results will be published in later phases.

4 Conclusion

Machine learning provides many helpful methods to look at problems that otherwise need manual solution. However, even now ML research could not totally overcome this problem. Many investigators give up into their private studies with a given dataset and work in separation to achieve correct algorithmic solution. Providing solutions to the ML community is the completion of this method. This method creates opportunities for the impact that are widely spread in the field of law, accounts, politics, medical science, education, and also to support systems that should learn, adapt, and take steps to achieve solutions. This paper explains certain methods and algorithms which can be considered as an amalgamation of basic concepts that a machine learning investigator must possess which could in turn make ML the best to make a difference.

Acknowledgments The authors would like to acknowledge Prof. Krishna Shastri (Ex-Joint Director, CIR, Amrita School of Engineering), Sureya Sathiamoorthi and Sree Harini of B.Tech (Computer Science and Engineering), 2010–2014 batch for their support extended in this study. This work was carried out as part of the open-source cloud lab set up by Dr. T. Senthil Kumar, established in Amrita CTS Lab (Amrita Cognizant Innovation Lab) in Amrita School of Engineering, Coimbatore.

References

1. Kohavi R, Provost F (1998) Glossary of terms. Mach Learn 30:271–274
2. Bishop CM (2006) Pattern recognition and machine learning. Springer, New York. ISBN: 0-387-31073-8
3. Mohri M, Rostamizadeh A, Talwalkar A (2012) Foundations of machine learning, the MIT Press ISBN: 9780262018258
4. Harrington P (2012) Machine learning in action: classifying with K-nearest neighbors. New York, pp 18–36. ISBN: 9781617290183
5. Breiman L (2001) Random forests. Mach Learn 45(1):5–32
6. Ali J, Khan R, Ahmad N, Maqsood I (2012) Random forests and decision trees. IJCSI Int J Comput Sci Issues 9(5):3
7. Horning N Introduction to decision trees and random forests, American Museum of Natural History's
8. Yadav SK, Bharadwaj BK, Pal S (2011) Data mining applications: a comparative study for predicting student's performance. Int J Innovative Technol Creative Eng (IJITCE) 1(12):13–19
9. Breiman L, Friedman J, Olshen R, Stone C (1984) Classification and regression trees. Wadsworth, New York
10. Quinlan JR (1986) Introduction of decision tree. In: Journal of machine learning, pp 81–106
11. Park, Hyeoun-Ae (2013) An introduction to logistic regression: from basic concepts to interpretation with particular attention to nursing domain. Korean AcudNurs 43(2):154–164
12. Everitt BS (1998) The Cambridge dictionary of statistics. Cambridge University Press, Cambridge
13. Zhang H, Jiang L, Su J (2005) Augmenting Naive Bayes for ranking. In: International journal of pattern recognition and artificial intelligence, 2005

14. Sheng S, Zhang H (2004) Learning weighted naive Bayes with accurate ranking. In: Proceedings of the fourth IEEE international conference on data mining (ICDM-04), pp 567–570
15. Pazzani M, Domingos P (1997) The optimality of the simple Bayesian classifier under zero-one loss. Mach Learn 29:103–130
16. McCallum A, Nigam K (2003) A comparison of event models for naïve Bayes text classification. J Mach Learn Res 3:1265–1287
17. Rennie JDM, Shih L, Teevan J, Karger DR (2003) Tackling the poor assumptions of Naive Bayes text classifiers. In: Proceedings of the twentieth international conference on machine learning (ICML-2003), Washington DC
18. Ghahramani Z (2004) Gatsby computational neuroscience unit, University College London, Sept 16
19. Hastie T, Robert T, Jerome F (2009) The elements of statistical learning: data mining, inference, and prediction. Springer, New York, pp 485–586. ISBN 978-0-387-84857-0
20. Fasulo D (1999) An analysis of recent work on clustering algorithms, April 26
21. Steinbach M, Karypis G, Kumar V (2000) A comparison of document clustering techniques. In: Proceedings of the KDD-2000 workshop text mining
22. Chavent M, Ding Y, Fu L, Stolowy H, Wang H (2005) Disclosure and determinants studies: an extension using the divisive clustering method (DIV). Eur Account Rev 15(2):181–218
23. Ackermann MR, Blömer J, Kuntze D, Sohler C (2012) Analysis of agglomerative clustering. Algorithmica
24. Skurichina M, Duin RPW (2002) Bagging, boosting and the random subspace method for linear classifiers. Pattern Anal Appl 5:121–135
25. Bauer E, Kohavi R (1999) An empirical comparison of voting classification algorithms: bagging, boosting and variants. Mach Learn 36:105–139
26. Optiz D, Maclin R (1999) Popular ensemble methods: an empirical study. J Artif Intell Res 11:169–198. doi:10.1613/jair.614
27. Breiman L (1996) Bias, variance and arching classifiers. Technical report. Retrieved 19 Jan 2015

Statistical Analysis of Low Back Pain—A Real Global Burden

M. Sajeer and M.S. Mallikarjunaswamy

Abstract Low back pain (LBP) is a serious health problem throughout the world. Almost all persons may experience LBP at any time during their lifetime. There are many reasons related with this global health burden. Age, profession, body weight, etc. are some of the reasons for LBP. This paper presents a statistical analysis of LBP and discusses how they are related with daily life. Here we collected the data of about 153 subjects of age between 20 and 70 years with LBP and analyzed the results. We completed our study by dividing the population into two groups based on age and sex, and concluded that the chances of LBP are almost equal in both males and females and is generally related with the daily routine of a person.

Keywords Herniated disc · Low back pain (LBP) · Lumbar region · Spinal degeneration · Statistical analysis

1 Introduction

Back pain is a serious health problem which causes pain in the back of the body. Usually, it may originate from muscles, nerves, bones, joints, and other structures in the spine. It is one of the common health problems, which causes financial burden in the world [1–4]. According to a study, almost 80 % of the adults experience back pain during their lifetime. It will affect men and women equally and is associated with the loss of quality of life [5]. An US study reveals that acute LBP is the fifth most reason for physician visits in the U.S.A.; about 50 % of the working adults were having LBP which causes about 40 % of the missed days of their work every year [6, 7]. It is the main reason for disability worldwide [8].

M. Sajeer (✉) · M.S. Mallikarjunaswamy
Department of Instrumentation Technology,
Sri Jayachamarajendra College of Engineering, Mysore, Karnataka, India
e-mail: sajeermuhammed@gmail.com

M.S. Mallikarjunaswamy
e-mail: ms_muttad@yahoo.co.in

L.P. Suresh and B.K. Panigrahi (eds.), *Proceedings of the International Conference on Soft Computing Systems*, Advances in Intelligent Systems and Computing 397, DOI 10.1007/978-81-322-2671-0_11

Back pain is generally related with aging process; however, sedentary life styles with less exercise also causes LBP in the young adults. It is more prevalent in people aged between 40 and 80. As the age increases, the risk of experiencing LBP due to disc disease or spinal degeneration also increases. Even though, most of the time the exact cause of pain cannot be found, the common cause is related with the daily routine of a person [9–11].

LBP may be acute (short term), subacute, and chronic based on the duration of pain. Acute back pain lasts from a few days to a few weeks. Subacute pain lasts between 4 and 12 weeks. But the chronic pain persists for 12 weeks or longer. Generally, acute pain and subacute pain can be relieved by nonsurgical treatments but in some cases chronic pain can be cured only with surgical treatment [11].

LBP mainly occurs in the lumbar region, which consists of five vertebrae (L1–L5) and supports most of the weight of the upper body. The space between adjacent vertebrae are maintained by intervertebral discs, filled with a jelly–like material called nucleus pulposus, which acts as a shock absorber keeping the vertebrae from rubbing together, surrounded by a fibrous ring called annulus fibrosus. It also functions as a ligament to hold the vertebrae together.

In case of older people, LBP is mainly due to spondylosis, which is the general degeneration of the spine associated with normal tear and wear.

Mechanical causes [11] of LBP include:

Sprains and strains Generally, acute pain is mostly due to sprains and strains. The main reason for sprain is the overstretching or tearing of ligaments and tears in tendons or muscles. They can occur when we lift something improperly or heavy objects.

Intervertebral disc degeneration It is a condition of lose of cushioning ability of intervertebral discs due to aging and is one of the common causes of LBP.

Herniated or ruptured discs Herniation is the phenomenon of bulging of intervertebral discs in outward direction due to compression, which causes LBP.

Radiculopathy Compression, inflammation, or injury to a spinal nerve root is responsible for the condition of radiculopathy, which causes pain and is radiated to other parts of the body.

Sciatica This is a type of radiculopathy, which is due to the compression of the static nerve—a nerve that passes through the buttocks and extends toward the back of the leg. This causes burning LBP combined with the pain in buttock and back of the leg.

Spondylolisthesis Another mechanical reason for LBP, in which a vertebra of the lower spine slips out of place which causes pinching of the nerves and exiting through the spinal column.

A traumatic injury It may occur from accidents, which can injure tendons, ligaments, or muscles, which causes LBP. In some cases, this injury causes over compression of spine resulting in intervertebral disc herniation or rupturing.

Spinal stenosis Here narrowing of spinal column occurs, which in turn puts pressure in the spinal cords and nerves resulting in LBP.

Skeletal irregularities Certain skeletal irregularities like scoliosis (curvature of the spine); lordosis (an abnormally accentuated arch in the lower back) also causes LBP.

LBP is also related to certain serious underlying conditions like infections in the vertebra, tumors, cauda equina syndrome (ruptured disc), abdominal aortic aneurysms (enlargement of legs), kidney stones, inflammatory diseases of the joints, osteoporosis (metabolic bone disease), endometriosis (enlargement of uterine tissue), fibromyalgia (muscle pain and fatigue), etc. [11].

Different studies have been carried out in different countries to investigate the characteristics of LBP. Astrand [12] in 1987 completed a study regarding the back pain of male employees in a Swedish pulp and paper industry. Battié and Bigos [13] have done a study about 'Industrial back pain complaints' at North America in 1991. In 2000, Punnett and Herbert [14] carried out a study about 'work-related musculoskeletal disorders' at Canada. Punnet et al. [15] have done a statistical analysis of LBP, related with occupational exposures in 2005. In United Kingdom, Dunn et al. [16] performed an analysis of LBP between 2001 and 2003, among 342 consulters and classified the patients based on pain intensity score. Sakai [17] conducted a study among 387 elderly persons of age between 60 and 81 of Japan, about the osteophyte formation in the lumbar spine and its relevance to the LBP. Murtezani et al. [18] recently completed a study among power plant workers in Kosovo in 2012 and examined the correlation between age and job tenure with LBP.

Also many studies have been carried out to find the effectiveness of different treatments for LBP. Cramer et al. [19] recently carried out a study about the meta-analysis of Yoga among LBP patients in 2013. Bhatikar and Magdum [20] investigated about the effectiveness of shortwave diathermy for management of chronic mechanical back pain in 2014.

We conducted a statistical analysis of LBP and the aim of our study was to investigate different factors related to them.

2 Materials and Methods

2.1 Study Design

This study was based on the data collected from the medical records of patients with LBP at Govt. medical college, Trivandrum, Kerala, India from January 2013 to December 2014.

2.2 Questionnaire

We also collected the data using a questionnaire which was filled by the LBP patients. The questionnaire comprised of personal information and details of diagnosed data.

2.3 Study Population

We investigated medical records of 153 patients with LBP. The population comprised of age ranged between 20 and 70 years of both male and female.

2.4 Study Variables

Patient data were collected from the hospital records and extracted information like age, sex, body mass index (BMI), occupation, smoking history etc. Also collected diagnosed data like presenting symptoms, duration of symptoms, causes of LBP, nature of pain, diagnosed discs and regions, nature of discomfort due to pain, recommended treatments etc.

2.5 Statistical Analysis

For analysis purpose, we have divided the patients into two groups based on their age and sex; one group is between the age of 20 and 45 years and the other between 46 and 70 years. For each group, we have collected the data and analyzed the results.

3 Results

3.1 Patient Characteristics and Symptoms

We collected about 153 patient's data with the help of medical records and questionnaire which consists of 88 male patients and 65 female patients with LBP. Table 1 gives the patient characteristics that we obtained from the analysis. BMI obtained for male patients between the ages of 20–45 was 24.4 ± 2.1 and for the age group 46–70, it was 24.8 ± 3. For female patients, the BMI was 24 ± 2.6 and 23.9 ± 2.8, respectively, for the above-mentioned age groups. About 18 % of office workers, 30 % uses two wheelers for their job, 15 % of drivers, 22 % of labors of

Table 1 Characteristics and symptoms of patients with low back pain (*n*-total number of subjects)

Parameters	Age in years (male)		Age in years (female)	
	(20–45) (n = 54)	(46–70) (n = 34)	(20–45) (n = 41)	(46–70) (n = 24)
Body mass index (BMI): (mean ± SD, kg/m^2)	24.4 ± 2.1	24.8 ± 3	24 ± 2.6	23.9 ± 2.8
Occupation				
Office worker	10 (18 %)	8 (24 %)	16 (39 %)	5 (21 %)
Uses 2 wheelers for job	16 (30 %)	3 (9 %)	–	–
Driver	8 (15 %)	9 (26 %)	–	–
Labor	12 (22 %)	10 (29 %)	5 (12 %)	2 (24 %)
Housewife	–	–	15 (37 %)	11 (46 %)
Others	8 (15 %)	4 (12 %)	5 (12 %)	6 (25 %)
Smoking habit				
Yes	16 (30 %)	20 (59 %)	–	–
No	38 (70 %)	14 (41 %)	–	–
Nature of pain				
No pain	2 (4 %)	–	3 (7 %)	–
Moderate pain	40 (74 %)	18 (53 %)	28 (68 %)	13 (54 %)
Chronic pain	12 (22 %)	16 (47 %)	10 (25 %)	11 (46 %)
Duration of pain				
Acute	34 (63 %)	18 (53 %)	27 (66 %)	14 (58 %)
Subacute	12 (22 %)	9 (26 %)	8 (19 %)	3 (13 %)
Chronic	8 (15 %)	7 (21 %)	6 (15 %)	7 (29 %)
Sleep disturbance due to pain				
Yes	14 (26 %)	20 (59 %)	14 (34 %)	16 (67 %)
No	40 (76 %)	14 (41 %)	27 (66 %)	8 (33 %)
Whether working because of pain				
Yes	48 (89 %)	15 (44 %)	35 (85 %)	10 (42 %)
No	6 (11 %)	19 (56 %)	6 (15 %)	14 (58 %)

male experienced back pain within the age group of 20–45 and for the age group of 46–70, the calculations were 24, 9, 26, and 29 % respectively. Around 30 % have a positive smoking history in the first age group and for second age group around 59 % were smokers, for male patients. Most of them had acute back pain (63 and 53 %, respectively) for the above two age group of male. Around 11 % were not working because of LBP in the first age group and for second age group it was about 56 %.

For female patients about 39 % of office workers, 37 % housewives experienced back pain within the age group of 20–45 and for the age group of 46–70, it was around 21 and 46 %, respectively. About 66 % of the female had symptoms of acute back pain, 19 % of sub-acute pain, and 15 % had chronic pain in the first age group, and for second age group it was 58, 13, and 29 %, respectively. Around 15 % of the women were not working because of LBP in the first category and for the second category it was 58 %.

3.2 MRI Findings and Treatment Modalities

MRI findings and treatment modalities obtained are shown in Table 2. Disc herniation was the main diagnosis investigated in the age group of 20–45 and for the age group of 46–70, spondilosis and degenerative disc disease were the causes of LBP in male and female. In all the cases, the diagnosed discs were mainly L4–L5

Table 2 MRI findings and treatment modalities (*n*-total number of subjects)

Parameters	Age in years (male)		Age in years (female)	
	(20–45) (n = 54)	(46–70) (n = 34)	(20–45) (n = 41)	(46–70) (n = 24)
Diagnosis				
Disc herniation	12 (22 %)	6 (18 %)	10 (24 %)	3 (13 %)
Disc extrusion	9 (17 %)	3 (9 %)	6 (15 %)	2 (8 %)
Disc protrusion	8 (15 %)	3 (9 %)	7 (17 %)	2 (8 %)
Spinal stenosis	6 (11 %)	2 (6 %)	4 (10 %)	2 (8 %)
Schmorl's node	2 (4 %)	2 (6 %)	3 (7 %)	1 (4 %)
Annular defects	4 (7 %)	3 (9 %)	2 (6 %)	2 (8 %)
Degenerative spondilosis	5 (9 %)	6 (18 %)	3 (7 %)	4 (17 %)
Intervertebral disc degeneration	6 (11 %)	5 (14 %)	3 (7 %)	3 (13 %)
Osteoporosis	2 (4 %)	4 (11 %)	3 (7 %)	5 (21 %)
Diagnosed disc				
T12–L1	1 (2 %)	1 (3 %)	1 (2 %)	1 (4 %)
L1–L2	4 (7 %)	2 (6 %)	3 (5 %)	1 (4 %)
L2–L3	4 (7 %)	3 (9 %)	3 (5 %)	3 (13 %)
L3–L4	6 (11 %)	4 (12 %)	3 (5 %)	4 (16 %)
L4–L5	22 (41 %)	11 (32 %)	16 (39 %)	7 (29 %)
L5–S1	17 (32 %)	13 (38 %)	15 (44 %)	8 (33 %)
Recommended treatment				
Surgical	9 (17 %)	12 (35 %)	8 (19 %)	10 (42 %)
Nonsurgical	45 (83 %)	22 (65 %)	33 (81 %)	14 (58 %)
Whether pain is relieved after treatment				
Yes	40 (74 %)	25 (73 %)	33 (80 %)	15 (62 %)
No	14 (26 %)	9 (27 %)	8 (20 %)	9 (38 %)
Main cause of low back pain is related with				
Aging	8 (15 %)	15 (45 %)	8 (20 %)	12 (50 %)
Profession	22 (41 %)	8 (23 %)	16 (39 %)	5 (21 %)
Obesity	10 (18 %)	7 (20 %)	9 (22 %)	4 (17 %)
Accidents	8 (15 %)	2 (6 %)	3 (7 %)	1 (4 %)
Genetics	4 (7 %)	1 (3 %)	3 (7 %)	1 (4 %)
Others	2 (4 %)	1 (3 %)	2 (5 %)	1 (4 %)

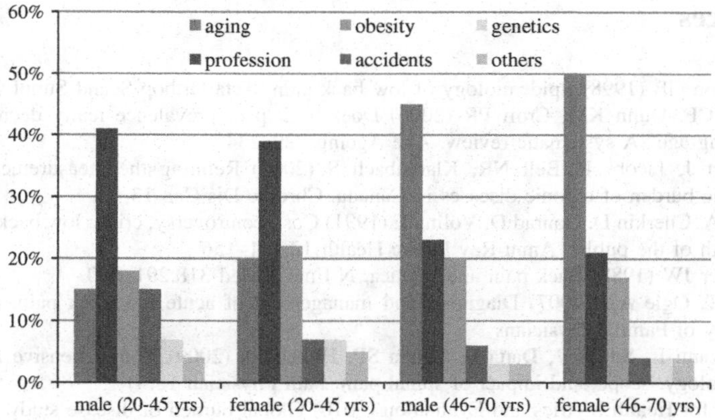

Fig. 1 Percentage variations of LBP in male and female with factors like aging, profession, obesity, etc

and L5–S1. Nonsurgical treatments like physiotherapy, meditation like Yoga, exercises, etc., were generally recommended and for some cases surgical treatments were performed. From the analysis we concluded that low back pain was mainly related with profession, age, obesity, etc.

Figure 1 represents a graph showing the variations of factors which causes LBP in both the age groups. From the graph, it can be seen that for both male and female patients of age between 20–45 years, LBP is mainly related with their profession, for the age group of 46–70 years it is related with aging.

4 Discussion

The investigated data showed that there are different factors which may be the cause of LBP. The factors like age and profession of a patient are directly related to back pain. Sedentary life styles and lack of exercises are also related to them. In most of the cases, LBP is mechanical in nature and may be acute. They can be overcome by changing the life style, doing exercises and taking rest between continuous works. Surgical treatments are needed only for few cases. From the presented data, we can see that there is almost equal chance for the occurrence of low back pain in both male and female.

Acknowledgments The authors wish to thank Dr. Shanavas, Assistant Professor, Department of Orthopedic, Govt. Medical College, Trivandrum, Kerala, India for his valuable help and suggestions during the preparation of this paper and Mr. Shibu, Managing Director, Metro Scan Centre, Kerala, who helped in getting the required MRI data.

References

1. Andersson GB (1998) Epidemiology of low back pain. Acta Orthop Scand Suppl 28–31
2. Dionne CE, Dunn KM, Croft PR (2006) Does back pain prevalence really decrease with increasing age? A systematic review. Age Ageing 229–234
3. Rapoport J, Jacobs P, Bell NR, Klarenbach S (2004) Refining the measurement of the economic burden of chronic diseases in Canada. Chronic Dis Can 13–21
4. Deyo RA, Cherkin D, Conrad D, Volinn E (1991) Cost, controversy, crisis: low back pain and the health of the public. Annu Rev Public Health 12:141–156
5. Frymoyer JW (1988) Back pain and sciatica. N Engl J Med 318:291–300
6. Patel AT, Ogle AA (2007) Diagnosis and management of acute low back pain. American Academy of Family Physicians
7. Manchikanti L, Singh V, Datta S, Cohen SP, Hirsch JA (2009) Comprehensive review of epidemiology, scope, and impact of spinal pain. Pain Physician 12(4)
8. Institute for Health Metrics and Evaluation (2010) Global burden of disease study
9. Hoy DG, Bain C, Williams G, March L, Brooks P, Blyth F, Woolf A, Vos T, Buchbinder R (2012) A systematic review of the global prevalence of low back pain. Arthritis Rheum 64(6):2028–2037
10. Medline Plus (2013) Low-back pain-chronic. http://www.nlm.nih.gov/medlineplus/ency/article/007422.htm. Accessed 11 Mar 2013
11. National Institute of Neurological Disorders and Stroke (2013) Low-back pain fact sheet. http://www.ninds.nih.gov/disorders/backpain/detail_backpain.htm. Accessed 11 Mar 2013
12. Astrand NE (1987) Medical, psychological, and social factors associated with back abnormalities and self reported back pain: a cross sectional study of male employees in a Swedish pulp and paper industry. Brit J Industr. Med 44:327–336
13. Battié MC, Bigos SJ (1991) Industrial back pain complaints: a broader perspective. Orthop Clin North Am 22:273–282
14. Punnett L, Herbert R (2000) Work-related musculoskeletal disorders: Is there a gender differential, and if so, what does it mean? In: Goldman MB, Hatch MC (eds) Women and health san diego. Academic Press, CA, pp 474–492
15. Punnet L et al (2005) Estimating the global burden of low back pain attributable to combined occupational exposures. Am J Ind Med
16. Dunn KM, Jordan K, Croft PR (2006) Characterizing the course of low back pain: a latent class analysis. Am J Epidemiol Adv Access
17. Sakai Y (2003) Osteophyte formation in the lumber spine and relevance to low back pain
18. Murtezani A et al (2012) Low back pain among Kosovo power plant workers-a survey. Ital J Public Health 9
19. Cramer H et al (2013) A systematic review and meta-analysis of yoga for low back pain. Clin J Pain 29:450–460
20. Bhatikar K, Magdum C (2014) Comparison of efficacy of shortwave diathermy (SWD) and high velocity low amplitude thrust manipulation (HVLATM) for management of chronic mechanical low back pain. Int J Res Health Sci 2

Robust Text Detection and Recognition in Blurred Images

Sonia George and Noopa Jagadeesh

Abstract In this paper, we propose a novel method for detecting and recognizing the text from the blurred images. Text detection in natural scenery images is an important issue in the processing stage. All the previously proposed methods use different algorithms to detect text in images; however, they suffer from poor performance while performing detection in blurred images. The proposed algorithm is capable of removing blur with an iterative deconvolution method and a linear invariant filter. The proposed method can achieve detection and recognition of the text with a time complexity of 4.53 s. Experiments show our method achieves a better text detection than the other existing methods.

Keywords Text detection · Text recognition · Image deblurring · Text segmentation

1 Introduction

A small piece of text in an image gives information regarding that item which helps to understand its objective more easily. Text detection is a technique, through which the text in an image can be identified and extracted. Nowadays, tremendous improvements are seen in the field of digital information over Internet. Browsing and finding pictures in large-scale and heterogeneous collections are important problems which have attracted much research within the information retrieval. Due to the complexity of background, font, and noise in images, text detection becomes a more difficult task.

Existing methods for text detection is presented in Fig. 1.

S. George (✉) · N. Jagadeesh
Computer Science and Engineering, Mahatma Gandhi University Federal Institute of Science and Technology, Angamaly, Kerala, India
e-mail: soniageorge111@gmail.com

L.P. Suresh and B.K. Panigrahi (eds.), *Proceedings of the International Conference on Soft Computing Systems*, Advances in Intelligent Systems and Computing 397, DOI 10.1007/978-81-322-2671-0_12

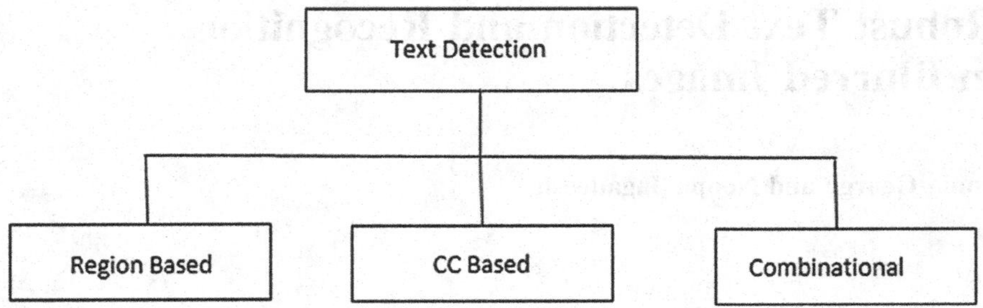

Fig. 1 Classification methods for text detection methods

The region-based method [1] uses a window in order to extract the text region from the image. The CC-based method [2] uses connected component analysis to find the connected character in the image. The combinational method [3] uses both region-based and CC-based method to extract the text line. More recently, the maximally stable extremal region (MSER)-based methods are used for text detection and obtain more promising results. Existing methods are only capable of detecting texts from images without blur. Chen [4] proposes a text detection algorithm capable of detecting the text from blurred images. It uses canny edge detection to reduce the effect of noise. Thillou et al. [5] which proposes a selective metric-based clustering to extract the textual information. Here, clustering metric and Log-Gabor filters are used parallelly to identify the text.

Wexler method [6] mainly focuses on the detection of text with the help of stroke width transform. It is a novel image operator to find the value of stroke width. Neumann et al. [7] propose a method which uses an effectively pruned exhaustive search. Chuci Yi et al. [8] have presented an algorithm for text localization, worked by the combined action of boundary clustering, stroke segmentation, and string fragment classification. The proposed framework is to localize the text regions under complex backgrounds and multiple text appearances. Jiri Matas et al. [9] performed text localization and recognition in real-time images. The method uses class-specific extremal regions. Here extremal regions are selected from the whole component tree of the image. Selection of extremal region is different from MSER.

Shi et al. [10] propose a graph model based on maximally extremal regions. It detects the MSER region and constructs an irregular graph with a MSER node. Yin et al. [11] propose a new method for text detection. First it explores the hierarchical structure of MSER and designs the MSER pruning algorithm. A self-training algorithm is used to cluster the text candidate. The algorithm uses a character classifier to extract text and nontext candidates and eliminates the nontext candidates having higher efficiency.

In this paper, we propose a novel method for text detection and recognition from the blurred images. Blur can be of either gaussian or motion type. Motion-type blur can be due to the movement of the camera. First, blurred images are fed into the system and then the images are enhanced during the blur removal phase. Enhanced images are input to text detection phase. MSER and stroke width of a character helps to detect the text. Detected text is given to the recognition phase. Finally, the above method is integrated to produce a robust text detection and recognition system. Our method shows a considerable improvement in detecting the output than Chen's method [4]. Moreover, experiments on multilingual images, street view images, and scene images show an improvement over traditional methods.

Rest of this paper is organized as follows: In Sect. 2, the materials and methods are discussed. The experimental result of the proposed system is described in Sect. 3. Section 4 discusses the comparison results. Finally, Sect. 5 summaries the conclusion of this paper.

2 Materials and Methods

By incorporating all the improvements over the traditional method, we have proposed a robust method for text detection from the blurred images. The flowchart of the proposed system is presented in Fig. 2. The proposed method includes the following stages:

A. Enhanced Input Image
   Blur in the input images are identified and corresponding blur removal techniques are applied. It either uses iterative deconvolution or linear time-invariant filter to deblur the input.
B. Text Extraction
   Characters are extracted using MSER. Remaining characters are identified by allowing the edges to grow in opposite to the edge direction. After segmentation, geometrical filtering is performed to extract the text.
C. Text Detection
   Distance weight and stroke width of each extracted character is calculated and the result is normalized. After stroke width calculation, a probability is calculated to eliminate nontext characters from text characters. Finally, detect the text from image.
D. Text Recognition
   Text recognition is performed by optical character recognition (OCR) method. It recognizes the input with a region of interest.

The proposed work as a whole is integrated and shown in Fig. 3.

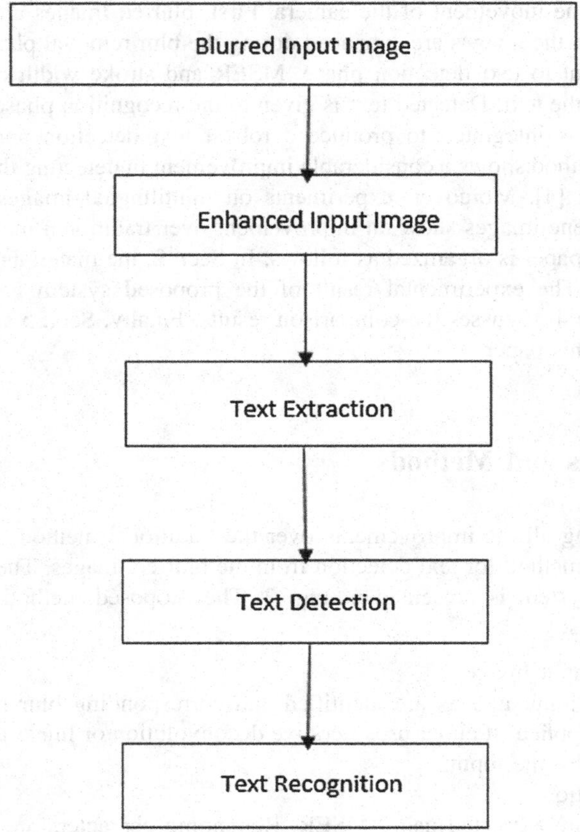

Fig. 2 Flowchart of the proposed method

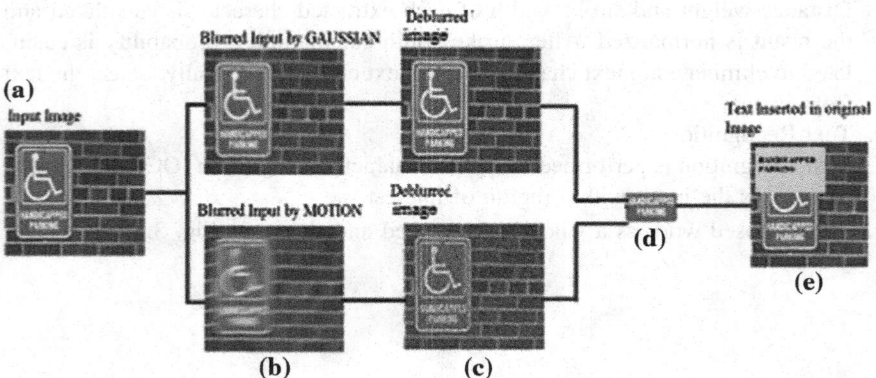

Fig. 3 The framework of robust text detection and recognition from blurred images. **a** Input image. **b** Blurred the image using either Gaussian or motion. **c** Deblur the input image. **d** Extracted text from input. **e** Recognition of text input

2.1 Blur Removal Phase

Presence of blur in an image is a pitfall to the text detection process. Several algorithms are available for the detection of blur. Blur can be of different types mainly gaussian and motion. Detection and removal of blur can be performed by deblurring techniques [12, 13]. Iterative deconvolution is applied to remove gaussian blur from an image. Linear time-invariant filter is used to deblur the motion blur from the input image. Figure 4 shows blurred and deblurred image example.

2.2 Text Extraction Phase

Enhanced image is given as an input to the extraction phase. Sort all pixels in the enhanced image according to their intensity. After sorting, find the pixels having higher or lower intensity than its outer boundary. Select the regions having stable local binarization over a large range of threshold that are candidates to the text detection. Find the thickness of the outer and inner boundaries of objects in an image. Eliminate the selected region outside the detected boundary. It will produce the text region. Geometrical filtering is applied to get noiseless output. Filtering is applied with the help of features like area, eccentricity, and solidity.

2.3 Text Detection Phase

Distance weight of each text region is calculated using Eucledian distance. The stroke width of each character is computed by averaging its pixel stroke width. As shown in Fig. 5, normalize the stroke width and filter the unwanted pixels. Determine the probability to eliminate nontext characters from text characters.

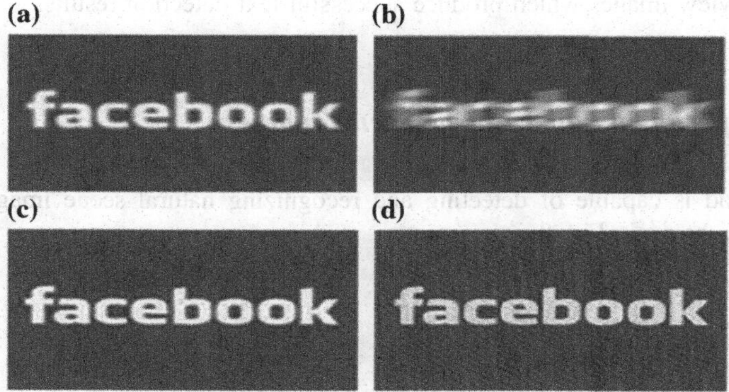

Fig. 4 **a** Gaussian-blurred image. **b** Motion-blurred image. **c** Deblur input a using iterative deconvolution method. **d** Deblur input *B* using linear time-invariant filter method

Fig. 5 Stroke width of the input image *A*

2.4 Text Recognition Phase

The detected text region is given to the recognition phase. OCR [14] is used to recognize the text. OCR uses a region of the input to recognize the text.

3 Results

Our system exhibits an average time of 4.53 s for processing an image. Proposed technology is profiled on a Windows laptop with a 2.00 GHZ processor. Figure 6 shows the successful and the failed samples.

The proposed method was applied on different types of image sets. They are:

3.1 Experiments on Street View Images

Our proposed method is capable of detecting street view images. We considered around 20 images, out of which 15 images are correctly classified. Figure 7 shows the street view images which produce successful text detection results.

3.2 Experiments on Natural Scene Images

Our method is capable of detecting and recognizing natural scene images. The results are shown in Fig. 8.

Fig. 6 Shows **a** for correctly detected samples. **b** Failed samples

Fig. 7 Correctly classified street view images

Fig. 8 Correctly classified natural scene images

3.3 Experiments on Multilanguage Images

The proposed system can detect and recognize the multilingual images. The results are shown in Fig. 9.

4 Discussion

In this section, we compare the proposed method with David Chen's method [4]. Chen's method fails to detect the text with a motion blur. Figure 10 shows the 3D graph obtained using our proposed method. Chen's method fails to generate the 3D graph of the detection output.

Figure 11 shows the contour graph of our method and David Chen's method [4]. The proposed method outperforms Chen's method. Chen's method failed to detect the text within a motion-blurred image. The area graphs obtained for both the methods are shown in Fig. 12.

Fig. 9 Correctly classified multilingual images

Fig. 10 3D graph of an output obtained after text extraction

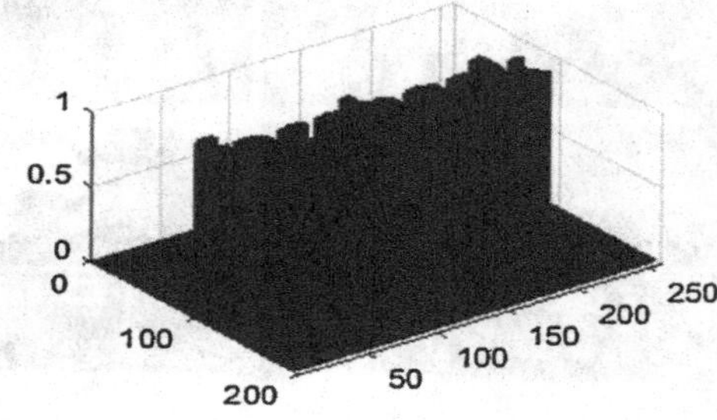

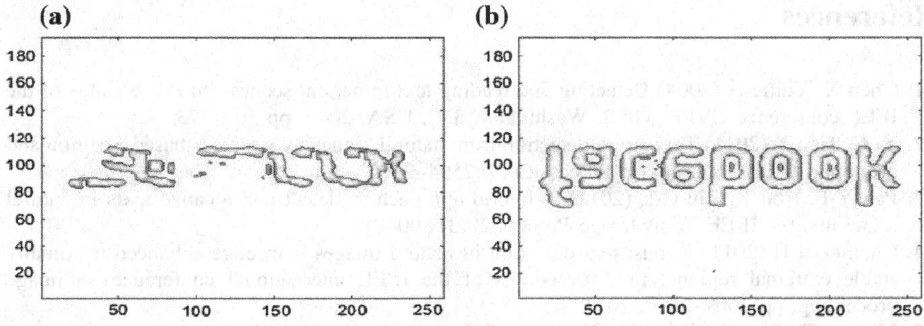

Fig. 11 **a** Contour graph for text detection by Chen's method. **b** Contour graph for text detection by the proposed method

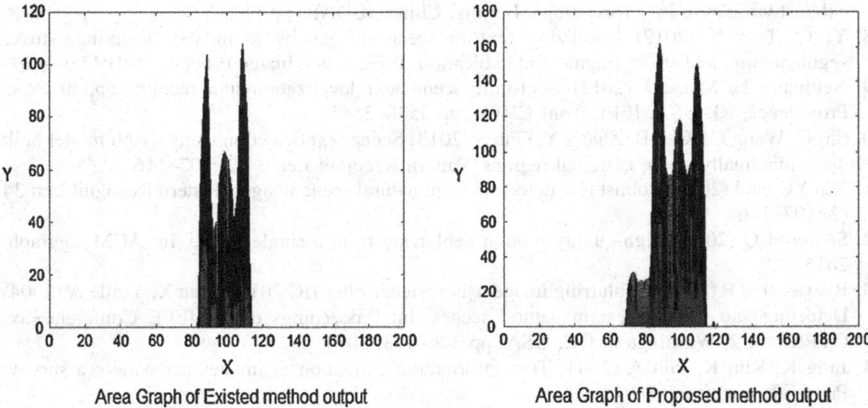

Fig. 12 Comparison of outputs of area graph for the proposed method and the existing method

5 Conclusion

We have proposed a fast and accurate algorithm for text detection from the blurred input image. The algorithm employs an image enhancement by deblurring to avoid the noise sensitivity of MSER. Text candidates are identified by intersection between the maximal stable extremal regions with edges. Stroke width computation helps to filter the nontext region and estimates a probability to eliminate it from the text regions. Integrate the above to build a robust scene text detection and recognition system which exhibits superior performance.

Acknowledgments I am thankful to Mrs. Noopa Jagdeesh, Assistant Professor of Computer Science Department, FISAT, Kerala for her keen interest and useful guidance in my paper.

References

1. Chen X, Yuille A (2004) Detecting and reading text in natural scenes. In: Proceedings of the IEEE conference CVPR, vol 2. Washington, DC, USA, 2004, pp 366–373
2. Yi C, Tian Y (2011) Text string detection from natural scenes by structure-based partition and grouping. IEEE Trans Image Process 20(9):2594–2605
3. Pan Y-F, Hou X, Liu C-L (2011) A hybrid approach to detect and localize texts in natural scene images. IEEE Trans Image Process 20(3):800–813
4. Chen et al H (2011) Robust text detection in natural images with edge enhanced maximally stable extremal regions. In: Proceedings of the IEEE international conference on image processing, pp 2609–2612, 2011
5. Mancas-Thillou C, Gosselin B (2007) Color text extraction with selective metric-based clustering. Comput Vis Image Und 107(1–2):97–107
6. Epshtein B, Ofek E, Wexler Y (2010) Detecting text in natural scenes with stroke width transform. In: Proceedings of the IEEE Conf CVPR. San Francisco, CA, USA, pp 2963–2970
7. Neumann L, Matas J (2011) Text localization in real-world images using efficiently pruned exhaustive search, In: Proceedings. Beijing, China, ICDAR, pp 687–691
8. Yi C, Tian Y (2012) Localizing text in scene images by boundary clustering, stroke segmentation, and string fragment classification. IEEE Trans Image Process 21(9):4256–4268
9. Neumann L, Matas J (2012) Real-time scene text localization and recognition, in Proc. Providence, RI, USA, IEEE Conf CVPR, pp 3538–3545
10. Shi C, Wang C, Xiao B, Zhang Y, Gao S (2013) Scene text detection using graph model built upon maximally stable extremal regions. Pattern Recognit Lett 34(2):107–116
11. Yin YC et al (2014) Robust text detection from natural scene images. Pattern Recognit Lett 34 (2):107–116
12. Sha et al Q (2015) High-quality motion deblurring from a single image. In: ACM siggraph, 2015
13. Biswas et al P (2015) Deblurring image using wiener filter IJC 2015. Chen X, Yuille A (2004) Detecting and reading text in natural scenes. In: Proceedings of the IEEE Conference on CVPR, vol 2. Washington, DC, USA, pp 366–373, 2004
14. Jung K, Kim K, Jain A (2004) Text information extraction in images and video: a survey. Pattern Recognit 37(5):977–997

Video Analysis for Malpractice Detection in Classroom Examination

T. Senthil Kumar and G. Narmatha

Abstract Malpractice in examinations is one of the banes in the education system of a nation. This paper aims in developing a robust face detection algorithm that tracks and analyzes multiple faces in a real-time video scene in a classroom examination. This work proposes automated face detection from a preprocessed surveillance video. First, the foreground object is extracted and the face region is detected using the Haar cascades. Second, the activity classification is performed to detect whether it is normal or suspicious, based on the face orientation, hand contact detection using background subtraction, and Gaussian mixture model (GMM). This system detects the commonly occurring suspicious activities like object exchange, peeping into others' answer sheet, and people exchange during examination. Automated suspicious activity detection would help in decreasing the error rate due to manual monitoring. Promising results were obtained using this approach which show significant improvement over the existing methods, especially in case of illumination and orientation.

Keywords Viola–Jones algorithm · Face detection · Integral image · Haar features

1 Introduction

Video analytics has proven to be a key factor that performs a variety of tasks ranging from real-time analysis of a video for fast detection of events of interest to analysis of a prerecorded video for the purpose of extracting events and information

T. Senthil Kumar (✉) · G. Narmatha
Department of CSE, Amrita School of Engineering, Amrita Vishwa Vidhyapeetham, Coimbatore 641112, India
e-mail: t_senthilkumar@cb.amrita.edu

G. Narmatha
e-mail: narmatha.gopal@gmail.com

L.P. Suresh and B.K. Panigrahi (eds.), *Proceedings of the International Conference on Soft Computing Systems*, Advances in Intelligent Systems and Computing 397, DOI 10.1007/978-81-322-2671-0_13

from the recorded video. It automatically monitors cameras and alerts for events of interest in many cases which are more effective than depending on a human operator, which is a costly resource with limited alertness and attention. Analyzing a recorded video is a need that can rarely be answered effectively by human operators. Examination malpractice is an illegal behavior by a candidate during the examination which is a cankerworm that threatens the nation. This unlawful behavior of students in the examination hall makes them idle and their seriousness toward education gets reduced.

Monitoring an examination hall is a very challenging task in terms of man power which may be prone to error during human supervision. Hence, a system should be designed to automatically detect suspicious activities and help in minimizing such malpractices in examination hall. Moreover, the probability of error must be lesser. This system will serve as a useful surveillance system for educational institutions. This paper aims in analyzing a real-time video used for monitoring human activities in an examination hall, and thus helping to identify whether the particular person's activity is suspicious or not. The system developed identifies abnormal head motions and hand contact detections, thereby prohibiting copying. It also identifies a student moving from one place to another. Finally, the system detects contacts between students and hence prevents passing incriminating material among students. This work proposes a thoughtful algorithm that can monitor and analyze the activities of students in an examination hall and can quickly alert the educational institution on occurrence of any such malpractices.

2 Literature Survey

Human behavior analysis involves stages such as motion detection with the help of background modeling and foreground segmentation, object classification, motion tracking, and activity recognition. For person identification from a surveillance video, there were several methods employed in the early systems. The proposed system uses a face recognition technique to identify students present in the examination hall and to locate features such as eyes, ears, nose, mouth, etc., from the detected face region image.

Turk and Pentland discovered that while using the Eigenface technique, the residual error could be used to detect faces in images, a discovery that enabled reliable real-time automated face recognition systems. PCA removes the correlations among the different input dimensions and significantly reduces data dimensions. But PCA is sensitive to scale and hence it should be applied only on data that have approximately the same scale. It may lose important information on discrimination between different classes.

Tsong-Yi Chen, Chao-Ho Chen, Da-Jinn Wang, and Yi-Li Kuo proposed skin color-based face detection approach which is based on the NCC color space, where the initial face of a candidate is obtained by detecting the skin color region and then the face feature of the candidate is analyzed to determine whether it is the candidate's real face or not. It allows fast processing and is robust to resolution changes but has higher probability of false positive detection.

The local binary pattern (LBP) introduced by Ojala et al. is one such operator which is robust to monotonic illumination variations [1]. Thus, various face detection systems have been proposed using LBP or its variants, such as improved local binary patterns (ILBP). It should be noted that the basic LBP features have performed very well in various applications, including texture classification and segmentation, image retrieval, and surface inspection. The most important properties of LBP features are their tolerance to monotonic illumination changes and their computational simplicity.

Haar features approach introduced by Viola–Jones can be used for face recognition. The strength of this methodology is that it gives low false positive rate. Hence, it has a very high accuracy while detecting the faces, it needs to be coupled with a classification algorithm like AdaBoost to give the best performance and hence has an extra overhead attached to it. It has high detection accuracy and minimizes the computational time. The approach was used to construct a face detection system which is approximately 15 times faster than any previously mentioned approaches.

3 Proposed Method

The proposed system identifies various kinds of malpractices. As seen in Fig. 3.1, the real-world video is fed into the system. Preprocessing consists of background subtraction and noise removal [2]. These preprocessed frames are used to identify suspicious activities by the two basic techniques: head motion detection and hand contact detection. The head motion detection is identified by detecting the human face (Fig. 1).

The proposed framework mainly consists of two parts: the first is to detect the face region and the second is to classify the activity of the person based on the orientation of the detected face and hand contact. When a frame is captured from a video sequence, it is preprocessed and a face detection method is performed to determine the position of the face. Next, the face region's local features are extracted to examine their activity whether it is a normal or a suspicious activity. The face region is detected using the extracted Haar feature from the given image and the GMM. Then, their behavior is monitored and the activity is classified using AdaBoost classifier [3]. The three major contributions/phases of algorithm are: feature extraction, classification using boosting, and cascading.

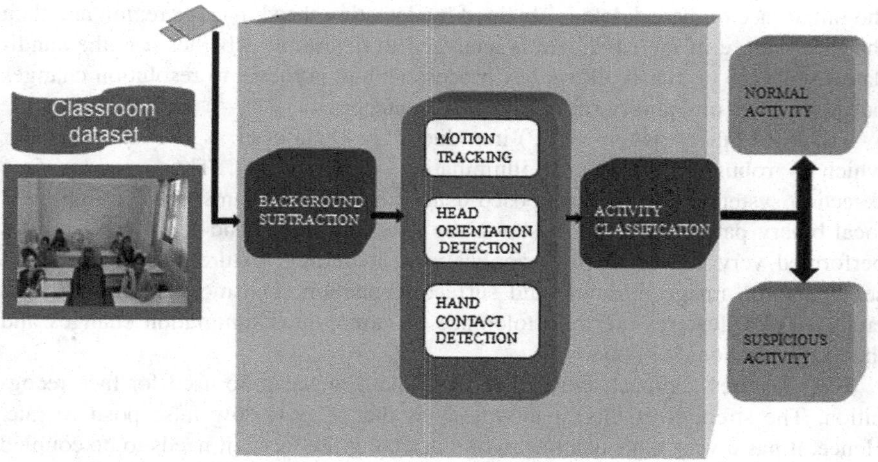

Fig. 1 Architecture diagram for malpractice detection in classroom examination

3.1 Face Detection

3.1.1 Integral Image

Integral image is computed by the summed area table in which every block is the summation of the previous block. The value at any point (*x*, *y*) in the summed area table is just the sum of all the pixels above and to the left of (*x*, *y*)

$$I(x,y) = \sum_{\substack{x' \leq x \\ y' \leq y}} i(x',y') \tag{1}$$

Moreover, the summed area table can be computed efficiently in a single pass over the image in Fig. 2a, using the fact that the value in the summed area table at (*x*, *y*) is just

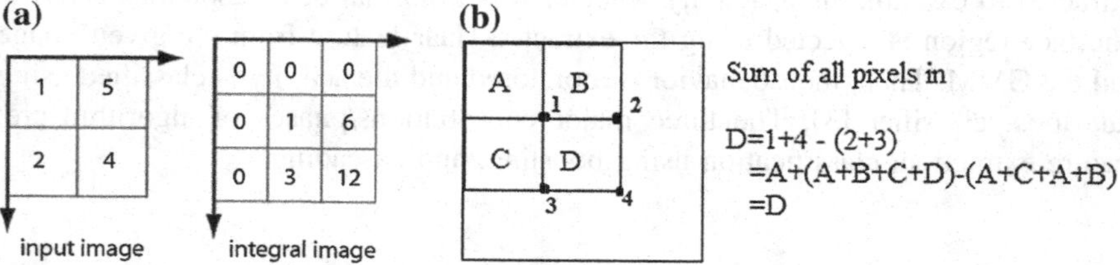

Fig. 2 **a** Integral image computation and **b** summed area table

$$I(x,y) = i(x,y) + I(x-1,y) + I(x,y-1) - I(x-1,y-1) \tag{2}$$

Once the summed area table has been computed, the task of evaluating any rectangle can be accomplished in constant time with just four array references as given in Fig. 2b.

3.1.2 Haar-Like Features

Haar-like features are digital image features used in object recognition. Each feature is the difference of the calculation of the white area minus the dark area. Simple Haar features provide much faster implementation and requires less data. A Haar-like feature considers adjacent rectangular regions at a specific location in a detection window, sums up the pixel intensities in each region, and calculates the difference between these sums. This difference is then used to categorize the subsections of an image. For example, in a database with human faces, the region of the eyes is darker than the region of the cheeks. Therefore, a common Haar feature for face detection is a set of two adjacent rectangles that lie above the eye and the cheek region.

Viola and Jones also defined three-rectangle features and four-rectangle features. The values indicate certain characteristics of a particular area of the image. Each feature type can indicate the existence (or absence) of certain characteristics in the image, such as edges or changes in texture. For example, a two-rectangle feature can indicate where the border lies between a dark region and a light region (Fig. 3)

$$\text{Sum} = I(C) + I(A) - I(B) - I(D) \tag{3}$$

where points A, B, C, and D belong to the integral image i. Each Haar-like feature may need more than four lookups, depending on how it was defined. Viola and Jones' two-rectangle features need six lookups, three-rectangle features need eight lookups, and four-rectangle features need nine lookups.

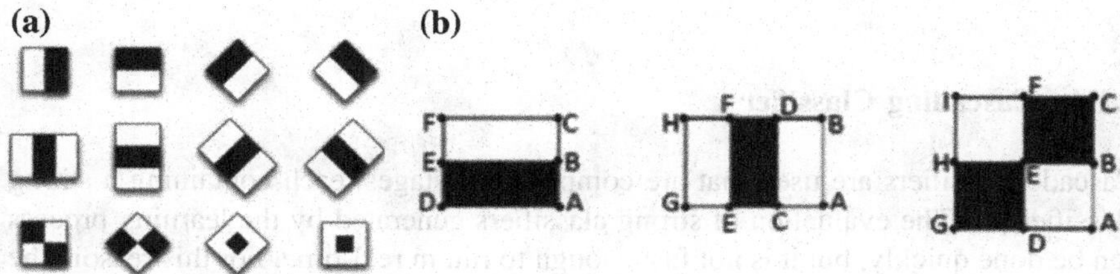

Fig. 3 **a** Haar-like and **b** Rectangular feature lookups

3.1.3 AdaBoost: The Boosting Algorithm

AdaBoost is a short form for Adaptive Boosting, which helps in finding only the best features among all other features. It constructs a strong classifier as a linear combination of weak classifiers.

$$\underbrace{F(x)}_{\text{Strong classifier}} = \underbrace{\alpha_1 f_1(x) + \alpha_2 f_2(x) + \alpha_3 f_3(x) + \ldots}_{\text{Weak classifiers}} \tag{4}$$

This classifier tells what the best features are and selects the best threshold and helps in combining them to a classifier. We represent our labeled training data by the set $\{(x_1, y_1), (x_2, y_2), \ldots, (x_m, y_m)\}$ with $x_i \in X$ and $y_i \in \{-1, 1\}$. The set X represents our training data and the set $\{-1, 1\}$ represents two class labels for the data elements. AdaBoost maintains a probability distribution over all the training samples. This distribution is modified iteratively with each selection of a weak classifier. Using the probability distribution $D_t(i)$, we conjure up a weak classifier for the training data. We apply weak classifier h_t to all of our training data. Then, estimate the classification error rate for the weak classifier. The trust factor for h_t is calculated by

$$\propto_t = \frac{1}{2} \ln \frac{1 - \epsilon_t}{\epsilon_t} \tag{5}$$

Finally, we update the probability distribution over the training data for the next iteration as

$$D_{t+1}(x_i) = \frac{D_t(x_i) e^{-\propto_t y_i h_t(x_i)}}{Z_t} \tag{6}$$

where the role of Z_t is to serve as normalize. At the end of T iterations, we construct the final classifier H as follows (Fig. 4):

$$H(x) = \text{sign}\left(\sum_{t=1}^{T} \alpha_t h_t(x) \right) \tag{7}$$

3.1.4 Cascading Classifier

Cascade classifiers are used that are composed of stages, each containing a strong classifier [4]. The evaluation of strong classifiers generated by the learning process can be done quickly, but it is not fast enough to run in real-time. For this reason, the strong classifiers are arranged in a cascade in the order of complexity, where each successive classifier is trained only on those selected samples which pass through the preceding classifier.

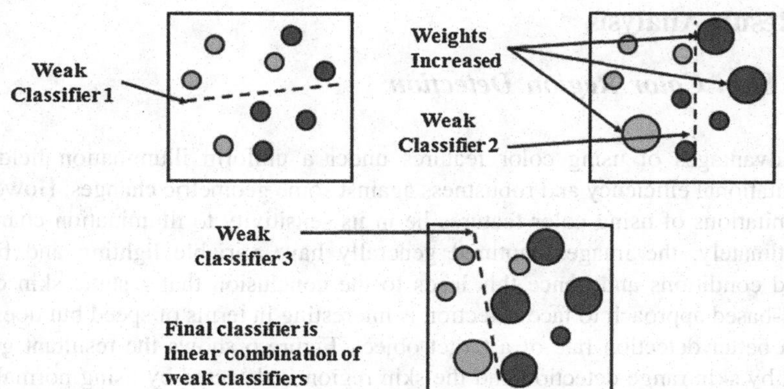

Fig. 4 AdaBoost classifier

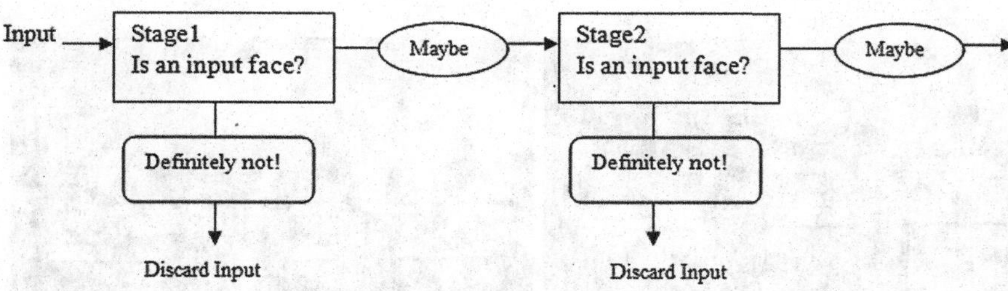

Fig. 5 Cascading classifier

If at any stage in cascade, a classifier rejects the subwindow under inspection, no further processing is performed and continues searching the next subwindow. The cascade therefore has the form of a degenerate tree (Fig. 5).

3.2 *Hand Movement Detection*

The hand movement from a given video frame is detected by constructing a grid for every student in an examination hall; and when there is an overlap of grid or any abnormal action occurs, an alert is given to the invigilators. Before which, background subtraction is performed to extract the foreground objects. The GMM [5] is used to monitor the human activity, where each activity is described by the combination of GMMs, with each GMM representing the distribution of a category feature vector (CFV).

4 Result Analysis

4.1 Skin Color Region Detection

The advantages of using color features under a uniform illumination field are computational efficiency and robustness against some geometric changes. However, the limitations of using color features lie in its sensitivity to illumination changes. Unfortunately, the images captured generally have variable lighting and background conditions and hence this leads to the conclusion that a static skin color model-based approach to face detection is interesting in terms of speed but does not give a better detection rate of a target object. Figure 6 shows the resultant group image by skin range detection and the skin region is detected by using normalized color space (Fig. 7).

Fig. 6 Grid formation for human activity detection

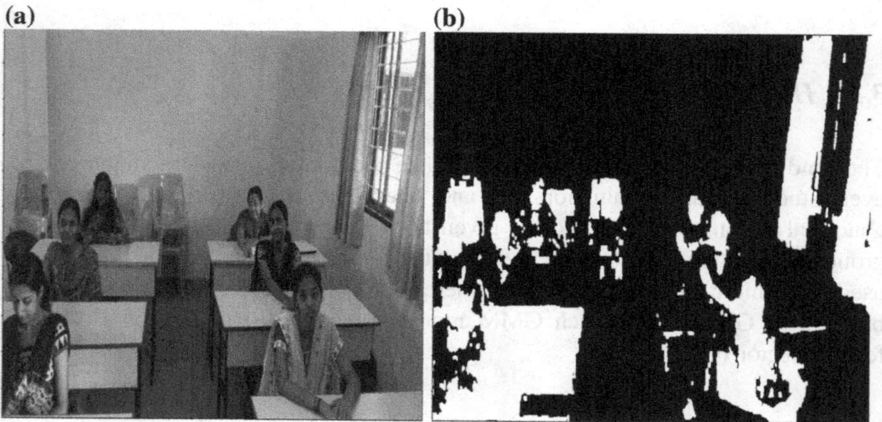

Fig. 7 **a** Input image and **b** Skin region-detected output

Fig. 8 **a** Input image and **b** Face-detected using LBP cascade

4.2 Face Detection Using Local Binary Pattern

Figure 4.2 shows the face-detected image using the LBP cascade. The main advantage of the present method is that it considers all the LBP features by reducing them into features based on the number of transitions (Fig. 8).

4.3 Face Detection Using Haar Cascade

The results given below were formed by running the Haar cascade in an input video stream of the classroom dataset.

When there is a head pose variation beyond the threshold value, then the respective person's face is detected separately by a head bounding box and its activity is monitored (Figs. 9 and 10).

4.4 Performance of the detectors on datasets

Algorithm	Total number of faces	Processing time (s)	True positive	False positive	Positive predictive value
Skin detection	50	0.721	45	12	0.789
LBP cascade	100	1.235	87	17	0.83
Haar cascade	100	1.467	92	12	0.88

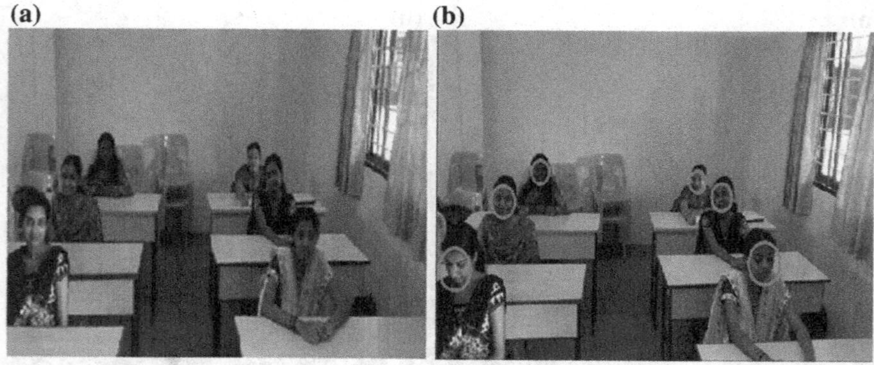

Fig. 9 **a** Input image and **b** Face-detected image using Haar cascade

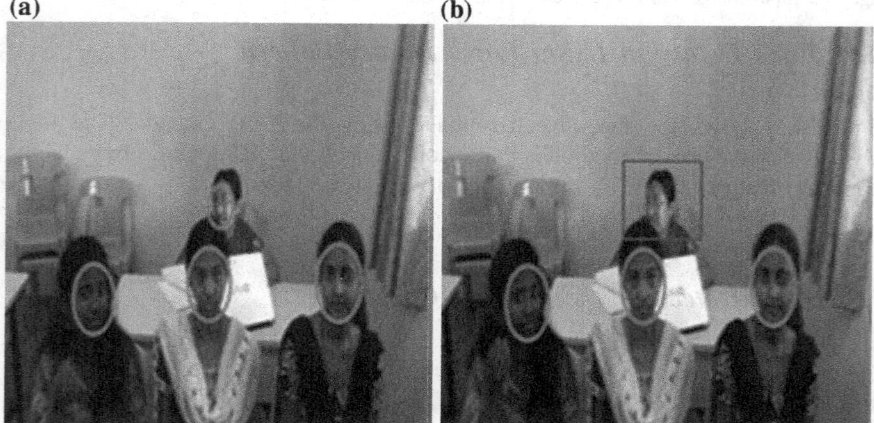

Fig. 10 **a** Input image and **b** Suspicious activity detection

4.5 Hand Movement Detection

Hand movement is detected by using background subtraction and the activity is monitored and the alert is given when there is a suspicious activity carried out (Fig. 11).

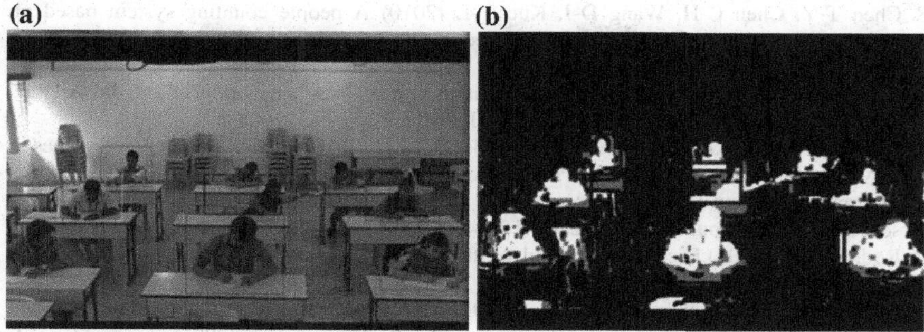

Fig. 11 **a** Input image and **b** Suspicious activity detected

5 Conclusion

The proposed work uses Viola–Jones algorithm which detects the human face in less time with high accuracy. Then, the result identifies the head motion detection through which the suspicious activity of the student is identified. When the faces cross a certain threshold, the system changes the box color from pink to red which indicates that a suspicious activity is being carried out. These days, having human invigilators in the examination halls requires a lot of man power and is inconvenient at times. To avoid this, we have provided a system which has the vision and ability to automatically invigilate in an examination hall during real-time situations. This system will certainly bring a revolution in the field of security surveillance in the education sector.

References

1. Ojala T, Pietikainen M, Harwood D (1996) A Comparative study of texture measures with classification based on Feature distributions. Pattern Recogn 29:51–59
2. Gowsikhaa D (2012) Suspicious human activity detection from surveillance videos. In: Int J Internet Distrib Comput Syst 2(2)
3. Khandagale P, Chaudhari A, Ranawade A (2013) Automated video survellience to detect suspicious human activity. Int J Emerg Technol Comput Appl Sci 13–128
4. Viola P, Jones MJ (2004) Robust real-time face detection. Int J Comput Vision 57(2):137–154
5. Lin W, Sun M-T, Poovandran R, Zhang Z (2008) Human activity recognition for video surveillance, 978-1-4244-1684-4/08 (2008) IEEE
6. Yan Z, Yang F, Wang J, Shi Y, Li C, Sun M (2013) Face orientation detection in video stream based on harr-like feature and LQV classifier for civil video surveillance. In: International communications satellite systems conferences (2013)
7. Borges PVK, Conci N, Cavallaro A (2013) Video-based human behavior understanding:a survey. IEEE Trans Circuits Syst Video Technol 23(11)

8. Chen T-Y, Chen C-H, Wang D-J, Kuo Y-L (2010) A people counting system based on face-detection using skin color detection technique. In: Fourth international conference on genetic and evolutionary computing, (2010)
9. Ko T (2008) A survey on behaviour analysis in video surveillance applications. In: Applied imagery pattern recognition workshop '08. 37th IEEE, pp 1–8, 2008
10. Lee GGK, Ka H, Kim BS, Kim WY, Yoon JY, Kim JJ (2010) Analysis of crowded scenes in surveillance videos. Can J Image Process Comput Vis 1(1):52–75
11. Maurin B, Masoud O, Papanikolopoulos N (2010) Camera surveillance of crowded traffic scenes. IEEE Comput Soc Press 22(4):16–44
12. Rodriguez Y (2006) Face detection and verification using local binary patterns. Ph.D. Thesis, École Polytechnique Fédérale de Lausanne, 2006
13. Schneiderman H, Kanade T A statistical method for 3D object detection applied to faces and cars. In: IEEE conference on computer vision and pattern recognition 2000 proceedings, vol 1. Digital Object Identifier 10.1109/CVPR.2000.855895
14. Kernighan BW, Ritchie DM, The C programming language, second edition, Prentice Hall Software Series, (1988)
15. Yang M-H, Kriegman DJ, Ahuja N (2002) Detecting faces in images: a survey. IEEE Trans Pattern Anal Mach Intell 24:34–58
16. Hadid A, Heikkil¨a JY, O. Silven & M. Pietik¨ainen,“ Face And Eye Detection For Person Authentication In Mobile Phones”, Machine Vision Group IEEE Transactions, 244-1354-0/07
17. Chellappa R, Vaswani (2006) Principal Components Space Analysis for Image and Video Classification. IEEE Trans. Image Processing 15(7):1816–1830
18. Fan XA, Lee TH, Xiang C (2006) Face Recognition using Recursive Fisher Linear Discriminent. IEEE Trans. Image processing 15(8):2097–2105
19. Jafri R, Arabnia HR (2009) A survey of face recognition techniques. J Inform Process Syst 5 (2)
20. Toygar O, Acan A (2003) Face recognition using Pca, Lda and Ica approaches on colored images. J Electr Electron Eng 3(1)
21. Tang X, Zongying O, Tieming S, Zhao P (2005) Cascade AdaBoost classifiers with stage features optimization for cellular phone embedded face detection system, ICNC 2005. LNCS 3612:688–697

Block-Based Forgery Detection Using Global and Local Features

Gayathri Soman and Jyothish K. John

Abstract Nowadays, many image-editing tools have emerged. So image authentication has become an emergency issue in the digital world, since images can be easily tampered. Image hash functions are one of the efficient methods used for detecting this type of tampering. Image hashing is a technique that extracts a short sequence from the image that represents the content of the image and thus can be used for image authentication. This method proposes an image hash that is formed using both the global and local features of the image. The Haralick texture features are used as the local feature. The global features are based on the Zernike moments of the luminance and the chrominance component. This robust hashing scheme can detect image forgery such as insertion and deletion of the objects. The features are extracted from the blocks of the image and so can detect forgery in small areas of the image also. The proposed hash is robust to common content-preserving modifications and sensitive to malicious manipulations.

Keywords Zernike moment · Haralick texture features · Image authentication · Image forgery

1 Introduction

In recent years, digital images and videos have gained more popularity because of its use in social networks. People can easily alter the content of the digital media by the use of various editing softwares. The technologies available on the Internet such as emails, social networks, etc. are interested on the systems which ensure the authenticity of the multimedia information received in a communication. Image

G. Soman (✉) · J.K. John
Computer Science and Engineering, Federal Institute of Science and Technology, Mahatma Gandhi University, Angamaly, Kerala, India
e-mail: gayathrisoman025@gmail.com

L.P. Suresh and B.K. Panigrahi (eds.), *Proceedings of the International Conference on Soft Computing Systems*, Advances in Intelligent Systems and Computing 397, DOI 10.1007/978-81-322-2671-0_14

authentication is the technique by which one can ensure that the image has not been altered during the transmission and the image is from the legal user.

Cryptographic hash functions have been used earlier, but such hashes are sensitive to small changes in the image. The image must be considered non-authentic even if only one pixel or even one bit of data has been changed. Content-preserving modifications such as compression and quantization, image format conversion, image enhancement, etc. should be tolerated by image hashing.

Image hashing can be used for image authentication because it extracts a short sequence from the image that represents the content of the image. In image hashing, image authentication is done by comparing the similarity of two image hashes that are extracted from the original image and the tampered image. Two different images would have a large hash distance between them.

An efficient image hash should have the following properties; it should be reasonably short, robust to ordinary image authentication manipulations, and sensitive to tampering. Unique in the sense that two different images should have different hash values and secure so that an unauthorized party cannot find the hash value. Robust image hashing is considered to be the desired authentication system for most practical cases. To meet all the requirements simultaneously is a difficult task.

At the sender side, the hash is calculated for the input image which is encrypted and sent along with the image to the receiver side. Receiver detaches the hash from the image and generates the hash for the received image in the same way as the sender did. Then at the receiver side, both the hashes are compared to check the authenticity of the image and to locate the tampered regions of the image.

A good hash generation technique should be perceptually robust while capable of detecting and locating content forgery. Most image forgery detection techniques use distance metrics indicating the degree of similarity between two hashes to detect the forgery.

This paper has the following structure: Sect. 2 explains about the related works, Sect. 3 presents the methodology employed for image hash construction and image authentication, Sect. 4 presents the experimental analysis, and Sect. 5 concludes the paper.

2 Related Work

Various image hashing methods have been proposed. Monga et al. [1] proposed a perceptual hashing technique. A perceptual image hash function captures the perceptual qualities of the image. In this method two steps are used; the first one is feature extraction followed by data clustering to obtain the final hash. In [2] Xiang et al. proposed a method which uses image histogram. The image is first filtered from which the histogram is extracted followed by hash generation. The hash function is simple to implement and robust to geometric attacks, which are the major advantages. However, the limitation is that it cannot distinguish images with similar histograms but with different contents.

Tang et al. [3] developed a global method using nonnegative matrix factorization. First, obtain the fixed-size pixel array from the image. Rearrange the pixels to obtain a secondary image and apply NMF to produce a coefficient matrix, which is then quantized. In [4], Aswin Swaminathan proposed a method which used Fourier transform of the preprocessed image, which is then converted into polar coordinates, from which the features are extracted. Gray coding is applied to obtain the binary hash sequence.

In [5], Lei proposed a method which first finds the DFT of significant random transform coefficients and then normalize/quantize the DFT coefficients to form image hash for content authentication. Khelfi et al. [6] proposed a secure perceptual image hashing technique based on virtual watermark detection.

In another work, Tang et al. [7] used a scheme in which a dictionary is constructed from the words which are actually large number of image feature vectors. Words are used to represent image blocks in generating the hash. Zhenjun Tang et al. [8] proposed a scheme in which the secondary image is transformed by the DFT, and robust salient features are extracted using nonuniform sampling. The extracted features are quantized into a binary string to form the final hash. In [9], Tang et al. proposed a hashing scheme which is achieved by converting the input image into a normalized image. The normalized image is divided into different rings and from each ring's entropies the hash is generated.

In another method, Ashwin Swaminathan et al. [10] used Fourier transform, and the image is then converted into polar coordinates. From the polar coordinates, the image features are extracted. Discretization is done followed by quantization to obtain the final hash value. In [11] Marc Schnand et al. used a method which extracts the content of interest. The content is then hashed using a hash function to produce a hash value. Then encryption is performed.

Chun-Shein Lu et al. [12] proposed a method in which the mesh generation algorithm is executed which extracts robust salient points. After that, a mesh normalization process is used to transfer the decomposed meshes to normalized meshes of fixed sizes. In [13], Yan Zhao et al. proposed that both the global and local features are considered. Zernike moments are taken as the global features and the local features include the position and texture information of the salient regions. These two features are combined to obtain the hash value (Table 1).

3 Methodology

The image hashing scheme which is proposed and the procedure of image authentication using the hash are described in this section.

Table 1 Advantages and disadvantages of different hashing techniques

Ref no	Advantages	Disadvantages
1	Good strength of the hash value	Significant feature points are not present in small regions
2	Hash function simple to implement	Hamming distance increases in some cases
3	Feature extraction process used for image hashing is easy	Considers only the *Y* component of the YCbCr representation
4	Performs very well for both bending and cropping	Suffers from some classical attacks such as additive noise
5	This can tolerate almost all the image processing manipulations	Not robust for high-frequency texture
6	Allows analytical determination of threshold	Need more storage for hash value
7	Very high robustness to jpeg compression	A security loop hole is created
8	Very low probability of collision of hashes	Considers only the luminance component of the image
9	Time complexity is less	Hash length depends on the number of rings
10	Performs better for cropping, filtering, shearing, etc.	It still suffers from some classical attacks such as additive noise
11	Can be used for more than just image authentication	Sometimes weakens the signature based on the type of hashing algorithm
12	A novel geometric distortion-invariant image hashing scheme	This scheme is somewhat complex
13	The probability of collision of hashes of two different images approaches zero.	Do not use features that best represent the image content and so cannot determine tampering in small area

3.1 Image Hash Construction

3.1.1 Image Hash Construction

The image is first rescaled to a fixed-size $K * K$. Large value of K leads to high computation complexity, while small value of K leads to loss of fine details. We select $K = 256$. Then the image is converted from its RGB representation into YCbCr representation. The Y component represents the luminance of the image and the |Cb–Cr| represents the chrominance component.

3.1.2 Global Feature Extraction

The global features are obtained by finding the Zernike moments of the luminance and the chrominance component of the image. We choose the order of the Zernike moment, $n = 5$. We get $Z' = [Z_y\ Z_c]$ which is a row vector with $11 \times 2 = 22$ elements. We randomly generate a vector $Q1$ of size 1×22 with values in [0, 255] using a secret key $K1$. Then the encrypted global feature vector Z is obtained as $Z = [(Z' + Q1) \bmod 256]$.

3.1.3 Local Feature Extraction

The image is then divided into blocks of size 32×32, and the local features are extracted from each of these blocks. The mean of the 13 Haralick features is extracted from each block. So we get local features as $L' = [H]$, where H denotes Haralick features, L' is a 1×64 sized vector. We randomly generate a vector $Q2$ of size 1×64 with values in [0,255] using a secret key $K2$. The encrypted local feature vector L is obtained as $L = [(L' + Q2) \bmod 256]$.

3.1.4 Final Hash Construction

The intermediate hash is constructed by concatenating the global and local feature vectors, we get $h' = [Z, L]$. We randomly generate a vector $Q3$ of size 1×86 with values with in [0, 255] using a secret key $K3$. Finally, the hash is generated as $h = [(h' + Q3) \bmod 256]$.

3.2 *Image Authentication*

In this step, the hash of a trusted image $H0$ is available and this is called the reference hash. The hash of the received image is calculated at the receiver end using the above method to obtain the hash $H1$, called test hash. Then these two hashes are compared to find whether the image is tampered or not.

3.2.1 Distance Calculation

We use a distance metric to calculate the distance between the hashes to find whether the image has been tampered. The euclidean distance between the reference and test image is given by

$$D \approx \|\mathbf{Z}_1 - \mathbf{Z}_1\| \triangleq D_G$$

If D is greater than a particular threshold, we can say that the image has been tampered.

3.3 Tamper Localization

The reference hash is decrypted and decomposed to get intermediate hash $h0' = [Z0', H0']$. At the receiver, we have $h1' = [Z1', H1']$ which is compared with intermediate reference hash. We compute the differences between, Haralick features as DH = $H0' - H1'$ If the total difference has all the values below 1, the image is recognized as the original image. Otherwise, the image blocks that have total difference greater than 1 are found as tampered blocks.

3.4 Tamper Classification

The Haralick texture feature of the reference and the test image is been compared. If the Haralick feature of any block of the reference image is more compared to same position block in the test image, then we can say that something has been deleted. If the Haralick feature of any block of the reference image is less compared to same position block in the test image, then we can say that something has been inserted (Figs. 1 and 2).

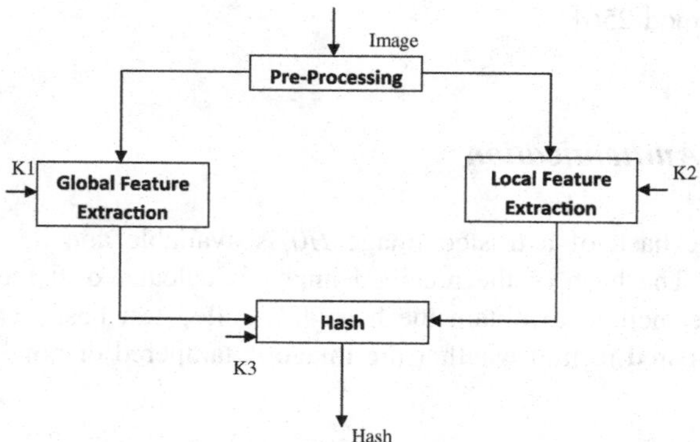

Fig. 1 Hash construction

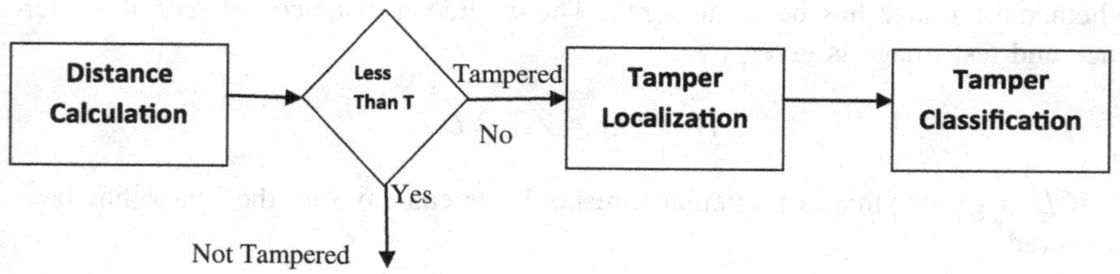

Fig. 2 Tamper detection

4 Experimental Results and Analysis

We have tested image pairs downloaded from the Internet and from the dataset CASIA4. All the tampered images are detected and are correctly localized. The success rate of tampering detection is 100 %. The proposed technique is tested on MATLAB R2012a, to justify the effectiveness of the proposed algorithm. Robustness of the hash is checked against various content-preserving modifications: JPEG coding, addition of noise, rotation, scaling, brightness and contrast adjustment, and slight cropping. The proposed method tolerates JPEG compression above a quality level of 15 which is an average value. Rotation tolerated by the proposed hash is up to 1 % above which the texture features of the image change, so that rotations above 1 % can be considered as a malicious manipulation. The proposed hash can tolerate contrast adjustment and brightness adjustment up to 20 and 10, respectively (Fig. 3).

4.1 Forgery Localization

Table 2 shows the tampering detection and tampering localization results of two hashing methods, viz., method based on Haralick and Zernike moments and method proposed in “robust hashing for image authentication using Zernike moments and local features.”

4.2 Sensitivity to Tampering

The percentage of the tampering is considered to study the sensitivity to tampering. Table 2 gives the results of tampering detection and the methods detect above the

Fig. 3 Result

Table 2 Tampering localization

Input			Tampered	*A*	*B*
Name	Size	Image			
Img1.jpg	384 * 256	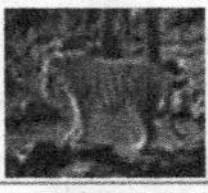		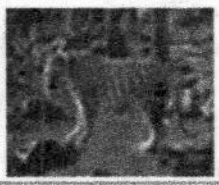	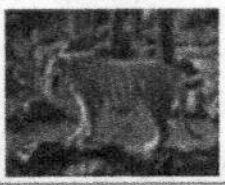
Img2.jpg	256 * 384	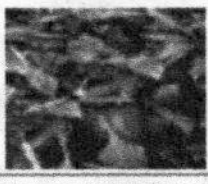	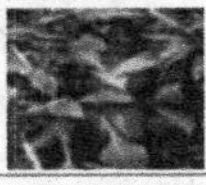	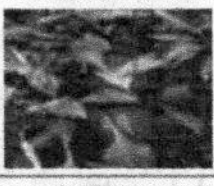	
Img3.jpg	2272 * 1704	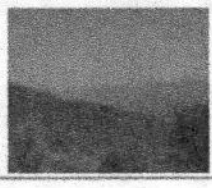			

Nomenclature
A. Method proposed in "Robust hashing for image authentication using Zernike moments and local features" [13]
B. Method based on Haralick and Zernike moments (proposed method)

Table 3 Sensitivity to tampering

Image	*A* (%)	*B* (%)
Img1.jpg	0.41	0.91
Img2.jpg	0.31	0.51
Img3.jpg	0.51	0.91

specified percentages in the table. However, by combining both local features and global features in **B**, the hash can be made sensitive to forgery (Table 3).

4.3 Robustness to Other Content-Preserving Modifications

The method B shows satisfying robustness to other content-preserving modifications such as image rotation, slight cropping, and addition of noise. The proposed hash tolerates rotation by 1 %, slight cropping below 1 % of the image. The proposed hash can tolerate contrast adjustment and brightness adjustment up to 20 and 10, respectively.

5 Conclusion

In this paper, a new modified algorithm for image hash generation is developed using both the global and local features. The global features are based on the Zernike moments of the luminance and the chrominance component. The local features include Haralick features extracted from the image block. The hash generated is robust to common content-preserving modifications such as JPEG compression, addition of noise, contrast and brightness adjustment, scaling, small angle rotation, and slight cropping, but sensitive to forgery. The Haralick features extracted from image blocks are highly sensitive to tampering. The proposed hash is robust to content-preserving manipulations and sensitive to malicious modifications.

The method explained here is used for image authentication. The hash can differentiate similar and forged images. It can also identify the type of forgery that is whether it is an insertion or deletion. The method uses block-based forgery detection, and hence can be used for finding tampering in small regions also.

Acknowledgments I am thankful to Mr. Jyothish K John, Senior Assistant Professor of Computer Science department, FISAT, Kerala for his keen interest and useful guidance in my paper.

References

1. Monga V, Banerjee A, Evans BL (2006) A clustering based approach to perceptual image hashing. IEEE Trans Inf Forensic Secur 1(1):68–79
2. Xiang S (2007) Histogram based image hashing scheme robust against geometric deformations. In: Proceedings of the ACM multimedia and security workshop, New York, pp 121–128, 2007
3. Tang Z, Wang S, Zhang X, Wei W, Su S (2008) Robust image hashing for tamper detection using non-negative matrix factorization. J Ubiquitious Convergenge Technol 2(1):18–26
4. Swaminathan A, Mao Y, Wu M (2006) Robust and secure image hash. IEEE Trans Inf Forensic Security 1(2):215–230
5. Lei Y, Wanga Y, Huang J (2011) Robust image hash in Randomn transform domain for authentication. Signal Process: Image Commun 26(6):280–288
6. Khelfi F, Jiang J (2010) Perceptual image hashing based on virtual watermark detection. Trans Image Process 19(4):981–994
7. Tang Z, Wang S, Zhang X, Wei W, Zahoe Y (2011) Lexiographical framework for image hashing with implementation based on DCT and NMF. Multimedia Tools Appl 52(2–3):325–345
8. Tang Z, Wang S, Zhang X, Wei W, Zahoe Y (2011) Lexiographical framework for image hashing with implementation based on DCT and NMF. Multimedia Tools Appl 52(2–3):325–345
9. Tang Z, Xianquan Z, Huang L, Dai Y (2013) Robust image hashing using ring based entropies. Sig Process 93(7):2061–2069
10. Swaminathan A, Mao Y, Wu M (2004) Image hashing resilient to geometric and filtering operations. In: IEEE workshop on multimedis signal processing, Siena Italy, 2004
11. Schnand M, Changeider SF (1996) A robust content based digital signature for image authentication. In: Proceedings of the IEEE international conference on image processing, vol 3, Lausanne, Switzerland, 1996, pp 227–230
12. Lu C-S, Hsu C-Y (2005) Geometric distortion resilent image hashing scheme and its application on copy detection and authentication. Multimedia Syst 11(2):159–173
13. Zhao Y, Wang S, Zhang X, Yao H Robust hashing for image authentication using Zernike moments and local features. IEEE Trans Inf Forensics Security 8(1)

3D Modelling of a Jewellery and Its Virtual Try-on

G. Radhika, T. Senthil Kumar, P. Privin Dolleth Reyner, G. Leela, N. Mangayarkarasi, A. Abirami and K. Vinayaka

Abstract Nowadays, everything is becoming automated. So, automation is indeed needed in the world of jewellery. The goldsmith or any jewellery vendor, rather than having all the real patterns of jewellery, can have the model of these jewellery, so that he can display them virtually on the customer's hand using Augmented Reality. 2D representation of an object deals only with the height and the width of an object. 3D representations include the third dimension of an image which is the depth information of an object. This paper presents an overall approach to 3D modelling of jewellery from the uncalibrated images. The datasets are taken from different viewing planes at different intervals. From these images, we construct the 3D model of an object. 3D model provides a realistic view for the users by projecting it on human hand using the augmented reality technique.

Keywords Image acquisition · Noise removal · Image segmentation · Depth estimation · 3D modelling · Augmented reality · GUI buttons

---

G.Radhika (✉) · T. Senthil Kumar · P. Privin Dolleth Reyner · G. Leela · N. Mangayarkarasi · A. Abirami · K. Vinayaka
Department of Computer Science and Engineering, Amrita School of Engineering, Amrita Vishwa Vidyapeetham (University), Coimbatore 641112, India
e-mail: g_radhika@cb.amrita.edu

T. Senthil Kumar
e-mail: t_senthilkumar@cb.amrita.edu

P. Privin Dolleth Reyner
e-mail: privindolleth3@gmail.com

G. Leela
e-mail: leelapatel120@gmail.com

N. Mangayarkarasi
e-mail: mangainehru@gmail.com

A. Abirami
e-mail: abi17anandan@gmail.com

K. Vinayaka
e-mail: vinayaka.rinku11@gmail.com

L.P. Suresh and B.K. Panigrahi (eds.), *Proceedings of the International Conference on Soft Computing Systems*, Advances in Intelligent Systems and Computing 397, DOI 10.1007/978-81-322-2671-0_15

1 Introduction

The increase in technological developments and the use of highly sophisticated techniques have created a greater impact on the human perception of viewing things. During earlier days, pictorial representation of an object was done by drawings. Later on, cameras came into existence, which gave pictures which were only 2D. To prove this quest of pictorial representation of an object and to improvise it, modern man introduced the concept of augmented reality which is the 3D representation of an object. With regards to the representation of an object, augmented reality is considered to be the state-of-the-art technology. The technique of augmented reality not only gives the 3D representation of an object but also the feeling of the real-time presence of the object.

Stereo vision is used to obtain the depth information from the images [1]. This method gives a 3D perception of an image by having two different images of the same scene [2, 3]. Using this concept, we extend our idea by taking multiple images of an object. The dataset is obtained by capturing the 360-degree view of a bangle followed by the depth estimation and finally 3D reconstruction of the jewellery [4]. Since the output which we got out of this was only 60 % accurate and the pattern over the bangle was not visibly clear enough, an alternate and quicker way of generating a 3D model of the jewellery would be using VisualSFM software with CMVS installed, which will give a high resolution model of the jewellery and this will serve our purpose of virtual try-on. This 3D model is given as an input to the UNITY software and we apply 3D transformations on the 3D model to align and resize it as required and we use Vuforia SDK to project the jewellery for the customers. For easier understanding, in our further discussion in this paper, bangle is used for demonstration. Parallely, a user interface is created by using the GUI, where eight interactive buttons are created specifying the standard size of the human hand. These measurements are used to resize the model of the bangle, as to fit in the customer's hand (Fig. 1).

The steps involved are as follows:

Camera view—The camera is fixed and images are captured for 360° in total by dividing the field of view into equal partitions of 20° each.

Obtain 2D images—The obtained 2D images are now grouped together in order to obtain a complete 360-degree image when they are stitched together [5].

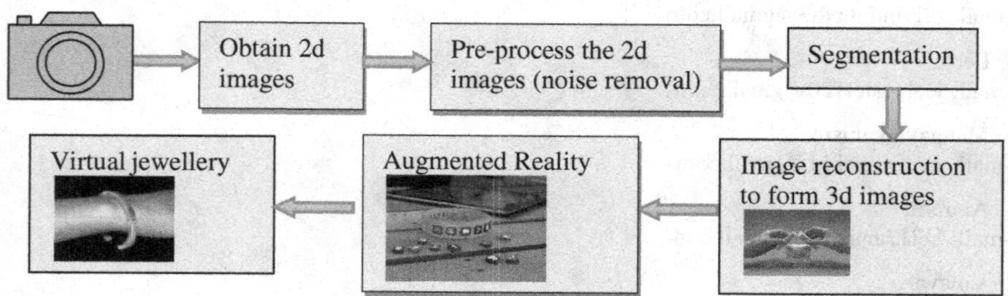

Fig. 1 Block diagram which gives the steps involved in the project

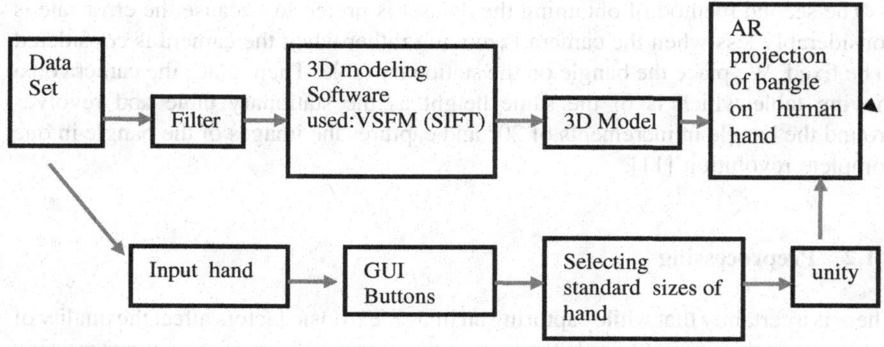

Fig. 2 The overall flow of the project when VSFM software is used for 3D reconstruction

Noise Removal—This is the preprocessing step for the 2D images which involves the noise removal [6].

A bilateral filter is used for the process of noise removal [7, 8]

Segmentation—Similar pixels are now clustered together to obtain the area of interest [9].

Reconstruction of 3D images—The 2D images are stitched together to obtain a 3D image of the jewellery. While constructing a 3D image from a 2D image, there is a reduction in accuracy and loss in minute details. 3D modelling is used to rectify these defects [10].

3D Modelling—Here, the modelling of a 3D image is done using the software Visual Structure from Motion System (VSFM). The segmented images are given as the input for this, which combines all the 2D images forming a 3D mesh.

Augmented Reality and Projection over the human hand—On giving the minimum and maximum measure of the wrist to the unity software, the software resizes the dimensions of the 3D image to best fit the human hand (Fig. 2).

2 Related Work

2.1 Work Accomplished

2.1.1 Image Acquisition

Multiple images of the jewellery are to be captured at different angles. This serves as the dataset. There are two common methods of obtaining the data set which includes

(a) A fixed camera with a rotating table on which the bangle is mounted
(b) The camera is allowed to rotate around a fixed table on which the bangle is mounted.

The second method of obtaining the dataset is preferred because the error rate is considerably less when the camera is moving rather when the camera is considered to be fixed. We place the bangle on the stationary table. Then, place the camera on a moving table which is of the same height as the stationary table and revolves around the bangle in increments of 20° and captures the images of the bangle in one complete revolution [11].

2.1.2 Preprocessing

There is a certainty that while capturing an image, extrinsic factors affect the quality of the image such as noise. In order to remove these noises, we perform a preprocessing step where filters are used to remove these unwanted noises [12]. When the Gaussian low-pass filtering is considered it computes an average weight of the neighbouring pixels whose weight decreases with respect to the distance from the centre of neighbourhood [12]. The noises are averaged and the signal values are preserved. However, when the low-pass filters are used the edges are blurred so, we prefer bilateral filtering as it preserves the edges by smoothing the image and combining the neighbouring pixels in a nonlinear manner [12, 13]. This method is considered to be non-iterative.

2.1.3 Segmentation

Segmentation generally means extracting the area of interest. Image segmentation classifies or clusters an image into several parts or regions according to the feature of an image. In *RGB threshold-based segmentation,* the RGB image is first converted to a binary image. Then we give a threshold value as required for our need [14]. Based on this threshold value, the pixels whose intensity values are above the threshold value will be white and the ones below the threshold value will be black. You get an intermediate image as output. Now, select a point on the image which is a part of the consideration. All the points connected to this point will be segmented from the rest of the image [15]. So, with this only the object of consideration from the rest of the image can be selected.

2.1.4 Depth Estimation

Maximum flow algorithm is used to detect the depth information. On detecting a point on the first image, a similar point is found to lie on the epipolar line formed by the point on the second image [16]. We thus consider two assumptions to calculate the second point:

(a) Two images are stereos.
(b) A disparity d [17]

To confirm that the point ($x1$, $y1$) and ($x2$, $y2$) are matching, we take into consideration that

- Since the images are considered to be lambertian, their corresponding images should match.
- Matching or same order of epipolar lines (ordering constraint) [16].

We now stack all the epipolar lines together and find the corresponding difference in pixel intensities which is named as P. We have to find a surface with minimum P such that we have to associate a capacity P' to each point (x, y, d) [18]. The quantities P' and P are considered to be proportional. The points with the minimum capacity are said to have the maximum flow. Thus, from such points for a desired isosurface, we get minimum P and an optimum disparity [19]. From this, you can reconstruct the 3D model using *marching cubes algorithm* [10]. Since we are dealing with jewellery objects, we expect clarity which cannot be achieved using this algorithm. So we moved on to VisualSFM which is discussed later.

2.1.5 3D Modelling

In VisualSFM, SFM stands for Structure from motion, is a GUI application used in 3D reconstruction. VisualSFM performs three major functions such as feature detection, feature matching and bundle adjustment. It estimates 3D structures or image from 2D image sequences combined with local motion signals. It is used in the field of 3D vision and visual perception. In biological field, SFM helps to recover the 3D structures from the projected 2D images. The output of the VSFM is given as an input to CMVS (Clustering View for Multiview Stereo). The CMVS software should always be used in combination with SFM software as it converts and forms image clusters from the input images. The formed clusters are then processed independently and reconstructed and then union of reconstructions should not have any defect of missing details. Steps involved are

1. Dataset is imported into the VisualSFM software.
2. Missing matches are computed. This creates SIFT for each image in the dataset and feature matching is done.
3. Sparse reconstruction for the dataset is generated
4. Dense reconstruction [20] using CMVS is generated and the output is stored in . nvm format.
5. The generated .nvm file is imported in Mesh lab and the required object is modelled.
6. The file is saved in .obj format.

2.1.6 Augmented Reality

The 3D technology has become more and more developed as it is being used in various areas. The display of these 3D models in the real world has lead to the invention of Augmented Reality (AR). AR helps to interact with the virtual images using the real objects. The applications of AR are increasing day-by-day in various fields. It has gained a wider importance in the gaming field. The 3D AR games are trending everywhere. Its application also includes virtual learning class rooms, medical fields, CAD, virtual dressing rooms, etc. This can be achieved in AR using the available software. The Vuforia SDK helps to build vision-based AR applications for both Android and iOS platforms. The details about Vuforia SDK will be discussed below

The Vuforia SDK used in this project is Vuforia 3.10 Image targets.

Image targets—There are three predefined targets in the SDK(tarmac, stones and chipset). We use stone images as our image target on which the reconstructed bangle is placed.

2.1.7 Virtual Jewellery

The main scope of this paper is to develop a 3D model of a bangle, to project it over the human hand using augmented reality. Though the prices of jewels are increasing day-by-day, the craze among people has never gone down. Each time a person has to wear a jewel, say bangle, he/she has to check if it suits them. To make it more interesting and ease for the customers, just imagine a shop where you find the models of bangles displayed on the monitor screen. You only have to select the bangle of your choice and drag and drop it over your hand; you will be able to view it virtually in the real world. This is where the augmented reality works. So, first we make 3D models of the bangles of different design by using the VisualSFM software as described above. By taking this 3D model as an input for the AR part using the Vuforia SDK for unity we create a scene where the 3D model is applied with various 3D transformations (i.e., translation, scaling and rotation) as per the standard sizes mentioned in the eight virtual buttons ranging from 1.3 to 2.7 in. with the increment of 0.2 in. and finally this model is placed over the image target (Vuforia stones), position it correctly in the 3D axis and then, when the scene is built and run the target stones are shown in front of AR camera, which tracks the target and places the bangle over it, Thus making augmented reality available in virtual environment by getting the real feel of the jewellery.

2.2 *Future Works*

2.2.1 Resizing of the Bangle

The size of the human hand varies with each user, thus to make the process more efficient we have included eight buttons which varies from 1.3 to 2.7 in. (standard human hand size range). The user can choose a virtual button according to his hand size. The size of the bangle to be projected over the human hand is correlated with the user's choice and thus the best fit bangle is projected over the human hand. The user can select the GUI buttons on the screen to resize the bangle.

3 Results and Discussions

The camera detects the Vuforia stone placed over the hand and skin classifier detects the skin and calculates the maximal and minimal width of the hand thus projected over the human hand as shown in Fig. 3. This paper deals only with the calculation of the minimal width. The output from the unity software used is instantaneous. It gives the users the privilege to choose a bangle of his/her own interest and it gets projected over the human hand. Using unity combined with VSFM preserves even minute details of the bangle. Our future work would be with respect to resizing the bangle with the virtual buttons to fit in the customer's hand.

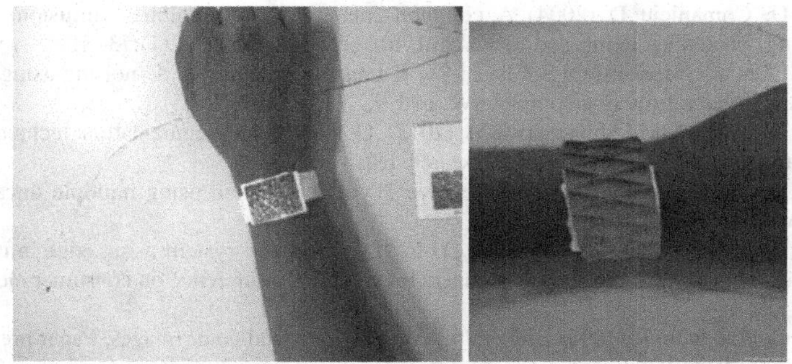

Fig. 3 Augmented reality using Vuforia image target (stones)

4 Conclusion

This paper has presented a detailed survey and findings of the algorithms and the steps that are involved in 3D reconstruction and AR. This virtual jewellery makes it more useful for the customers both in online shopping through the Internet and offline shopping through the stores. As the size of the hand varies from person to person, the model should adapt accordingly. The constraints faced by the project include though the size of the human hand varies from person to person, designing a well-fitted AR-based 3D model of a bangle is our scheduled future work.

Acknowledgments Gowtham, Ph. D. Scholar, Amrita School of Engineering, Coimbatore. Cognizant Technology Systems Lab, Amrita School of Engineering, Coimbatore.

References

1. Chandran S (2014) novel algorithm for converting 2D image to stereoscopic image with depth control using image fusion. Lect Notes Inf Theory 2(1)
2. Seitz S, Curless B, Diebel J, Scharstein D, Szeliski R (2006) A comparison and evaluation of multi-view stereo reconstruction algorithms. In: Proceedings IEEE conference on computer vision and pattern recognition, vol 1, pp 519–528. IEEE
3. Zach C (2000) Fast and high quality fusion of depth maps. In: Proceedings international symposium on 3D data processing, visualization and transmission (3DPVT), vol 1
4. François ARJ, Medioni GG (2001) Interactive 3D model extraction from a single image. Image Vis Comput 19:317–328
5. Gosta M, Grgic M (2010) Accomplishments and challenges of computer stereo vision. In: 52nd international symposium ELMAR-2010, Zadar, Croatia, 15–17 Sept 2010
6. Tomasi C, Manduchi R (1998) Bilateral filtering for gray and color images. In: IEEE international conference on computer vision, 1998
7. Barash D, Comaniciu D (2004) A common framework for nonlinear diffusion, adaptive smoothing, bilateral filtering and mean shift. Image Vis Comput 22(1):73–81
8. Jyoti S, Ekta M, Shreelakshmi S, Shashi Raj K Image sharpening & de-noising using bilateral & adaptive bilateral filters-a comparative analysis. IJAREEIE 2(8)
9. Sharma N, Mishra M, Shrivastava M (2012) Colour image segmentation techniques and issues: an approach. Int J Sci Tech Research 1(4)
10. Mohr R, Quan L, Veillon F (1995) Relative 3D reconstruction using multiple uncalibrated. Int J Robot Res 14(6):619–632
11. Cheng CC, Li CT, Chen LG (2010) A 2D-to-3D conversion system using edge information. In: Proceedings of digest of technical papers international conference on consumer electronics, pp 377–378
12. Tomasi C, Manduchi R (1998) Bilateral filtering for gray and color images. Paper presented at proceeding international conference on computer vision, pp 839–846
13. Elad M (2002) On the origin of the bilateral filter and the ways to improve it. IEEE Trans Image Process 11(10):1141–1151
14. Al-amri SS, Kalyankar NV, DImage KS (2010) Segmentation by using threshold techniques. J Comput 2(5). ISSN 2151-9617
15. Singh P (2013) A new approach to image segmentation. Int J Adv Res Comput Sci Software Eng 3(4)

16. Lorensen WE, Cline HE (1987) Marching cubes a high resolution 3d surfacing algorithm. Comput Graph 21(4)
17. Wang S, Wang X, Chen H (2008) A stereo video segmentation algorithm combining disparity map and frame difference. In: 3rd international conference on intelligent system and knowledge engineering, vol 1, pp 1121–1124
18. Newman TS, Yi H (2006) A survey of the marching cubes algorithm. Comput Graph 30:854–879
19. Lorensen W, Cline H (1987) Marching cubes a high resolution 3D surface construction algorithm. Comput Graph 21(4):163–169
20. Newcombe RA, Davison AJ (2010) Live dense reconstruction with a single moving camera. In: Proceedings of the IEEE computer society conference on computer vision and pattern recognition (CVPR), pp 1498–1505
21. Kim Y, Kim W (2014) Implementation of augmented reality system for smartphone advertisements. Paper presented at the international journal of multimedia and ubiquitous engineering, vol 9, issue 2, pp 385–392

Fuzzy-Based Multiloop Interleaved PFC Converter with High-Voltage Conversion System

P. Elangomenan and G. Nagarajan

Abstract This paper presents the implementation of an interleaved boost converter controlled by fuzzy logic control, combined with conventional multiloop control and interleaved pulse-width modulation. The conventional method has the drawback of increased current and voltage ripple. The generated modified proposed system will reduce the voltage and current ripple and improve PFC correction. The average behavior of the interleaved three-level switch-mode rectifier (SMR) is similar to the voltage stress and can be reduced using ZVS and ZCS soft switching method. The proposed method is simulated using MATLAB software with embedded controller in the hardware; both the results are verified by this system.

Keywords Fuzzy logic · Interleaved boost converter (ILBC) · PI controller · PWM technique · Switch-mode rectifier (SMR)

1 Introduction

The multiloop interleaved boost converter improves the power factor correction (PFC); the drawback of this converter is increased MOSFET heating. As shown in the block diagram, AC supply is converted to DC supply, the interleaved boost converter boosts more voltage ratio, and then the converter operates at high frequency as it reduces the voltage ripple and the current ripple and losses should be minimized by zero voltage switching (ZVS) [8, 9] and zero current switching (ZCS). The converter's operating frequency is 20 kHz if multiloop-controlled technique is used to compare the different controllers used. The conventional method using PI

P. Elangomenan (✉) · G. Nagarajan
Department of Electrical and Electronics Engineering, Sathyabama University, Chennai, Tamil Nadu, India
e-mail: elangomenan@gmail.com

G. Nagarajan
e-mail: nagarajanme@yahoo.co.in

L.P. Suresh and B.K. Panigrahi (eds.), *Proceedings of the International Conference on Soft Computing Systems*, Advances in Intelligent Systems and Computing 397, DOI 10.1007/978-81-322-2671-0_16

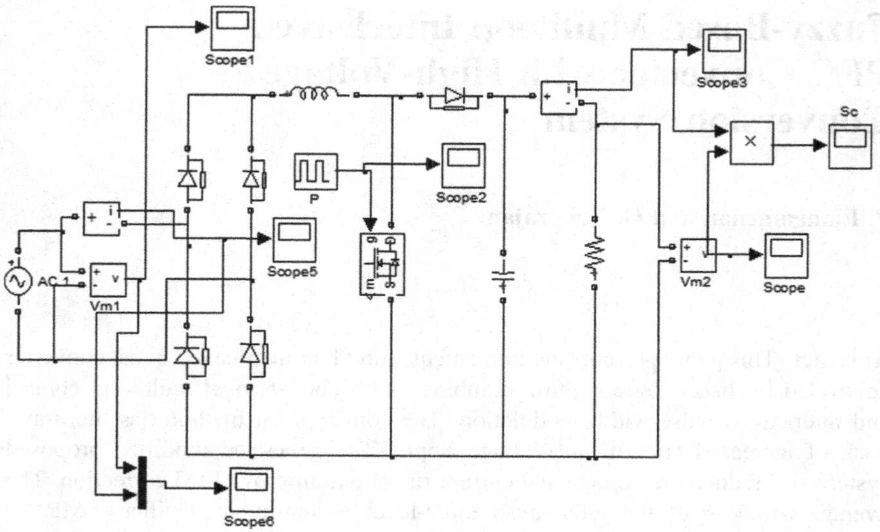

Fig. 1 Conventional circuit diagram

controller has more steady-state errors and the settling time is also very high. The modified proposed method fuzzy logic controller used when compared to PI and fuzzy controller reduces steady-state errors and the voltage settling time is also very fast. The interleaved boost converter operates switching mode at 180° switch and conducts switching sequence. As shown in Fig. 7, the drain-to-source voltage (Vds) is reduced to the MOSFET conduction losses and also reduces the drain current conduction and diode conduction losses Shown in Fig. 5 is the conventional loop; Fig. 1 shows increased current ripple and high voltage ripple poor PFC. The proposed method is shown in Fig. 8. Multiloop two-stage interleaved boost converter good PFC and voltage and current ripple are also reduced.

2 Block Diagram

In this project multiloop-interleaved boost converter's PFC corrections are improved by fuzzy logic control scheme. PWM technique used in [4–7]. Sinusoidal pulse-width modulation technique is employed for the control of ILBC.

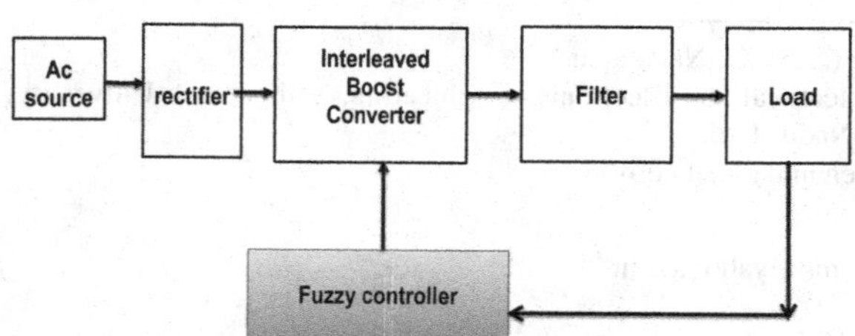

3 Design of the Existing Single-Stage ILBC

This cut converter model of the circuit is simulated with the input voltage of 15 V and an output voltage of 60 V. This system shows high current ripple and voltage ripple of the output simulation results. Figure 2 shows the input voltage and current wave form of the conventional method's circuit, Fig. 4 shows the output current, and Fig. 5 shows the output voltage wave form.

The conventional circuit diagram is shown in Fig. 1. In the conventional circuit we can use only one switch. In single switching operation, we observe more ripple currents and switching losses are also high. The output efficiency is low. Improved efficiency is necessary to get the additional auxiliary active switches to for this version conventional circuit for PI controller.

Figure 2 shows the input current and input voltage wave form for the conventional circuit. Our input voltage is an AC source. We gave a 15 V AC supply to the system. We took the input current as 2 A as shown.

Figure 3 shows the output current wave form of the conventional system. In this graph, *x*-axis represents the time while *y*-axis represents the output current. The output current is taken as 0.175 A, which contains more ripples of the conventional method.

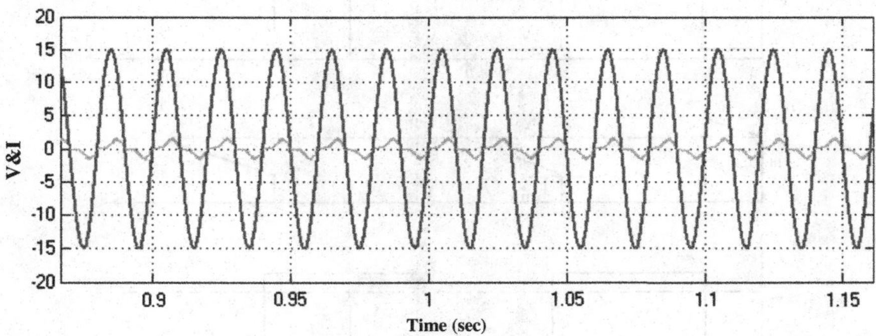

Fig. 2 Input voltage and input current

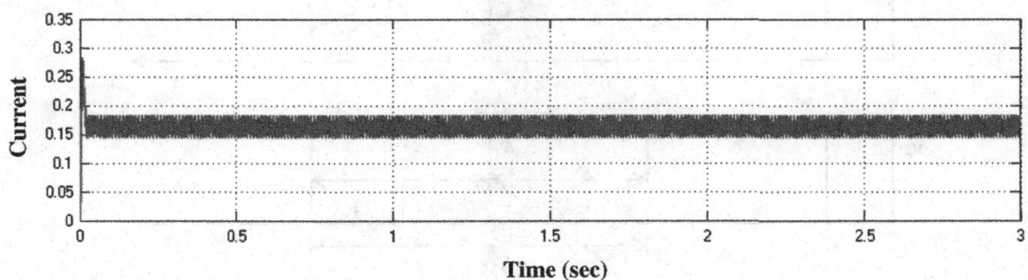

Fig. 3 Output current

Figure 4 shows the output power wave form of the conventional system. In this graph, *x*-axis represents the time while *y*-axis represents the output power. The output power is measured as 7 W which contains more ripples than the conventional method. Figure 4 represents the output power of a single stage ILBC, *x*-axis represents the time, and *y*-axis represents the power (Fig. 5).

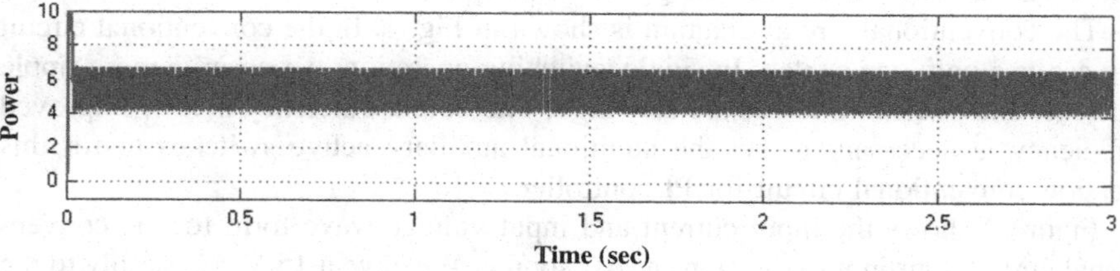

Fig. 4 Output power

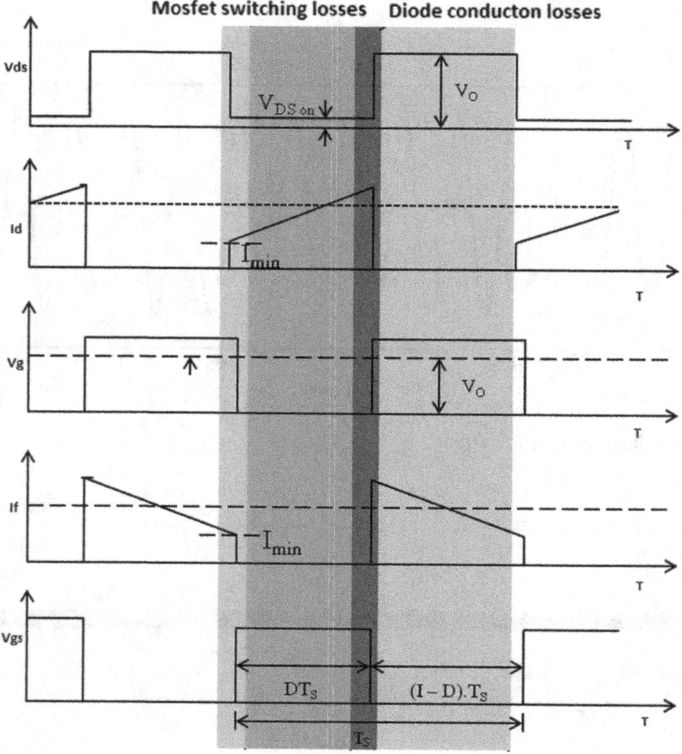

Fig. 5 Typical waveforms of one leg of the ILBC at the rated power

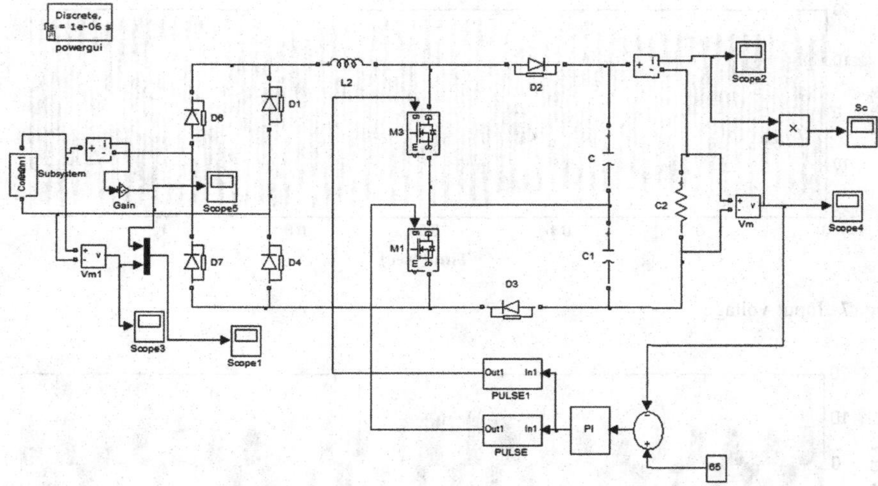

Fig. 6 Closed loop with PI controller

4 Design of the Proposed Multiloop ILBC

The proposed system of the multiloop ILBC simulation circuit diagram is shown in Fig. 8. This is the improved method. The proposed circuit's better PFC correction current ripple and voltage ripple ratio is compared in Table 2. The phase angle difference between the gate signals of the two MOSFETs V_{g1} and V_{g2}, is kept at 180° in the ILBC.

The open-loop method is shown in Fig. 8 along with the input voltage and current wave form results and output voltage and current wave form results. ILBC conducting mode, the open-loop method control, the pulse-width modulation technique used in this paper compared with the open loop system and closed loop system simulation results are given in the figures.

5 Closed Loop with PI Controller

Figure 6 shows the closed-loop system with PI controller. The three-level boost converter is shown in Fig. 7, where two capacitors are connected across the switches, respectively. Thus, each switch needs to withstand only a half output voltage. In addition, the inductor voltage in the three-level boost converter has three levels, but the inductor voltage in the conventional boost converter has only two levels. Therefore, the three-level boost converter is able to yield smaller inductor current ripple than the conventional boost converter. It follows that the three-level converters are often used in the applications, such as high-voltage ratio DC/DC

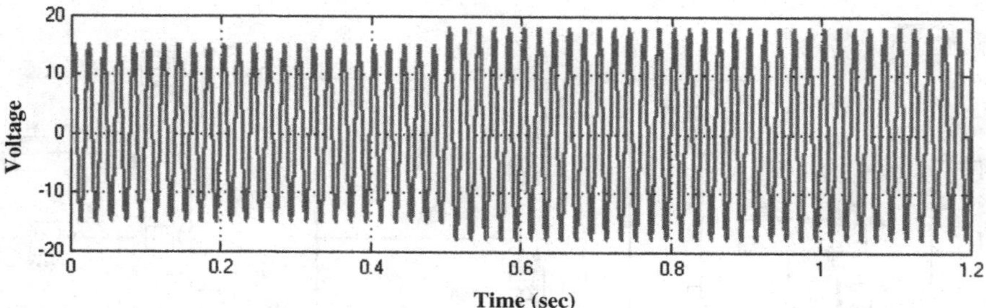

Fig. 7 Input voltage

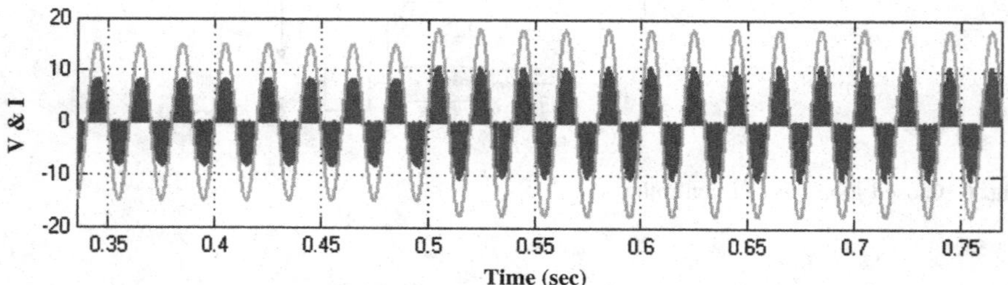

Fig. 8 Input voltage and input current

conversion. Additionally, the high-withstanding-voltage semiconductor switches often have larger drain-source resistances than the low-withstanding-voltage ones. Thus, the three-level converter has the advantages of low voltage stress, small inductor current ripple, and low switching loss.

Figure 7 shows the input voltage of the closed system with a disturbance voltage creating 0.5 s and *x*-axis represents time while *y*-axis represents voltage

Figures 7 and 8 show the input current and input voltage wave form to the closed loop with PI controller. The proposed system was studied by simulating the single stage fly back PFC using MATLAB/Simulink. The output from the FLC is taken as reference signal in Fig. 5 which is compared with ramp signal to produce the gate pulse to MOSFET gate which is shown. The gate pulses are decided by the duty cycle of the MOSFET. The output voltage and current waveforms are shown in Figs. 7 and 8.

Figure 9 shows the output voltage of proposed system, *x*-axis represents, the time and *y*-axis represents the voltage. The output voltage is near to 60 V.

Figure 10 shows the output current of proposed system, *x*-axis represents the time, while *y*-axis represents the current. The ripple current is reduced compared to the conventional method.

Figure 11 shows the output power of proposed system, *x*-axis represents the time and *y*-axis represents the power. The output power is also improved when compared with the conventional system. We have improved the better power factor correction using the fuzzy logic system.

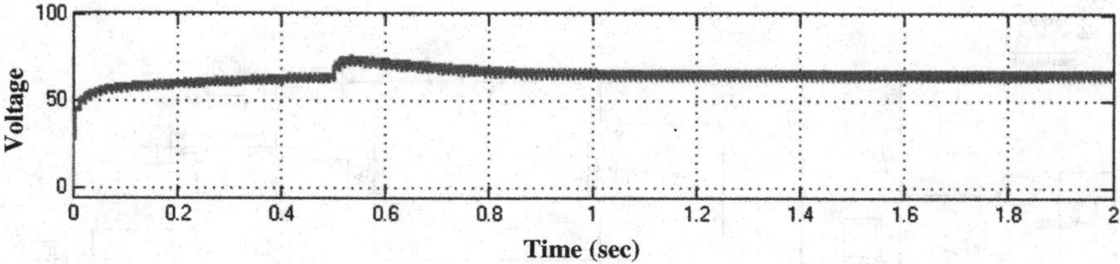

Fig. 9 Output voltage

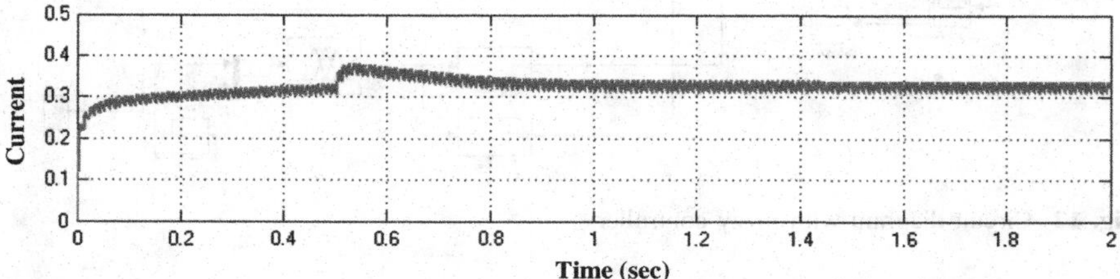

Fig. 10 Output current

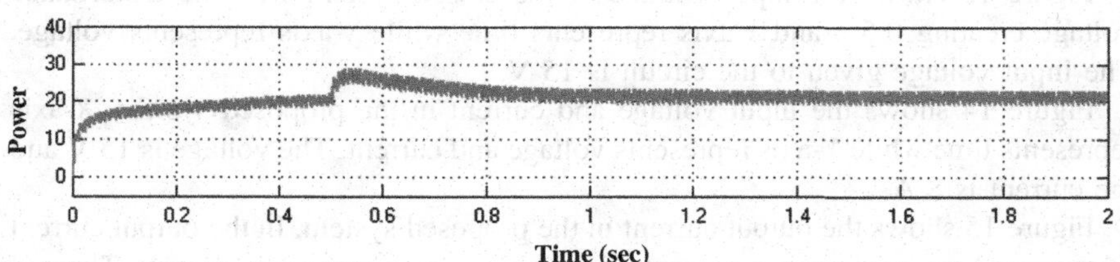

Fig. 11 Output power

6 Circuit Diagram with Fuzzy Controller

Figure 12 shows the circuit diagram with fuzzy logic controller. The conventional method using PI controller method produces more steady-state error and the settling time is also very high. The modified method's fuzzy logic controller compared to PI and fuzzy controller reduces the steady-state error and voltage settling time is also very fast and the interleaved boost converter operates at a switching mode of 180° switch which will operate and conduct switching sequence. Figure 14 shows the drain-to-source voltage (Vds) to reduce the MOSFET conduction losses, and also reduces the drain current conduction losses diode conduction losses; shown in Fig. 16 is the conventional loop's increased current ripple generating and high-voltage ripple's poor PFC The proposed method as shown in Fig. 15 is the

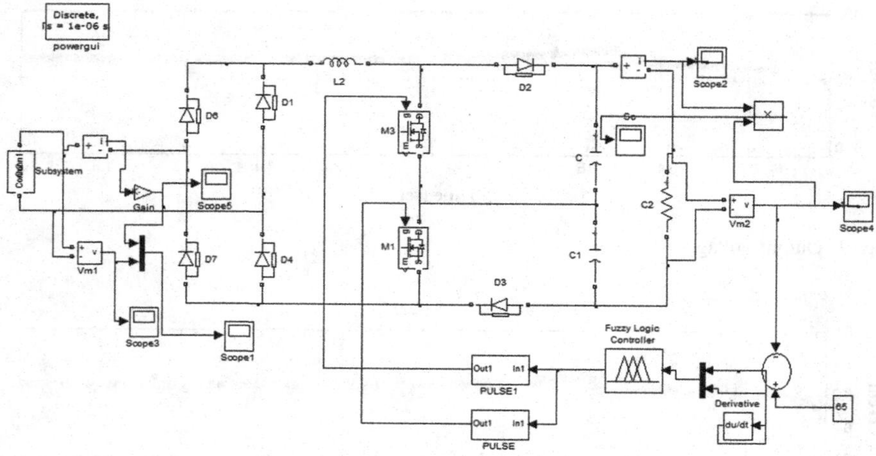

Fig. 12 Circuit diagram with fuzzy controller

multiloop two-stage-interleaved boost converter that has good PFC and voltage and current ripples are also reduced.

Figure 13 shows the input voltage of the closed system with the disturbance voltage creating 0.5 s and *x*-axis represents time while *y*-axis represents voltage. The input voltage given to the circuit is 15 V.

Figure 14 shows the input voltage and current in the proposed system. *X*-axis represents time while *Y*-axis represents voltage and current. The voltage is 15 V and the current is 8 A.

Figure 15 shows the output current in the proposed system. In the output current of proposed system, *x*-axis represents time while *y*-axis represents current. There is no ripple current in this graph.

Figure 16 shows the output power of proposed system, *x*-axis represents time, and *y*-axis represents power. From this FFT analysis, the THD is reduced to 38.60 % and the power factor is 0.93. When the same circuit is simulated for the

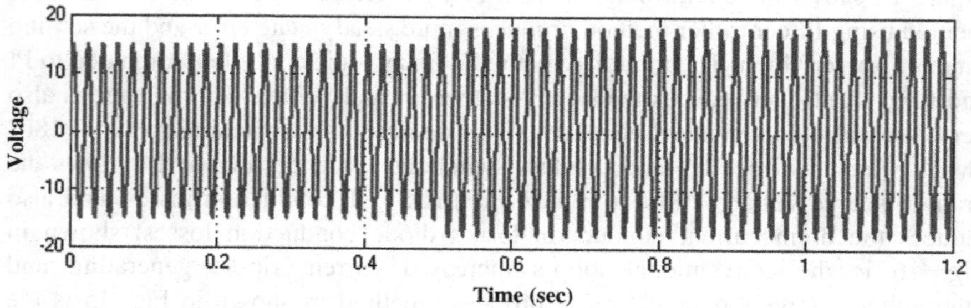

Fig. 13 Input voltage

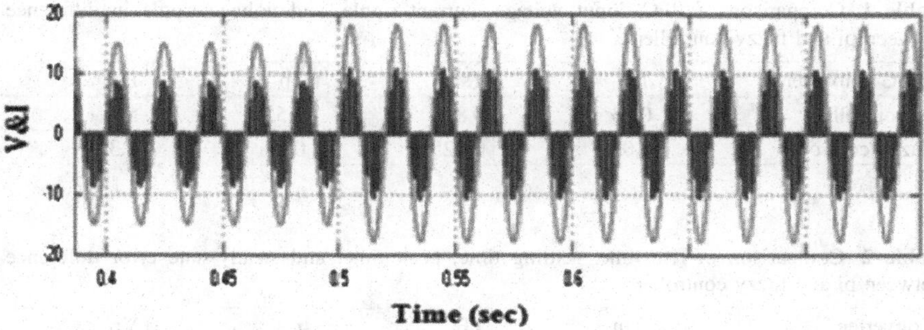

Fig. 14 Input voltage and input current

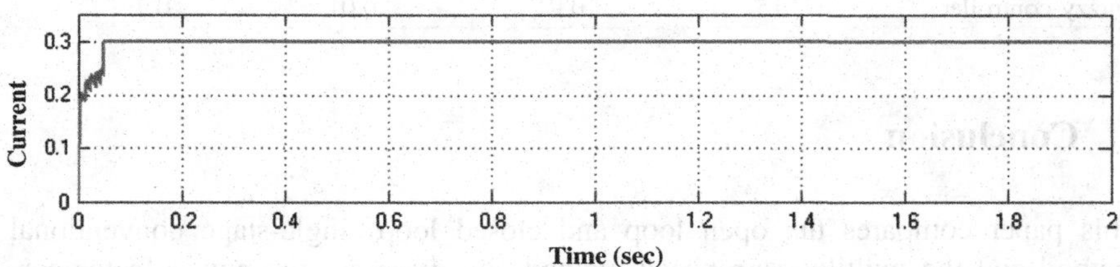

Fig. 15 Output current

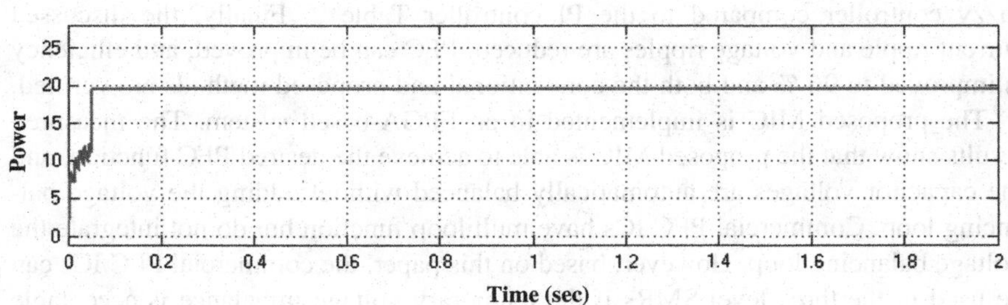

Fig. 16 Output power

same output voltage by MATLAB 2010 a with DC current feedback from the uncontrolled rectifier to generate the control pulses without fuzzy logic controller, the THD was 136.38 % (Table 1).

Thus, the efficiency curve as 96 %, improved the output of the proposed method.

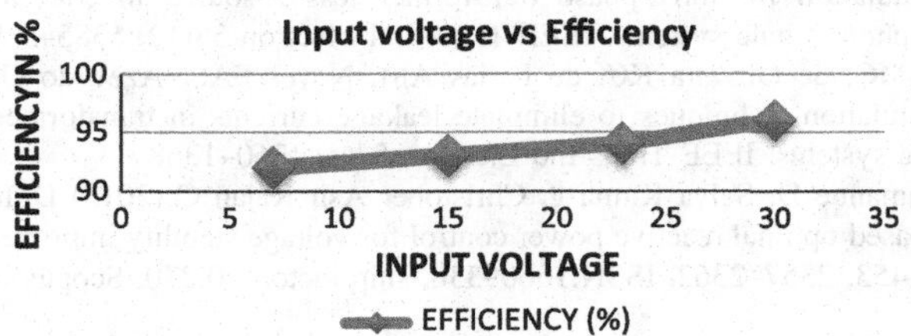

Table 1 Comparison of PFC, input voltage, current ripple, and voltage ripple in difference between pi and fuzzy controllers

ILBC converter	Tr	Ts	Tp	Ess
PI controller	0.03	0.8	0.52	5
Fuzzy controller	0.02	0.025	0.0	0.3

Table 2 Comparison of rise time, settling time, peak time, and steady-state error difference between pi and fuzzy controller

Converter	Vin	PFC	Ir	Vr
Open loop	15	0.81	0.05	8
PI controller	15	0.87	0.03	5
Fuzzy controller	15	0.91	0.01	0.3

7 Conclusion

This paper compares the open loop and closed loop single-stage conventional method and the multiloop proposed method and discusses the power factor correction of conventional method which is 0.82 while that of proposed method is 0.91. It compares PI controller and fuzzy controller; the timing response is better for fuzzy controller compared to the PI controller Table 2. Finally, the discussed current ripple and voltage ripples are reduced. PFC can be improved, and efficiency is improved to 96 % and both the conventional and proposed methods are verified.

The proposed MIC is implemented in an FPGA-based system. The measured results show that the proposed MIC is able to achieve the desired PFC function, and the capacitor voltages are automatically balanced without adding the voltage balancing loop. Commercial PFC ICs have multiloop function but do not integrate the voltage balancing loop. However, based on this paper, the commercial PFC ICs can be used in the three-level SMRs if the temporary voltage imbalance is acceptable during the system operation.

References

1. Bradaschia F, Cavalcanti MC, Ferraz PEP, Neves FAS, dos Santos EC Jr, da Silva JHGM (2011) Modulation for three-phase transformer less Z-source inverter to reduce leakage currents in photovoltaic systems. IEEE Trans Ind Electron 58(12):5385–5395
2. Cavalcanti MC, de Oliveira KC, de Farias AM, Neves FAS, Azevedo GMS, Camboim F (2010) Modulation techniques to eliminate leakage currents in transformer less three-phase photovoltaic systems. IEEE Trans Ind Electron 57(4):1360–1368
3. Godwin Immanue D, Selva Kuma g, Christober Asir Rajan C (2013) Differential evolution algorithm based optimal reactive power control for voltage stability improvement. Appl Mech Mater 448–453, 2357–2362. ISSN:16609336. snip factor—0.270, Scopus indexed

4. Godwin Immanue D, Selva Kuma G, and Christober Asir Rajan C (2014) A multi objective hybird differential evolution algorithm assisted genetic algorithm approach for optimal reactive power and voltage control. Int J Eng Technol 6(1):199–203. ISSN:0975-4024, Scopus indexed
5. Lopez O, Freijedo FD, Yepes AG, Fernandez-Comesana P, Malvar J, Teodorescu R, Doval-Gandoy J (2010) Eliminating ground current in a transformer less photovoltaic application. IEEE Trans Energy Convers 25(1):140–147
6. Nagarajan G, Thanigaivel K (2013) An Implementation os SSSC-based cascade H-bridge model series compensation scheme. IEEE Power Comput Technol ICCPCT 147–151
7. Ravi CN, Selvakumar G, Christober Asir Rajan C (2013) Hybrid real coded genetic algorithm-differential evolution for optimal power flow. Int J Eng Technol 5(4):3404–3412
8. Samuel Rajesh Babu R, Joseph H (2012) Embedded controlled ZVS Dc-Dc converter for electrolyzer application. Int J Intell Electron Syst 5(1):6–10; Xiao H, Xie S (2012) Transformer less split-inductor neutral point clamped three-level PV grid-connected inverter. IEEE Trans Power Electron 27(4):1799–1808
9. Sivachidambaranathan V, Dash SS (2011) A novel soft switching high frequency Ac to Dc series resonant converter. Nat J Electron Sci Syst 2(2):30–36
10. Shen (2012) Novel transformer less grid-connected power converter with negative grounding for photovoltaic generation system. IEEE Trans Power Electron 27(4):1818–1829

A Voice Recognizing Elevator System

D. Meenatchi, R. Aishwarya and A. Shahina

Abstract The paper proposes a voice controlled, simulated elevator system for the benefit of differently abled persons, such as those who are visually impaired or are paraplegics. The proposed approach for an eight floor elevator system uses a speaker independent automatic speech recognition system to recognize spoken words which includes the floor numbers, directions and door commands. Sphinx4 is used for this purpose. To handle emergencies, the recognition of ten digit cellular phone numbers is incorporated with a text to speech system that gives voice feedback for verification before the telephone call is placed. The mean recognition accuracy for floor numbers is 97 %, while it is 90 % for directions and door operations.

Keywords Automatic speech recognition (ASR) · Sphinx 4 · Text to speech (TTS) · Dynamic time warping (DTW) algorithm

1 Introduction

In this rapid world of technology where voice begins its era of domination to replace the touch screens from smart phones to huge computer systems, bringing voice in day-to-day affairs becomes significant. Elevators being one such system used in daily life serves this purpose of making future generations hands free which also becomes a boon for the disabled.

D. Meenatchi (✉) · R. Aishwarya · A. Shahina
Department of Information Technology, SSN College of Engineering, Chennai, India
e-mail: meenatchidau@gmail.com

R. Aishwarya
e-mail: r.aishwarya5294@gmail.com

A. Shahina
e-mail: shahinaa@ssn.edu.in

L.P. Suresh and B.K. Panigrahi (eds.), *Proceedings of the International Conference on Soft Computing Systems*, Advances in Intelligent Systems and Computing 397, DOI 10.1007/978-81-322-2671-0_17

The basic working principle of elevator is based on the elevator algorithm, where an elevator can decide to stop based on two conditions. The first one being the direction and second one based on the current floor and destination floor. The elevator is generally made up of rotors, cables, pulleys based on traction, climbing or hydraulic model. To serve laboratory purposes it can also be designed by connecting the elevator system to a desktop or microprocessor to accept input voice.

Voice control option is attractive for several reasons. It is potentially appropriate for a large number of elevator users since the system can be used by any individual capable of consistent and distinguishable vocalization [1]. Voice control also reduces physical requirements. However, the recognition accuracy of automatic speech recognition (ASR) system is a constraint in the deployment of many voice controlled system in real-world application [2].

In this paper, a voice controlled elevator system is proposed where the input commands to stimulate the movement of the elevator system are kept convenient for the users. The commands include voice input for the floor operations, directions, elevator car's door operation and a special option to place a call of speaker's choice in case of any unexpected event that requires immediate action.

2 Related Work

Speech recognition had been effectively contemplated since 1950s, however, late improvements in personal computer (PC) and telecommunication technology have enhanced speech recognition abilities [3]. In a useful sense, speech recognition has tackled issues, enhanced benefits and bought a greater revolution in current scenario [3]. Voice control can replace the function of a push-button efficiently [4]. Speech recognition is a very complex issue. It includes numerous calculations which oblige high computational necessities [5]. Automatic speech recognition is used in various areas ranging from medical transcription to game control, from call centre dialogue systems to data recovery [6, 7].

This automatic speech recognition process is used in most of the voice controlled systems. The voice controlled wheelchair for the physically challenged was proposed in 2002 [8]. This paper had described an experiment that compared the performance of abled and disabled people using voice control to operate a power wheelchair both with and without navigation assistance, where the navigations were assisted by the sensors to identify and avoid obstacles in wheelchair's path [1, 8].

The voice controlled home automation system was proposed in 2008 [9]. This paper had discussed the speech recognition and its application in control mechanism [9]. Speech recognition can be used to automate many tasks that usually require hands-on human interaction, such as turning on lights or shutting a door or driving a motor [9, 10].

The intelligent lift control model that uses voice recognition was proposed in 2003 [11]. This proposed model had been controlled by voice and sensor panel [11]. The modification of the well-known dynamic time warping (DTW) algorithm was used [11]. The set of voice commands for the model consisted of eight Lithuanian words [11]. The model was specifically designed for domestic use (smart houses). In contrast to the intelligent lift control model [11], this paper proposes a simulated model that utilizes speech recognition to implement an elevator system that could be of help to visually or physically challenged person where the system generated voice gives assistance. Also it brings in the concept of immediate safety measures for emergency.

3 Adapting Sphinx4 into This Project

3.1 Overview of Sphinx4

Sphinx4 is a Java speech recognition library where speech recognition used is an open-source recognition platform from CMU [12]. It gives a quick API to change the speech recordings into text with the help of "CMUSphinx" acoustic models. It can be utilized on servers and as a part of desktop applications and it is highly portable. Beside speech recognition Sphinx4 serves to distinguish speakers, adapt models, allows highly flexible user interfacing and more [12]. Sphinx4 supports US English and many other dialects.

3.2 Using Sphinx4 into This Project

The recognition platform of Sphinx4 helps to add library files into dependencies of the project and there are few high level recognition interfaces such as LiveSpeechRecognizer [12], StreamSpeechRecognizer [12], SpeechAligner [12] and SpeechResult [12]. These interfaces along with the acoustic model, dictionary, grammar/language model and source of speech are executed for the task of speech recognition [12].

3.3 Using Text-to-Speech Conversion in This Project

Text-to-speech (TTS) synthesizer is a computer-based system that reads any text aloud, whether it was introduced in the PC by an administrator or checked and submitted to an optical character recognition (OCR) system [13]. It is a speech synthesis application that is used to make a spoken sound version of the text in a computer report, such as a help document or a web page [14]. It can empower the reading of computer display information for the visually challenged person or may just be utilized to enlarge the perusing of an instant message.

4 Elevator Execution

4.1 Database

The experiment (refer Sect. 5) is performed under laboratory conditions with the use of an Audio Technica pro37 condenser microphone. The experimental conditions are impervious to superfluous noise. One speaker at a time is allowed to give any of the following input commands.

The system generated voice greets the user, guides the user about the current floor and also instructs them to feed in the input voice commands (Table 1).

4.2 Working

The input voice to the automatic speech recognition (ASR) system deduces a command for the control device. This device controls entire elevator operations including its speed and traction. It also handles any emergency situation. In the event of unexpected situations, to avoid panicking, calling methods has been improvised where the user can verify the number using a TTS system. Any cellular number of speaker's choice can be called; otherwise the normal lift operations are resumed.

The different parts of the block diagram are explained in the following sub Sects. 4.3 (ASR), 4.4 (Emergency Situations) and 4.5 (Control Device) (Fig. 1).

4.3 Automatic Speech Recognition

The acoustic model detects the relationship between the input speech signal and phonetic units in the language. The speech output utilizes the search algorithm to provide the recognition results. The search takes place by mapping the words to the phones in the lexicon (dictionary), where the mapped words undergo search space restriction of the language model to filter out the results (Fig. 2).

Table 1 List of voice commands

	Commands	Description
Floor numbers	One, two, three, four, five, six, seven, eight, nine	The simulated elevator moves to the corresponding floor numbers
Direction	Up, down	The simulated elevator moves along the corresponding floor numbers
Door operations	Open, close	The opening and closing of elevator cars door is controlled
Cellular numbers	Sequence of ten digit numbers	Call for help is placed to this number

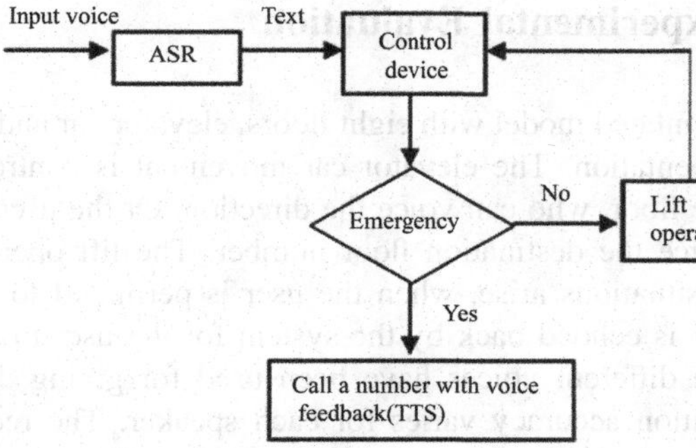

Fig. 1 Block diagram of voice controlled lift

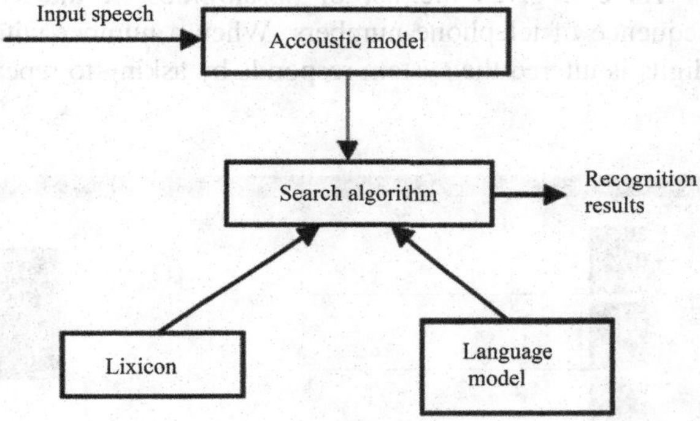

Input voice
ASR
Text
Control device
Emergency
No
Lift operations
Yes
Call a number with voice feedback(TTS)

Fig. 2 Block diagram of ASR [5]

Input speech
Accoustic model
Search algorithm
Recognition results
Lixicon
Language model

4.4 Control Device

The general movement of the elevator between different floors including the acceleration and time of travel is determined by the control device. The directions sent to the control device are from ASR or the feedback from normal lift operations. The operation done here includes the aspect of safety, performance and perfect coordination.

4.5 Emergency Situation

The effective and unique handling of emergency situations is a significant part of the proposed model. The ten digit phone number of speaker's choice is accepted. The TTS conversion system replays the number that is ratified by the user. The call is placed to that number which can bring immediate help.

5 Experimental Evaluation

The simulated model with eight floors, elevator car and request pool is shown in the implementation. The elevator car movement is controlled either by the user at a specific floor, who can voice the direction. Or the user inside the elevator car, who can voice the destination floor number. The lift operations proceed unless emergency situations arise, when the user is permitted to call a cellular number. This number is echoed back by the system for the user to verify (Fig. 3).

Five different voices have been used for giving the input commands and the recognition accuracy varies for each speaker. The recognition of both digits and words varies according to the pronunciation. The results are discussed in (Table 2).

Table 2 gives the list of possibilities of the system response for different sequence of telephone numbers. When a number with less than or more than ten digits is uttered the system responds by asking to repeat the ten digit number (Case

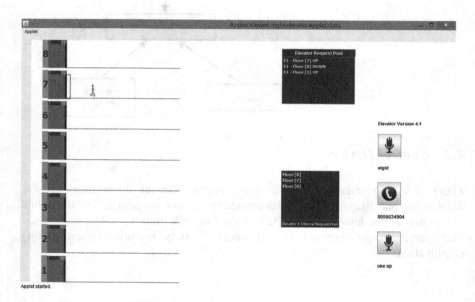

Fig. 3 Implementation of the voice controlled lift

Table 2 List of voice responses for emergency situations

Case	Phone number	Voice response
A	805603490	Please say a ten digit phone number
B	9789941254	Nine seven eight nine nine four one two five four
C	9441424821	Nine four four one four two four eight two one
D	74166947211	Please say a ten digit phone number

Table 3 Confusion matrix for floor number recognition

Digits	1	2	3	4	5	6	7	8
1	9	0	0	0	1	0	0	0
2	2	6	0	0	0	0	0	2
3	0	0	7	0	0	1	0	1
4	0	0	0	8	0	1	0	1
5	1	0	0	0	8	0	0	0
6	0	0	0	2	0	8	0	0
7	1	0	0	0	0	0	9	0
8	0	0	0	0	0	0	0	10

Table 4 Confusion matrix for lift operation recognition

Lift operations	Open	Close	Up	Down
Open	8	1	1	0
Close	0	7	3	0
Up	0	0	10	0
Down	0	0	1	9

A & D). When an exact ten digit phone number is uttered the system replays the number (Case B).

Tables 3 and 4 shows the confusion matrix for the recognition of the floor numbers and lift operations, respectively. For a scale of one to ten, the number *eight* is recognized all the ten times as *eight*, whereas the number *two* and *three* are recognized as themselves for six and seven times, respectively. And the word *up* is recognized all the ten times as *up*, whereas the word *close* is recognized as *close* for seven times.

The graph in Figs. 4 and 5 show the percentage accuracy of the digits and words for each of the five speakers. The average recognition accuracy is calculated for each of the floor numbers (90.13 %), door operations (85.2 %) and directions (93.6 %) (Table 5).

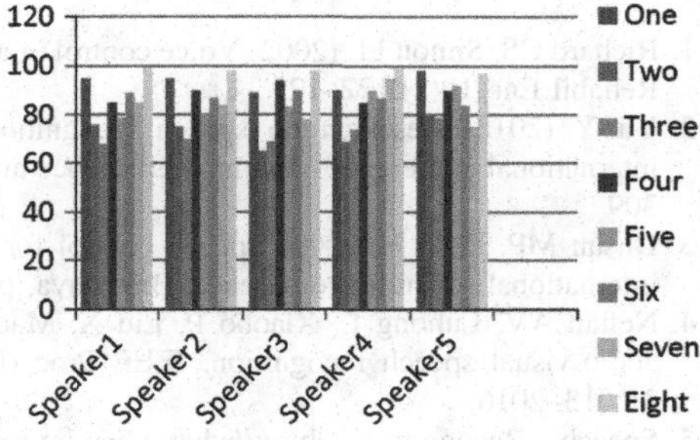

Fig. 4 Recognition accuracy for floor numbers (one-eight) depending upon the speaker

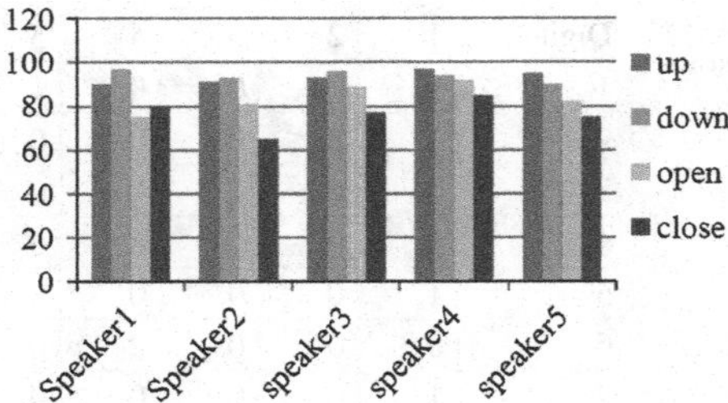

Fig. 5 Recognition accuracy for directions (*up*, *down*) and door operations (*open*, *close*) depending upon the speaker

Table 5 Percentage accuracy for lift operations

Commands	% Accuracy
Floor numbers	90.13
Directions	93.6
Door operation	85.2

6 Conclusion

This paper explains how voice control can become a boon in the future in everyday life by means of an elevator simulation. The proposed work shows the feasibility of developing an elevator system based on voice control. It also incorporates a voice feedback system in case of placing emergency calls, which helps the user to verify the correctness of the number. In future, the size of the experiment can be improvised to make this model a real-time system. Also the recognition accuracy can be improved by making the speech recognition system speaker dependent and by including the aspect of robustness to noise.

References

1. Richard CS, Simon PL (2002) Voice control of a powered wheelchair. IEEE Trans Neural Syst Rehabil Eng 10(2):122–125
2. Yu Y (2012) Research on speech recognition technology and its application. In: 2012 international conference on computer science and electronics engineering (ICCSEE), pp 306–309
3. Dušan MP, Peter F (2000) Speech control for car using the TMS320C6701 DSP. In: 10th international scientific conference, Bratislava, pp 97–100
4. Nefian AV, Luhong L, Xiaobo P, Liu X, Mao C, Murphy K (2002) A coupled HMM for audio-visual speech recognition. IEEE Proc (ICASSP '02) Acoust Speech Signal Process 2:2013–2016
5. Speech Recognition. http://what-when-how.com/video-search-engines/speech-recognition-audio-processing-video-search-engines/. Accessed 9 Mar 2015

6. Thomas H, Asmaa EH, Stuart NW, Vincent W (2008) Automatic speech recognition for scientific purposes—webASR. In: 9th annual conference of the international speech communication association, pp 504–507
7. Suzuki H, Zen H, Nunkuku Y, Miyajima C, Tokuda K, Kitumuru I (2013) Speech recognition using voice-characteristics dependent acoustic models. In: Acoustics, speech, and signal processing, 2003. Proceedings. (ICASSP '03). 2003 IEEE international conference, pp 740–743
8. Muhammad TQ, Syed AA (2009) Voice controlled wheelchair using DSK TMS320C6711. In: International conference on signal acquisition and processing, pp 217–220
9. Jeong H-DJ, Ye S-K, Lim J, You I, Hyun W (2013) A remote computer control system using speech recognition technologies of mobile devices. In: 2013 seventh international conference on innovative mobile and internet services in ubiquitous computing, pp 595–600
10. Haleem MS (2008) Voice controlled automation system. In: Multitopic conference, INMIC, pp 508–512
11. Cernys P, Kubilius V, Macerauskas V, Ratkevicius K (2003) Intelligent control of the lift model. In: Proceedings of the second IEEE international workshop, pp 428–431
12. Sphinx4 dependencies. http://cmusphinx.sourceforge.net/wiki/tutorialsphinx4. Accessed 9 Mar 2015
13. FreeTTS. http://freetts.sourceforge.net/docs/index.php#what_is_freetts. Accessed 9 Mar 2015
14. Zhu J, Gao X, Yang Y, Li H, Ai Z, Cui X (2010) Developing a voice control system for zigbee-based home automation networks. In: Network infrastructure and digital content 2010, 2nd IEEE international conference, pp 737–741

Application of DSTATCOM for Loss Minimization in Radial Distribution System

P. Shanmugasundaram and A. Ramesh Babu

Abstract Power system consists of three parts, first one is generation, second one is transmission, and third one is distribution. The distribution system provides electric power to end consumer. The distribution line has three configurations. They are radial, two end fed, and ring distribution system. Among them radial distribution is most commonly used which is economical and rugged system. This radial distribution system is suffers by loss. This loss can be reduced by including capacitor in the system which is discrete size and not suitable for fine adjustment and it is a passive device and has slow response. Distributed static compensator (DSTATCOM) is flexible AC transmission system (FACTS) device which is fast and continuous adjustable device. The DSTATCOM provide adequate reactive power support and reduce losses in the radial distribution system. To demonstrate the effect of DSTATCOMIEEE 33 Bus test system is considered.

Keywords DSTATCOM · Loss minimization · VAR injection · FACT · Sensitive bus

1 Introduction

Nowadays the distribution system play important role of electrical utility, due to its power quality and planning. The power system loss in India has been in the range of 25–30 % of its power production in that distribution systems contribute more compared other systems. This major distribution loss is due to the voltage sag and limitation of stability. These distribution losses are reduced by application of distribution networks, such as dynamic voltage restorer (DVR), unified power flow controller (UPFC), and distribution STATic COMpensator (DSTATCOM). Among

P. Shanmugasundaram (✉) · A. Ramesh Babu
Department of EEE, Sathyabama University, Chennai, India
e-mail: shanmugam12101960@gmail.com

A. Ramesh Babu
e-mail: rameshbabuaa@gmail.com

L.P. Suresh and B.K. Panigrahi (eds.), *Proceedings of the International Conference on Soft Computing Systems*, Advances in Intelligent Systems and Computing 397, DOI 10.1007/978-81-322-2671-0_18

them DSTATCOM is normally used due to its parallel connection, compensate power quality issues like voltage sag and over loading [1]. The DSTATCOM placement problem involves determining the location, and size to be installed in the buses of the radial distribution system which requires reactive power. Hence reduces the power losses which results in the maximization of annual savings. A numerous work has been done to solve the optimal capacitor placement problem. Though various methods have been investigated by the researchers' in the area of loss minimization, placement, and sizing of reactive power injection, it is found that FACTS device gives the better results. Many papers have dealt with the problem of capacitor allocation using intelligent techniques in the distribution system for the purpose of power loss reduction. The reactive power injection improves the bus voltage magnitude and thereby it reduces transmission line losses [2]. The sensitive buses for reactive power injection are identified and reactive power of DSTATCOM is injected into the buses to reduce the loss. DSTATCOM is fast acting reactive power support as compared to capacitor. In this paper, loss minimization of both capacitor and DSTATCOM is considered and compared.

Baran and Wu [3], considered the capacitor placement in the radial distribution system. Their objective of the work is to find best capacitor location and size of the capacitor in the radial system. They used mixed integer technique for the problem solving. They decompose the problem into master and slave, to simplify the execution. Gallego et al. [4], introduce reactive power injection using capacitor in discrete size. These capacitor size and position in the radial distribution system is vital for better control and to reduce loss. Their mathematical model uses resistance for the calculation of real power and reactance for calculating reactive power loss. The current injection in the bus should be enough to supply the demanded current in behind buses. Haque [5] developed a procedure to minimize cupper loss of the distribution line. His method is to introduce capacitor in the buses which will reduce reactive current thereby total current in the branch. Reduction of current leads to reduction cupper loss in the distribution line. He ordered the buses to be compensated by the capacitor for reactive power for location and then size of capacitor is decided based on the requirement in the bus. Baghzouz and Ertem [6] used capacitor to compensate non-sinusoidal harmonic distortion. The objective of the work is to minimize total harmonic distortion in the radial system. The capacitor shunted radial distribution line improves the quality of the power and reduces the total harmonic distortion. Baghzouz [7] used nonlinear load for the radial distribution line for loss minimization and minimization harmonics. Harmonic currents due to nonlinear load are reduced by connecting capacitor in the radial distribution line. Current injection model is considered for the work. Voltage variation in the bus is used to decide the size of capacitor. Sultana and Roy [8] used teaching learning optimization for placing and sizing of capacitor. It uses teacher teaching technique and student learning process for the implementation of algorithm for location and sizing of capacitor. They used standard test buses for the implementation and the results are compared to other intelligent technique. Huang [9], proposed tabu search technique for find optimal positioning and sizing of the capacitor placement. For finding best location for giving more impact sensitive analysis method is used. This analysis provides set of buses for the placement of capacitor so only size is decided by the search algorithm for loss minimization. Das and Verma [10] used artificial neural network for the decision of connecting discrete size of capacitor in the radial line for reduction loss. Aman [11] proposed parallel capacitor for loss minimization and to increase the load-ability of the considered radial distribution lines. They used six different approaches for the placement of capacitor. The capacitors are connected weakest bus. Devi and Geethanjali [12] used DSTATCOM for loss minimization in the radial distribution system. They used loss sensitive factor to identify sensitive bus for location of DSTATCOM and its size or amount of reactive power injection is find using PSO algorithm [13, 14]. It is the efficient population-based intelligent algorithm for solving optimization problem. In this paper, bus sensitive factor is used to find the location of DSTATCOM and the result of DSTATCOM is compared to capacitor compensation.

2 Problem Formulation

The objective of the problem is to minimize the power loss in the radial distribution line and subject to power balance equality constraint and voltage level of inequality constraint. The problem formulation is given below

$$\min \cdot F = \sum_{i=1}^{\text{nbus}} P_{\text{Loss}(i,i+1)} \tag{1}$$

$$P_{\text{Loss}(i,i+1)} = \frac{(P_i + Q_i)}{V_i^2} * R_{(i,i+1)} \tag{2}$$

Subjected To:
Equality constraint,

$$P_{\text{T,Gen}} = P_{\text{T,Load}} + P_{\text{T,Loss}} \tag{3}$$

Inequality constraint,

$$V_{\min} \le V_i \le V_{\max} \tag{4}$$

where,

$P_{\text{Loss}(i,i+1)}$ = Loss between bus i and $i + 1$
P_i = Real power at bus i
Q_i = Reactive power at bus i
V_i = Voltage magnitude at bus i
$R_{(i,i+1)}$ = Resistance between bus i and $i + 1$
$P_{\text{T,Gen}}$ = Total real power generation

$P_{T,Load}$ = Total real power Demand or Load
$P_{T,Loss}$ = Total real power loss in the system
V_{min}, V_{max} = Minimum and Maximum voltage limit

2.1 Sensitivity Analysis

In order to determine the candidate location for placing the capacitors in the radial distribution system, sensitivity analysis [9] is employed. The evaluation of these candidate locations basically helps in reducing the search space during optimization procedures. It is used to select locations that reduce system real power losses when we place capacitors at those locations. Find out the potential buses for the capacitor placement following steps to be given below,

Calculate the loss sensitivity factor

$$S = \frac{\partial P_{Loss}}{\partial Q} \tag{5}$$

Arrange the value of loss sensitivity factor of all buses in descending order and then find the normalized voltage magnitude using following Eq. (6)

$$\text{norm}(i) = \frac{V(i)}{0.95} \tag{6}$$

The placement of capacitor or DSTATCOM bus selected, if value of norm is less than 1.01.

3 Procedure for Solving Problem

The common forward and backward load flow is used to find loss in the system to find objective function value given in Eq. (1) and to satisfy equality constraint given by Eq. (3). The outcome of the load flow provides voltage magnitude of all bus is used to check the constraint of Eq. (4). Then the sensitive buses are finding using Eqs. (5) and (6). The algorithm for solving this problem is listed below.

Step 1: Get radial distribution line data.
Step 2: Set initial voltage value for the flat start and use backward sweep and forward sweep to know load flow.
Step 3: Find P_{Loss} and voltage magnitude of the radial system.
Step 4: Find sensitive bus using Eqs. (5) and (6).
Step 5: Add capacitor in the sensitive bus.

Step 6: Run forward backward load flow to know P_{Loss} and V.
Step 7: Place DSTATCOM in selected bus.
Step 8: Run forward backward load flow to know P_{Loss} and V.

4 Results and Discussion

For the simulation IEEE 33 bus is considered as given in Fig. 1. Using sensitive factor analysis best buses for compensation is selected. In this research work three capacitors and then 3 DSTATCOM is used to connect in the sensitive bus to reduce the loss. In the simulation forward and backward load flow is used. To calculate the cost for loss 168 rupees per kw is considered hence when loss is reduced the cost for the loss is reduced as discussed below.

Base case result is given below in the Table 1 as follows.

Base case total real power loss in kw: 199.102

The sensitive buses for reactive power compensation are:

6	28	29	30	9	13	10	8	27	31	26	14	7	12	17	16	15	11	32	18	33	1

After 3 Capacitor placed in the first three sensitive buses are given in the Table 2.

For capacitor total real power loss in kw: 168.456

After 3 DSTATCOM placed in first three sensitive buses are given in Table 3

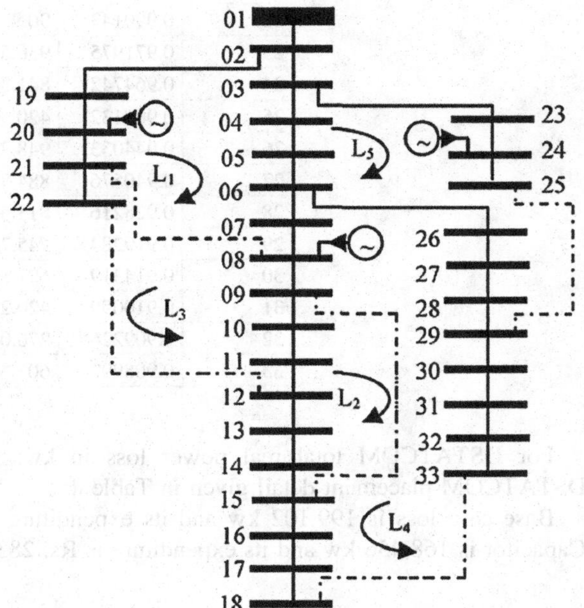

Fig. 1 Single line diagram of IEEE 33 bus system

Table 1 Base case result

Bus no.	Voltage (PU)	Cumulative real Power load (kvar)	Cumulative reactive Power load (kvar)
1	1.00334	3917.91	2458.88
2	0.999673	3905.67	2452.55
3	0.984357	3392.61	2205.02
4	0.977662	2342.89	1697.63
5	0.968257	2204.06	1608.04
6	0.94823	2105.85	1528.09
7	0.947799	1093.12	528.131
8	0.941539	888.345	419.836
9	0.934628	684.217	316.87
10	0.928186	620.701	294.377
11	0.927081	560.154	274.196
12	0.925168	514.283	243.908
13	0.918582	451.649	206.835
14	0.916507	390.93	170.888
15	0.91505	270.578	90.5747
16	0.913568	210.3	80.372
17	0.911711	150.052	60.041
18	0.911059	90	40
19	0.995094	360.979	160.928
20	0.991394	270.145	120.176
21	0.990724	180.044	80.0579
22	0.990143	90	40
23	0.971975	936.531	455.148
24	0.964742	841.308	401.023
25	0.961132	420	200
26	0.944033	948.222	972.35
27	0.940576	884.888	945.651
28	0.928216	813.569	910.671
29	0.919284	745.729	883.839
30	0.914319	621.823	811.848
31	0.910099	420.226	210.269
32	0.909226	270.013	140.021
33	0.908992	60	40

For DSTATCOM total real power loss in kw: 138.656.The capacitor and DSTATCOM placement detail given in Table 4.

Base case loss is 199.102 kw and its expenditure is Rs. 33449.211, Loss for Capacitor is 168.456 kw and its expenditure is Rs. 28300.688 and hence saving is

Table 2 Result after 3 capacitor

Bus no.	Voltage	Cumulative real	Cumulative reactive
	(PU)	Power load (kw)	Power load (kvar)
1	1.00251	3886.42	1984.09
2	0.999132	3875.51	1978.44
3	0.984998	3368.89	1734.2
4	0.979101	2322.74	1228.61
5	0.970919	2187.41	1140.81
6	0.95362	2096.44	1070.31
7	0.953449	1092.94	377.951
8	0.947231	888.213	419.738
9	0.940366	684.126	316.802
10	0.933965	620.644	294.334
11	0.932867	560.103	274.155
12	0.930966	514.241	243.87
13	0.924423	451.632	206.817
14	0.922362	390.921	170.879
15	0.920914	270.572	90.5691
16	0.919441	210.297	80.3684
17	0.917597	150.052	60.0406
18	0.916949	90	40
19	0.995374	360.979	160.927
20	0.991675	270.145	120.176
21	0.991005	180.044	80.0578
22	0.990425	90	40
23	0.973723	936.508	455.13
24	0.966504	841.303	401.02
25	0.9629	420	200
26	0.949804	939.934	665.774
27	0.94696	877.617	639.593
28	0.936905	809.954	607.837
29	0.929713	744.805	583.35
30	0.925321	621.787	661.811
31	0.921152	420.222	210.264
32	0.92029	270.013	140.02
33	0.920059	60	40

Rs. 5148.52. Loss for DSTATCOM is 138.656 kw and its Expenditure is Rs. 25654.171 and hence savings is Rs. 7795.04. Figures 2 and 3 shows the comparison of losses and savings of the system (Tables 3 and 4.)

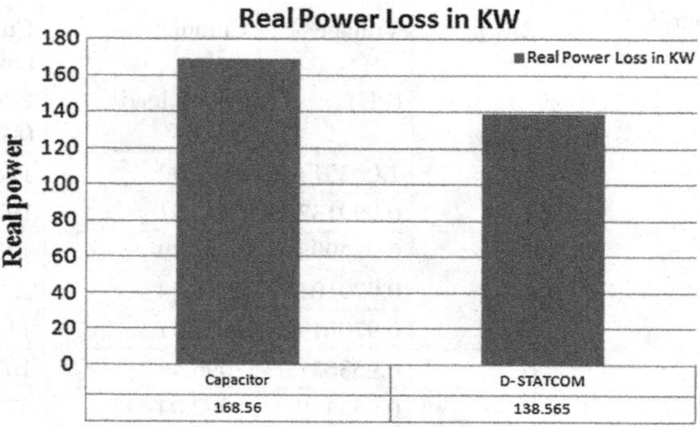

Fig. 2 Comparison of real power losses

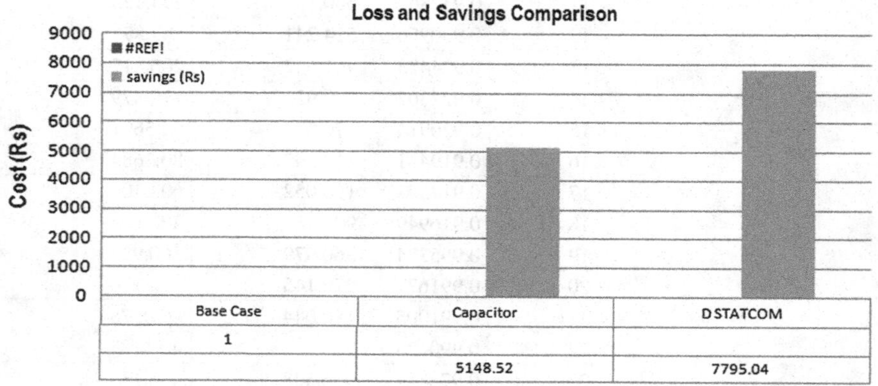

Fig. 3 Comparison of savings

Table 3 Result after 3 DSTATCOM

Bus no.	Voltage (PU)	Cumulative REAL Power Load(kw)	Cumulative REACTIVE Power Load(kvar)
1	1.00221	3876.34	1785.71
2	0.998947	3865.9	1780.31
3	0.985318	3361.5	1537.19
4	0.979767	2316.51	1032.19
5	0.97209	2182.3	944.953
6	0.955919	2093.61	877.448
7	0.955834	1092.87	304.968
8	0.949635	888.156	419.696
9	0.942788	684.087	316.773
10	0.936405	620.62	294.315
11	0.93531	560.081	274.137

(continued)

Table 3 (continued)

Bus no.	Voltage (PU)	Cumulative REAL Power Load(kw)	Cumulative REACTIVE Power Load(kvar)
12	0.933413	514.223	243.853
13	0.926889	451.626	206.81
14	0.924833	390.917	170.876
15	0.923389	270.57	90.5667
16	0.92192	210.296	80.3668
17	0.920081	150.052	60.0405
18	0.919435	90	40
19	0.99549	360.978	160.927
20	0.991791	270.145	120.176
21	0.991121	180.044	80.0578
22	0.990541	90	40
23	0.97445	936.498	455.122
24	0.967236	841.301	401.018
25	0.963636	420	200
26	0.952282	937.47	546.237
27	0.949674	875.442	520.204
28	0.940507	808.809	489.356
29	0.933987	744.409	465.522
30	0.929885	621.772	583.305
31	0.925737	420.22	210.262
32	0.924878	270.013	140.02
33	0.924648	60	40

Table 4 Capacitor and DSTATCOM location and size

Bus no.	Cap. size (kvar)	DSTATCOM size (kvar)
6	150	222.906
28	150	189.128
29	150	228.491
Tt. loss (Kw)	168.456	138.656
Saving (Rs.)	5148.52	7795.04

5 Conclusion

In this research work, radial distribution system IEEE 33 bus is considered to find effect of capacitor and DSTATOM for the reduction of loss. Capacitor gives reactive component of the current, and hence, it compensates the reactive current partially and reduces total load current. This reduction current reduces losses in the radial distribution line. DSTATCOM is a FACTS device which is fast and better reactive power support device. Connection of this DSTATCOM reduces distribution system loss further and saves cost better than capacitor compensation.

References

1. Taher SA, Afsari SA (2014) Optimal location and sizing of DSTATCOM in distribution systems by immune algorithm. Electri Power Energy Syst 60:34–44
2. Sundarsingh Jebaseelan SD, Raja Prabu R (2013) Power quality improvement of fourteen bus system using STATCOM. Nat J Electron Sci Syst 15–23. ISSN:0975-7325, SathyabamaUniversity
3. Baran ME, Wu FF (1989) Optimal capacitor placement on radial distribution systems. IEEE Trans Power Delivery 4(1):725–734
4. Gallego RA, Monticelli AJ, Romero R (2001) Optimal capacitor placement in radial distribution networks. IEEE Trans Power Syst 16(4):630–637
5. Haque MH (1999) Capacitor placement in radial distribution systems for loss reduction. IEE Proc Gener Transm Distrib 146(5):501–505
6. Baghzouz Y, Ertem S (1990) Shunt capacitor sizing for radial distribution feeders with distorted substation voltages. IEEE Trans Power Delivery 5(2):650–657
7. Baghzouz Y (1991) Effects of nonlinear loads on optimal capacitor placement in radial feeders. IEEE Trans Power Delivery 6(1):245–251
8. Sultana S, Roy PK (2014) Optimal capacitor placement in radial distribution systems using teaching learning based optimization. Int J Electr Power Energy Syst 54:387–398
9. Huang Y-C, Yang H-T (1996) Solving the capacitor placement problem in a radial distribution system using Tabu search approach. IEEE Trans Power Delivery 11(4):1868–1873; Das BP, Verma K (2001) Artificial neural network-based optimal capacitor switching in a distribution system. Int J Electr Power Syst Res 60(2):55–62
10. Aman MM et al (2014) Optimum shunt capacitor placement in distribution system—a review and comparative study. Renew Sustain Energy Rev 30:429–439
11. Devi S, Geethanjali M (2015) Placement and sizing of D-STATCOM using particle swarm optimization. Power Electron Renew Energy Syst. 941–951 (Springer India)
12. Sundarsingh Jebaseelan SD, Immanuel DG (2014) Reactive power compensation using STATCOM with PID controller. Int J Appl Eng. 9(21):11281–11290. ISSN:0973-4562
13. Sundarsingh Jebaseelan SD, Raja Prabu R (2013) Reactive power control using FACTS devices. Ind Stream Res J 3(2):45–59. ISSN:2230-7850, Impact Factor 0.2105
14. Ravi CN, Christober Asir Rajan C (2012) Optimal power flow solutions using constraint genetic algorithm. Nat J Adv Comput Manag 3(1):48–54. Sathyabama University

Automated Health Monitoring Through Emotion Identification

S. Ananda Kanagaraj, N. Kamalakannan, M. Devosh, S. Uma Maheswari, A. Shahina and A. Nayeemulla Khan

Abstract Emotional health refers to the overall psychological well-being of a person. Prolonged disturbances in the emotional state of an individual can affect their health and if left unchecked could lead to serious health disorders. Monitoring the emotional well-being of the individual becomes a vital component in the health administration. Speech and physiological signals like heart rate are affected by emotion and can be used to identify the current emotional state of a person. Combining evidences from these complementary signals would help in better discrimination of emotions. This paper proposes a multimodel approach to identify emotion using a close-talk microphone and a heart rate sensor to record the speech and heart rate parameters, respectively. Feature selection is performed on the feature set comprising features extracted from speech-like pitch, Mel-frequency cepstral coefficients, formants, jitter, shimmer, and heart beat parameters like heart rate, mean, standard deviation, root mean square of interbeat intervals, heart rate variability, etc. Emotion is individually modeled as a weighted combination of speech features and heart rate features. The performance of the models is evaluated. Score-based late fusion is used to combine the two models and to improve recognition accuracy. The combination shows improvement in performance.

S. Ananda Kanagaraj
Department of Computer Science and Engineering, University at Buffalo, New York, United States

N. Kamalakannan · M. Devosh
Zoho Corporation, Chennai, India

S. Uma Maheswari (✉) · A. Shahina (✉)
Department of Information Technology, SSN College of Engineering, Chennai, India
e-mail: umamaheswaris@ssn.edu.in

A. Shahina
e-mail: shahinaa@ssn.edu.in

A. Nayeemulla Khan
School of Computing Sciences and Engineering, Vellore Institute of Technology, Chennai, India

L.P. Suresh and B.K. Panigrahi (eds.), *Proceedings of the International Conference on Soft Computing Systems*, Advances in Intelligent Systems and Computing 397, DOI 10.1007/978-81-322-2671-0_19

Keywords Emotional health · Emotion recognition · Physiological signals · Health monitoring · PAD model · Heart rate monitoring

1 Introduction

Positive emotions are beneficial in building good physical health. People who experience more positive emotions live longer and healthier lives [1]. Similarly, negative emotions can cause adverse effects on health. Stress and depression are associated with substantial risks of heart diseases, stroke, obesity, and sleep disorders [2]. The World Health Organization (WHO) estimates that globally more than 350 million people of all ages are affected by depression [3]. Manual analysis of physiological signals like finger temperature and heart rate with respect to various emotions resulted in the following cues [4]. The heart rate of a person is usually high when the emotions of anger, fear, and sadness are felt rather than when happiness is felt. Similarly, the heart rate is relatively high for surprise than for disgust [4]. The skin temperature is relatively high for the emotion anger than for happiness and high for sadness than for fear, surprise, and disgust [4]. A computer game with interfaces to incorporate emotion in game play via voice and physiological signals is demonstrated in [5]. It was suggested that the emotion recognition can be improved by the combined analysis of physiological signals such as skin conductance and heart rate which are good indicators for arousal and voice harmonics which may assist in predicting positive and negative emotions [5]. There are a variety of models available for speech-based emotion recognition. The pleasure–arousal–dominance (PAD) emotional model is reported as a widely used efficient model [6].

We propose a system to monitor the mental or emotional health of an individual. The system periodically monitors the emotions of the individual and analyzes them. The emotions of a person are identified based on the speech inputs and the physiological signals recorded from the person during everyday activity. Features extracted from these signals are used to model a person's emotional characteristics under a supervised learning paradigm.

The remainder of the paper is organized as follows: Sect. 2 discusses the speech-based emotion recognition using EmoMeter [7] and PAD model. Emotion recognition using physiological signals is presented in Sect. 3. Sections 4 and 5 illustrate the combination of the results from the models and the experimental study. Finally, Sect. 6 concludes the paper with future improvements.

2 Voice-Based Emotion Recognition

2.1 *Using EmoMeter*

The individual's speech is continuously recorded daily for a period of 30 days. Whenever there is a long pause in the speech, the speech is processed to extract emotions from it. Emotions are identified using the EmoMeter [7], which uses a weighted combinational model for measuring mixed emotions. Preprocessing of speech is carried out by denoising and amplitude normalization. The features such as pitch, formants, intensity, Mel-frequency cepstral coefficients (MFCCs), jitter, and shimmer are extracted from speech inputs (254 features) and feature selection is performed using forward feature selection algorithm. The selected features (24 features) are used in a combinational model consisting of the classifiers such as k-nearest neighbors (k-NN), neural network, Gaussian mixture model (GMM), naïve Bayesian classifier, and support vector machines (SVM). The different emotions present in the speech are identified by each model (classifier). Based on the accuracy of each classifier, weights are assigned to each model and the speech-based combinational model is formed as a summation of the weighted scores of individual models. The combinational model has an overall accuracy of 73 %. The emotion anger is recognized 100 % of the time in the EmoMeter [7].

2.2 *Using PAD (Pleasure, Arousal, Dominance) Emotional State Model*

The performance of the EmoMeter is enhanced in combination with the pleasure–arousal–dominance (PAD) emotional state model for speech [8]. The PAD emotional state model is a psychological model consisting of three dimensions to represent most emotions. The pleasure–displeasure scale is used to measure how pleasant an emotion may be. The arousal–nonarousal scale is used to measure how intense an emotion may be. The dominance–submissiveness scale is used to measure how dominant an emotion may be.

Samples are taken from Surrey audio-visual expressed emotion (SAVEE) database [9] which consists of 480 sample recordings from four males in seven emotions like anger, disgust, fear, joy, neutral, sadness, and surprise. The emotion discriminating features like pitch, intensity, formants, MFCCs, jitter, and shimmer for the speech inputs are extracted as described previously [7]. The extracted features are used to classify the emotions based on the PAD scales using a GMM. The outputs of the GMM are pleasure score and arousal score ranging from 1 to 4 (low to high). The GMM is used to determine how pleasant and intense (arousal) the emotion is. Then the emotion is identified based on the PAD emotional state model.

The results of the testing data of 48 samples (10 % of the sample database) for pleasure and arousal are shown as a confusion matrix in Tables 1 and 2. It can be

Table 1 Confusion matrix for pleasure ranging from 1 to 4 (Low to High)

Class	1	2	3	4
1	**15**	2	0	0
2	1	**14**	0	1
3	1	0	**7**	0
4	0	0	0	**7**

Table 2 Confusion matrix for arousal ranging from 1 to 4 (Low to High)

Class	1	2	3	4
1	**6**	1	0	0
2	2	**15**	0	1
3	0	4	**15**	1
4	0	1	1	**1**

seen that the model can classify pleasure correctly at 89.6 % and arousal correctly at 77 %. Based on the pleasure and arousal scores, the combinational model classifies the emotions with an efficiency of 81.25 %.

3 Emotion Recognition Using Physiological Signals

Physiological signals like heart rate, skin conductance, skin temperature, etc. can help to identify emotions. Different emotional behaviors produce different changes in the physiological signals. For example, anger causes increased heart rate and skin temperature, fear causes increased heart rate and decreased skin temperature, and happiness causes decreased heart rate and no change in skin temperature. The main advantage of identifying emotions using physiological signals is that the emotions can be measured even when the individual is silent. For this study, physiological signals like heart rate, interbeat intervals (IBIs), and heart rate variability (HRV) are measured simultaneously along with the individual's speech for the purpose of emotion recognition. The circuit consists of an Arduino microcontroller board and a heartbeat sensor used to measure the heart rate parameters, in order to detect emotion.

For the experimental setup, the heart beat sensor consists of a light source and a light detector. When the finger is placed between the light source and sensor, the diffused light falls on the sensor and is measured. Changes in the measured values correspond to changes in blood flow, which in turn corresponds to the heart beat of the person. The heart beat sensor can be used to measure the heart rate (HR) in beats per minute (bpm), IBI, and HRV. These features play a vital role in discriminating the emotions.

3.1 Database

A database of samples is necessary for training classifiers. The heart beat sensor is used to collect samples from 10 healthy subjects (5 male and 5 female) aged between 20 and 40 years. Since it is difficult to insist a subject to feel or express a particular mood, an alternate approach of inducing emotion is used. The subjects were exposed to various stimuli such as images, music, noise, and videos. For example, in order to induce joy, a comedy video of 5 min is played and samples are collected for a period of 1 min without the knowledge of the user. Similarly, for inducing anger, an appropriate video is played. The advantage of this approach is that no professional actors are needed. Thus, a database of 70 samples for seven different emotions like anger, disgust, fear, joy, neutral, sadness, and surprise is collected.

3.2 Feature Extraction

The parameters relevant for detecting emotions are extracted from the sensor signals. The features like mean, standard deviation, variance, and root mean square values of the IBIs and the heart rate are determined. In the proposed system, a total of five features are used for identifying emotion from the heart rate.

3.3 Classifiers

The five-dimensional feature vector extracted as described above is used to train different classifiers. In the voice-based emotion recognition, a combination of various models such as k-nearest neighbors (k-NN), neural network, GMM, naïve Bayesian classifier, and SVM is used for classifying the emotions based on the PAD scale using the physiological signals [10, 11]. The heart beat features are used to measure how pleasant and intense the emotions are. The classification of emotions is carried out using each model. Based on the accuracy of each model, a weighted score is assigned to each classifier. The proposed system consists of a combinational model which is the summation of the weighted scores of the individual models. The output of the combined model will be the pleasure and arousal scores similar to the voice-based PAD classifier discussed in Sect. 2.2. The emotion is identified based on the PAD scale.

For testing the model, the emotions are induced for the subjects, the heart rate parameters are measured as previously, and the model is tested with the data. This is repeated for 20 samples and the results of the testing data for pleasure and arousal are shown as a confusion matrix in Table 3 and 4.

It can be seen that the model has an efficiency of 70 % in classifying pleasure correctly and an efficiency of 80 % to find arousal correctly. Based on the pleasure and arousal scores, the model is able to find the emotions correctly with an efficiency of 75 %.

Table 3 Confusion matrix for pleasure ranging from 1 to 4 (Low To High)

Class	1	2	3	4
1	**8**	0	0	0
2	2	**3**	0	1
3	2	0	**2**	0
4	1	0	0	**1**

Table 4 Confusion matrix for arousal ranging from 1 to 4 (Low To High)

Class	1	2	3	4
1	**6**	1	0	0
2	1	**5**	0	1
3	0	0	**3**	1
4	0	0	0	**2**

4 Combining the Results

In order to compute the emotion based on the voice and physiological signals, the results from various models should be combined. In the proposed system, the weightage of each emotion is computed from each model by summing up the product of the efficiency of the model and the score returned by the model as shown in the formula:

$$A(x) = \sum_{i=1}^{N} E(i) \times P_x(i) \tag{1}$$

where A is the weightage given to the emotion x by all the models, E is the efficiency of the model i, P_x is the score returned by the model i for the emotion x, and N is the total number of individual models to be combined. The efficiency of the PAD model is calculated using the testing data from the samples collected while the efficiencies of the other five models are calculated as described in EmoMeter [7]. Finally, the weightage given by the models for all the emotions is normalized as percentage.

5 Experimental Analysis

Testing is performed for all the individual models and the proposed combinational model with a random testing set created using the 10 % of the sample database. Since the testing set is generated at random, there is no fixed number of samples taken for each emotion. For real-time testing, the system records the voice and processes it to identify the mixed emotions present in it. The testing can be done with or without the physiological signals integrated to it.

Once the live recording is done, the emotion identification process is carried out. It includes the feature extraction, feature selection, and classification using the proposed combinational model. After the classification, the results of emotion recognition by individual classifiers are shown separately.

When the heart rate parameters are combined with the voice-based combination model, the result of a testing sample is 0 % anger, 0 % disgust, 0 % fear, 96 % joy, 4 % neutral, 0 % sad, and 0 % surprise. Similarly, the results of three other testing samples are found and analyzed. Emotions like joy and sad, fear and joy, and anger and joy are complementary and the results are appropriate for the testing samples as shown in Table 5.

The results of testing the individual models and the proposed voice and physiological signals combinational model are determined and their performance is computed.

A comparative study is made of the accuracy of the emotion detection with individual models and the proposed combinational model. The comparison of their performance is plotted as graphs shown in Figs. 1, 2, and 3. It can be seen from Fig. 1 that the voice-based combined model has the maximum accuracy for neutral and minimum accuracy for disgust. Using physiological signal heart rate alone, the combinational classifier has a higher accuracy for anger and joy and also a better

Table 5 Testing results for four samples

Case	Anger	Disgust	Fear	Joy	Neutral	Sad	Surprise
A	0	0	0	**96**	**4**	0	0
B	**89**	**5**	0	0	0	**6**	0
C	0	0	**85**	0	**8**	**7**	0
D	**58**	**23**	0	0	0	**19**	0

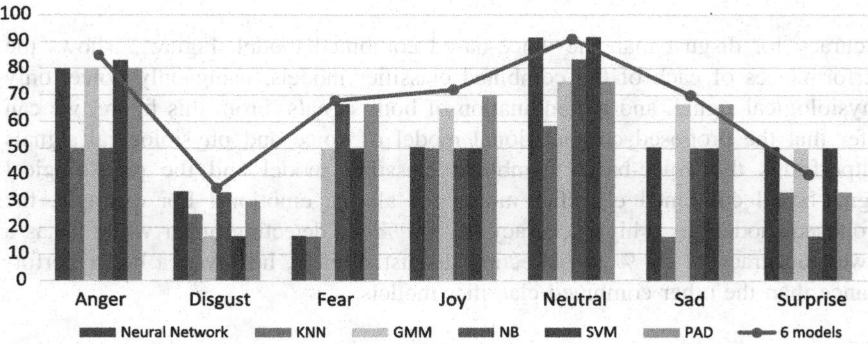

Fig. 1 Comparison of performance of the six classifiers and combined model for voice-based emotion recognition

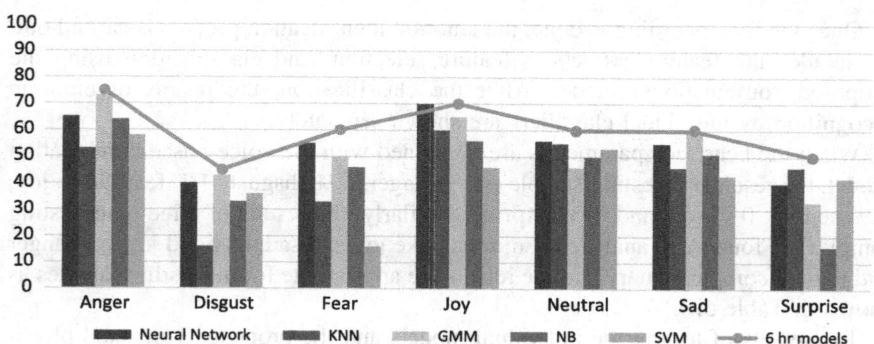

Fig. 2 Comparison of performance of the combined classifier model against the individual classifiers using only the physiological signals (Heart rate)

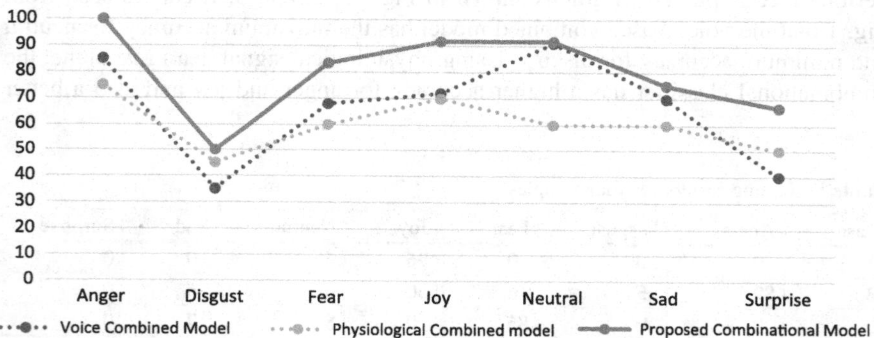

Fig. 3 Comparison of the proposed voice and physiological combined classifier models against the individual voice and physiological signal-based combined classifier models

accuracy for disgust than the voice-based combined model. Figure 3 shows the performances of each of the combined classifier models, using only voice, only physiological signal, and a combination of both signals. From this figure, we can infer that the proposed combinational model of voice and physiological signals outperforms the voice-based combined classifier model and the physiological signal-based combined classifier model for all the emotions. For example, the proposed model has a high accuracy of 100 % for detecting anger while it has a lowest accuracy of 50 % for detecting disgust. This is, however, a better performance than the other combined classifier models.

6 Conclusion and Future Improvements

Thus the automated health monitoring system based on the emotion identification is presented. The results of emotion recognition from the individual classifiers based on voice and physiological signals are analyzed and found that the accuracy of the proposed combinational model surpasses the other individual models. The circuit measuring heart rate can be made as a wearable component so that a person can use the device to monitor the emotions continuously even in outdoors.

Further improvements and extensions of the system include the addition of other physiological signals like skin conductance and skin temperature to obtain the maximum accuracy of the system, building a compact human wearable device for this system.

References

1. Kok BE, Coffey KA, Cohn MA et al (6 May 2013) How positive emotions build physical health: perceived positive social connections account for the upward spiral between positive emotions and vagal tone, psychological science. doi:10.1177/0956797612470827
2. Unhappiness by the Numbers: 2012 Depression Statistics. [URL:http://www.healthline.com/health/depression/statistics-infographic]. Accessed 25 March 2013
3. Depression Fact sheet N°369 October 2012, WHO. [URL:http://www.who.int/mediacentre/factsheets/fs369/en/]. Accessed 25 March 2013
4. Levenson RW, Ekman P (2002) Difficulty does not account for emotion-specific heart rate changes in the directed facial action task. Psychophysiology 39:397–405. doi:10.1017.S0048577201393150
5. Kim J, Bee N, Wagner J, André E (September 2004) Emote to win: affective interactions with a computer game agent. In: Lecture Notes in Informatics (LNI)—Proceedings, vol P-50, pp 159–164
6. Gunes H, Schuller B, Pantic M, Cowie R (21–25 March 2011) Emotion representation, analysis and synthesis in continuous space: a survey. In: 2011 IEEE international conference on automatic face and gesture recognition and workshops (FG 2011), pp 827, 834. doi:10.1109/FG.2011.5771357
7. Ananda Kanagaraj. S, Shahina. A, Devosh M, Kamalakannan N (10–12 April 2014) EmoMeter: measuring mixed emotions using weighted combinational model. Presented at the fourth international conference on recent trends in information technology, Chennai, India
8. Mehrabian A (1996) Pleasure-arousal-dominance: a general framework for describing and measuring individual differences in Temperament http://dx.doi.org/10.1007/BF02686918, Journal Article 0737-8262 Current Psychology 14 4 10.1007/BF02686918T, Springer-Verlag 1996-12-01 261-292
9. Haq S, Jackson PJB (2010) Multimodal emotion recognition. In: Wang W (ed) Machine audition: principles, algorithms and systems, chapter 17. IGI Global Press, pp 398–423. ISBN 978-1615209194
10. Costa T, Boccignone G, Ferraro M (2012) Gaussian mixture model of heart rate variability. doi:10.1371/journal.pone.0037731
11. Picard RW, Vyzas E, Healey J (2001) Toward machine emotional intelligence: analysis of affective physiological state. IEEE Trans Pattern Anal Mach Intell 23:1175–1191

A Network Model of GUI-Based Implementation of Sensor Node for Indoor Air Quality Monitoring

Siva V. Girish, R. Prakash, S.N.H. Swetha, Gargi Pareek, T. Senthil Kumar and A. Balaji Ganesh

Abstract This paper describes a wireless sensor network-based indoor air quality monitoring system. The indoor air quality defines the quality of the environment where people live. Here wireless sensor network serves as the tool for estimating the indoor air quality. The WSN comprises sensor nodes and a coordinator node which communicates using the IEEE 802.15.4 wireless standard ZigBee wireless module. The indoor air quality estimation is done by interfacing CO_2, temperature and RH (Relative humidity) sensors with the sensor node. The sensor node gathers the sensor data and reports it to the coordinator for real-time monitoring using a GUI (Graphical user interface) developed in Java NetBeans to run on windows PC. The collected data can be used to maintain the environment parameters by interfacing it to a HVAC (Heating, Ventilation and Air Conditioning) system.

Keywords Indoor air quality · Sick building syndrome · Wireless sensor networks · GUI · HVAC

S.V. Girish (✉) · R. Prakash · A. Balaji Ganesh
Electronic System Design Laboratory, TIFAC-CORE, Velammal Engineering College, Chennai, India
e-mail: sivagirish1@gmail.com

R. Prakash
e-mail: prakash.rama121@gmail.com

A. Balaji Ganesh
e-mail: abganesh@live.in

S.N.H. Swetha · G. Pareek · T. Senthil Kumar
Computer Science and Engineering Department, Amrita University, Coimbatore, India
e-mail: swetha.199444@gmail.com

G. Pareek
e-mail: pareek23gargi@gmail.com

T. Senthil Kumar
e-mail: senthan111@gmail.com

L.P. Suresh and B.K. Panigrahi (eds.), *Proceedings of the International Conference on Soft Computing Systems*, Advances in Intelligent Systems and Computing 397, DOI 10.1007/978-81-322-2671-0_20

1 Introduction

Air quality is a major factor that affects the health of those who resides most of their times inside the buildings such as office, home, etc. The changes in physical parameters like carbon dioxide, temperature and humidity become the most important factors which affect the indoor air quality directly or indirectly by providing a nourishing environment for the growth of microbial. These pollutants can induce adverse health effects to the building occupants. The sick building syndrome is one of the best examples of the health effects caused due to poor indoor air quality in a building [1]. Some of the sources of carbon monoxide inside a house are cooking activities and smoking of tobacco products or incomplete combustion of any material. As the pollutants are colourless, odourless and tasteless, it is impossible to detect with our sensory system. It affects the victim by reducing the oxygen-carrying capacity of the blood as it can bind easily with haemoglobin than oxygen-forming carboxyhaemoglobin (COHb) [2]. Likewise, carbon dioxide is also a colourless and odourless gas which is a normally available in the atmosphere at 330-440 PPM. It affects the victim inducing headache, fatigue, burning and irritating sensation in eye and respiratory tract. To avoid such adverse effects, an air quality monitoring system is required. The indoor air quality monitoring device which is available to be used for home is desktop-type monitoring gadgets.

The professional monitoring systems come with a mesh of nodes which is capable of collecting information regarding indoor air quality and makes the manager to view and control the environment accordingly. However, they are high in cost to be used by common people. Each sensor node comprises a microcontroller interfaced with ZigBee wireless communication module and a mounting stand. The basic system costs around $856 and additional cost for extended sensor support. In order to overcome all these constraints, a cost-effective system which can give early warning about the indoor air quality should be developed.

As wireless communication systems have achieved a rapid development in every aspect of day-to-day life, deployment of wireless sensor network for monitoring situations has become so common. So it will be very suitable for the development of indoor air quality monitoring system.

2 Related Work

Choi et al. [3] presented an air pollutant monitoring system titled Micro-sensor node for air pollutant monitoring system (APOLLO). The system incorporates a collection of MEMS-based micro-gas sensors which collect information about the various constituents of air to evaluate the air quality information and forward the data to the host system.

Lozano et al. [4] proposed an indoor air quality monitoring system which comprises WSN using ZigBee standard IEEE 802.15.4 integrated with sensors to measure temperature, humidity and light. The sensed parameters were monitored by a specific program developed with Labview.

Li et al. [5] proposed a WSN-based indoor air quality (IAQ) system. The sensor node in the system is equipped with sensors for gathering information about the concentration of CO_2, CO, VOC, temperature, RH and dust. The sensor boards were designed to be interfaced with Arduino board which is configured to be a sensor node. The sensor data from the individual nodes were collected using a gateway node which is interfaced with the PC.

Khedo et al. [6] presented an air pollution monitoring system which utilizes the air quality index (AQI). The AQI represents the amount of air pollutants present in the air, thereby defining the air quality. It presents a data aggregation algorithm named recursive converging quartiles (RCQ) in order to improve the efficiency of WAPMS.

Washimkar [7] proposed a system for evaluating air quality by developing a hardware unit which monitors the concentration of gases such as CO, NO_2 and SO_2 using semiconductor sensors and transmits the sensed data to the central server via ZigBee network. A high-end personal computer application server acts as the central server for the system.

Noh [8] proposed a wireless sensor module that had a ZigBee communication module and a sensor module for monitoring a room or an office environment using WSN. The sensor module has a humidity sensor, temperature sensor, and O_2 and CO_2 sensors. Using ZigBee technology of 2.4 GHz industrial, scientific, and medical (ISM) band, it could monitor the information from terminal PC modules

3 Proposed System

The proposed architecture involves monitoring of the environmental parameters using individual sensor nodes which are capable of transferring the sensed data in a preassigned ZigBee wireless channel to a coordinator node.

The coordinator node is interfaced with a PC using UART communication through which collected sensor data from various sensor nodes were logged and monitored in a GUI developed using Java NetBeans. The coordinator node can also be made to communicate with remote machines by establishing GSM-based mobile data network (Fig. 1).

3.1 Sensor Node

The sensor node comprises carbon dioxide, temperature, and relative humidity sensors interfaced with the microcontroller. The microcontroller fetches the sensor data through the ADC and places the sensor data on the payload of the packet according to the designed frame format.

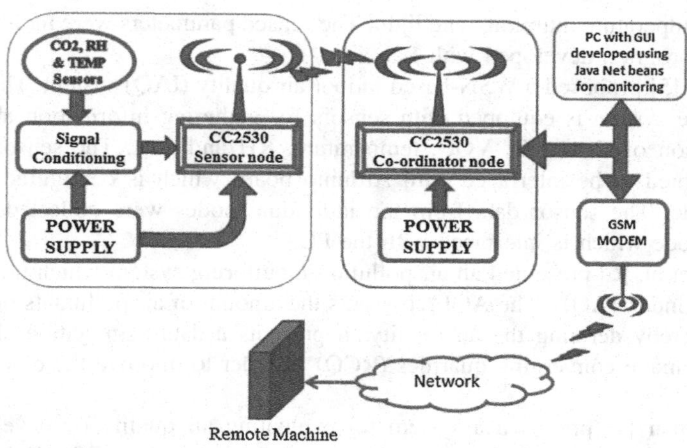

Fig. 1 Wireless sensor network model

The frame structure of the packet of the ZigBee protocol is shown above [9]. The physical packet field starts with a preamble sequence. The preamble sequence serves the synchronization sender and receiver. The 32-bit sequence of preamble is generated to make the receiver detect the start of next packet. It is followed by a 8-bit sequence of bits to insist the end of preamble (Fig. 2).

The microcontroller in the sensor node places the 3x8-bit sensor data (Temperature, CO_2 and Relative humidity sensors-8 bit each) on the PSDU of the packet and sends the data to the coordinator node.

The sensor node sends the report periodically and enters the low power mode (sleep) to reduce the power consumption. The above figure shows the sensor node designed with CC2530SOC of 2.4 GHz ZigBee transceiver with extended 8051 microcontroller which is very optimal to work as a low power device. The sensor node system comprises a temperature, relative humidity and CO_2 sensing elements. CC2530 system on chip has its own in-built temperature sensor for monitoring the core temperature. There were sensors of various types such as metal oxide, optical, MEMS and thin film [10] which can be utilized for these applications. The proposed sensor node is embedded with ChipCap2 sensor which is an integrated solution for humidity, temperature, and TGS4161 sensor for sensing the concentration of CO_2 (Fig. 3).

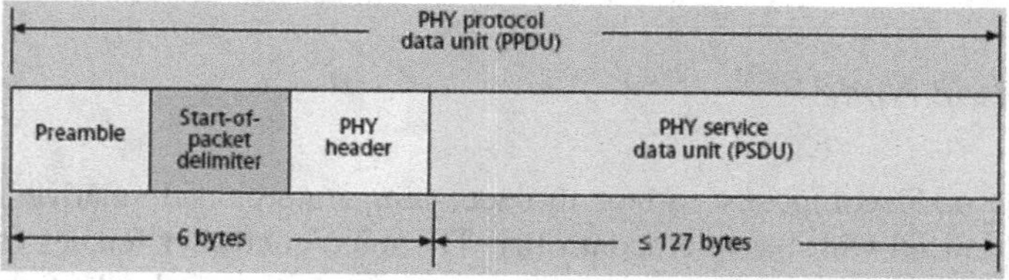

Fig. 2 Frame structure of 802.15.4

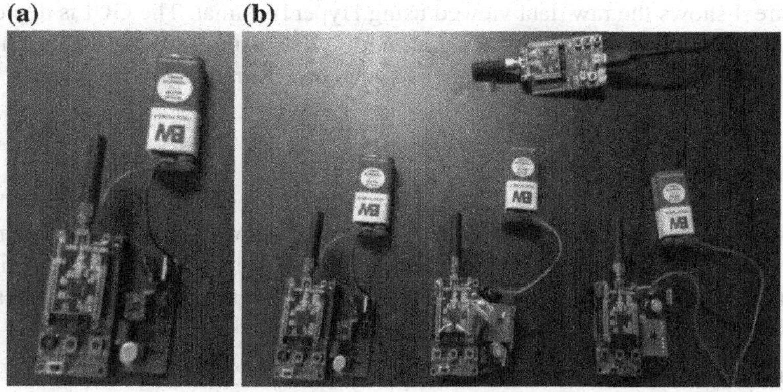

Fig. 3 **a** Sensor node with the sensors **b** Sensor nodes with a coordinator node

3.2 *Coordinator Node*

The coordinator node acts as a central node which commands and controls its child nodes (sensor node). It collects data from the sensor nodes and computes them for evaluating meaningful data of indoor air quality.

The coordinator node evaluates the collected data and sends it to the PC to which it is interfaced using UART. The data from the UART can be viewed using serial port monitoring tools like HyperTerminal. The raw data comprises the node address from which the data is received and their corresponding parameter concentration as shown in Fig. 4.

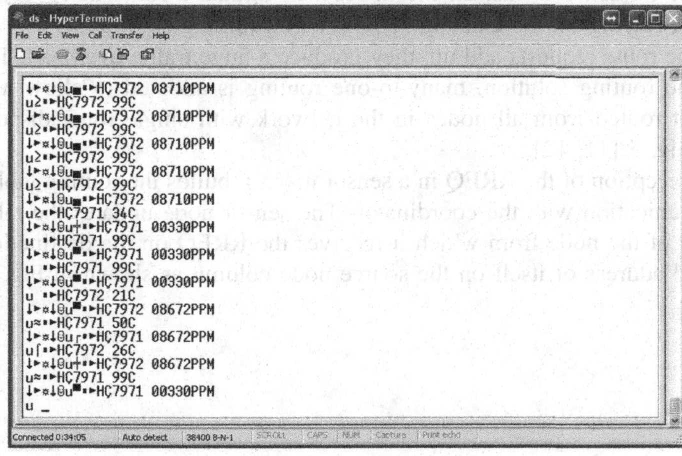

Fig. 4 Raw data from the coordinator node

Figure 4 shows the raw data viewed using HyperTerminal. The GUI is deployed on the PC to monitor the individual sensor node's data using Java NetBeans making it easy to be deployed in any computer having Java Runtime Environment.

3.2.1 Addressing Schemes

The coordinator node broadcasts its PAN-ID to the network. When a sensor node receives the beacon, it gives a binding request with a request for new address to join with the coordinator. The ZigBee 2007 uses two different addressing schemes for assigning network address:

(a) Tree addressing
(b) Stochastic addressing

The coordinator node assigns the address for all the sensor nodes once they requests for binding. The addressing scheme ensures that all assigned network addresses are unique throughout the network [7]. This avoids the ambiguity about which device a particular packet should be routed to. The tree addressing algorithm also ensures that the device only has to communicate with its parent device to receive its network address. This helps in scalability of the network.

3.2.2 Routing Protocol

The network follows a many-to-one routing protocol. The ZigBee pro standard adopts the many-to-one routing protocol to minimize traffic particularly when centralized nodes were involved. The centralized node may be a gateway or a data concentrator. All the nodes in the network will have at least one valid route to the coordinator. To achieve this, all the nodes will initiate a route discovery for the coordinator, relying on the existing AODV-based routing solution.

When the route requests add up, they produce a huge traffic overhead. To better optimize the routing solution, many-to-one routing is adopted to allow a coordinator to get routed from all nodes in the network with single route discovery as shown in Fig. 5 [11, 12].

On the reception of the RREQ in a sensor node, it builds the routing table for its data communication with the coordinator. The sensor node updates the table with the address of the node from which it receives the RREQ on the destination node column and address of itself on the source node column as shown in Fig. 6.

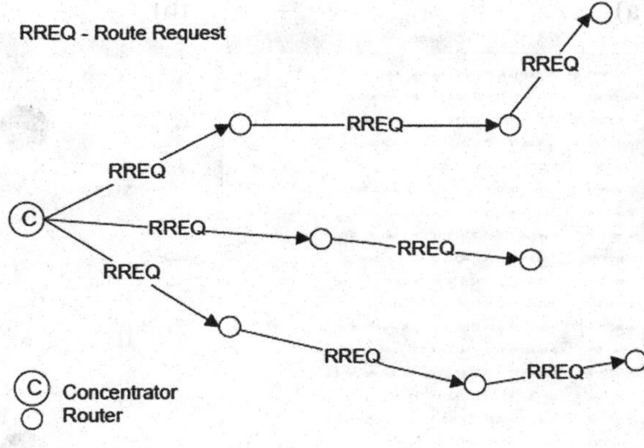

Fig. 5 Many-to-one route discovery illustration (ROUTING REQUEST)

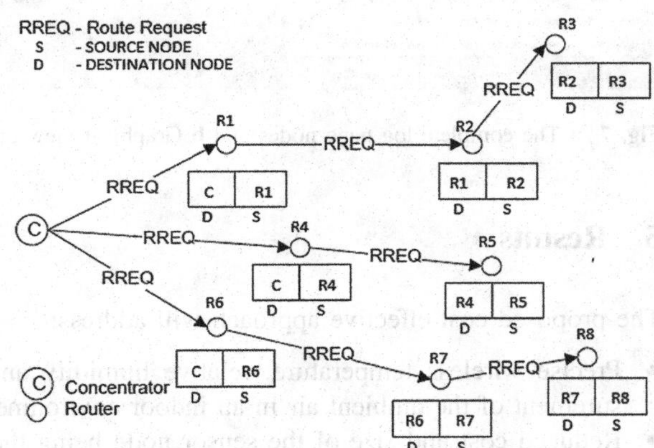

Fig. 6 Routing table determination

4 Graphical User Interface

As soon as the application starts, it will ask the user to enter the filename (with correct path specified) in which the data logging takes place.

When the user presses "OK" button, a screen called "values" is displayed that displays different node data. It includes node value, date and respective sensor values. This screen has two buttons: "Individual sensors" and "Show Graph".

i. Individual Sensors—This helps the user to view the details of an individual sensor node. When this button is pressed, a window listing the available nodes will be displayed as shown in Fig. 7a. Choose the desired node to get its details. Temperature node displays temperature, likewise humidity and CO_2 nodes show their respective values.
ii. Show Graph—This helps the user to view the nodes connected to the coordinator node graphically as shown in Fig. 7b. This graph changes from time to time as the node connects and disconnects.

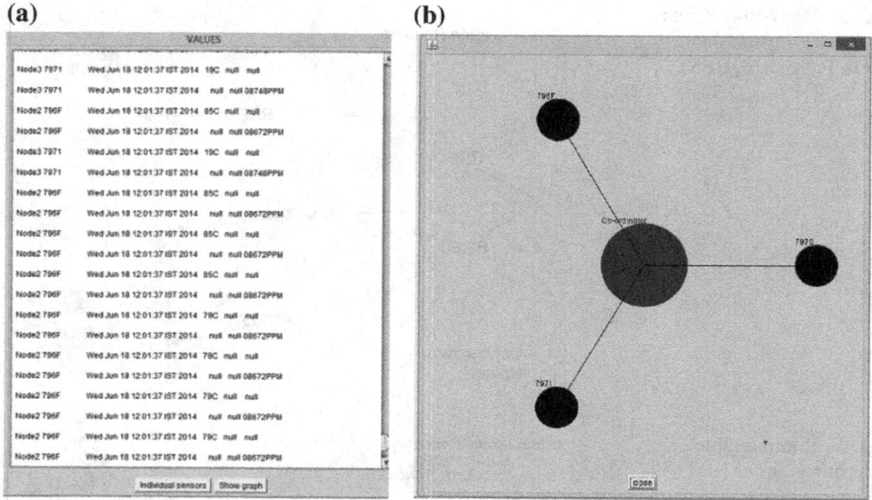

Fig. 7 **a** The complete log page nodes and **b** Graphical view of connected nodes

5 Results

The proposed cost-effective approach will address:

- Precise wireless temperature, relative humidity and CO_2 concentration measurement of the ambient air in an indoor environment.
- Reduced cost and size of the sensor node being the controller and transceiver integrated as an SOC.
- A real-time monitoring is possible using GUI which is developed using Java NetBeans.

Acknowledgments The authors gratefully acknowledge the financial support from Department of Science and Technology, New Delhi, through Instrumentation Development Program Division No: IDP/IND/2012/22 to Velammal Engineering College, Chennai.

References

1. Hedge A, Erickson WA (1997) The study of indoor air environment and sick building complaints in air-conditioned offices: benchmark for facility performance. Int J Facil Manage 4:185–192
2. Brauer M, Hirtle R (2000) Assessment of indoor fine aerosol contributions from environmental tobacco smoke and cooking with a portable nephelometer. J Expo Anal Environ Epidemiol 136–144

3. Choi S, Kim N, Cha H, Ha R (12 October 2009) Micro sensor node for air pollutant monitoring: hardware and software issues. Sensors 9:7970–7987
4. Lozanoa J, Suárezb JI, Arroyoa P, Ordialesa JM, Álvareza F (2012) Wireless sensor network for indoor air quality monitoring. The Italian Association of Chemical Engineering 30:319–324
5. Xinrong L (2012) Applications of wireless sensors in monitoring Indoor Air Quality in the classroom environment, Research Experiences for Teachers Program 2012
6. Khedo KK, Perseedoss R, Avinash Mungur (May 2010) A wireless network air polluting monitoring system. Int J Mobile Netw 2(2)
7. Chourasia NA, Washimkar SP (2012) Implementation of Zigbee based wireless air pollution monitoring system. Int J Eng Res Appl (IJERA), ISSN, pp 761–766
8. Noh S-K, Kim K-S, Ji Y-K, (2013) Design of a room monitoring system for wireless sensor networks. Int J Distrib Sens Networks 2013:7 p, Article ID 189840
9. Z-Stack developer's guide. http://www.ti.com/tool/z-stack
10. Chung WY, Lim JW (2003) Patterning of thin tin oxide filmwith nano-size particle for two-dimensional micro-gas sensorarray. Curr Appl Phys 3(5):413–416
11. Method for Discovering Network Topology. http://www.ti.com/tool/z-stack
12. Z-Stackdeveloper's guide. http://www.ti.com/tool/z-stack

Automated Segmentation of Skin Lesions Using Seed Points and Scale-Invariant Semantic Mathematic Model

Z. Faizal Khan

Abstract A color image-based segmentation method for segmenting skin lesions is proposed in this paper. This proposed methodology mainly includes two parts: First, a combination of scale-invariant and semantic mathematic model is utilized to classify different pixels. Second, a strategy based on skeleton corner point's extraction is proposed in order to extract the seed points for the skin lesion image. By this method, the skin slices are processed in series automatically. As a result, the lesions present in the skin can be segmented clearly and accurately. The proposed algorithm is trained and tested for 360 skin slices in order to evaluate the accuracy of segmentation. Overall accuracy of the proposed method is compared with existing conventional techniques. An average missing pixel rate of 3.02 % and faulting pixel rate or 2.36 % has been obtained for segmenting the skin lesion images.

Keywords Semantic mathematic model · Color image segmentation · Skin lesion · Seed points

1 Introduction

Melanoma is the most deadly form of skin cancer, with an estimated 76,690 people being diagnosed with melanoma and 9480 people dying of melanoma in the United States in 2013 [1]. In the United States, the lifetime risk of getting melanoma is 1 in 49 [1]. Melanoma accounts for approximately 75 % of deaths associated with skin cancer [2]. It is a malignant tumor of the melanocytes and usually occurs on the trunk or lower extremities [3]. Recent trends found that incidence rates for non-Hispanic white males and females were increasing at an annual rate of approximately 3 % [4]. If melanoma is detected early, while it is classified at

Z. Faizal Khan (✉)
Department of Computer and Network Engineering, College of Engineering, Shaqra University, Al Dawadmi, Kingdom of Saudi Arabia
e-mail: faizalkhan@su.edu.sa

L.P. Suresh and B.K. Panigrahi (eds.), *Proceedings of the International Conference on Soft Computing Systems*, Advances in Intelligent Systems and Computing 397, DOI 10.1007/978-81-322-2671-0_21

Stage I, the 5-year survival rate is 96 % [5]; however, the 5-year survival rate decreases to 5 % if the melanoma is in Stage IV [5]. With the rising incidence rates in certain subsets of the general population, it is beneficial to screen for melanoma in order to detect it early. To reduce costs of screening melanoma in the general population, development of automated melanoma screening algorithms have been proposed.

Human body visualization research includes cutting the sequences of body slices, developing the computerized methods with the overall goal of reconstructing the 3D virtual models of the human body. With the help of these 3D virtual models, the human organs can be hidden selectively and viewed from any side. The virtual human models can be implemented for risky experiments instead of real human body. In summary, the visible human project (VHP) has a lot of applications and significant value for human society. This ambitious scheme has caused attentions of many countries. By the end of 2012, the United States, South Korea, China, and Japan had all begun their own Visible Human projects. Researchers all over the world are now planning to append physical and chemical attributes to the 3D Visible. If new technology by which the live human body can be scanned to build personalized 3D virtual models is invented, the Virtual Human Project will have a more promising future [1–3].

Image segmentation is a critical stage of the cancer detection. That is because errors at this stage will inevitably lead to later bigger problems. In order to actualize the precise and accurate descriptions of inner structure of the human body, immense volumes of raw data are collected. Therefore, a rapid and accurate method should be utilized to segment the cross-sectional photographs of the Visible Human.

Currently, a wide variety of color image segmentation methods have been proposed. Among them, the k-means method and fuzzy c-means method are accurate and practical [6, 7]. However, considering the large data set of Visible Human, their high computational complexity will lead to considerable time cost. Color image segmentation methods based on region-growing can be implemented easily. The result is also acceptable. However, the seed points should be manually picked. It is hard to imagine picking out seed points for each picture. So does the graph cut method whose manual mode will be tedious for the large-scale image datasets [8, 9]. The color structure code (CSC) segmentation algorithm can tolerate the sensitivity of threshold, but it is easy to be affected by the noise [10–13]. The segmentation algorithm based on fuzzy connectedness can well mark the target region which is not easy to be accurately defined due to the fuzzy edge, and this algorithm is not sensitive to the noise. But there is a lot of iterative calculation algorithm which will seriously reduce the speed of the segmentation [14–17]. The support vector machine (SVM)-based segmentation algorithm can achieve the automatic segmentation in some degree, but it need too much time for acquiring the accurate space information during the training. What is more, the segmentation speed is not fast enough. So it cannot realize the real-time and efficient segmentation [18]. Another kind of segmentation methods is the interactive image segmentation methods which can manually extract the regions of interest. Most importantly, these methods are designed only for a single-layer color image, and they cannot be directly utilized as the serialized slice image segmentation method for the immense datasets of the Visible Human. The cross-sectional photographs of Visible Human are serialized, coherent, and slowly changing. Taking these features into account, an automatic-serialized color image segmentation method is proposed in this paper. The purpose of sequential and automatic segmentation in 3D color image spaces with lower time cost and higher accuracy can be achieved.

The rest of the paper is organized as follows: Sect. 2 presents the computation method of color similarities between two pixels and the automatic seed points extraction method for serialized slices. Section 3 presents experimental segmentation results for the primary organs of Visible Human. Concluding remarks are given in Sect. 4.

2 Materials and Methods

2.1 Selection of Color Feature

There are many color similarity measure methods. For example, Minkowski distance measure, Canberra distance measure, and cylindrical distance metric [19]. They all regard the color as feature and try to group the pixels with a similar color in the image. However, too much time is wasted on the calculations during the color space transformation to decouple brightness and hue. These calculations will be a bad influence when I am dealing with thousands of images in the Virtual Human Project.

The most widely used color space is RGB model. RGB is an abbreviation of red-green-blue. The three fundamental components of RGB color space are all influenced by brightness, which make it not perfectly good at color discrimination. Besides RGB color space, many other color spaces are designed based on RGB color space. By some calculations, the RGB color space can be turned into other spaces, such as HSI color space [20]. Some of these spaces are specially designed for color image processing. They mimic human visual perception in a more accurate way. HSI space is a typical representative of them. H means hue (the difference of color). S means saturation (the gradation of color). I represents intensity (the degree of light and shade of color). Human is more sensitive to brightness rather than hue. HSI space shows the brightness and hue explicitly and separately. So, HSI space is more common-used on color image processing. Although HSI space and RGB space are two expressions of the same thing, the transformation between them is very complicated. Therefore, it will surely improve the speed of the segmentation method if it does not need the color space transformation procedure (RGB → HIS) but can accord with the color vision perception characteristics of human eyes (HSI space).

2.2 Computation of Color Similarities Between Two Pixels

In this paper, a color similarity based segmentation algorithm which was proposed by Shikai Wang was utilized. This method is accurate and robust with low computational complexity [21]. A scale-invariant and semantic mathematic model is utilized to classify different pixels. This method can compute the color similarity between two RGB pixels of a color image in real time. Color space transformation is avoided in this method, so the computation time will be further reduced. However, this similarity reflects the characteristics of the HSI color space, which is what the most notable feature of this method.

Given two pixels *f* and *g* in the RGB color space: (R_0, G_0, B_0), (R_1, G_1, B_1). Let

$$(R', G', B') = \left(\frac{R_1}{R_0}, \frac{G_1}{G_0}, \frac{B_1}{B_0}\right)$$

Then the proposed algorithm is computed as follows:

$$AM = \frac{R' + G' + B'}{3} \tag{1}$$

$$HM = \frac{3}{\frac{1}{R'} + \frac{1}{G'} + \frac{1}{B'}} \tag{2}$$

$$\delta_M = \frac{AM}{HM} \tag{3}$$

$$\delta_M = \frac{\frac{3}{\frac{1}{R'} + \frac{1}{G'} + \frac{1}{B'}}}{\frac{R' + G' + B'}{3}}$$

$$= \frac{9}{(R' + G' + B') \times \left(\frac{1}{R'} + \frac{1}{G'} + \frac{1}{R'}\right)} \tag{4}$$

The proposed model considers both brightness and hue and does not perform color space transformation. The proposed method, which represents the level of similarity, ranges from 0 to 1. When the scale-invariant and semantic mathematic model is equal to 1, it means two color pixels are same. When the proposed model is equal to 0, it means two color pixels are absolutely different. After the seed point is picked out, (R_0, G_0, B_0) is obtained. So a threshold should be chosen to help us determine whether the ambient pixels in the slice are similar to the seed point or not. Experiment results indicate that this threshold can be found by several tests. Once confirmed, the threshold will be valid and invariant in the whole segmentation process of an image sequence.

2.3 Automatic Seed Points Extraction for Serialized Slices

In the traditional image segmentation algorithm based on region-growing, the seed points must be clicked manually. To pick out seed points in each slice image is exhausting. In this paper, the research goal is achieving an automatic and serialized segmentation of skin lesions in the image sequence. Automatic image segmentation requires less manual intervention to some extent. The automation degree of the segmentation algorithm is limited by setting the seed points. In other words, before the next image is to be segmented, the corresponding seed points should be provided automatically rather than manually.

The next image to be segmented must be different from the current image. However, the changes are small and predictable. For instance, current target regions are merged into one region or split into more regions. In these cases, more seed points should be provided in order to make sure at least one seed point in each separate region. Because of the characteristics of the visible skin dataset, the contours of the target regions in adjacent images are similar. So after extracting the target regions of the current image, the skeletons of target regions can be computed and stored. And then, an effective method to further extract the feature points of the skeletons to generate the seed points is needed. As a typical feature point extraction method, Harris operator has the advantage of simple calculation, uniform distribution, and stable performance. Therefore, the Harris corners of skeleton will be utilized as the seed points of the next image. The detailed process is as follows:

The current image which has already been segmented is regarded as a binary image. The pixel belongs to the target region is stored as 1, while the others are stored as 0. Zhang-Suen Thinning algorithm is utilized to extract the skeleton of the segmented region [22]. Because the skeleton can reflect topological features and is close to the center of the region, the seed points can be obtained from the skeleton pixels instead of checking each pixel. This new strategy is better than my previous method [23].

Harris corners are the points where the curvature has a sudden change [24]. The edges of the segmented regions are frequently changing. However, the Harris corners of the skeleton are not. These corner points are characteristic of the most unchangeable local features owned by the segmented regions. In other words, the extracted corner points will land in the target regions of next slice for certain. So Harris corners can be seed points of the adjacent image. As a consequence, the automatic seed points extraction method proposed in this paper is possible.

Algorithm for the proposed serialized segmentation method is as follows:

(1) Set the color similarity threshold T;
(2) Select a number of pixels manually in the target region, and set each of them as an initial seed pixel (x_0, y_0);
(3) Set pixel (x_0, y_0) as the center, and obtain the eight neighborhood pixels (x_i, y_i) $(i = 0, 1, \ldots., 10)$;
(4) Compute the color similarity between (x_i, y_i) and (x_0, y_0) by the combination of scale-invariant and semantic mathematic model;

(5) If the color similarity between pixel (x_i, y_i) and pixel (x_0, y_0) is larger than threshold T, pixel (x_i, y_i) and pixel (x_0, y_0) can be grouped into a same region. Pixel (x_i, y_i) will be pushed into a stack;
(6) If the stack is not empty, iterate each pixel and regard one as the pixel (x_0, y_0). Return to step (3). When the stack is empty, the growing process for the target region segmentation of the current color slice ends. If current slice is the last image, the whole segmentation process ends;
(7) In the obtained target regions, compute the seed pixels (number ≥ 1) for the next slice by utilizing the skeleton corner point method;
(8) For the next slice, each seed pixel is regarded as (x_0, y_0). Then turn to step (3).

3 Results and Discussion

Accuracy of segmentation has been calculated in order to determine the performance of the proposed segmentation algorithm. In this section, the results of this segmentation methodology in terms of quality evaluation through image display and experimental analysis are presented. The system is validated through accuracy metrics for finding satisfactory segmentation results. The performances of various segmentation techniques have been analyzed and compared with the results obtained for segmenting skin lesion images. The segmented outputs of the proposed model are shown in Table 1.

In this paper, methods of calculating the missing pixel rate P_m and faulting pixel rate P_f are used to evaluate the segmentation quality of different algorithms.

$$P_m = |G \cap \bar{S}| / |G| \tag{5}$$

$$P_f = |\overline{G} \cap S| / |G| \tag{6}$$

where G is the pixel set of reference segmentation result (Ground truth) and S is the set of pixels present in the original image. The experimental results are shown in Table 2. From the tables, it can be seen that the missing pixel rate of the color structure code method is most serious and the faulting pixel rates of the first three segmentation methods are similar. The missing pixel rate and the faulting pixel rate of this method are all lower than other three methods.

In terms of time, the proposed method shows satisfactory results. The experiments were performed on an ordinary dual-core personal computer. The time cost of different methods is shown in Table 3. From the table, it can be observed that the overall speed of this proposed method is faster than other methods.

Table 1 Segmentation results for different images using proposed methodology

Image	Original image	Segmented result
Image 1		
Image 2		
Image 3		
Image 4		

Table 2 Quality evaluation of the skin lesion segmentation

	CSC segmentation (%)	Fuzzy connectedness (%)	SVM (%)	Proposed method (%)
P_f	12.60	8.87	4.22	3.02
P_f	8.21	6.08	4.97	2.36

Table 3 Time taken for different segmentation methods (unit: ms)

	CSE segmentation	Fuzzy connectedness	SVM	Proposed method
Image 1	1369	3972	1380	462
Image 2	1017	2962	1449	447
Image 3	1289	2367	1489	349
Image 4	1980	2678	1540	430
Image 5	1456	2120	1562	413
Image 6	1678	3249	1329	389

4 Conclusions

In this paper, a novel lesion segmentation algorithm using the concept of combination of scale-invariant and semantic mathematic model is proposed. A scale-invariant and semantic mathematic model is introduced based on the seeded points of normal skin and lesion textures. The images are divided into numerous smaller regions and each of those regions are classified as lesion or skin. The entire proposed framework is tested by using various skin lesion images as the input to the proposed segmentation algorithm. The proposed algorithm is compared with other conventional segmentation algorithms, including three algorithms designed for color images. The proposed framework produces the highest segmentation accuracy using manually segmented images as ground truth. From the experimental results, it can be observed that the proposed method is very effective for the segmentation of skin lesions with relatively consistent color.

References

1. Howlader N, Noone AM, Krapcho M, Garshell J, Neyman N, Altekruse SF, Kosary CL, Yu M, Ruhl J, Tatalovich Z, Cho H, Mariotto A, Lewis DR, Chen HS, Feuer EJ, Cronin KA (2013) SEER cancer statistics review, 1975–2010. National cancer institute, Bethesda, MD, USA, technology reports
2. Jerants AF, Johnson JT, Sheridan CD, Caffrey TJ (2000) Early detection and treatment of skin cancer. Am Family Phys 62(2):1–6
3. Public Health Agency of Canada (2013) Melanoma skin cancer. http://www.phac-aspc.gc.ca/cd-mc/cancer/melanomaskincancer-cancerpeaumelanome-eng.php
4. Jemal A, Saraiya M, Patel P, Cherala SS, Barnholtz-Sloan J, Kim J, Wiggins CL, Wingo PA (2011) Recent trends in cutaneous melanoma incidence and death rates in the united states, 1992–2006. J Am Acad Dermatol 65(5):S17.e1–S17.e11
5. Freedberg KA, Geller AC, Miller DR, Lew RA, Koh HK (1999) Screening for malignant melanoma: a cost-effectiveness analysis. J Am Acad Dermatol 41(5, pt. 1):738–745
6. Lim YW, Lee SU (1990) On the color image segmentation algorithm based on the thresholding and the fuzzy c-means techniques. Pattern Recogn 23(9):935–952
7. Siang Tan K, Mat Isa NA (2011) Color image segmentation using histogram thresholding–fuzzy C-means hybrid approach. Pattern Recogn 44(1):1–15
8. Shi J, Malik J (2000) Normalized cuts and image segmentation. IEEE Trans Pattern Anal Mach Intell 22(8):888–905
9. Felzenszwalb PF, Huttenlocher DP (2004) Efficient graph-based image segmentation. Int J Comput Vision 59(2):167–181
10. Priese L, Sturm P. Introduction to the color structure code and its implementation [[EB/OL]. doi:10.1.1.93.3090. http://citeseerx.ist.psu.edu/viewdoc/summary?
11. Lia H, Gua H, Hana Y, Yang J (2010) Object-oriented classification of high-resolution remote sensing imagery based on an improved colour structure code and a support vector machine. Int J Remote Sens 31(6):1453–1470
12. Priese L, Rehrmann V, Schian R, Lakmann R, Bilderkennen L (1993) Traffic sign recognition based on color image evaluation [C]. In: Proceedings IEEE intelligent vehicles symposium '93, pp 95–100

13. von Wangenheim A, Bertoldi RF, Abdala DD, Richter MM, Priese L, Schmitt F (2008) Fast two-step segmentation of natural color scenes using hierarchical region-growing and a color-gradient network. J Braz Comput Soc 14(4):29–40
14. Udupa JK, Samarasekera S (1996) Fuzzy connectedness and object definition: theory, algorithms, and applications in image segmentation. Graph Models Image Process 58(3):246–261
15. Udupa JK, Saha PK (2003) Fuzzy connectedness and image segmentation. Proc IEEE 91 (10):1649–1669
16. Saha PK, Udupa JK, Odhner D (2000) Scale-based fuzzy connected image segmentation: theory, algorithms, and validation. Comput Vis Image Underst 77:145–174
17. Yu Z, Bajaj CL (2002) Normalized gradient vector diffusion and image segmentation [C]. In: Proceedings of the 7th European conference on computer vision (ECCV 2002), pp 517–530
18. Cyganek B (2008) Color image segmentation with support vector machines: applications to road signs detection. Int J Neural Syst 18(4):339–345
19. Ikonomakis N (2000) Color image segmentation for multimedia applications. J Intell Robot Syst 28:5–20
20. Tao W, Jin H, Zhang Y (2007) Color image segmentation based on mean shift and normalized cuts. IEEE Trans Syst Man Cybern 37(5):1382–1389
21. Wang S (2009) Color image segmentation based on color similarity [C]. In: IEEE international conference on computational intelligence and software engineering, pp 1–4
22. Zhang TY, Suen CY (1984) A fast parallel algorithm for thinning digital patterns. Commun ACM 27(3):236–239
23. Liu B, Li H, Xianyong Jia X, Zhao ZL, Zhao Q, Zhang H (2014) A simple method of rapid and automatic color image segmentation for serialized Visible Human slices. Comput Electr Eng 40(3):870–883
24. Harris C, Stephens MJ (1988) A combined corner and edge detector [C]. Alvey vision conference, pp 147–152

Prospective Bio-Inspired Algorithm-Based Self-organization Approaches for Genetic Algorithms

M. Ilamathi, R. Raju and P. Victer Paul

Abstract The genetic algorithm (GA) is a population-based meta-heuristic global optimization technique for dealing with complex problems with very large search space. Nature plays a grand and enormous starting place of inspiration for solving NP completeness problems in biology, mathematics, and computer science. In the interim, it is the use of computers to represent the living phenomena, and concurrently the study of life to progress the usage of computers. Biologically inspired computing is a most important subset of computation. It constantly finds the optimal solution to explain its problem maintaining faultless steadiness among its components. Bio-inspired algorithms are meta-heuristics that imitate the nature for solving optimization problems in computation. Self-organization is the mechanism to improve the fitness and diversity among the population in the optimization algorithms. This paper presents a broad overview to examine several types of bio-inspired algorithm that can be used as a self-organization technique in genetic algorithms to improve its overall performance.

Keywords Genetic algorithm (GA) · Grey wolf optimizer (GWO) · Self-organization blending with genetic algorithm (SOGA) · Prospective Bio-inspired algorithm

1 Introduction

Genetic algorithm (GA) is an exploration technique used in computing to find proper or estimated solutions to optimization and search problems. Genetic algorithm is an exacting class of evolutionary algorithms that use techniques motivated

M. Ilamathi (✉) · R. Raju · P. Victer Paul
Department of Information Technology, Sri Manakula Vinayagar Engineering College, Puducherry, India
e-mail: ilamathimanjini@gmail.com

P. Victer Paul
e-mail: victerpaul.ap@gmail.com

L.P. Suresh and B.K. Panigrahi (eds.), *Proceedings of the International Conference on Soft Computing Systems*, Advances in Intelligent Systems and Computing 397, DOI 10.1007/978-81-322-2671-0_22

by evolutionary biology such as inheritance, mutation, selection, and crossover [1]. The evolution usually initiates [2] from a population of erratically generated individuals and happens in generations. In every generation, the fitness of each individual in the population is estimated; numerous individuals are selected from the existing population and personalized to form a fresh population. The fresh population is used in the subsequent iteration of the algorithm. The algorithm gets terminated when either a greatest numbers of generations have been created, or an agreeable fitness level has been reached for the population. The advantage of genetic algorithm is the effortlessness [3] with which it can handle subjective kinds of restrictions and objectives, and all such things can be handled as subjective components of the fitness function, making it easy to acclimatize the genetic algorithm scheduler to the exacting requirements of a very wide range of probable objectives. Genetic algorithms have been used for problem solving and for modeling. Genetic algorithms are applied to many scientific, engineering problems in business and entertainment including optimization, automatic programing, machine and robot learning, economic models, etc.

Hybrid genetic algorithms are based on the corresponding view of search methods. Genetic and other search methods can be seen as corresponding tools that can be brought collectively to attain an optimization goal. In these hybrids, genetic algorithm incorporates one or more methods to improve the performance of the genetic search. Integrating a search method within a genetic algorithm can advance the search performance on the situation that their roles assist to attain the optimization goal. There is a prospect in hybrid optimization to confine the finest of both schemes. Most of hybrid genetic algorithms that patch up chromosomes to gratify confines are Lamarckian and the technique has been predominantly effective in solving TSP. The hybrid genetic algorithm should thump stability between exploration and exploitation, in order to be able to solve global optimization problems. According to hybrid presumption, solving an optimization problem and accomplishing a solution of desired quality can be attained in one of the two ways. Moreover, the global search method alone reaches the solution or the global search method directs the search to sink of attraction from where the local search method can persist to lead to the preferred solution.

Self-organization is a vibrant process which is not obligatory by the peripheral power. In self-organizing, individuals are interacting directly to fabricate the widespread pattern which influences the performance of low-level individual in the population. It is also a well-organized stochastic search method for dealing multifarious problems with very large space. In SO, individuals are interacting directly to produce common model which is based on the social behavior of individuals in the population. Self-organizing and genetic algorithm are shared to evade the premature convergence and get trapped from the local optimum. The advantage of amalgamating the perception of self-organization enhances the working competence of other techniques to discover a solution of huge search problem.

Self-organizing genetic algorithm requires a complete knowledge of various parameters of self-organization and its relationships. The inherent property of self-organization and the method of amalgamating genetic algorithm develop a self-organizing genetic algorithm (SOGA). The endeavor of self-organizing genetic algorithm is to generate a computerized program that solves the problem with modest or no information from the user. The self-organizing genetic algorithm is used to reduce the number of external parameter. The genetic algorithm with self-organizing coding, operator, and parameter ethics is capable and effortless to use.

2 Recent and Best Working on Bio-Inspired Algorithms

Although there are number of bio-inspired algorithms survive in novel to solve optimization problems, until now there is constantly a need of innovative algorithm which can search for optimum solution in least time. This paper proposes three different new optimization algorithms like grey wolf algorithm, krill herd algorithm, and bull eye algorithm for solving optimization problems.

Grey wolf optimizer (GWO) is a population-based meta-heuristics algorithm that simulates the leadership hierarchy and hunting mechanism of grey wolves in nature proposed by Mirijalili et al. in 2014. Grey wolves are considered as apex predators, which they are at the top of the food chain [4]. Grey wolves prefer to live in groups each group containing 5–12 members on average. All the members in the group have a very strict social dominant hierarchy. The grey wolf hierarchy consists of four levels such as alpha, beta, delta, and omega. Alpha wolves are responsible for making decisions about hunting, time to walk, sleeping place, and so on. The alpha wolf is considered the dominant wolf in the group and all his/her orders should be followed by the pack members. The pack members have to dictate the alpha decisions and they acknowledge the alpha by holding their tails down. The beta wolves are considered as the best candidate to be the alpha, and when the alpha passes away or become very old the beta reinforces the alpha's commands throughout the pack and gives the feedback to alpha (Fig. 1).

The delta wolves called as subordinates. They have to submit to the alpha and beta but they dominate the omega. The omega wolves are considered as the scapegoat in the pack, and they have to submit to all the other dominant wolves [4]. They may seem are not important individuals in the pack and they are the last allowed wolves to eat. In the grey wolf algorithm, we consider the fittest solution as alpha, beta, and delta. The rest of the solutions are considered as omega.

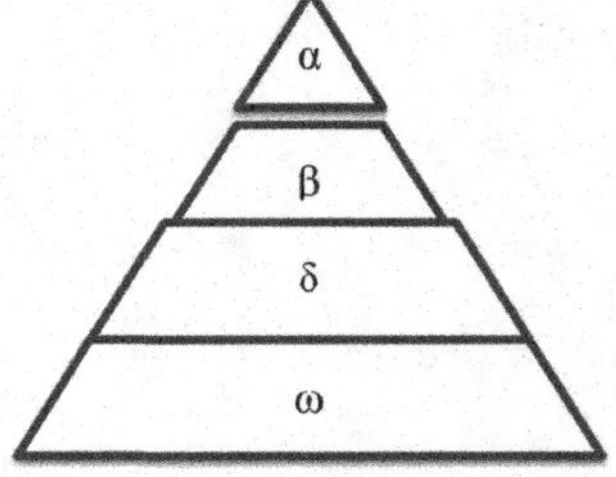

Fig. 1 Grey wolf representations

Krill herd algorithm is proposed by Gandomi et al. in 2012, inspired by the herding behavior of krill individuals. It mimics the krill's enslavement on the herd density while trying to forage for food. In krill herd algorithm, the time dependency position of krill individuals involves three main components like movement led by other individuals, foraging motion, and random physical diffusion. The advantage of krill herd algorithm is that it uses random search instead of a gradient search because derivative information is unnecessary. The fitness of each krill individual is defined as its distance from food and highest density of the swarm [5]. Bull eye approach is the center of the target which is a model that perfectly predicts the correct values. As we move away from the bull eye, our predictions get worse and worse. Imagine that we can repeat the entire model process to get separate hits on the target [1]. Each hit represents an individual's realization of the model, given the chance variability in the training data that are gathered. Sometimes it will get a good distribution of training data that predict very well and we are close to that target. In the bull eye approach, even though the moves are away from the target they would not result in the increased scatter of estimates. The low samples result in the wide scatter of estimate.

The major drawback of genetic algorithm is that it has certain optimization problems which occur due to poorly known fitness functions where generated bad chromosomes block in spite of the fact that only good chromosomes block crossover [6]. Like other artificial intelligence techniques, the genetic algorithm cannot assure constant optimization response times.

Even more, the difference between the shortest and the longest optimization response time is much longer than with conventional gradient methods [7]. The proposed genetic algorithm technique mainly lags in the premature convergence and computation time. This drawback can be overcome using self-organization blending with genetic algorithm (SOGA). The aim of SOGA is used to produce an automated computer program to solve the problem with no information from the user. Blending self-organization with genetic algorithm [8] avoids premature convergence and gives the better result for the individuals in the population (Fig. 2).

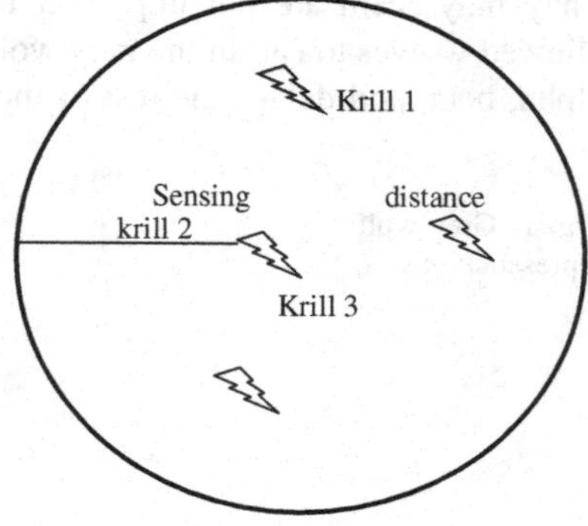

Fig. 2 Krill herd representation

3 Integrating Bio-Inspired Algorithm as Self-organization in Genetic Algorithm

Self-organization is a self-motivated methodology and adaptable procedure where the frameworks gain and keep up the structure without anyone else present, without any outside control. It is very confirm that it gives more prominent profits to take care of the complex problems with capable effectiveness levels. In self-arranging, individuals are connecting specifically with the hopeful answers for producing the common pattern which impacts the conduct of low-level individual in the population. The recent genetic algorithm like grey wolf algorithm, krill herd algorithm, and bull eye algorithm can be integrated with self-organization technique to yield better performance in producing a common pattern and make the individuals come under the capacity.

Grey wolf algorithm mimics the leadership hierarchy of grey wolves in nature. As stated earlier the grey wolves are classified as alpha, beta, delta, and gamma. In grey wolf algorithm, we consider the fittest solution as the alpha, and the second fittest solution as beta, and third fittest sólution as delta. In grey wolf algorithm, the hunting process is guided by the hunting process which is usually guided by the alpha [4]. The beta and delta might participate in hunting occasionally. In mathematical model of hunting behavior of grey wolves, we assumed the alpha, beta, and delta have better knowledge about the potential location of the prey. The first three best solutions are saved and the other agents are obliged to update their positions according to the position of the best search agent. The exploration process in grey wolf algorithm is applied according to the position and that diverge from each other to search for the prey and converge to attack the prey. In the population set after finding the best three individuals solution like alpha, beta, and delta, the other individuals in the generation are arranged based on the three fittest solutions using self-organization method. Self-organization technique [9] is done by identifying the common pattern in the three best fittest solutions. The common patterns are recognized by the calculating weights of the each fittest individual and the intermediate table is formulated to arrange the other individuals in the generation. Self-organize the individuals such that the position can be alerted to replicate the element position based on intermediate table. By replicating the position of the individuals, each generation is used to enhance the performance of the genetic algorithm [8].

Krill herd algorithm is based on the simulation of the herding behavior of krill individuals. The minimum distances of each individual krill from food and from highest density of the herd are considered as the objective function for the krill movement. The defined motions frequently change the position of the krill individual toward the best fitness. The time-dependent positions of the individual are movement induced by the presence of the individuals, foraging activity, and random diffusion. Each of the first two actions contains a global and a local optimizer [10]. The two global and two local strategies work in parallel resulting in notable efficiency of the krill herd algorithm. To improve the searching process, self-organization [11] methodology can also be added as the third action. As mentioned earlier, the major advantage of the krill herd algorithm is that it uses stochastic random search instead of a gradient search because derivate information is unnecessary [12]. For the self-organization process, if we use random search process, the infeasible solutions in the population are randomly altered to get a feasible solution in each generation based on the fittest solution. This approach is very effective in solving several constrained optimization problems [11].

Bull eye approach uses the center of the target model that perfectly predicts the correct value. As stated earlier, the low sample size results in a wide scatter of estimates. Increasing the sample size would make the estimate clump closer together. Increasing the sample size is used to reduce the variance [1]. After predicting the perfect value from the target, the worst individual in the generation is altered using nearest neighbor algorithm. The nearest neighbor algorithms [13] use a flexible machine learning technique. After finding fittest from the target, the nearest neighbor algorithm is performed. The nearest neighbor [14] is plotted on the plane with the other individuals in generation [13]. The nearest other individuals will be found using the geographic measure of distance. If the nearest individual is less than or equal to the target, then it is considered as the next fittest solution.

4 Discussions

Nature-inspired algorithms such as particle swarm algorithm, cuckoo search, and firefly algorithm which are integrated with genetic algorithm have become popular and they are widely used in current years in many application. Apart from these algorithms, there are three more recent algorithms such as grey wolf algorithm, krill herd algorithm, and bull eye algorithm which are versatile, well-organized, and effortless to implement. The above section summarizes the most recent hybridization algorithms and their numerous applications. In this paper, various hybrid algorithms are discussed, for the global search, genetic algorithm is used, and for the local search, different hybrid optimization techniques are used. This type of integration is used to improve the convergence rate in the algorithm and also to prove the efficiency of the computational time.

5 Conclusions

The hybrid genetic algorithm with self-organizing coding, operators, and parameter value is capable and effortless to use. If we tend to run the genetic algorithm many times, it will converge, probably at total different optimum chromosomes. The schemata that promise convergence are literally indicative of the regions within the search house wherever sensible chromosomes could also be found. In this paper, the overall idea of the recent algorithm is integrated with self-organization and is stated to yield the best performance. Typically, genetic algorithm is in addition to an area search mechanism to seek out the optimum chromosomes in an exceeding region. Genetic algorithm aren't sensible at distinctive the optimum price of a body for a retardant however do fine in distinctive the regions whenever those optima lie.

References

1. Gandomi AH, Yang X-S, Alavi AH Cuckoo search algorithm: a metaheuristic approach to solve structural optimization problems. Engineering with Computers (in press). doi:10.1007/s00366-011-0241-y
2. Murphy EJ, Morris DJ, Watkins JL, Priddle J (1988) Scales of interaction between Antarctic krill and the environment. In: Sahrhage D (ed) Antarctic Ocean and resources variability. Springer, Berlin, pp 120–130
3. Object Group Management (2003) Light weight CORBA component model. Revised submission, edn. OMG document realtime/03-05-05
4. Hoseini P, Shayesteh MG (2012) Efficient contrast enhancement of images using hybrid ant colony optimisation, genetic algorithm, and simulated annealing. Digit Signal Proc 23 (2013):879–893
5. Gandomi AH, Alavi AH (2012) Krill herd: a new bio-inspired optimization algorithm. Commun Nonlinear Sci Numer Simulat 17(2012):4831–4845
6. Miller DGM, Hampton I (1989) Krill aggregation characteristics: spatial distribution patterns from hydroacoustic observations. Polar Biol 10:125–134
7. Roy N, Shankaran N, Schmidt DC (2006) Bulls-eye—a resource provisioning service for enterprise distributed real-time and embedded systems. LNCS 4276:1843–1861
8. Baskaran R, Victer Paul P, Dhavachelvan P (2012) Ant colony optimization for data cache technique in MANET. In: International conference on advances in computing (ICADC 2012), advances in intelligent and soft computing series, vol 174. Springer, pp 873–878. ISBN:978-81-322-0739-9
9. Baskaran R, Victer Paul P, Dhavachelvan P (2012) Algorithm and direction for analysis of global replica management in P2P network. In: IEEE international conference on recent trends in information technology (ICRTIT), Chennai, pp 211–216. ISBN:978-1-4673-1599-9
10. Beasley DR, Bull R, Martin R (1993) An overview of genetic algorithms: part 1, fundamentals. Univ Comput 15:58–69
11. Baskaran R, Victer Paul P, Dhavachelvan P (2012) Analytical inspection for replica management in WANET using distributed spanning tree. In: IEEE international conference on recent trends in information technology (ICRTIT), Chennai, pp 297–301. ISBN:978-1-4673-1599-9
12. Kallel L, Schoenauer M (1997) Alternative random initialization in genetic algorithm. In: Proceedings of the 7th international conference on genetic algorithms
13. Victer Paul P, Ramalingam A, Baskaran R, Dhavachelvan P, Vivekanandan K, Subramanian R (2014) A new population seeding technique for permutation-coded genetic algorithm: service transfer approach. J Comput Sci (5):277–297. (Elsevier). ISSN:1877-7503
14. Victer Paul P, Ramalingam A, Baskaran R, Dhavachelvan P, Vivekanandan K, Subramanian R, Venkatachalapathy VSK (2013) Performance analyses on population seeding techniques for genetic algorithms. Int J Eng Technol (IJET) 5(3):2993–3000. ISSN:0975-4024
15. Price HJ (1989) Swimming behavior of krill in response to algal patches: a mesocosm study. Limnol Oceanogr 34:649–659
16. Object Management Group (2003) Deployment and configuration. Adopted submission, edn. OMG document ptc/03-07-08

17. Deng G, Balasubramanian J, Otte W, Schmidt D, Gokhale A (2005) DAnCE: a QoS-enabled component deployment and configuration engine. In: Proceedings of the 3rd working conference on component deployment, Grenoble, France
18. Muro C, Escobedo R, Spector L, Coppinger R (2011) Wolf-pack (Canis lupus) hunting strategies emerge from simple rules in computational simulations. Behav Process 88:192–197
19. Kirkpatrick S, Gelatt CD, Vecchi MP (1983) Optimization by simulated annealing. Science 220:671–680
20. Han K-H, Kim J-H (2002) Quantum-inspired evolutionary algorithm for a class of combinatorial optimization. IEEE Trans Evol Comput 6:580–593
21. Fort JC (1988) Solving combinatorial problem via self-organizing process: an application of the Kohonen algorithm to the traveling sales-man problem. Biol Cybern 59(1):33–40
22. Soltoggio (2005) An enhanced GA to improve the search process reliability in tuning of control systems. In: Proceedings of the 2005 conference genetic and evolutionary computation, GECCO'05, Washington, DC, pp 2165–2172
23. Meng W, Han XD, Hong BR (2006) Bee evolutionary genetic algorithm. Acta Electronica Sinica 34:1294–1300

Dual Converter Multimotor Drive for Hybrid Permanent Magnet Synchronous in Hybrid Electric Vehicle

G. Nagarajan, C.N. Ravi, K. Vasanth, D. Godwin Immanuel and S.D. Sundarsingh Jebaseelan

Abstract Electric vehicles (EV) and hybrid electric vehicles (HEV) are operated by battery power, which has to be optimized. HEV is the combination of an electric machine and an internal combustion engine (ICE) which is a promising means of reducing emissions and fuel consumption without compromising vehicle functionality and driving performance. The hybrid electric vehicle can be classified according to the way in which power is supplied to drive the train as parallel hybrid electric vehicle (PHEV), series hybrid electric vehicle (SHEV), dual-mode hybrid electric vehicle (DMHEV). In this proposed work, dual-mode hybrid electric vehicle (DMHEV) is used. One of the motors in the drive is a permanent magnet synchronous motor (PMSM). This PMSM serves two operations: As a parallel motor with an existing motor and the winding of PMSM acts as a boost inductor. A dual converter is proposed which acts as a DC-to-AC converter when PMSM is working as a motor. The same converter acts as a boost converter which uses the PMSM inductance in the boost mode. In this work, a total automobile HEV system is simulated using MATLAB. This system has an electrical subsystem which contains a generator, a battery and motors. The dual converter and PMSM are included in the electrical subsystem to improve the performance of the automobile system.

G. Nagarajan (✉) · C.N. Ravi · K. Vasanth · D. Godwin Immanuel
S.D. Sundarsingh Jebaseelan
Department of Electrical and Electronics Engineering, Sathyabama University, Chennai, Tamil Nadu, India
e-mail: nagarajanme@yahooo.co.in

C.N. Ravi
e-mail: dr.ravicn@gmail.com

K. Vasanth
e-mail: vasanthecek@gmail.com

D. Godwin Immanuel
e-mail: dgodwinimmanuel@gmail.com

S.D. Sundarsingh Jebaseelan
e-mail: sundarjeba@yahoo.co.in

L.P. Suresh and B.K. Panigrahi (eds.), *Proceedings of the International Conference on Soft Computing Systems*, Advances in Intelligent Systems and Computing 397, DOI 10.1007/978-81-322-2671-0_23

Keywords Car · Dual converter · Permanent magnet synchronous motor (PMSM) · Generator and motor drives

1 Introduction

Vehicles are essential for the transportation of people and goods from one place to other place. In ancient days, animals were used instead of the vehicles, and later in the modern world fossil fuels are used to propel the vehicles. These fossil-fuelled vehicles pollute the environment by producing emissive gases and noise. Thus it produces noise and air pollutions. In recent years, environmental awareness pushes the vehicle propels using electric energy. Electric cars are good from the pollution point of view but earlier conform given by fossil fuel vehicle may not be afforded. This setback is overcome by using hybrid electric vehicles. Almost all leading car manufacturing industries are researching and developing these types of hybrid electric cars. The advantage of these hybrid electric vehicle is driving good things both fossil fuel and electric car. The advantage in a fossil fuel car is its high speed and a comfortable journey. The advantages of an electric car are pollution-free atmosphere and sound. The power transmission in hybrid electric vehicle (HEV) is classified into three types. The three types of classification are (1) series HEV, (2) parallel HEV and (3) dual-mode HEV. In series HEV, internal combustion (IC) engines are used to rotate the electric generator and the power of this generator is given to electric motor which will propel the vehicle. Parallel HEV has an IC engine and its shaft is connected to the same shaft of electric motor shaft. This common shaft is connected to the wheels of the vehicle which propel it. In dual-mode HEV, the IC engine is connected to the alternator and the wheels of the vehicle. The electric power generated by the alternator is used to charge the battery. In this work, the dual-mode configuration is used. The dual-mode hybrid vehicle is similar to parallel configuration, but the difference is that an alternator (generator) is coupled to the IC engine that charges the battery [1]. The multimotor drive system will use two or more motors to boost the torque, especially under low speed and high-torque region. The proposed circuit allows the permanent magnet synchronous motor (PMSM) to operate in motor mode or acts as a boost inductor by using the dual converter. It is used for charging the battery at a rated DC voltage level [2, 3]. In the parallel hybrid electric vehicle (HEV) the drivetrain is connected to the electric motor engine through a mechanical coupling or an angle gear. These vehicles require an alternator (generator). The generator is a permanent magnet synchronous generator as it generates the power at a rated speed. The HEV initially runs on a battery, if load is increased in the internal combustion engine will be running and also an alternator will start to run, the generator generates the power at a rated speed. Habib Ullah [2] designed a hybrid electric vehicle which is a dual-mode HEV, he used a DC motor and eliminates the use of converter which in turn reduces the converter loss [4], and this concept is considered in the proposed work. He used a gasoline IC engine for parallel operation with a DC motor. For electric power management, he used PIC16F877A microcontroller. Pulse width modulation (PWM) is used to control the electric drive required for the DC motor [5]. Omar Hegazy [3] implemented multiple interleaved converters for DC-to-DC conversion. The motor used by him is a three-phase induction motor and it needs one more DC-to-AC converter [6]. Instead the battery he used is the fuel cell for the electric energy backup. Wei Qian [1] used 3X DC-to-DC converter which is an advantage than the four-level converter using capacitors. In the proposed converter, he used a small size inductor and achieved the same efficiency of four-level converters. Motor generator set and additional traction inverters are used for the HEV [7–10].

2 HEV Classifications

This section gives an overview and the working procedures of different types of HEV. The three main classifications based on power transfer among IC engine and electric motor are explained below.

2.1 *Series HEV*

The vehicle is run by the electric motor. The rating of the electric motor should be large enough to propel the vehicle. The electric motor gets its electric power either from a battery or an alternator. To run the alternator, an IC engine is used, which gives the required mechanical power. Since IC engine needs to run the alternator its size is small. This IC engine is optimized in such a way that it is running only when electric power is demanded. The alternator supplies electric power for driving the electric motor and the battery. When the battery is fully charged the alternator and IC engine are stopped, then the motor runs from the battery's charge. When the battery's charge is reduced to its minimum charge level, the IC engine and the alternators are switched on to get the electric power. The vehicle runs by an electric motor so there is no emission for propulsion and very less noise for operation. IC engine is used only to run the alternator so emission by the IC engine is very less as compared to conventional fossil-fuelled vehicle [11]. Advantages of this series HEV are less emission, less noise, the electric motor is directly connected to vehicle drives hence smooth operation, freedom IC engine and alternator location in vehicle. Disadvantage of this series HEV is it is suitable for short trip, large size of electric motor is required, need more battery backup and not suitable for high-speed applications.

2.2 Parallel HEV

The vehicle is run by the electric motor and the IC engine. The electric motor and the IC engine are connected to a common shaft which propels the vehicle. It has the freedom of running through an electric motor or an IC engine or by both. For less speed and short distance travel, electric motor may be used to run the vehicle. For long distance and high-speed applications IC engines may be used to propel the vehicle. Hence it can be a pure electric vehicle or a pure fossil fuel vehicle. In some situations such as hill climbing, IC engine's power may not be enough to run the vehicle at this time, in addition to IC engine the electric motor is used to boost the power to propel the vehicle. In this case, a separate alternator is not required as in the case of series HEV. The same machine acts as motor/alternator based on the command given by electric control system. When IC engine is working to run the vehicle, the electric machine which is connected to the same shaft works as an alternator and charges the battery. When the IC engine is not operated the electric machine acts as a motor which consumes power from the battery and propels the vehicle [12].

Advantage of this parallel HEV is that the battery size is small as compared to the series HEV, performance is better than the series HEV and good like conventional fossil fuel vehicles, space required for the battery is reduced and weight of the battery is also reduced. Disadvantage of this parallel HEV is that the battery power is not enough when it runs as a pure electric vehicle. Power transmission from electric motor as well as from IC engine creates complexity in the power transmission system [13].

2.3 Dual-Mode HEV

The vehicle is run either by the electric motor or IC engine or by both. This dual HEV is similar to parallel HEV but the common electric machine which acts as motor/alternator is replaced. Instead of a single machine, in dual HEV there are two separate machines, one is always electric motor and another one which is always alternator is employed. The alternator in the configuration is always connected to the IC engine. Hence, whenever the IC engine runs the alternator rotates and gives electric power and it is stored in the battery. The motor is used when the vehicle runs as an electric vehicle. During this time electric power is taken from the battery. When the vehicle is required to run at its maximum power, at that time IC engine also runs at its maximum power and electric motor is utilized to boost the power. So the requirement of a large-power IC engine is reduced, since electric motor may provide the peak power required by the vehicle. Latest hybrid electric vehicles use this concept and it becomes the dominating among the three classifications. The electric power generated by the alternator is used for online charging of the batteries always. When the batteries are fully charged the motor may switch on and reduce the burden of the IC engine until the battery discharges to a minimum level. This cyclic charging and discharging of battery may extend the life and efficiency of the battery.

It has the advantage of both series and parallel HEV. Its performance is good like the conventional fossil fuel vehicles. Requirement of the IC engine power is reduced. The size of the battery requirement is reduced. Optimal use of battery, IC engine and electric motor drive is ensured. Disadvantages of this dual HEV are it needs two electric machines, a motor and an alternator. Complex electronic control scheme is required for dual HEV.

3 Existing System

The hybrid electric car working on dual HEV mode and having a DC motor and an IC engine is considered for the work. In the existing model, it has a DC motor and an associated electrical system, energy management system, IC engine, gear system and vehicle dynamics system.

Energy management system is the main controlling system which decides the operation of the IC engine, electrical subsystem and vehicle dynamics system. For low-speed operation and when battery is charged enough, the vehicle operates as an electric vehicle and it is run by electric motor. When battery charge is not enough, the IC engine is switched on and the vehicle operates as fossil-fuelled vehicle. When the vehicle runs above the rated speed of the alternator coupled to the shaft, energy management system gives command to generate electric power. The generated power is stored in the battery for the operation of the motor.

Electrical subsystem has one motor, one generator, one DC-to-DC converter and a battery [5]. This motor is used for the electric drive. This drive motor is separately exited DC motor. DC-to-DC converter conditions the DC power of battery to serve the requirements of the DC motor. The same converter is also used to charge battery when the alternator generates electric power.

IC engine operates based on the energy management system's command. When electric motor is not able to run the vehicle IC engine throttles and runs the vehicle. Gear system connects electric motor shaft and IC engine shaft to run the vehicle. It is a complex mechanical system for mechanical power transmission. The common shaft is connected with a vehicle wheel to propel it.

Vehicle dynamics is the mechanical structure and mass of the vehicle which has to be propelled. It has a wheel, mechanical fixtures and seats of the vehicle. When luxurious accessories are added to the vehicle, its mass and the need for mechanical thrust are increased. Figure 1 shows the block diagram of existing system.

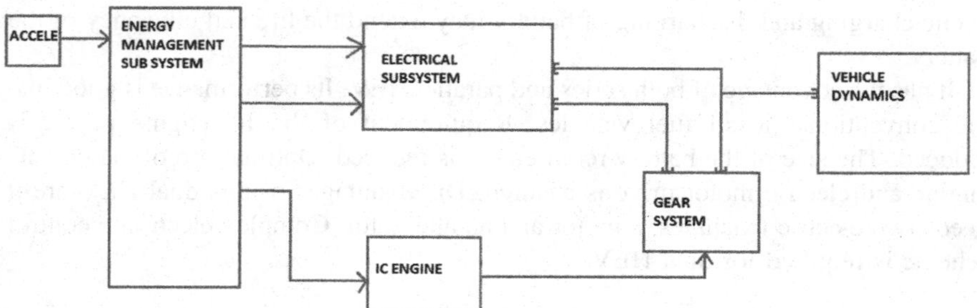

Fig. 1 Block diagram of hybrid electric vehicle in the existing system

3.1 Electrical Subsystem

Figure 2 shows the electrical subsystem. In this electrical subsystem, the battery is connected to the DC–DC converter, the converter is connected to the motor and generator. The DC–DC converter is a bidirectional converter. If motor is running, the power will be supplied to the motor from the battery, and the generator rotates at the rated speed which will the generate power to charge the battery. The initial condition of the motor is operating.

3.2 Energy Management Subsystem

Figure 3 shows the energy management subsystem. In this energy management subsystem, the primary input is the accelerator input. Based on the acceleration and

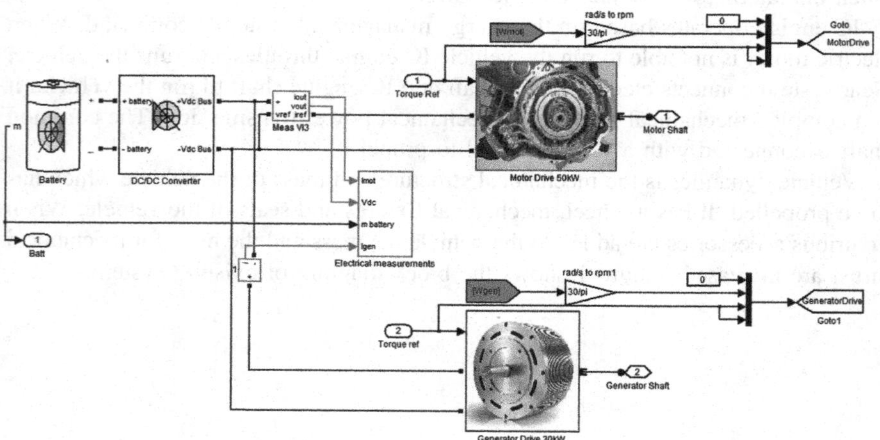

Fig. 2 Electrical subsystem

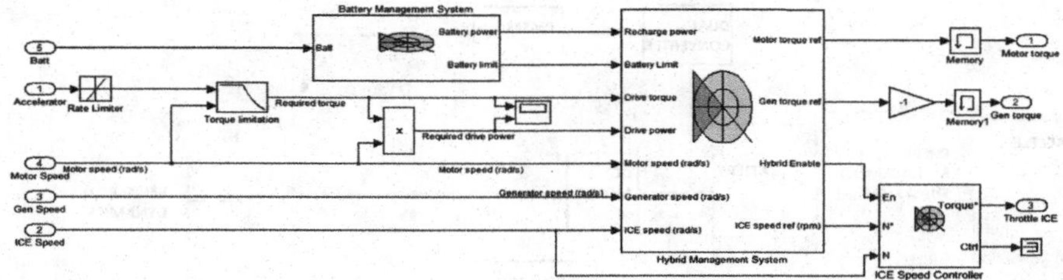

Fig. 3 Energy management system

the energy availability in the battery, this system decides the switching on of either the motor or the IC engine or both.

The energy available in the battery is determined using its state of charge (SOC). Speed is measured and if it is above the rated speed of the alternator then the alternator is switched on. It gives reference torque for the motor and generator operation and throttles input for the IC engine

3.3 Internal Combustion Engine

The internal combustion engine generates the mechanical motive power by the burning of petrol, oil or other with the air inside the engine, the hot gases produced being used drive a piston or do other work they expand.

3.4 Separately Excited DC Motor

In this system, separately excited DC motor is used for the drive. Based on the energy management system's command the motor is switched on and gives required mechanical power. The motor is switched on if enough electric power is available to it. This electric power is derived from the battery as well as from the alternator which is run by the IC engine. Alternator power is fed to the common DC bus which is connected to the motor. During peak mechanical thrust requirement along with IC engine DC motor works to relive IC engine from excess stress.

4 Proposed System

Multimotor is applied to improve the electrical drive system. The same dual HEV model in the previous section is considered, in addition permanent magnet synchronous motor (PMSM) is used for multimotor implementation. To feed electric

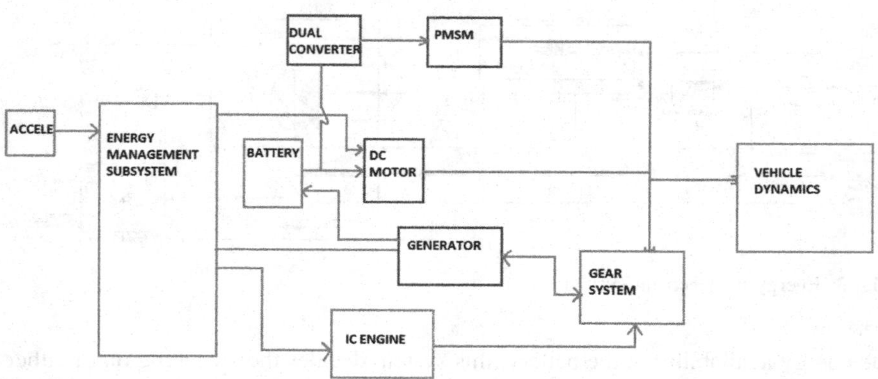

Fig. 4 Block diagram of the proposed system

power to PMSM dual converter is used. It serves as DC-to-AC converter when PMSM run as motor in parallel with existing DC motor. In dual operation, it utilizes PMSM stator windings as boost inductor to boost DC bus voltage. DC bus is common for DC motor, PMSM motor drive and alternator. Figure 4 shows the proposed model's block diagram. It consists of energy management system, IC engine; gear system, vehicle dynamic system and electrical system consists of generator, battery, DC motor, dual converter and PMSM.

The interleaved control idea, a boost control technique using motor windings as boost inductors for the proposed integrated circuit will be proposed. Under light load, the integrated circuit acts as a single-phase boost converter for not invoking additional switching and conduction losses, and functions as the two-phase interleaved boost converter under heavy load to significantly reduce the current ripple and thereby reducing the losses and thermal stress. Therefore, the proposed control technique for the proposed integrated circuit under boost converter mode can increase the efficiency.

The dual converter has two modes of operations. Mode-1: the converter acts as normal DC-to-AC conversion for PMSM. This electric power given to PMSM is used to run it in parallel with existing DC motor connected in common shaft. Mode-2: converter acts as interleaved boost converter to boost common DC bus voltage. Stator windings of PMSM act as interleaved inductor and used to boost common DC bus voltage. Since the DC bus voltage is increased the existing DC motor speed is increased to serve the vehicle dynamics. During high-torque requirement the energy management system directs the PMSM to run as motor. During high-speed requirements, the energy management system directs the dual converter to act as interleaved boost converter to boost the common DC bus voltage and it is used to boost the speed of existing DC motor.

5 Simulation Results

This section overviews the simulation result of single motor and multimotor dual converter HEV system. First subsection gives the single or existing system's simulation result and the next section overviews the multimotor dual converter's simulation result.

5.1 Existing System Results

This section briefs the simulation result obtained by single-motor dual HEV system. The input given to the simulation is accelerator position which is the intern speed requirement of the vehicle. This input is given to energy management system and decides the operation of the IC engine and the motor drive. Figure 5 shows the simulation output. First waveform shows the acceleration position, initially it is 0.25 and it is increased to 1 at 10 s. Second waveform shows the actual speed of the vehicle. Up to 10 s the speed is gradually increased since the speed requirement is less. During this time period the electric motor is operated and the vehicle acts like electric vehicle. At 10 s acceleration is increased to 1 and requires high speed. Hence IC engine is throttled by the energy management system and vehicle run by both IC engine and motor. During this time period the vehicle speed is above the rated speed of the alternator. Third waveform shows drive torque of the vehicle till 10 s motor gives driving torque and after 10 s IC engine gives the driving torque. Fourth part of Fig. 5 shows three waveforms, in which the blue-coloured waveform

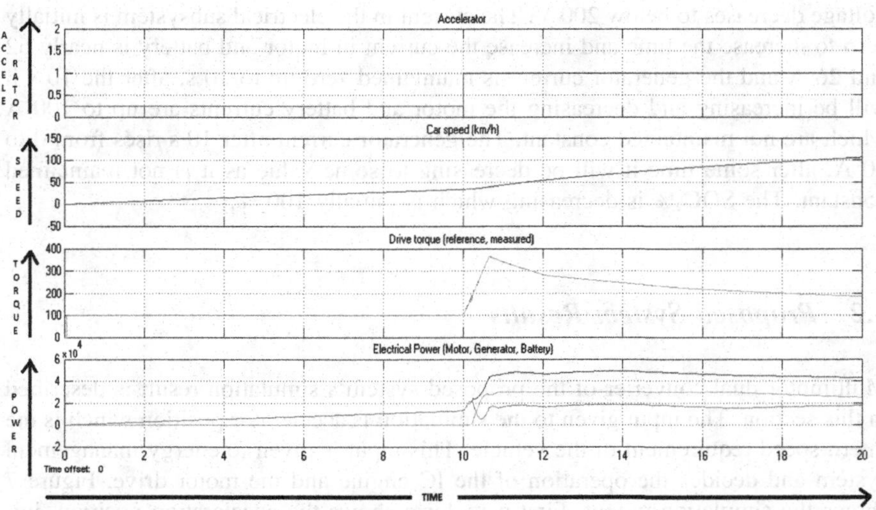

Fig. 5 Existing system's simulation output results

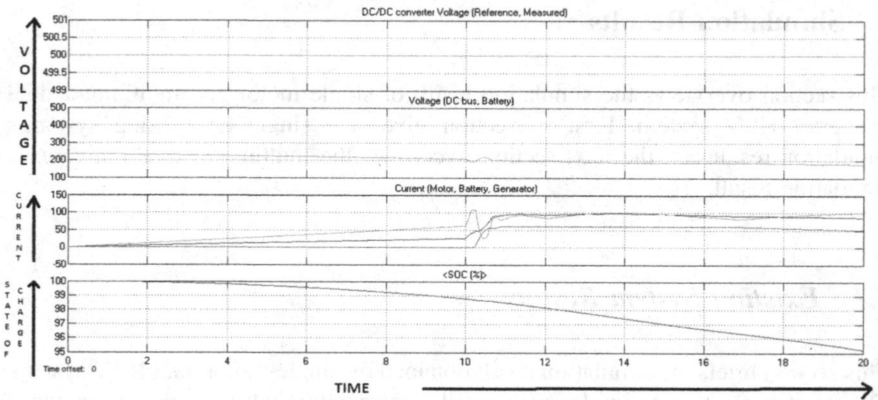

Fig. 6 Electrical subsystem waveform for the existing system

shows the motor speed. It is gradually increased till 10 s and sudden rise at 10 s. Green-coloured waveform shows the generator output. Generator output is zero till the vehicle speed is less than its rated speed and command from energy management system. After this it generates constant electric power and is supplied to the common DC bus. Red-coloured waveform shows battery energy availability in terms of state of charge. The SOC of battery get increased after the generator starts supplying electric power

Electrical Subsystem Figure 6 shows the simulation results of the electrical subsystem. In the electrical subsystem, the DC/DC converter's output voltage is maintained at 500 V and the voltage of the battery output is 230 V which is given to the converter. It does not maintain the constant voltage as the output of the battery voltage decreases to below 200 V. The current in the electrical subsystem is initially zero to increase the time and increase the current in motor and battery is nearly 62 and 26 A and the generator current is maintained zero up to 10 s, after the 10 s it will be increasing and decreasing the motor and battery currents are up to 100 A which are not maintained constant. The generator current after 10 s rises from 0 to 60 A, after some time it will be decreasing to some value as it is not maintained constant. The SOC % is decreasing which is initially 100 %.

5.2 Proposed System Results

Multimotor dual converter of the proposed system's simulation result is described in this section. The input given to the simulation is accelerator position which is the intern speed requirement of the vehicle. This input is given to energy management system and decides the operation of the IC engine and the motor drive. Figure 7 shows the simulation output. First waveform shows the acceleration position, initially it is 0.25 and it is increased to 1 at 10 s.

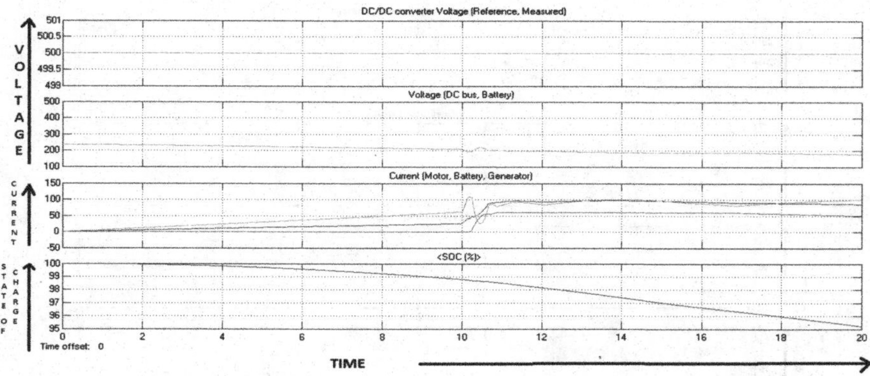

Fig. 7 Proposed system's simulation results

Second waveform shows the actual speed of the vehicle. Up to 10 s the speed gradually increases since the speed requirement is less. During this time period the electric motor is operated and the vehicle acts like electric vehicle. At 10 s acceleration is increased to 1 and requires high speed. Hence IC engine is throttled by energy management system and the vehicle is run by both IC engine and motor. During this time period vehicle speed is above the rated speed of an alternator. Third waveform shows the drive torque of the vehicle. Till 10 s motor gives the driving torque after 10 s the IC engine gives the driving torque. Fourth part of Fig. 7 shows three waveforms, in which the blue-coloured waveform shows the motor speed. It gradually increases till 10 s and suddenly rises at 10 s. Green-coloured waveform shows the generator output. Generator output is zero till the vehicle speed is less than its rated speed and command from energy management system. After this it generates constant electric power and supplied to the common DC bus. Red-coloured waveform shows the battery energy availability in terms of state of charge (SOC). The SOC of battery get increased after the generator starts supplying electric power. In this system we give the acceleration of the amplitude as 0.25–1 and time as 0–10. Initially, during 0–10 s the speed of the motor is 0 to nearly 66 km/h. The vehicle after 10 s is going to rise and after 20 s nearly to 140 km/h. Up to 20 s to give the run time.

Electrical Subsystem Figure 8 shows the simulation results of the electrical subsystem. In the electrical subsystem, the DC–DC converter output voltage is maintained as 500 V and the voltage of the battery output is 230 V which is given to the converter. It does not maintain constant voltage when the output of the battery voltage decreases below 200 V. The current in the electrical subsystem is initially zero and when we increase the time and increase the current in motor and battery to nearly 62 and 26 A and the generator current maintains zero up to 10 s, after the 10 s it will be increasing and decreasing as the motor and battery currents are up to 100 A not maintaining constant. The generator current is after 10 s rises from 0 to 60 A after some time it will be decreasing to some value, it is not maintained constant. The SOC % decreases which is initially 100 %.

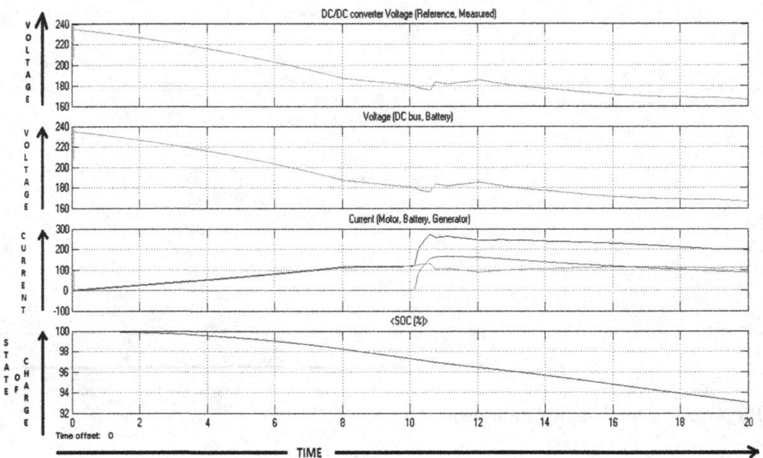

Fig. 8 Electrical subsystem waveform for the proposed system

6 Conclusion

Electric and hybrid electric vehicles will be the promising future vehicles. Electric vehicle suffers from low speed and short distance coverage. This drawback may be overcome by the hybrid electric vehicles. It has both an IC engine and an electric motor. There are three major classifications, namely series, parallel and dual HEV. Dual HEV is considered since it is most versatile for the proposed work. The existing system has single motor for electric drive. In the proposed model, two electric motor and dual converter is used to serve the electric drive. One of the motor is PMSM which acts as a motor for high-torque requirement and it provides its stator winding as interleaved boost inductor for dual converter operation. The system is simulated for a vehicle having energy management system, IC engine system, gear system and vehicle dynamic system. The simulation waveform for the existing and the proposed HEV is obtained. In future, for multimotor application more than two motors may be used for heavy vehicle.

References

1. Ullah MH, Gunawan TS, Sharif MR, Muhida R (2012) Design of environmental friendly hybrid electric vehicle. In Computer and Communication Engineering (ICCCE), 2012 International Conference on, IEEE 2012, pp 544–548
2. Lai YS, Lee WT, Lin YK, Tsai JF (2014) Integrated inverter/converter circuit and control technique of motor drives with dual-mode control for EV/HEV application. Power Electron. 29(3):1358–1365

3. Hegazy O, Van Mierlo J, Lataire P (2012) Analysis, modeling, and implementation of a multidevice interleaved DC/DC converter for fuel cell hybrid electric vehicles. IEEE Trans Power Electron 27(11):4445–4458
4. Jayaprakash S, Ramakrishnan V (2014) Simulation of solar based DC-DC converter for armature voltage controlled separately excited motor. In: 2014 international conference on advances in electrical engineering (ICAEE), IEEE 2014
5. Sundar Rajan GT, Christober Asir Rajan C (2011) Input stage improved power factor of three phase diode rectifier using hybrid unidirectional rectifier. In: 2011 international conference on nano science, engineering and technology (ICONSET), IEEE 2011, pp 589–592
6. Puttalakshmi GR, Paramasivam S (2014) Mathematical modeling of BLDC motor using two controllers for electric power assisted steering application. In: Proceedings of Springer international conference on artificial intelligence and evolutionary algorithms in engineering systems 2014 (ICAEES2014). Noorul Islam University, pp 1437–1444, 978-81-322-211-97. Published in Springer LINEE series of power electronics and renewable energy systems, April 2014
7. Qian W, Cha H, Peng FZ, Tolbert LM (2012) 55-kw variable 3X DC-DC converter for plug-in hybrid electric vehicles. IEEE Trans Power Electron 27(4):1668–1678
8. Maggetto G, Van Mierlo J (2011) Electric and electric hybrid vehicle technology: a survey. In: Proceedings of IEEE seminar electric, hybrid fuel cell vehicles, April 2000, pp 1/1–111
9. Yilmaz M, Krein PT (2013) Review of battery charger topologies, charging power levels, and infrastructure for plug in electric and hybrid electric vehicles. IEEE Trans Power Electron 28 (5):2151–2169
10. Emadi A, Lee YJ, Rajashekara K (2008) Power electronics and motor drives in electric, hybrid electric, and plug-in hybrid electric vehicles. IEEE Trans Ind Electron 55(6):2237–2245
11. Elankurisil SA, Dash SS (2012) Compare the performance of controllers in non-isolated D.C to D.C. converters for DC motor. Natl J Electron Sci Syst 3(1):1–12
12. Samuel Rajesh Babu R, Joseph H (2011) Embeded controlled ZVS DC-DC converter for electrolyzer application. Int J Intell Electron Syst 5(1):6–10
13. Sasi Kumar M, Pendian SC (2011) Modeling and analysis of cascaded H bridge inverter for wind driven isolated self excited induction generators. J Electr Eng Inf 3(2):132–145

Performance Enhancement of Minimum Volume-Based Hyperspectral Unmixing Algorithms by Empirical Wavelet Transform

Parvathy G. Mol, V. Sowmya and K.P. Soman

Abstract Hyperspectral unmixing of data has become one of the essential processing steps for crop classification. The endmembers to be extracted from the data are statistically dependent either in the linear or nonlinear form. The primary focus of this paper is on the effect of empirical wavelet transform (EWT) on hyperspectral unmixing algorithms based on the geometrical minimum volume approaches. The proposed method is experimented on the standard hyperspectral dataset, namely Cuprite. The performance analysis of proposed approach is evaluated based on the standard quality metric called root mean square error (RMSE). The experimental result analysis shows that our proposed technique based on EWT improves the performance of hyperspectral unmixing algorithms based on the geometrical minimum volume approaches.

Keywords Hyperspectral unmixing (HU) · Abundance map · Empirical wavelet transform (EWT) · Endmember signature

1 Introduction

The advancement of hyperspectral sensors and the development of softwares led to a significant growth in hyperspectral image analysis. A large number of wavelength bands have been measured, hence the name 'hyper', means 'too many'. Hyperspectral remote sensing imagery contains detailed information [1, 2]. Hyperspectral unmixing (HU) is defined as the separation of spectral content

---

P.G. Mol (✉) · V. Sowmya · K.P. Soman
Centre for Excellence in Computational Engineering and Networking, Amrita Vishwa Vidyapeetham, Amritanagar PO, Coimbatore 641112, Tamil Nadu, India
e-mail: parvathy.g.mol@gmail.com

V. Sowmya
e-mail: sowmiamrita@gmail.com

K.P. Soman
e-mail: kp_soman@amrita.edu

L.P. Suresh and B.K. Panigrahi (eds.), *Proceedings of the International Conference on Soft Computing Systems*, Advances in Intelligent Systems and Computing 397, DOI 10.1007/978-81-322-2671-0_24

of each pixel into various constituent spectral signatures, termed as endmembers [3]. The important components of hyperspectral unmixing are the endmembers and the abundances. The pure pixel data vectors are known as the endmembers and the abundances correspond to the amount of each endmember in a pixel vector.

Mixing models can either be linear or nonlinear [4, 5]. The limitation of linear spectral unmixing is overcome by the unsupervised endmember extraction method such as vertex component analysis (VCA) [6]. Geometrical approaches have the assumption that they work under linear mixing models and here scattering does not take place [3]. These geometrical-based unmixing algorithms are widely used for hyperspectral unmixing. It is divided into two subclasses, namely pure pixel-based approaches and minimum volume-based approaches [3]. Relevant pure pixel-based algorithms are vertex component analysis (VCA) [6], alternating volume maximization (AVMAX) [7], successive volume maximization (SVMAX) [7], alternating decoupled volume max–min (ADVMM) [8], successive decoupled volume max–min SDVMM [8], N-finder [9], etc. Pure pixels are not available in all the scenes. To avoid this issue, minimum volume-based approaches are used, which works violating the pure pixel assumption. The examples for this class are, RMVES [10], MVSA [11], and SISAL [12] based on Craig's criterion, which are known as popular methods for unmixing.

In this paper, a method based on EWT is proposed to unmix the hyperspectral data followed by minimum volume-based unmixing algorithms. In order to extract the relevant details, the well-known reconstruction technique, EWT is applied to the hyperspectral dataset. The reconstructed data is given as the input to the minimum volume unmixing approaches such as, RMVES, MVSA, and SISAL. Results of the EWT-based proposed method enhance the performance of unmixing techniques. The paper is structured as follows. The overview of EWT is described in Sect. 2. Section 3 presents the minimum volume-based geometrical approaches such as RMVES, MVSA, and SISAL. Based on the experimental results, the analysis of the proposed technique is described in Sect. 4. The concluding statements derived from this paper and future research opportunities are presented in Sect. 5.

2 Overview of Empirical Wavelet Transform (EWT)

Empirical wavelet transform (EWT) can be explained as the combination of wavelets formalism and the adaptability of empirical mode decomposition (EMD). Wavelets and their geometric extensions are very efficient in image processing. Making use of this same character, the improvement of unmixing has been done here. In EWT, a set of wavelets are made by adapting from the processed signal like in the Fourier method, i.e., building a set of band-pass filters. For the adaption process, the location of information in the spectrum is identified with frequency,

$\omega \in [0, \pi]$. As this is to make the filters support, partition plays a significant role. As an initial step, the Fourier transformed signal is partitioned into N segments. Each segment will have the boundary limits, denoted as ω [13]. Each partition is denoted as $\wedge_n = [\omega_{n-1}, \omega_n]$, $\bigcup_{n=1}^{N} \wedge_n$.

Around each ω_n, a small area of width $2\tau_n$ is defined. This denotes a transition phase. The empirical wavelets are defined on each of the $\wedge_n$. It is a band-pass filter constructed using Littlewood–Paley and Mayer's wavelets [13]. Through this filtering operation the subbands are extracted. Each partition in the spectrum is considered as modes, which contains a central frequency with certain supports. If there are N partitions, then there will be $N + 1$ boundary limits. Since 0 and π are used as the limits to the spectrum, the number of boundary limits required will be $N - 1$ [13]. The boundaries are calculated in two steps:

- Finding local maxima's in the spectrum.
- Sorting it in the decreasing order by excluding and selecting the M boundary values.

Therefore the partition boundaries, n comprises of 0, selected maxima and π. Two possibilities for the selection of the boundaries are $M \geq N$ and $M \leq N$. For the function f, the detail coefficients are obtained by taking the inverse of the convolution operation between f and the wavelet function ψ_n which is explained in [10]. In spectrum domain, the convoluted output is obtained through simple multiplication of the two functions. The approximate coefficients are obtained by taking inverse of the convolution operation between f and the scaling function ϕ_n. Therefore the signal $f(t)$ can be reconstructed. The mathematical derivations in detail are given in [10].

3 Geometrical-Based Hyperspectral Unmixing Approach

Geometrical approach aims to identify the vertices which are equivalent to the extraction of endmembers, under the linear mixing model [14]. So identifying the vertices is equivalent to finding out the endmembers. Robust minimum volume enclosed simplex analysis (RMVES) and minimum volume simplex analysis (MVSA) are the non-pure-pixel-based hyperspectral unmixing algorithms proposed by Ambikapathi et al. and Bioucas Dias et al. respectively [10, 11]. MVSA has better performance than MVES due to the utilization of the VCA technique [6]. In [12], an enhanced version of the MVSA algorithm for hyperspectral unmixing, simplex identification via split augmented lagrangian (SISAL) is proposed. In this method, to solve the hard non-convex optimization problem, i.e., to obtain a constraint formulation, variable splitting is used followed by an augmented Lagrangian technique. All the three algorithms are very efficient in the hyperspectral image unmixing area.

4 Experimental Result and Analysis

The hyperspectral image data used in this experiment is the Cuprite data, acquired by airborne visible/IR imaging spectrometer (AVIRIS) sensor over NEVADA, U.S.A. in 1997 [15]. In this work, a dataset containing 250 × 191 pixels and 188 bands in the spectral range of 400–2500 nm, which excludes the bands with low SNR are used to experiment the proposed method [16]. The effectiveness of the proposed method is evaluated based on the well-known quality metric, RMSE [17].

The experiment is performed on the AVIRIS Cuprite dataset. A characteristics extraction technique called EWT is applied prior to unmixing techniques. In EWT, adaptive wavelet frames are used for the reconstruction. To enhance the performance of proposed method, the design parameter $\eta = 0.001$ and regularization parameter $\lambda > 0$ is set. Table 1 shows the performance comparison of our proposed methods (EWT + RMVES, EWT + MVSA, EWT + SISAL) with RMVES, MVSA, and SISAL based on the RMSE values calculated for different numbers of endmembers (P) and SNR = 40. From the results, it is evident that the RMSE obtained after applying EWT (proposed method) is much better than the result without EWT. Using our proposed method, the average RMSE obtained is reduced from 0.007892 to 0.00773 when $P = 3$ and from 0.003459 to 0.0032 when $P = 8$ (RMVES). Among the three hyperspectral unmixing algorithms (RMVES, MVSA, and SISAL), our proposed method enhances the performance of RMVES algorithm for various endmembers and SNR = 40 dB. These results highlight the quality improvement of unmixing techniques by our proposed method based on EWT.

Table 1 Performance evaluation of our proposed method based on RMSE values

P	RMVES	Proposed method	MVSA	Proposed method	SISAL	Proposed method
3	0.0079	0.0077	0.8427	0.3132	1.0209	1.0501
5	0.0061	0.0059	0.5110	0.4464	0.7391	0.7059
6	0.0047	0.0045	2.2762	0.6875	1.0079	0.9551
8	0.0035	0.0032	0.6043	1.0844	1.0740	1.2556

5 Conclusion

This paper presents a characteristics extraction method called EWT to improve the performance of minimum volume-based unmixing techniques. Unmixing algorithms used in this work are RMVES, MVSA, and SISAL which have been tested on the standard hyperspectral data. The standard metric, RMSE, is used to validate the results. The tabulated quality metric for the dataset used in this experiment shows that the performance of the empirical wavelet transform-based proposed method is better than the existing minimum volume hyperspectral unmixing methods. From our experimentation, we have observed that the application of EWT enhances the performance of the endmember identification algorithms. As a future work, EWT can be used to improve the performance of other existing geometrical spectral unmixing algorithms.

References

1. Iordache MD, Bioucas-Dias JM, Plaza A (2011) Sparse unmixing of hyperspectral data. IEEE Trans Geosci Remote Sens 49(6):2014–2039
2. Shippert P (2003) Introduction to hyperspectral image analysis. Online J Space Commun 3
3. Bioucas Dias JM, Plaza A, Dobigeon N, Parente M, Du Q, Gader P, Chanussot J (2012) Hyperspectral unmixing overview: geometrical, statistical, and sparse regression based approaches. IEEE J Sel Top Appl Earth Observations Remote Sens 5(2):354–379
4. Keshava N, Mustard JF (2002) Spectral unmixing. Sig Process Mag IEEE 19(1):44–57
5. Liangrocapart S, Petrou M (1998) Mixed pixels classification. In: Remote sensing international society for optics and photonics, 1998 Oct, pp 72–83
6. Nascimento JM, Bioucas Dias JM (2005) Vertex component analysis: a fast algorithm to unmix hyperspectral data. IEEE Trans Geosci Remote Sens 43(4):898–910
7. Chan TH et al (2011) A simplex volume maximization framework for hyperspectral endmember extraction. IEEE Trans Geosci Remote Sens 49(11): 4177–4193
8. Chan TH et al (2013) Robust affine set fitting and fast simplex volume max-min for hyperspectral endmember extraction. IEEE Trans Geosci Remote Sens 51(7):3982–3997
9. Winter ME (1999) N-FINDR: an algorithm for fast autonomous spectral end-member determination in hyperspectral data. In: International symposium on optical science, engineering, and instrumentation. International Society for Optics and Photonics (1999)
10. Ambikapathi A, Chan TH, Ma WK, Chi CY (2011) Chance-constrained robust minimum volume enclosing simplex algorithm for hyperspectral unmixing. IEEE Trans Geosci Remote Sens 49(11):4194–4209
11. Li J, Bioucas-Dias JM (2008 July). Minimum volume simplex analysis: fast algorithm to unmix hyperspectral data. In: Proceedings of the IEEE international geoscience and remote sensing symposium (IGARSS), vol 3, pp. III250–III253, July 2008
12. Bioucas-Dias JM (2009) A variable splitting augmented Lagrangian approach to linear spectral unmixing. In: IEEE workshop on hyperspectral image and signal processing: evolution in remote sensing, (WHISPERS'09), IEEE, pp 1–4
13. Gilles J, Tran G, Osher S (2014) 2d empirical transforms. wavelets, ridgelets, and curvelets revisited. SIAM J Imaging Sci 7(1):157–186

14. Chan TH, Chi CY, Huang YM, Ma WK (2009) A convex analysis based minimum volume enclosing simplex algorithm for hyperspectral unmixing. IEEE Trans Signal Process 57(11): 4418–4432
15. AVIRIS Free standard data products. http://aviris.Jpl.nasa.gov/html/aviris.freedata.html
16. Bijitha SR, Geetha P, Nidhin Prabhakar TV, Soman KP (2013) Performance analysis of minimum volume based geometrical approaches for spectral un- mixing. Int J Sci Eng Technol Res (IJSETR) 2(7). ISSN: 2278-7798, July 2013
17. Hendrix EMT et al (2012) A new minimum volume enclosing algorithm for end-member identification and abundance estimation in hyperspectral data. IEEE Trans Geosci Remote Sens 50(7):2744–2757

Empirical Wavelet Transform for Multifocus Image Fusion

S. Moushmi, V. Sowmya and K.P. Soman

Abstract Image fusion has enormous applications in the fields of satellite imaging, remote sensing, target tracking, medical imaging, and much more. This paper aims to demonstrate the application of empirical wavelet transform for the fusion of multi-focus images incorporating the simple average fusion rule. The method proposed in this paper is experimented on benchmark datasets used for fusing images of different focuses. The effectiveness of the proposed method is evaluated across the existing techniques. The performance comparison of the proposed method is done by visual perception and assessment of standard quality metrics which includes root mean squared error, relative average spectral error, universal image quality index, and spatial information. The experimental result analysis shows that the proposed technique based on the empirical wavelet transform (EWT) outperforms the existing techniques.

Keywords Image fusion · EWT · DWT · FIHS · MSVD · Simple average · Quality metric evaluation

1 Introduction

Image fusion is the technique of extracting the details from different images from different sensors into a single image. The fused output image contains all relevant information present in the input images. The goal of multifocus image fusion is to combine various images of different focuses into single image for which different

S. Moushmi (✉) · V. Sowmya · K.P. Soman
Center for Excellence in Computational Engineering and Networking, Amrita Vishwa Vidyapeetham, Amritanagar, Coimbatore 641112, Tamil Nadu, India
e-mail: moushmi24@gmail.com

V. Sowmya
e-mail: sowmiamrita@gmail.com

K.P. Soman
e-mail: kp_soman@amrita.edu

L.P. Suresh and B.K. Panigrahi (eds.), *Proceedings of the International Conference on Soft Computing Systems*, Advances in Intelligent Systems and Computing 397, DOI 10.1007/978-81-322-2671-0_25

approaches are available in the literature. Image fusion plays a vital role in navigation guidance, satellite imaging for remote sensing, object detection and recognition, military and civilian surveillance, etc. [1]. Image fusion algorithms can be classified into different levels: pixel, feature, and decision levels [2].

Image fusion methods can be divided into two groups such as spatial domain and transform domain [3]. Fast intensity hue saturation (FIHS), discrete wavelet transform (DWT), and multiresolution singular value decomposition (MSVD) based image fusion methods are few popular image fusion techniques. FIHS-based image fusion falls under the spatial domain method. DWT- and MSVD-based image fusion methods lie under the transform domain fusion method.

Fast intensity hue saturation-based image fusion method is one of the commonly used techniques for various applications in the field of remote sensing [4]. In this method, the analysis of image is done in intensity hue saturation (IHS) color space followed by linear transformations. Intensity can be defined as the amount of light that an eye can attain. Hue is the ascendant wavelength of a color, and saturation is the clarity of total amount of white light of a color [5]. The key advantage of FIHS method is its simplicity and good visual effects in the fused image, however, it has a disadvantage that this method suffers from artefacts and noise which tends to higher contrast. Details about fast intensity hue saturation technique for fusion are available in [4–6].

The wavelet transform has become a convenient method in image fusion after the advent of multiresolution analysis in image processing. The quality of wavelet-fused image is better than the fast intensity hue saturation method in terms of spatial and spectral content [7]. Multiresolution singular value decomposition is similar to wavelet transform, where the signal is filtered by high-pass and low-pass finite impulse response filters separately. In multiresolution singular value decomposition, the finite impulse response filters are replaced with singular value decomposition (SVD) [8].

In this paper, we focused on empirical wavelet transform (EWT) based image fusion scheme in which the input images are of different focuses. The experimental result is compared with other existing techniques, namely fast intensity hue saturation, discrete wavelet transform, and multiresolution singular value decomposition technique for image fusion. The comparative assessment is done based on standard image fusion quality metrics. In our proposed method, the input images are decomposed into number of modes and the desired modes from each source images are merged together to reconstruct the fused image.

2 Empirical Wavelet Transform

EWT decomposes a signal or an image on wavelet tight frames which are built adaptively. The key advantage of this empirical approach is to keep together some information that should be spitted in the case of dyadic filters. This property of EWT is used in image fusion to improve the worth of the fused image.

In EWT, a set of wavelets are built by adaption from the processed signal. The main idea behind EWT is to extract different modes of a signal by designing the appropriate wavelet filter banks. This algorithm was first proposed by Gilles in [9]. Similar approach is used in the Fourier method for forming band-pass filters. For the adaption process, the location of information in the spectrum is identified with frequency, $\omega \in [0, \pi]$. This is used to make the support of the filter. Initially, the Fourier transformed signal is partitioned into N segments. Each segment will have the boundary limits denoted as ω_n [9].

Each partition is denoted as ¤n. Around each ω_n, a small area of width $2\tau_n$ is defined. This denotes a transition phase. The empirical wavelets are defined on each of the ¤n empirical wavelets which are band-pass filters constructed using Littlewood–Paley and Mayer's wavelets [9]. The subbands are extracted through these filtering operations. Each partition in the spectrum is considered as modes which contains a central frequency with certain supports. If there are N partitions, there will be $N + 1$ boundary limits. Since 0 and π are used as the limits to the spectrum, the number of boundary limits required will be $(N - 1)$ [9, 10]. The boundaries are calculated by finding local maxima in the spectrum and sorting it in the decreasing order (excluding 0 and π). Assume that the algorithm found M maxima. Then two possibilities for the selection of the boundaries are $M \geq N$ and $M \leq N$. The mathematical derivation for empirical wavelet transform is given in [9].

3 Proposed Method

We propose a method to build the composite image from two multifocus input images decomposed using EWT. Primarily, we define the empirical wavelet transform parameters such as length of the filter, degree for polynomial interpolation, detection method, and maximum number of bands for detection method for the source images. EWT decomposes the source images into several modes. The decomposition of modes is in the increasing order of frequency from mode 1 to mode N. Simple average [11], [12] is the fusion rule used in this experiment. The fusion rule is applied on the mode 1 of both input images and the same process is repeated for all N modes. Finally, the inverse empirical wavelet transform (IEWT) is performed on the resultant modes to reconstruct the fused image. The flow diagram for the proposed method is shown in Fig. 1. In the proposed method, the maximum number of modes are two, hence each input image is decomposed into two modes horizontally and two modes vertically.

4 Experimental Results and Analysis

The proposed method was experimented on five benchmark multifocus datasets. The multifocus datasets contain clock, saras, book, toy, and radiolarian images. The experimental results of the proposed method are compared with the existing techniques (DWT, FIHS, and MSVD) based on the well-known image quality metrics.

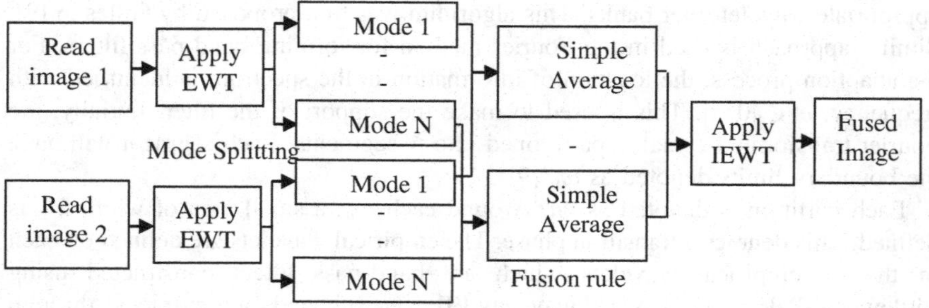

Fig. 1 Flow diagram of the proposed method

4.1 Metrics Performance Evaluation

Quality metrics include root mean squared error (RMSE) [4], relative average spectral error (RASE) [4], universal image quality index (UIQI) [13], and spatial quality of the fused image [4]. For a better quality image, the value of RMSE and RASE should be very low. The value of UIQI and spatial quality should be high for the output fused image with maximum information.

4.2 Results and Discussion

This section gives the comparison of the proposed method with the existing fusion methods by visual perception and also based on the computation of quality metrics. The output of the proposed EWT-based image fusion methods and other existing methods are shown in Figs. 2, 3, 4, 5 and 6. The quality metrics calculated for the proposed EWT-based image fusion method and other existing methods are presented in Tables 1, 2, 3, 4, and 5.

The universal image quality index value for the clock image is 0.9524 for the proposed EWT-based method, whereas the value is low for other existing methods which show that the quality of output of the EWT-based fusion method is better. The spatial data metric of our proposed method is in the range of 0.81–0.99 for all datasets. The quality of the fused image will be high when the value for the relative average spectral error is minimum. In case of saras image, the value of RASE for the proposed method is 2.8562 which is very much less than the value obtained for the other existing methods. Even though it is hard to find much difference through visual perception, the tabulated quality metric calculated for all datasets used in this experiment shows that the EWT-based method performs better than the other existing methods. Therefore it can be concluded that the proposed empirical wavelet transform-based multifocus image fusion is able to preserve the detail features of the original image.

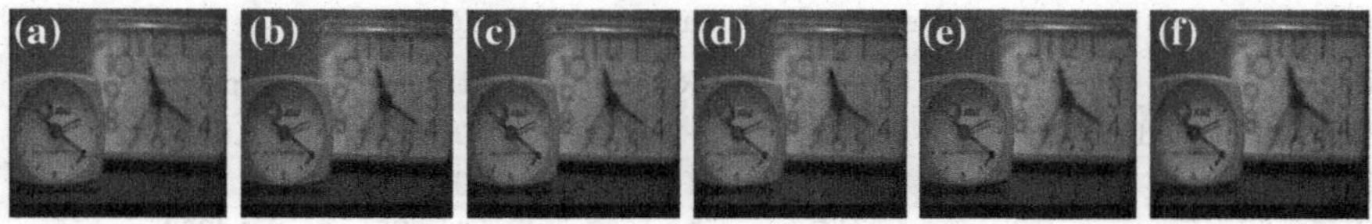

Fig. 2 Fused output for clock image. **a** Input image 1. **b** Input image 2. **c** MSVD-based fusion. **d** FIHS-based fusion. **e** DWT-based fusion. **f** Output of the proposed method (EWT)

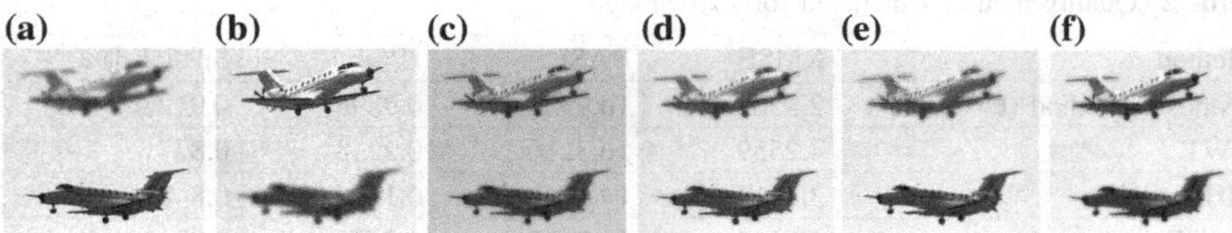

Fig. 3 Fused output for saras image. **a** Input image 1. **b** Input image 2. **c** MSVD-based fusion. **d** FIHS-based fusion. **e** DWT-based fusion. **f** Output of the proposed method (EWT)

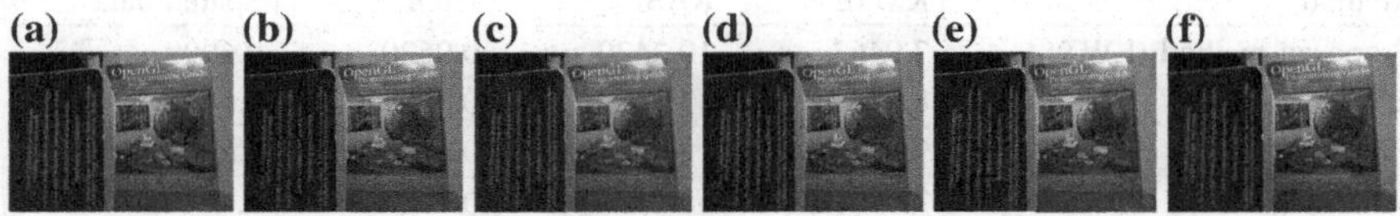

Fig. 4 Fused output for the book image. **a** Input image 1. **b** Input image 2. **c** MSVD-based fusion. **d** FIHS-based fusion. **e** DWT-based fusion. **f** Output of the proposed method (EWT)

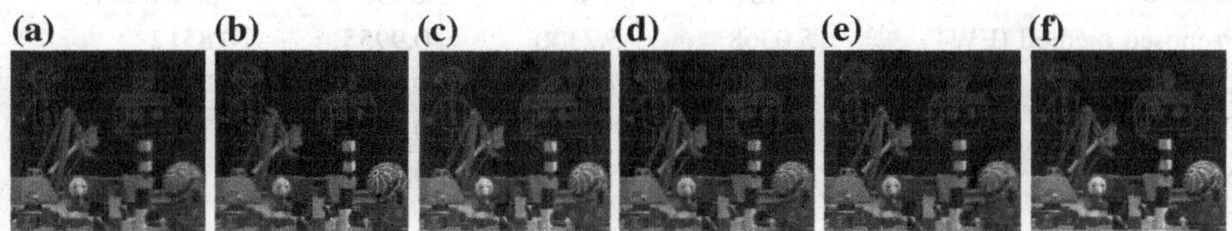

Fig. 5 Fused output for the toy image. **a** Input image 1. **b** Input image 2. **c** MSVD-based fusion. **d** FIHS-based fusion. **e** DWT-based fusion. **f** Output of the proposed method (EWT)

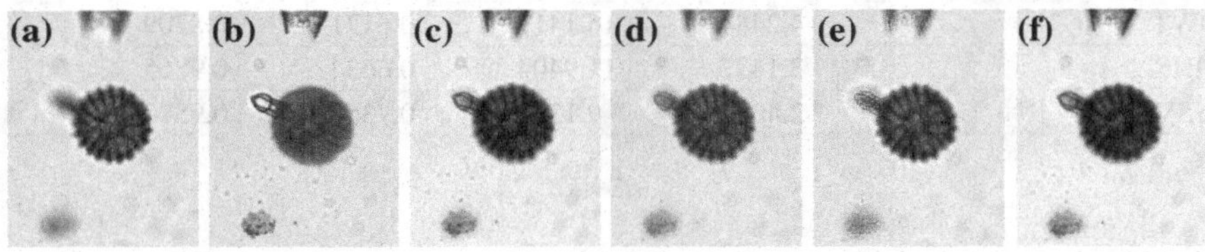

Fig. 6 Fused output for radiolarian image. **a** Input image 1. **b** Input image 2. **c** MSVD-based fusion. **d** FIHS-based fusion. **e** DWT-based fusion. **f** Output of the proposed method (EWT)

Table 1 Quality metrics calculated for clock image

Method	RMSE	RASE	UIQI	Spatial data
Proposed method (EWT)	4.5168	7.5539	0.9524	0.8130
DWT	7.4990	12.5412	0.6295	0.3810
FIHS	6.0415	10.1037	0.6099	0.4834
MSVD	4.7713	7.9795	0.6877	0.4059

Table 2 Quality metrics calculated for saras image

Method	RMSE	RASE	UIQI	Spatial data
Proposed method (EWT)	2.8562	10.9535	0.9896	0.9122
DWT	4.2559	16.3210	0.2218	0.8271
FIHS	3.8113	14.6163	0.1296	0.8486
MSVD	3.0537	11.7109	0.3109	0.8215

Table 3 Quality metrics calculated for book image

Method	RMSE	RASE	UIQI	Spatial data
Proposed method (EWT)	7.0463	10.3420	0.9520	0.6999
DWT	8.8769	13.0287	0.6688	0.6673
FIHS	9.5364	13.9967	0.5906	0.5981
MSVD	7.6581	11.2400	0.7084	0.4944

Table 4 Quality metrics calculated for toy image

Method	RMSE	RASE	UIQI	Spatial data
Proposed method (EWT)	5.0368	5.2300	0.9955	0.8512
DWT	7.0205	7.2898	0.8542	0.7246
FIHS	6.7154	6.9730	0.7679	0.7461
MSVD	5.3956	5.6026	0.8386	0.7105

Table 5 Quality metrics calculated for radiolarian image

Method	RMSE	RASE	UIQI	Spatial data
Proposed method (EWT)	2.3533	8.9411	0.9885	0.9921
DWT	4.2484	16.1417	0.6171	0.8709
FIHS	3.1427	11.9404	0.6634	0.9865
MSVD	2.4035	9.1320	0.7354	0.9570

5 Conclusion

In the present work, a new fusion method based on empirical wavelet transform is proposed. Experiments are done with standard multifocus image datasets. The effectiveness of the proposed fusion method is evaluated with the well-known quality metrics. Experimental analysis proves that the fusion method based on EWT outperforms the existing techniques.

References

1. Cyril Prasanna Raj P, Venkateshappa SBS (2014) High speed low power DWT architecture for image fusion. In: International conference on electrical, electronics and communications-ICEEC, pp 156–158
2. Abass HK (2013) A study of digital image fusion techniques based on contrast and correlation measures, Phd Thesis. Al-Mustansiriyah University
3. Rani K, Sharma R (2013) Study of different image fusion algorithm. Int J Emerg Technol Adv Eng 3(5):288–291
4. Strait M, Rahmani S, Merkurev D (2008) Evaluation of pan-sharpening methods. UCLA Department of Mathematics
5. Patel R, Rajput M, Parekh P (2015) Comparative study on multi-focus image fusion techniques in dynamic scene. Int J Comput Appl
6. Rahmani S, Strait M, Merkurjev D, Moeller M, Wittman T (2010) An adaptive IHS pan-sharpening method. Geosci Remote Sens Lett IEEE 7(4):746–750
7. Huang SG (2010) Wavelet for image fusion. National Taiwan University, Graduate Institute of Communication Engineering and Department of Electrical Engineering
8. Naidu VPS (2011) Image fusion technique using multi-resolution singular value decomposition. Defence Sci J 61(5):479–484
9. Gilles J (2013) Empirical wavelet transform. IEEE Trans Signal Process 61(16):3999–4010
10. Gilles J, Tran G, Osher S (2014) 2D empirical transforms. wavelets, ridgelets, and curvelets revisited. SIAM J Imaging Sci 7(1):157–186
11. Sahu DK, Parsai MP (2012) Different image fusion techniques a critical review. Int J Modern Eng Res (IJMER) 2(5):4298–4301
12. Jagruti V Implementation of discrete wavelet transform based image fusion. IOSR J Electron Commun Eng (IOSR-JECE) 9(2):107–109
13. Wang Z, Bovik AC (2002) A universal image quality index. Signal Process Lett IEEE 9(3):81–84

Extraction of Photovoltaic (PV) Module's Parameters Using Only the Cell Characteristics for Accurate PV Modeling

A. Vijayakumari, A.T. Devarajan and N. Devarajan

Abstract This paper presents a method for extraction of the model parameters pertaining to a commercially available PV module, which can be used to develop an accurate PV model that can serve as a source for simulation studies of both grid-connected and stand-alone PV systems. All the physical parameters required for the PV model are determined only from the published V–I characteristics of the module and the standard cell equation. A curve fitting based extraction technique is proposed to obtain the parameters pertaining to a particular PV module available in the market. The extracted parameters are used and a simulation model of the PV module is developed in MATLAB/Simulink. The inputs to the model are irradiance, ambient temperature, number of series and parallel modules in an array as required in an application. A 115 W commercial module S115 is taken for verification, and its physical parameters are extracted and its simulation model is developed. The characteristics of the developed model with the extracted parameters, for several G and T_C are presented and compared with the published data of S115 and compliance confirmed. The model has been tested for sudden changes in irradiance and temperature using a maximum power point tracker.

Keywords Photovoltaic model · Physical parameters of PV module · PV array · Irradiance · Maximum power point tracking

A. Vijayakumari (✉) · A.T. Devarajan
Department of Electrical and Electronics Engineering, Amrita Vishwa Vidyapeetham University, Amrita Nagar, Coimbatore 641112, India
e-mail: a_vijayakumari@cb.amrita.edu

A.T. Devarajan
e-mail: at_devarajan@cb.amrita.edu

N. Devarajan
Department of Electrical and Electronics Engineering, Govt. College of Technology, Coimbatore 641013 India
e-mail: Profdevarajan@yahoo.com

L.P. Suresh and B.K. Panigrahi (eds.), *Proceedings of the International Conference on Soft Computing Systems*, Advances in Intelligent Systems and Computing 397, DOI 10.1007/978-81-322-2671-0_26

1 Introduction

Power generation using fossil fuels generate green house gas emission and they also are getting depleted fast. Time has come to utilize the abundantly available wind and solar energy extensively.

Photovoltaic (PV) cell converts the light energy into electric energy. Many cells are connected in series and parallel making modules. The DC power from the PV modules is to be conditioned, so as to make it usable efficiently by the practical loads. The resulting PV system can be of two types, the one based on the interconnection with the utility grid, i.e., grid-connected PV systems and the other stand-alone PV systems.

It is necessary to use an interfacing converter system between PV system and load/utility because the output of a PV module varies widely with the operating temperature and the irradiance level. For the design of the interfacing power conditioning schemes, the PV module's V–I characteristics is essential.

Various simulation models available in the literature [1–7] adapt various methods for obtaining the V–I relationship based on the Schockley equation representing the PN junction of the solar cell. In these models the physical parameters which depend on the semi conducting materials and the fabrication processes are assumed, but they are not generally available for accurate modeling.

This paper focuses on the modeling of a typical PV module, available in the market by extracting the physical parameters pertaining to that particular module from the manufacturers' data sheet. Thus the proposed model makes it possible to obtain the V–I characteristics of any commercially available PV module. This will enable the power electronic engineers and students to develop simulation model for the interfacing converters, for the PV system, and study the converters under all possible G and T_C conditions both in the steady state and transient conditions.

2 Model Description

In a solar cell, when photons of the incident radiation collide with the valence electrons of the silicon they will be absorbed, releasing electrons and holes into the crystal lattice. On open circuit a direct EMF or voltage appears across its terminals. When an external electrical circuit is connected to its terminal a direct electric current flows.

2.1 Equivalent Circuit of a Solar Cell

A solar cell is basically a current generator, whose output current and voltage depend on G and T_C. Figure 1 shows the equivalent circuit of a single solar cell.

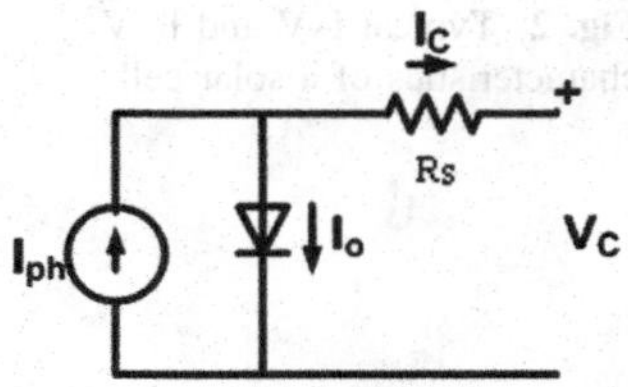

Fig. 1 Equivalent circuit of a solar cell

The V–I characteristics of the cell are determined by the diode. The output voltage of the cell is expressed [1] as,

$$V_C = \frac{AkT_C}{e} \ln\left(\frac{I_{\text{ph}} + I_O - I_C}{I_O}\right) - R_S I_C \tag{1}$$

where A = the diode ideality factor, e = electron charge, k = Boltzmann constant, I_C = cell output current (A), I_{ph} = photocurrent (A), I_O = reverse saturation current (A), R_S = series resistance of the cell (Ω), T_C = cell operating temperature in K, V_C = cell output voltage (V).

2.2 *Development of the Cell Model*

There are four manufacture-dependent parameters present in Eq. (1), viz., I_{ph}, I_O, A, and R_S. The value of these parameters is unique for a module and depends on the type of semiconducting materials used, type, and amount of doping and the processing technology used. For an accurate modeling these parameters are to be known accurately. In the proposed work these parameters are estimated from the published V–I characteristics at standard test condition (STC) of the selected PV module as explained in Sect. 2.3.

The value of V_C is evaluated for any current I_C at STC from Eq. (1). Then correction in voltage ΔV_C due to change in irradiance, the consequential change in cell current ΔI_C, and temperature are introduced as explained in Sect. 2.4 to get cell voltage V_C at any irradiance to complete the model.

2.3 *Estimation of the Process-Dependent Parameters in Cell Equation*

The cell current under short circuit will be equal to I_{ph} because the diode will not be conducting under short circuit and the same can be obtained from Eq. (1). If the data sheet values of short circuit current I_{SC} and the number of parallel branches are known then the I_{ph} can be found by dividing the module SC current by the parallel

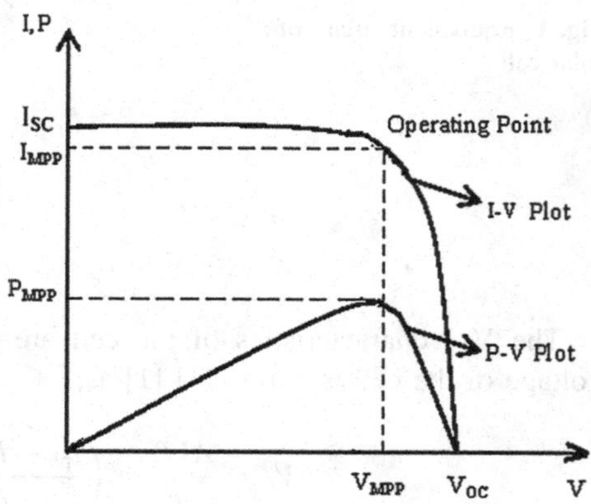

Fig. 2 Typical I–V and P–V characteristics of a solar cell

branches. The rest of the three unknown configuration parameters can be obtained through curve fitting method from the published I–V characteristics [8].

Figure 2 presents the I–V and P–V characteristics of a typical PV module. The curve fitting is based on selecting multiple points and trying to replicate or reproduce the same trend in the aimed model also. Three points are selected on the published characteristic curve at STC for the module to be modeled with the first point selected as MPP, the second and the third points selected as a point on the left of MPP and at the right of MPP. The coordinates of these points are (V_1, I_1) at P_1, (V_2, I_2) at P_2, (V_3, I_3) at P_3 are obtained from the published V–I characteristics. These points are converted to single cell values as by dividing the voltages V_1, V_2, and V_3 by the number of series cells. Let they be (V_{C1}, I_{C1}) at P_{C1}, (V_{C2}, I_{C2}) at P_{C2}, (V_{C3}, I_{C3}) at P_{C3}. These values of voltages are substituted in Eq. (1) and are solved for the three unknown configuration parameters I_O, A, and R_s.

2.4 *Account of Variations in V–I Characteristics Due to* G *and* T_C

The change in cell voltage due to change in temperature is expressed as [8],

$$\Delta V_C = -\alpha_{VOC}\Delta T_C - R_S \Delta I_C \tag{2}$$

where α_{VOC} is the temperature coefficient of voltage from PV module data sheet. The values of ΔI_C and ΔT_C can be obtained as,

$$\Delta I_C = \beta_{I_{SC}}\left(\frac{G}{G_{STC}}\right)\Delta T_C + \left(\frac{G}{G_{STC}} - 1\right) I_{SC,STC} \tag{3}$$

$$\Delta T_C = T_C - T_{\text{STC}} \tag{4}$$

where β_{ISC} is the temperature coefficient of current from PV module data sheet and T_C is the cell temperature at any irradiance and T_{STC} is the cell temperature at STC. The cells operate at a higher temperature than the ambient [9] and the cell temperature at any G is found as,

$$T_C = T_A + \left(\frac{T_{\text{NOCT}} - 20}{G_{\text{NOCT}}}\right) G \tag{5}$$

where T_A is the ambient temperature, T_{NOCT} is the normal operating T_C and G_{NOCT} is G at NOCT available from the PV module data sheet. Finally, the cell output voltage and current for at a required G and T_C are found as,

$$V_C = V_{C@\text{STC}} + \Delta V_C \tag{6}$$

$$I_C = I_{C@\text{STC}} + \Delta I_C \tag{7}$$

The cell model is implemented using Eqs. (1)–(7) and gives the cell voltage for a given I_C at any T_A and G.

3 Development of Simulink Model for the Array

PV Array model is developed in Simulink based on Sects. 2.3 and 2.4. The various stages of the array model are shown in Figs. 3, 4 and 5, wherein the inputs are T_{STC}, $I_{\text{sc,STC}}$, T_{A} and G. The array model can have N_{pa} and N_{sa} number of parallel and series modules respectively. The estimated unknown parameters A, I_O, R_S are supplied as constant values for the model. The output voltage obtained for a single cell is altered to result module output voltage. The model can give output voltage at any G and T_A with N_S, number of modules in series.

A model for implementing Eq. (1) is shown in Fig. 3. Implementation of Eqs. (2), (3), and (4) are shown in Fig. 4. The model gives the change in module current ΔI_C and module voltage ΔV_C as a function of T_A and G making use of the values G_{STC}, $I_{\text{sc,STC}}$, and $T_{C,\text{STC}}$. In Fig. 4 *DIm* and *DVm* refers the change in current and change in voltage of the module, as the $I_{\text{SC,STC}}$ used is for the module as found from the data sheet. The array voltage V_A is given by Eq. (8)

$$V_A = (V_C \times N_{\text{sm}} \times N_{\text{sa}}) + DVm \tag{8}$$

where N_{sm} is the number of series cells in a module and N_{sa} is number of modules forming the array. To obtain the output voltage from the model, a cell current I_C need to be given as input. For any given load, I_C is obtained by dividing the array

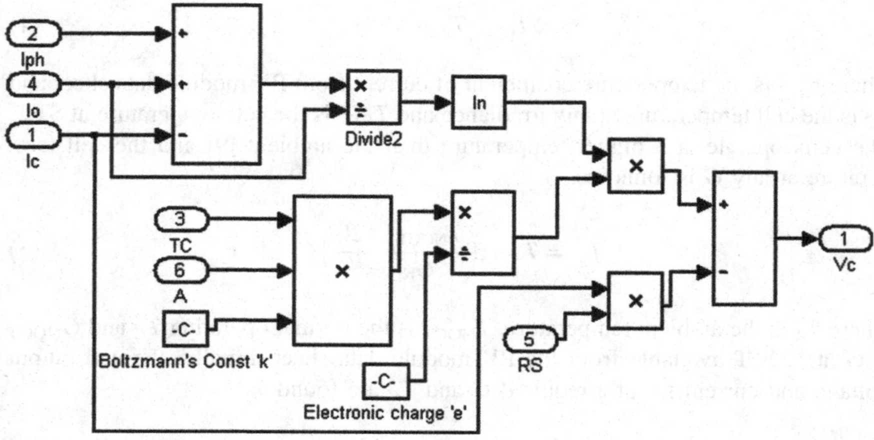

Fig. 3 Model of cell equation

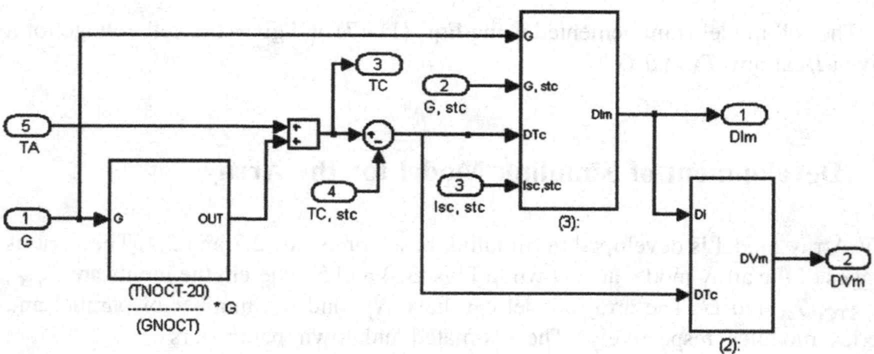

Fig. 4 Model to include the effect of temperature and irradiance

current by the product $N_{sm} \times N_{sa}$. This is included in the modeling stage of Fig. 5, where the required load is to be connected between points 1 and 2 as shown.

4 Model Verification

To verify the applicability of the general array model, Shell Solar S115 module is taken as a specific case and the array developed is applied to it. As an example, data [10] of Shell Solar's commercial module S115 is used to determine the cell model. The published data for S115 at STC are given in Table 1 also it contains a single string of 54 series connected cells [10].

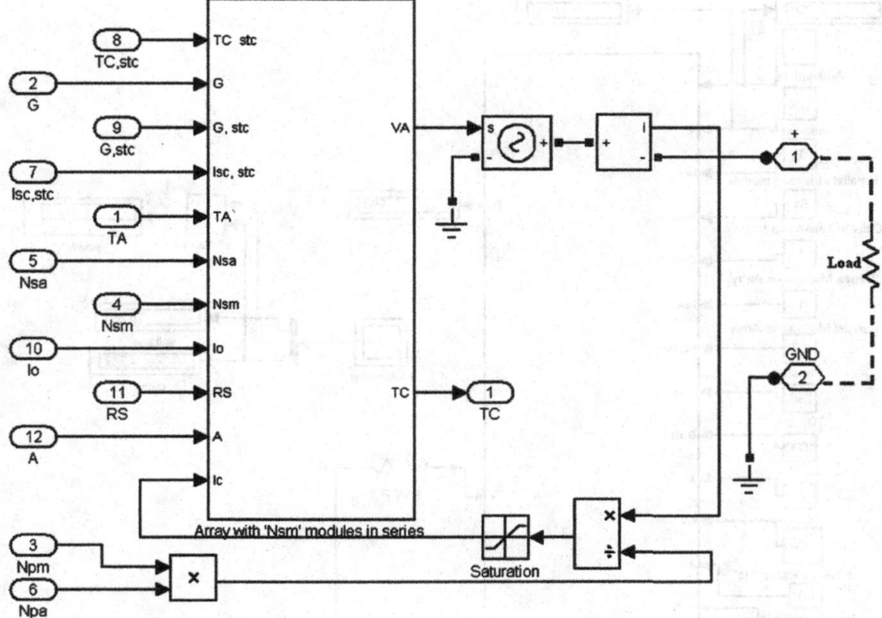

Fig. 5 Model of an array to get V_C for any load

Table 1 Typical electrical characteristics of shell S115 PV module

Parameter	Value
Peak power* P_{MPP}	115 W
Peak power voltage V_{MPP}	26.8 V
Open circuit voltage V_{OC}	32.8 V
Short circuit current I_{SC}	4.7 A
Coefficient of voltage α_{Voc}	–115 mV/°C
Coefficient of current α_{Isc}	+2 mA/°C

The point corresponding to MPP condition is obtained from Table 1 and other two points are taken from the V–I characteristics from datasheet [10]. The points are P_1 (26.80 V, 4.29 A), P_2 (25.74 V, 4.43 A) and P_3 (27.93 V, 4.06 A). Substituting the coordinates of P_1, P_2, and P_3 suitably scaled down to a single cell value, in Eq. (1) and eliminating A and R_S, a nonlinear equation in terms of I_O is obtained as Eq. (9)

$$3.054\ln\left[\frac{0.64}{I_o}+1\right]-7.086568\ln\left[\frac{0.41}{I_o}+1\right]+4.06\ln\left[\frac{0.27}{I_o}+1\right]=0 \qquad (9)$$

Equation (9) is solved using Newton–Raphson method for the value of I_O when the equation becomes zero. The solution of Eq. (9) gives the value of I_O as

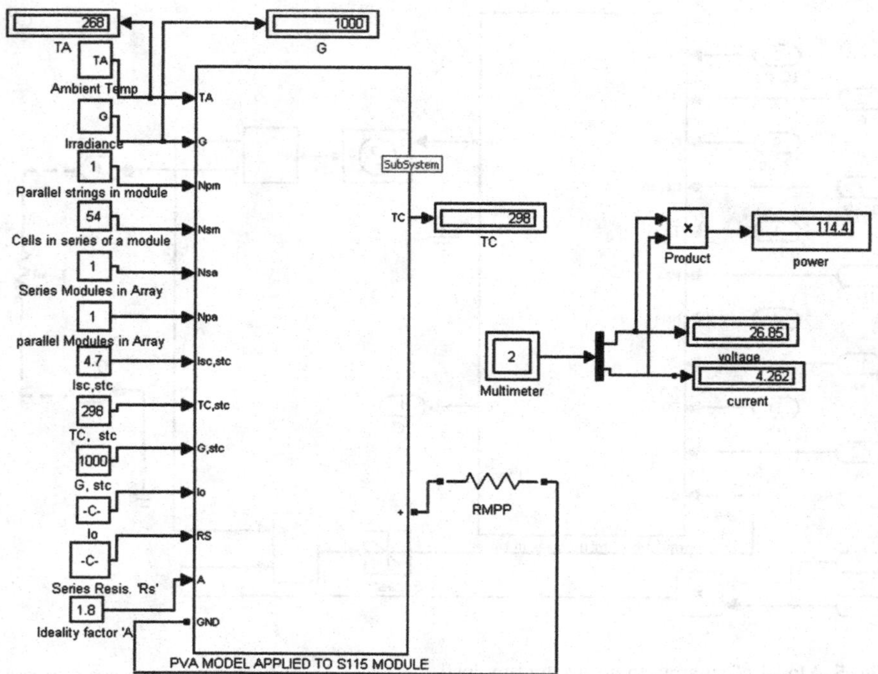

Fig. 6 Model of Shell 115 PV array under open circuit condition

9.25×10^{-6} A. Using this value of I_O the other two unknown parameters are found as $A = 1.8$, $R_S = 0.000192\ \Omega$.

The choice $N_{sm} = 54$ and $N_{sa} = N_{pa} = N_{pm} = 1$ reduces the model of array to the model of S115 module. Figure 8 shows the PV model applied to S115 commercial module.

The displayed values in Fig. 6 are obtained by simulation corresponding to the characteristics of S115 at MPP. The data sheet MPPT values are given at T_C of 25 ° C corresponding to a T_A of−5 °C.

The V–I and P–V characteristics obtained with the developed model are given in Fig. 7a, b, respectively, for varying G and constant T_C. The same sets of characteristics are presented for varying T_C and constant G conditions in Fig. 7c, d.

On comparison with data sheet it is seen that the steady state characteristics of the developed model in Fig. 7 are in close agreement with the published data for S115 at all irradiance levels and temperatures.

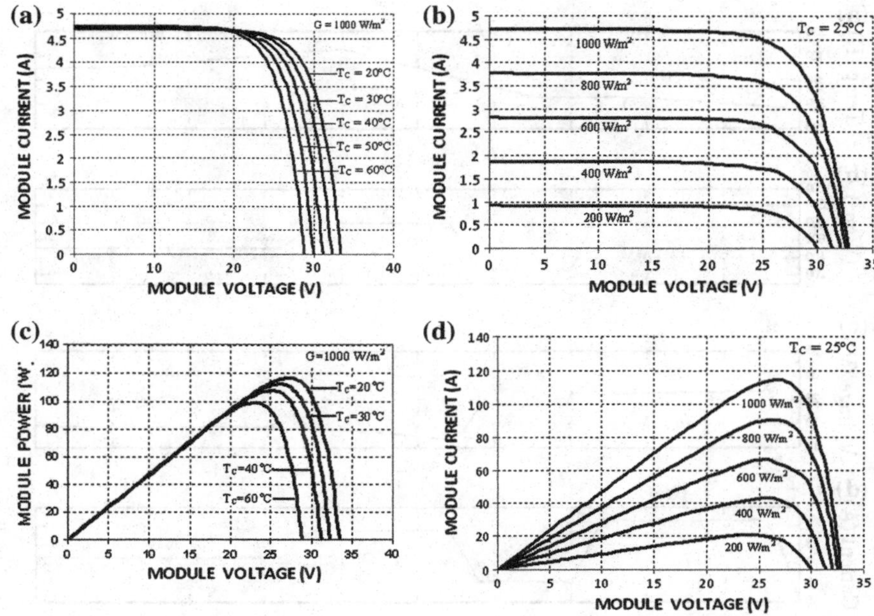

Fig. 7 V–I and P–V characteristics obtained for S115 by simulation **a** V–I at constant T_C and different G, **b** V–I constant G and different T_C, **c** P–V at constant G and different T_C, **d** P–V at constant T_C and different G

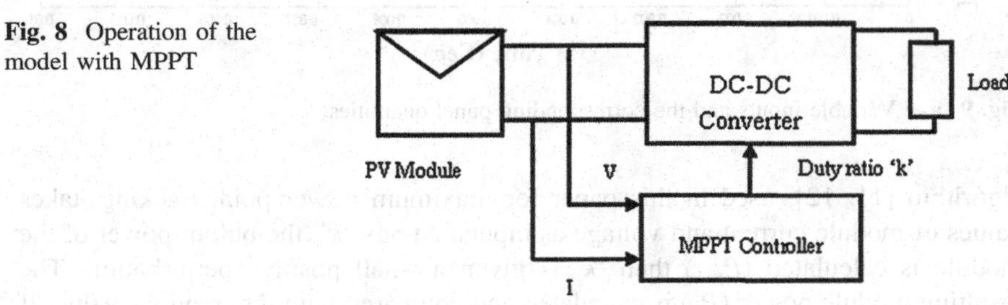

Fig. 8 Operation of the model with MPPT

5 Operation of the Model at MPP

To test the model under dynamic input conditions a DC–DC converter is used with an MPPT controller. Figure 8 shows the block diagram of the MPPT scheme. The time response of the system for rapidly changing irradiance and temperature is tested with variable input signals.

A fixed load is made to draw different powers from the panel by varying the duty ratio (k) of the DC–DC converter. MPPT controller sets the duty ratio 'k' at value so that maximum power is drawn from the panel. The perturb and observe (P&O)

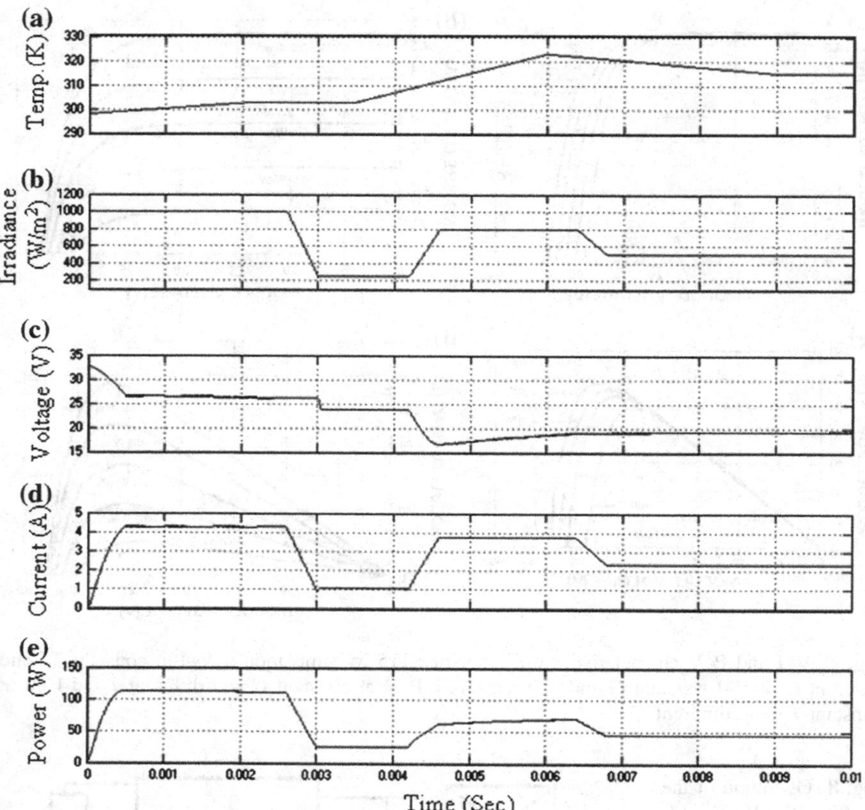

Fig. 9 **a–e** Variable inputs and the corresponding panel quantities

algorithm [11, 12], used in this paper for maximum power point tracking, takes values of module current and voltage as inputs. At any '*k*', the output power of the module is calculated ($P_{n\text{-}1}$) then '*k*' is given a small positive perturbation. The resulting module power (P_n) is calculated and compared with the previous value. If $P_n > P_{n-1}$ then '*k*' is given another positive perturbation and the P&O process is continued till P_{MPP} is reached. At any time if $P_n < P_{n-1}$, a negative perturbation is given to '*k*'.

5.1 Simulation Results for Dynamic Input Conditions

At 1000 W/m^2 and 25 °C, the V_{MPP}, I_{MPP}, and P_{MPP} are found from the developed model with MPPT are 26.65 V, 4.296 A, and 114.5 W, respectively. They compare well with the data sheet values at MPP, viz., 26.8 V, 4.29 A, 115 W. The operation

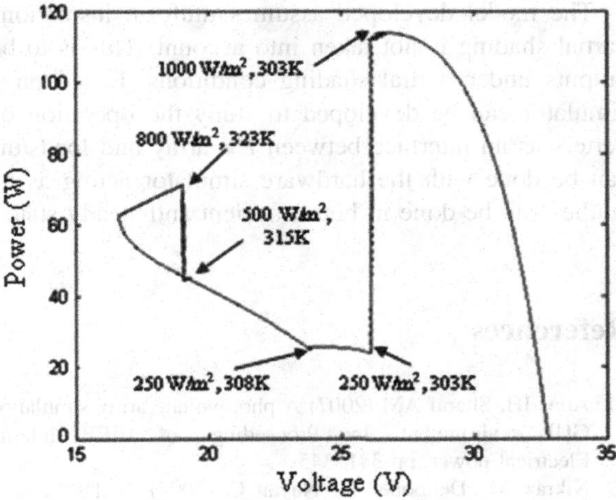

Fig. 10 P–V trajectory while tracking

of the MPPT is verified for sudden variations in irradiance and temperature, by applying a variable irradiance and temperature using signal builder blocks. The variable input signals and the resulting panel voltage, panel current, and the panel power are shown in Fig. 9a–e. The P–V trajectory obtained during the dynamic tracking with the variable inputs is shown in Fig. 10. The DC–DC converter in the system is operated at a switching frequency of 20 kHz and the values of L-C-R used in the converter are $L = 2$ mH, $C = 30$ μF and $R = 4\ \Omega$.

6 Conclusion

A MATLAB/Simulink model of a PV array, using only the published V–I characteristics of a PV module at Standard Test Conditions (STC), is developed. It outputs the V–I characteristics at any irradiance G and temperature T_C. The model uses the values of temperature coefficient of open circuit voltage α_{Voc} and short circuit current α_{Isc} from the data sheet of modules. A method of selection of operating points on the V–I characteristics for determination of the unknown parameters in the cell equation, so that the model can accurately give the characteristics for almost all insolation conditions has been suggested. None of the values required for the cell equation has been assumed as done often in the literature [1, 2]. The model developed for the array has been applied to a specific module (S115). The V–I characteristics obtained from the model developed for different load are found to be in good agreement with the published characteristics of S115 for all insolation. The model developed for the array has been tested for MPP tracking with a DC–DC converter using perturb and observe algorithm. Maximum power point tracking has been observed for step changes in the irradiance and temperature and the results are presented.

The model developed assumes uniform insolation for all cells. The effect of partial shading is not taken into account. This is to be provided to obtain correct outputs under partial shading conditions. Based on this model, a hardware PV simulator can be developed to study the operation of the power electronic converters as an interface between PV array and loads/utility. Hardware experiments can be done with the hardware simulator acting as a PV source, and the system studies can be done in both transient and steady state conditions.

References

1. Atlas IH, Sharaf AM (2007) A photovoltaic array simulation model for MATLAB-Simulink GUI environment. In: Proceedings of IEEE International conference on Clean Electrical power, pp 341–345
2. Nikraz M, Dehbonei H, Nayar C (2004) A DSP-controlled photovoltaic system with maximum power point tracking. Curtin University of Technology
3. Shanthi T, Ammasai Gounden N (2007). Power electronic interface for grid-connected PV array using boost converter and line-commutated inverter with MPPT. In: ICIAS 2007. international conference on intelligent and advanced systems, 2007, pp 882–886. IEEE
4. Ropp ME, Gonzalez S (2009) Development of a MATLAB/Simulink model of a single-phase grid-connected photovoltaic system. IEEE Trans Energy Conv 24(1):195–202
5. Esram T, Chapman PL (2007) Comparison of Photovoltaic array maximum power point tracking techniques. IEEE Trans Energy Convers 22(2):439–449
6. Tsai HL, Huan Liang (2010) Insolation oriented model of photovoltaic module using Matlab/Simulink. Sol Energy 84:1318–1326
7. PonVenkatesh R, Edward Rajan S (2011) Investigation of cloud less solar radiation with PV module employing Matlab/Simulink". Sol Energy 85:1727–1734
8. Vijayakumari A, Devarajan AT, Devarajan N (2012) Design and development of a model-based hardware simulator for photovoltaic array. Int J Electr Power Energy Syst 43(1):40–46
9. Messenger RA, Venture J (2003) Photovoltaic systems engineering, 2nd edn. CRC Press, New York
10. Shell Solar S115 Photovoltaic Solar Module, SAP ref: 400353
11. Femia N, Petrone G, Spagnuolo G, Vitelli M (2005) Optimization of perturb and observe maximum power point tracking method. IEEE Trans Power Electron 20(4):963–973
12. Femia N, Petrone G, Spagnuolo G, Vitelli M (2005) Optimization of perturb and observe maximum power point tracking method. IEEE Trans Power Electron 20(4):963–973

Effect of Nanoparticles on the Dielectric Strength and PD Resistance of Epoxy Nanocomposites

D. Kavitha, T.K. Sindhu and T.N.P. Nambiar

Abstract The dielectric strength and partial discharge resistance of epoxy material are enhanced by the addition of nanomaterial as fillers. The enhancement depends on the type of nanofiller and percentage loading. Epoxy nanocomposites with a good dispersion of nanoparticles in the base epoxy were prepared and experiments were conducted to measure the dielectric strength and partial discharge resistance with PDIV. In addition, the influence of water molecule on dielectric strength and PD resistance in base epoxy and epoxy nanocomposites were studied. This paper deals with the experimental analysis of dielectric strength and PD resistance of the nanocomposites with four types of insulating nanofillers, viz., $CaCO_3$, TiO_2, Al_2O_3, and layered silicate are analyzed at different filler concentration by weight. Results are given for both dry and wet (water absorbed) samples.

Keywords Epoxy nanocomposites · Dielectric strength · Partial discharge (PD) · Partial discharge inception voltage (PDIV)

1 Introduction

Epoxy resin is widely used for insulation of high voltage switchgear, bushings, cables, and transformers. Epoxy has found its widespread use of its good thermal characteristics, excellent adhesion, excellent antipollution performance, suitable for high polluted area free of cleaning, economical maintenance and suitable for difficult maintenance area. It is light weight, easy, and economical to transport and install. Epoxy insulators are made with epoxy resin because it is self-curing,

D. Kavitha (✉) · T.N.P. Nambiar
Department of Electrical and Electronics Engineering,
Amrita Vishwa Vidyapeetham University, Coimbatore 641 112, Tamil Nadu, India
e-mail: Kavi_viswa15@yahoo.co.in

T.K. Sindhu
Department of Electrical Engineering, National Institute of Technology, Calicut, India

L.P. Suresh and B.K. Panigrahi (eds.), *Proceedings of the International Conference on Soft Computing Systems*, Advances in Intelligent Systems and Computing 397, DOI 10.1007/978-81-322-2671-0_27

non-weathering and resistant to repetitive heating. Thus, epoxy has perfect electrical and mechanical properties for the manufacturing of bushings and insulators [1, 2]. Polymer nanocomposites have become popular and have been attracting attention of researchers for electrical insulation in high voltage applications [3].

In the study of AC breakdown characteristics of epoxy nanocomposites by Preetha et al. [4], it was shown that dielectric strength of unfilled epoxy was lower than epoxy nanocomposites. It was observed that there was a certain value of threshold weight percentage of nanofiller above which the dielectric strength decreases and the nanofiller has no considerable effect on the dielectric strength. In a similar research carried out by Sarathi et al. [5], it was shown that the breakdown strength of epoxy nanoclay composite increased to a great amount when compared to unfilled base epoxy. Also it was concluded that breakdown voltage increased with increase in nanoclay content up to 5 wt% loading, under AC and DC voltages. Another study done by Imai et al. [6] showed that epoxy resin containing nanoscale silica particles showed superior partial discharge resistance and breakdown strength when compared to unfilled base epoxy [7]. There are also a lot of studies which contradict the advantages of epoxy nanocomposites [8, 9]. According to IEC (International Electrotechnical Commission) Standard 60270, partial discharge is a localized electrical discharge that only partially bridges the insulation between conductors and which may or may not occur adjacent to a conductor [10]. It is observed that with the increase in applied voltage the PD activity also increases [11–13]. Once the PD is initialized in the insulation of high voltage power equipment, it continues for a long time if it is not taken care of and finally dielectric properties of such materials degrade. Because of the above reason PD detection and measurement is necessary [14].

Kozako et al. performed experiments to study the partial discharge resistance of polyamide layered silicate nanocomposites and observed that with 5 wt% fillers, the nanocomposite had better resistance to discharge when compared with the unfilled and other wt% [15]. Tanaka et al. [16] also studied an improvement in the partial discharge resistance in the epoxy clay nanocomposites as compared to the unfilled epoxy. In all these studies the influence of water absorption was not taken up. A detail study of water absorption by epoxy nanocomposites were taken up in this paper [17]. The objective of this paper is to characterize the dielectric strength and PD resistance of epoxy nanocomposites both dry and wet samples based on the filler type and percentage of filler loading. The AC breakdown experiments were performed on epoxy nanocomposites and studied the effect of breakdown strength on the percentage of filler loading, the filler type and the filler shape (nanomer is a layered silicate whereas the other fillers are spherical in shape). With the help of PD measurement system obtained PDIV and studied the PD distribution on both wet and dry samples.

2 Sample Preparation and Methodology

Bisphenol A epoxy resin along with hardener is used for present study. Four different types of nanofillers like TiO_2, Al_2O_3, $CaCO_3$, and layered silicate (Nanomer) were weighed according to the weight percentage of sample to be prepared and it was mixed with the calculated amount of epoxy resin [18]. The mixture was ultra sonicated for 1 h so that the nanofiller uniformly disperses in the epoxy resin. The calculated amount of hardener was added and it was manually stirred for 5 min. This epoxy nanocomposite solution was poured into the mold of dimension 40 mm × 40 mm × 2.5 mm. This was allowed to dry for 20 h in room temperature and was kept in a hot air oven for 1 h at 60 °C. The sample of epoxy nanocomposite was cured.

2.1 SEM Analysis of Samples Under Study

The dispersion of the nanofillers and also inter particle distance between nanofillers in the base epoxy was analyzed using a scanning electron microscope (SEM). Samples of approximately 1 mm × 1 mm × 1 mm are prepared by cutting epoxy nanocomposite at different locations and analyzed for microscopy. The epoxy nanocomposites of alumina (Al_2O_3) of 5 wt% and nanoclay of 5 wt% are studied under SEM. In Al_2O_3, each nanofiller is of spherical shape and has a diameter of 50 nm. The SEM studies of same are shown in Fig. 1. The voids present in the sample can also be seen. Nanomers are nanofillers of layered mineral silicates having a diameter of 40 nm. The layered structure can be seen in the SEM analysis in Fig. 2.

Fig. 1 SEM analysis of Al_2O_3 of 5 wt%

Fig. 2 SEM analysis of nanomer of 5 wt%

3 AC Breakdown Studies

3.1 Experimental Setup

The AC breakdown studies were performed on unfilled epoxy and epoxy nanocomposites under both dry and wet (water-absorbed samples) using a high voltage test kit as per ASTM D 149 [19]. The experiment setup and electrode arrangement for dielectric strength studies is shown in Fig. 3. All tests were carried out at room temperature. The samples were placed between the rods to plane electrodes configuration of the kit filled with transformer oil. The top electrode used for the experiment had an edge radius of 3.2 mm and ground electors is plane one. The kit was connected to supply and then the voltage was slowly applied to the samples step by step until puncture or breakdown occurred. Care was taken to ensure external flashover did not occur.

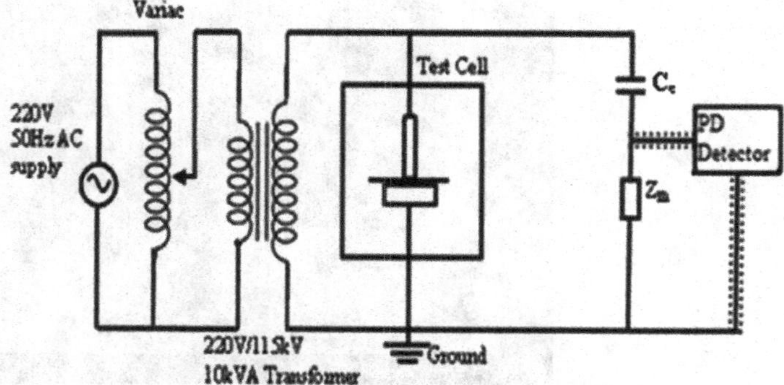

Fig. 3 Experimental setup for BDV

4 Result and Discussion

4.1 AC Dielectric Strength for Dry Samples

The experiment was performed on both plain epoxy and epoxy nanocomposites. The nanofillers used were alumina (Al_2O_3), titanium oxide (TiO_2), calcium carbonate ($CaCO_3$), and layered silicate (nanomer). There were four samples for each filler type but with different filler loadings of 2.5, 5, 7.5,10, and 15 %. All the tests were carried out at room temperature and all the samples were identical, that is, of same dimension of 40 mm × 40 mm and thickness of 2.5 mm. Six set of breakdown voltage of unfilled dry epoxy and epoxy nanocomposites were measured and breakdown strength was evaluated using the equation $E = V/d$, where V is the breakdown voltage and d is the thickness of the sample. The dielectric strength of four different dry epoxy nanocomposites are compared with unfilled epoxy is shown in Fig. 4.

4.2 Water Absorbed by Epoxy Nanocomposites

Epoxy has a tendency to absorb moisture when it is exposed to an environment with high water vapor content or humidity. This has adverse effects on the properties of epoxy, both mechanical as well as electrical properties. The epoxy material tends to swell up because the water gets accumulated in the cavities or voids of the material. Also some of the nanomaterial may leads or restrict to the absorption of water under various conditions. The dry weights of the epoxy nanocomposite samples were taken and the samples were immersed in water for 30 days. After 30 days, the wet weights were observed. From the wet weights, the percentage change in the weight were calculated by the formula given below [20],

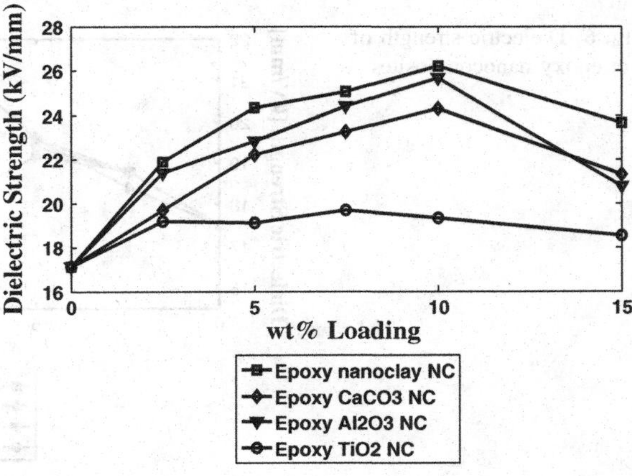

Fig. 4 Dielectric strength of dry epoxy nanocomposites

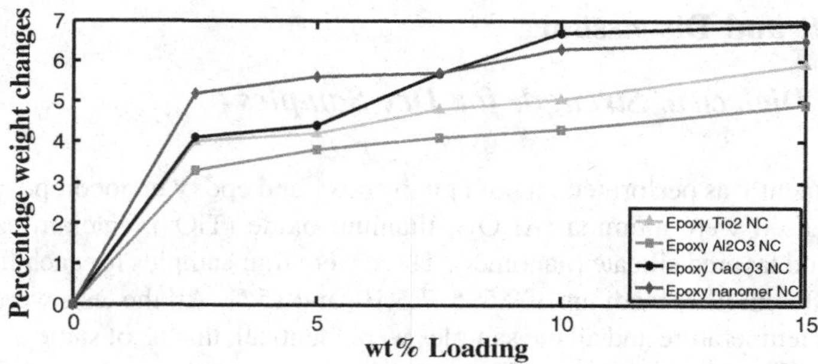

Fig. 5 Percentage weight changes for different epoxy nanocomposites

$$\text{Percentage Weight Change} = \frac{M_\text{w} - M_\text{d}}{M_\text{w}} \quad (1)$$

where M_w is the wet weight and M_d is the dry weight. Figure 5 shows the percentage weight changes for different epoxy nanocomposites.

4.2.1 AC Dielectric Strength of Water-Absorbed Samples

The samples that were kept in water for 30 days were used for testing. It was observed that there was significant decrease in breakdown voltage values of these samples. Six set of breakdown voltage of unfilled water-absorbed epoxy and epoxy nanocomposite samples were measured and breakdown strength was evaluated using the equation $E = V/d$, where V is the breakdown voltage and d is the thickness of the sample. There is an approximate of 8 % decrease in breakdown strength of plain epoxy when it is exposed to water. The dielectric strength of wet epoxy nanocomposites compared with unfilled is shown Fig. 6. It was observed that the

Fig. 6 Dielectric strength of wet epoxy nanocomposites

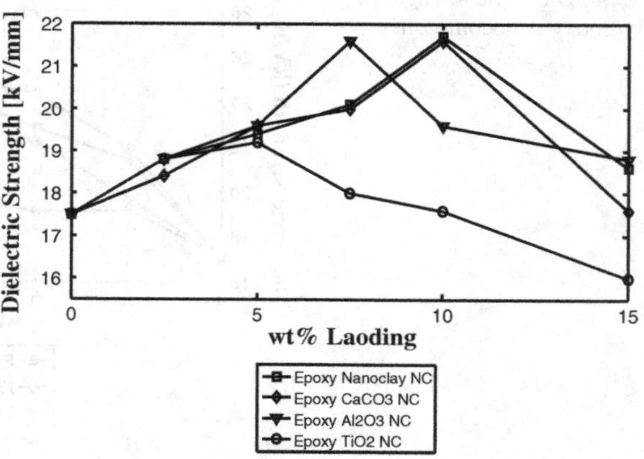

breakdown voltage and hence the breakdown strength of epoxy was increased by adding nanofillers. Among the different nanofillers, nanomer (layered silicate) showed highest resistance to breakdown on application of AC voltage. It could be attributed to the layered structure which prolonged the formation of conduction path and acted as obstacles for the flow of current. Also it can be inferred that nanomer has the least permittivity value compared with Al_2O_3 and $CaCO_3$; maybe this was another reason for epoxy nanomer composites to have good dielectric strength.

It was seen that as the filler concentration increased, the breakdown voltage increased up to a certain limit and which reduced later. This showed that there was an optimum weight percentage of nanofiller that had to be added. After this optimum weight percentage value, even if the filler loading is increased, the presence of nanofiller had no significant effect on the breakdown strength of the samples. Five percentages (5 wt%) by weight of nanomer acted as the optimum value and it exhibited the highest dielectric strength and there is a 5 % increase in dielectric strength when compared to the sample with 2.5 wt% nanomer loading. At higher filler loading, the nanofillers tend to interact with one another and it leads to coalescence of the particles [6]. These coalesced particles do not act as obstacles for the current thus allowing the current to flow easily leading to reduction in breakdown voltage. Epoxy Al_2O_3 nanocomposite exhibited high breakdown strength for higher filler loading (7.5 wt%). So, 7.5 weight percentage (7.5 wt%) of alumina acts as the optimum filler loading in this case. At this filler loading, there was a 7.2 % increase in breakdown strength when compared to unfilled epoxy. Thus nanofillers enhanced the breakdown strength of epoxy, when subjected to humidity absorbed water. This reduced the performance of epoxy as an insulation material of electrical apparatus. Epoxy Al_2O_3 nanocomposite absorbed least amount of water. Water absorption deteriorated the dielectric strength of epoxy nanocomposites and unfilled epoxy. There was a significant decrease in dielectric strength of wet samples when compared to the dry samples. It was noted that, the dielectric strength decreased by 8 % for wet unfilled epoxy when compared to dry unfilled epoxy. Here again epoxy filled with 5 wt% nanomer and 7.5 wt% nano alumina showed the best performance.

5 PDIV Measurement on Dry and Wet Epoxy Nanocomposites

The measurement of partial discharges is a nondestructive test on electrical apparatus or equipment. When dry and wet epoxy nanocomposites are exposed to PD measurement, their internal surface will suffer to physical damage by the impact of charged particles on it. Due to discharge the reactive oxygen produced and further which leads oxidize the internal surface thus enhancing the degradation. To understand and monitoring PDIV the ϕ-q-n distribution of the PD pulse for dry and wet unfilled epoxy and epoxy nanocomposites are shown in Fig. 7a–f. It was observed that, there is a reduction in number and magnitude of pulse in both the cycle for dry unfilled and epoxy nanocomposites when compared with that of wet samples.

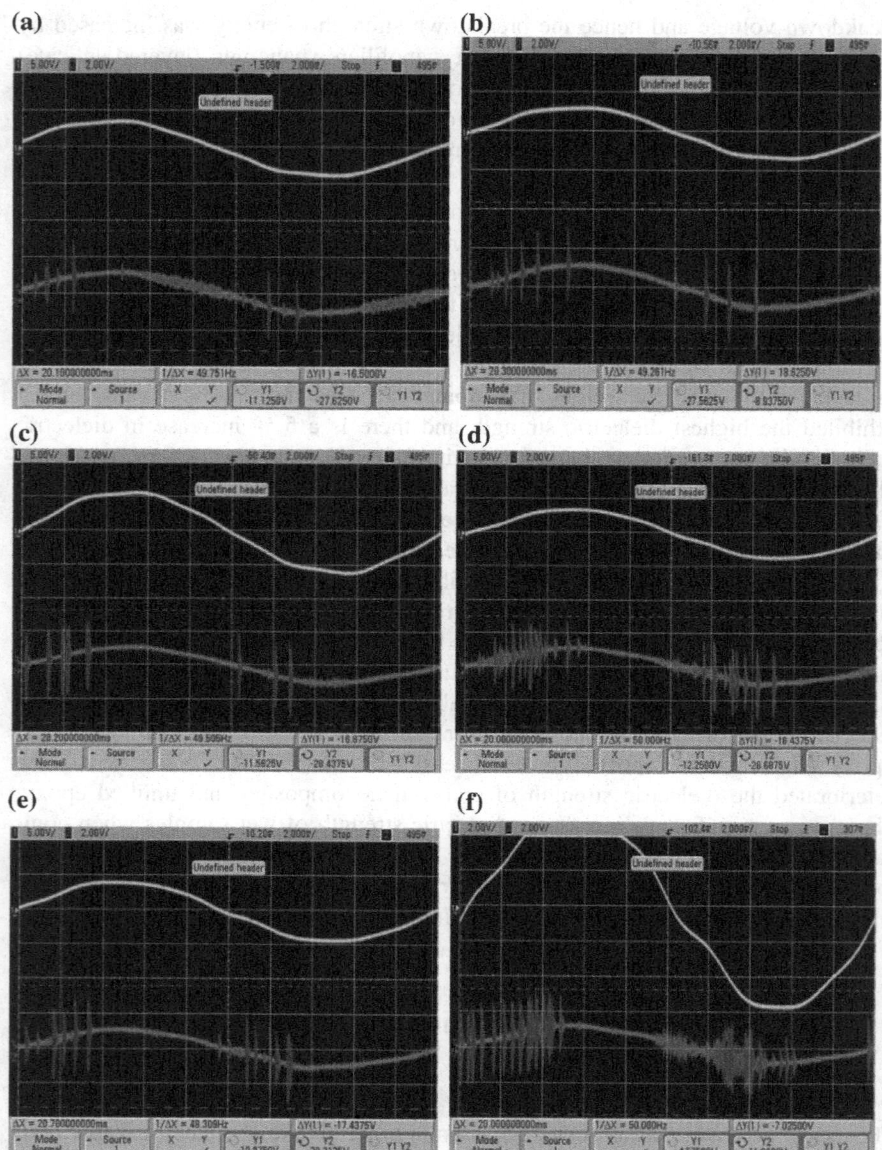

Fig. 7 Typical PD patterns **a** unfilled dry epoxy, **b** unfilled wet epoxy, **c** dry epoxy 5 wt% TiO_2 nanocomposite, **d** wet epoxy 5wt% TiO_2 nanocomposite, **e** dry epoxy 5 wt% $CaCO_3$ nanocomposite, **f** wet epoxy 5wt% $CaCO_3$ nanocomposite

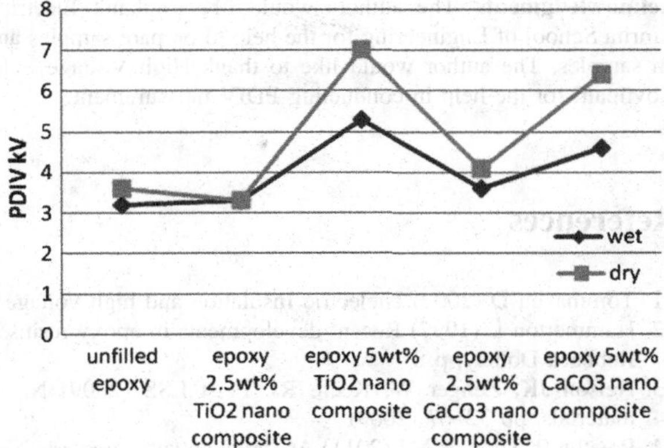

Fig. 8 Partial discharge inception voltage for wet and dry nanocomposites

It was found that partial discharge inception voltage (PDIV) is considerably increased with higher wt% loading of both wet and dry samples. The results are shown in Fig. 8 indicate that PDIV of wet samples was lower than the dry samples. According to percentage weight change due to water absorption in wet samples with dry samples, the PDIV was about 7 kV in the dry epoxy 5 wt% TiO_2 nanocomposite and about 5.3 kV in the wet sample of same loading. Same way the PDIV was about 6.4 kV in the dry epoxy 5 wt% $CaCO_3$ nanocomposite and about 4.6 kV in the wet sample of same loading. When compared epoxy 2.5 wt% of TiO_2 and $CaCO_3$ under wet and dry conditions the PDIV is higher in $CaCO_3$ than TiO_2 nanocomposites. This due to nano TiO_2 has high value of inherent permittivity than other nanofillers. It was found that the PDIV is mainly depended upon the wt% loading of nanoparticles, nanoparticle size and percentage change in weight due to water absorption.

6 Conclusions

The experiment analysis is made to detect the promising nanofiller and their optimum wt% of loading among four different nanofillers under dry and wet conditions. For this purpose epoxy nanocomposite is made with different wt% loading and subjected to AC breakdown test under dry and wet conditions. It was observed that the epoxy nanomer (layered silicate) with 5 wt% and epoxy nano Al_2O_3 with 7.5 wt % had maximum dielectric strength under dry and wet conditions. And for PD resistance, it was observed that the epoxy 2.5 wt% of TiO_2 had minimum PDIV and lesser number of pulses than that of epoxy 2.5 wt% $CaCO_3$ under dry and wet condition. Same way, epoxy 5 wt% of $CaCO_3$ had minimum PDIV and lesser number of pulses than that of epoxy 5 wt% TiO_2 under dry and wet conditions.

Acknowledgments The authors would like to thank Electrical workshop, EEE Department, Amrita School of Engineering for the help to prepare samples and conduct dielectric strength test on samples. The author would like to thank High Voltage Lab, National Engineering College, Kovilpatti for the help in conducting PDIV measurement.

References

1. Tommasini D (2009) Dielectric Insulation and high voltage issues, pp 16–25
2. Hammerton L (1997) Recent developments in epoxy resins. RAPRA Review Reports, 1997, Rebecca Dolbey, p 8
3. Nelson JK, Zenger W, Keefe RJ, Feist LSS (2009) Nano structured dielectic composite materials, pp 32–67 (2009)
4. Preetha P, Thomas MJ (2011) AC breakdown characteristics of epoxy nanocomposites. IEEE Trans Dielectr Electr Insul 18(5):1526–1534
5. Sarathi R, Sahu RK, Rajeshkumar P (2007) Understanding the electrical, thermal and mechanical properties of epoxy nanocomposites. Mater Sci Eng, A 445–446:567–578
6. Imai T, Sawa F, Ozaki T, Shimizu T, Kido R, Kozako M, Tanaka T (2005) Evaluation of insulation properties of epoxy resin with nano-scale silica particles. Electr Insul Mater 1:239–242
7. Singha S, Thomas MJ (2008) Dielectric properties of epoxy nanocomposites. IEEE Trans Dielectr Electr Insul 15(1):12–23
8. Shin JY, Park HD, Choi KJ, Lee KW, Lee JY, Hong JW (2009) Electrical properties of the epoxy nano-composites. Trans Electric Electr Mater 10(3)
9. Nelson JK (2003) Electrical insulation and dielectric phenomena. Annual Report, pp 19–22
10. Samat SS, Musirin I, Kusim AS (2012) The effect of supply voltage on partial discharge properties in solid dielectric. In: IEEE international power engineering and optimization conference (PEOCO2012), 2012
11. Fasil VK, Karmakar S (2012) Modeling and simulation based study for on-line detection of partial discharge of solid dielectric. In: IEEE 10th international conference on the properties and applications of dielectric materials, 2012
12. Sabat A, Karmakar S (2011) Simulation of partial discharge in high voltage power equipment. Int J Electric Eng Inform 3(2)
13. Kavitha D, Alex N, Nambiar TNP (2013) Classification and study on factors affecting partial discharge in cable insulation. J Electric Syst 9:346–354
14. Kozako M, Kido R, Imai T, Ozaki T, Shimizu T, Tanaka T (2005) Surface roughness change of Epoxy/TiO_2 nanocomposites due to partial discharges. Electric Insul Mater 3:661
15. Kozako M, Fuse N, Ohki Y, Okamoto T, Tanaka T (2004) Surface degradation of polyamide nanocomposites caused by partial discharges using IEC (b) electrodes. IEEE Trans Dielectr Electr Insul 11(5):833–839
16. Tanaka T, Tatsuya Y, Ohki Y, Mitsukazu O, Miyuki H, Imai T (2007) Frequency accelerated partial discharge resistance of epoxy/clay nanocomposite prepared by newly developed organic modification and solubilization methods. IEEE international conference on solid dielectrics, 2007, pp 337–340
17. Chen Z, Rowe SW (2008) The effect of water absorption on the dielectric properties of epoxy nanocomposites. vol 1, pp 235–239 (2008)
18. Hsueh HB, Chen C-Y (2003) Preparation and properties of LDHs/epoxy nanocomposites. Polymer 44:5275–5283
19. ASTM-D149-97a (reapproved 2004) Standard test method for dielectric breakdown voltage and dielectric strength of solid electrical insulating materials at commercial power frequencies
20. Zhao H, Robert Li KY (2008) Effect of water absorption on the mechanical and dielectric properties of nano-alumina filled epoxy nanocomposites. Appl Sci Manufact 39(4):602–611

Performance Analysis of Artificial Neural Network and Neuro-Fuzzy Controlled Shunt Hybrid Active Power Filter for Power Conditioning

Jarupula Somlal and M. Venu Gopala Rao

Abstract Harmonics are developed in the power systems at various stages with the increased role of power electronic converters. Harmonics reduces the quality of power systems results in instability and voltage distortion. Several filtering techniques with different controllers have been proposed earlier for reducing the harmonics, but accurate and fast controllers are needed. This paper presents different intelligent control techniques such as artificial neural network (ANN) and neuro-fuzzy controllers for shunt hybrid active power filter (SHAPF), based on feed forward-type (trained by a back propagation algorithm) ANN and mamdani-type neuro-fuzzy method for mitigating the harmonics in the distribution system. In SHAPF, the active power filters (APF) mainly uses the energy of the capacitor in order to maintain its DC-link bus voltage and thus reduces the time of the transient response when there is abrupt variation in the load. The suggested control techniques are usually appropriate for any type of other APF. The proposed control strategies for SHAPF have been constructed in MATLAB/SIMULINK environment. In this paper, simulation results of both the methods are presented, it is observed that there is a considerable reduction in harmonics with both controllers.

Keywords Feed forward ANN · Shunt hybrid active power filter · Total harmonic distortion (THD) · Mamdani neuro-fuzzy controller · Distribution system

J. Somlal (✉)
Department of EEE, K L University, Guntur 522502, Andhra Pradesh, India
e-mail: jarupulasomu@kluniversity.in

M. Venu Gopala Rao
Department of Electrical & Electronics Engineering,
Prasad V Potluri Siddhartha Institute of Technology, Kanuru, Vijayawada 520007, Andhra Pradesh, India
e-mail: venumannam@kluniversity.in

L.P. Suresh and B.K. Panigrahi (eds.), *Proceedings of the International Conference on Soft Computing Systems*, Advances in Intelligent Systems and Computing 397, DOI 10.1007/978-81-322-2671-0_28

1 Introduction

Power conditioning is one of the most important and major criteria in electrical industry. In this, nonlinear loads are known to be the major sources of power quality problems. This nonlinear load imposes not only current harmonics but also demands reactive power at the supply mains [1]. Traditionally, passive filtering is the simplest feasible technique for compensating the current harmonics and improving the reactive power, but it is suffering from many disadvantages such as harmonic current absorption, bulk in size, resonance problem, its performance depending on impedance of the system, number of passive filters to be installed depends on which harmonic current to be compensated [2]. All such problems can be effectively overcome by active power filters (APF), but the active filters are not cost effective for high power rating applications due to their large rating and high frequency switching requirement of the inverter [3].

All the problems which are mentioned above can be easily and effectively overcome by hybrid active power filter (HAPF), which is a device in which the passive and active filters are used together and both are connected in shunt with the electrical system [4]. This filter can successfully overcome the problem of a passive and active filter. It provides cost effective harmonic compensation, particularly for high power nonlinear loads [5]. In literature, several hybrid-filter configurations are proposed by researchers for compensating reactive power and current harmonics. One of the most common configuration is series HAPF, in which active filter is connected in series with the passive filter [6–9] and the other is shunt configuration [10, 11]. Both the configurations have certain drawbacks such as Voltage source Converter (VSC) has small voltage rating in the first configuration and VSC has small current rating in second configuration, but it should to be designed for the nominal voltage. An overview of all other configurations of hybrid filters can be observed in [12–15].

In literature, different control techniques such as instantaneous power (p-q) theory, synchronous reference frame (SRF) transformation, etc., are used for obtaining current harmonic components of the source from the fundamental components. A large number of modern methodologies have been developed, offering solutions to several challenging control problems in industry and manufacturing areas [16]. One is fuzzy logic, implemented for shunt active power filter to improve power quality by compensating reactive power and current harmonics. Instantaneous p-q theory is used with the inclusion of neural network filter for reference current generation and fuzzy logic controller for DC voltage control to improve the dynamic behavior of a shunt active power filter. This is suitable for both steady state and transient conditions of load [17, 18].

In this paper, artificial intelligence (AI) control methods have been used for APF, namely artificial neural networks, and neuro-fuzzy based control strategy. These are having a major impact on the power electronics applications. The main objective of designing the controller for the SHAPF is for effective controlling, to obtain the reliable control algorithms, and quicker response to create the output signals.

Neural-network-based controllers offer rapid dynamic response in order to maintain the stability of the converter system over a wide operating range and are counted as a novel instrument to design control circuits for power quality devices. Neuro-fuzzy based control strategy is used in order to improve the APF dynamics to minimize the harmonics for wide range of variations of load current under various conditions, it is simple and also capable of maintaining the compensated line currents balanced, irrespective of unbalancing in source voltages. Unlike their conventional counterparts, these intelligent controllers can learn, remember, and make conclusions. The present research work emphasis on the comparability of the transient time of DC voltage and THD factor based on the ANN and neuro-fuzzy.

This paper proceeds as follows. Section 2 gives an overview of SHAPF and estimation of compensating current. In Sect. 3, proposed control strategies for APF based on ANN and neuro-fuzzy are discussed. Results and discussions of proposed methods for SHAPF are explained in Sect. 4. The main conclusions of this paper are presented in Sect. 5.

2 Overview of SHAPF and Estimation of Compensating Current

Figure 1 shows the configuration of SHAPF (a combination of both passive power filter (PPF) and active power filter (APF)) is connected in shunt with a three-phase source feeding power to the nonlinear load (universal bridge) through a distribution

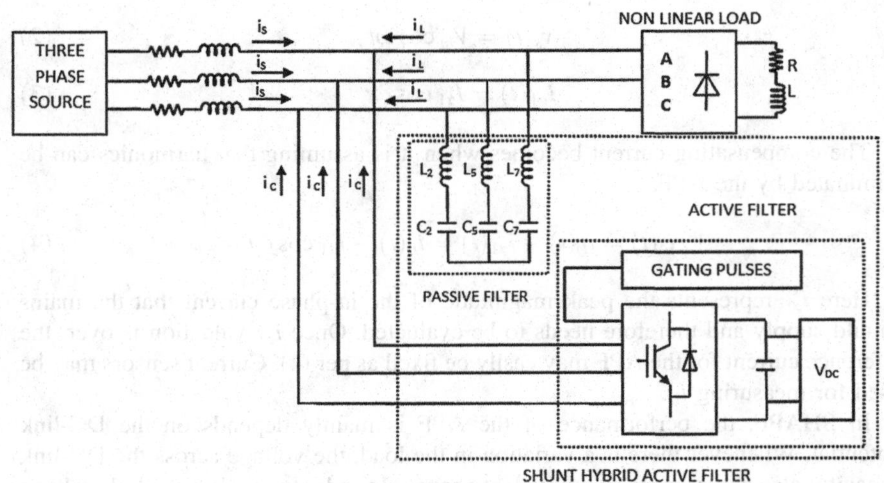

Fig. 1 Configuration of three-phase distribution system with SHAPF

system. The PPF is generally designed for specific order of harmonics for reducing the burden of APF. In SHAPF, the main function of PPF is to reduce the lower and higher order harmonics, whereas the other order harmonics are taken care of by the APF. In this work, the PPF is designed for compensating 2nd, 5th, and 7th order harmonics and are represented as C2, L2; C5, L5; and C7 and L7 in Fig. 1, while the APF is normally utilized to increase the working performance of PPF. The shunt APF takes a 3-Φ voltage source inverter (VSI) as the main circuit and it uses capacitor (C) as the voltage storage element on the DC side to maintain the DC bus voltage V_{DC} constant [17]. The VSI of APF utilizes two artificial intelligent control strategies, namely ANN and neuro-fuzzy control strategy. The main features of these controllers is both accounts for DC voltage control and THD, but the added advantage of neuro-fuzzy controller is that it provides excellent dynamic response in case of load current deviations. The proper operation of these two controllers results in the production of gate signals for three-phase VSI which in turn is responsible for production of compensating currents. Harmonic reduction and reactive power compensation can be done by injecting compensating currents through three-phase inverter.

A general formulation for the load current corresponding to Fig. 1 is

$$i_L(t) = i_{\alpha 1}(t) + i_{\beta 1}(t) + i_h(t) \tag{1}$$

where, $i_{\alpha 1}$ is in-phase component of the phase current at the fundamental frequency and $i_{\beta 1}$ is quadrature component of the phase current at the fundamental frequency. i_h includes all other harmonics.

The per-phase source voltage and the corresponding in-phase component of the load current may be conveyed as

$$v_s(t) = V_m \cos \omega t \tag{2}$$

$$I_{\alpha 1}(t) = I_{\alpha 1} \cos \omega t \tag{3}$$

The compensating current becomes when it is assuming that harmonics can be eliminated by the APF,

$$i_c(t) = i_L(t) - i_{\alpha 1}(t) = i_L(t) - I_{\alpha 1} \cos \omega t \tag{4}$$

Here $i_{\alpha 1}$ represents the peak magnitude of the in-phase current that the mains should supply and therefore needs to be evaluated. Once $i_{\alpha 1}$ valuation is over, the reference current for the APF may easily be fixed as per (4). Current sensors may be used for measuring i_{L}.

In SHAPF, the performance of the APF is mainly depends on the DC-link potential. Whenever there is a variation in the load, the voltage across the DC-link capacitor also experiences an equivalent change. In order to regulate the link voltage at a required value, a controller is utilized. An easy investigation of the dynamics of the DC-link voltage is first carried out. Parameters that regulate the dynamics are recognized, following which a procedure is improved to assessment the compensating current of the APF. In-phase (i.e., in phase with the source voltage) current i_{sa} is drawn by the capacitor to keep the DC bus voltage to a required magnitude. This is in addition to the compensating current i_c.

From equation of power balance

$$P_{\mathrm{dc}} = C_{\mathrm{dc}} V_{\mathrm{dc}} \frac{\mathrm{d}V_{\mathrm{dc}}}{\mathrm{d}t} \tag{5}$$

where, P_{dc} is the power required to keep the voltage V_{dc} across the DC link.

From equation of power balance

$$\begin{aligned} &\sum_{i=a,b,c} v_{si}(t) i_{s\alpha}(t) - \sum_{i=a,b,c} R_f\left(i_{s\alpha}^2(t) + i_{ci}^2(t)\right) - \frac{1}{2} \sum_{i=a,b,c} L_f \frac{\mathrm{d}}{\mathrm{d}t}\left(i_{s\alpha}^2(t) + i_{ci}^2(t)\right) \\ &\quad = i_{\mathrm{dc}}(t) v_{\mathrm{dc}}(t) = P_{\mathrm{dc}} \end{aligned} \tag{6}$$

where R_f and L_f represent the resistance and inductance of the inductor, respectively. In between the point of common coupling, the VSI is linked by the inductor of the inductance and resistance. Note that α delivers the system loss at the steady state and the capacitor charges/discharges during transient to maintain the DC-link voltage.

Considering scalar quantity for the power system (6), for a balanced 3-Φ system may be represented as

$$3v_s(t) i_{s\alpha}(t) - 3R_f\left(i_{s\alpha}^2(t) + i_c^2(t)\right) - \frac{3}{2} L_f \frac{\mathrm{d}}{\mathrm{d}t}\left(i_{s\alpha}^2(t) + i_c^2(t)\right) = i_{\mathrm{dc}}(t) v_{\mathrm{dc}}(t) = P_{\mathrm{dc}} \tag{7}$$

The new set of variables can be derived by applying small perturbations in each i_c, i_{sa}, V_{dc}, and V_s, around an operating point

$$i_c(t) = I_c + \Delta i_c \tag{8}$$

$$i_{s\alpha}(t) = I_{s\alpha} + \Delta i_{s\alpha} \tag{9}$$

$$v_{\mathrm{dc}}(t) = V_{\mathrm{dc}} + \Delta v_{\mathrm{dc}} \tag{10}$$

$$v_s(t) = V_s + \Delta v_s \tag{11}$$

where I_c, I_{sa}, and V_s are rms and V_{dc} is the DC value of the equivalent quantities at the operating level. Again, in steady state

$$3V_sI_{s\alpha} - 3R_f\left(I_{s\alpha}^2 + I_c^2\right) = 0 \tag{12}$$

Substituting (8)–(12) in (7), the equivalent equation is obtained

$$3(\Delta v_sI_{s\alpha} + V_s\Delta i_{s\alpha}) - 6R_f(I_{s\alpha}\Delta i_c) - 3L_f\left(I_{s\alpha}\frac{\mathrm{d}\Delta i_{s\alpha}}{\mathrm{d}t} + I_c\frac{\mathrm{d}\Delta i_c}{\mathrm{d}t}\right) = C_{\mathrm{dc}}V_{\mathrm{dc}}\frac{\mathrm{d}\Delta v_{\mathrm{dc}}}{\mathrm{d}t} \tag{13}$$

Transforming the variables to s-domain and after readjusting, (13) may be derived as

$$\begin{aligned}\Delta V_{\mathrm{dc}}(s) = {} & \frac{KG_s(s)G_1(s)G_2(s)}{1 + KG_s(s)G_1(s)G_2(s)}\Delta V_{\mathrm{dc}}^*(S) \\ & - \frac{G_2(s)G_3(s)}{1 + KG_s(s)G_1(s)G_2(s)}\Delta I_c(s) + \frac{G_2(s)G_4(s)}{1 + KG_s(s)G_2(s)G_4(s)}\Delta V_s(s)\end{aligned} \tag{14}$$

where K denotes the small-signal gain.

Equation (14) proves that ΔV_{dc}, ΔI_c, and ΔV_s effect on the ripples of DC-link voltage. In our present problem, the distortion in reference DC bus voltage and source voltage is not counted.

Therefore, (14) further modifies to

$$\Delta V_{\mathrm{dc}}(s) = \frac{G_2(s)G_3(s)}{1 + KG_s(s)G_1(s)G_2(s)}\Delta I_c(s) \tag{15}$$

Equation (15) researches the possibility of taking out an assessment of the compensating current from the change in V_{dc}.

3 Control Strategies Used in APF in MATLAB/SIMULINK Environment

3.1 *Artificial Neural Network (ANN) Controller Used in APF*

Instantaneous abc_to_dq0 transformation is used for reference current calculation. This transformation block (in Simulink) computes the direct-axis, quadrature-axis, and zero sequence quantities in a two-axis rotating reference frame for a three-phase sinusoidal signal. The Eqs. (16), (17), and (18) are used for reference current calculation.

$$I_d = \frac{2}{3}(I_a \sin \omega t) + I_b \sin\left(\omega t - \frac{2\pi}{3}\right) + I_c \sin\left(\omega t + \frac{2\pi}{3}\right) \quad (16)$$

$$I_q = \frac{2}{3}(I_a \cos \omega t) + I_b \cos\left(\omega t - \frac{2\pi}{3}\right) + I_c \cos\left(\omega t + \frac{2\pi}{3}\right) \quad (17)$$

$$I_0 = \frac{1}{3}(I_a + I_b + I_c) \quad (18)$$

where ω = rotation speed (rad/s) of the rotating frame.

The rapid spotting of the disturbance signal with high accuracy, fast processing of the reference signal, and high dynamic response of the controller are the prime prerequisites for desired compensation in case of APF. The conventional controller fails to achieve satisfactorily under parameter variations nonlinearity load disturbance, and so forth [19].

A Back propagation training algorithm based single layer feed forward network is used for optimizing the performance of the APF with ANN technique. This network consists of two layers with its corresponding neuron interconnections. For input layer to receive the inputs, two neurons are used. Hidden layer comprises of 21 neurons to which each of the processed input is fed. One neuron is used for output layer, whose output is to be calculated as reference current. The tan-sigmoidal function and pos-linear functions are given as activation functions to input layer and output layer, respectively, which are assigned for each of the layers in order to train them.

The large data of the DC-link voltage for 'n' and '$n - 1$' intervals from the conventional method are gathered and are stored in the MATLAB workspace and is used for training the ANN. The data which is stored in workspace is being retrieved using the following training algorithm. The neurons in the input and output layers is almost a fixed quantity to obtain the provided input. The accuracy of the ANN operation mostly depends on the number of hidden neurons.

$$y = \sum_{n=1}^{21} w_n \cdot i_n + b \quad (19)$$

3.2 *Algorithm Used in ANN Controller*

Step 1 Normalize the inputs and outputs with respect to their maximum values. It is shown that the neural networks work better if the inputs and outputs lie between 0 and 1. There are two inputs given by {P} 2 × 20 and one output {O} 1 × 20 in a normalized form.

Step 2 Enter the number of inputs for a fed network.

Step 3 Enter the number of layers.

Step 4 Create a new feed forward network with 'tansig and poslin' transfer functions.
Step 5 Train the network with a learning rate 0.02.
Step 6 Enter the number of epochs.
Step 7 Enter the goal.
Step 8 Train the network for given input and targeted output.
Step 9 Generate simulation of the given network with a command 'gensim'.

The neural network in each layer is created with the set number of neurons by using the algorithm given in Sect. 3.2. 500 iterations are done in each and every training session, and six similar validation checks are taken out in order to minimize the scope of error occurrence. The learning rate is a very important factor to be considered in the training process of ANN (change of interconnection weights). The learning rate should not be too small that the training becomes too delayed or it should not be too high to avoid the oscillations occur about the target values which leads to the training gets delayed. In this controller, neural network is trained at a learning rate of 0.02.

Figure 2 shows the block diagram representation of proposed ANN controller. Initially, the load currents, point of common coupling (PCC) voltages and DC bus voltage of APF are sensed in this controller. Maintaining of constant DC bus voltage is so important for effective compensation of harmonics and it can be effectively done by DC voltage loop in the controller. The difference between V_{dc} and reference value (400) is given as an input to ANN controller, whereas the output of ANN controller is responsible for mitigating the harmonics in the system. The main function of PLL (phase-locked loop) is to synchronize the positive-sequence component of the current. The output of PLL (angle $\theta = \omega t$) is used to compute the direct-axis and quadrature-axis components of the three-phase currents. The output signals of ANN controller and direct-axis and quadrature-axis component of currents from d-q-o transformation block are compared by using the error detector and it produces d-q-o component of reference signal. The signals from d-q-o frame are again converted to a-b-c frame and are compared with a filter

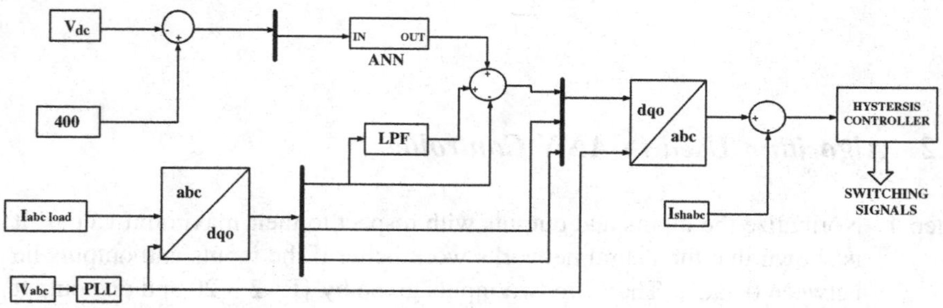

Fig. 2 Block diagram representation of ANN controller

current (Ishabc). It results in generation of reference compensation current and is given as input to the hysteresis band controller, which results in production of triggering signals required for switching ON/OFF of IGBT's of APF. The main objective of hysteresis band controller is to control the compensation currents by forcing it to follow the reference ones. The switching strategies of the three-phase inverter will keep the currents into the hysteresis band. The real load currents are sensed and their non-active components are compared with the reference compensation currents. The hysteresis comparator output signals are used to turn on the inverter power switches.

3.3 Neuro-Fuzzy Controller

Neural fuzzy systems are characterized by the use of neural network to provide fuzzy systems with a kind of tuning by the specified fuzzy set rules, but without altering their functionality. In the training process, a neural network adjusts its weights in order to minimize the mean square error between the output of the network and the desired output. Fuzzy controller is connected to the output of the ANN to improve the output further. Figure 3 shows the internal architecture used in neuro-fuzzy controller.

The fuzzy controller represent the parameters of the fuzzification function, fuzzy word membership function, fuzzy rule confidences and defuzzification function, respectively. In this sense, the training of this neural network results in automatically adjusting the parameters of a fuzzy system and determining their optimal values. Neural networks identify patterns and adjust themselves to cope with

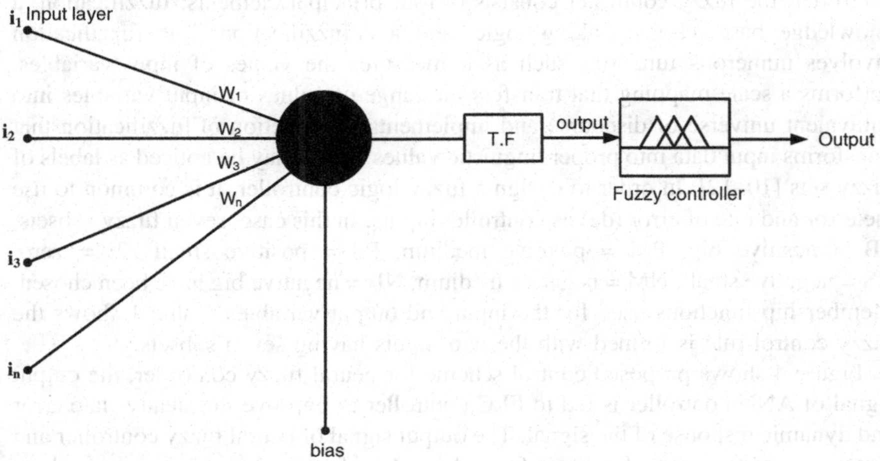

Fig. 3 Internal architecture of neuro-fuzzy system

Table 1 Fuzzy control rule

de	E						
	NB	NM	NS	ZE	PS	PM	PB
NB	NB	NB	NB	NB	NM	NS	ZE
NM	NB	NB	NB	NM	NS	ZE	PS
NS	NB	NB	NM	NS	ZE	PS	PM
ZE	NB	NM	NS	ZE	PS	PM	PB
PS	NM	NS	ZE	PS	PM	PB	PB
PM	NS	ZE	PS	PM	PB	PB	PB
PB	ZE	PS	PM	PB	PB	PB	PB

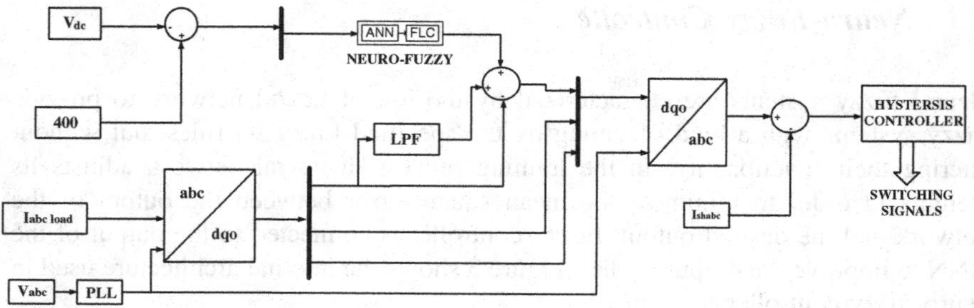

Fig. 4 Block diagram representation of mamdani neuro-fuzzy controller

varying environment and fuzzy inference systems that integrate human knowledge and achieve inferencing and decision-making. The incorporation of these two correlative approaches together with certain derivative free optimization methodologies results in a novel discipline called neuro-fuzzy. In mamdani neuro-fuzzy controller, the fuzzy controller consists of four principal elements: fuzzification, a knowledge base, choice-making logic, and a defuzzification. The fuzzification involves numerous functions such as it measures the values of input variables, performs a scale mapping that transfers the range of values of input variables into equivalent universe of discourse and implements the function of fuzzification that transforms input data into proper linguistic values which may be noticed as labels of fuzzy sets [10–12]. In order to design a fuzzy logic controller, it is common to use the error and rate of error (de) as controller inputs. In this case, seven fuzzy subsets, PB = positive big, PM = positive medium, PS = positive small ZE = zero, NS = negative small, NM = negative medium, NB = negative big have been chosen. Membership functions used for the input and output variables. Table 1 shows the fuzzy control rule is formed with the two inputs having seven subsets.

Figure 4 shows proposed control scheme for neural fuzzy controller, the output signal of ANN controller is fed to FLC controller to improve the steady state error and dynamic response of the signal. The output signal of neural fuzzy controller and direct-axis component of current from d-q-o transformation are compared which produces direct-axis component of reference signal. The signals from d-q-o frame are again converted to a-b-c frame are compared with a filter current (Ishabc), which results in generation of reference compensation current, which is given as input to the hysteresis controller.

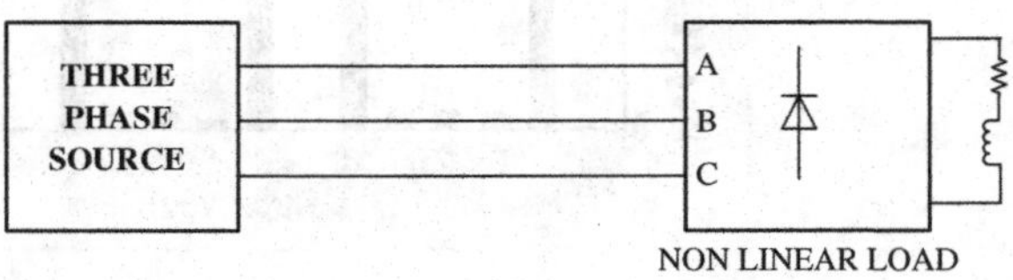

Fig. 5 Uncompensated system

Table 2 System parameters

System parameters	Values used
Source impedance	L = 0.01e−3 mH
Load	R = 10 Ω, L = 30e−3 mH
Active filter	R = 0.1 Ω, L = 3e−3 mH

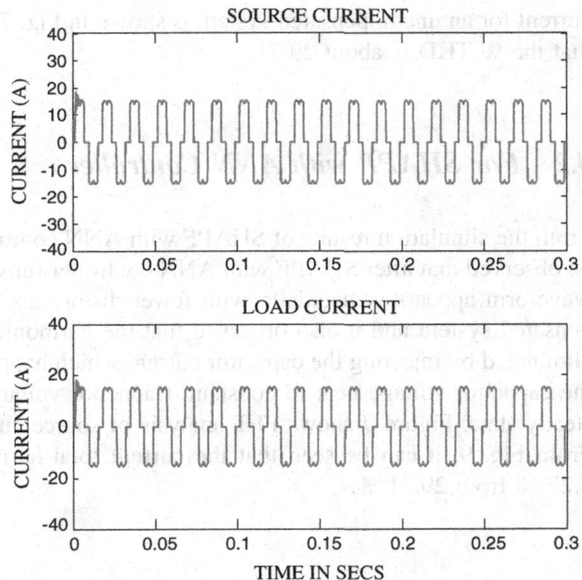

Fig. 6 Waveforms of load current and source current of uncompensated system

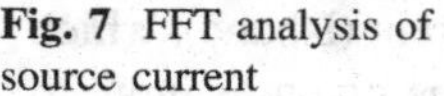
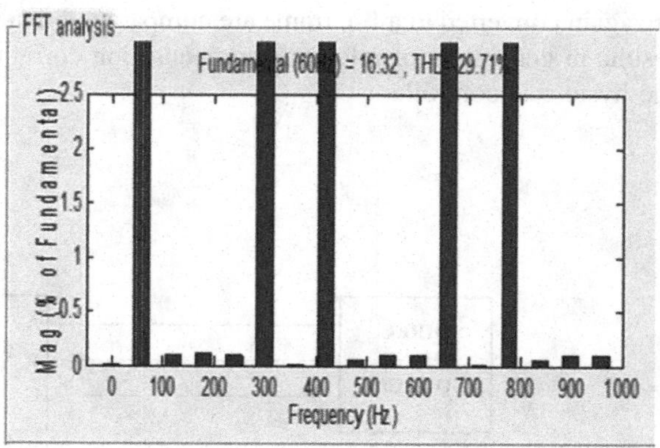

Fig. 7 FFT analysis of source current

4 Results and Discussions

4.1 *For Uncompensated System*

Any nonlinear load connected to the source produces harmonics in the system. A simulation circuit for an uncompensated system is shown in Fig. 5. The parameters of the connected system are framed in Table 2. The source current waveform and the response of the load are distorted by the effect of nonlinear load and the plots are shown in Fig. 6. It can be observed that a huge distortion in the source current instead of the actual sinusoidal waveform. FFT analysis of source current for an uncompensated system is shown in Fig. 7. From which it is observed that the % THD is about 29.71.

4.2 *For SHAPF with ANN Controller*

From the simulation results of SHAPF with ANN controller shown in Fig. 8, it can be observed that after SHAPF with ANN controller runs, it reduces much delay and waveform appears sinusoidally with fewer distortions when compared to uncompensated system and it also observed that the harmonics of the source current are eliminated by injecting the capacitor current which happens because of maintaining the capacitor voltage near to constant. Capacitor voltage takes 0.08 s to reach the steady state. Figure 9 shows FFT analysis of source current with ANN controller. From Fig. 9, it can be seen that the current total harmonic distortion reduces to 2.27 % from 29.71 %.

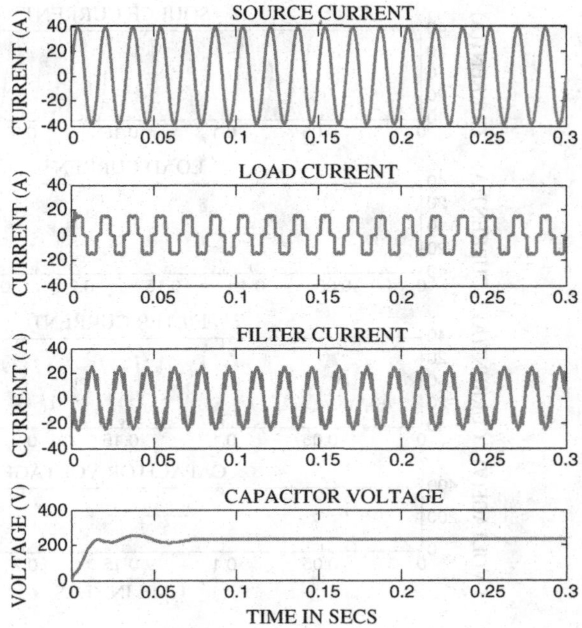

Fig. 8 Simulation results of SHAPF with ANN controller

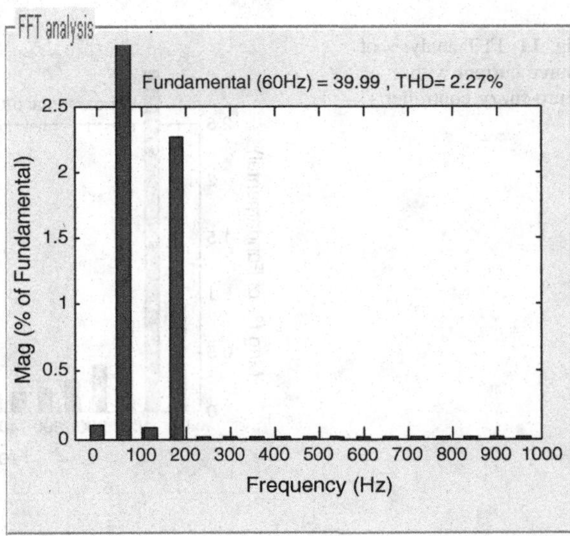

Fig. 9 FFT analysis of source current with ANN controller

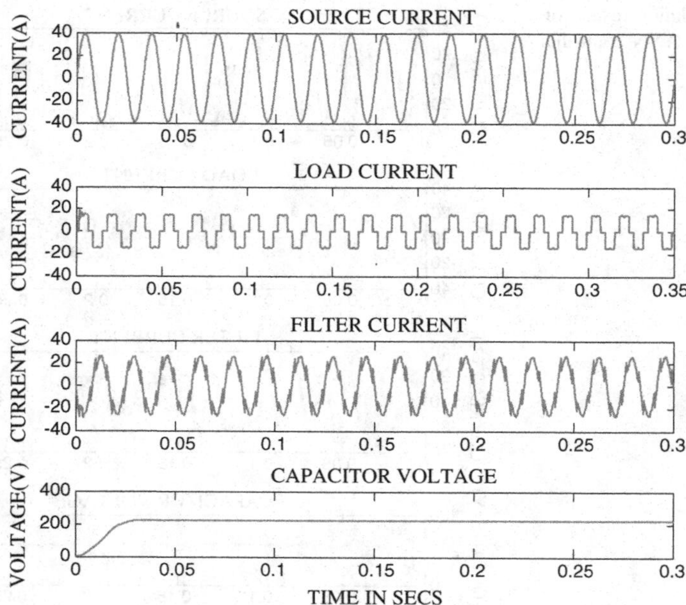

Fig. 10 Simulation results of SHAPF with neuro-fuzzy controller

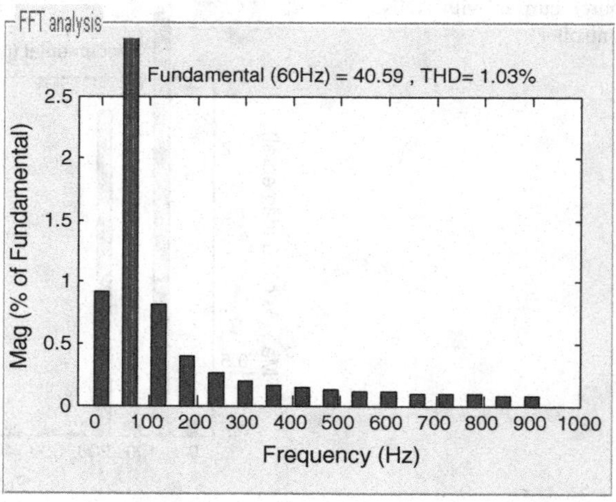

Fig. 11 FFT analysis of source current with neuro-fuzzy controller

Table 3 Comparison of ANN and neuro-fuzzy controller

	HAPF with type of controller		
	Uncompensated system	ANN	Neuro-fuzzy
Settling time (V_{DC}) in seconds	–	0.08 s	0.03 s
%THD	29.71	2.27	1.03

4.3 *For SHAPF with NEURO-FUZZY Controller*

From the simulation results of SHAPF with neuro-fuzzy controller as shown in Fig. 10, it can be observed that after neuro-fuzzy controller runs, it reduces the much delay and waveform appears pure sinusoidal with no distortions when compared to ANN controlled system and it also observed that the harmonics of the source current are eliminated by injecting the capacitor current which happens because of maintaining the capacitor voltage exactly constant. Here, capacitor voltage takes 0.03 s to reach the steady state where the capacitor voltage took 0.08 s to reach the steady state in ANN controlled system. Figure 11 shows FFT analysis of source current with neuro-fuzzy controller. From Fig. 11, it can be seen that the current total harmonic distortion reduces to 1.03 % from 2.27 %. Table 3 shows the performance comparison of ANN and neuro-fuzzy controller based shunt hybrid active power filter. So it can be observed that the neuro-fuzzy based current controller exhibits much better performance in terms of improving dynamic performance and reducing steady state error than the ANN controller.

5 Conclusion

In this paper, SHAPF with ANN and neuro-fuzzy controller has been proposed to mitigate harmonics and to increase the distribution power capacity of the three-phase system.

The obtained result shows the simplicity and the effectiveness of the two proposed intelligent controllers under nonlinear load conditions. From the results, it can be observed that the current total harmonic distortion reduces much better with neuro-fuzzy controlled filter than ANN controlled hybrid filter. The simulation and experimental results also show that the new control methods are not only easy to be calculated and implemented, but also very effective in reducing harmonics.

Acknowledgments This work was supported in part by the SERB under Grant SB/EMEQ-321/2014.

References

1. Dougan RC, Beaty HW (2002) Electrical power systems quality. McGraw-Hill, New York
2. Das JC (2004) Passive filters; potentialities and limitations. IEEE Trans Indus Appl 40:232–241
3. Singh B, Al-Haddad K, Chandra A (1999) A review of active filters for power quality improvement. IEEE Trans Indus Electron 46(5):960–971
4. Somlal J, Mannam VGR (2012) Analysis f discrete and space vector PWM controlled hybrid active filters for power quality enhancement. Int J Adv Eng Technol (IJAET) 2(1):331–341. ISSN: 2231-1963
5. Fujita H, Akagi H (1991) A practical approach to harmonic compensation in power systems; series connection of passive and active filters. IEEE Trans Indus Appl 27:1020–1025
6. Fujita H, Akagi H (1991) A practical approach to harmonic compensation in power systems series connection of passive and active filters. IEEE Trans Indus Appl 27(6):1020–1025
7. Bhattacharya S, Cheng P-T, Divan DM (1997) Hybrid solutions for improving passive filter performance in high power applications. IEEE Trans Indus Appl 33(3):732–747
8. Singh B, Verma V (2006) An indirect current control of hybrid power filter for varying loads. IEEE Trans. Power Del 21(1):178–184
9. Inzunza R, Akagi H (2005) A 6.6-kV transformerless shunt hybrid active filter for installation on a power distribution system. IEEE Trans Power Electron 20(4):893–900
10. Corasaniti VF, Barbieri MB, Arnera PL, Valla MI (2009) Hybrid active filter for reactive and harmonics compensation in a distribution network. IEEE Trans Indus Electron 56(3):670–677
11. Chen Z, Blaabjerg F, Pedersen JK (2005) Hybrid compensation arrangement in dispersed generation systems. IEEE Trans Power Del 20(2)pt. 2:1719–1727
12. Herman L, Papic I, Blazic B (2014) A proportional-resonant current controller for selective harmonic compensation in a hybrid active power filter. IEEE Trans Power Delivery 29 (5):2055–2065
13. Chen L, Xie Y, Zhang Z (2008) Comparison of hybrid active power filter topologies and principles. In: Proceedings of international conference on electronic machine system, Oct 17–20, 2008, pp 2030–2035
14. Luo A, Tang C, Shuai ZK, Zhao W, Rong F, Zhou K (2009) A novel three-phase hybrid active power filter with a series resonance circuit tuned at the fundamental frequency. IEEE Trans Ind Electron 56(7):2431–2440
15. Sathya Priyanka A, Satheesh A (2014) Harmonic compensation and reactive power support using ultracapacitor with shunt active filter in distribution system. IOCR J Electric Electr Eng (IOCR-JEEE) 9(5):60–65
16. Asiminoaei L, Aeloiza E, Enjeti PN, Laabjerg FB (2008) Shunt active-power- filter topology based on parallel interleaved inverters. IEEE Trans Ind Electron 55(3):1175–1189
17. Somlal J, Mannam VGR (2014) FUZZY logic based space vector PWM controlled hybrid active power filter for power conditioning. WSEAS Trans Power Syst 9, Art. #24:242–248
18. Jain SK, Agrawal P, Gupta HO (2002) Fuzzy logic controlled shunt active power filter for power quality improvement. In: IEE Proceedings in electrical power applications, vol 149, no. 5, Sept. 2002
19. Somlal J, Mannam VGR, Narsimha Rao V (2014) Performance analysis of artificial neural network based shunt active power filter. Int J Appl Eng Res 9(19):5697–5708

Design of Reversible Logic Based ALU

R. Dhanabal, Sarat Kumar Sahoo, V. Bharathi, V. Bhavya, Patil Ashwini Chandrakant and K. Sarannya

Abstract Nowadays reversible logic circuits are gaining fascinating attraction in many fields. Main aim of using reversible gate is that we can easily get input from output. In this paper, we are using some reversible gates which help us to perform logical and arithmetic operations. Reversible gates namely TSG gate performs 1-bit addition with carry. This is the first reversible gate which alone can acts as full adder. PV gate, Fredkin gate is used to perform logical operations like AND, OR. So here we are designing 1-bit alum using pass transistor with virtuoso tool of cadence. Based on analysis of the result, this design using reversible gates is better than that using the irreversible gates.

Keywords Reversible gate · ALU · Fredkin gate · TSG gate · PV gate

---

R. Dhanabal (✉)
VLSI Division, SENSE, VIT University, Vellore, India
e-mail: rdhanabal@vit.ac.in

S.K. Sahoo
School of Electrical Engineering, VIT University, Vellore, India
e-mail: sksahoo@vit.ac.in

V. Bharathi
GGR College of Engineering, Vellore, India
e-mail: bharathiveerappan@yahoo.co.in

V. Bhavya · P.A. Chandrakant · K. Sarannya
SENSE, VIT University, Vellore 632014, TN, India
e-mail: patil2892ashwini@gmail.com

P.A. Chandrakant
e-mail: vbnanu@gmail.com

K. Sarannya
e-mail: sarannyakadhir61@gmail.com

L.P. Suresh and B.K. Panigrahi (eds.), *Proceedings of the International Conference on Soft Computing Systems*, Advances in Intelligent Systems and Computing 397, DOI 10.1007/978-81-322-2671-0_29

1 Introduction

Year-by-year size of the VLSI circuits are reducing which results increase in complexity of the circuit, hence power dissipation is increasing. As Landauer says irreversible gate causes to loss of the energy. And afterwards Bennett showed that reversible gates help us to prevent this loss of energy. As if we use irreversible OR gate giving two 1-bit inputs there will be loss of 1-bit information and we will get only 1-bit output. This energy loss can be recovered by using reversible gates. In some applications we need some input data for further operations but using irreversible gate it is not possible to recover the lost input. But here reversible gates allow us to get inputs from output as it has separate output for each input.

In this project we are using reversible TSG gate, Fredkin gate, PV gate to perform some boolean functions like addition, XOR, OR, AND. All these sub circuits are simulated in cadence.

2 Literature Survey

The real problem in present day innovation is the bender of energy due to the data destitution in technological circuits which was constructed using irreversible hardware. A device is called as irreversible when the input can not retrace from the output. This was found by Landauer in the year 1960 [1]. Landauer's principle says that the loss of 1 bit of information will blow out KT ln 2 J of energy where the Boltzmann constant and k is $=1.380 \times 10^{-23}$ J/K, T is the absolute heat temperature in Kelvin.

The primitive combinational circuits disperse the heat energy for all the data that is lost amid operation. This on the account of, once the datum is lost can never be re-coupled by any routines as indicated by second law of thermodynamics.

In 1973, Bennett made a circuit from reversible gates in order to avert KT ln 2 J of dissipation of energy in that circuit [2]. His reversible Turing machine is a specific three-tape Turing machine which works as follows:

1. (Progressive estimation):
   The machine does the obliged reckoning, by sparing the historical backdrop of that in the second tape and utilizing the first tape as workspace.
2. The yield of the processing is duplicated into the third tape.
3. The forward computation is followed back by utilizing the history tape and emptying the first tape.

In this way, at last, the first and the second tapes come back to their starting setup, and the third contains the result. Thus the energy dissipation can be diminished if the calculation gets to be message lossless. Reversible gates uphold the procedure of running the framework both backward and forward.

3 Concept of Reversible Gates

A reversible logic gate is l × m logic gate where l and represents the number of inputs and outputs. As reversible gate has equal number inputs and outputs so that we can easily get the inputs from outputs. It has some outputs which we are not using called as garbage outputs. These outputs will be used while recovering inputs from outputs. The l × m reversible gate is shown as

$$i_V = (i1, i2, i3, .il) \tag{1}$$

$$ov = (o1, o2, o3, om) \tag{2}$$

There are different types of reversible gates. Among which NOT gate is the simplest 1 × 1 gate and Fredkin gate, Toffoli gate, Peres gate, and TR gate are 3 × 3 gates, where CNOT is the 2 × 2 gate [4]. There will be a great recuperation of energy dissipation if the circuit is developed by reversible gates. Hence we are using different reversible logic gates to perform various boolean functions, delay and power analysis results, comparison of existing 1 bit ALU and Proposed 1-bit ALU and its graphical representation are shown below in Tables 1, 2, 3, 4 and figures.

Table 1 Truth table of logical and arithmetic function

S3	S2	S1	S0	Output	Operation
0	0	0	0	A XOR B	XOR
0	1	0	0	A+B	ADD
1	0	0	0	Cout	CARRY
1	1	0	0	B	TRANSFER B
1	1	0	1	A AND B	AND
1	1	1	0	A OR B	OR
1	1	−1	1	A	TRANSFER A

Table 2 Delay and power analysis

Parameters	Power	Avg delay (ps)
TSG gate	48.08 μW	93.5
Fredkin gate	16.99 μW	28.53
PV gate	958.24 μW	10.375
ALU design	66.02 μW	132.40

Table 3 Comparison of existing 1-bit ALU and proposed 1-bit ALU

Parameter	Conventional CMOS based 4-bit ALU [5]	HybridSET CMOS based 4-bit ALU	Proposed ALU
Rise time delay(s)	1.9982e−7	1.0046e−8	365.2e−12
Fall time delay(s)	1.0002e−7	4.6238e−9	4.1019e−9

Table 4 Comparison of existing 1-bit ALU and Proposed 1-bit ALU

Terms to be compared	Existing 1-bit ALU	Existing 1-bit ALU [10]	Proposed ALU
Gate count	22	10	7
Constant input	10	4	3

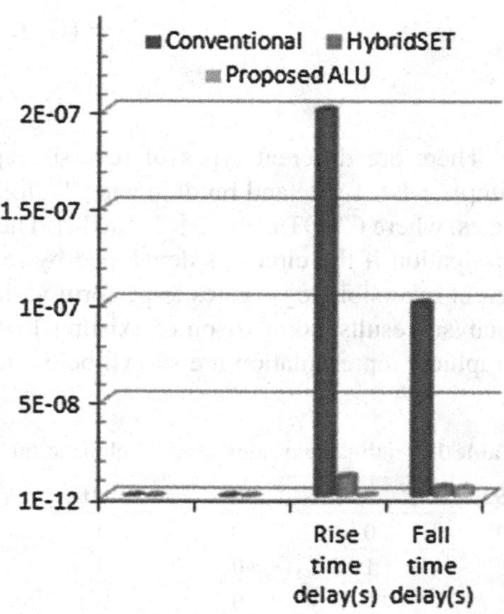

Fig. 1 Graphical representation of comparison of delay

4 Design of Proposed ALU Using Revrsible Gate

In this proposed design we are using two reversible gates TSG AND Fredkin gate for performing ALU functions. We are using two 2:1 PV mux to select different logical operations performed by Fredkin gate and one 4:1 PV mux to select particular operation of ALU. Here for TSG gate we are giving third input as zero to reduce the calculation complexity. It will give addition of 1-bit inputs with carry at third and fourth output terminal, respectively. It will give XOR operation of first and second inputs at second output terminal. First output we are considering as garbage output. Fredkin gate can give various functions. First PV gate helps us to pass B or 0 two the second input terminal. Second 2:1 PV gate helps us to pass B or 1 to the third input terminal so that gate can perform AND OR TRANSFER operation accordingly. The 4:1 PV mux will select specific function to be performed based on control signals.

4.1 Block Diagram

See Fig. 2.

5 Design of Reversible Gates

5.1 Fredkin Gate

Fredkin gate has three inputs and three outputs and hence it is said as (3*3) gate. In above block diagram of Fredkin gate we are using only second output terminal. First and third output terminal gives garbage outputs. If we pass B and 1 to second and third terminal, respectively, we get OR operation (A+B) of first (A) and second (B) input at the middle output terminal. If we give 0 and B as input to second and third terminal we will get AND operation (A.B) of first (A) and second (B) inputs. For passing appropriate input to this gate we are using two 2:1 PV mux. If we keep second and third input as 0 and 1, respectively, we get transfer of first input at second output terminal. Implementation of this gate in cadence has shown below (Figs. 3, 4 and 5).

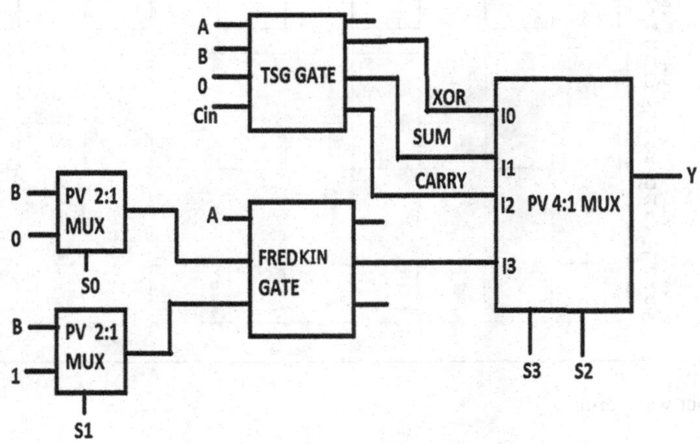

Fig. 2 Block diagram

Fig. 3 Fredkin gate [3]

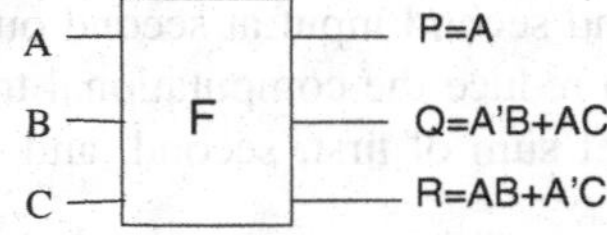

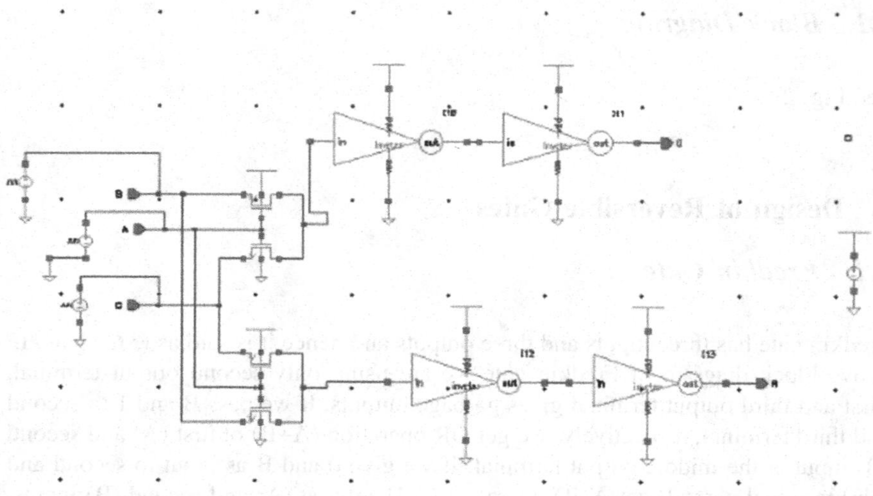

Fig. 4 Schematic of Fredkin gate

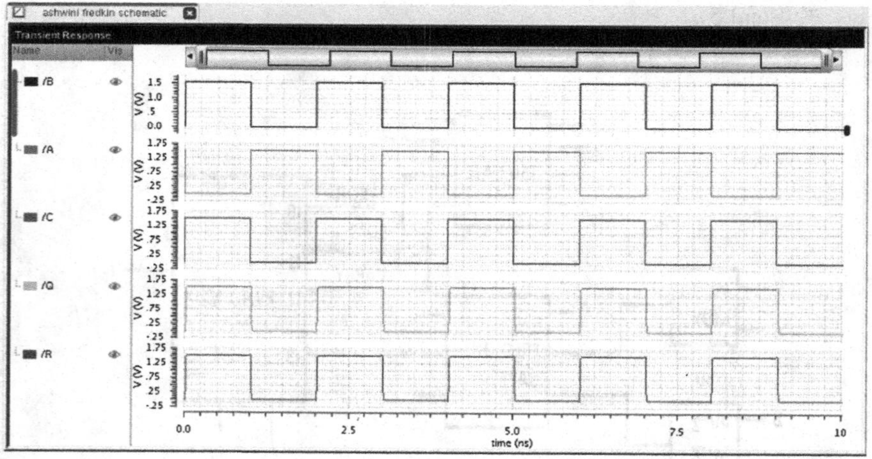

Fig. 5 Output waveform

5.2 *TSG Gate*

TSG gate is used to perform addition with carry which also provide us XOR of first and second input at second output terminal. We are keeping zero at the third input to reduce the computational toughness [6–8]. At the third output terminal, we will get sum of first, second, and fourth input and carry at the fourth output terminal.

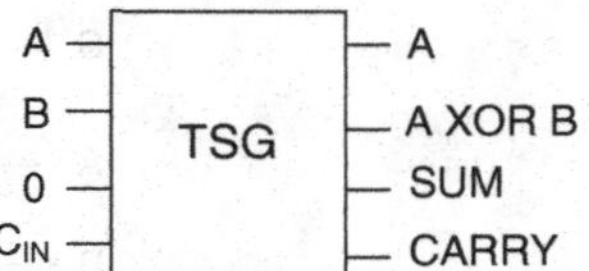

Fig. 6 TSG gate

Here we will get one garbage output at the first output terminal. Last three outputs of this gate are connected to 4:1 PV mux for selection of required function.

This gate alone acts as full adder which help us to reduce the complexity of circuits and also power and area will be reduced (Figs. 6 and 7).

5.3 PV Gate

The above diagram shows the implementation of 2:1 multiplexer using PV gate and the performance as follows. It is a reversible gate which works as multiplexer [7]. The 4:1 mux has 4 inputs and 1 output. Last three outputs of TSG gate is given to

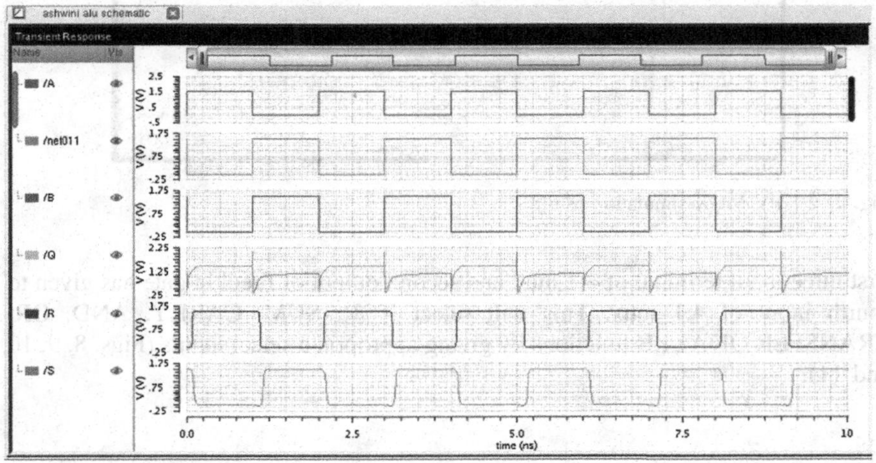

Fig. 7 Output of TSG gate

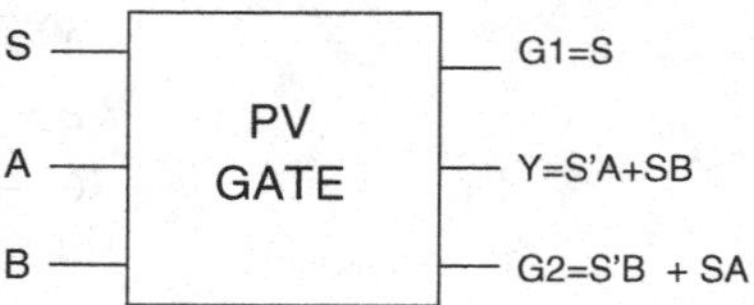

Fig. 8 PV gate

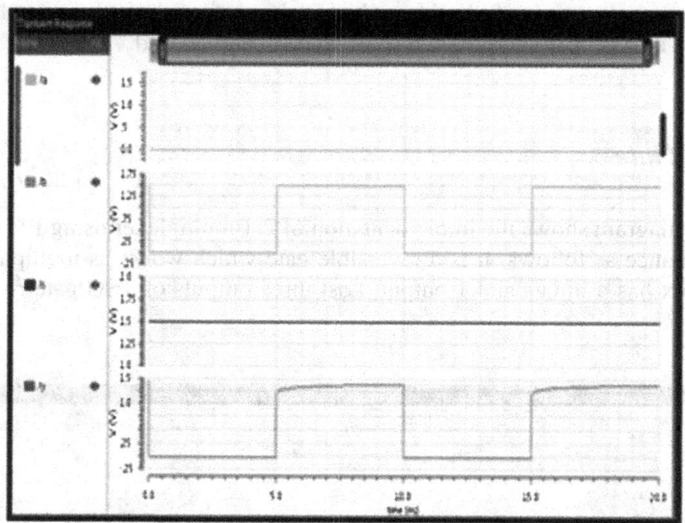

Fig. 9 2:1 PV MUX simulation result

first three input terminal of 4:1 mux and second output of Fredkin gate has given to fourth input of 4:1 mux. This will select XOR, SUM, CARRY, AND, OR, TRANSFER OF A or B functions by giving appropriate select inputs (Figs. 8, 9, 10 and 11).

6 Simulation Results of Proposed ALU

The above result shows function of sum of inputs A, B, and C in Figs. 1, 2, 3, 4, 5, 6, 7, 8, 9, 10 and comparison of results are provided in Tables 1, 2, 3, 4. First is output waveform which shows the addition of last three waveforms A, B, and C_{in}, respectively. Four waveforms after output waveform are s3, s2, s1, s0.

For select line 0100, output will be sum of A, B, and C_{in}.

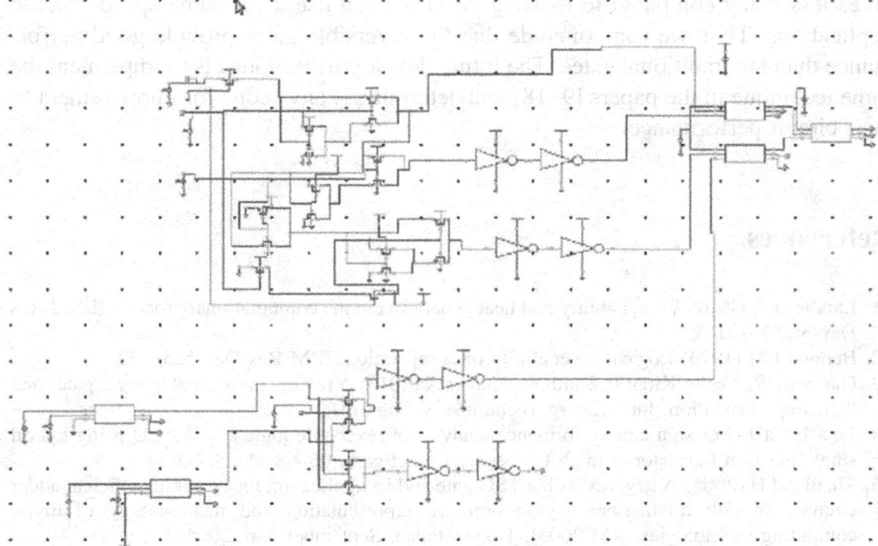

Fig. 10 Schematic of proposed ALU

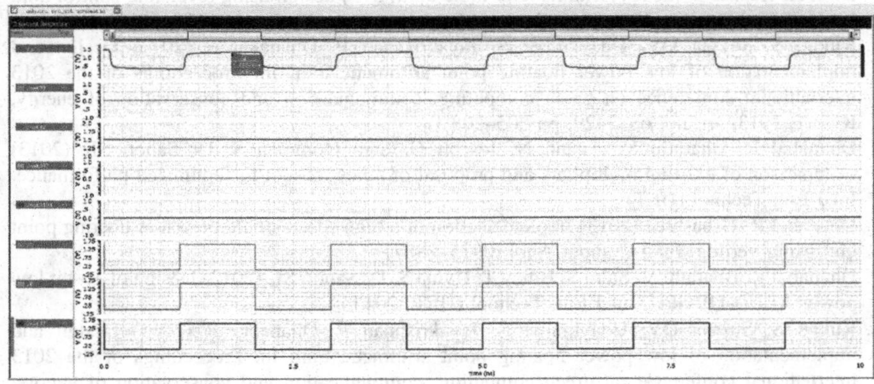

Fig. 11 Simulation result of proposed ALU

7 Conclusion

We have designed an ALU using reversible logic gates like Fredkin, TSG, and PV gate instead of normal traditional gates. By using these reversible gates the number of inputs bits and the information loss is reduced. And also the power dissipation and computational complexity of the circuits is reduced to a certain extent. As it gives less delay compared to existing ALU we can use it for some speed-specific applications. Thus we can conclude that the reversible gates provide good performance than the traditional gates. The future development of idea is to implement the same technique in the papers [9–18] and determine is any scope for improvement in their circuit performance.

References

1. Landauer R (1961) Irreversibility and heat generation in the computational process. IBM J Res Dev 3:183–191
2. Bennett CH (1973) Logical reversibility of computation. IBM Res Dev 525–532
3. Garipelly R, MadhuKiran P, Santhosh Kumar A (2013) A review on reversible logic gates and their implementation. Int J Emerg Technol Adv Eng 3(3)
4. Jana B (2014) Design and performance analysis of reversible logic based ALU using hybrid single electron transistor. Punjab University Chandigarh, 06–08 March 2014
5. Thapliyal H (2005) A new reversible TSG gate and its application for designing efficient adder circuits. In: 7th international symposium on representation and methodology of future computing technologies (RM 2005), Tokyo, Japan, September 5–6, 2005
6. Praveen B, Vinay Kumar SB 2n:1 Reversible multiplexer and its transistor implementation. Int J Comput Eng Res 2(1):182–189. ISSN: 2250–3005
7. Thapiyal H, Vinod AP (2006) Transistor realisation of reversible TSG gate and reversible adder architecture. 1-4244-0387-1/2006 IEEE
8. Dixit A, Kapase V (2012) Arithmatic & logic unit(ALU) design using reversible control unit. ISSN:2277-3754 ISO 9001:2008 Certified Int J Eng Innov Technol (IJEIT) 1(6)
9. Kukati S, Sujana DV, Udaykumar S, Jayakrishnan P, Dhanabal R (2013) Design and implementation of low power floating point arithmetic unit. In: Proceedings of the 2013 international conference on green computing, communication and conservation of energy, ICGCE 2013, art. no. 6823429, pp 205–208
10. Dhanabal R, Bharathi V, Anand N, Joseph G, Sam Oommen S, Dr Sahoo SK (2013) Comparison of existing multipliers and proposal of a new design for optimized performance. Int J Eng Technol (IJET)
11. Dhanabal R, Ushashree (2013) Implementation of a high speed single precision floating point unit using verilog. Int J Comput Appl (0975–8887)
12. Dhanabal R, Bharathi V, Saira S, Bincy T, Hyma S, Dr Sahoo SK (2013) Design of 16-bit low power ALU—DBGPU. Int J Eng Technol (IJET) 2013
13. Kukati S, Sujana DV, Udaykumar S, Jayakrishnan P, Dhanabal R (2013) Design and implementation of low power floating point arithmetic unit. In: Proceedings of the 2013 international conference on green computing, communication and conservation of energy, ICGCE 2013, art. no. 6823429, pp 205–208 (2013)
14. Dhanabal R, Bharathi V, Samhitha NR, Pavithra S, Prathiba S, Eugine J (2014) An efficient floating point multiplier design using combined booth and dadda algorithms. J Theoret Appl Inform Technol 64(3):819824
15. Dhanabal R, Sahoo SK, Barathi V, Samhitha NR, Cherian NA, Jacob PM (2014) Implementation of floating point MAC using residue number system. J Theoret Appl Inform Technol 62(2):458463
16. Dhanabal R, Bharathi V, Sri Chandrakiran G, Bharath Bhushan Reddy M (2014) VLSI design of floating point arithmetic and logic unit. J Theoret Appl Inform Technol 64(3):703709

17. Dhanabal R, Srivastava D, Bharathi V (2014) Low power and area Efficient half adder based carry select adder design using common boolean logic for processing element. J Theoret Appl Inform Technol 64(3):718723
18. Dhanabal R, Gunerkar R, Bharathi V (2014) Design and implementation of improved area efficient weighted modulo 2n+1 adder design. ARPN J Eng Appl Sci 9(12):25692575

Design of Basic Building Blocks of ALU

R. Dhanabal, Sarat Kumar Sahoo, V. Bharathi, Asha Devi, Rikta Sarma and Divya Chowdary

Abstract There were no limits for speed of operation of arithmetic/logical circuits. One can always try to increase their speed. There were many proposed algorithms, which would work fast to specified arithmetic operations. So, there is the need for the implementation of a faster design by putting these fastest algorithms in a single ALU. The carry-select adder with K–S algorithm is found to be one of the fastest algorithms for addition and Urdhva-Tiryagbhyam Karastuba algorithm for multiplication, which are the most important operations in any central processing unit. We have used QUARTUS-II software. This design can be used where high speed computation is needed. This design would work for unsigned, fixed point, 8-bit operations. We have taken the different adder circuits and compared their performance. These circuits are the basic elements or building blocks of an ALU. The circuits have been simulated using 90 nm technology of Cadence and Quartus II EP2C20F484C7. Adders can be implemented using EX-OR/EX-XNOR gates, transmission gates, HSD (High Speed Domino) technique, domino logic. Parallel feedback carry adder, ripple carry adder, carry look ahead adder, carry-select adder are some of the adders that been implemented using Cadence and Quartus-II. We found that 10T PFCA is efficient compared to 11 T PFCA to some extent. Adders

R. Dhanabal (✉)
VLSI Division, SENSE, VIT University, Vellore, India
e-mail: rdhanabal@vit.ac.in

S.K. Sahoo
School of Electrical Engineering, VIT University, Vellore 632014, India
e-mail: sksahoo@vit.ac.in

V. Bharathi
GGR College of Engineering, Anna University, Vellore, India
e-mail: bharathiveerappan@yahoo.co.in

A. Devi · R. Sarma · D. Chowdary
SENSE, VIT University, Vellore 632014, TN, India
e-mail: Asha.devi2014@vit.ac.in

R. Sarma
e-mail: rikta.sarma2014@vit.ac.in

L.P. Suresh and B.K. Panigrahi (eds.), *Proceedings of the International Conference on Soft Computing Systems*, Advances in Intelligent Systems and Computing 397, DOI 10.1007/978-81-322-2671-0_30

based on XOR and XNOR gates have the least delay compared to the other adders that we have used.

Keywords K–S algorithm · Urdhva-tiryagbhyam karastuba algorithm · QUARTUS-II · ALU

1 Introduction

It is not uncommon in mathematics to perform operation of a particular stage with one of the inputs depending on previous stages. A simple example of it is a convolving circuit, where amplitude of any impulse depends upon the intermediately sums and products. This type of problem, calculating the value of a particular (last) stage is called recurrence problem and is a set of recurrence function. Likewise multiplication operation has to face a constraint, i.e., speed.

So there is need for implementing a design which works fast to perform these arithmetic operations. This paper provides solution for such initial value based processing's (recurrence function). This paper uses carry-select adder with K–S algorithm for addition operation and Urdhva-Tiryagbhyam Karastuba algorithm for multiplication. These two modules were integrated onto same ALU with global reset/enable input. Equality operation is also provided in its design. The global reset/enable input would drive a 2 × 4 active high decoder, which provides enable inputs for individual modules. The ALU has 8-bit combination as input and 16-bit result in which the lower nine bits are multiplexed for adder/substractor operation and multiplication, remaining seven higher ordered bits are given for multiplication alone.

2 Carry-Select Adder with K–S Algorithm

The most commonly used adder in ALU is ripple carry adder and carry look head adder. But in ripple carry adder for getting the final carry we have to wait until its succeeding stage propagates carry and which in turn has to wait for its preceding stage's propagated carry and finally the MSB stage depends on the initial carry. Similarly in the carry look ahead adder the initial carry has to travel longer distances.

So it takes much time to get the final result, and it will lead to poor performance of ALU. So to avoid the delay we have to reduce the path length. This can be done by computing the present stage carry independent of the previous carry. For doing the same we have used carry-select adder with the Kogge-Stone adder circuit.

2.1 Carry-Select Adder

The carry-select adder [1, 2] given in Fig. 1 computes the carry of present stage by assuming the previous carry as both one and zero. When these are given to a multiplexer with the carry from the previous stage as selection line, the actual carry would do the same for its succeeding stages. Computing the present stage carry independent to the previous carry reduces the delay and makes the circuit speed.

2.2 Kogge-Stone (K–S) Adder

Kogge-Stone adder is parallel prefix form of carry look head adder. In parallel prefix forms the whole operations can divide into three types such as preprocessing stage, carry generation, and post-processing stage as given in Fig. 2.

$$CP_0 = P_i \text{ and } P_j \tag{1}$$

$$CG_0 = (P_i \text{ and } G_j)|G_i \tag{2}$$

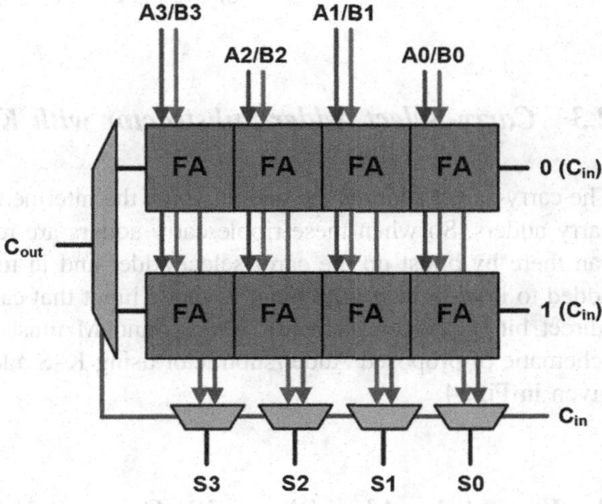

Fig. 1 Structure of carry-select adder

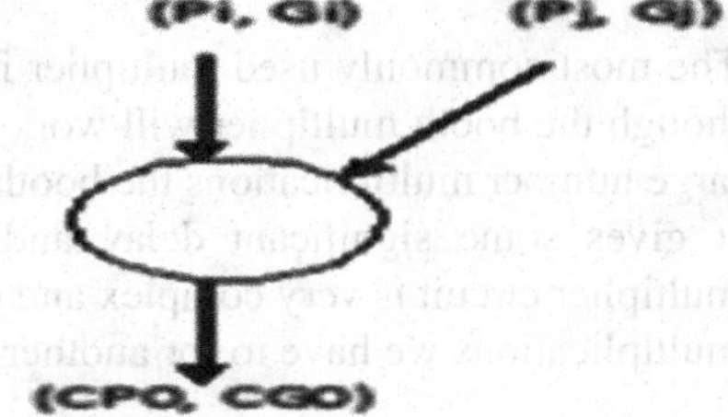

Fig. 2 K–S Operator [2]

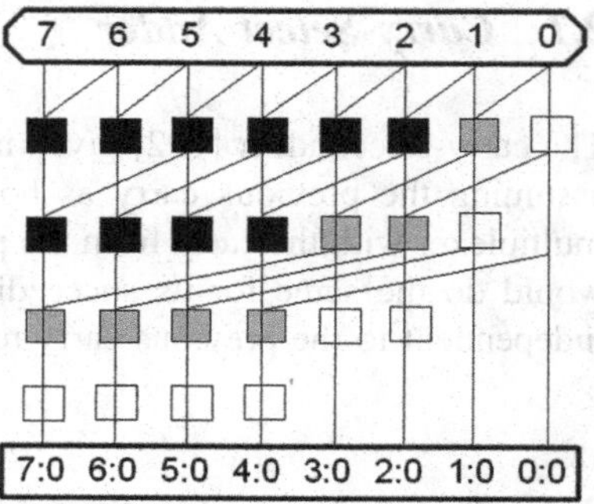

Fig. 3 Structure of 8-bit K–S adder

In preprocessing stage we generate both propagate and generate by taking the inputs. After completion of (propagate, generate) we will compute the intermediary carries. By using propagate, generate in the preprocessing stage we compute the final sum bits.

The sum bits can be calculated by using the following Boolean expressions. The structural block diagram of 8-bit K–S adder is provided in Fig. 3.

$$C_i - 1 = (P_i \text{ and } C_{\text{in}})|G_i \tag{3}$$

$$S_i = P_i \wedge C_i - 1 \tag{4}$$

2.3 *Carry-Select Adder/Substractor with Kogge–Stone*

The carry-select adder is the one in which the intermediate stage contains the ripple carry adders. So when these ripple carry adders are replaced with K–S adders we can there by boost up the carry-select adder and in turn when the XOR gates are added to it with one of the input as mode input that can control the input of adders (direct bit or its complement). The c_{in} and M must be given logic-1. The RTL schematic of proposed Adder/subtractor using K–S adder obtained from the tool is given in Fig. 4.

3 Karastuba Algorithm with Carry-Select K–S Adders

The most commonly used multiplier in ALU architecture is booth multiplier. Even though the booth multiplier will work efficiently on small numbers, when we go for large number multiplications the booth multiplier is not sufficient for large numbers. It gives some significant delay and power consumption. Moreover, the booth multiplier circuit is very complex and occupies more area. So, for the large numbers multiplications we have to for another algorithm. Both Vedic, Karastuba algorithms

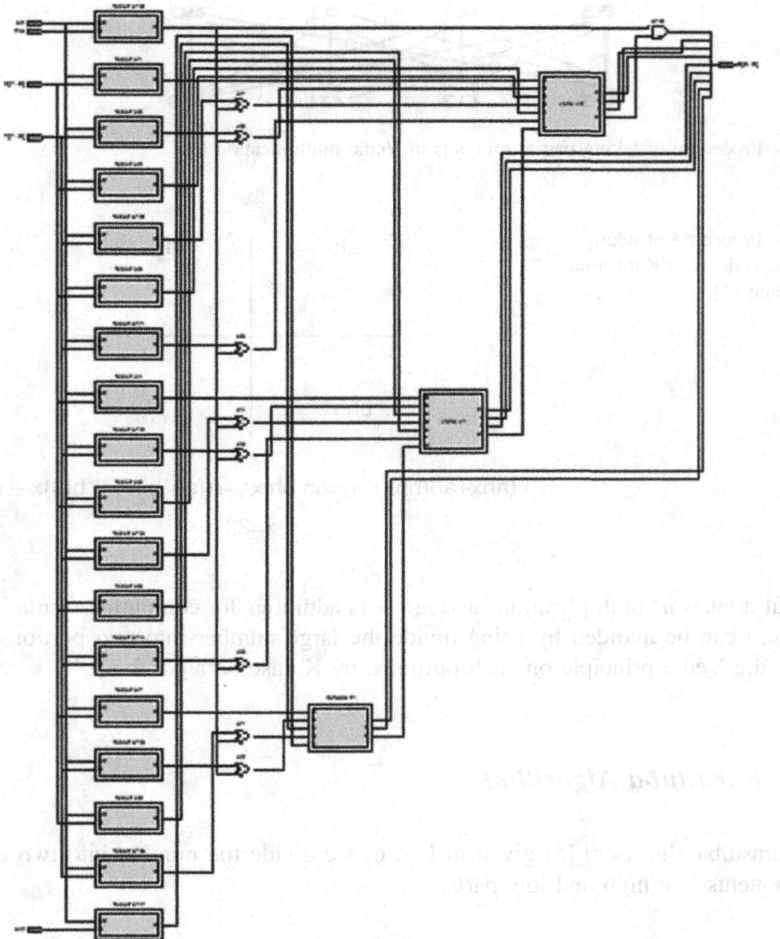

Fig. 4 RTL schematic of proposed Adder/Subtractor using K–S adder

will deal with area and power specifications. So for large numbers we prefer both Vedic and Karastuba.

3.1 Vedic Multiplier

In Vedic multiplier we use Urdhva–Tiryagbhyam principle [3, 4] for getting results faster. By using this principle the multiplication will look like Fig. 5.

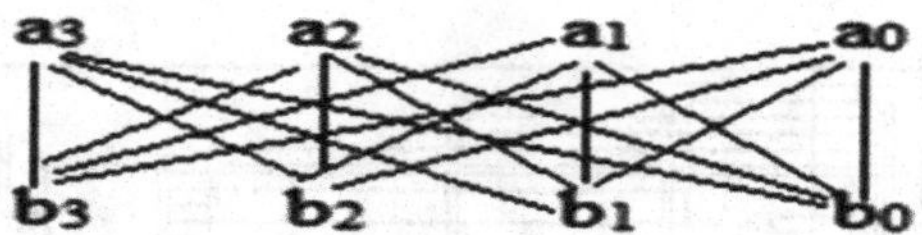

Fig. 5 Procedure of taking partial products in Vedic multiplication [4]

Fig. 6 Procedure of taking partial products in Karastuba algorithm [5]

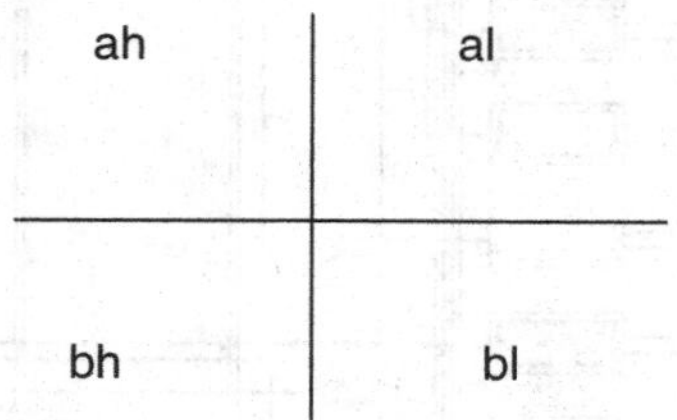

$$(ahx+al)(bhx+bl)=ah\ bh\ x^2\ +((ah*bl)+(al*bh))x + al*bl$$

$$x=2^{n/2}$$

But it takes n^2 multiplications and $n(n + 1)$ additions for computing results. This problem can be avoided by using divide the large number into two portions and apply the Vedic principle on each portions, by Karastuba algorithm.

3.2 *Karastuba Algorithm*

In Karastuba algorithm [5] given in Fig. 6, we divide the number into two equal components like high and low parts.

$$A = ah + al \tag{5}$$

$$B = bh + bl \tag{6}$$

Even though each one has its own advantages and disadvantages, so we are going for Vedic multiplier with Karastuba algorithm.

So, in our project we used both Karastuba and Vedic multiplier principles. We break the number into two components by using Karastuba and applying Vedic multiplications on individual blocks. The proposed multiplier structure is given in Fig. 7.

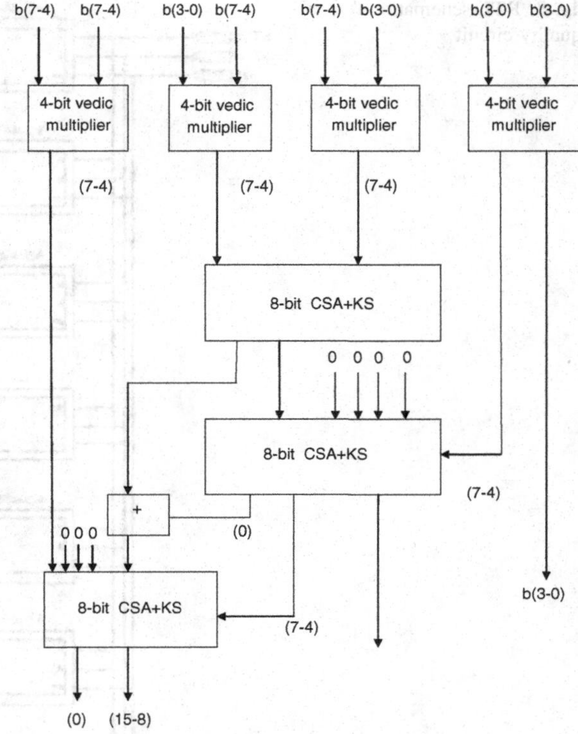

Fig. 7 Proposed multiplier structure

4 Other Operations

The other operations include equality operation. This also would get reset when the global reset pin goes low. This contains a bank of eight XNOR gates which will simultaneously give output as 1 when both of its inputs are same. All these ones when given to AND gate, it would give output as logic 1, indicating that the inputs are same. The structure of equality can be shown as in Fig.8 obtained from Altera Quartus II design tool.

The global reset/enable input would drive a 2 × 4 active high decoder, which provides enable inputs for individual modules as given in Fig. 9.

5 Power-Number of Elements-Delay

The delay is found to be less in CSA+KS type of adder when compared to other types of adders. But the requirement that this design is not meeting is the power consumption and logic elements. So this type of designs can be used mostly where

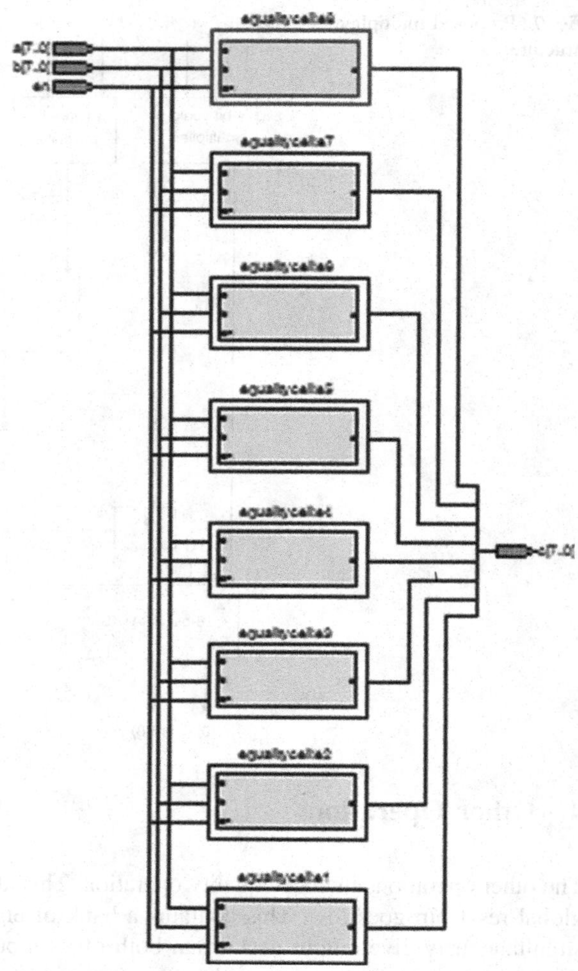

Fig. 8 RTL schematic of equality circuit

high speed of operation is the main point. Multiplier is fast in comparison but at the cost of less power. The schematic of full adder based on EX-OR and EX-XNOR gate is given in Fig. 10.

5.1 Simulation Result

Proposed Circuit: In the 10T PFCA, there is voltage degradation, so when we add buffers to both sum and c_{out} in the form of inverters. This results in waveforms free from voltage degradation and glitches as given in Figs. 11 and 12. Comparison of power consumption for various adders implemented in Altera Quartus II design tool

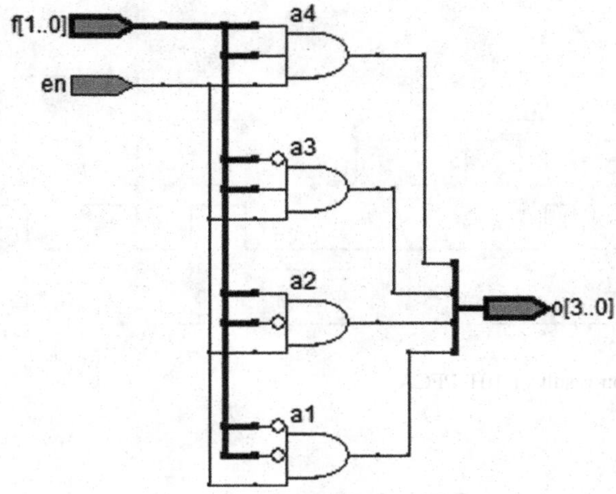

Fig. 9 ALU decoder which generates enables inputs for various modules

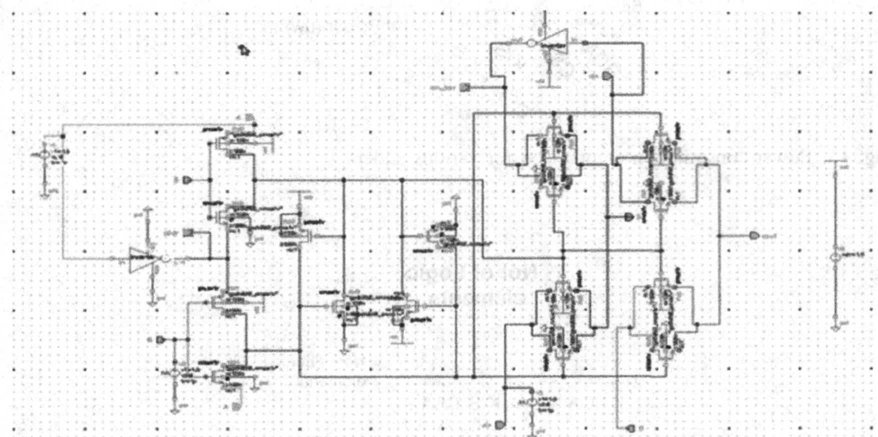

Fig. 10 Schematic of FA based on EX-OR and EX-XNOR

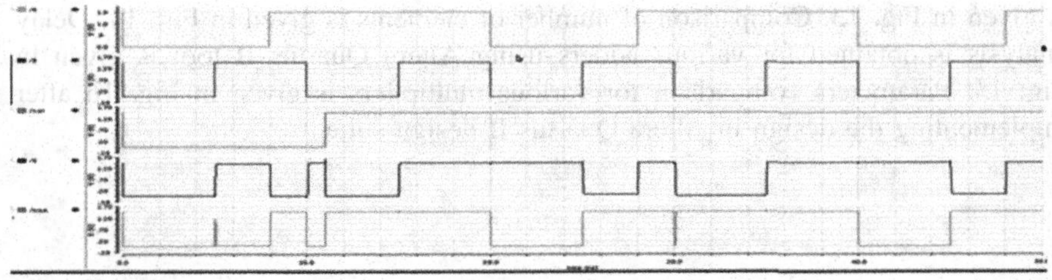

Fig. 11 Input and output waveforms of adder based on XOR and XNOR gates

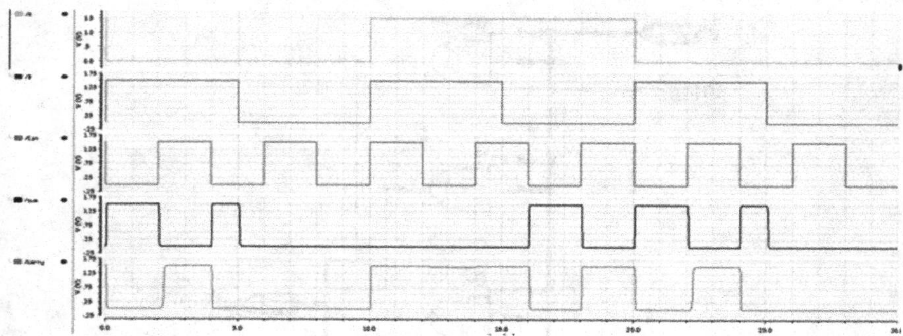

Fig. 12 Simulation result of 10T PFCA

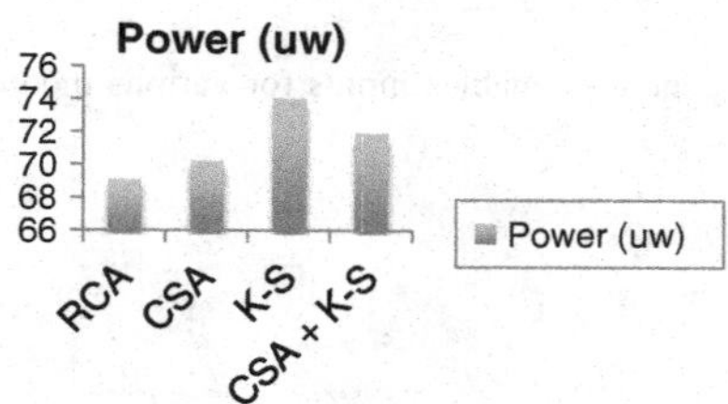

Fig. 13 Power consumption comparison for various adders

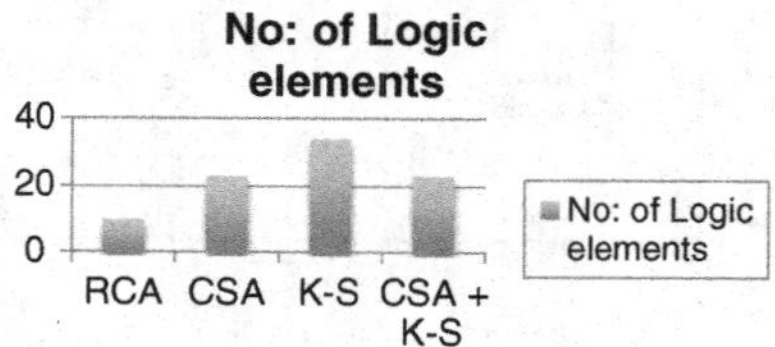

Fig. 14 Comparison of number of elements

is given in Fig. 13. Comparison of number of elements is given in Fig. 14. Delay analysis is obtained for various adders using Altera Quartus II tool is given in Fig. 15. Parameters comparison for various multipliers is given in Fig. 16 after implementing the design in Altera Quartus II design suite.

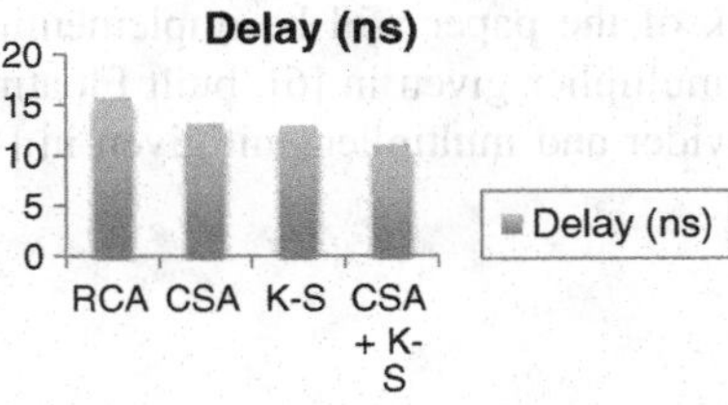

Fig. 15 Delay analysis for various adders

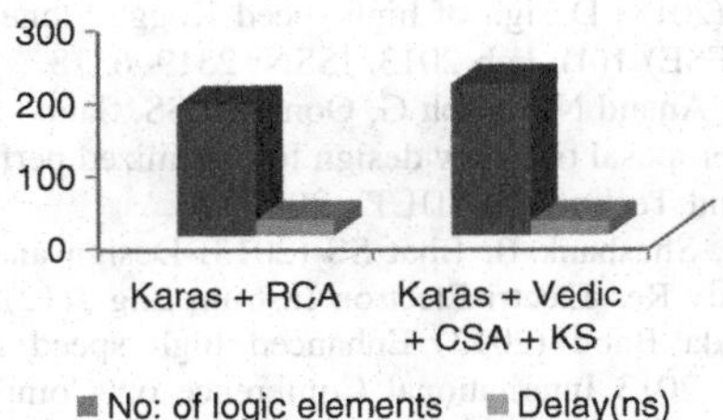

Fig. 16 Parameters comparison for various multipliers

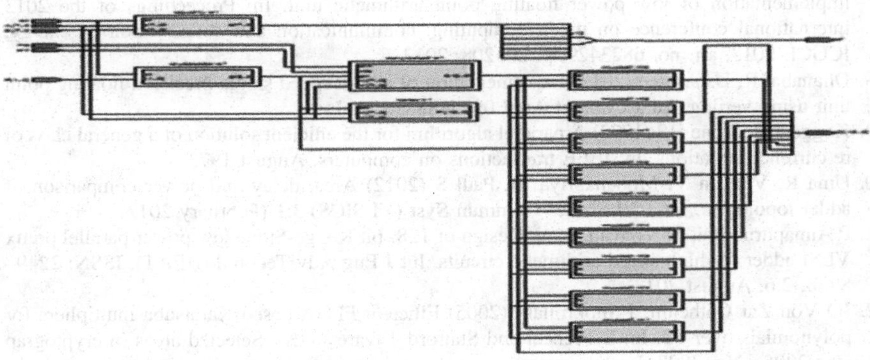

Fig. 17 RTL schematic for proposed ALU

6 Summary and Conclusion

Hence, if the aspect of area/power has not been considered this type of design would give better performance when compared to all the other adders and multipliers. There would be some applications where such high speed processing is needed like satellite image processing, enemy reflected radar signal processing. The RTL schematic for proposed ALU is obtained from Altera Quartus-II is given in Fig. 17. Future work of the paper will be implementing the proposed architecture in Floating point multiplier given in [6], built Floating point ALU given in [7] and Floating point divider and multiplier unit given in [8] with meeting IEEE 754 standard.

References

1. Shekhar S, Pandey, Bakshi A, Sharma V (2013) 128 Bit low power and area efficient carry select adder. Int J Comput Appl 69(6):0975–8887, May 2013
2. Chakali P, Patnala MK (2013) Design of high speed Kogge-Stone based carry select adder. Int J Emerg Sci Eng (IJESE) 1(4), Feb 2013. ISSN: 2319–6378
3. Dhanabal R, Bharathi V, Anand N, Joseph G, Oommen SS, Sahoo SK (2013) Comparison of existing multipliers and proposal of a new design for optimized performance. In: International journal of engineering and Technology (IJET), 2013
4. Premananda BS, Pai SS, Shashank B, Bhat SS (2013) Design and implementation of 8-bit vedic multiplier. Int J Adv Res Electri Electron Instrum Eng 2(12), Dec 2013
5. C Prema, CS Mainkanda Babu (2013) Enhanced high speed modular multiplier using Karatsuba algorithm. In: 2013 International Conference on Computer Communication and Informatics (ICCCI-2013), Jan 04–06, 2013, coimbatore, India
6. Dhanabal R, Bharathi V, Samhitha NR, Pavithra S, Prathiba S, Eugine J (2014) An efficient floating point multiplier design using combined booth and dadda algorithms. J Theor Appl Inf Technol 64(3):819824
7. Kukati S, Sujana DV, Udaykumar S, Jayakrishnan P, Dhanabal R (2013) Design and implementation of low power floating point arithmetic unit. In: Proceedings of the 2013 international conference on green computing, communication and conservation of energy, ICGCE 2013, art. no. 6823429, p. 205208, 2013
8. Dhanabal R, Ushashree (2013) Implementation of a high speed single precision floating point unit using verilog. Int J Comput Appl 0975–8887, 2013
9. Kogge PM, Stone HS (1973) A parallel algorithm for the efficient solution of a general class of re currence equations. In: IEEE transactions on computers, August 1973
10. Uma R, Vijayan V, Mohanapriya M, Paul S (2012) Area, delay and power comparison of adder topologies. Int J VLSI Des Commun Syst (VLSICS) 3:1 (February 2012)
11. P Annapurna Bai, M Vijaya Laxmi, Design of 128- bit Kogge-Stone low power parallel prefix VLSI adder for high speed arithmetic circuits. Int J Eng Adv Technol (IJEAT), ISSN: 2249–8958, 2:6, August 2013
12. JO Von Zur Gathenm, J Shokrollahi (2005) Efficient FPGA-based Karatsuba multipliers for polynomials over F2. In: B Preneel and Stafford Tavares (eds), Selected areas in cryptograp (sac 2005), No. 3897. Lecture notes in computer science. Springer, Kingston
13. Ratna Raju B, Satish DV (2013) A high speed 16*16 multiplier based on Urdhva Tiryakbhyam Sutra. Int J Sci Eng Adv Technol IJSEAT 1:5, Oct 2013
14. Dhanabal R, Bharathi V, Salim S, Thomas B, Soman H, Sahoo SK (2013) Design of 16-bit low power ALU—DBGPU. In: International journal of engineering and technology (IJET) 2013
15. Dhanabal R, Sahoo SK, Barathi V, Samhitha NR, Cherian NA, Jacob PM (2014) Implementation of floating point MAC using residue number system. J Theor Appl Inf Technol 62(2):458463
16. Dhanabal R, Bharathi V, Sri Chandrakiran G, Bharath Bhushan Reddy M (2014) VLSI design of floating point arithmetic and logic unit. J Theor Appl Inf Technol 64(3):703709

17. Dhanabal R, Srivastava D, Bharathi V (2014) Low power and area Efficient half adder based carry select adder design using common boolean logic for processing element. J Theor Appl Inf Technol 64(3):718723
18. Dhanabal R, Gunerkar R, Bharathi V (2014) Design and implementation of improved area efficient weighted modulo 2n + 1 adder design. ARPN J Eng Appl Sci 9(12):25692575

Implementation of Low Power and Area Efficient Floating-Point Fused Multiply-Add Unit

R. Dhanabal, Sarat Kumar Sahoo and V. Bharathi

Abstract In this paper, a modified architecture for Floating-Point Fused Multiply-Add (FMA) unit for low power and reduced area applications is presented. FMA unit is the one which computes a floating-point $(A \times B) + C$ operation as a single instruction. In this paper a bridge unit has been used, which connects the existing floating-point multiplier (FMUL) and the FMUL's add-round unit in the co-processor to perform FMA operation. The main objective of this modified FMA unit is to reuse as many components as possible to allow parallel floating-point addition and floating-point multiplication or floating-point fused multiply-add functionality by addition of little hardware into the FMUL's add-round unit. In this paper each unit is designed using Verilog HDL. The design is simulated using Altera ModelSim and is synthesized using Cadence RTL compiler in 45 nm. All the floating-point arithmetics are implemented in IEEE-754 double precision format. It is found that the proposed FMA architecture achieved 17 % improvement in power and 6 % improvement in area when compared to the existing Bridge FMA unit.

Keywords Fused multiply-add · Floating-point arithmetics · IEEE-754 double precision standard

R. Dhanabal (✉)
VLSI Division, SENSE, VIT University, Vellore, India
e-mail: rdhanabal@vit.ac.in

S.K. Sahoo
SELECT, VIT University, Vellore 632014, India
e-mail: sksahoo@vit.ac.in

V. Bharathi
VLSI Division, CSE, GGR College of Engineering,
Anna University, Vellore, India
e-mail: bharathiveerappan@yahoo.co.in

L.P. Suresh and B.K. Panigrahi (eds.), *Proceedings of the International Conference on Soft Computing Systems*, Advances in Intelligent Systems and Computing 397, DOI 10.1007/978-81-322-2671-0_31

1 Introduction

In digital signal processing applications the floating-point fused multiply-add (FMA) operation has become one of the fundamental operations. Many of the commercial processors like IBM PowerPC, Intel Itanium have included the FMA unit in its floating-point units to execute double precision fused multiply-add operation [1]. FMA unit improves the accuracy of the floating-point $(A \times B) + C$ operation as it performs single rounding instead of two. FMA operation is very useful when a floating-point multiplication is followed by a floating-point addition.

Floating-point fused multiply-add implementation has two advantages over implementation of floating-point addition (FADD) and floating-point multiplication (FMUL) separately: (1) The FMA operation is performed with only one rounding instead of two (one for floating-point adder and other for floating-point multiplier) reducing overall error due to rounding. (2) There will be a reduction in delay and hardware required by sharing components [2, 3].

In some designs the existing FMUL unit and FADD unit is entirely replaced with a FMA unit. It performs single FMUL operation by making $C = 0$ and single FADD operation by making $A = 1$ (or $B = 1$) in $(A \times B) + C$, e.g., $(A \times B) + 0.0$ for single multiplier and $(A \times 1.0) + C$ for single adds. But due to the insertion of constants, the latencies of stand-alone FMUL, and FADD operations increase due to the complexity of FMA unit. In such designs there will not be any possibility to perform parallel FMUL and FADD instructions [3].

The first floating-point FMA unit was introduced on IBM RISC System/6000 in 1990 for single instruction execution of $(A \times B) + C$ operation as an indivisible operation [2, 4]. Executing parallel FMUL and FADD operations is not possible in basic FMA unit. In [5] the Concordia FMA architecture is designed, which uses alignment blocks before the multiplier array. So multiplier tree input range widens. Due to this larger variable multiplier tree is required. A few possible solutions have been identified in the Lang/Bruguera fused multiply-add architecture, which is designed for reduced latencies [3, 6]. But it did not reach the latency of a common FADD/FMUL instruction. A bridge FMA design is introduced in [7] to avoid the stand-alone FMUL and FADD latencies due to the insertion of constants by adding extra blocks between existing FMUL and FADD components in the processor. But the cost added to this architecture is increase in area and power consumed when compared to the basic FMA architecture.

The main objective of this work is to design a low power, area efficient FMA unit which performs FMA operation or parallel FMUL and FADD operations based on requirement. In this paper a modified add-round block is designed, which supports add-round for FMUL as well as FMA. Common add-round unit for both FMA and FMUL instructions is used to save chip area.

All the floating-point arithmetic operations here are done using IEEE-754 double precision format. The standard IEEE-754 double precision format [8] consists of 64 bits which are divided into three sections as shown in Fig. 1. To represent any

Sign (1-bit)	Exponent (11-bit)	Mantissa (52-bit)

Fig. 1 IEEE-754 double precision format [8]

floating-point number, all the three sections have to be combined. The double precision floating-point number is calculated as shown in Eq. (1).

$$A = (-1)^{\mathrm{sign}_A} \times 1 \cdot \mathrm{fraction}_A \times 2^{\exp_A - \mathrm{bias}} \tag{1}$$

2 Architecture of Proposed FMA Unit

Block diagram for proposed floating-point fused multiply-add unit is shown in Fig. 2. The FMA unit starts with the common multiplier and adder units which can perform single stand-alone operations. The main components in this design are:

1. Floating-Point Multiplier
2. Floating-Point Adder
3. Bridge Unit
4. Add-Round unit for FMA/FMUL
5. Add-Round Unit for FADD

Our FMA unit performs parallel floating-point addition and multiplication or floating-point fused multiply-add operations based on the requirement. Suppose when a FMA operation is to be performed, this bridge architecture is connected between the existing FMUL and FMUL's add-round unit. When FMA operation is

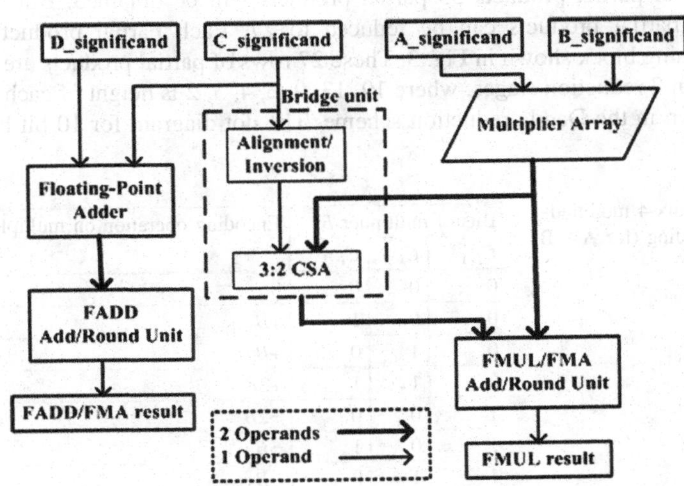

Fig. 2 Block diagram for proposed FMA unit

not needed stand-alone FMUL and FADD operation can be performed without using the intermediate bridge unit.

2.1 Multiplier

Efficient double precision floating-point multiplier using radix-4 modified booth algorithm (MBE) and Dadda algorithm has been implemented. This hybrid multiplier is designed by using the advantages in both the multiplier algorithms. MBE has the advantage of reducing partial products to be added. Dadda scheme has the advantage of adding the partial products in a faster manner [9, 10]. Our main objective is to combine these two schemes to make the multiplier design power efficient and area efficient. Finally obtained two rows (sums and carries) are added using an efficient parallel prefix adder [11].

MBE generates at most $\lfloor \frac{N}{2} \rfloor + 1$ partial products, where N is the number of bits. Radix-4 recoding is done with the digit set $\{-2, -1, 0, 1, 2\}$ is shown in Table 1. Each three consecutive bits of the multiplier B represents the input to the booth recoding block. This block selects the right operation on multiplicand A which can be shift or invert $(-2B)$ or invert $(-B)$ or zero or no operation (B) or shift $(2B)$. Figure 3 shows the generation of one partial product using MBE.

In Dadda scheme, the reduction of obtained partial products is done in stages using half adders and full adders. The reduction in size of each stage is obtained by working back from the final stage. Each preceding stage height must be not greater than $\lfloor 3 \cdot \text{successorheight}/2 \rfloor$ [10].

For a double precision floating-point multiplication two 53-bit (1 hidden bit + mantissa 52 bits) numbers are to be multiplied. If normal method is used for generation of partial products 53 partial products will be obtained. But by using MBE the partial products can be reduced to 27. Each partial product can be obtained using block shown in Fig. 3. These 27 rows of partial products are reduced to 2 rows in 7 reduction stages, where 19, 13, 9, 6, 4, 3, 2 is height of each stage as we go down in the Dadda reduction scheme. The dot diagram for 10 bit by 10 bit

Table 1 Radix-4 modified booth's recoding (for A × B)

Bits of multiplier B			Encoding operation on multiplicand A
C_{i+1}	C_i	C_{i-1}	
0	0	0	0
0	0	1	$+B$
0	1	0	$+B$
0	1	1	$+2B$
1	0	0	$-2B$
1	0	1	$-B$
1	1	0	$-B$
1	1	1	0

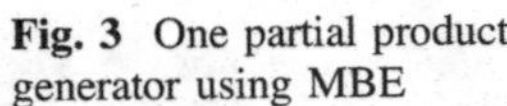

Fig. 3 One partial product generator using MBE

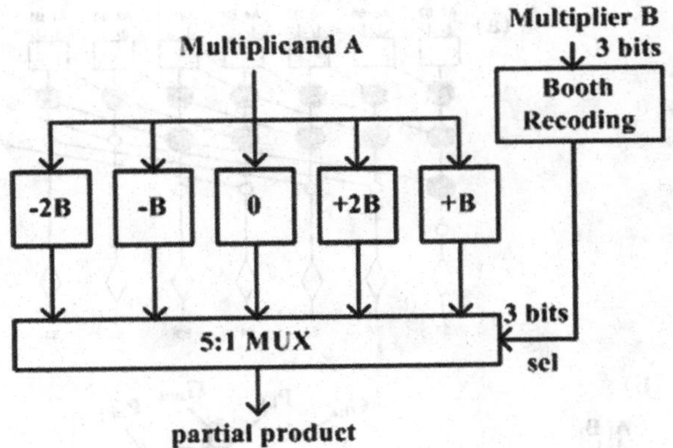

booth encoding with Dadda reduction is shown in Fig. 4. The same method in Fig. 4 is extended to 53 bit by 53 bit.

The final sums and carries are added using parallel prefix adders as it offer a highly efficient solution to the binary addition and suitable for VLSI implementations [11, 12]. Among the parallel prefix adders, Kogge–Stone architecture is the widely used and the popular one. Kogge–Stone adder is considered as the fastest

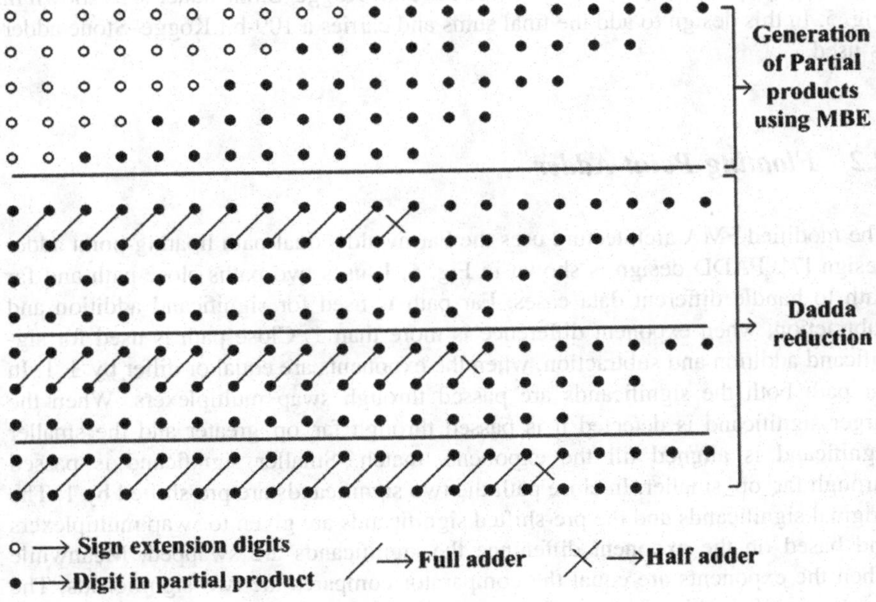

Fig. 4 10 bit by 10 bit booth encoding with Dadda reduction

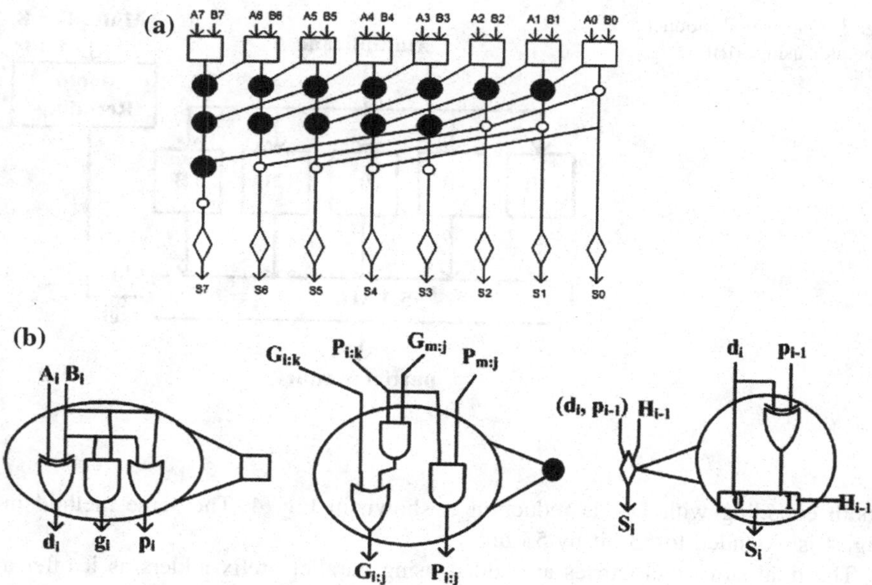

Fig. 5 **a** 8-bit Kogge–Stone adder. **b** Logic implementation of each block of Kogge–Stone adder [11]

adder design possible [11]. Architecture for 8-bit Kogge–Stone adder is as shown in Fig. 5. In this design to add the final sums and carries a 109-bit Kogge–Stone adder is used.

2.2 *Floating-Point Adder*

The modified FMA architecture uses the Farmwald's dual-path floating-point adder design [7]. FADD design is shown in Fig. 6. It uses two paths close path and far path to handle different data cases. Far path is used for significand addition and subtraction, when exponent difference is more than 1. Close path is used for significand addition and subtraction, when the exponents are equal or differ by $\pm$ 1. In far path both the significands are passed through swap multiplexers. When the larger significand is detected it is passed through far_op_greater and the smaller significand is aligned till the exponents match. Smaller significand is passed through far_op_smaller. In close path the two significands are pre-shifted by 1. The original significands and the pre-shifted significands are given to swap multiplexers and based on the exponent difference the significands are swapped. Meanwhile when the exponents are equal the comparator compares the two significands. The greater significand in close path is passed through close_op_greater and the smaller significand is passed through close_op_smaller.

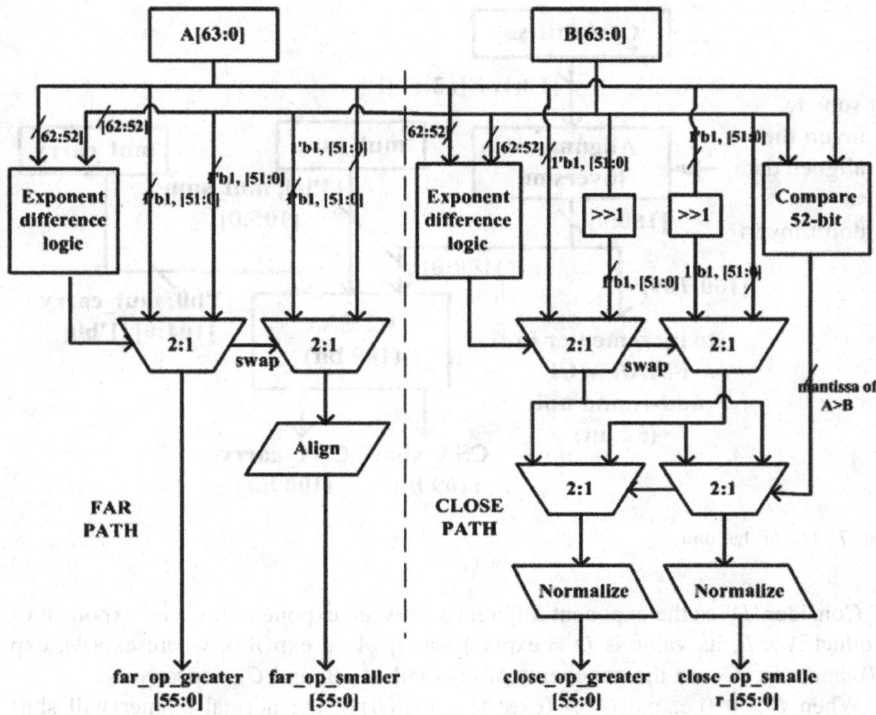

Fig. 6 Floating-point adder unit [7]

2.3 Bridge Unit

The bridge unit is as shown in Fig. 7. This bridge unit is capable of carrying data from multiplier array to FMUL's add-round unit to perform FMA operation $((A \times B) + C))$ efficiently. Inputs to this unit are mantissa of the operand C and the carry save format product of $A \times B$ from multiplier array. The operand C is aligned based on the exponent difference between exponent of C and exponent of the product. After alignment, the select line 'sub' decides whether to perform inversion or not. This inversion provides effective 2's compliment for effective subtraction. If sub = 1, it performs inversion on the aligned data else the aligned data is buffered.

Bridge unit adds the product (i.e., mul_sum and mul_carry) $A \times B$ along with a part ([108:0]) of pre-aligned 161-bit addend (operand C) using 3:2 CSA as shown in Fig. 7. The remaining 52 ([161:109]) bits of the 161 added is given to the incrementer in FMA/FMUL's add-round unit. The 109-bit sum and carry obtained from 3:2 CSA is given to multiplexer stage in FMA/FMUL's add-round unit.

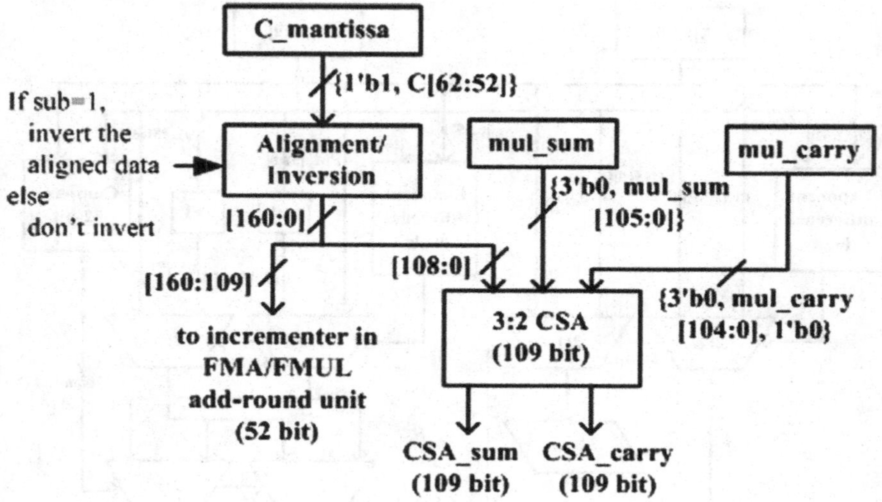

Fig. 7 The bridge unit

Consider 'D' as the exponent difference between exponent of C and exponent of product $A \times B$, its value is $D = \exp(C) - (\exp(A) + \exp(B))$, where exp($A$), exp ($B$), and exp ($C$) are the exponents of operands A, B, and C, respectively.

When $D \geq 0$ (i.e.,$\exp(C) > (\exp(A) + \exp(B))$), the normal aligner will shift exponent of $A \times B$ right by D bits or shift 'C' left by D bits until the exp ($A \times B$) = exp (C). When $D \geq 56$, the sum and carry are placed right of LSB of operand 'C'

When $D < 0$, the operand 'C' will be shifted right by D bits until exp ($A \times B$) = exp (C). For the right shift greater than 105 (i.e., $D < -105$), the operand C is placed to the right of the LSB of the sum and carry (product).

To avoid bidirectional shifter, the alignment is totally implemented as right shift by placing operand 'C' left to that of sum and carry and by placing two extra bits (guard bit and round bit) between the two. Combining both the cases the shift amount will be in the range of 161-bit right shifter. Figure 8 shows how to align operand 'C' in different cases in detail.

- In case of $D \geq 0$, the shift amount is shift amount = max $\{0, 56 - D\}$
- In case of $D < 0$, the shift amount is shift amount = min $\{161, 56 - D\}$

2.4 FMA/FMUL Add-Round Unit

FMA/FMUL add-round unit is shown in Fig. 9. This same add-round unit is used for both FMA and FMUL operation. When a stand-alone FMUL is required it acts

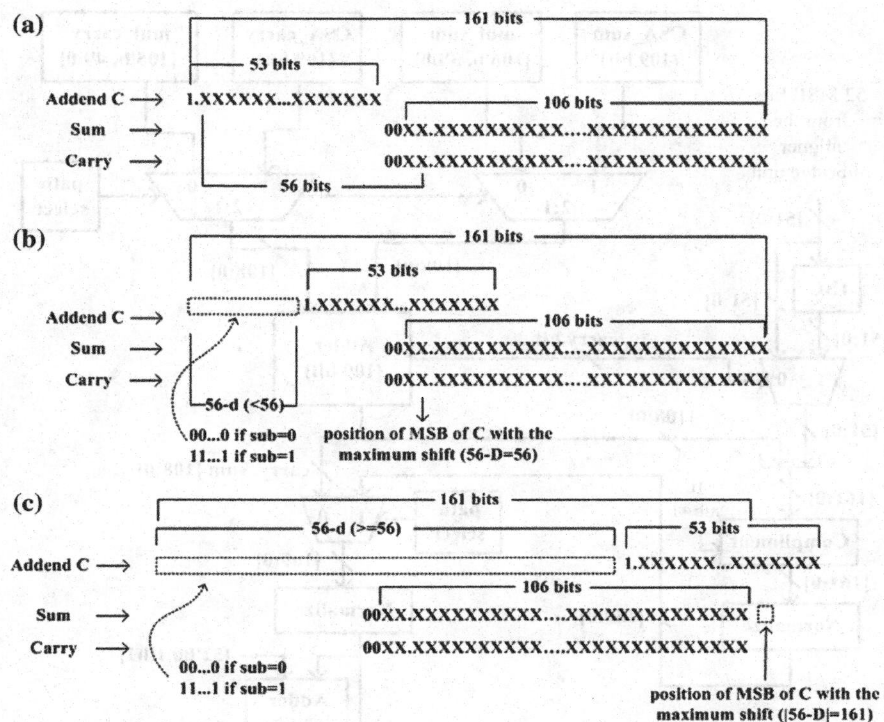

Fig. 8 Alignment of operand C. **a** Before alignment. **b** Alignment with $D \geq 0$. **c** Alignment with $D < 0$

as FMUL add-round unit and when FMA is required it acts as FMA add-round unit. Multiplexer stage is used to select FMA or FMUL. 109 bit Kogge–Stone adder is used to add the data from the mux stages. In parallel to this part of aligner output (52 MSB's) from the bridge unit is given to incrementer. Based on the carry from 109-bit adder the 2:1 mux will select the aligner output or the incrementer output. Compliment the output if necessary. After normalizing the data is sent to perform rounding.

Basically three bits after the LSB decides the rounding. The three bits next to LSB are guard bit (g), round bit (r), and sticky bit (s), respectively. Sticky bit is the logical OR of all bits beyond the guard bit. In the Fig. 9, $R[2:0]$ represents $\{g, r, s\}$, respectively. Round-up method which is in [13] is used for rounding purpose, result and result + 1 need to be generated for rounding up. By using the rounding table given in [13] the result is rounded. Finally mantissa of the FMA/FMUL output will be obtained. In parallel to this the exponent is to be adjusted accordingly whenever the normalization or shifting is done.

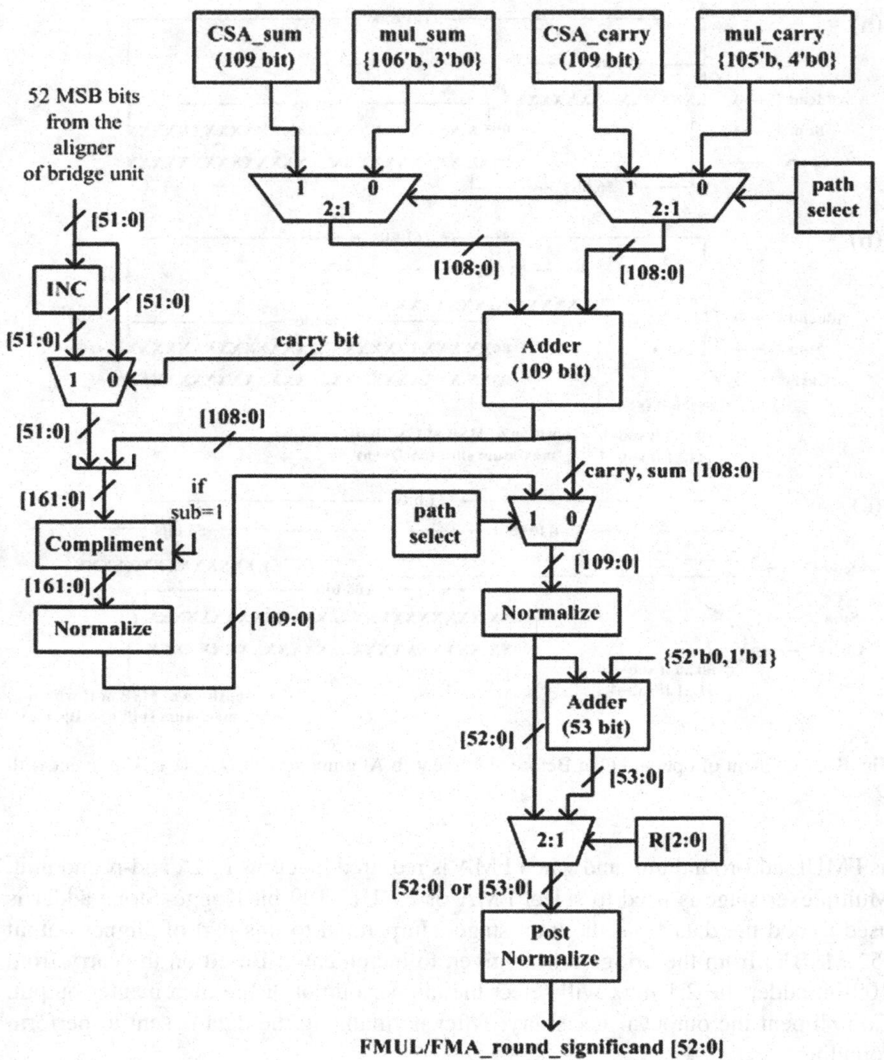

Fig. 9 FMA/FMUL add-round unit

2.5 *FADD Add-Round Unit*

The add-round unit which is shown in Fig. 10 is exclusively used for FADD operation. The far path and close path operands from floating-point adder are given to FADD add-round unit. The two selected inputs are passed through 56 bit Kogge–Stone adder and the 56-bit 3:2 CSA. In order to perform round-up we are taking third input of the CSA as {55′b0, 1′b1}. The rounding is done in the same way as

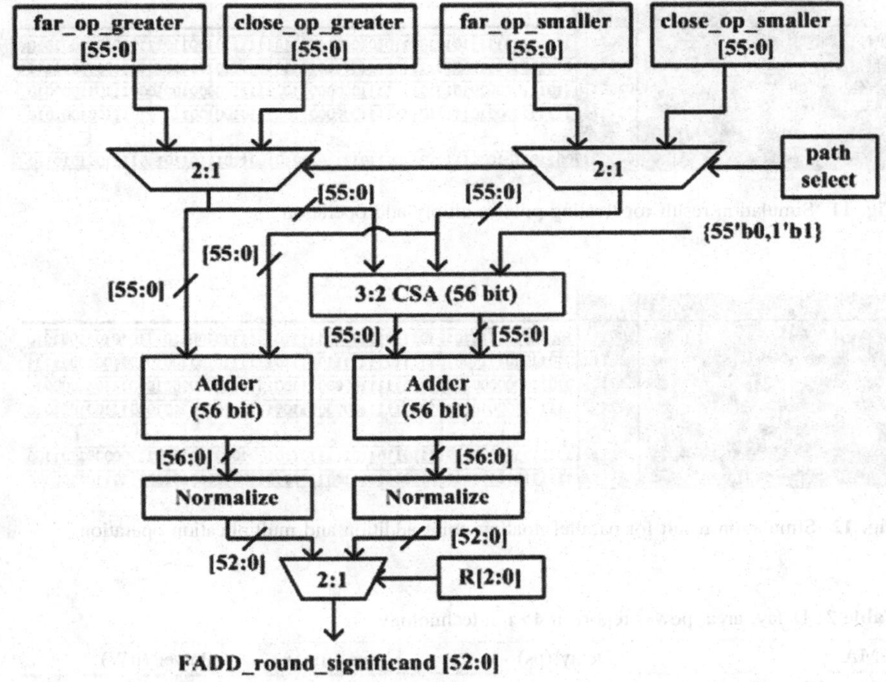

Fig. 10 FADD add-round unit

the multiplier. Then sum and carry from CSA is added with one more 56-bit Kogge–Stone adder. Finally mantissa of the adder output will be obtained. In parallel to this the exponent is to be adjusted accordingly whenever the normalization or shifting is done.

3 Results and Discussion

Simulation results for floating-point multiply-add operation is shown in Fig. 11. Simulation result for parallel floating-point addition and multiplication operation is shown in Fig. 12. Synthesis report for proposed FMA design and FMA design in [7] using Cadence RTL Compiler in 45 nm technology is given in Table 2.

From Table 2 we found that delay and power consumed for stand-alone FADD and FMA operation decreased. The comparison charts for delay, power, and area is shown in Figs. 13, 14 and 15 respectively.

The proposed FMA unit has achieved 7 and 18 % improvement in delay for FADD and FMA instructions respectively, 19 and 17 % improvement in power consumption for FADD and FMA instructions, respectively, and 6 % improvement

Messages		
/FMA_unit_test/A	100111	1001111011111010101110101010111111011111101001101110101011001010
/FMA_unit_test/B	010110	0101101011110010111110111111101111011111101011001010110010011111
/FMA_unit_test/C	010111	0101110000001010101111111100000001100111110000010010000111011000010
/FMA_unit_test/D	111111	11111110001011101011111110000000110011101001001101010111001001010
/FMA_unit_test/sel	0	
/FMA_unit_test/FMA_or_FMUL_out	001101	0011011010011111101101110111101101010011011101101101111011010110

Fig. 11 Simulation result for floating-point multiply-add operation

Messages		
/FMA_unit_test/A	100111	1001111011111010101110101010111111011111101001101110101011001010
/FMA_unit_test/B	010110	0101101011110010111110111111101111011111101011001010110010011111
/FMA_unit_test/C	010111	0101110000001010101111111100000001100111110000010010001110110000010
/FMA_unit_test/D	111111	11111110001011101011111110000000110011101001001101010111001001010
/FMA_unit_test/sel	1	
/FMA_unit_test/FMA_or_FMUL_out	101110	1011101000000101111011011011111011010010001111000111000001101100
/FMA_unit_test/FADD_out	111111	1111110111010100000001111110011000101101100101010001101101100000

Fig. 12 Simulation result for parallel floating-point addition and multiplication operation

Table 2 Delay, area, power report in 45 nm technology

FMA	Delay (ps)	Area (μm²)	Power (μW)
Bridge FMA [7]	FADD: 7498.90	34271.51	FADD: 522.22
	FMUL: 6998.90		FMUL: 3510.21
	FMA: 9156.10		FMA: 4582.01
Proposed FMA design	FADD: 6995.20	32542.21	FADD: 422.40
	FMUL: 7499.00		FMUL: 3564.82
	FMA: 7499.00		FMA: 3823.64

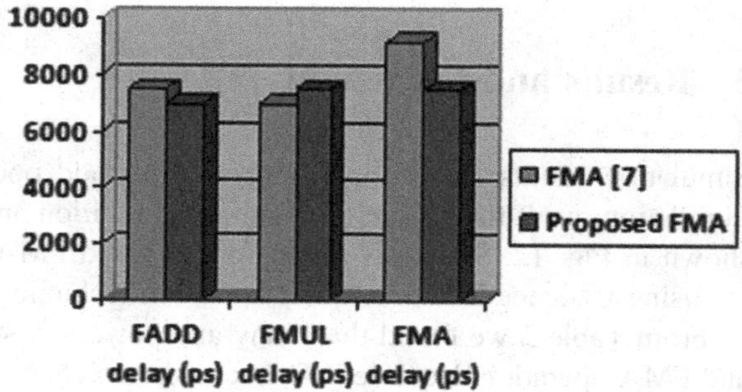

Fig. 13 Comparison chart for delay of proposed FMA with FMA in [7]

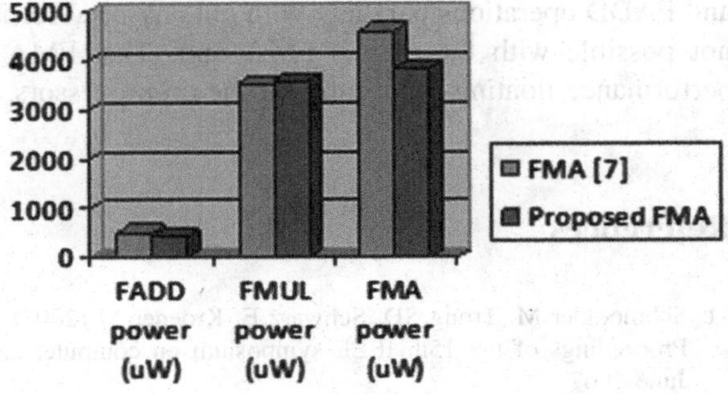

Fig. 14 Comparison chart for power of proposed FMA with FMA in [7]

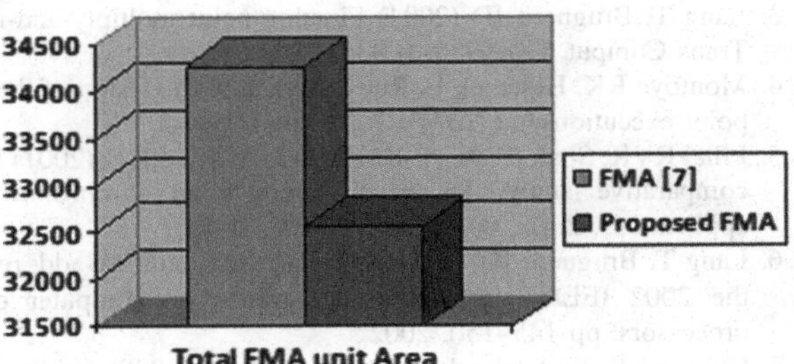

Fig. 15 Comparison chart for delay of proposed FMA with FMA in [7]

in total area when compared to FMA in [7]. Proposed FMA unit for FMUL instruction consumes almost same power as that of the FMA in [7]. But the drawback of proposed FMA unit is that, it has 7 % degradation in timing for FMUL instruction when compared to FMA in [7].

The stand-alone FMUL and FADD operations in existing floating-point units and ALUs [14, 15, 16] can be replaced by floating-point fused multiply-add, if a floating-point addition is followed by a floating-point multiplication. Further this FMA design can be extended and implemented using Residue Number System as it is gaining popularity for fast arithmetic operations [17].

4 Conclusion

This paper presents a low power and area efficient double precision floating-point fused multiply-add unit. The use of common add-round unit for FMUL and FMA instruction is the main reason for reduction in area occupied by the unit. By this the overall power consumption of the unit also decreased. The design has been compared with existing bridge FMA and it is found to be efficient in terms of power and area. But the only drawback is the degradation in timing for FMUL instruction. The proposed FMA can perform FMA operation or it can perform stand-alone FMUL and FADD operations parallely with out any need for insertion of constants. This is not possible with the classic FMA unit. This FMA design is suitable for high performance floating-point units of the co-processors.

References

1. Schmookler M, Trong SD, Schwarz E, Kroener M (2007) P6 Binary floating-point unit. In: Proceedings of the 15th IEEE symposium on computer arithmetic, Montpellier, pp 77–86, June 2007
2. Hokenek E, Montoye R, Cook PW (1990) Second-generation RISC floating point with multiply-add fused. IEEE J Solid-State Circuits 25(5):1207–1213
3. Lang T, Bruguera JD (2004) Floating-point multiply-add-fused with reduced latency. IEEE Trans Comput 53(8):988–1003
4. Montoye RK, Hokenek E, Runyon SL (1990) Design of the IBM RISC System/6000 floating point execution unit. IBM J Res Dev 34:59–70
5. Pillai RVK, Shah SYA, Al-Khalili AJ, Al-Khalili D (2001) Low power floating point MAFs-a comparative study. In: Sixth international symposium on signal processing and its applications, 2001, vol 1, pp 284–287, 2001
6. Lang T, Bruguera JD, Floating-point fused multiply-add: reduced latency. In: Proceedings of the 2002 IEEE international conference on computer design: VLSI in computers and processors, pp 145–150, 2002
7. Quinnell E, Swartzlander EE, Lemonds C (2008) Bridge floating-point fused multiply-add design. IEEE Trans Very Large Scale Integr (VLSI) Syst 16(12):1727–1731
8. IEEE Standard for Binary Floating-Point Arithmetic (1985) ANSI/IEEE Standard 754–1985, Reaffirmed 6 Dec 1990, 1985
9. Dadda L (1964) Some schemes for parallel multipliers. IEEE Trans Comput 13:14–17
10. Waters RS, Swartzlander EE (2010) A reduced complexity wallace multiplier reduction. IEEE Trans Comput 59(8):1134–1137
11. Dimitrakopoulos Giorgos, Nikolos Dimitris (2005) High-speed parallel-prefix VLSI ling adders. IEEE Trans Comput 54(2):225–231
12. Anitha RV, Bagyaveereswaran (2012) High performance parallel prefix adders with fast carry chain logic. Int J Adv Res Eng Technol 3(2):01–10
13. Quach N, Takagi N, Flynn M, (1991) On fast IEEE rounding, Stanford University, Stanford, CA, Technical Report CSL-TR-91-459, Jan 1991
14. Dhanabal R, Bharathi V, Shilpa K, Sujana DV, Sahoo SK (2014) Design and implementation of low power floating point arithmetic unit. Int J Appl Eng Res 9(3):339–346, 2014. ISSN: 0973-4562
15. Ushasree G, Dhanabal R, Sahoo SK (2013) VLSI implementation of a high speed single precision floating point unit using verilog. In: Proceedings of IEEE conference on information and communication technologies (ICT 2013), pp 803–808, 2013
16. Dhanabal R, Bharathi V, Salim S, Thomas B, Soman H, Sahoo SK (2013) Design of 16-bit low power ALU-DBGPU. Int J Eng Technol 5(3):2172–2180
17. Dhanabal R, Sarat Kumar Sahoo, Barathi V, Samhitha NR, Cherian NA, Jacob PM (2014) Implementation of floating point mac using residue number system. J Theor Appl Inf Technol 62(2), April 2014

Implementation of MAC Unit Using Reversible Logic

R. Dhanabal and V. Bharathi

Abstract Reversible quantum logic plays an important role in quantum computing. This paper proposes implementation of MAC unit using reversible logic. We have discussed all the elementary reversible logic gates which are used in the design. Here, 4 × 4 irreversible MAC has been compared with reversible MAC and it has been found that, there is 25.6 % reduction in the power consumption. The design has been simulated using ModelSim and synthesized using Cadence RTL compiler.

Keywords Reversible logic · MAC · Quantum logic

1 Introduction

A Reversible logic does not lose energy which motivates to use the logic in various applications. In the reversible design there are three parameters in determining performance of the circuit. The reversible multiplier and adders are used for designing MAC unit which gives better performance compare to existing logics. Other than reversible logic if Vedic algorithms are adapted in reversible logic cell design it may provide much more sophisticated circuits for supreme performance.

R. Dhanabal (✉)
VLSI Division, SENSE, VIT University, Vellore, India
e-mail: rdhanabal@vit.ac.in

V. Bharathi
CSE, GGR College of Engineering, Anna University, Vellore, India
e-mail: bharathiveerappan@yahoo.co.in

L.P. Suresh and B.K. Panigrahi (eds.), *Proceedings of the International Conference on Soft Computing Systems*, Advances in Intelligent Systems and Computing 397, DOI 10.1007/978-81-322-2671-0_32

2 Methodology

Any reversible gate has to be physically reversible [1, 2] and logically reversible. So any NxN reversible [3] gate has N inputs and N outputs with few inputs getting transferred to output unaltered. All the inputs are uniquely mapped to their corresponding outputs. The basic reversible gates are (Figs. 1 and 2).

1 × 1 Not gate has only one input and the output is the not of it.

2 × 2 Feynman gate has two inputs and one of its input is unaltered at the output and the second output is the XOR operation. B is the target signal and A is the control signal which controls the transfer of A to the output.

2 × 2 Controlled V gate has a V operator in between as shown in the Fig. 3. When A is 1, the output is V(B) otherwise B (Fig. 4).

2 × 2 Controlled V^+ gate is operated in the same way according to the value of A. There are few more gates like Fred kin gate, Peres gate [5], etc. (Figs. 5 and 6).

The 3 × 3 Fredkin gate [6] is shown above. It has one input unaltered and the other outputs implement some function.

The 3 × 3 Peres gate also has one of its inputs unaltered while the remaining two outputs can be used to implement half adder if C is made zero (Fig. 7).

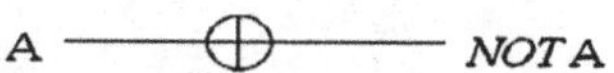

Fig. 1 1 × 1 NOT gate [4]

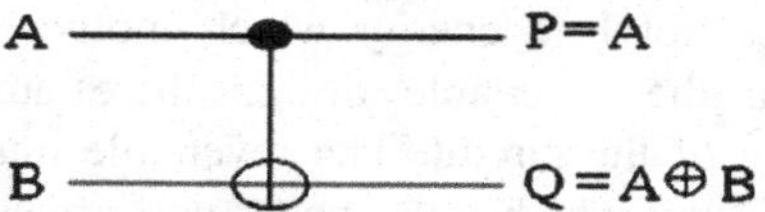

Fig. 2 2 × 2 Feynman gate [4]

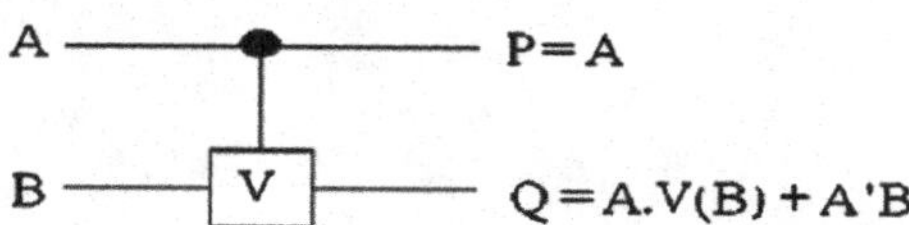

Fig. 3 2 × 2 Controlled V gate [4]

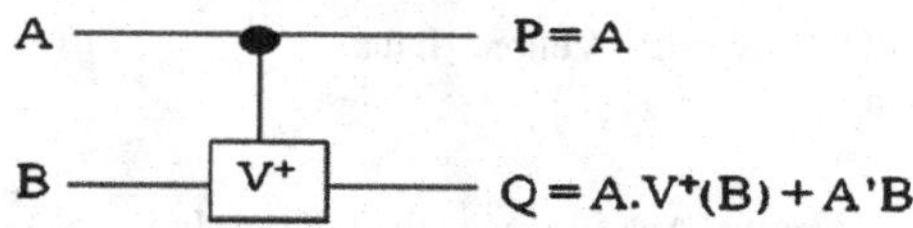

Fig. 4 2 × 2 Controlled V^+ gate [4]

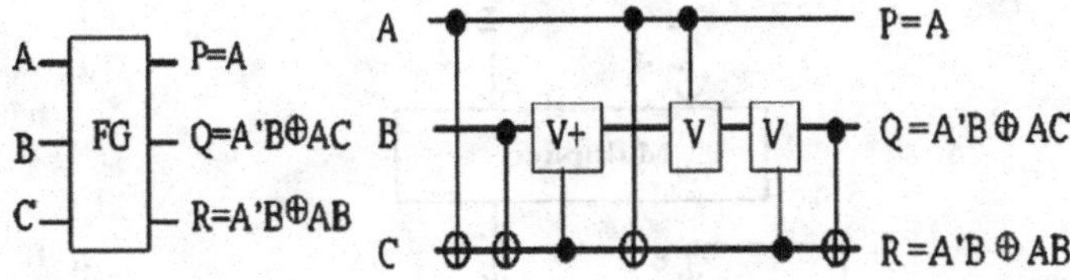

Fig. 5 Fredkin gate [4]

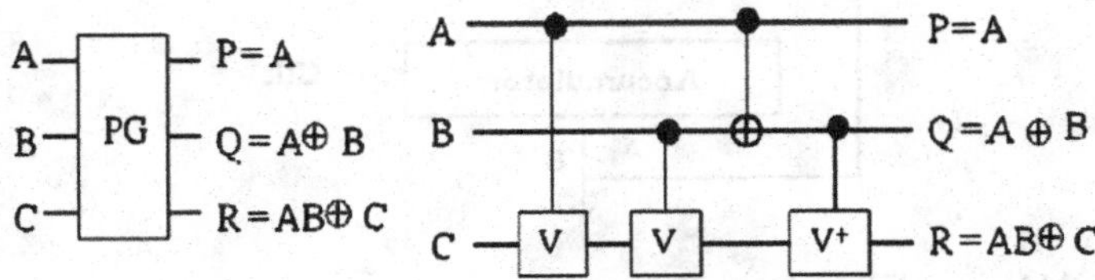

Fig. 6 Peres gate [4]

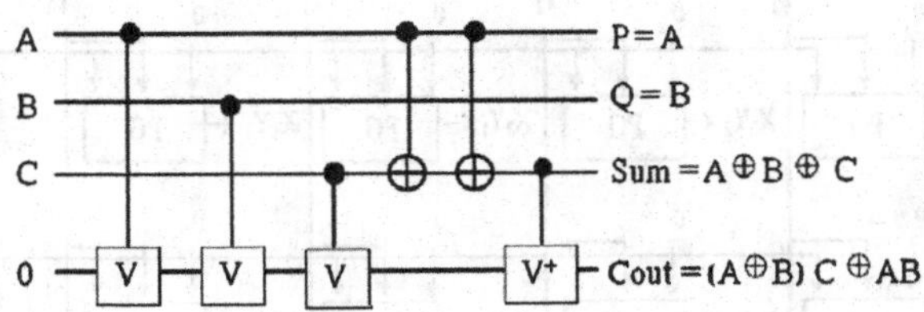

Fig. 7 Full adder [4]

3 Full Adder

The full adder circuit is shown above. As shown, the fourth input is made zero to implement the full adder functionality.

4 Proposed Architecture

The MAC Unit preforms repetitive, multiplication, and summation and of the operands, an adder to add the result to the previously stored value and an accumulator to store the values. As shown in Fig. 8, the MAC unit [4] shown here has two 4 bit inputs that have to be multiplied and an 8 bit output has to be accumulated in the accumulator. This has to be done within one clock cycle.

The two inputs are multiplied using the partial product generator formed by the interconnection of Peres gates like shown below can be used to implement proposed circuit (Fig. 9).

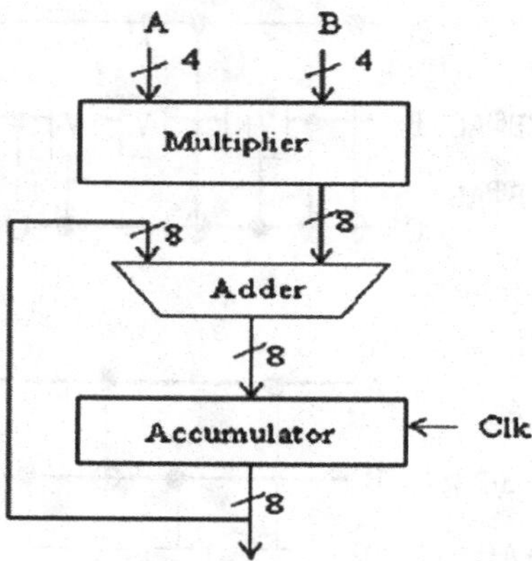

Fig. 8 Basic MAC unit [7]

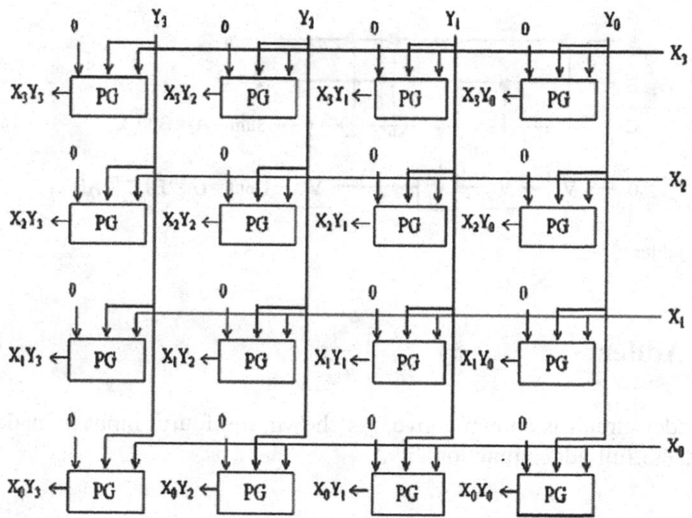

Fig. 9 Partial product generator [4]

The partial products generated must be added to get the final product. So here the adder is shown for that which is made up of half adders and full adders. These half adders and full adders are further made up of basic reversible logic gates, i.e., half adder can be implemented using Peres gate and full adder also has a special circuitry as already mentioned (Fig. 10).

The 8 bit adder for adding the products to the already accumulated result is shown below. It is a combination of full adders and half adder built of reversible logic gates (Fig. 11).

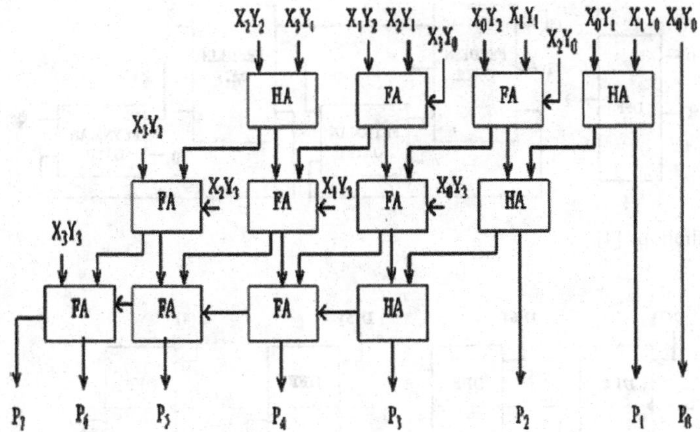

Fig. 10 Partial product adder [4]

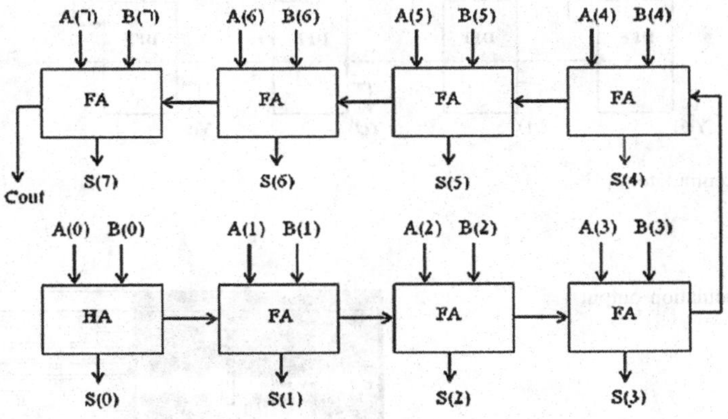

Fig. 11 Adder [4]

The Accumulator circuitry can be formed by the interconnection of D-flipflops [8]. Each D-flipflop can be built of basic Fredkin and Feynman gates to get the functionality implemented in reversible logic (Figs. 12 and 13).

5 Simulation

Here as shown in Fig. 14, in the first cycle the inputs are binary values of 3 and 6 which when multiplied give 18 and it is stored in the accumulator. For the second cycle, inputs are binary values of 5 and 3 which when multiplied give 15 as result and it is added to the previously stored 18 to get the final accumulator's result.

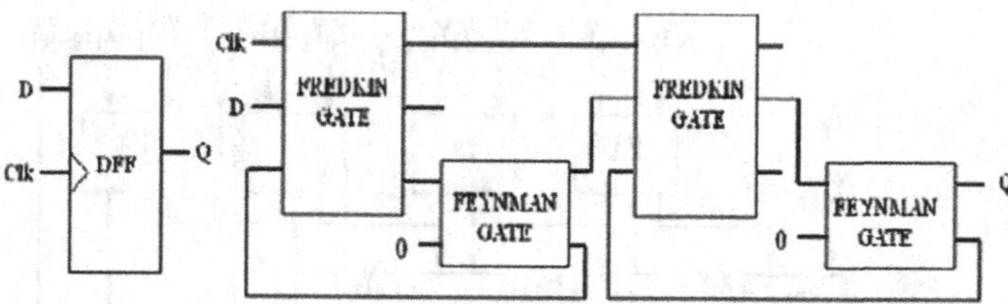

Fig. 12 D-flipflop [4]

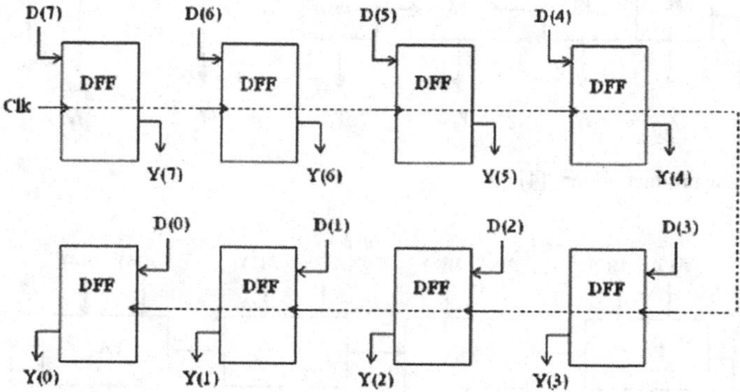

Fig. 13 Accumulator [4]

Fig. 14 Simulation output

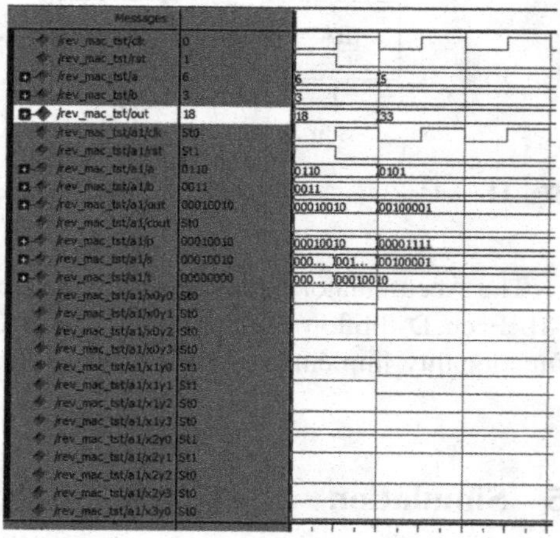

Table 1 Comparison table

Architecture	Area (μm^2)	Power (MW)	Timing (ps)
With reversible logic	270	108143.317	1766
Without reversible logic	300	145,360	1890

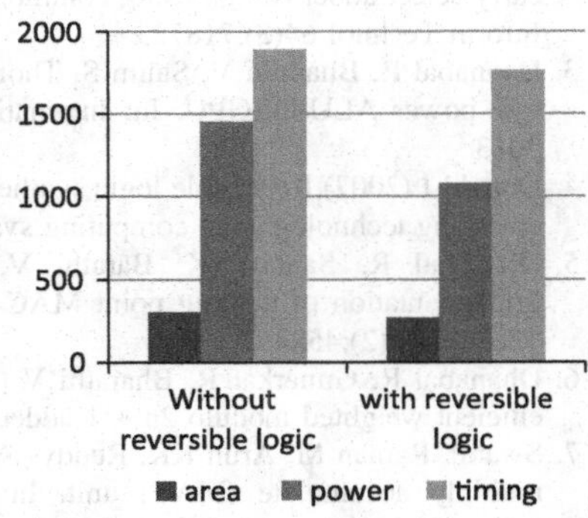

Graph 1 Comparison of area, power and speed/timing analysis graph of MAC unit proposed with MAC without reversible logic

6 Results and Comparison

Here, the results of synthesis are tabulated for both MAC with reversible logic and without reversible logic in terms of power, area, and timing as shown in Table 1. The reversible logic has lower power and area compared to the other (Graph 1).

7 Conclusion

The reversible MAC which is proposed in this paper has 25.6 % of power reduction compared to normal 4 × 4 bit MAC [9]. Hence, it can be concluded that the reversible logic consumes lesser power, and therefore efficient to use for low power circuits for portable processing device elements. The proposed architecture can be used for low power portable applications like mobile phones, tablets, etc. where low power dominates more for enabling battery operated device modeling compare to high-speed device modeling. Future enanchement for this paper will be implementing the reversible logic technique in [10–12] and compare the performance with the existing ciruicts in the papers [10–12], modify the circuit in [13] with reversable logic technique and improve its capabilities.

References

1. Kukati S, Sujana DV, Udaykumar S, Jayakrishnan P, Dhanabal R (2013) Design and implementation of low power floating point arithmetic unit. In: Proceedings of the 2013 international conference on green computing, communication and conservation of energy, ICGCE 2013, art no. 6823429, pp 205208, 2013
2. Dhanabal R, Srivastava D, Bharathi V (2014) Low power and area efficient half adder based carry select adder design using common boolean logic for processing element. J Theor Appl Inform Technol 64(3):718723
3. Dhanabal R, Bharathi V, Salim S, Thomas B, Soman H, Sahoo SK (2013) Design of 16-bit low power ALU-DBGPU. In: International journal of engineering and technology (IJET), 2013
4. Donald J (2007) Reversible logic synthesis with Fredkin and Peres gates. In: ACM journal on emerging technologies in computing systems, 3/1/2008
5. Dhanabal R, Sahoo SK, Barathi V, Samhitha NR, Cherian NA, Jacob PM (2014) Implementation of floating point MAC using residue number system. J Theor Appl Inform Technol 62(2):458463
6. Dhanabal R, Gunerkar R, Bharathi V (2014) Design and implementation of improved area efficient weighted modulo 2n + 1 adder design. ARPN J Eng Appl Sci 9(12):2569–2575
7. Swaraj, Raman M, Arun KK, Reddy, Srinivas K (2012) Reversible implementation of novel multiply accumulate (MAC) unit. In: 2012 international conference on communication information and computing technology (ICCICT), 2012
8. Dhanabal R, Bharathi V, Sri Chandrakiran G, Bharath Bhushan Reddy M (2014) VLSI design of floating point arithmetic and logic unit. J Theor Appl Inform Technol 64(3):703–709
9. https://www.ece.ucsb.edu/its/bluespec/training/…/Lec12_Conclusion.pdf
10. Dhanabal R, Ushashree (2013) Implementation of a high speed single precision floating point unit using verilog. In: International journal of computer applications (0975–8887), 2013
11. Dhanabal R, Bharathi V, Samhitha NR, Pavithra S, Prathiba S, Eugine J (2014) An efficient floating point multiplier design using combined booth and dadda algorithms. J Theor Appl Inform Technol 64(3):819824
12. Kukati S, Sujana DV, Udaykumar S, Jayakrishnan P, Dhanabal R (2013) Design and implementation of low power floating point arithmetic unit. In: Proceedings of the 2013 international conference on green computing, communication and conservation of energy, ICGCE 2013, art. no. 6823429, pp 205208, 2013
13. Dhanabal R, Bharathi V, Anand N, Joseph G, Oommen SS, Sahoo SK (2013) Comparison of existing multipliers and proposal of a new design for optimized performance. In: International journal of engineering and technology (IJET), 2013

3D Modelling Using Depth Sensors Present in Kinect

G. Radhika, S. Ramkumar and K. Narasimha Prasath

Abstract Presently, animation and augmented reality has taken a leap into the boundary of various industries such as gaming, entertainment and architecture. To make the works more amiable, three-dimensional models are constructed normally by hands and by technological invasions, and scanners are being employed. 3D models have become more eminent for they are flexible and accurate. But modelling a 3D object manually can be arduous and engaging scanners are lavish. Thus, in this paper, we propose an approach of using Kinect's depth sensor. Kinect is an add-on for the Xbox 360 (gaming console). The depth image is obtained using the depth sensor where segmentation of a part to be 3D modelled is the exclusivity.

Keywords 3D modelling · Kinect · Depth sensor · Pre-processing · Transformation matrix · Segmentation and bilateral filter

1 Introduction

The 3D modelling process is a set of structured steps by which a real scene is modelled, within a virtual space using a computer. This can be done in two different ways. It can be designed manually using softwares such as SketchUp and

G. Radhika (✉) · S. Ramkumar · K. Narasimha Prasath
Department of Computer Science and Engineering, Amrita School of Engineering, Amrita Vishwa Vidyapeetham (University), Coimbatore 641112, India
e-mail: g_radhika@cb.amrita.edu

S. Ramkumar
e-mail: ramkumar.sokkalingam@gmail.com

K. Narasimha Prasath
e-mail: k.narasimhaprasath@gmail.com

L.P. Suresh and B.K. Panigrahi (eds.), *Proceedings of the International Conference on Soft Computing Systems*, Advances in Intelligent Systems and Computing 397, DOI 10.1007/978-81-322-2671-0_33

AutoCAD. In this method, human has to start designing of models from the scratch and every minute details has to be designed manually. Scanners can be used to scan the shape of the object that has to be modelled. Using those scanned data of the object, the 3D model can be created. This paper gives a step by step procedure for 3D modelling of objects using Kinect depth sensors. This paper will explain modelling of rigid real-world objects using Kinect depth sensors. The design of the proposed system includes pre-processing, finding transformation matrix, segmentation and creating 3D model. First, the 3D data and the image of the object have taken from different directions. The noise in the input data has to be removed using filters. From the image, the features are detected and matched with the image taken from different direction. The transformation matrix is found between the two consecutive images. The 3D data and images are segmented. The 3D segmented data is combined based on the transformation matrix and the 3D model of the object is formed. The detailed design of the proposed system is explained in the following topic.

2 Related Works

3D modelling can be done in many ways. The following are the works done on the field of 3D modelling.

Jurgen Sturm, Erik Bylow, Fredik Kahl and Daniel Cremers describe an approach for 3D modelling of a human body. They print the 3D model that is formed using a 3D model to create miniatures using the 3D printer. There work totally focuses on creating a 3D model of a person. The colour image and depth image of the person is captured continuously by rotating the swivel chair on which the person is seated. They use signed distance function to represent the person's model. The model is optimized to reduce the printing cost.

Zhang [1] has worked on re-implementing the paper from Microsoft Research to reconstruct the cloud points of the whole indoor scene using Kinect. In this work, they have given a approach to create a 3D model of the whole scene. The work done, by Jiakai Zhang, gives an approach to generate a 3D model of the whole scene. The work done by Jurgen Sturm, Erik Bylow, Fredik Kahl and Daniel Cremers gives a specific approach to generate a 3D module of a person and they has included algorithms to optimize the model for printing. We have given a novel approach to create a 3D model of the specified real-world object that can be used in entertainment and gaming industries.

3 Proposed Model

The proposed model is the slight variation of the approach given by Zhang [1]. We have included a step for segmenting the specified project to be modelled. Our approach toward 3D modelling is a five-step process which will generate a 3D model of the object. They are

1. Acquiring images using Kinect
2. Pre-processing
3. Finding transformation matrix
4. Segmentation
5. Creating 3D model

After execution of these 5 steps, the 3D model of the object will be formed.

4 Acquiring Image Using Kinect

The depth image (3D) and the colour image (2D) of the object are taken using Kinect with the help of Kinect SDK. The depth image is taken using Kinect depth sensors and this image is used to form the 3D model of the Object. The colour image is taken using Kinect camera and this image is used to colour the 3D model. The images have to be taken around the object. The object should be placed at a distance within the range of 3–12 feet because the visible depth range of Kinect is 2.5–13 feet. However, a recommended usage range is 3–12 feet (Figs. 1 and 2).

Fig. 1 Depth image on *left*

Fig. 2 Colour image on *right*

5 Pre-processing

The 3D depth image obtained using Kinect sensors are very noisy. The noise should be removed using bilateral filter. Bilateral filter is used because this filter will smooth.

$$I^{\text{filtered}}(x) = \sum_{xi \in \Omega} I(x_i) fr(||I(x_i) - ||) g_s(||(x_i) - ||)$$

$$W_p = \sum_{xi \in \Omega} fr(||I(x_i) - I(x)||) g_s(||x_i - x||)$$

$$w(i,j,x,y) = e\left(-\frac{(i-k)^2 + (j-l)^2}{2\sigma_d^2} - \frac{||I(i,j) - (k,l)||^2}{2\sigma_r^2}\right)$$

$$I_d(i,j) = \frac{\sum_{k,l} I(k,l) * w(i,j,k,l)}{\sum_{k,l} w(i,j,k,l)}$$

- I^{filtered} is the filtered image
- I is the original input image to be filtered.
- x are the coordinates of the current pixel to be filtered.
- Ω is the window centred in.
- f_r is the range kernel for smoothing differences in intensities. This function can be a Gaussian function.
- g_s is the spatial kernel for smoothing
- differences in coordinates. This function can be a Gaussian function.
- (i,j) is the pixel location that has to denoised.
- (k,l) is one of the neighbouring pixels.
- I_D is the denoised intensity of pixel (i,j).
- σ_d and σ_r are the smoothing parameters. $w(i, j, k, l)$ gives the weight assigned for pixel (k,l) to denoise the pixel (i,j).

6 Finding Transformation Matrix

The transformation matrix provides the 6 degrees of freedom (DOF) of the camera. The 6DOF of the camera are 3 directions of movement along the principle axis (*x*, *y* and *z*) and rotation about the 3 principle axis. It shows displacement details and angular rotation made by the camera to take the second image from the first one. The transformation matrix of the two images can be obtained using many methods. In this approach, we have made use of iterative closest point (ICP) algorithm. We give the 2 consecutive depth image of the object as point cloud and the normal vector to the ICP algorithm [1]. The following formulas are used to find the normal vector of the 2 depth images.

$$u = (x, y) \tag{1}$$

$$n(u) = v_i(u) x v_j(u) \tag{2}$$

$$v(\mathrm{u}) = (\mathrm{u})K^{-1}(\mathrm{u}, 1) \tag{3}$$

n(*u*) is the normal vector that has to be passed to ICP algorithm. *K* is the intrinsic calibration matrix. Normally *K* should be $\begin{matrix} 1/f & 0 & 0 \\ 0 & 1/f & 0 \\ 0 & 0 & 0 \end{matrix}$ for kinect camera the formula is revised as follows [1]

$$v \cdot x = (u \cdot x - c \cdot x) * D(u)/f \cdot x \tag{4}$$

$$v \cdot y = (u \cdot y - c \cdot y) * D(u)/f \cdot x \tag{5}$$

$$v \cdot z = D(u) \tag{6}$$

D(*u*) gives the depth value at *u*.

Using the above formulas, normal vector is calculated for each pair of consecutive images. The pseudo code of ICP algorithm is as follows [2] principle axis (*x*, *y* and *z*) and rotation about the 3 principle axis. It shows displacement details and angular rotation made by the camera to take the second image from the first one. The transformation matrix of the two images can be obtained using many methods. In this approach, we have made use of ICP algorithm. We give the 2 consecutive depth image of the object as point cloud and the normal vector to the ICP algorithm [1]. The following formulas are used to find the normal vector of the 2 depth images.

a) Reconstruct projection rays from the image points of Image 1
b) For each projection ray R:
c) For each point in Image 2
    (c1) Estimate the nearest point P1 of ray R to a point on the contour
    (c2) if (n==1) choose P1 as actual P for the point – line correspondence (c3)
    else compare P1 with P: if
        distance(P1,R) is smaller than distance(P,R) then
    choose P1 as new P
d) Use (P,R) as correspondence set.
e) Estimate pose with this correspondence set
f) Transform Image 1 using the estimated pose
g) Goto (b)

Using the above algorithm the transform matrix is found for each pair of depth image.

7 Segmentation

To obtain the depth data of particular object, segmentation is applied to each image taken using the camera. For segmentation, histogram-based method is used [3]. Histogram-based image segmentation uses histogram to select the grey levels for grouping pixels into regions. Before segmentation, the images have to be pre-processed. The steps of pre-processing are as follows [3]

- Histogram equalization.
- Histogram smoothing.
- Thresholding.
- Region Growing.

For histogram-based segmentation, we have used adaptive histogram technique. This [3] technique uses the peaks of the histogram in a first pass and adapts itself to the objects found in the image in a second pass.

8 Creating a 3D Model

To generate a 3D model of the object, we have to combine all the depth images. To do that, first, we select one of the depth data to be fixed and we have to transform the other depth data using the transform matrix obtained in the third step. Take 2 consecutive depth images. Transform the second image with the transform matrix obtained from step 3 and combine the points in depth data 2 with depth data 1. Find the transform matrix of the next depth data with reference to the first depth data. Transform it with reference to the first image and combine its depth data to the first depth data. This operation is to be performed for each depth data. While combining the depth data, there will be some irregularities in the combining process. So the point cloud formed after combining the depth data of the object is again smoothed using bilateral filter. After completing this process, the point cloud of the object will be formed. Triangular mesh is applied for the point cloud of the object. This process is carried out with the help of Mesh Lab libraries.

9 Conclusion

3D modelling of real-world objects plays a important role in many industries. In this paper, we have explained a novel approach towards 3D modelling of an object using Kinect depth sensors. The approach is a five-step process. First, we have to capture point cloud of the object from different angles using the Kinect sensors. Next, we have to smooth the point clouds with a bilateral filter which will smooth the data while preserving the edges. From the smoothed point clouds, first, we find the transform of 2 consecutive point clouds using ICP algorithm. From the point cloud, the depth data of the object is alone segmented out using the histogram-based segmentation method. From the point cloud, one of the point clouds is made as a reference. The transform matrix found using the ICP algorithm is changed with reference to this point cloud. Other point clouds taken from different direction are combined with this point cloud by applying a transform of the transform matrix with reference to the first depth data found in step 3. The output of step 4 will be a point cloud of the object. Finally, a triangular mesh is formed using the point cloud with the help of Mesh Lab. The output of the final step will be a 3D model of the object.

10 Future Enhancement

This paper discusses about how the 3D model of an object is formed using the Kinect sensors. This paper explains a five-step process to create a 3D model. This project can be extended in future using the concept of augmented reality where a user can interact with the 3D model as if it is a real object. Augmented reality is a technology that superimposes a computer-generated image on a user's view of the real world, thus providing a composite view. The user should be able to perform actions on 3D model displayed using augmented reality. The action can include rotate, slice, move, scale and pane. Many research and projects are being performed on this field. If the 3D model obtained is combined with the concept of augmented reality, then the 3D model can be used in the gaming industry where augmented reality is used.

11 Result

We took the depth data of the object for which the 3D model is to be formed. The object is placed stationary. The depth data is taken by moving the camera around the object with a constant interval. To form a better 3D model of the object, the depth data is taken with an approximate interval of 12 degrees around the object. Each data is taken by pointing the Kinect sensor toward the object. We took totally 30 images around the object. The depth data is stored as a point cloud. The point cloud obtained using Kinect cameras were very noisy so we applied a bilateral filter which smoothes the image while preserving the edges. The bilateral filter is applied with $\sigma d = 3$ and $\sigma r = 50$. The filter is used to smooth the depth values. The filter will smooth the noise present in the point cloud. Next, we found the transformation matrix of each pair of consecutive point clouds. The transformation matrix is obtained with the help of ICP algorithm. Totally there will be 30 pairs of consecutive depth data where the last 30th data and 1st depth data are given to the ICP algorithm to find the last transformation matrix. Next we apply histogram-based segmentation to the depth image to segment out the object out from the rest of the depth details. Next we fixed the first point cloud to be reference and found the transformation matrix of each image with respect to the first point cloud. The first point cloud is made fixed and other point clouds are transformed and combined with the first point cloud. The result of this step will produce the point cloud of the 3D object that has to be modelled. The result will be containing noise because the combining of point cloud from different direction did not set together so the result of this step is smoothed again using the bilateral filter with the same parameter ($\sigma d = 3$ and $\sigma r = 50$). At the end of the 4th step, the point cloud of the object is formed. A mesh of the point cloud is formed using the Mesh Lab tool which is an open source tool. When a triangular mesh is formed using this tool the final output of this step is the 3D model of the object.

References

1. Zhang J (2012) Real-time 3D reconstruction using Kinect, Sep 2012. http://jiakaizhang.com/project/real-time-3dreconstruction/
2. Rosenhahn B (2005) Pose estimation of 3D freeform contours. Int J Comput Vis, May 2005
3. Phillips D (1993) Image processing, part 9: histogram-based image segmentation (Tutorial). C Users J, Feb 1993 Issue

Closed-Loop Control of Facts Devices in Power System

S.D. Sundarsingh Jebaseelan, C.N. Ravi, G. Nagarajan and S. Marlin

Abstract This paper presents the improvement of reactive power and voltage in fourteen bus system using TCTC and STATCOM. Flexible AC transmission systems (FACTS) are the option to mitigate the problem of overloaded lines due to increased electric power transmission by controlling power flows. In this work, combined controller based on optimal power flow (OPF) with multiple objectives is derived in order to provide secure transmission and reduces transmission losses. Static compensator and thyristor-controlled tap changer are two such compensators belonging to FACTS devices are used in this work. The fourteen bus system with closed-loop-controlled STATCOM and TCTC is modeled and simulated.

Keywords Flexiable AC transmission system · Static compensator · Thyristor-controlled tap changer

S.D. Sundarsingh Jebaseelan (✉) · C.N. Ravi · G. Nagarajan
Sathyabama University, Chennai 600119, Tamil Nadu, India
e-mail: sundarjeba@yahoo.co.in

C.N. Ravi
e-mail: dr.ravicn@gmail.com

G. Nagarajan
e-mail: nagarajanme@yahoo.co.in

S. Marlin
T.J. Institute of Technology, Chennai 600119, Tamil Nadu, India
e-mail: sagayarajmarlin@gmail.com

L.P. Suresh and B.K. Panigrahi (eds.), *Proceedings of the International Conference on Soft Computing Systems*, Advances in Intelligent Systems and Computing 397, DOI 10.1007/978-81-322-2671-0_34

1 Introduction

A lot of energy is used for commercial, home, space, and military applications with the application of power electronics [1]. Power electronics technology has advanced a lot over the last two decades. High-power handling electronic devices are used for FACTS devices [2].

FACTS, which are power electronic-based devices can change parameters like impedance, voltage, and phase angle [3]. The important feature of FACTS is that they can vary the parameters rapidly and continuously, which will allow a desirable control of the system operation.

The most important power quality parameter is system voltage. Voltage is important for proper operation of electrical power system. This voltage depends on reactive power [4–7]. Decreasing reactive power causes voltage to fall and increasing reactive power causes voltage to rise. Active power is the energy supplied to electrical equipment. Reactive power provides the important function of regulating voltage [8].

2 FACTS Controller

Additionally, available generating plants are often not situated near the load centers and power must consequently be transmitted to many kilometers. Due to ever increase in load demand, existing transmission system has to restructure or FACTS devices need to be used [9].

The first group employs reactive impedances or tap changing transformers with thyristor switches as controlled elements. The second group employs self-commutated voltage source switching converters. The static var compensators (SVC), thyristor-controlled series capacitor (TCSC), and thyristor-controlled tap changing transformer (TCTC) belong to the first group of controllers, while static synchronous compensator (STATCOM), static synchronous series compensator (SSSC), unified power flow controllers (UPFC), and interline power flow controllers (IPFC) belong to the other group [10–13]

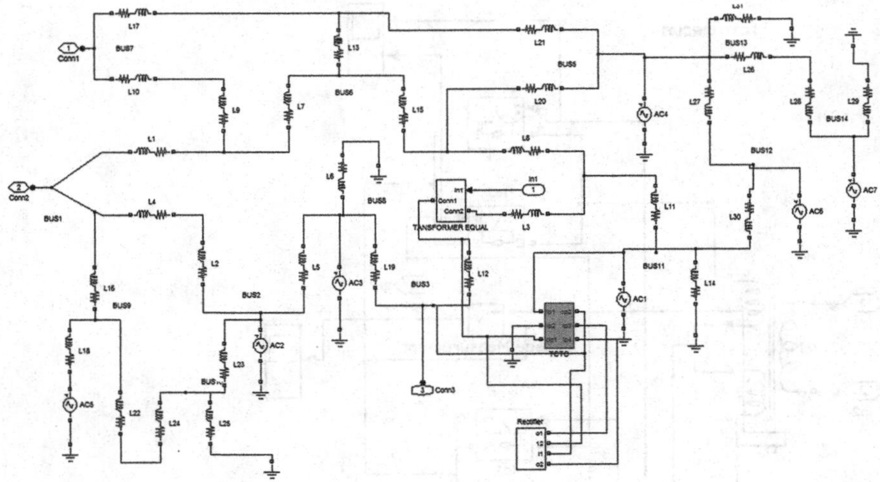

Fig. 1 Closed-loop model for fourteen bus system with STATCOM and TCTC

3 Results and Discussion

A. Closed-Loop Controller

In the fourteen bus system, the STATCOM is installed at bus 4 and TCTC is installed at bus 11. The Matlab Simulink diagram for closed-loop controller fourteen bus system with STATCOM and TCTC is shown in Fig. 1. The TCTC and STATCOM model for closed-loop system is shown in Figs. 2 and 3.

When STATCOM is connected at bus 4 and TCTC is connected in bus 11 the real and reactive power increases. The real and reactive power at bus 4 is shown in Fig. 4. The real and reactive power at bus 11 are shown in Fig. 5. Table 1 shows the summary of real and reactive powers for fourteen bus system with STATCOM and TCTC using with and without PI controller. Table 2 shows the summary of reactive powers for fourteen bus systems with open loop and closed loop using combined STATCOM and TCTC. The Fig. 6 shows the comparison graph reactive power with and without PI controller for 14 bus system. The Fig. 7 shows the comparison graph open loop and closed loop system for 14 bus system.

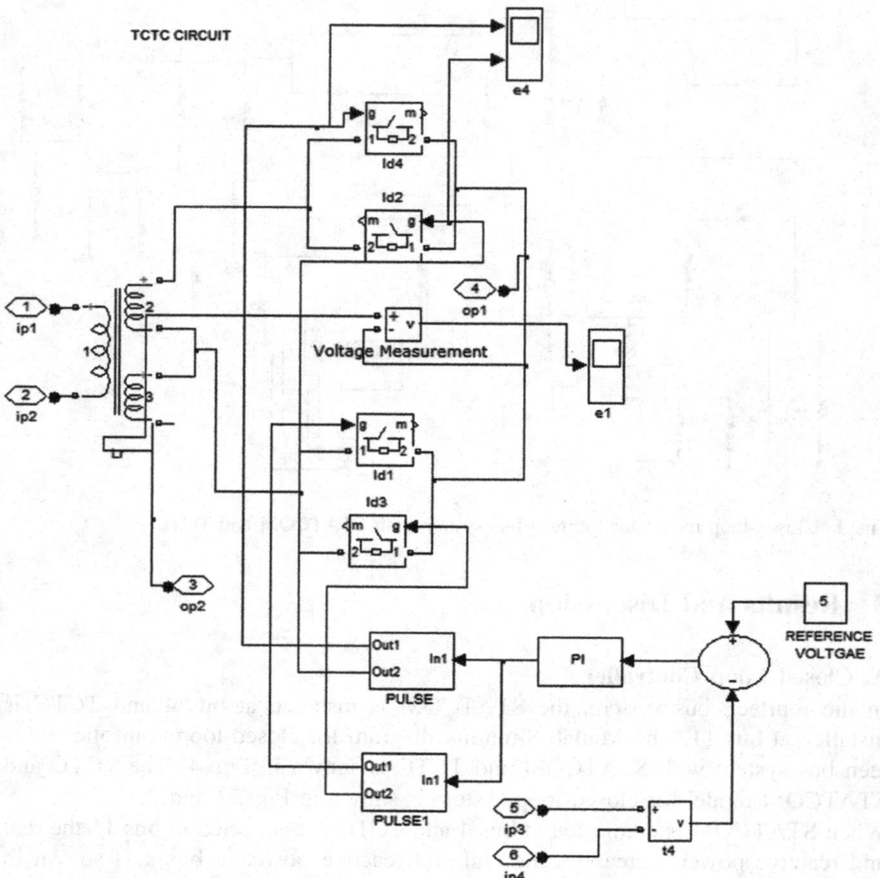

Fig. 2 Closed-loop model for TCTC

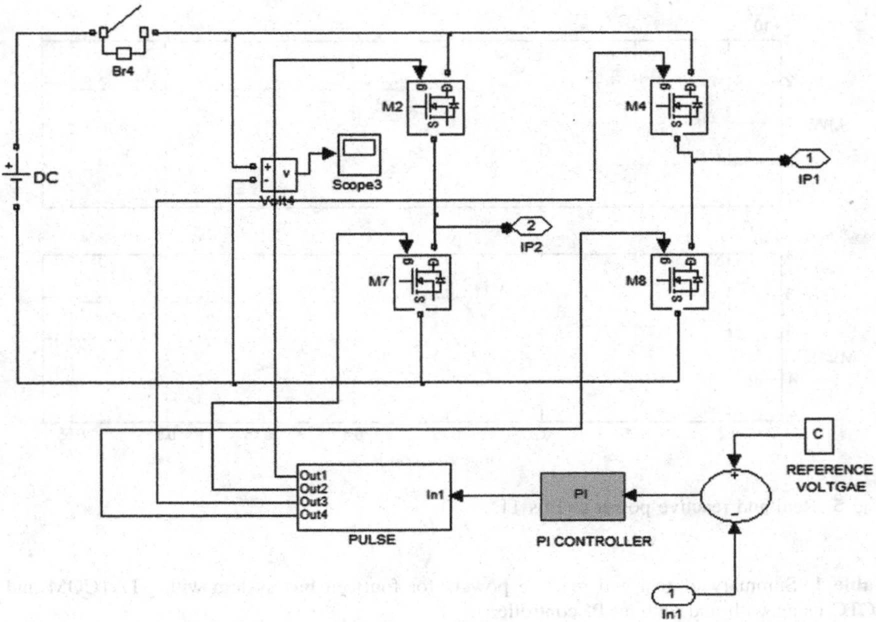

Fig. 3 Closed-loop model for STATCOM with PI controller

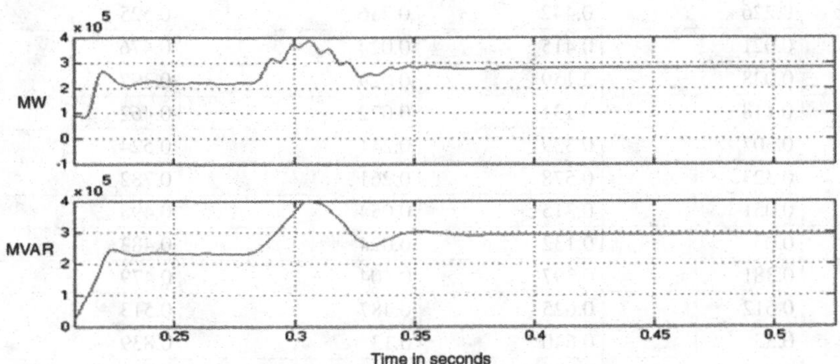

Fig. 4 Real and reactive power at Bus 4

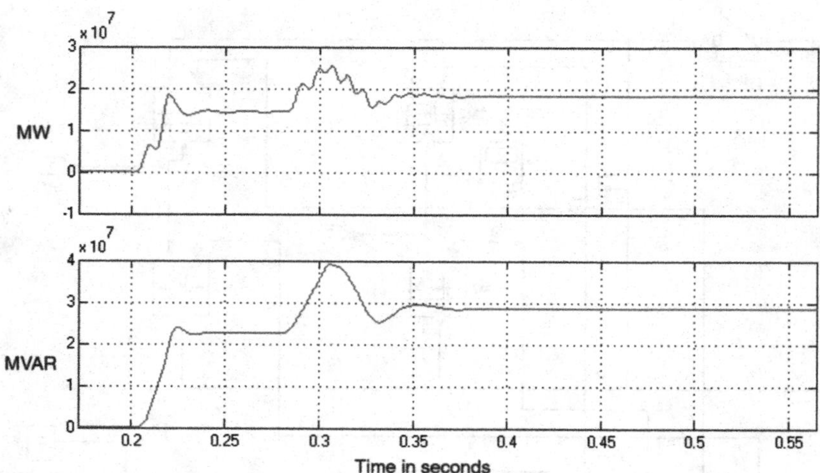

Fig. 5 Real and reactive power at Bus 11

Table 1 Summary of real and reactive powers for fourteen bus system with STATCOM and TCTC using with and without PI controller

Bus no.	Real power (MW) without PI	Real power (MW) with PI	Reactive power (MVAR) without PI	Reactive power (MVAR) with PI
1	0.199	0.146	0.208	0.641
2	0.226	0.442	0.236	0.525
3	0.021	0.415	0.024	0.476
4	0.028	2.159	0.029	0.397
5	0.118	0.131	0.072	0.462
6	0.307	0.339	0.241	0.524
7	0.323	0.578	0.261	0.782
8	0.051	0.315	0.054	0.498
9	0.05	0.132	0.024	0.483
10	0.381	0.397	0.304	0.479
11	0.612	0.625	0.487	0.513
12	0.33	0.640	0.13	0.839
13	0.432	0.434	0.169	0.673
14	0.296	0.397	0.076	0.678

Table 2 Summary of reactive powers for fourteen bus systems with open loop and closed loop using combined STATCOM and TCTC

Bus no.	Open loop with STATCOM and TCTC	Closed loop with STATCOM and TCTC
1	0.169	0.641
2	0.18	0.525
3	0.457	0.476
4	0.379	0.397
5	0.438	0.462
6	0.402	0.524
7	0.495	0.782
8	0.478	0.498
9	0.426	0.483
10	0.474	0.479
11	0.491	0.513
12	0.478	0.839
13	0.484	0.673
14	0.486	0.678

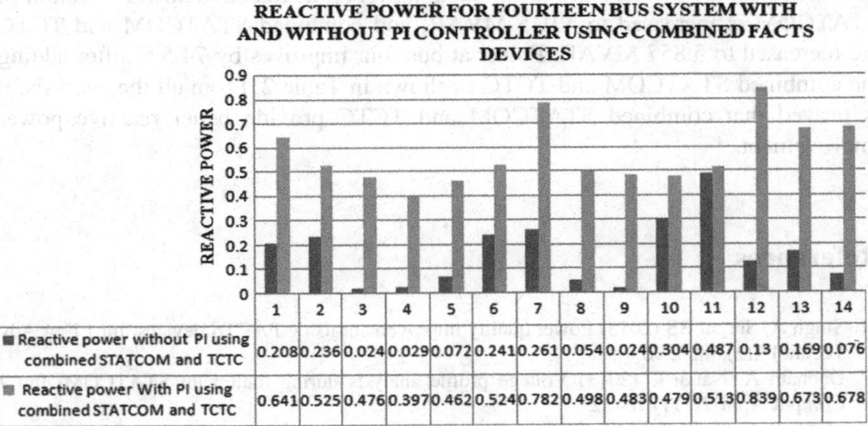

Fig. 6 Comparision of reactive power for fourteen bus system with and without PI controller using combined FACTS devices

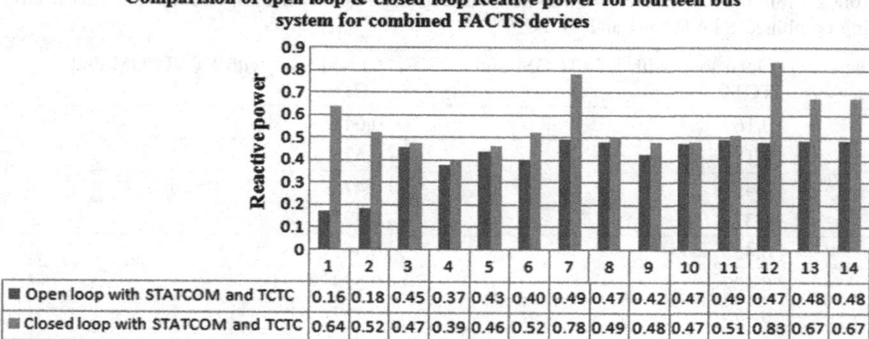

Fig. 7 Comparision of open-loop and closed-loop reactive power for fourteen bus system for combined FACTS devices

4 Conclusion

FACTS devices are analyzed and the circuit model for the fourteen bus system is developed using the blocks available in MATLAB. Fourteen bus systems with TCTC and STATCOM are simulated and the results are presented. Reactive power increases with the increase in the conduction of TCTC. Reactive power injection of STATCOM is increased to 4.945 MVAR, and combined STATCOM and TCTC are increased to 5.857 MVAR. Power at bus four improves by 74.5 % after adding the combined STATCOM and TCTC as shown in Table 2. From all the analysis, it is proved that combined STATCOM and TCTC provide better reactive power improvement.

References

1. Singh A, Surjan BS (2013) Power quality improvement using FACTS devices. Int J Eng Adv Technol 3(2):383–390
2. Debnath A, Nandi C (2013) Voltage profile analysis during fault with STATCOM. Intl J Comput Appl 72(11):16–22
3. Mehta Balwant K, Patel PJ (2013) Static voltage stability improvement in power system using STATCOM FACTS controller. J Inf Knowl Res Electr Eng 2(2):312–316
4. Tooraji HE, Abdolamir N (2013) Improving power quality parameters in AC transmission systems using unified power flow controller. Res J Recent Sci 2(4):84–90
5. Masdi H, Mariun N, Bashi SM, Mohamed A (2004) Design of a prototype D- STATCOM for voltage sag mitigation. In: IEEE conference, proceedings of power and energy, pp 45–49
6. Murali D, Rajaram M, Reka N (2010) Comparison of FACTS devices for power system stability enhancement. Int J Comput Appl 8(4):30–35
7. Quamruzzaman Md, Hossain Miraj R (2013) Voltage level and transient stability enhancement of a power system using STATCOM. Int J Adv Res Comput Sci Electron Eng 2(1):32–37

8. Ahmad A Prof, Shehzad A Dr, Shrivastava VB (2013) Power quality improvement using STATCOM in IEEE 30 bus system. Adv Electron Electr Eng 3(6): 727–732
9. Usha N, Vijaya Kumar M (2013) Enhancement of power quality in thirty bus system using ZSI based STATCOM. Int J Modern Eng Res 3(2):773–778
10. Murugan A, Nagarajan G, Sundarsingh Jebaseelan SD, Ravi CN (2014) Performance analysis of voltage profile, power, angle of injection using combined FACTS device. Int J Appl Eng 9 (21):10303–10316. ISSN 0973-4562
11. Marlin S, Sundarsingh Jebaseelan SD, Padmanabhan B, Nagarajan G (2014) Power Quality improvement for thirty bus system using UPFC and TCSC. Indian J Sci Technol 7(9):1320–1324, Sept 2014
12. Sundarsingh Jebaseelan SD, Godwin Immanuel D (2014) Reactive power compensation using STATCOM with PID controller. Int J Appl Eng 9(21):11281–11290. ISSN 0973-4562
13. Kalai Murugan A, Rajaprabu R Dr (2013) Improvement of transient stability of power system using solid state circuit breaker. Am J Appl Sci 10(6):563–569

A Study on Face Recognition Under Facial Expression Variation and Occlusion

Steven Lawrence Fernandes and G. Josemin Bala

Abstract Two well-known problems are recognizing faces in the presence of facial expression variation and in the presence of occlusion. Humans depict their feelings through facial expressions, and this is an effective way of nonverbal communication. Facial expressions are dynamic, and recognizing faces under varying facial expressions thus are the challenging task. The ability to recognize human affective state through an intelligent machine will empower to interpret, understand, and respond to human emotions, moods, and possibly intentions which is similar with one person to another. On the other hand, occlusion in an image refers to obstructions in the view of an object. Face recognition systems in real-world applications need to manage an extensive variety of obstructions, and faces can occluded by facial accessories (e.g., sunglasses, scarf, cap, cloak), objects before the face (e.g., hand, food, cellular telephone), extreme illumination (e.g., shadow), self-occlusion (e.g., non-frontal pose), or poor picture quality (e.g., blurring). In this paper, we perform a comparative study on various state of the art techniques to recognize faces under varying facial expressions and in the presence of occlusion.

Keywords Face recognition · Facial expression variation · Occlusion

1 Introduction

Face recognition is a critical component in numerous machine vision applications, for example, access control, video surveillance, and public security [1–5]. With the expanding requirements for security-related applications, for example, computational forensics and antiterrorism, face recognition has been a dynamic theme for

S.L. Fernandes (✉) · G. Josemin Bala
Department of Electronics & Communication, Karunya University, Coimbatore, India
e-mail: steva_fernandes@yahoo.com

G. Josemin Bala
e-mail: josemin@karunya.edu

L.P. Suresh and B.K. Panigrahi (eds.), *Proceedings of the International Conference on Soft Computing Systems*, Advances in Intelligent Systems and Computing 397, DOI 10.1007/978-81-322-2671-0_35

researchers in the regions of machine vision and image processing [6–10]. The primary focus of any expression-invariant face recognition algorithm is to mitigate the changes related to facial expression [11]. Automatic facial expression analysis systems are also aiming toward the application of computer vision techniques in the medical care via space mapping between the continuous emotion and a set of discrete expression categories [12, 13]. Face recognition under occlusion is utilized to enhance the execution level such that it will perceive any suspect, disguised in an alternate way, for example, face covered by hand, helmet, goggles, hair scarf, and so forth [14]. The approaches to recognize faces under varying facial expressions and occlusions can be broadly classified into are: feature-based techniques, model-based techniques, and correlation-based techniques. Section 2 contains feature-based techniques. Section 3 contains model-based techniques. Section 4 contains correlation-based techniques. Section 5 contains result and discussion, and Sect. 6 draws the conclusion.

2 Feature-Based Techniques

Authors in [15] proposed a geometric system for breaking down 20 facial appearances with the particular objectives of contrasting, matching, and averaging their shapes. Here it is pointed out to facial surfaces by spiral bends emanating from the nose tips and use elastic shape investigation of these curves. To create a Riemannian system for investigation of full facial surfaces, the systems utilized takes into consideration formal factual impedances. Local feature-based algorithms [16] exploits curvelet change in two routes, i.e., as a key point finder to extract salient points on the face region. And as a multi-scale local surface descriptors that can separate very unique peculiarities around located key points. The main step is extracting sub patches. Tests have demonstrated that 5 × 5 sub patches are perfect. At that point, the patches are reordered utilizing circular shifts. Sparse representation-based face expression recognition (FER) strategy [17] lessen the intra-class variety while accentuating the facial expression in a query face image. To that end, it introduces another system for creating an intra-class variation image of every representation by utilizing preparing training expression image. Authors in [18] propose that a solitary image of a face will have high measure of instability in it. So this could not be an exceptionally precise representation, so different picture of this face is integrated for more precision. First, a set of virtual pictures is integrated utilizing the given image. At that point, an ideal image is chosen among the original and integrated pictures; finally, l2 standard-based representation calculation is connected to arrange test specimen. Authors in [19] proposed face Recognition using ensemble string matching. This is a compact syntactic string face representation that can perform nonsequential string matching between 2 strings face invariant to sequential order and direction of each one string.

3 Model-Based Techniques

The authors in [20] propose a deformable 3D facial expression model and D-Isomap-based classification. Different strategies like finding key fiducial point inborn geometries are used. An enhancement of was proposed by authors in [21]. In this method, a novel multi-view facial expression recognition technique is displayed. Here multi-scale sizes are utilized to parcel every facial image into a set of subregions and completed the extraction in each sub regions. Here, to tackle the optimization problem of GSRRR, an effective algorithm is proposed utilizing inexact augmented Legrangian multiplier approach (ALM) [22]. Authors in [23] proposed another functional framework for the arrangement of facial expressions emoted by people in a spontaneous environment seen by video cameras. The key destination of this technique was prepared an alternate database from the databases utilized for testing. Matching pursuit-based method proposes facial expression recognition utilizing dictionary based on component separation algorithm. This methodology sees the expressive face as superimposition of neutral segment with expression segments. Authors in [24] proposed 3D face recognition under occlusion using masked projection. In this approach, authors propose a completely automatic 3D face recognizer. This model is equipped for attaining to an effective parts-based representation for robust face recognizable proof.

4 Correlation-Based Techniques

Correlation-based technique [25] proposed for facial expression recognition frame work utilizes audio visual information analysis. A model was proposed to cross-modality data correlation while permitting them to be dealt as asynchronous streams. This frame work can enhance the recognition performance while essentially decreasing the computational cost by isolating repetitive frame processing. This is done by consolidating auditory data. A one-to-one binary classification and multi-class order classification performance is assessed utilizing both subject-dependent and independent strategies. Authors in [8] propose a novel geometric frame work for 3D faces, with the particular objectives of comparing, matching and averaging their shapes, and represent facial surfaces by spiral bends to create a riemannian structure for analyzing shapes of full facial surfaces. Authors in [26] propose regularized robust coding (RRC) which vigorously relapse a given sign with regularized relapse coefficients to automatically locate specific facial features such as eyes, nose, and mouth based on known distances between them. Both error correction and error recognition is performed.

5 Results and Discussions

To analyze the face recognition rate, under facial expression variation various standard face databases are considered. The databases considered are GEMEP-FERA [27], BU-3DFE [28], CK+ [29], Bosphorous [30], MMI [31], JAFFE [32], LFW [33], FERET [34], CMU-PIE [35], Georgia tech [36], AR [37], eNTERFACE 05 [38], and FRGC [39]. Face recognition rate obtained using feature-based techniques is tabulated in Table 1, from the analysis we have found that curvelet-based feature extraction technique gives the best face recognition rate of 97.83 % on FRGCv2 database. Face recognition rate obtained model-based techniques is tabulated in Table 2; from the analysis, we have found that automated spontaneous random forest classifiers technique gives the best face recognition rate of 94.70 % on JAFFE database. Face recognition rate obtained correlation-based techniques is tabulated in Table 3; from the analysis, we have found that regularized robust coding gives the best face recognition rate of 97.50 % on AR database.

Table 1 Face recognizing using feature-based technique

Author	Method	Dataset	Recognition rate	Remarks
Drira et al. [15]	Statistical shape analysis of facial surfaces [15]	FRGCv2.0	97.7 %	Novel geometric system for breaking down facial appearances with the particular objectives of contrasting, matching, and averaging their shapes
		Gavab DB	96.99 %	
		Bosphorous	89.25 %	
Elaiwat et al. [16]	Curvelet-based feature extraction [16]	FRGCv2	97.83 %	The use of local feature-based algorithms as multi-scale local surface descriptors that separates unique peculiarities around located key points
Lee et al. [17]	Sparse representation [17]	CK+DB and MMI	65.47	To lessen the intra-class variety while accentuating the facial expression in a query face image
		JAFFE	94.70	
		MMI	78.72	
		BU-3DFE	87.85	
Yong et al. [18]	Data uncertainty in face recognition [18]	ORL	35.385	Proposes that image of a face will have high measure of instability
		AR	73.535	
		FERET	80.915	
		GEORGIA TECH	75.84	
		LFW	35.385	
Wang et al. [19]	Manifold regularization [19]	AR	94.17	MRLSR model can hold more group sparsity

Table 2 Face recognition using model-based techniques

Author	Method	Dataset	Recognition rate	Remarks
Tie et al. [20]	Automatic emotion recognition method [20]	RMC emotional dataset	80.3	An automatic emotion recognition technique from video sequences
		Mind reading DVD	80.0	
Zheng [21]	Group sparse reduced-rank regression using GSRRR and ALM [21]	BU-3DFE	66.0	Novel multi-view facial expression recognition technique
		BU-3DFE	78.9	
		BU-3DFE	72.4	
		Multi-PIE	81.7	
El Meguid and Levine [22]	Automated, Spontaneous, Random Forest Classifiers [22]	BU-4DEF	73.10	A functional framework for the arrangement of facial expressions emoted by people in a spontaneous environment seen by video cameras
		FEED	53.66	
		AFEW	44.53	
		JAFFE	90.17	
		CK-CK	90.26	
Taher et al. [23]	Utilizes a dictionary based component separation algorithm [23]	DCS-S1	81.6	It does not support some facial features such as the shape of facial components
		DCS-S2 selected subset	86.8	
		DCS-S2 whole dataset	89.21	
He et al. [24]	HQ framework [24]	AR	55	HQ-based robust sparse representation shows that outliers significantly affect the estimation of sparse coding
		Yale B	85	

Table 3 Face recognition using correlation-based techniques

Author	Method	Dataset	Recognition rate	Remarks
Tavari et al. [25]	Cross Modal Data Association [25]	eNTERFACE '05	82.2	Rules used might not be suitable for the particular emotion class
Juefei-Xu et al. [8]	Discrete Transform —Local Binary Pattern [8]	FRGC v2	96.5	Re-confirmed the robustness and efficacy of the proposed DTLBP feature for face recognition
Yang et al. [26]	Regularized Robust Coding [26]	AR	97.5	RRC with *l*2-norm regularization could achieve very high recognition rate

6 Conclusion

Face recognition is a critical component for security-related applications. Recognizing faces under facial expression variation and occlusion are still challenging task. In this paper, we perform a comparative study on various state of the art techniques to recognize faces under varying facial expressions and in the presence of occlusion. The approaches to recognize faces under varying facial expressions and occlusions can be broadly classified: feature-based techniques, model-based techniques, and correlation-based techniques. Among various feature-based techniques, curvelet-based feature extraction technique gives the best face recognition rate of 97.83 % on FRGCv2 database. Among model-based technique, automated spontaneous random forest classifiers technique gives the best face recognition rate of 94.70 % on JAFFE database. Finally, among correlation-based techniques, regularized robust coding gives the best face recognition rate of 97.50 % on AR database.

Acknowledgments The proposed work was made possible because of the grant provided by Vision Group Science and Technology (VGST), Department of Information Technology, Biotechnology and Science and Technology, Government of Karnataka, Grant No. VGST/SMYSR/GRD-402/2014–15 and the support provided by Department of Electronics and Communication Engineering, Karunya University, Coimbatore, Tamil Nadu, India.

References

1. Steven L, Fernandes G, Bala J (2014) 3D and 4D face recognition: a comprehensive review. Recent Pat Eng 8(2):112–119
2. Steven L, Fernandes G, Bala J (2014) Development and analysis of various state of the art techniques for face recognition under varying poses. Recent Pat Eng 8(2):143–146
3. Yang M, Zhang L et al (2013) Regularized robust coding for face recognition. IEEE Trans Image Process 22(5):1753–1766
4. He R, Zheng W-S, Tan T et al (2014) Half-quadratic-based iterative minimization for robust sparse representation. IEEE Trans Pattern Anal Mach Intell 36(2): 261–275
5. Steven L, Fernandes G, Bala J (2015) Recognizing faces when images are corrupted by varying degree of noises and blurring effects. Adv Intell Syst Comput 337(1):101–108
6. Steven L, Fernandes G, Bala J (2015) Low power affordable, efficient face detection in the presence of various noises and blurring effects on a single-board computer. Adv Intell Syst Comput 337(1):119–127
7. Fernandes SL, Bala GJ (2014) Recognizing facial images in the presence of various noises and blurring effects using gabor wavelets, DCT-neural network, hybrid spatial feature interdependence matrix. In: 2nd IEEE International Conference on Devices, Circuits and Systems
8. Juefei-Xu F, Savvides M (2014) Subspace- based discrete transform encoded local binary patterns representations for robust periocular matching on nist's face recognition grand challenges. IEEE Trans Image Process 23(8)
9. Wei C-P, Chen C-F, Wang Y-CF (2014) Robust face recognition with structurally incoherent low-rank matrix decomposition. IEEE Trans Image Process 23(8)

10. Fernandes SL, Bala GJ (2014) Recognizing facial images using ICA, LPP, MACE gabor filters, score level fusion techniques. IEEE Int Conf Electron Commun Syst
11. Fernandes SL, Bala GJ et al (2013) Robust face recognition in the presence of noises and blurring effects by fusing appearance based techniques and sparse representation. IEEE Int Conf Adv Comput Networking Secur
12. Fernandes SL, Bala GJ et al (2013) A comparative study on score level fusion techniques and MACE gabor filters for face recognition in the presence of noises and blurring effects. IEEE Int Conf Cloud Ubiquit Comput Emerg Technol
13. Fernandes SL, Bala GJ (2013) A comparative study on ICA and LPP based face recognition under varying illuminations and facial expressions. IEEE Int Conf Signal Proc, Image Proc Pattern Recognit
14. Alyuz N, Gokberk B et al (2013) 3-D face recognition under occlusion using masked projection. IEEE Trans Inf Forensics Secur 8(5)
15. Drira H, Amor BB et al (2013) 3D face recognition under expressions, occlusions, and pose variations. Pattern Anal Mach Intell 35:2270–2283
16. Elaiwat S, Bennamoun M et al (2014) 3-D face recognition using curvelet local features. Biometrics Compendium 21(2):172–175
17. Lee SH, Plataniotis KNK et al (2014) Intra-class variation reduction using train in expression images for sparse representation based facial expression recognition. Affective Comput 5:340–351
18. Yong X, Fang X et al (2014) Data uncertainty in face recognition. Biometrics Compendium 44:1950–1961
19. Wang L, Wu H, Pan C Manifold regularized local sparse representation for face recognition. IEEE Trans Circuits Syst Video Technol
20. Tie Y, Cuan L et al (2013) A deformable 3-D facial expression model for dynamic human emotional state recognition. Biometrics Compendium 23:142–157
21. Zheng W (2014) Multi-view facial expression recognition based on group sparse reduced-rank regression. Affective Comput 5:71–85
22. El Meguid MKA, Levine MD (2014) Fully automated recognition of spontaneous facial expressions in videos using random forest classifiers. Affective Comput 5:418–431
23. Taheri S, Patel VM et al (2013) Component based recognition of faces and facial expressions. Affective Comput 4(4):360–371
24. He R, Zheng W-S, Tan T, et al (2014) Half-quadratic-based iterative minimization for robust sparse representation. IEEE Trans Pattern Anal Mach Intell 36(2)
25. Tavari A, Trivedi MM et al (2013) Face expression recognition by cross modal data association. Multimedia 15:1543–1552
26. Yang M, Zhang L et al (2013) Regularized robust coding for face recognition. IEEE Trans Image Process 22(5)
27. GEMEP-FERA SSPNE, http://sspnet.eu/2011/05/gemep-fera/
28. Facial expression database, http://www.cs.binghamton.edu/~lijun/Research/3DFE/3DFE_Analysis.html
29. Facial expression public database, http://www.pitt.edu/~emotion/ck-spread.html
30. 3D face database, http://bosphorus.ee.boun.edu.tr/
31. Facial expression database, http://mmifacedb.eu/
32. Japnees femail expression database, http://www.kasrl.org/jaffe.html
33. LFW database, http://vis-www.cs.umass.edu/lfw/
34. The gray scale FERET database, www.itl.nist.gov/iad/humanid/feret/feret_master.html
35. CMU pose, illumination, and expression(PIE) Database, http://vasc.ri.cmu.edu/idb/html/face/
36. FEI face database, http://www.anefian.com/research/face_reco.htm
37. AR face database, http://www2.ece.ohio-state.edu/~aleix/ARdatabase.html
38. Emotional speech database, http://www.enterface.net/enterface05/main.php?frame=emotion
39. FRGC v2 database, http://blendmetrics.com/frgc-v2-database-1e636-free

A Novel Technique to Recognize Human Faces Across Age Progressions

Steven Lawrence Fernandes and G. Josemin Bala

Abstract Individual's appearance changes as age progresses, this shows immeasurable potential uses of programmed face recognition over ages. Since different individuals age in different way, a "sufficient and complete" dataset ought to contain the complete aging examples of the number of individuals as are important to speak to the entire populace. In any case, age progression cannot be artificially controlled. The collection of the aging images hence normally obliges extraordinary exertion in looking for photos taken years back, and future images cannot be procured. The two methodologies considered to recognize faces across age variations are discriminative-based methodology and generative-based methodology. In this paper, we propose a novel technique to recognize faces across age progressions. We also analyze different best in class systems accessible in discriminative-based methodology and generative-based methodology; this analyzes are performed on different standard age varying face databases.

Keywords Face recognition · Age progressions · Age variation

1 Introduction

Faces experience progressive varieties because of aging [1]. Human aging and age estimation have gotten to be dynamic exploration themes in computer vision [2]. Here, our center is the aging impact on expression recognition, which is connected, however, not quite the same as age estimation [3–5]. It is also different from our recent study on age estimation under expression changes. A multitude of age-invariant face recognition approaches employs an initial transformation into a

S.L. Fernandes (✉) · G. Josemin Bala
Department Electronics & Communication, Karunya University, Coimbatore, India
e-mail: steva_fernandes@yahoo.com

G. Josemin Bala
e-mail: josemin@karunya.edu

L.P. Suresh and B.K. Panigrahi (eds.), *Proceedings of the International Conference on Soft Computing Systems*, Advances in Intelligent Systems and Computing 397, DOI 10.1007/978-81-322-2671-0_36

general representation space before performing further processing [6]. To wipe out the false positive, the frames experience second detector. The frames in which no eyes are detected are disposed of. Amid, second goes as not containing face [7–13]. This methodology is effective since offers the detection of eye directions from the face pictures and recognizes the false positives. The two methodologies considered to recognize faces across age variations are discriminative-based methodology and generative-based methodology.

The rest of the papers organized as follows: Sect. 2 describes discriminative-based approach for age-invariant face recognition, Sect. 3 describes generative-based approach for age-invariant face, Sect. 4 describes the proposed system, Sect. 5 describes the results and discussions, and Sect. 6 draws the conclusion.

2 Discriminative-Based Approach for Age-Invariant Face Recognition

Age estimation utilizing unfiltered faces methodology considers a novel dataset of face images, marked for age and gender, obtained by advanced mobile phones and other versatile gadgets, and transferred without manual filtering to online picture archives. The dropout-SVM methodology is utilized for face characteristic estimation, to evade over-fitting [14]. At long last, a robust face alignment method is considered which explicitly considers the instabilities of facial feature detectors. As opposed to considering every face image as an occurrence, authors in [15] consider every face image as an occasion connected with a label distribution. The label covers a specific number of class labels; speaking to the extent that every label depicts the case. With the increment in age, there are changes in skeletal structure, muscle mass, and body fat. For recognizing faces with age variations, authors in [16] have concentrated on the skeletal structure and muscle mass. This additionally fuses weight data to enhance the execution of face recognition with age variations. At last, this uses neural network and irregular choice decision forest to encode age variations over different weight classifications [17].

3 Generative-Based Approach for Age-Invariant Face Recognition

Authors in [18] proposed dimensionality reduction, after which it catches topological features to lead pattern recognition. Therapeutically changed face transformation considers a dataset of more than 1.2 million face images was developed from videos acquired from YouTube who are experiencing hormone replacement therapy (HRT) for age transformation over a period of several months to 3 years. HRT accomplishes age transformation by extremely adjusting the balance of sex hormones, which reasons of changes in the physical appearance of the face and body. This work investigates the effect of face changes due of hormone control and its capacity to disguise the face and henceforth, its capacity to impact match rates [19]. Face disguise is attained organically as hormone control causes pathological changes to the body resulting in modification of face appearance. This work investigates and assesses face segments versus full face algorithms trying to distinguish locales of the face that are flexible to the HRT process [20].

4 Proposed System

The proposed system has 6 modules: preprocessing for train images, sparse representation via L1-minimization, classification based on sparse representation, load train images, test image preprocessing, identified image, and calculate accuracy. The block diagram of the proposed system is shown in Fig. 1.

4.1 Preprocessing for Train Images

In this module, noise removal process for training images is done. Removing noise process is done by median filter. Median Filtering is to replace each pixel value in an image by the median of its neighborhood. Procedure of median filtering sort the pixel values: find the median and replace the pixel value by the median.

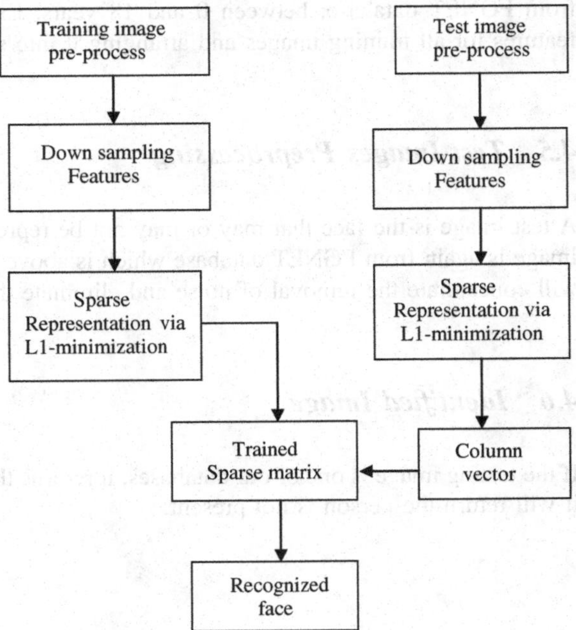

Fig. 1 Block diagram of the proposed system to recognize faces across age progressions

4.2 *Sparse Representation via L1 Minimization*

Recent developments in sparse representation say that if x0 solution is very sparse, then the solution of 0-minimization problem is same as the solution of 1-minimization problem.

$$\hat{}X1 = \arg\min ||x||1 \text{ where } Ax = y \quad (1)$$

This problem is solved using standard linear programming methods. If the solution is very sparse, then we get even more efficient methods.

4.3 *Classification Based on Sparse Representation*

We compute sparse representation, i.e., ^x1 using an input image y. Nonzero elements of ^x1 will be associated with columns of train images of a single image i and the test image y can be assigned to that class easily.

4.4 *Load Train Images*

In this module, we will load the training images. Training images consist of images from FGNET database, between 0 and 18 years. Extracting the down sampling features for all training images and arranging it into sparse representations.

4.5 *Test Images Preprocessing*

A test image is the face that may or may not be represented in our database. This image is again from FGNET database which is above 18 years. In this module, we will concentrate the removal of noise and eliminate the background information.

4.6 *Identified Image*

If the testing image is one of our databases, it returns the identified image otherwise it will return the person is not present.

4.7 Calculate Accuracy

Here we compare our proposed sparse representation technique with other techniques such as higher verifications. And also partial features are trained by sparse and give the best results for face recognition.

5 Result and Discussions

Face recognition rate are investigated by considering the following databases: LFW, YTF, FGNET, MORPH, WhoIsIt, Gallagher, Pubfig 83, and Twin days dataset [21–28]. Experimental results on two aging face databases show remarkable advantages of label distribution learning algorithms over the compared single-mark learning algorithms, either extraordinarily intended for age estimation or for general purpose. The correlation with existing state-of-the-art algorithms and commercial system on FGNET databases demonstrates that the proposed sparse representation beats existing algorithm significantly. Recognition rate is classified in Tables 1 and 2.

Table 1 Recognition rate for discriminative-based technique for age-invariant face recognition

Author	Method	Database	Recognition set
Geng [14]	IIS-LLD	FGNET	5.77 %
		MORPH	5.67 ± 0.15 %
	CPNN	FGNET	4.76 %
		MORPH	4.87 ± 0.31 %
Paone et al. [15]	Baseline	Twin days Dataset	1.0 %
Singh et al. [17]	Neural network and random decision forest-based classification	WhoIsIt	28.532 ± 1.03 %
		FGNET	20.34 ± 0.47 %

Table 2 Recognition set for generative-based techniques for age-invariant face recognition

Author	Method	Dataset	Recognition set
Bouchaffra et al. [18]	NTCA	FGNET	48.96 %
		MORPH1	71.30 %
		MORPH2	83.30 %
Chiachia et al. [20]	SVM	Pubfig 83	92.8 ± 0.28 %
Fernandes et al.	Sparse Representation (Proposed System)	FGNET	98.90 %

6 Conclusion

Face recognition under age variation recognition techniques are classified into two categories—discriminative-based techniques and generative-based technique. In this paper, we propose a novel sparse representation-based age-invariant face recognition; this technique was tested on FGNET database and it gives excellent face recognition rate of 98.90 %.

Acknowledgments The proposed work was made possible because of the grant provided by Vision Group Science and Technology (VGST), Department of Information Technology, Biotechnology and Science and Technology, Government of Karnataka, Grant No. VGST/SMYSR/GRD-402/2014–15 and the support provided by Department of Electronics and Communication Engineering, Karunya University, Coimbatore, Tamil Nadu, India.

References

1. Eidinger E, Enbar R et al (2014) Age and gender estimation of unfiltered faces. Inf Forensics Security 9(12):2170–2179
2. Mahalingam G, Ricanek K Jr et al (2014) Investigating the periocular-based face recognition across gender transformation. Inf Forensics Security 9(12):2180–2192
3. Bruyer R, Scailquin J-C Person recognition and ageing: the cognitive status of addresses-an empirical question. Int J Psychol 29(3):351–366
4. Li L, Su H, Xing E, Fei-Fei L (2010) Object bank: a high-level image representation for scene classification & semantic feature sparsification. Proc Adv Neural Inf Process Syst (NIPS) (2010)
5. Li L, Su H, Lim Y, Fei-Fei L (2010) Objects as attributes for scene classification. Proc. European Conf computer vision (ECCV). In: First Int'l workshop parts and attributes
6. Park U, Tong Y, Jain AK (2010) Age-invariant face recognition. IEEE Trans Pattern Anal Mach Intell 32(5):947–954
7. Hoffmann H (2007) Kernel PCA for novelty detection. Pattern Recognit 40(3):863–874
8. IARPA Broad Agency Announcement: BAA-13-07, Janus Program. http://www.iarpa.gov/Programs/sc/Janus/solicitation_janus.html (Nov. 2013)
9. Belhumeur P, Hespanha J, Kriegman D (1997) Eigenfaces vs. fisherfaces: recognition using class specific linear projection. IEEE Trans Pattern Analysis Mach Intell 19(7):711–720
10. Vasilescu MAO, Terzopoulos D (2002) Multilinear image analysis for facial recognition. In: IEEE international conference on pattern recognition
11. Wolf L, Hassner T, Taigman Y (2009) The one-shot similarity kernel. In: IEEE international conference on computer vision
12. Wright J, Yang A, Ganesh A, Sastry S, Ma Y (2009) Robust face recognition via sparse representation. IEEE Trans Pattern Anal Mach Intell 31(2):210–227
13. Bolme DS, Draper BA, Beveridge JR (2009) Average of synthetic exact filters. Comput Vision Pattern Recognit
14. Geng X, Yin C, Zhou Z-H et al (2013) Facial age estimation by learning from label distributions. Pattern Anal Mach Intell 35(10):2401–2412
15. Paone JR, Flynn PJ, Jonathon Philips P et al (2014) Double trouble: differentiating identical twins by face recognition. Inf Forensics Security 9(2):285–295
16. Chen H, Gallagher AC et al (2014) The hidden sides of names-face modeling with first name attributes. Pattern Anal Mach Intell 36(9):1860–1873

17. Singh M, Nagpal S, Singh R, Vatsa M et al (2014) On recognizing face images with weight and age variations. Access 2:822–830
18. Bouchaffra D (2014) Nonlinear topological component analysis: application to age-invariant face recognition. Neural Networks Learn Syst 99:1
19. Lacey Best-Rowden H, Han C et al (2014) Collection unconstrained face recognition: identifying a person of interest from a media. Inf Forensics Security 9(12):2144–2157
20. Chiachia G, Falcao AX, Pinto N et al (2014) Learning person-specific representations from faces in the wild. Inf Forensics Security 9(12):2089–2099
21. Kong AW, Zhang D, Lu G (2006) A study of identical twins' palmprints for personal verification. Pattern Recognit 39(11):2149–2156
22. Prabhakar JS, Pankanti S (2002) On the similarity of identical twin fingerprints. Pattern Recognit 35(11):2653–2663
23. Hollingsworth K, Bowyer K, Flynn P (2010) Similarity of iris texture between identical twins. In: Proceedings IEEE Computer Society Conference CVPRW, pp. 22–29
24. Ariyaeeiniaa A, Morrison C, Malegaonkara A, Black B (2008) A test of the effectiveness of speaker verification for differentiating between identical twins. Sci Justice 48(4):182–186
25. Fernandes Steven L, Josemin Bala G (2014) 3D and 4D face recognition: a comprehensive review. Recent Patents Eng 8(2):112–119
26. Fernandes SL, Josemin Bala G (2014) Development and analysis of various state of the art techniques for face recognition under varying poses. Recent Patents Eng 8(2):143–146
27. Fernandes Steven L, Josemin Bala G (2015) Recognizing faces when images are corrupted by varying degree of noises and blurring effects. Adv Intell Syst Comput 337(1):101–108
28. Fernandes SL, Josemin Bala G (2015) Low power affordable, efficient face detection in the presence of various noises and blurring effects on a single-board computer. Adv Intell Syst Comput 337(1):119–127

Control of a Time-Delayed System Using Smith Predictive PID Controller

Rahul Vijayan, P.P. Praseetha, Dwain Jude Vaz and V. Bagyaveereswaran

Abstract In this paper, a theoretical framework for controlling a time-delayed system using Smith predictive PID controller based on fuzzy logic is proposed. The conventional PID controller is not effective if the system is nonlinear and has high overshoot. A combination of fuzzy logic controller and Smith predictor is used to get optimum response of the system. Also the PID parameters are optimized using LQR and the system response is compared with fuzzy logic-based controllers. Simulation results showed that proposed control algorithm has the advantages of strong adaptive ability and noise immunity.

Keywords Smith predictive controller · PID · Fuzzy logic controller · LQR

1 Introduction

The controlled variable is always characterized by the volume lag of three-tank system as described in [1]. Hence the conventional PID controller cannot maintain the requirement of the variable. In [2], intelligent fuzzy controller is used for the implementation of three-tank system. Also they analyzed the effect on its efficiency which resulted in the increase of robustness for the disturbance rejection under different circumstances.

R. Vijayan (✉) · P.P. Praseetha · D.J. Vaz
MTech in Control and Automation, VIT University, Vellore, India
e-mail: rahul.nmra@rediffmail.com

P.P. Praseetha
e-mail: praseethaprakash12@gmail.com

D.J. Vaz
e-mail: dwainvaz@gmail.com

V. Bagyaveereswaran
SELECT, VIT University, Vellore, Tamil Nadu, India
e-mail: bagyaveereswaran@gmail.com

L.P. Suresh and B.K. Panigrahi (eds.), *Proceedings of the International Conference on Soft Computing Systems*, Advances in Intelligent Systems and Computing 397, DOI 10.1007/978-81-322-2671-0_37

There are simple techniques for the control of time-delayed system. Three-tank system offers a great time delay to the output. This paper gives a brief description about the Smith predictor and how it rejects the system time delay. Also it analyzed the robustness of this control scheme and identified that it offers poor robustness [3, 4].

Rather than the conventional PID tuning methods such as Ziegler–Nichols method and Cohen-Coon method, now several researches are going on the modern techniques such as particle swarm optimization, genetic algorithm, and fuzzy controller. In [5] soft computing technique, fuzzy controller is used for the tuning of PID controller which gave the better characteristics in the output. In [6], combination of PID controller and fuzzy logic is used. It shows that with the use of this time-delayed system becomes more efficient.

A three-tank system is used to generate a data and develop a linear model using neuro-fuzzy [7]. Immunity of a three-tank system to noise and other performance while using Smith predictive technique along with fuzzy is illustrated in this paper [8]. Tuning a PID controller using fuzzy inference system and certain rules for designing is shown here [9]. In [10], optimal tuning of PID for a fractional-order system is done using LQR. Fractional-order systems having slow, sluggish and oscillatory response is tuned with LQR controller. The weighting matrices are optimized using genetic algorithm (GA) [11]. In [11], several controllers like internal model controller, PID controller, etc., have been tuned using GA.

Here in this paper, several soft computing techniques are used, and their performances and efficiency are compared along with the conventional methods. This paper mainly focuses on the performance variations happening when Smith PID controller is changed to fuzzy adaptive Smith predictor and also concentrates on the differences of the response of system with conventional PID controller and the response with Smith PID controller. Also this paper focuses on the optimization of system using linear quadratic regulator (LQR).

2 System Description

The system used here is a three-tank noninteracting system with a large time delay shown in Fig. 1. The time delay may be due to mass transportation phenomena. Time delay is also caused due to processing time or accumulation of time lags of a number of systems connected in series.

A system is noninteracting because response of one system does not affect the other. In this case, level of fluid in first tank does not depend on the level of fluid in the second tank. Same is the case with second and third tanks. The system dynamic behavior can be given as:

$$A_1 \frac{\mathrm{d}h_1(t-\tau/3)}{\mathrm{d}t} = q_0(t) - q_1(t-\tau/3) \tag{1}$$

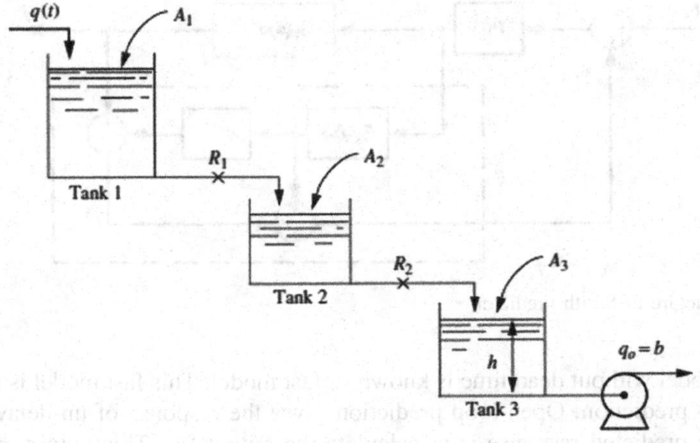

Fig. 1 Schematic diagram of three-tank system

$$A_2 \frac{\mathrm{d}h_2\left(t-\frac{2\tau}{3}\right)}{\mathrm{d}t} = q_1\left(t-\frac{\tau}{3}\right) - q_2\left(t-\frac{2\tau}{3}\right) \tag{2}$$

$$A_3 \frac{\mathrm{d}h_3(t-\tau)}{\mathrm{d}t} = q_2(t-2\tau/3) - q_3(t-\tau) \tag{3}$$

where A_i is the cross sectional area of tank, h_i is the height of the tank, and q_i is the flow rates.

The overall transfer function of the system is:

$$G(s) = \frac{2e^{-2s}}{125s^3 + 75s^2 + 15s + 1} \tag{4}$$

3 Smith Predictive Controller

Here Smith predictive controller scheme is the first control loop test used. This scheme deals with the time delay of the system. It is configured in a particular manner in order to achieve effective time delay compensation. Hence it is also known as dead time compensator.

The structure of smith predictor is shown in Fig. 2.

The model is divided into two parts namely, primary controller and predictive controller. Predictive controller structure consist of process model without dead time, $G_{\mathrm{pa}}(s)$ and a model of dead time, e^{-sT_d}. Basically the model's linear transfer function part is separated from the dead time part.

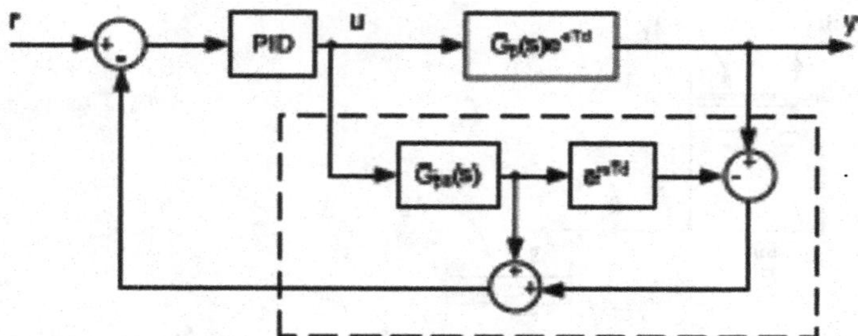

Fig. 2 Structure of Smith predictor

The model without dead time is known as fast model. This fast model is used for open-loop prediction. Open-loop prediction gives the response of un-delayed process. This predicted response is to calculate the output '*u*.' This output is further used to find out the actual un-delayed output '*y*.' This actual un-delayed output and predicted delayed output is compared and given as the feedback which will compensate the difference. By this way, the structure will compensate the dead time existing in the system.

4 LQR Algorithm for Optimisation of PID Controller Using the Template

The LQR method is used to optimize the cost function

$$J = \int (X^T QX + U^T RU)\mathrm{d}t \tag{5}$$

where X is the state variables and U is the control action, Q and R is the weighting matrices. Q and R should be chosen such that it is positive semi-definite and positive definite, respectively. Minimizing the above cost function gives us the continuous time Ricatti Eq. (2) which is used to find out the state feedback control law (3)

$$A^T P + PA - PBR^{-1}B^T P + Q = 0 \tag{6}$$

$$U = -KX \tag{7}$$

where $K = R^{-1}B^T P$

The PID controller optimization for the noninteracting three-tank system using LQR is shown in the above figure. The state variables are chosen as the error signals and its differentials. In case of feedback control design, the external set point does not affect and result hence $r(t) = 0$.. Hence $y(t) = e(t)$.

$$x_1(t) = D^{-1}e(t) \tag{8}$$

$$x_2(t) = e(t) \tag{9}$$

$$x_3(t) = De(t) \tag{10}$$

where $D = \frac{\mathrm{d}}{\mathrm{d}t}$, the above equations can be obtained from Fig. 9.

The transfer function of the system is given by

$$\frac{Y(S)}{U(S)} = \frac{2}{125S^3 + 75S^2 + 15S + 1} \tag{11}$$

As $Y(S) = E(S)$

$$\frac{E(S)}{U(S)} = \frac{2}{125S^3 + 75S^2 + 15S + 1} \tag{12}$$

The above equation reduces to

$$\dot{x}_4 = 0.016U(s) - 0.6x_4 - 0.12x_3 - 0.008x_2 \tag{13}$$

where $\dot{x}_3 = x_4$. The state space representation can be formulated from the above equations

$$A = \begin{matrix} 0 & 1.0000 & 0 & 0 \\ 0 & 0 & 1.0000 & 0 \\ 0 & 0 & 0 & 1.0000 \\ 0 & -0.0080 & -0.1200 & -0.6000 \end{matrix}$$

$$B = \begin{matrix} 0 \\ 0 \\ 0 \\ 0.0160 \end{matrix}$$

The state feedback gain matrix can be obtained using the lqr () function in the control system toolbox in MATLAB. The state feedback gain matrix is obtained from the equation,

$$K = R^{-1}B^{T}P \tag{14}$$

The state feedback gain matrix gives the optimal tuning values for the PID controller tuning parameters K_i, K_p, K_d.

5 Fuzzy Controller

The fuzzy control is an intelligent control technique which can overcome many shortcomings of traditional PID controller. Fuzzy is a nonlinear control strategy which imitates human logical thinking and accurate mathematical model of controlled object is not needed. Here we use fuzzy logic to tune a Smith PID controller for getting better static and dynamic performance.

Controller design includes three stages:

(1) *Fuzzification*—Here the crisp values of inputs are transformed to fuzzy sets. It gives a quantitative idea of the range of inputs, each membership function will contribute.
(2) *Evaluating rules*—The user creates a rule base which tells the system what to do in a particular case. The set of rules are connected by if-then statements. To compute input, we use Mamdani method. It contains the following steps. Determining a set of rules, fuzzifying the inputs and combining it according to the rules. Rule strength and output membership function give the consequences of rules. Combining these consequences, we get output distribution and defuzzification is the final step.
(3) *Defuzzification*—It is the reverse process of fuzzification which gives an extend of taking action to a particular rule and is a resolution of conflict between competing rule strength. Fuzzy sets corresponding to the control outputs are transformed to crisp values.

(a) Design of Fuzzy Adaptive Controller

The inputs to the fuzzy controller are the error 'e' and rate of change of error 'e_c' and based on the inputs at different time the demand of self-tuning of PID controller is met. The controller finds the relationship between e, $e_{c,}$ and PID parameters.

(b) Fuzzy inputs

The error e, rate of change of error $e_{c,}$ proportional parameter k_p, integral parameter k_{i}, and differential parameter k_d are fuzzed by 7 fuzzy sets "NB," "NM," "NS," "Z," "PS," "PM," and "PB." Error 'e' and rate of change of error 'e_c' take the discrete universe $\{-3, -2, -1, 0, 1, 2, 3\}$ discrete universe of outputs are $k_d\{-3, -2, -1, 0, 1, 2, 3\}$, $k_p\{-0.3, -0.2, -0.1, 0, 0.1, 0.2, 0.3\}$, and k_{i} $\{-0.06, -0.04, -0.02, 0\ 0.02, 0.04, 0.06\}$ (Table 1).

(c) Fuzzy rules

The value of K_p determines the adjusting accuracy and speed of response. If it is high response is quicker and smaller value of K_p will affect both static and dynamic performance of the system.

The value of K_i determines how fast the steady state error is eliminated and the output becomes saturated. Higher value of K_i means steady state error is quickly eliminated and if its value is low it will affect the adjusting accuracy of the system.

Table 1 Fuzzy rules set

e_c / e	NB	NM	NS	Z	PS	PM	PB
NB	PB/NB/PS	PB/NB/NS	PM/NM/NB	PM/NM/NB	PS/NS/NB	Z/Z/NM	Z/Z/PS
NM	PB/NB/PS	PB/NB/NS	PM/NM/NB	PS/NS/NM	PS/NS/NM	Z/Z/NS	NS/Z/Z
NS	PM/NB/Z	PM/NM/NS	PM/NS/NM	PS/NS/NM	Z/Z/NS	NS/PS/NS	NS/PS/Z
Z	PM/NM/Z	PM/NM/NS	PS/NS/NS	Z/Z/NS	NS/PS/NS	NM/PM/NS	NM/PM/Z
PS	PS/NM/Z	PS/NS/Z	Z/Z/Z	NS/PS/Z	NS/PS/Z	NM/PM/Z	NM/PB/Z
PM	PS/Z/PB	Z/Z/NS	NS/PS/PS	NM/PM/PS	NM/PM/PS	NM/PB/PS	NB/PB/PB
PB	Z/Z/PB	Z/Z/PM	NM/PS/PM	NM/PM/PM	NM/PM/PS	NB/PB/PS	NB/PB/PB

K_d improves the dynamic characteristics of the system. If K_d is too large the adjusting time and noise immunity of the system will be decreased.

5.1 Fuzzy Adaptive Smith PID Controller

The delay part of the system can be compensated using smith control technique. For large time-delayed system, if pure fuzzy control is implemented the adjusting time will be prolonged and we will not get effective response. Here if we add Smith predictive controller the delay time will be compensated and the control effect obtained will be ideal (Fig. 3).

Here the fuzzy controller adjusts all the PID parameters.

The output of the controller $G_c(s)$ is:

$$U(k) = U(k-1) + kp[e'(k) - e'(k-1)] + k_i[e'(k)] + k_d[e'(k) - 2e'(k) + e'(k-2)]$$

$U(k)$ is controller output at kth sampling time.

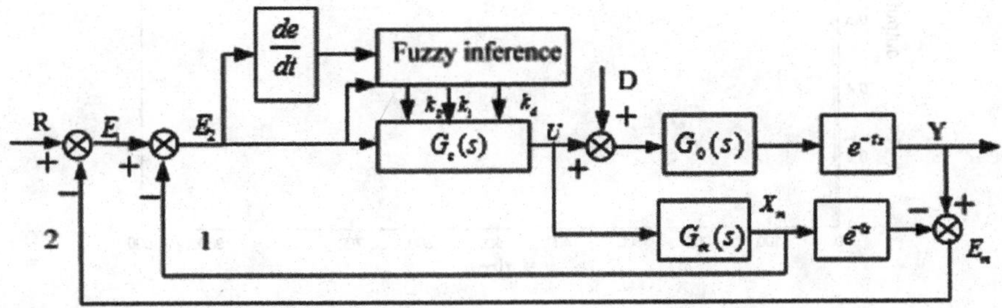

Fig. 3 Fuzzy adaptive Smith PID

6 Simulation and Results

The Simulink model of system with conventional PID controller is given in Fig. 4 and its response is shown in Fig. 5 (Figs. 6).

The Simulink model and response of the system with Smith PID controller with zero disturbance is shown in Fig. 7 (Figs. 8 and 9).

The response of system under LQR optimization is shown below (Fig. 10),

Simulation results show that the proposed LQR optimization works well for a three-tank system. The response is oscillatory but the settling time is less compare

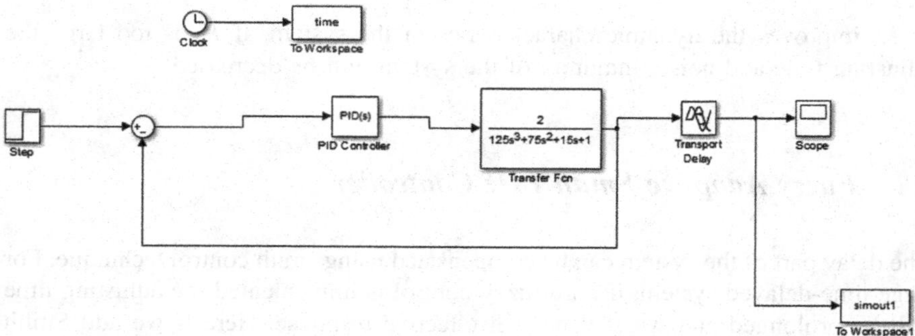

Fig. 4 Simulink model of PID controller

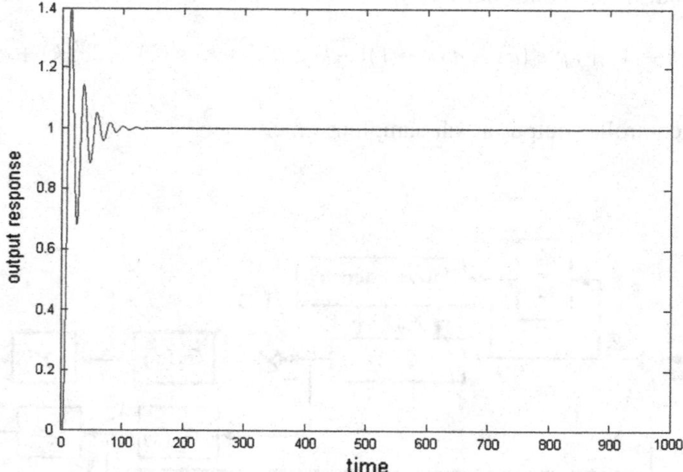

Fig. 5 Response of system with conventional PID controller

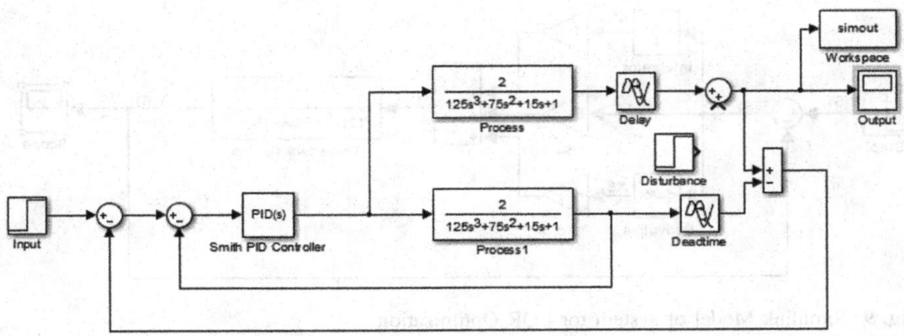

Fig. 6 Simulink model of Smith PID controller

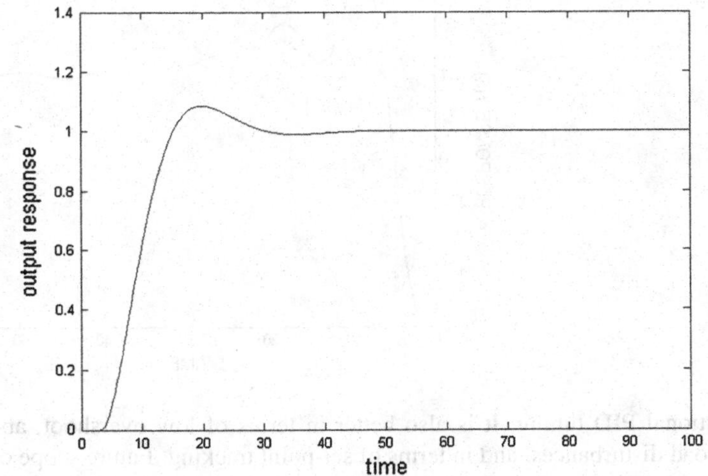

Fig. 7 Response of system with Smith PID controller

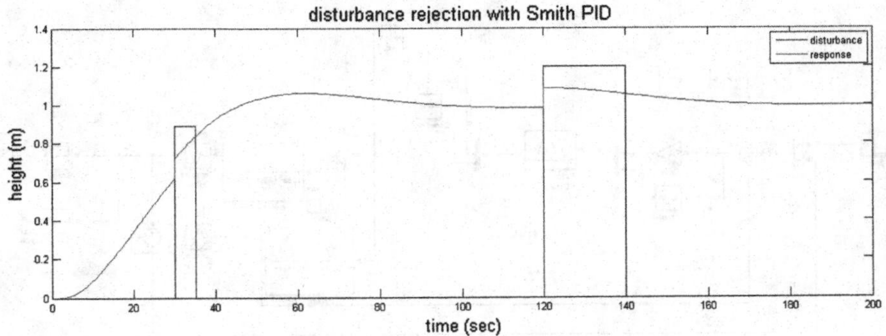

Fig. 8 Disturbance rejection with Smith PID

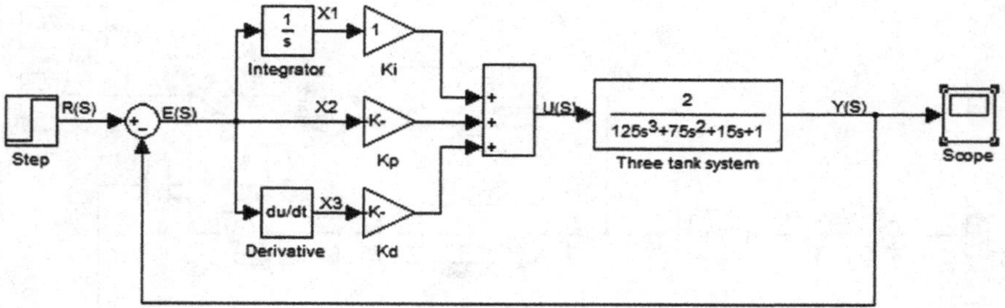

Fig. 9 Simulink Model of system for LQR Optimization

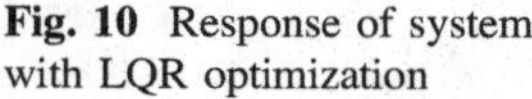

Fig. 10 Response of system with LQR optimization

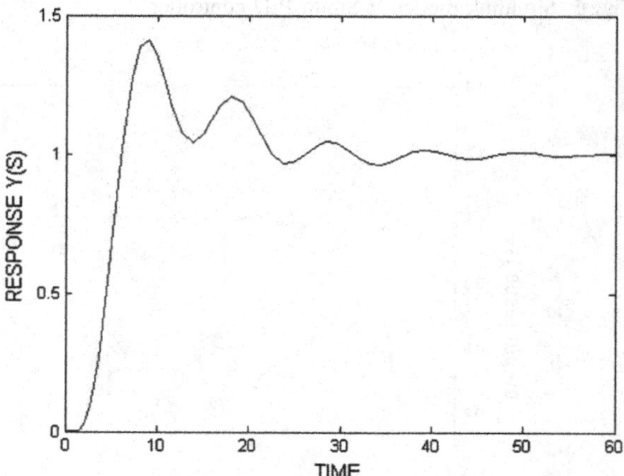

to conventional PID tuning. It is also better in terms of low overshoot, ability to suppress load disturbances, and in terms of set-point tracking. Future scope of work may include LQR-based FOPID controller tuning (Figs. 11, 12, 13, 14, 15, 16, 17 and 18).

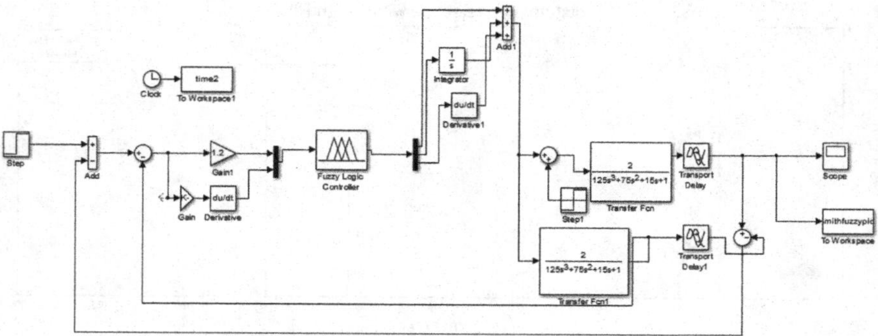

Fig. 11 Simulink model of fuzzy adaptive Smith PID controller

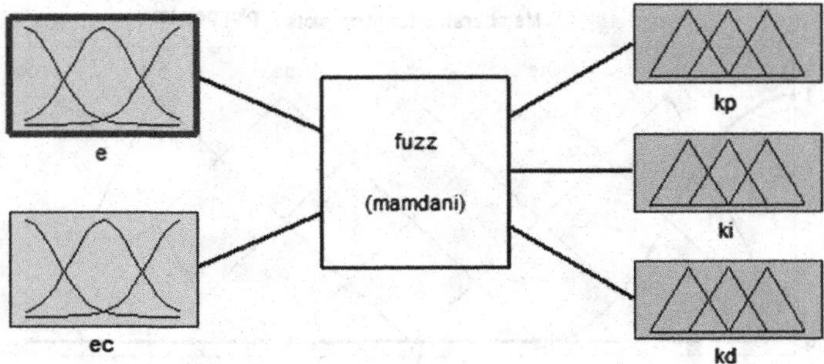

Fig. 12 Fuzzy inference system

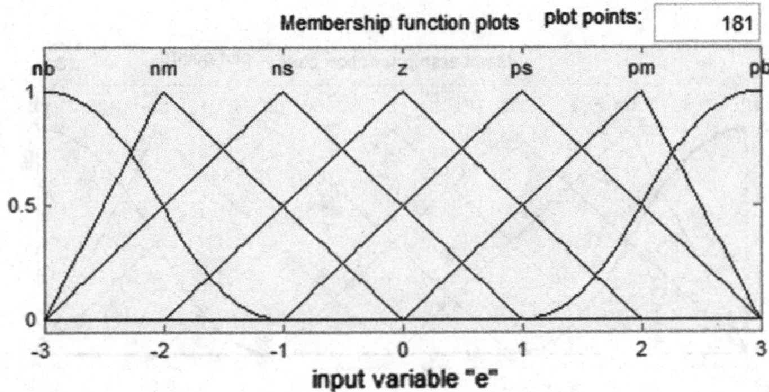

Fig. 13 Membership function of input e

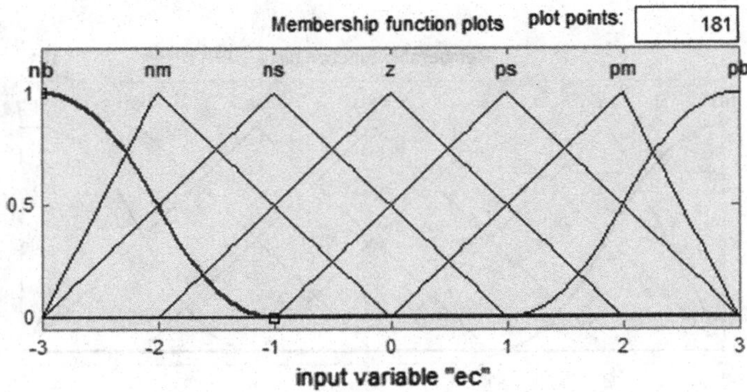

Fig. 14 Membership function of input e_c

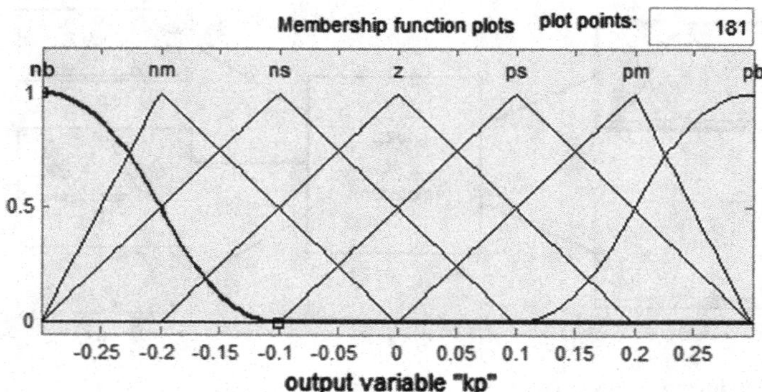

Fig. 15 Membership function of K_p

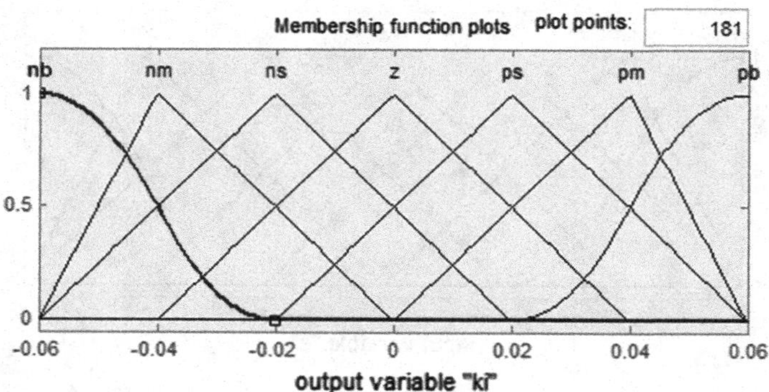

Fig. 16 Membership function of K_i

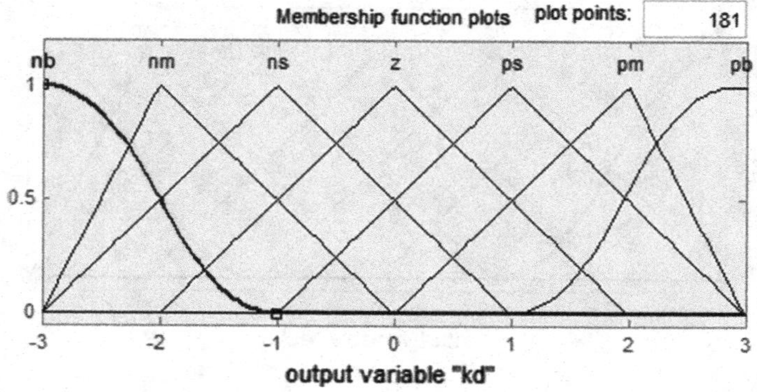

Fig. 17 Membership function of K_d

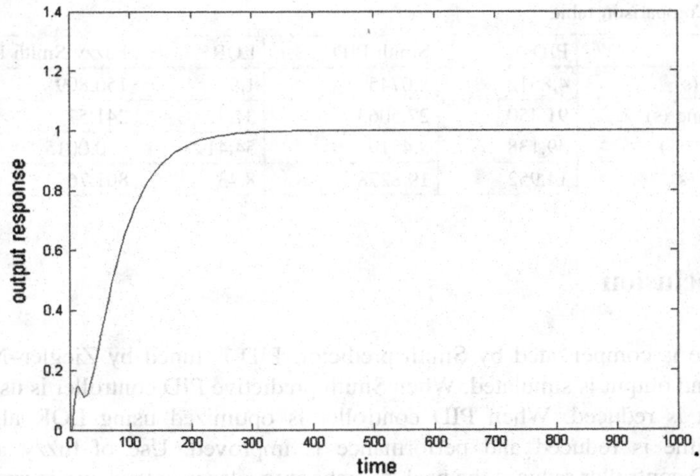

Fig. 18 Output response of system using fuzzy adaptive Smith PID

Peak overshoot is reduced to a large extend and output settles with less oscillations. Fuzzy PID part can improve the adaptive as well as resistive ability. Smith predictive part overcome the delay. Peak over shoot is reduced to 0.0015 which almost negligible (Fig. 19).

The output tracks set point effectively. It is clear that the system rejects the disturbance very effectively.

A comparative study on all models is done, and the result is tabulated in Table 2.

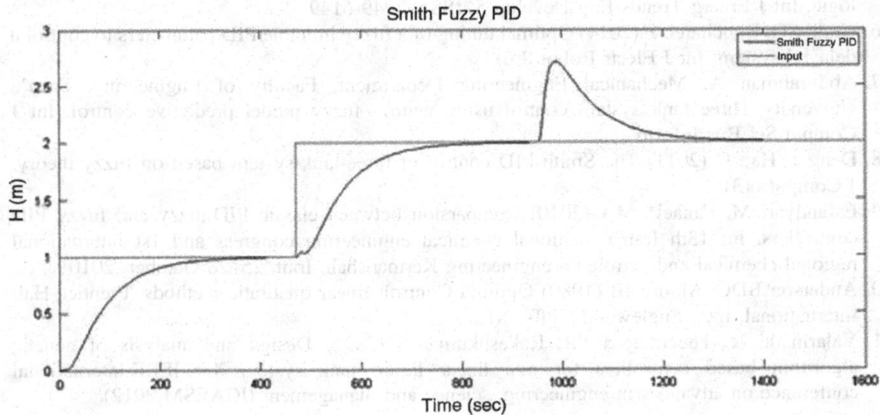

Fig. 19 Set point tracking with fuzzy Smith PID

Table 2 Comparison table

	PID	Smith PID	LQR	Fuzzy Smith PID
Rise time (s)	4.8512	8.0745	1.9	150.809
Settling time (s)	91.450	27.5063	31.42	241.57
Overshoot (%)	39.138	8.4319	54.41	0.0015
Peak time (s)	14.952	19.8278	8.43	801.76

7 Conclusion

Dead time is compensated by Smith predictor. PID is tuned by Ziegler–Nichol's method and output is simulated. When Smith predictive PID controller is used peak overshoot is reduced. When PID controller is optimized using LQR algorithm settling time is reduced and performance is improved. Use of fuzzy adaptive Smith PID controller reduces the peak overshoot to a large extend and increased the overall performance effectively.

References

1. Wang Q-I, Xie X-J (2008) Smith control of three tank water system based on C++ builder. J. Qufu Normal University (in Chinese) 1:33–35
2. Suresh M, Srinivasan GJ, Hemamalini R Integrated Fuzzy Logic based Intelligent Control of three tank system. Serbian J Electric Eng
3. Smith OJ (1959) A controller to overcome deadtime. ISA J
4. Terry Bahil A A simple adaptive smith predictor for controlling time delay systems
5. Venugopal P, Ganguly A, Singh P Design of tuning methods of PID Controller using Fuzzy logic. Int J Emerg Trends Eng Develop 5. ISSN 2249-6149
6. Herbi SG, Bouchareb F (2014) Optimal tuning of a fuzzy immune PID parameters to control a delayed system. Int J Electr Robot **8**(6)
7. Abdurahman A, Mechanical Engineering Department, Faculty of Engineering, Sana'a University Three tank system control using neuro—fuzzy model predictive control. Int J Comput Sci Bus Inform
8. Deng J, Hao C (2011) The Smith-PID control of three-tank-system based on fuzzy theory. J Comput 6(3)
9. Esfandyari M, Fanaei* MA (2010) Comparsion between classic PID,fuzzy and fuzzy PID controllers. In: 13th Iranian national chemical engineering congress and 1st international regional chemical and petroleum engineering Kermanshah, Iran, 25–28 October, 2010
10. Anderson BDO, Moore JB (1989) Optimal Control: linear quadratic methods. Prentice-Hall International, Inc., Englewood Cliffs, NJ
11. Valarmathi R, Theerthagiri PR, Rakeshkumar S (2012) Design and analysis of genetic algorithm based controllers for non linear liquid tank system. In: IEEE-international conference on advances in engineering, science and management (ICAESM-2012)

Optimal Tuning of Fractional-Order PID Controller

M.K. Arun, U. Biju, Neeraj Nair Rajagopal and V. Bagyaveereswaran

Abstract The paper proposes a comparative study of the performance of the fractional-order PID(FOPID) controller with the conventional controllers. The parameters of the FOPID controller is estimated using particle swarm optimization (PSO) algorithm, and the values are compared with the values obtained through conventional optimization techniques. This paper aims to find out optimal settings for a fractional $PI^{\alpha}D^{\beta}$ controller in order to achieve five different design specification for a closed system and also presents identification of fractional-order system using PSO. By obtaining optimal settings, fractional-order PID controller can enhance system control performance and robustness.

Keywords FOPID controller · PSO algorithm

1 Introduction

Proportional integral derivative controllers are commonly used in industries for process control applications because of its simplicity of design and good performance including low percentage overshoot and small settling time. The performance can be further enhanced by adding fractional-I and fractional-D actions [1]. In addition to set the proportional, integral and derivative constants K_P, T_i, T_d,

---

M.K. Arun (✉) · U. Biju · N.N. Rajagopal · V. Bagyaveereswaran
SELECT, VIT University, Vellore, India
e-mail: arunnambiar043@gmail.com

U. Biju
e-mail: helloubiju@gmail.com

N.N. Rajagopal
e-mail: neeraj_chembra@hotmail.com

V. Bagyaveereswaran
e-mail: bagyaveereswaran@gmail.com

L.P. Suresh and B.K. Panigrahi (eds.), *Proceedings of the International Conference on Soft Computing Systems*, Advances in Intelligent Systems and Computing 397, DOI 10.1007/978-81-322-2671-0_38

respectively, we have two more parameters which mean α and β used as power of 's' in integral and derivative part of the controller. The optimization on a five-dimensional space is done by finding optimal solution of [K_P, T_i, T_d, α, $_\beta$] to the given process. Due to roughness of objective function, other classical optimization techniques could not be implemented. Compared to its integer counterpart, the performance of optimal fractional PID controller is better. Hence it can be used in many applications.

For the better performance of PID controller, fractional calculus is used extensively. Fractional calculus is a branch where it uses fractional derivative or integral of the function which helps to generalize a concept using noninteger orders instead of integer order.

The five-dimensional parameters of the FOPID controller is estimated using particle swarm optimization (PSO) technique [2]. This is a newly accepted optimization technique put forwarded by Kennedy and Eberhart [3] in 1995. The method is based on bird's predation where the optimal value of the solution is found out from a solution space using swarm intelligence. The method is widely used in various applications nowadays because of its easy implementation and fast convergence rate [4, 5].

2 Integer-Order and Fractional-Order PID Controllers

When the parameters K_p, T_i, T_d of a PID controller takes integer orders, it is called an integer-order PID controller. It follows the transfer function as:

$$K_p + T_i S^{-1} + T_d S. \tag{1}$$

But in reality, fractional powers may provide more precise understanding of the process. Since the computation of the fractional order is difficult, the integer-order controllers are widely used [6]. Integer-order controllers are a generalized form of the fractional-order controllers. A fractional-order PID controller thus takes the form as:

$$K_p + T_i S^{-\alpha} + T_d S^{\beta} \tag{2}$$

This is the more general form of PID controller. When the values of α and β are unity it becomes the integer-order PID controllers. The number of parameters to be evaluated in a fractional-order PID controller is five where it was only three in the case of integer-order controller.

One of the major advantages of using the fractional-order controller is that it increases the region of stability [7]. The conventional stability region was only the left hand side of the s-plane. The region can be further enlarged to the right side for a particular angular length using the fractional-order controllers (Figs. 1 and 2).

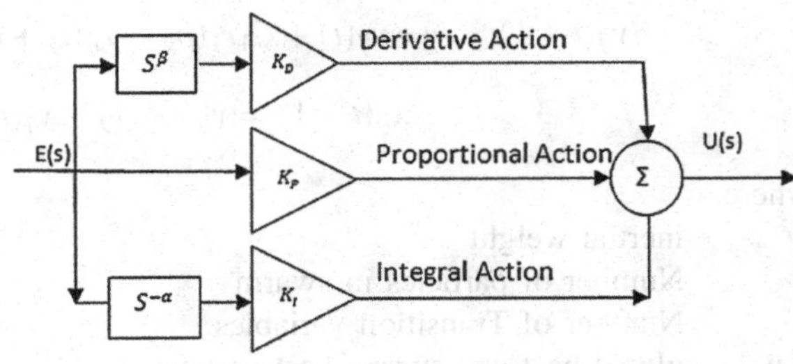

Fig. 1 Fractional PID controller block diagram

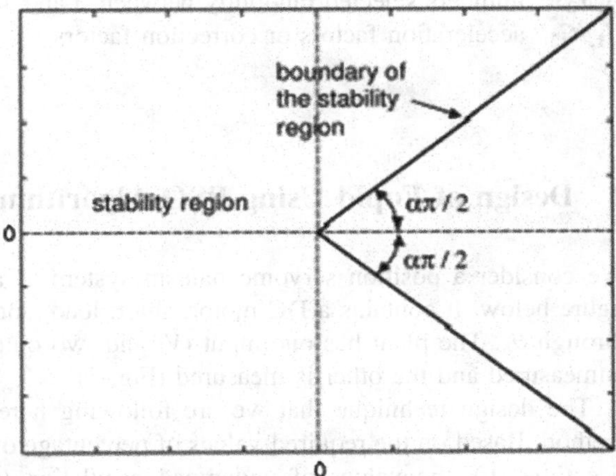

Fig. 2 Improved stability region by fractional-order controllers

3 Particle Swarm Optimization (PSO) Algorithm

PSO is a generic form of population-based stochastic optimization algorithm in swarm intelligence. It accounts for its simplicity of implementation and ability to converge to an optimal solution quickly. Similar to other typical optimization methods, in PSO initial, population is initiated with random solutions and algorithm searches for global optimum in course of time. Unlike other evolutionary algorithms, PSO do not have crossover and mutation operations. While following the present iterations, optimal values of other solutions flow through search space and global optimum is identified up to present iteration. In PSO, particles keep track of their previous coordinates associated with best solutions achieved so far. Best value of any particle obtained so far and fitness value are stored, and the best value is termed as gbest. This swarm of particles for each iterations varies its velocity and acceleration towards its pbest and lbest locations, where accelerations specially varies for every iteration. The velocity and position in PSO is updated using the formulas given.

$$v_{ij}(t+1) = w * v_{ij}(t) + C_1 r_1 (p_g - r_{ij}(t) + C_2 r_2 (g_j - x_{ij}(t)) \quad (3)$$

$$x_{ij}(t+1) = v_{ij}(t+1) + x_{ij}(t) \quad (4)$$

where

w	inertial weight
i	Number of particles in swarm
j	Number of Transition variables
g,p	global best and swarm best particle
l	local best particle
r_1, r_2	numbers selected randomly between 0 and 1
C_1, C_2	acceleration factors or correction factors

4 Design of Fopid Using PSO Algorithm

We consider a position servomechanism system as an illustration. It is given in figure below. It contains a DC motor, shaft, load, and a gearbox. The feedback is through θ_L. The plant has one input (V) and two outputs (T and θ_L) where one is unmeasured and the other is measured (Fig. 3).

The design technique that we are following here is based on the root locus method. Based on the required values of percentage of peak overshoot (% M_p) and rise time (t_r), the values of undamped oscillation (ξ) and natural frequency of oscillation (ω_0) are calculated. The pole locations are the values obtained by the given formulae:

$$-\zeta\omega_0 - j\omega_0\sqrt{1-\zeta^2}, -\zeta\omega_0 + j\omega_0\sqrt{1-\zeta\omega^2} \quad (5)$$

Let the closed-loop transfer function of the system is

$$\frac{G(S)}{1+G(S)H(S)} \quad (6)$$

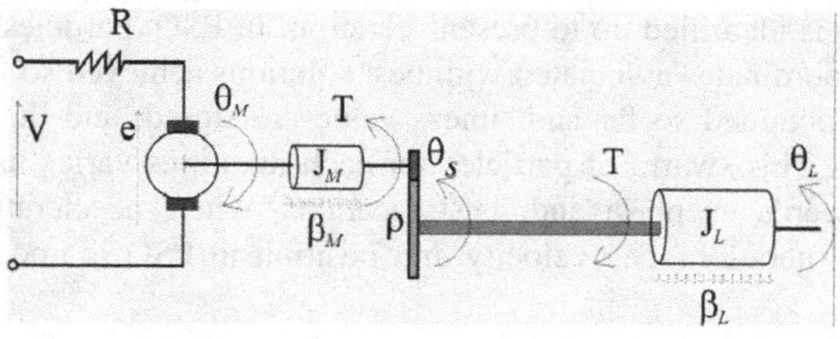

Fig. 3 Position servomechanism diagram

where $G(s) = G_c(s)G_p(s)$
$G_c(s)$ transfer function of the controller
$G_p(s)$ transfer function of the process

$G_c(s) = K_p + T_i s^{-\alpha} + T_d s^{\beta}$
Let the poles are located at $p_1 = -a + ib$ and $p_2 = -a - ib$, then at pole p_1,

$$1 + G(p_1)H(p_1) = 0 \tag{7}$$

$$1 + G_c(p_1)G_p(p_1)H(p_1) = 0 \tag{8}$$

$$1 + (K_p + T_i S^{-\alpha} + T_d S^{\beta})G_p(p_1)H(p_1) = 0 \tag{9}$$

Let $H(s) = 1$

$$1 + (K_p + T_i(-a+ib)^{-\alpha} + T_d(-a+ib)^{\beta}G_p(-a+ib) = 0 \tag{10}$$

There are five unknown parameters in this complex equation. $\{K_p, T_i, T_d, \alpha, \beta\}$. Solving of these 5 values cannot be done using normal methods. So PSO technique to be adopted for getting the five-dimensional solution space.

We define
R_e real part of the complex equation
I_m imaginary part of the complex equation
P_h phase of the complex equation
and $f = |R_e| + |I_m| + |P_h|$

As an illustration, we take the following transfer function for the position servomechanism model obtained from specifications of 10 % maximum overshoot and 0.3 s rise time.

$$1 + \frac{K_p + T_i(-1.44+j1.66)^{\alpha} + T_d(-1.44+j1.66)^{\beta}}{0.8(-1.44+j1.66)^{2.2} + 0.5(-1.44+j1.67)^{0.9} + 1} \tag{11}$$

As a practical assumption, we take K_p in the range of 1 and 1000, T_i, T_d in the range of 1 and 500 and α, β in the range of 0 and 2. From this, we will be getting,

$$R_e = (K_p + 1) + (T_i/2.2^{\alpha}) * \cos(2.2788\alpha) + T_d * 2.2^{\beta} * \cos(2.2788\beta) + 0.875 \tag{12}$$

$$I_m = -(T_i/2.2^{\alpha}) * \sin(2.2788\alpha) + T_d * 2.2^{\beta} * \sin(2.2788\beta) - 3.425 \tag{13}$$

$$P_h = \tan^{-1}(I_m/R_e) \tag{14}$$

Table 1 Comparison of parameters of integer and fractional PID

Type of controller	K_p	T_i	T_d	α	β
Integer PID	15.93	9.65	5.46	1	1
FOPID	25.4	24.11	1.988	0.54	0.47

We minimize $f = |R_e| + |I_m| + |P_h|$ according to the mentioned ranges and constraints. Now we have to minimize f using PSO technique to get the five solution set such that $f = 0$.

We also take inertia factor $w = 0.729$ and $c1$, $c2 = 1.494$. After running the PSO algorithm, the set of five solutions are obtained. After PSO computation, we have got the optimal values of K_p, T_i, T_d, α, β are 25.4, 24.11, 1.988, 0.54, and 0.47. The values are used as input to a simulator diagram in SIMULINK which shows the effect of both integer-order controller and fractional-order controller. The parameter values used in both cases are summarized as follows (Table 1):

Figures depicting this are illustrated as follows: (Fig. 4)

The comparison of the results with a conventional PID controller is made which shows that the performance of the fractional-order PID is better in terms of overshoot as well as rise time (Fig. 5).

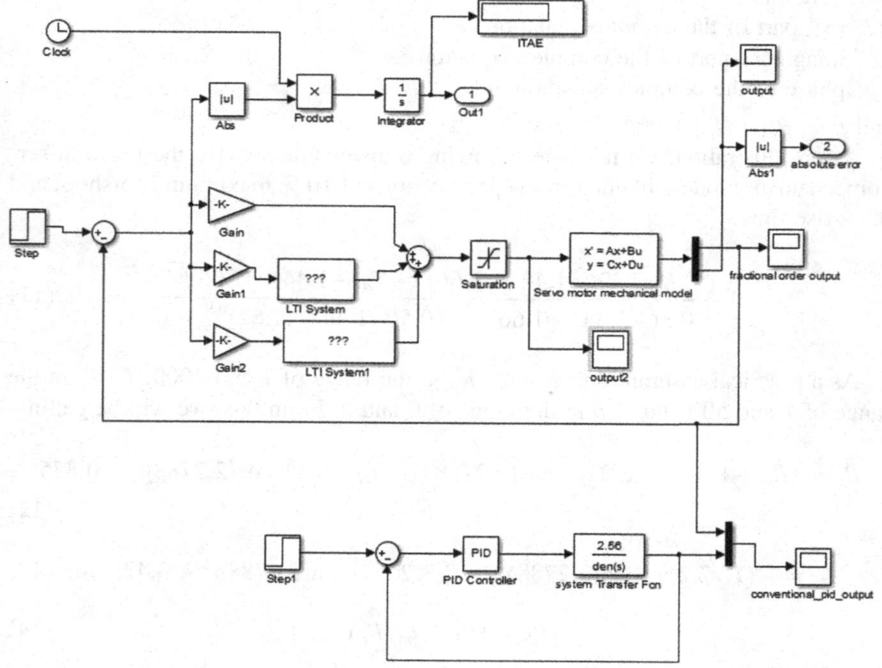

Fig. 4 SIMULINK diagram of the PID controller

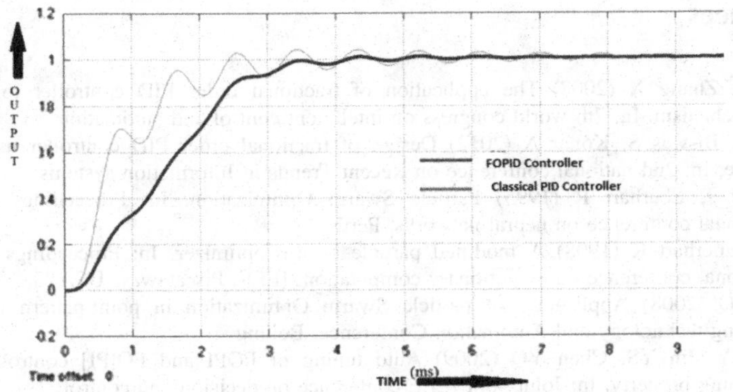

Fig. 5 Comparison of results

Table 2 Comparison of time-domain specifications

Controller	Rise time (t_r)	Settling time (t_s)	Percentage overshoot (M_p)
Classical PID	2.5 ms	5 ms	10 %
FOPID	3.5 ms	9 ms	1 %

The time-domain specifications of the integer-order and fractional-order PID controllers can be summarized as follows (Table 2):

5 Conclusion

The simulations for the fractional order have been very promising. Fractional-order controllers always had the upper hand due to its preciseness in mathematically explaining the system. But the problem lied in computing the fractional order which was made possible by fractional calculus. Now with the advent of optimal tuning for fractional order, controller has become more sophisticated and accurate, although there still problems lie with reliability of values. This can be covered by more robust techniques for optimal tuning. This can be taken a step further by introducing automatic optimal tuning for such controllers. It helps in using such controllers in real-time systems where system characteristics and its parameters changes dynamically with time.

References

1. Zhao C, Zhang X (2008) The application of fractional order PID controller to Position Servomechanism. In: 7th world congress on intelligent control and automation
2. Maiti D, Biswas S, Konar A (2012) Design of fractional order PID controller using PSO technique. In: 2nd national conference on Recent Trends in Information systems
3. Kennedy J, Eberhart R (1995) Particle Swarm Optimization. In: Proceedings of IEEE international conference on neural networks, Perth
4. Shi Y, Eberhart R (1998) A modified particle swarm optimizer. In: Proceedings of IEEE international conference on evolutionary computation. IEEE, Piscataway, USA
5. Deng XQ (2008) Application of particle Swarm Optimization in point-pattern matching. Computing technology and Automation Conference, Beijing
6. Wang CY, Jin YS, Chen YQ (2009) Auto tuning of FOPI and FO[PI] controllers with Iso-damping property. In: Joint 48th IEEE conference on decision and control
7. Cao JY, Cao BG (2000) Design of fractional order PID controllers based on Particle Swarm Optimization

Automatic Steering Using Ultracapacitor

Nurul Hasan Ibrahim, R. Saravana Kumar and Siddharth Kaul

Abstract In today's scenario, parking a vehicle is big challenge due to space availability, lack of driver attention, etc. In developed countries such as North America and Europe, several car manufacturers such as Ford, Audi, BMW, and GM are in market introduction of auto parking cars. In emerging market such as Indian subcontinent, still this feature is not introduced due to traffic congestion and affordability by buyers. This proposal is to develop a retro fit auto parking assist which can be a module and fitted in any vehicle manufacturer. Once the driver enables auto parking mode, the steering wheel rotates automatically based on the parking trajectory required. The speed, angle and direction of rotation of steering column will be sensed by Hall effect-based multi turn steering position sensor which will be powered by ultracapacitor to provide power back up for recording the turn when main power goes off. Based on the sensor information, the controller based on Labview/PLC will take effective action on the motor drive which in turn controls the speed, angle, and direction of motor. The real-time location of the car can be viewed in GUI developed in PC/tablet. Indian road conditions such as pedestrian crossing and neighboring vehicles can be simulated and based on that the parking trajectory will vary. The parking trajectory can be parallel parking, perpendicular parking, reverse parking, etc. The acceleration and speed location of the vehicle can also be simulated and shown in tablet/PC GUI. Later this project can be extended with wheels, transmission, and chassis to show entire car in auto parking mode. In this paper, a Simulink model of the concept is also described. This concept is filed for a patent in India patent office.

Keywords Ultracapacitor · Steering position sensor · Auto parking · GUI · Labview

---

N.H. Ibrahim (✉) · R. Saravana Kumar · S. Kaul
SELECT, VIT University, Vellore 632014, India
e-mail: nurulhasan_i@yahoo.com

R. Saravana Kumar
e-mail: rsaravanakumar@vit.ac.in

S. Kaul
e-mail: siddharth.kaul.k10@gmail.com

L.P. Suresh and B.K. Panigrahi (eds.), *Proceedings of the International Conference on Soft Computing Systems*, Advances in Intelligent Systems and Computing 397, DOI 10.1007/978-81-322-2671-0_39

1 Introduction

Ultracapacitors are modern electric energy storage devices, designed similarly to conventional electrolytic capacitors. Their advantages are high specific energy [5] and high specific power capability along with high cycle life which makes them ideal for peak power applications. Ultracapacitors application in electric vehicles can improve their performance, making them more interesting for automotive market by operating range increase and operating costs cut. In this paper, their application as energy storage device for steering position sensor is discussed.

Ultracapacitors, as shown in Fig. 1, consist of two electrodes, separated by separator soaked in electrolyte. As a result, we get two capacitors connected in series. Capacitance of each is proportional to the electrode area and inversely proportional to the distance between charges. The difference between conventional electrolytic capacitor and ultracapacitor is in electrode material. Ultracapacitors use activated carbon; so achievable electrode area is approximately 2000 $m^2 g^{-1}$, so far allowing production of single cells with high capacitance, up to thousands of farads. This could make ultracapacitors similar to batteries, but the most important difference between batteries and ultracapacitors is that in ultracapacitors the ions form electrolyte approach electrode material, but do not react with it, so there is no electrode structural degradation. This highly improves cycle life of an ultracapacitor. Another concept is asymmetrical electrochemical capacitor, which uses two different electrochemical processes on different electrodes in one cell. Typical example is C–NiOOH system with KOH electrolyte. Advantage of such device is higher specific energy capability. Today's commercial symmetrical cells reach capacity up to 5 kF, with rated voltage 2.5 V and rated current up to 600 A. The highest available capacity of asymmetrical cells is 80 kF, with operating voltage window (0.8–1.7) V. For higher voltages, there are modules with cells connected in series. Parameters of conventional capacitors, batteries and ultracapacitors are shown in Table 1.

The top level block diagram is as shown below (Fig. 2).

At end of this project, we will realize an entire auto parking assist which can be retrofitted in any car/truck. To realize an automatic steering system for auto parking is the objective. This will use a Hall effect-based steering position sensor which

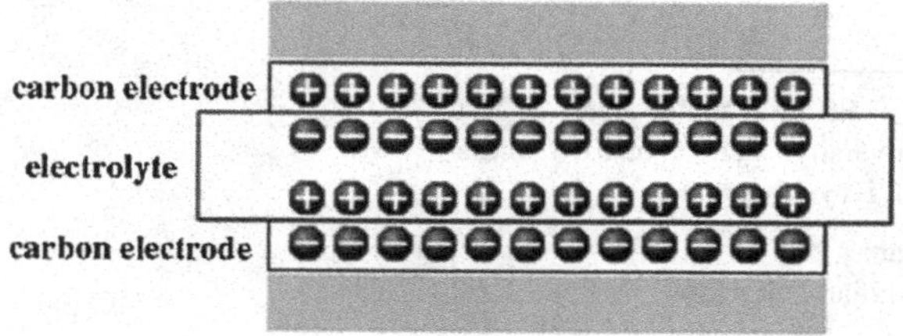

Fig. 1 Double-layer capacitor (charged state)

Table 1 Energy storage devices parameter comparison

Characteristic	Lead-acid battery	Conventional capacitor	Ultracapacitor
Charging time	(1 ÷ 5) h	(10^{-3} ÷ 10^{-6}) s	(0.3 ÷ 30) s
Discharging time	(0.3 ÷ 3) h	(10^{-3} ÷ 10^{-6}) s	(0.3 ÷ 30) s
Energy density	(10 ÷ 100) Wh kg^{-1}	<0.1 Wh kg^{-1}	(1 ÷ 10) Wh kg^{-1}
Cycle life	1,000 cycles	>1,000,000 cycles	>500,000 cycles
Power density	<1 kW kg^{-1}	<100 kW kg^{-1}	<10 kW kg^{-1}
Efficiency	(70 ÷ 85) %	>95 %	(85 ÷ 98) %

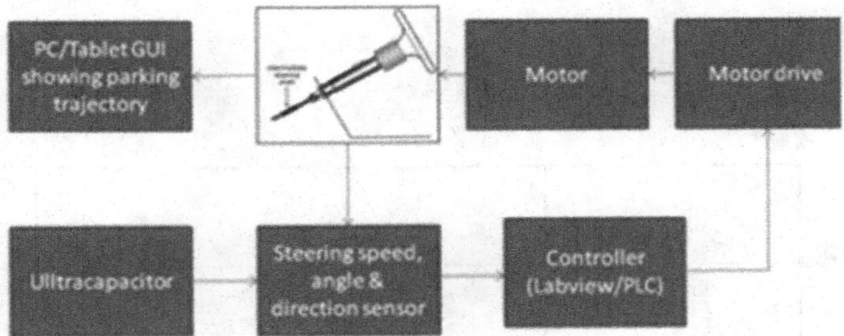

Fig. 2 Top level block diagram

senses speed, angle, and direction of steering column with ultracapacitor powered and feeds a controller (lab view/PLC). The controller will drive a motor which will create automatic steering as per parking trajectory required. When vehicle is stopped, the steering will align the wheels inside the vehicle periphery.

2 Simulation Model in Matlab/Simulink

Total resistance will be magnetic encoder resistance. Charging/discharging time will be based on ultracapacitor configuration. The load current profile is as shown below. The simulation for the ultracapacitor is done with taking in the specification of magnetic encoder as from datasheet of AS5040. This specifies the typical voltage to be 3 V and typical current to be 16 mA (Fig. 3).

The voltage is regulated through a regulator so our ultracapacitor can have values greater that 3 V but the challenge is that with all the resistance in the magnetic encoder, the ultracapacitor needs to maintain the at least 16 mA of current supply in the circuit.

This same thing has been shown in the graph. Where one can see that the load current gradually reduces to 16 mA hence we can conclude that the ultracapacitor can power the magnetic encoder for at least 800–1200 s (Fig. 4).

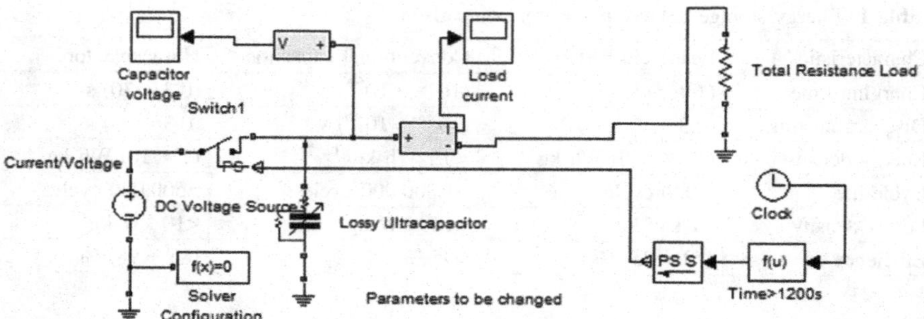

Fig. 3 Simulink model

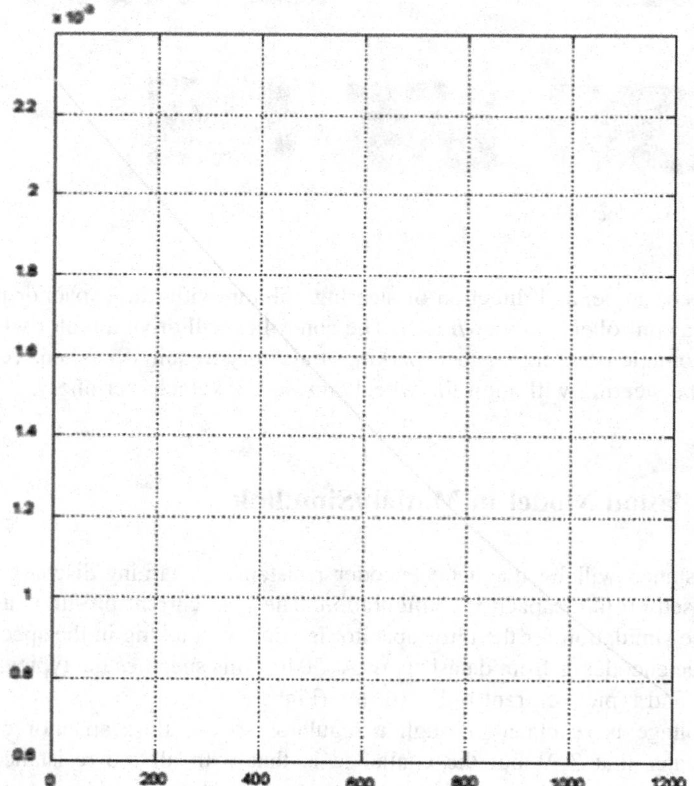

Fig. 4 Load current versus time

3 Steering Position Sensor

This Hall effect-based steering position sensor to sense speed, angle, and direction will be used. The sensor will be mounted at the end of steering column. A diametrically magnetized permanent magnet will be fixed to the rotating column. A hall effect IC either from Austria Micro systems will be placed at an air gap from the magnet. When the magnet rotates, the Hall effect IC will provide the speed, angle, and direction. The no of turns will be captured by indexing in IC. When main power goes off, the current turn information will be stored with ultracapacitor back up.

4 Controller, Motor Drive, and Motor

The sensor output will be sent to the controller through data acquisition card from NI/Siemens. The controller knows the standard information of the trajectory required for parking. Based on the base values with a look up table, it compares current values and takes necessary action to motor drive. The motor drive from parker or any other equivalent will be used. The motor can be either stepper or servo.

5 GUI

The location of car w.r.t. parking trajectory will be shown real time in a PC/tablet gui with labview/plc interface. In the same GUI, user can select type of parking profile and same will be sent to controller. The controller will drive the motor as per parking trajectory required. The conditions such as pedestrian crossing and adjacent vehicle collision can be simulated and parking trajectory can be varied accordingly. Below are the images of model GUI's (Figs. 5 and 6).

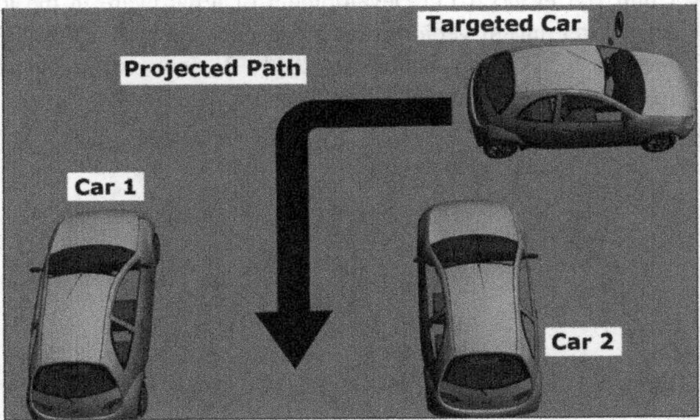

Fig. 5 Reverse parking

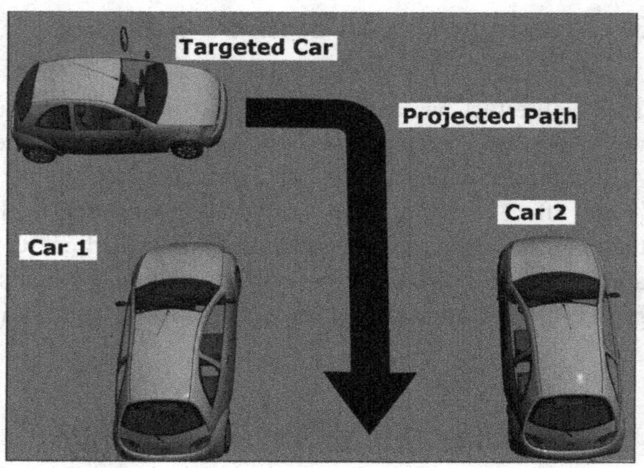

Fig. 6 Perpendicular parking

6 Conclusion

In this paper, the concept for an automatic steering was described. The simulation of the entire model is also explained. A retro fit auto parking assist was explained briefly.

As a future work, the entire vehicle with chassis, transmission, etc., can be considered with collision sensors around the vehicle. The entire vehicle parking can be dealt.

References

1. AMS 5040 datasheet
2. Sivaraj D et al (2011) Automatic Steering Control and Adaptive Cruise Control for a Smart Car. International Conference on VLSI, Communication and Instrumentation (ICVCI) 2011 Proceedings published by International Journal of Computer Applications® (IJCA)
3. Li X, Zhao X-P, Jie (2009) Chen controller design for electric power steering system using T-S fuzzy model approach. Int J Autom Comput 6(2):198–203
4. Chun HH et al (2011) Development of a hardware in the loop simulation system for electric power steering in vehicles. Int J Automot Technol 12(5):733–744
5. Mallika S, Saravana Kumar R (2011) Review on ultracapacitor—battery interface for energy management system. Int J Eng Technol 3(1):37–43

3D Multimodal Medical Image Fusion and Evaluation of Diseases

Nithya Asaithambi, R. Kayalvizhi and W. Selvi

Abstract In this paper, registration-based multimodal medical image fusion process implemented using 3D shearlet transform. Here, 3D MRI simulated slices with three different sets like T1-weighted, T2-weighted, proton density (Pd) are fused to view the abnormalities. As well as magnetic resonance imaging (MRI) and single-photon emission computed tomography (SPECT) images are fused to obtain more useful information and diseases like Cavernous angioma, Hungtinton's chorea are evaluated. 3D shearlet coefficients of the high-pass subbands are highly non-Gaussian and they are modeled into generalized Gaussian distribution (GGD) using heavy-tailed phenomenon. Kullback–Leibler distance is used to obtain local and global information between two high-pass bands and the images are fused by registration. By fusing memory required for storage is also reduced.

Keywords 3D image fusion · Shearlet transform · Generalized gaussian distribution · Registration · Disease evaluation

1 Introduction

Multimodal medical image fusion technologies facilitate better applications of medical imaging to provide an easy access for doctors to recognize the lesion structures and functional change by studying the data of anatomical and functional modalities. CT image is used to visualize a dense structure and it is not able to view soft tissues, the MRI image provides visualization of tumors and tissue abnormalities.

N. Asaithambi (✉) · R. Kayalvizhi
Annamalai University, Chidambaram, India
e-mail: info.nithi83@gmail.com

W. Selvi
Arunai College of Engineering, Tiruvannamalai, Tamil Nadu, India

L.P. Suresh and B.K. Panigrahi (eds.), *Proceedings of the International Conference on Soft Computing Systems*, Advances in Intelligent Systems and Computing 397, DOI 10.1007/978-81-322-2671-0_40

PET image provide the information of blood flow to the body. Magnetic resonance imaging plays an intensifying role on appraisal of multiple sclerosis. For clinical pinpointing work, current conventional MRI has many series of image acquisition which depends on available pulse sequences developed to provide optimal tissue contrast. Proton density and T2 weighted images are generated with long repetition time. The four parameters proton density, T1 relaxation time, T2 relaxation time, and flaw are responsible for signal intensity in MRI. For MR images the essential determines of contrast-based signal intensity are T1 and T2 relaxation time. If an deformity is found, an additional scan helps to distinguish the lesion. Noncontrast T1-weighted images are considered necessary only if the prelude scans suggest blood loss, lipoma, or dermoid. Or else, contrast-enhanced scans are suggested.

Gadolinium-based contrast agents for MR imaging are paramagnetic and have established outstanding biologic forbearance. No important complications or side effects have been reported. The gadolinium-based contrast agents do not annoy the undamaged blood–brain barrier (BBB). If the BBB is disrupt by a syndrome process, the gadolinium-based contrast agent diffuse into the interstitial space which reduces T1 relaxation time of the tissue and increase the T1-weighted images signal intensity. The scans should be captured between 3 and 30 min post vaccination for finest results. Image fusion is a subdivision of digital image processing where more sequences of images are combined to form a single fused image which is highly informative and reduces the storage memory capacity. Application of data fusions are computer vision, military surveillance, medical, and in forensic field. For image fusion data's can be 2D (still image), 3D (video sequences in the form of spatiotemporal volumes or volumetric still images), or of higher dimensions. Image fusion techniques are broadly categorized into multimodal, multi-view, multi-temporal, and multi-focus. Use of the discrete wavelet transform (DWT) in image fusion was approximately and concurrently proposed by Li et al. and Chipman et al. Computed tomography provides dense intense structure information and fails to provide information regarding soft tissues. Magnetic resonance image gives better information of tumor and abnormalities in human body. Positron emission tomography (PET) act as a nuclear radioactive medical imaging. Single-photon emission computed tomography (SPECT) gives the information of how blood flows to the tissues and other chemical reaction metabolism. Table 1 shows the MR signal intensities for different materials of internal organs.

Table 1 MR signal intensities

	T2WI	PD/FLAIR	TIWI
Solid mass	Bright	Bright	Dark
Cyst	Bright	Dark	Dark
Sub acute blood	Bright	Bright	Bright
Acute and chronic blood	Dark	Dark	Gray
Fat	Dark	Bright	Bright

2 Related Work

Many different types of transforms are used for image fusion in both 2D and 3D images. Cao et al. [1] appraise the performance of all levels of multi-focused image fusion of using discrete wavelet transform, stationary wavelet transform, lifting wavelet transform, multi-wavelet transform, dual tree DWT, and dual tree complex wavelet transform in terms of different parametric measures. Daneshvar et al. [2] proposed the method of image fusion based on hue, saturation, and intensity method. Texture retrieval is done using wavelet transform with generalized Gaussian distribution and Kullback–Leibler Distance [3, 4] describes the detailed multiscale decomposition of 2D shearlet Transform. Goshtasby et al. [5] presents a curvelet transform advance method for the fusion of magnetic resonance brain image (MRI) and computed tomography brain image (CT). Easley et al. [4] proposed a new discrete multiscale directional illustration called the discrete shearlet transform for reducing the noise application. For human visualization nonlinear multi-resolution signal disintegration gives better salient features [6]. To visualize tumor by knowing the anatomical and physiological uniqueness is done by fusing CT and positron emission tomography (PET) images [7]. Source images can be decomposed into different levels of low-pass subbands and high-pass subbands. Laplacian pyramid and wavelet transform are important MSD tools in image fusion [8]. Different fusion technologies available are based on three methods which are pixel level, decision level, and feature level [9].

3 Materials and Methods

3.1 Multiscale Decomposition

Figure 1 shows the overall proposed block diagram. The pixel level fusion methods can be roughly classified into (MSD) Multiscale decomposition based or non-MSD-based methods. The MSD coefficients only know the local relationship in a small region but not any of the global relationship between the two

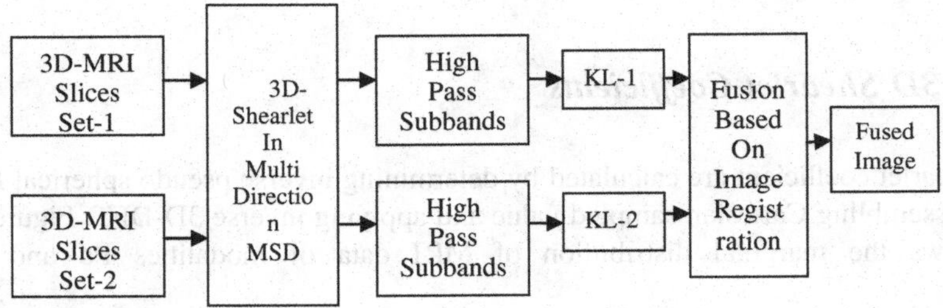

Fig. 1 Overall block diagram

corresponding high-pass subbands. MSD-based medical image methods outperform the other methods, for the source images can be decomposed into the low-pass subbands and the high-pass subbands in different levels to be more suitable to the mechanism of the human vision.

3.2 3D Shearlet Transform

It has both advantages of curvelet and surfacelet transform. By using DWT image can be decomposed into more high-pass subbands with each level of vertical, horizontal, and diagonal subbands. By using 3D shearlet transform high-pass subbands, more than three directions can be obtained; with four directions are first level and six directions in second level.

3D shearlet transform consists of two functions (1) multiscale partition using Laplacian filter (2) directional localization using pseudo spherical Fourier transform. By having $D = 3$, pyramidal regions are obtained by P_1, P_2, P_3 partition of Fourier space R3 [10].

$$\begin{aligned} P_1 &= \left\{ (\varepsilon_1, \varepsilon_2, \varepsilon_3) \in R^3 : \left|\frac{\varepsilon_2}{\varepsilon_1}\right| \leq 1, \left|\frac{\varepsilon_3}{\varepsilon_1}\right| \leq 1 \right\} \\ P_2 &= \left\{ (\varepsilon_1, \varepsilon_2, \varepsilon_3) \in R^3 : \left|\frac{\varepsilon_1}{\varepsilon_2}\right| < 1, \left|\frac{\varepsilon_3}{\varepsilon_1}\right| \leq 1 \right\} \\ P_3 &= \left\{ (\varepsilon_1, \varepsilon_2, \varepsilon_3) \in R^3 : \left|\frac{\varepsilon_1}{\varepsilon_3}\right| < 1, \left|\frac{\varepsilon_2}{\varepsilon_3}\right| < 1 \right\} \end{aligned} \tag{1}$$

3D shearlet transform is a collection of coarse scale shearlets, interior shearlets, and boundary shearlets with parameters l_1 and l_2 which control the orientations of support regions in 3D shearlet systems

$$\begin{aligned} &\{\tilde{\psi}_{-1,k} :\in k \in Z^3\} \cup \cup \{ \tilde{\psi}_{j,l,k} ; j \geq 0, l_1, l_2 = \pm 2^j, k \in Z^3\} \\ &\{\bar{\psi}_{j,l,k,d} : j \geq 0, |l_1| < 2^j; |l_2| \leq 2^j, k \in Z^3, d = 1,2,3\} \end{aligned} \tag{2}$$

3.3 3D Shearlet Coefficients

3D shearlet coefficient are calculated by determining inverse pseudo spherical DFT by reassembling Cartesian sampled value and applying inverse 3D-DFT. Figure 2a, b shows the marginal distribution of MRI data of modalities Pd and T2,

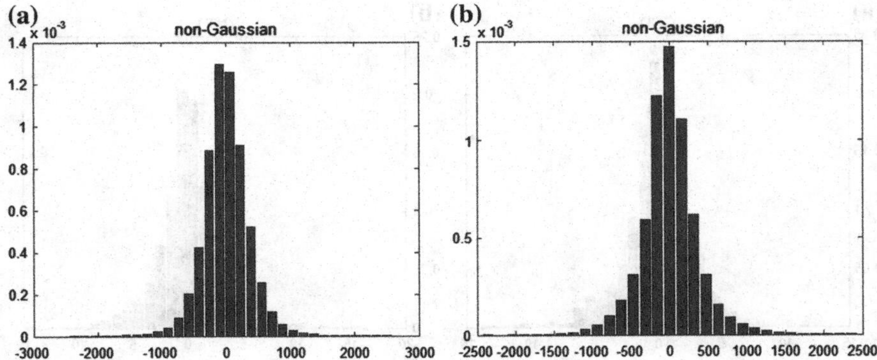

Fig. 2 **a** and **b** Marginal distribution for MRI—Pd weighted and marginal distribution for MRI—T2 weighted

respectively. Probability density distribution are characterized by very shape peak at zero amplitude and extended tails in both sides of the peak (also called heavy-tailed phenomenon). The marginal distributions of 3D high-pass subbands coefficient are highly non-Gaussian. And the graph is between histogram and shearlet coefficient amplitude.

3.4 *GGD for the 3D Shearlet Coefficients*

Generalized Gaussian distribution (GGD) is also called exponential power distribution, which is a widely used as parametric distribution. To obtain good probability density function (PDF) approximation for marginal density of shearlet coefficients at particular subband produced by varying two parameters is given by

$$P(x, \alpha, \beta) = \frac{\beta}{2\alpha\Gamma\left(\frac{1}{\beta}\right)} \mathrm{e}^{-\left(\frac{|x|}{\alpha}\right)^{\beta}} \tag{3}$$

where $\Gamma(.)$ is gamma functions (i.e.,) α models the width of PDF peak (standard deviation), β is inversely proportional to the decreasing rate of peak. α is scale parameter, β is shape parameter. Figure 3a, b shows the GGD of MRI data of modalities Pd and T2, respectively. And the graph is between probability distribution function and 3D shearlet Coefficient. α and β value of Fig. 3a is 0.9461 and similarly for Fig. 3b is 0.9462.

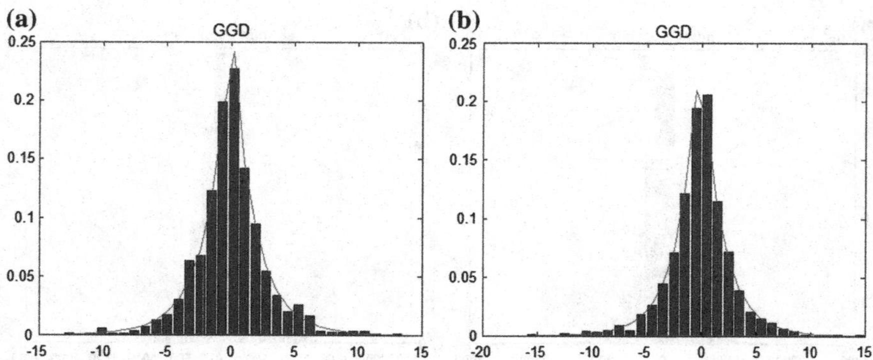

Fig. 3 **a** and **b** GGD for high-pass subbands (Pd) and GGD for high-pass subbands (T2)

3.5 Similarity Measurement and Image Registration

Global relationship or similarity measurement between two high-pass subbands can be described by the Kullback–Leibler distance (KLD) [10] as

$$D(p(.;\alpha_1,\beta_1)\|p(.;\alpha_2,\beta_2)) = \log\left(\frac{\beta_1\alpha_2\Gamma(1/\beta_2)}{\beta_2\alpha_1\Gamma(1/\beta 1)}\right) + \left(\frac{\alpha_1}{\alpha_2}\right)^{\beta_2}\frac{\Gamma((\beta_2+1)/\beta_1)}{\Gamma(1/\beta 1)} - \frac{1}{\beta 1} \tag{4}$$

Determining the point-by-point pixel correspondence between two slices is called registration and by registering two slices, the fusion of multimodality information becomes possible, the depth, changes in slices, and objects are detected. Figure 4a, b shows the KLD between two high-pass subbands of Pd weighted and T2 weighted, respectively.

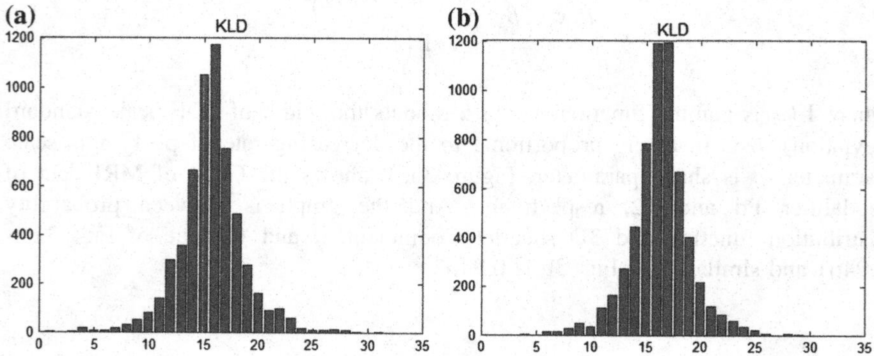

Fig. 4 **a** and **b** KLD between two subbands (Pd) and KLD between two subbands (T2)

Registration algorithm is given below:

```
[opt, met] = imregconfig('monomodal');
opt.MaximumIterations=300;
opt.MinimumStepLength=5e-4;
fc=cell(1,3);
fg=zeros(217,181,10);
 for j=1:3
 xx=real(KL1{j});
 yy=real(KL2{j});
 for i=1:10
 fs=imregister(xx(:,:,i),yy(:,:,i),'rigid',opt,met);
 fg(:,:,i)=fs;
end
 fc{j}=fg;
end
```

4 Results and Discussions

For the evaluation purpose, simulated 3-D MRI brain are obtained from Brain Web, which have three sets Pd, T1 and T2. The scans in each set were spatially registered for the simulation. Each scan has 181 × 217 × 181 voxels with 12-bit precision and the thickness is 1 mm. Simulation results are obtained using MATLAB 7.14 Version R2012 with Windows 07 Operating system and 2 GB RAM. Totally 181 slices are considered in this experiment in each sets of T1, T2, and Pd. But in this paper work only first 10 slices in each set are considered since to reduce length. Quantitative metrics considered here are mutual information, entropy, peak signal-to-noise ratio (PSNR), and structural similarity index measure. Figure 5

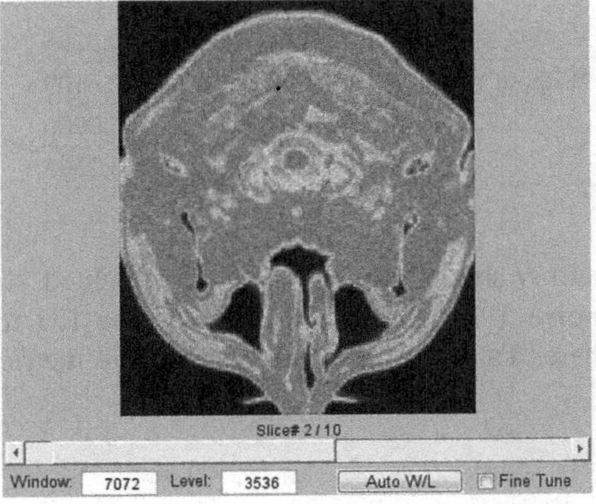

Fig. 5 3D MRI Pd weighted (1–10) slices

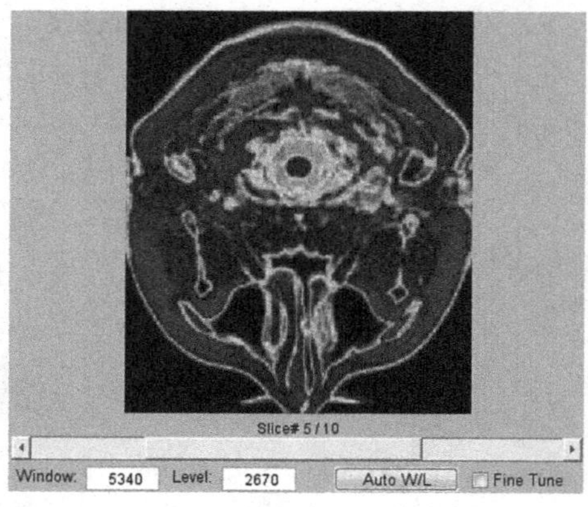

Fig. 6 3D MRI Pd weighted (1–10) slices

Table 2 Evaluation Of MRI Images

Noise	Modality	MI	Entropy	PSNR	SSIM
MRI noise 1 %	T1-T2	2.4946	37.1667	28.0353	0.6753
	T1-pd	4.0169	37.1667	92.8474	0.200
	T2-pd	4.0475	37.1667	47.0862	0.7264
MRI noise 3 %	T1-T2	2.2024	37.1667	28.1871	0.6707
	T1-pd	4.4462	37.1667	38.9057	0.6542
	T2-pd	2.7349	37.1667	47.1808	0.6569
MRI noise 5 %	T1-T2	1.9782	37.1667	28.5225	0.6557
	T1-pd	4.6440	37.1667	39.0281	0.6308
	T2-pd	2.0246	37.1667	47.3226	0.6153
MRI noise 7 %	T1-T2	2.1351	37.1667	29.0339	0.6375
	T1-pd	4.4784	37.1667	39.2059	0.6195
	T2-pd	1.7638	37.1667	47.5146	0.5816
MRI noise 9 %	T1-T2	1.6779	37.1667	29.6710	0.6173
	T1-pd	3.9796	37.1667	39.4410	0.5519
	T2-pd	1.8295	37.1667	47.7432	0.4036
Without noise	T1-T2	2.6109	37.1667	28.0257	0.6629
	T1-pd	4.0906	37.1667	38.8409	0.6337
	T2-pd	3.2393	37.1667	47.0730	0.7840

and 6 shows the 3D-MRI pd weighted slices and MRI T2 weighted slices with noise 5 %, respectively. Among the 181 total slices first ten of each set is shown here. Table 2 shows the quantitative metrics with MRI slices having noise 1, 3, 5,

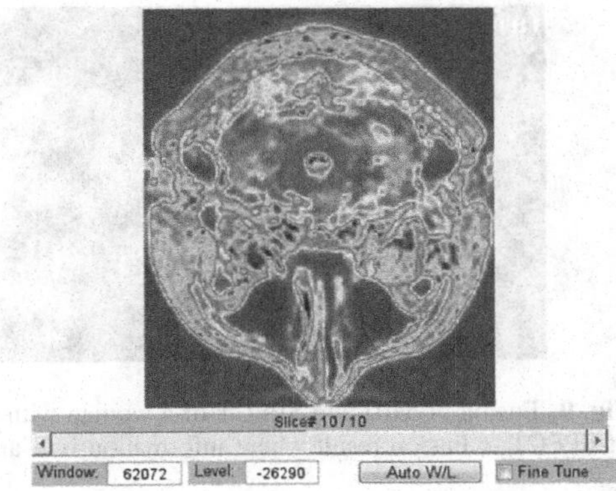

Fig. 7 Fused output image

9 % and MRI slices without noise. Figure 7 shows the output of the fused image which is a gray scale image using color map converted to color for highlighting purpose.

4.1 Evaluation of Diseases

By fusing MRI and SPECT images diseases like cavernous angioma and degenerative disease like Huntington's chorea are evaluated. Huntington disease is caused by changes in central area of the brain which affect the person's movement, mood and thinking skills. By fusing MRI—SPECT brain imaging, basal ganglia (BG) perfusion in patients can be well evaluated which is shown in Fig. 9. Cavernous angioma is a blood vessel abnormality characterized by large, adjacent capillaries with little or no intervening brain. The blood flow through these vessels is slow and their fused image is shown in Fig. 8.

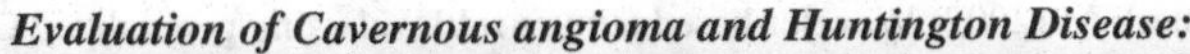

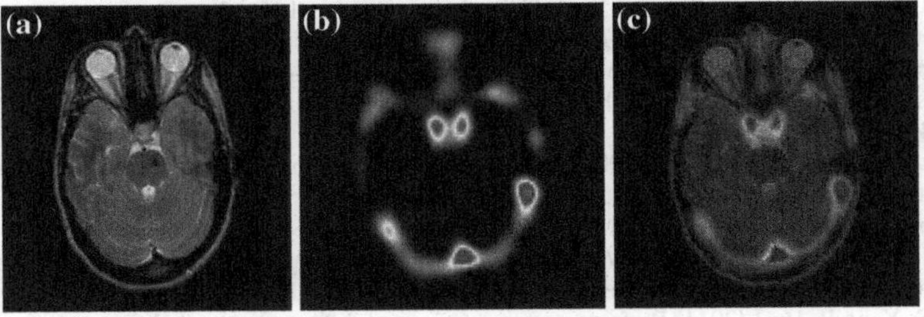

Fig. 8 Fusion of MRI and SPECT of a woman with Cavernous angioma. **a** Slice of MRI. **b** Slice of SPECT. **c** Fusion result where information in **a** and **b** are combined together

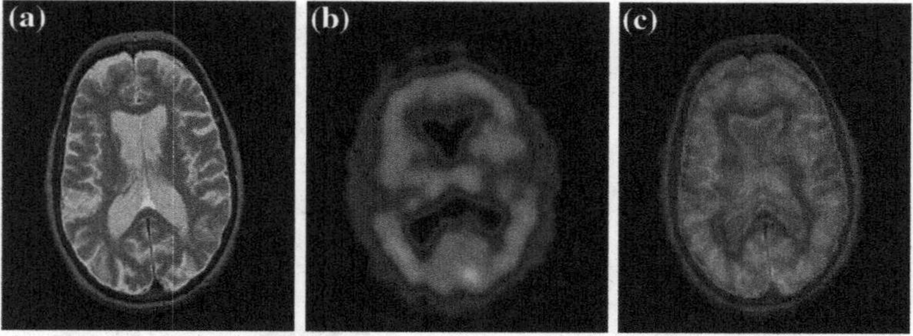

Fig. 9 Fusion of MRI and SPECT of a woman with Huntington disease. **a** Slice of MRI. **b** Slice of SPECT. **c** Fusion result where information in **a** and **b** are combined together

5 Conclusion

In this paper, we have developed a novel medical image fusion method based on registration with the 3D shearlet transform space. The 3D shearlet transform provided better image representations. The proposed fusion rule was stimulated from the asymmetry of the KLD, which could guarantee the validity of the proposed method in the information theory. Experiments on synthetic data and real-data demonstrated the effectiveness of the proposed method by the fusion results of higher quality. In the future, this method can be extended to 4D medical image fusion and with different software.

References

1. Cao Y, Li Sh, Hu J (2011) Multi-focus image fusion by nonsubsampled shearlet transform. Proc 6th Int Conf Image Graphics 17–21
2. Daneshvar S, Ghassemian H (2011) MRI and PET image fusion by combining IHS and retina-inspired models. Inf Fusion 11:114–123
3. Do MN, Vetterli M (2002) Wavelet-based texture retrieval using generalized Gaussian density and Kullback–Leibler distance. IEEE Trans Image Process 146–158
4. Easley G, Labate D, Lim W-Q (2008) Sparse directional image representations using the discrete shearlet transform. Appl Comput Harmon Anal 25–46
5. Goshtasby AA, Nikolov S (2007) Image fusion: advances in the state of the art. Inf Fusion 114–118
6. Goutsias J, Heijmans HJAM (2000) Nonlinear multi-resolution signal decomposition schemes. Part I: Morphological pyramids. IEEE Trans Image Process 1862–1876
7. Grosu AL, Weber WA, Franz M (2005) Reirradiation of recurrent highgrade gliomas using amino acid PET (SPECT)/CT/MRI image fusion to determine gross tumor volume for stereotactic fractionated radiotherapy. Int J Rad Oncol Biol Phys 511–519
8. Li S, Yang B, Hu J (2011) Performance comparison of different multiresolution transforms for image fusion. Inf Fusion 74–84
9. Piella G (2003) A general framework for multi-resolution image fusion: from pixels to regions. Inf Fusion 4(4):259–280

10. Wang L, Li B, Tian L (2014) Multimodal medical volumetric data fusion using 3D. Discrete sharlet transform and global to local rule. IEEE Trans Biomed Eng
11. Do MN, Vetterli M (2003) Framing pyramids. IEEE Trans Signal Process 2329–2342
12. Heijmans HJAM, Goutsias J (2000) Nonlinear multi-resolution signal decomposition schemes. Part II: morphological wavelets. IEEE Trans Image Process 1897–1913
13. Kwan RKS, Evans AC, Pike GB (1999) MRI simulation based evaluation of image processing and classification methods. IEEE Trans Med Imaging 1085–1097
14. Miao Q, Shi Ch, Xu P, Yang M, Shi Y (2011) A novel algorithm of image fusion using shearlets. Opt Commun 1540–1547
15. Negi P, Labate D (2012) 3D discrete shearlet transform and video processing. IEEE Trans Image Process 2944–2954
16. Pajares G, De la Cruz DM (2004) A wavelet-based image fusion tutorial. Pattern Recogn 1855–1872
17. Sharifi K, Leon-Garcia A (1995) Estimation of shape parameter for generalized Gaussian distributions in subband decompositions of video. IEEE Trans Circuits Syst Video Technol 5 (1):52–56
18. Shen R, Cheng I, Basu A (2013) Cross-scale coefficient selection for volumetric medical image fusion. IEEE Trans Biomed Eng 1069–1079
19. Yang L, Guo B, Ni W (2008) Multimodality medical image fusion based on multi-scale geometric analysis of contourlet transform. Neurocomputing 203–211
20. Johnson KA, Becker JA. The whole brain altas. http://www.med.harvard.edu/aanlib/

10. Wang L, Li [illegible] (2014) Multimodal medical volumetric [illegible] using 3D [illegible] transform and [illegible]
11. [illegible] (2013) [illegible]
12. [illegible] (2000) [illegible] morphological [illegible] IEEE [illegible]
13. [illegible] (1999) [illegible] processing [illegible] IEEE Trans [illegible]
14. [illegible] (2011) A novel [illegible] image fusion using [illegible]
15. [illegible] (2013) [illegible] shearlet [illegible] and [illegible] Image Process [illegible]
16. [illegible] based [illegible] Pattern Recognition [illegible]
17. [illegible] image fusion [illegible]
18. [illegible] (2013) [illegible] image fusion [illegible] IEEE Trans Biomed Eng [illegible]
19. Yang L, Guo [illegible] Multimodality medical image fusion [illegible] contourlet transform [illegible]
20. [illegible]

Automatic Generation of Multimodal Learning Objects from Electronic Textbooks

Aksheya Suresh and V. Mercy Rajaselvi

Abstract Technology-supported learning systems (TSLS) (e-learning, educational software, electronic books) require an appropriate representation of the knowledge to be learned that is the domain module. Domain module enables to guide the students in the learning process or to learn by themselves. The authoring of the domain module is cost- and labor intensive. Achieving greater accuracy in the generation of domain module is an issue. The proposed system considers the source of electronic textbooks that are documents in different formats—pdf, doc as input. The system is designed to efficiently handle the multimodal learning objects (images and captions) that are present in electronic textbooks. Also, the proposed system is designed to improve the generation of learning domain ontology (LDO), which employs ontology construction using natural language processing (NLP) techniques. Thus, improvisation in LDO guarantees to achieve a higher accuracy.

Keywords Domain module · E-learning · Ontology · Multimodal

1 Introduction

1.1 Motivation

The current system of education requires a transformation for today's generation in its speedy revolution. This is most important for children who have limited access to education, both in rural and urban India. In today's world, students need to equip themselves to enter into the job market. The only field that has not yet transformed is the influence of a teacher on the learning phase of a student's life. The advent of

A. Suresh (✉) · V. Mercy Rajaselvi
Computer Science and Engineering, Easwari Engineering College, Chennai, India
e-mail: aksheya.suresh@gmail.com

V. Mercy Rajaselvi
e-mail: mercyeec@gmail.com

L.P. Suresh and B.K. Panigrahi (eds.), *Proceedings of the International Conference on Soft Computing Systems*, Advances in Intelligent Systems and Computing 397, DOI 10.1007/978-81-322-2671-0_41

technology can help the education system through e-learning hubs. This creates a pool of learning and sharing knowledge across borders. It can also help to upgrade the approach to education for differently abled children, which is the priority for the government under the 12th 5-Year Plan.

E-learning has gained fame in the education technology revolution, which has substantial benefits. It offers unique opportunities for people with confined access to education and training. It blends innovative and inventive strategy and thus provides unlimited access to resources and information.

Theoretically, e-learning can also be known as instructional technology, information and communication technology (ICT) in education, EdTech, learning technology, multimedia learning, technology-enhanced learning (TEL), computer-based instruction (CBI), computer managed instruction, computer-based training (CBT), computer-assisted instruction or computer-aided instruction (CAI), Internet-based training (IBT), flexible learning, web-based training (WBT), online education, virtual education, virtual learning environments (VLE) (which are also called learning platforms), m-learning, and digital education [1].

1.2 *Mankato's Hierarchy of Educational Technology Needs*

The following diagram is Mankato's hierarchy [2] which depicts the impact on student's performance, and the difficulties the institutions may face if the following hierarchy is not met.

Established Infrastructure This bottom level of hierarchy is for ensuring that reliable, adequate, cost-effective, and secure technology infrastructure is being provided and it supports learning, teaching, and thereby achieves the administrative goals of the institution.

Effective Administration In administrative level of hierarchy, the technology to improve its administrative effectiveness through improvised business practices, effective communication, planning, and record keeping is to be provided.

Extensive Resources This hierarchy manages the resources effectively. It provides the most recent, accurate, and extensive information resources possible to all learners in a cost-effective and reliable manner. Figure 1 shows the Mankato's hierarchy of educational technology.

Enhanced Teaching This level of hierarchy is to guarantee that all mentors are equipped with technology training, skills, and resources which are needed to ensure students' success and assess their progress.

Empowered Students If all the below hierarchies are satisfied, then all students will demonstrate the mastered use of technology to access, process, organize, and communicate.

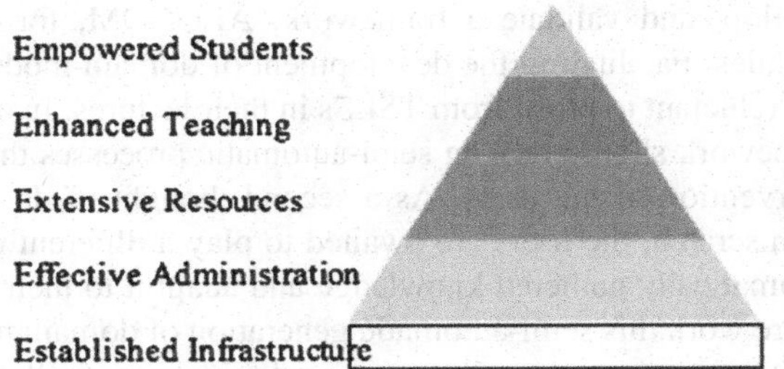

Fig. 1 Mankato's hierarchy of educational technology [2]

1.3 *Domain Module*

The domain module enables advantages in both exploratory learning and instructive learning as well. In exploratory learning, it enables the students to learn by themselves, and in instructive learning provides technology supported learning systems (TSLSs) to plan the learning process. For example, an ITS relies on the domain module to determine the level of class interaction, the preference of examples, questions, and statements, and to assess the potential of the students. However, building TSLSs, especially their domain modules, is cost- and labor intensive.

2 Related Work

The work in the construction of domain module can be elaborated as Calvo et al. [3] have introduced a framework—ErauzOnt, for identifying and gathering learning objects from electronic documents using natural language processing. This framework groups the learning objects based on identification of commonly used patterns from the document containing the educational material. Building of learning objects considers pieces of texts and images which are corresponding or appropriate to a particular topic. This framework is designed in such a way that it handles only pdf documents in Basque language. As a future work it can be extended to support various documents and can be developed for other international languages like—English and Spanish. Also, the result obtained by collecting learning objects can be used for the construction of domain module.

Olagaray et al. [4] introduced a framework to semi-automate the generation of TSLSs. More specifically, it aims at reproducing the way teachers prepare their lessons and the learning material they are about to use for the entire course. Teachers incline to prefer one or more textbooks that describe the contents of their subjects, diagnose the topics to be addressed, and discover the parts of the textbooks which may be beneficial for the students. Textbooks provide resources with educational purposes, e.g., definitions, theories, examples, or exercises, which might be effective for learning. Given many electronic textbooks, choosing appropriate sources of information is vital, this work aims to analyze, design, develop, and validate a framework, ALOCOM, for the construction of domain modules. Facilitating the development of domain modules may also make teachers less reluctant to profit from TSLSs in their lectures. In order to achieve that goal, the framework should rely on semi-automatic processes that minimize the need for the intervention of the users. As a second thought of developing the domain module from scratch, the users are awaited to play a different role. They will supervise the automatically gathered knowledge and adapt it to their preferences and needs. As a future work this semi-automatic generation of domain module can be done in a fully automatic way and can be compared with a manually generated one.

Elorriaga et al. [5] developed a framework—DOM Sortze which constructs automatic generation of the domain module. The domain module generation process entails three main tasks: preparing the document for knowledge extraction, building the ontology that describes the domain to be learned, and the generation of the learning objects (LOs). In this approach, they have considered electronic documents without images and have tested them using an electronic textbook and compared the automatically generated elements with the domain module manually developed by instructional designers. As a future work, the paper addresses to improve the generation of LDO as the accuracy achieved is 80.09 % and handle images which has not been done here. Also, it can be improved to support multilingual issues.

The extraction of figures from papers provides lots of information. The academic papers and research papers comprise of multiple figures (information graphics) representing significant findings and experimental results. The extraction of data automatically from those figures and classifying information graphics is a problematic solution. Choudhury et al. [6] have used indexing, classification, and data extraction of figures in PDF documents. Images and captions from the pdf file are extracted using Xpdf5 API and captions can also be extracted using regular expressions and filters. A figure caption would be a paragraph or a line starting with the term "Fig." or "Figure" immediately below the figure. However, the issue here is that images and captions are separately captured in different folders. Hence, it is necessary to have a mapping mechanism that maps the respective images and captions accordingly.

Clustering search results problem can be defined as grouping of similar documents which is done automatically from a search hits list that are returned from a search engine. Osiński et al. [7] proposed an approach that used open directory project that acts as a source of high-quality narrow topic document references and mixed them into several multitopic test sets. It is done by discovering groups of related documents and describes the subject of these groups in an understandable way to a human. Lingo uses a combination of phrases and singular value decomposition (SVD) techniques. SVD and the related dormant semantic indexing were presented for information retrieval for disarticulating concepts from textual data. It first attempts to discover descriptive names for future clusters and only then proceeds to assigning each cluster with matching documents. The process employed here is the reverse of the conventional clustering algorithm.

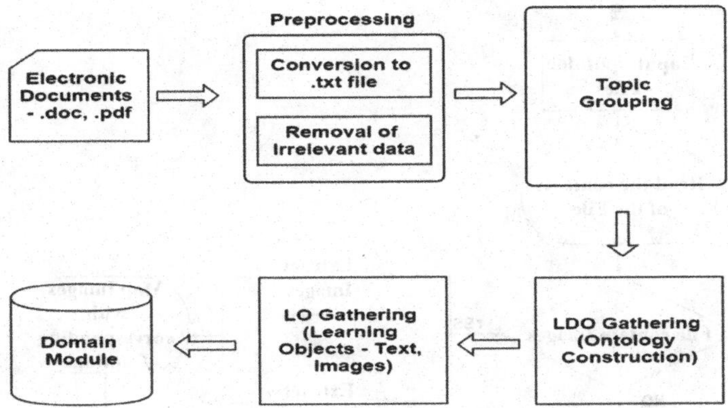

Fig. 2 Overall architecture of automatic generation of multimodal learning objects

3 Proposed System

The system is designed for automatic generation of domain module by handling electronic documents with images as input. Initially preprocessing of the documents is done and written into txt file. Irrelevant words, which are present in the documents, are removed using stop-word removal algorithm. The document may contain multimodal objects like images and those images and captions are extracted using Xpdf5 API and a mapping function is used to map the respective images with the corresponding captions. Finally the maximum number of occurrences of a topic is determined by clustering using Lingo Algorithm. Based on the occurrence, topics are grouped. To improve the efficiency of grouping, ontology construction and natural language processing techniques are used. Figure 2 shows the overall architecture of the automatic generation of multimodal learning objects.

4 Implementation

4.1 Preprocessing

In preprocessing the electronic documents with different formats—pdf, doc are accepted and preprocessed using preprocessing algorithms. Figure 3 shows the activity diagram for preprocessing of electronic documents.

Conversion to .txt File Here input is electronic documents of different formats—pdf, doc. To have a common format, conversions of these formats are done. These are converted to .txt files for easier handling. The conversions to .txt files are done using XPWF—Apache poi jar. XWPF has a fairly substantial core API, that provides read and write access to the main parts of various formats of documents like —.doc, .pdf.

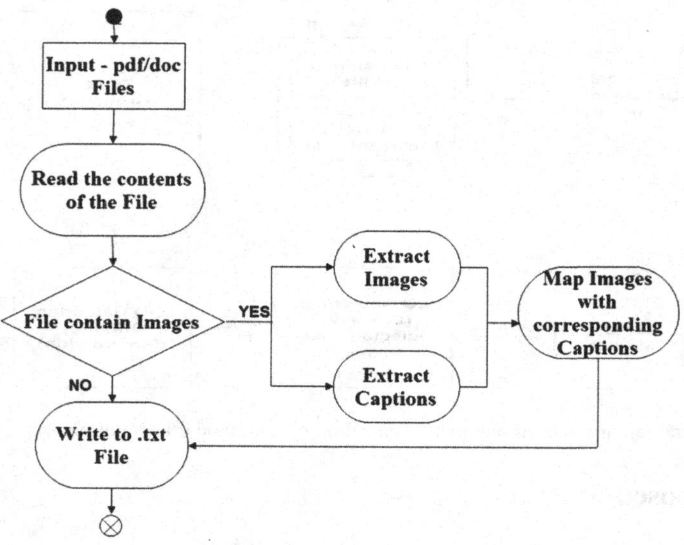

Fig. 3 Activity diagram for preprocessing

Removal of Irrelevant Words The removal of irrelevant words is an important step in preprocessing. Removing such words is done using stop-word removal algorithm, thereby increasing efficiency in searching, grouping, etc.

Extraction of Multimodal Objects Electronic documents contain multimodal objects—Images. The images from the documents are extracted using Xpdf5 and the captions from the documents that describe about the images are extracted using regular expressions.

Now the images and captions are stored as separate folders. In order to ensure that every image and caption coincides we apply mapping function between list of images and list of captions. Figure 3 depicts the activity diagram for preprocessing of documents with and without images.

4.2 Topic Grouping

Topic grouping is done using Lingo Clustering Algorithm. Lingo consists of five phases. In phase one, input snippets (document fragments) are preprocessed using stemming and stop-word removal algorithms. In phase two, frequent terms and phrases are discovered by combining all documents and storing it in suffix-arrays. In phase three, the actual group label induction is done. Orthogonal vectors [7] are assembled from previously extracted common phrases by using a vector space model and the score is calculated for each phrase against the SVD-decomposed matrix. To represent distinct topic from the input data, these orthogonal vectors have to be extracted and it is done using SVD. The fourth phase of the algorithm consists of cluster content discovery. Vector space model is used against input document set that is applied to group labels as artificial queries. In this scenario, the highest scoring documents for each cluster are assigned as the corresponding cluster's content. The last phase is applying a score function to all clusters for sorting them.

Figure 4 shows the activity diagram for topic grouping module that uses Lingo algorithm for grouping of paragraphs and lines.

4.3 LDO Gathering—Ontology Construction

To enhance the accuracy in grouping of sentences under each topic, construction of ontology is vital. Ontology is constructed for a specific domain, which includes all the topics and subtopics related to that domain. It is constructed in such a way that the relationships between the topics are clearly understood. Example: If a topic "A" is to be studied, what are all the prerequisite topics that are to be known before studying about topic "A".

Figure 5 is a sample for an ontology tree [8]. Likewise for any input a dynamically ontology tree is constructed using web scrapping technique.

Web Scrapping Technique An input keyword is taken and passed to Wikipedia link. From Wikipedia, various related words are taken for the respective keywords.

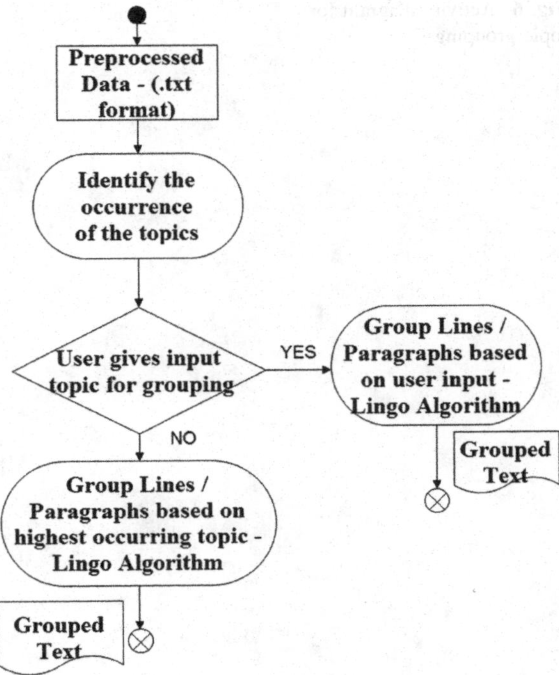

Fig. 4 Activity diagram for topic grouping

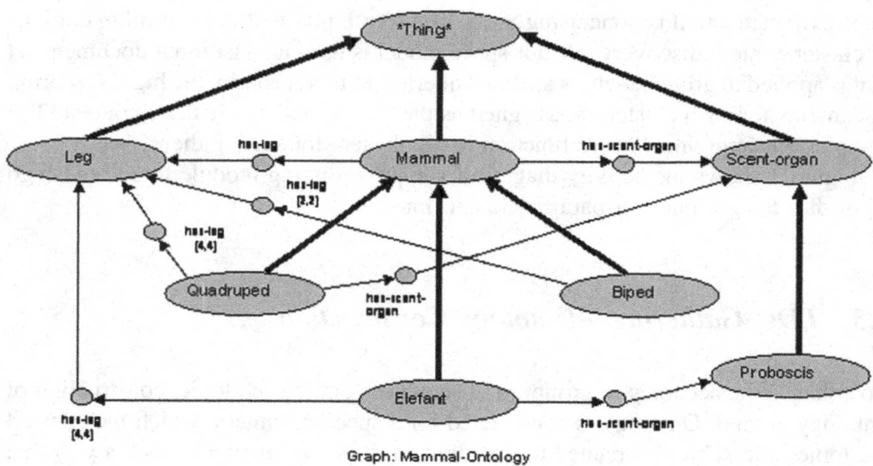

Fig. 5 Sample—ontology tree

The collected words are again passed to Wikipedia to get the next set of related words. This iteration takes place leading to the construction of ontology tree. Once the ontology tree is constructed dynamically, some more topics can be added manually to improve the efficiency and topic grouping is shown in Fig. 6.

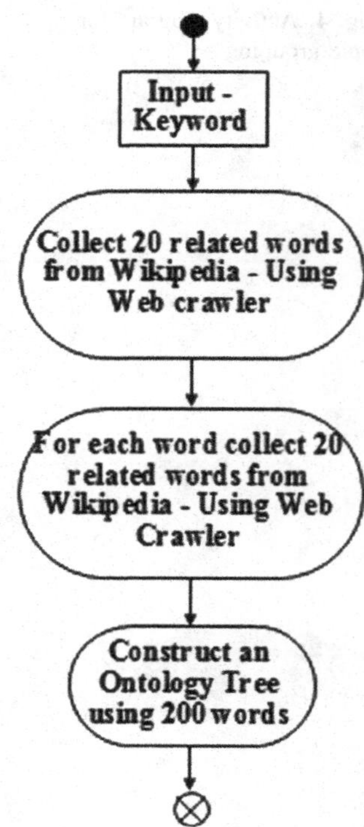

Fig. 6 Activity diagram for topic grouping

4.4 LO Gathering—Learning Objects

Based on the relationships gathered from the ontology in the previous step corresponding learning objects—text, images are gathered from the document.

5 Experimental Results

Handling various types of electronic documents is a hard task. To handle them efficiently the system takes input as electronic documents like—.doc/.pdf formats. These documents are written into a text file. Irrelevant words are removed using stop-word removal algorithm. Then pattern matching is performed using Lingo Algorithm and finally ontology is constructed to further enhance the accuracy. The system also handles multimodal objects from electronic documents like—images and captions. A total of 24 electronic documents are preprocessed of which 14 are

Table 1 Table showing the information about the processed documents

File formats	No. of docs containing multimodal objects	No. of docs not containing multimodal objects
PDF format	12	2
Doc format	7	3

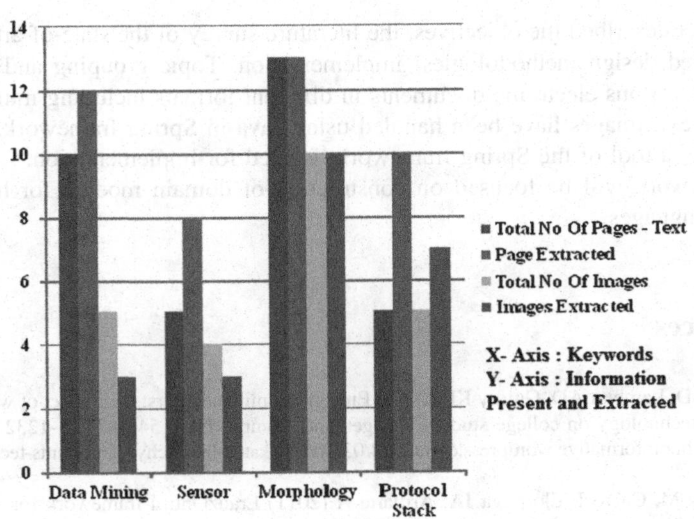

Fig. 7 Graph represents the keywords, information present, and information extracted

Table 2 Table showing the information about the amount of information gathered

	Amount of information present (%)	Amount of information gathered (%) (domain module)	Amount of information gathered (%) (ontology)
Outline	28.95	26.40	27.87
Document body	71.05	50.99	60.13
Total	100.00	77.39	88.00

in pdf format and 10 in.doc format. Also, 20 documents contain multimodal objects —images. The above information is tabulated in Table 1.

Figure 7 depicts four keywords namely—data mining, sensor, morphology, and protocol stack. For each keyword we have the number of pages and images extracted. The graph compares it with the total number of pages and images got using manual method and it is shown in Table 2.

The above table shows the total amount of information present, amount of information gathered using domain module [4], and amount of information gathered through ontology construction. The above table is a summarized one which was tested across various electronic books like—Principles of Data Mining, Digital Image Processing: An Algorithmic Introduction using Java. Thus the percentage of information gathered is increased compared to the previous work due to the dynamic construction of ontology.

6 Conclusion and Future Work

The project described the objectives, the literature survey of the state-of-art works, and detailed design methodologies' implementation. Topic grouping and preprocessing of various electronic documents in different formats including multimodal objects—text, images have been handled using Java in Spring framework. Spring Tool Suite, a tool of the Spring framework, is used for implementation.

Future work will be focused on construction of domain module for handling various languages.

References

1. Chen P-SD, Lambert AD, Guidry KR (2010) Engaging online learners: the impact of web-based learning technology on college student engagement. Comput Educ 54(4):1222–1232
2. https://technoinformative.wordpress.com/2014/03/10/mankatos-hierarchy-of-students-technology-use/
3. Larranaga M, Calvo I, Elorriaga JA, Arruarte A (2011) ErauzOnt: a framework for gathering learning objects from electronic documents. In: Proceedings of the 11th IEEE ICALT '11, pp 656–658, 2011

4. Olagaray ML (2012) Semi-automatic generation of learning domain modules for technology supported learning systems. University of the Basquev, 2012
5. Larranaga M, Elorriaga JA, Arruarte A (2014) Automatic generation of the domain module from electronic textbooks: method and validation. IEEE Trans Knowl Data Eng 26(1)
6. Choudhury SR, Mitra P, Kirk A, Szep S, Pellegrino D, Jones S, Giles CL (2013) Figure metadata extraction from digital documents. In: 12th international conference on document analysis and recognition, 2013
7. Osńiski S, Weiss D (2008) Conceptual clustering using lingo algorithm: evaluation on open directory project data. Institute of Computing Science, Pozna´n University of Technology, ul. Piotrowo 3A, 60–965 Poznań, Poland, (2008)
8. https://www.wikipedia.org/

Detection of Macular Degeneration in Retinal Images Based on Texture Segmentation

J. Jayasakthi and V. Mercy Rajaselvi

Abstract Age-related Macular Degeneration (AMD) is a kind of retinal disease that contains deposits affecting the macula. Macula occupies only 4 % of the retinal region but its effect is severe when it is affected. For many patients, the visual impairment associated with AMD means a loss of independence, depression, and increased financial concerns. Currently, these problems can be resolved using image processing techniques. In the existing system, the algorithm does not address the presence of artifacts in image. Also, it takes more time for the processing of the retinal images. Hence, the proposed system masks the input retinal image during preprocessing and keeps only the retinal part, thereby reducing the processing time. Image processing techniques such as generic quality indicators can be applied to assess the quality of the retinal image and the Gabor filter for detecting the retinal images containing macular degeneration (which leads to AMD). Such a retinal image containing artifacts in macula is identified and used for the medical application. Here, the processing time can be reduced effectively using an optimized method. Finally, the outcome is nature of the image, whether the image's quality is applicable for medical purpose and whether the retinal image is subjected to AMD.

Keywords Assessment · Macular degeneration · Texture segmentation

1 Introduction

Fundus photography is a technology in which the images of the retina or fundus are captured and used for diagnostic purposes by medical experts and through this the retinal diseases such as age-related macular degeneration (AMD), diabetic

J. Jayasakthi (✉) · V. Mercy Rajaselvi
Computer Science and Engineering, Easwari Engineering College, Chennai, India
e-mail: jairam.jj@gmail.com

V. Mercy Rajaselvi
e-mail: mercyeec@gmail.com

L.P. Suresh and B.K. Panigrahi (eds.), *Proceedings of the International Conference on Soft Computing Systems*, Advances in Intelligent Systems and Computing 397, DOI 10.1007/978-81-322-2671-0_42

retinopathy (DR), glaucoma, and cataract. Automated software tools are used nowadays for fundus image evaluation in order to give additional support for the diagnosis. Thus, for implementing the automation in the retinal image, the images must have good quality. The quality also means for more sharpness that separates the various components of the retina. Based on the clustering, thresholding, and texture feature techniques, the retinal images can be classified and their quality can be assessed.

2 Related Work

The work in Miguel Pires Dias et al. [1] proposed retinal image quality assessment using generic image quality indicators. The generic image quality indicators are used to assess the quality, which includes the four features, quantifying the image color, focus, contrast, and illumination. The overall quality using these indicators allows reliable classification. The quality of those individual indicators is evaluated partially. These quality indicators are also combined and classified to evaluate the image suitability for diagnostic purposes. It includes the retinal images from the DRIVE, Messidor, ROC, and STARE datasets. However, it has evaluated the quality of the normal retinal images and the images containing the diabetic retinopathy (DR). The computational complexity of the algorithm increases with the increase in image pixels.

Minhas et al. [2] proposed an efficient algorithm for focus measure computation in constant time. Gradient detectors and the neighborhood support are some of the significant components that influence the focus computation. This technique provides an efficient algorithm for focus measurement, in which a sequence of images is considered to measure the depth map of the image at different foci. The SFF based on steerable filters is used for the focus measurement. The performance analysis shows that the technique is reliable and durable, even in the presence of shadows, haze, and varying illumination and focus conditions. However, the 3-D images need more computational resource and time comparatively with the 2-D images.

Marrugo et al. [3] proposed retinal image restoration by means of blind deconvolution. The retinal images are unclear and are subjected to uneven illumination, which are the major drawbacks in the acquisition process. These drawbacks have a serious impact on the diagnosis. To overcome this problem, color retinal image restoration by means of multichannel blind deconvolution is used. In the multichannel blind deconvolution, the preprocessing includes segmentation, image restoration, and adjustment of uneven illumination. Then, a pair of retinal images that is acquired within a period of time is captured. The results show that the pair of retinal images obtained contains necessary information for the restoration of images. However, the variability in contrast and restoration affects the color matching.

Prabhu et al. [4] proposed the study of macular degeneration with respect to artifacts on retinal images. The segmentation process separated the foreground and background of the image. The filters are used to highlight the edges for its clear visibility. Thus, ridges are a kind of filter which separates the various components of the retinal image by providing high contrast areas. At the time of segmentation, the retinal images can be easily classified as diseased or not. If the non-diseased images are classified as diseased, then the resulting impact is severe. In this approach, several techniques are used for handling the evaluation of assessment of retinal images.

Carmen et al. [5] proposed retinal vessel segmentation using AdaBoost. It provides a method for automated vessel segmentation in retinal images. In AdaBoost approach, which is a machine learning technique, a number of images are trained. Then, the AdaBoost algorithm was tested for extraction of vessel with the test images obtained from datasets such as DRIVE and Messidor. The results are well documented and are compared with those of the other algorithms. In performance analysis, the test images show a sharp contrast between the vessel and the retina.

3 Proposed System

The system is designed to assess the quality of the image and to detect the lesions or drusen present in the retinal image, which leads to the age-related macular degeneration (AMD). Initially, masking is applied to crop the input retinal image, keeping only the retinal part, thereby reducing the processing time. Then the generic quality indicators such as color assessment, focus assessment, illumination assessment, and contrast assessment algorithms are employed to assess the image quality individually with each factor. Then these quality indicators are combined and classified to evaluate the overall image quality and suitability for the diagnostic purposes in the medical field.

The resulting gradable images are further tested for the presence of lesions and drusen in the image by detecting the artifacts which lead to AMD. The system identifies and detects visual artifacts and AMD-related lesions in the retinal image using Gabor filter (Figs. 1 and 2).

4 Implementation

4.1 Generic Image Quality Indicators

The generic image quality indicators include color assessment, focus assessment, illumination assessment, and contrast assessment algorithm to evaluate the overall quality of the retinal image.

Initially, preprocessing of the image is performed, in order to focus only on the retinal part. Each image is first subjected to a pre-processing phase, which removes unimportant information by masking and cropping the image to include only pixels

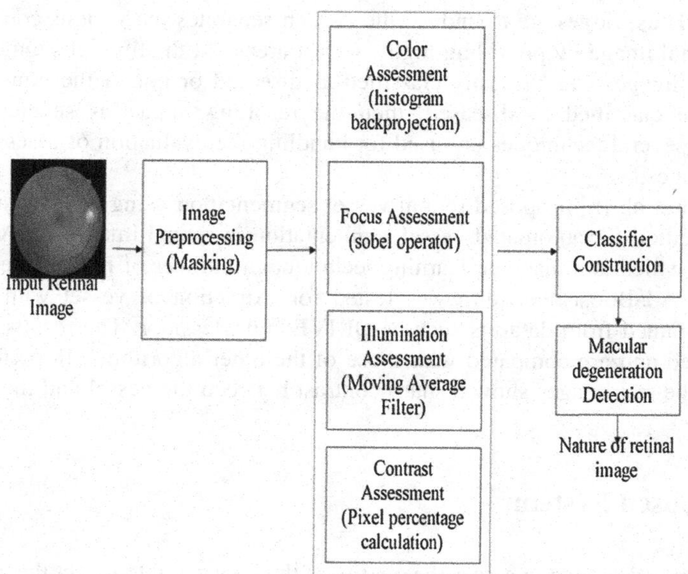

Fig. 1 Overall process of quality assessment and AMD detection

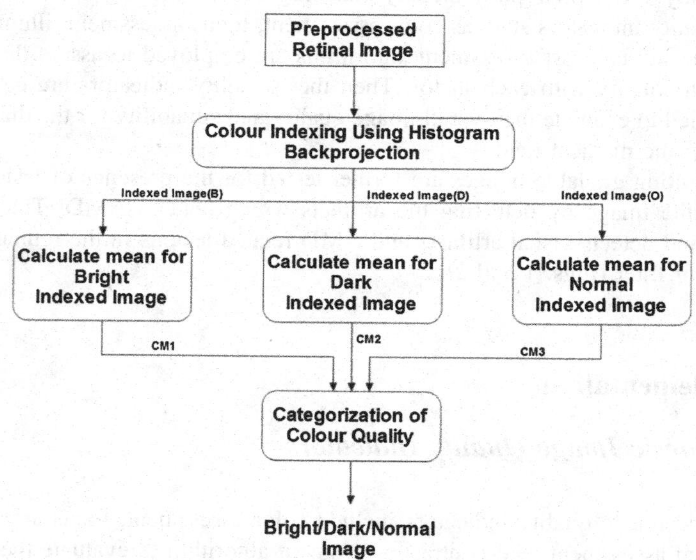

Fig. 2 Flowchart of color assessment algorithm

showing retinal data. The cropping step reduces the retinal image size and thus decreases processing time.

Color Assessment. The color assessment algorithm classifies the image into "bright", "dark," and "normal" images. Each of these color maps is used to perform color indexing of the retinal image being processed, with the creation of three different indexed images. Then the average values of the three indexed images are computed, yielding three color measures (CM1, CM2, and CM3).

Finally since each image color class occupies specific regions in the feature space, the image class ("bright", "dark," or "normal") can be determined using clustering or classification techniques applied to the image feature vector (CM1, CM2, CM3).

$$\mathrm{CM1} = (1/n)\sum B_i,\ B_i \text{ is the } i\text{th pixel of } B. \quad (1)$$

$$\mathrm{CM2} = (1/n)\sum D_i. \quad (2)$$

$$\mathrm{CM3} = (1/n)\sum O_i. \quad (3)$$

Focus Assessment. The focus assessment algorithm classifies the retinal image into "focused", "blurred," or the "borderline" images. Focus is measured through the application of the Sobel operator to the retinal image after conversion to grayscale, followed by a multi-focus-level analysis. The first stage corresponds to the application of the Sobel operator to the original grayscaled retinal image, and the resulting focus measure (FM1) corresponds to the mean of the respective gradient maps (*O*). Then, the grayscaled image is low-pass filtered with a 3 × 3 moving average filter. The Sobel operator is applied and the focus information (fm2) gathered through the means of the resulting gradient map (L1). The second focus measure (FM2) corresponds to the difference between FM1and fm2. In the final stage, the original grayscaled image is again low-pass filtered with a 5 × 5 moving average filter, the Sobel operator is applied, and the focus information (fm3) gathered through the means of the resulting gradient map (L2). The third focus measure (FM3) is obtained through the difference between fm2 and fm3. These three measures are computed according to the following equations:

$$\mathrm{FM1} = (1/n)\sum_{i=1}^{n} O_i, \quad O_i \text{ is the } i\text{th pixel of } O. \quad (4)$$

$$\mathrm{fm2} = (1/n)\sum_{i=1}^{n} \mathrm{L1}i.$$

$$\mathrm{FM2} = \mathrm{FM1} - \mathrm{fm2}. \quad (5)$$

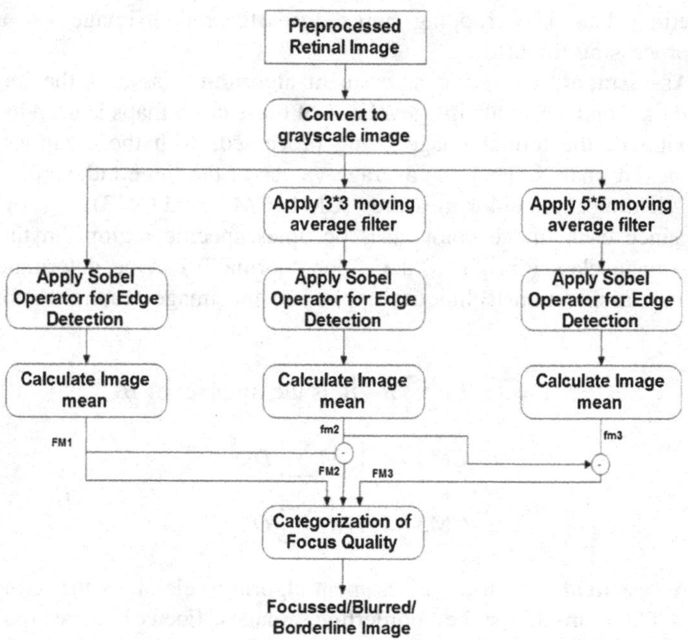

Fig. 3 Flowchart of focus assessment algorithm

$$\mathrm{Fm3} = (1/n)\sum_{i=1}^{n} \mathrm{L2}i.$$

$$\mathrm{FM3} = \mathrm{fm2} - \mathrm{fm3}. \tag{6}$$

The rationale behind the use of these differences to measure focus is that a focused retinal image is more affected by low-pass filtering than the one which is blurred and these equations quantify this effect. The flowchart for focus assessment algorithm is depicted in Fig. 3.

Illumination Assessment. The illumination assessment algorithm processes the retinal image and classifies whether the retinal image is evenly or unevenly illuminated.

Figure 4 shows the flowchart for the illumination assessment algorithm.

Thus, classifying the image as either "even" or "uneven," the four measures are computed as in the following equations:

$$\mathrm{IM1} = (1/n)\sum_{i=1}^{n} O_i, \quad O_i \text{ is } i\text{th pixel of } O. \tag{7}$$

$$\mathrm{IM2} = \mathrm{var}\{O_i\ \mathrm{s.t.} O_i < \mathrm{IM1}\}. \tag{8}$$

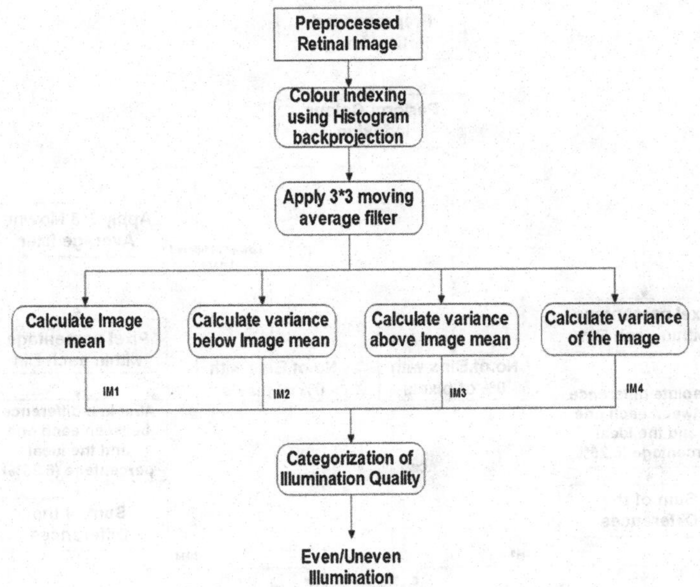

Fig. 4 Flowchart of illumination assessment algorithm

$$\mathrm{IM3} = \mathrm{var}\{O_i \, \mathrm{s.t.} O_i > \mathrm{IM1}\}. \tag{9}$$

$$\mathrm{IM4} = \mathrm{var}\{O_i, \quad \forall i\}. \tag{10}$$

Contrast Assessment. The retinal image contrast is classified as "low" or "high" contrast images. Since contrast is related to color perception, an easier way of solving the contrast measurement problem is to perform color indexing, just like in the color assessment algorithm (Fig. 5).

4.2 Quality Classifier

The final retinal image classification as "ungradable" or "gradable" relies on the measures computed by the four algorithms. Both the global quality classification (gradability classification) and the four partial quality classifications (color, focus, contrast, and illumination) are performed using machine learning techniques (Fig. 6).

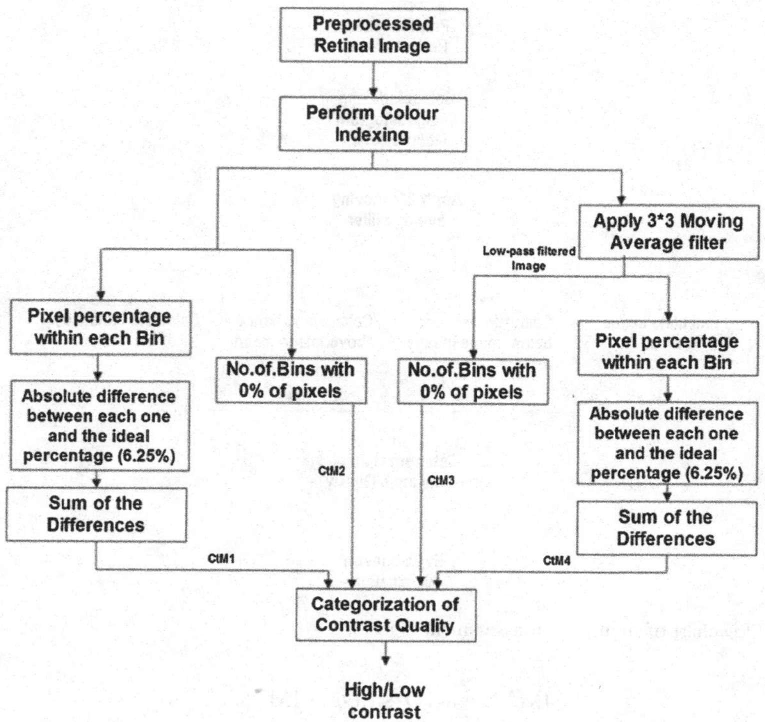

Fig. 5 Flowchart of contrast assessment algorithm

4.3 Detection of Macular Degeneration

The blood vessels and the degenerated macular areas in the retinal images are detected using Gabor filters [4]. The images whose quality is assessed as gradable are tested for the presence of artifacts, leading to the AMD (Fig. 7).

5 Experimental Result

AMD is very common among aged people which sometimes leads to complete blindness. The system evaluates the quality of the image and the Gabor filter identifies the artifacts in retinal images that lead to AMD. A total of 300 fundus images from the MESSIDOR, DRIVE, and the STARE datasets are evaluated for the color, focus, and the illumination quality. The datasets include the retinal images containing diabetic retinopathy (DR), AMD, and the non-diseased normal ones.

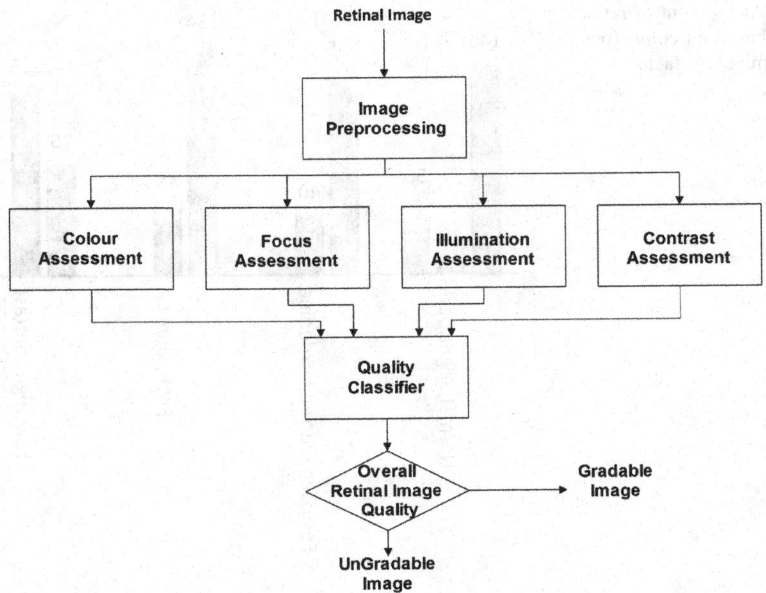

Fig. 6 Quality assessment of retinal images

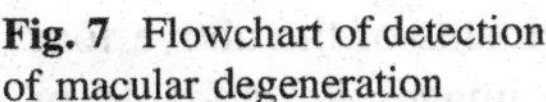

Fig. 7 Flowchart of detection of macular degeneration

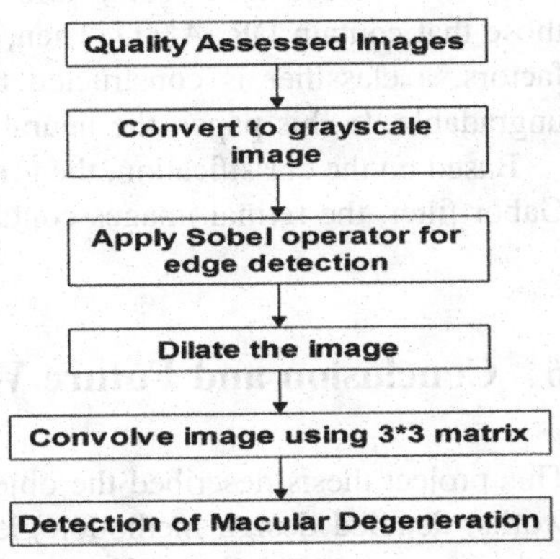

From a total of 10 distorted images, the system has classified two images as "bright", "Focused", "even illumination", and "Low contrast". From a total of 200 normal images, the system has classified 135 images as "normal", "Focused", "even illumination," and "High contrast". A total of 20 Images containing artifacts (AMD) are evaluated in which five are classified as "dark", "focused", "uneven", "High contrast" and 6 images as "Dark", "Blurred", "even illumination", "Low contrast". Thus, the retinal images that are graded by experts match with the classification.

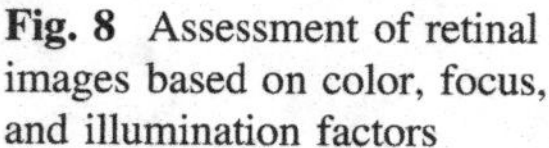

Fig. 8 Assessment of retinal images based on color, focus, and illumination factors

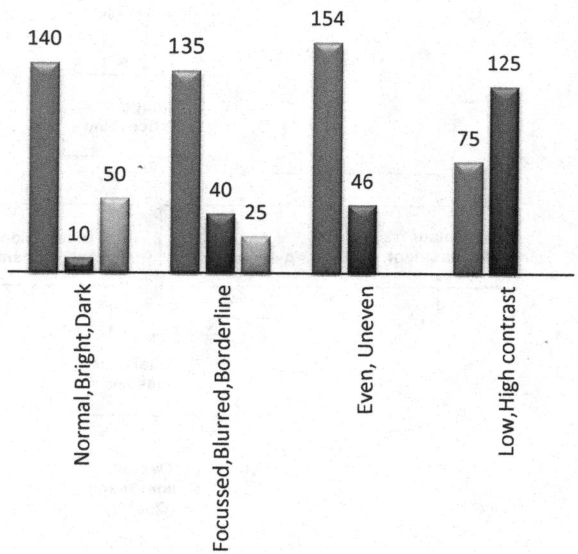

Figure 8 describes the number of images tested on each of the quality factors. It includes the retinal images of aged people more than 50 years and young people below 30 years without being affected by any retinal disease and the images of those that contain DR, AMD. Then, based on the measurements of the above four factors, a classifier is constructed to classify the retinal images as gradable or ungradable. In this paper, the neural network classifier is used.

Based on the classification, the gradable images are taken into account and using Gabor filter, the retinal images containing AMD are identified.

6 Conclusion and Future Work

This project thesis described the objectives, the literature survey of the state-of-art works, detailed design methodologies, and implementation of quality assessment techniques using the OpenCV using java interface. The evaluation of the quality of images is tested for the images containing age-related macular degeneration, diabetic retinopathy (DR), and the non-diseased ones from the datasets of ROC, MESSIDOR, and DRIVE. The retinal images, which are considered as gradable, can be tested for the presence of drusens or lesions, generally referred to as the macular degeneration in the retina (leads to AMD), and the areas of the macular degeneration are detected.

In future, the paper work can be extended for detecting the retinal images with glaucoma or diabetic retinopathy (DR).

References

1. Miguel Pires Dias J, Manta Oliveira C, da Silva Cruz LA (2014) Retinal image quality assessment using generic image quality indicators. Inf Fusion 19
2. Minhas R, Mohammed AA, Wu QMJ (2012) An efficient algorithm for focus measure computation in constant time. IEEE Trans Circuits Systems Video Technol 22
3. Marrugo AG, Šorel M, Šroubek F, Millán MS (2011) Retinal image restoration by means of blind deconvolution. J Biomed Optics 16
4. Prabhu S, Chakraborty C, Banerjee RN, Ray AK (2012) Study of macular degeneration with respect to artifacts on retinal images. Special Issue of Int J Comput Appl
5. Carmen AL, Domenico T, Emanuele T (2010) Retinal vessel segmentation using AdaBoost. IEEE Trans Inf Technol Biomed 4
6. Jelinek H, Cree MJ (2009) Automated image detection of retinal pathology. CRC Press
7. Abramoff MD, Niemeijer M, Suttorp-Schulten MS, Viergever MA, Russell SR, van Ginneken B (2008) Evaluation of a system for automatic detection of diabetic retinopathy from color fundus photographs in a large population of patients with diabetes. Diabetes Care 31

Power Angle Control of a Single Phase Grid Connected Photovoltaic Inverter for Controlled Power Transfer

A. Vijayakumari, A.T. Devarajan and S.R. Mohanrajan

Abstract This paper focuses on the design and development of a 500 W, single phase single stage low-cost inverter for the transfer of direct current (DC) power from the solar photovoltaic (SPV) panel to the grid while meeting the standards for interconnection. The power transfer from SPV to grid is facilitated by controlling the angle between the inverter voltage and the grid voltage i.e. the power angle or phase angle (δ). Simulation of the single stage inverter for a power rating of 500 W is implemented in MATLAB/Simulink. The synchronization of the inverter with the grid was carried out by monitoring the grid frequency continuously and updating it. The power transferred to the grid is made equal to the available power by monitoring and updating δ. A pulse-width modulation (PWM) control is implemented using PIC16F877 microcontroller for the inverter so as to deliver the available power from the solar panel to the grid, as well as, make the frequency of the output voltage of the inverter equal to the grid frequency for sustaining the synchronization.

Keywords Photovoltaic panel · Power angle · Phase angle · Grid synchronization

A. Vijayakumari (✉) · A.T. Devarajan · S.R. Mohanrajan
Department of Electrical and Electronics Engineering, Amrita Vishwa Vidyapeetham University, Amrita Nagar, Coimbatore 641112, India
e-mail: a_vijayakumari@cb.amrita.edu

A.T. Devarajan
e-mail: at_devarajan@cb.amrita.edu

S.R. Mohanrajan
e-mail: sr_mohanrajan@cb.amrita.edu

L.P. Suresh and B.K. Panigrahi (eds.), *Proceedings of the International Conference on Soft Computing Systems*, Advances in Intelligent Systems and Computing 397, DOI 10.1007/978-81-322-2671-0_43

1 Introduction

Global electricity consumption is expected to increase 7 % annually. With the increased awareness on environmental issues like Global Warming and Greenhouse Effect, researchers have turned their attention to renewable resources, as they are clean forms of energy [1]. Solar energy is available abundantly almost in every part of the globe and is yet to be commercialized. The transfer of power from the solar panel to mains is a challenging task, as the power from the panel is environment dependent. Also, the voltage and frequency of the utility also vary within a specified range. The DC output from the solar PV panel should be power conditioned and fed to the grid meeting the grid connection standards. So the use of power electronic interface becomes a must in the grid connected PV applications [2–5].

Commercialization of low power photovoltaic (PV) systems demands the control circuit and power circuit of the converter to be less complex, robust, compact and economical. Hence, appropriate converter configuration and control mechanism has to be designed and implemented. Commonly followed topologies are single stage or multistage. In the first stage of a typical two stage conversion, the amplitude of current injected into the mains is controlled followed by the control of the phase of the current injected in the second stage. The control action is divided between the two converters making it simple, compared to both controls done in the DC/AC converter itself [6, 7] in a single stage topology. The main disadvantage with the multistage technique is that the number of power devices required is more, hence higher power loss and reduced efficiency. In contrast when a single stage converter is opted for, there is advantage in that it has higher efficiency and reliability compared to the multistage type besides also being economical and compact [6–9]. Various control strategies are used for grid synchronization and power control of single stage inverters [10–12]. The common methods adapted for power control are by the control of amplitude of inverter output voltage, current delivered by the inverter and the power angle [10–14]. The power delivered is more sensitive to the power angle rather than the inverter output voltage [2, 6–14]. This paper uses the control of phase angle for power control in a grid connected inverter system feeding power from the solar PV panel to the grid. A methodology is presented for the estimation of phase angle with sustaining synchronization.

2 Concept of Power Angle Control

Figure 1 is an equivalent circuit of a voltage source inverter (VSI) feeding power to grid. The power transferred is expressed as,

$$P = \frac{V_{\text{inv}} V_g}{X} \sin \delta \tag{1}$$

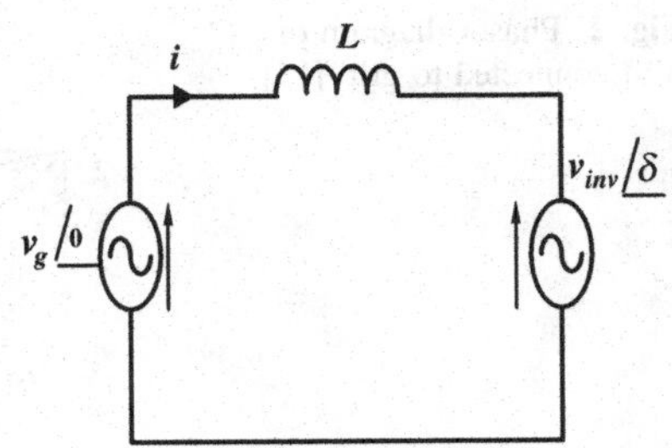

Fig. 1 Equivalent circuit of VSI connected to grid

where,

P is the power delivered by the inverter to the grid.
X is the reactance ($X = 2\pi fL$).
V_{inv} is the inverter output voltage.
V_g is the grid voltage.
δ is the power or phase angle.

The exchange of power between two voltage sources can be achieved either by varying the magnitude of the source voltages or by varying the load angle between the two output voltages or the reactance value. The decoupling inductor (L) is an essential part of such systems, as they cause a precise power flow control by adjusting its drop for producing currents with different phases and magnitudes as shown in the phasor diagram of Fig. 2. Using one of the sources as a VSI, the power flow is controlled by adjusting the magnitude and phase of the inverter output voltage with respect to the grid voltage. This generates an appropriate voltage drop across the decoupled inductor (V_{L}), to force the reference power to be delivered through a desired magnitude of current. Another important role of the decoupled inductor is prevention of large circulating currents between the two voltage sources due to the instantaneous mismatch in their voltages.

The concept of phase angle control is better explained using the phasor diagram of Fig. 2. It shows that, the locus of the output voltage phasor of VSI remains constant but the magnitude of the current delivered varies when the power angle "δ" is varied under different grid voltages. The power angle could be either leading or lagging, which facilitates the power flow to happen either from the grid to the VSI or vice versa. For a given power reference the phase angle (α), between the grid voltage and the current and the amplitude of the current, is determined by the magnitude and phase of the decoupled inductor voltage (V_{L}), when the power angle is varied.

Fig. 2 Phasor diagram of VSI connected to grid [15]

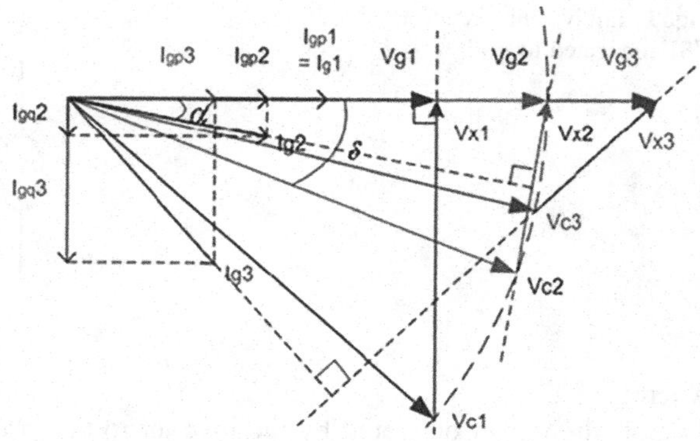

3 System Description

The proposed single phase single stage inverter feeding the grid from PV panel is represented in Fig. 3a. A single phase single stage inverter acts as an interface between the SPV and grid. A zero crossing detector is used to obtain the synchronization points of the grid voltage. The duration between two consecutive zero cross points is used to estimate the frequency of the grid voltage by the frequency estimator programmed in the microcontroller. The current and voltage at the point of common coupling (PCC) are sensed and the average power delivered is estimated. This is compared with the available power from any maximum power point tracking (MPPT) algorithm of the PV panel and the δ update block estimates the magnitude and the direction of the "δ". This block either increments or decrements the value of δ so as to match the power transferred to grid and the power available. The new values of frequency and δ are used to generate the sine reference for the PWM generator. The sinusoidal pulse width modulation (SPWM) pulses are so produced with the new reference to make the inverter voltage lead the grid voltage by δ at the estimated grid frequency.

Incorporating the power angle into the voltage reference of SPWM is shown in Fig. 3b. A proportional resonant (PR) controller is used to force the grid voltage magnitude to follow the reference. The transfer function of the PR controller is given as

$$H_{\mathrm{PR}}(s) = K_P + \frac{K_I s}{s^2 + \omega^2} \tag{2}$$

As PR controllers can work with AC quantities, no co-ordinate transformation is required in the control loop.

A phase locked loop grabs the phase and form of the grid voltage. With these values viz. magnitude, shape, frequency, phase and power angle, a reference sine wave is generated by mathematical manipulations in the reference signal generator.

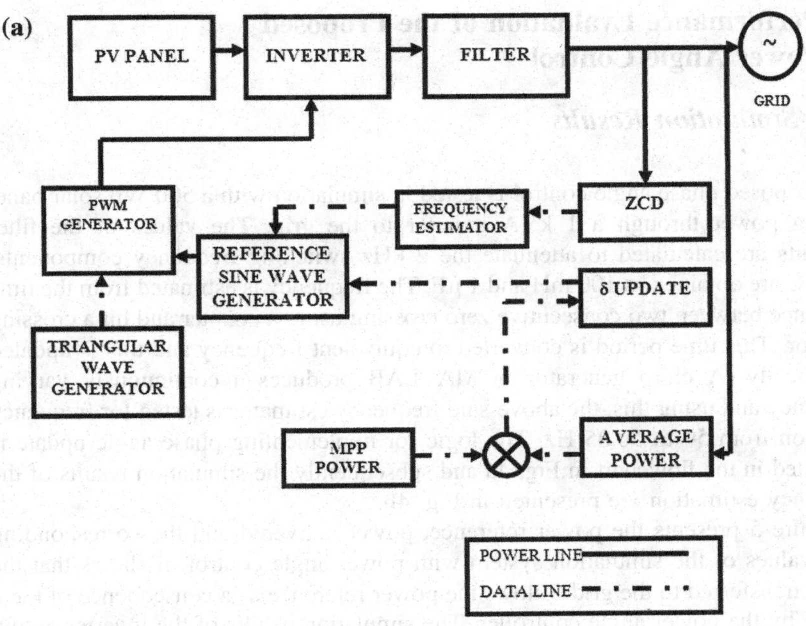

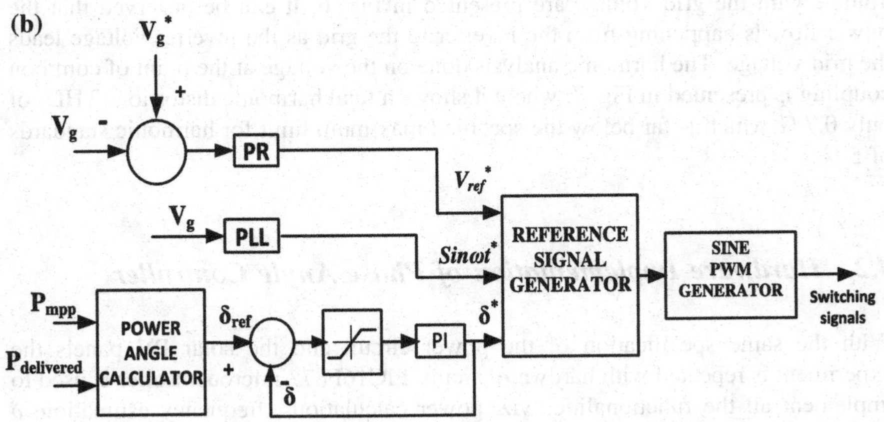

Fig. 3 a Block diagram of the proposed phase angle control. **b** Power angle control implementation

4 Performance Evaluation of the Proposed Power Angle Control

4.1 *Simulation Results*

The proposed phase angle control is tested in simulation with a 500 Wp solar panel feeding power through a 1 kVA inverter to the grid. The values of the filter elements are calculated to attenuate the 2 kHz switching frequency components. L and C are obtained as 100 mH and 1 μF. The frequency is estimated from the time deference between two consecutive zero crossing using a counter and hit a crossing detector. This time period is converted to equivalent frequency and this is updated periodically. A chirp generator in MATLAB produces a continuously varying frequency and using this, the above-said frequency estimator is tested for frequency variation from 50 to 50.05 Hz. The logic for implementing phase angle update is presented in the flowchart in Fig. 4a and subsequently the simulation results of the frequency estimation are presented in Fig. 4b.

Figure 5 presents the power reference, power delivered and the corresponding delta values of the simulation system with power angle control. It shows that the power transferred to the grid follows the power reference as a consequence of the δ update by the power angle controller. The simulation results of the inverter output voltage with the grid voltage are presented in Fig. 6. It can be observed that the power flow is happening from the inverter to the grid as the inverter voltage leads the grid voltage. The harmonic analysis done on the voltage at the point of common coupling is presented in Fig. 7, where it shows a total harmonic distortion (THD) of only 0.7 % which is far below the specified maximum limit for harmonic standards of 5 %.

4.2 *Hardware Implementation of Phase Angle Controller*

With the same specification of the power circuit and the solar PV panels the experiment is repeated with hardware circuit. PIC16F877 microcontroller is used to implement all the functionalities viz. power calculations, frequency estimation, δ update, phase locked loop, PR controller and reference sine wave generation. The snapshot of the hardware setup is shown in Fig. 8. The inverter voltage and the voltage at the point of common coupling are captured using the storage oscilloscope. The reference power is assumed to be available from MPPT, and for the testing of the hardware setup it is given through electrical equivalent signals. The grid voltage is sensed using a Hall effect potential transducer. The grid voltage's phase and form i.e. a unit sine wave is generated through a phase locked loop (PLL). The algorithm for PLL implementation is also coded in the same microcontroller.

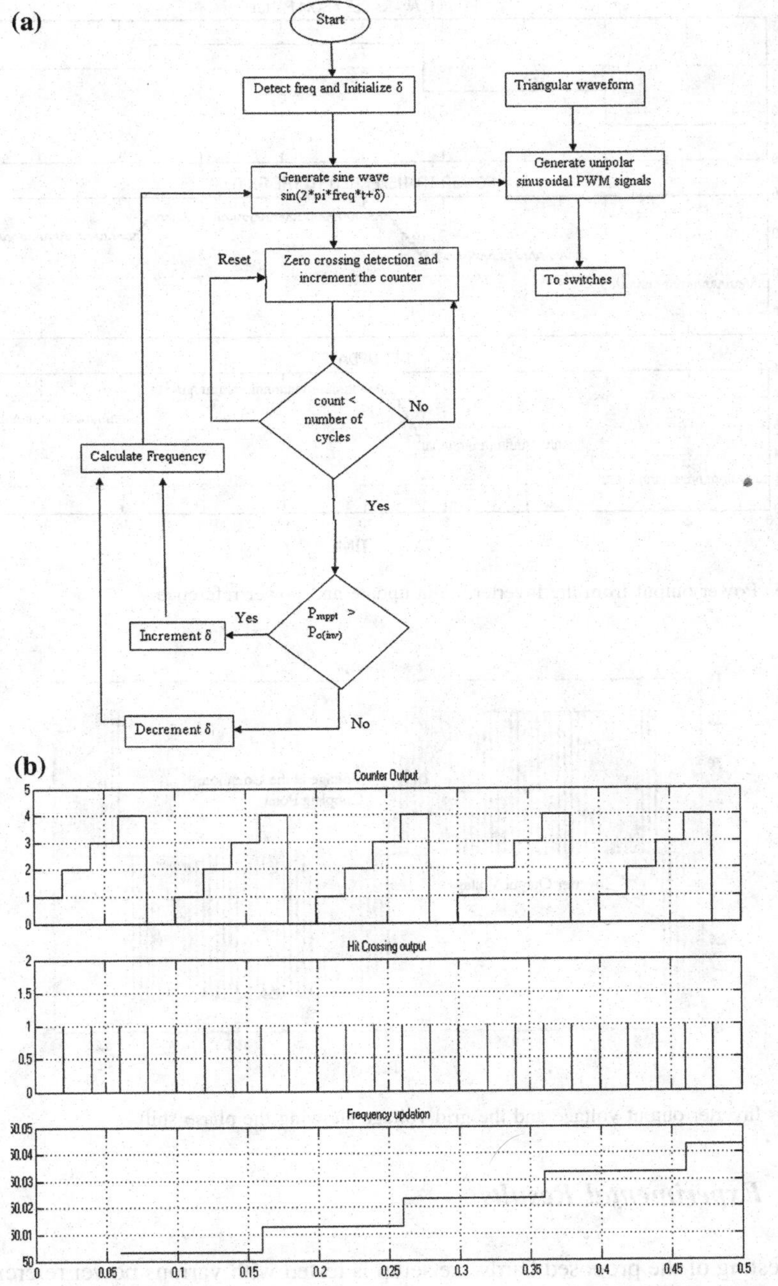

Fig. 4 **a** Logic for δ update. **b** Output from the counter, hit crossing blocks in frequency measurement

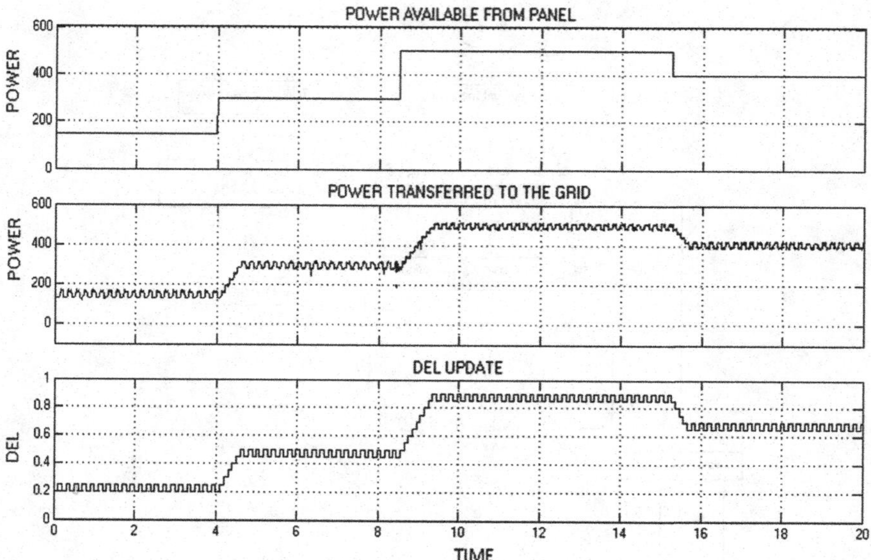

Fig. 5 Power output from the inverter, delta update and power reference

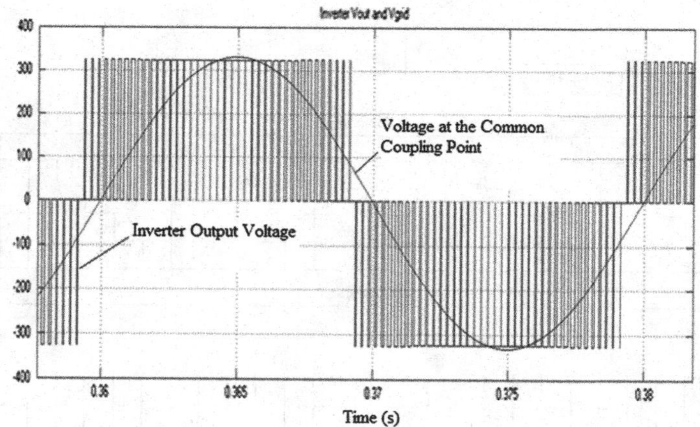

Fig. 6 Inverter output voltage and the grid voltage showing the phase shift

4.3 Experimental Results

The testing of the proposed hardware setup is tested with various power references. Figure 9a, b presents the output voltage of the inverter and the grid voltage for a δ of 0.49 corresponding to a power output of 300 W. Different power references were given and the power angle is observed and mentioned in Table 1. The voltage and

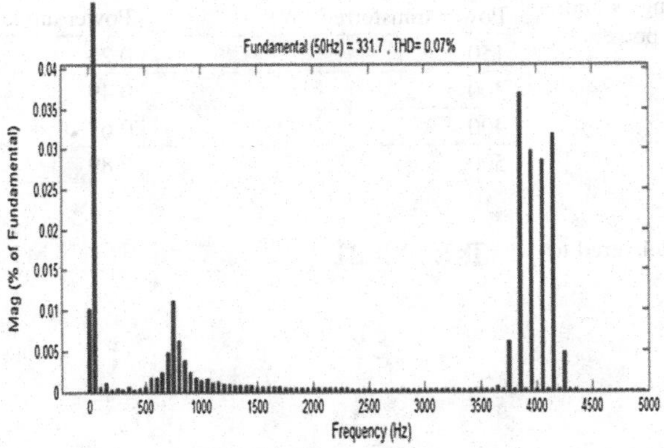

Fig. 7 Fast Fourier transform (FFT) analysis of voltage at PCC

Fig. 8 Hardware setup of the proposed system

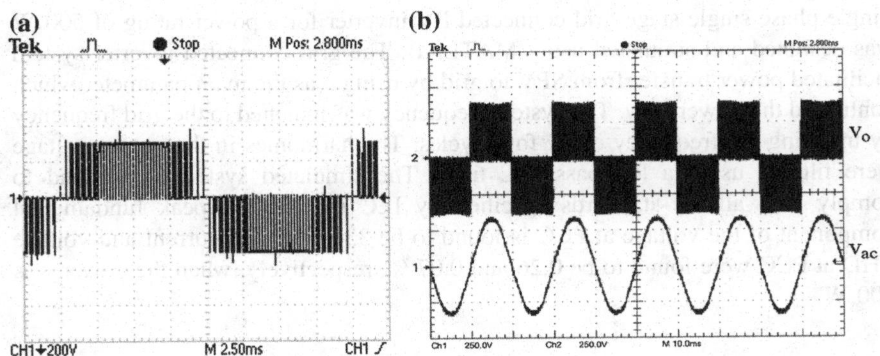

Fig. 9 **a** Inverter voltage. **b** Inverter voltage and grid voltage

Table 1 Power angles and the corresponding powers

Power transferred (W)	Power angle (rad)
150	0.23
300	0.49
400	0.67
500	0.89

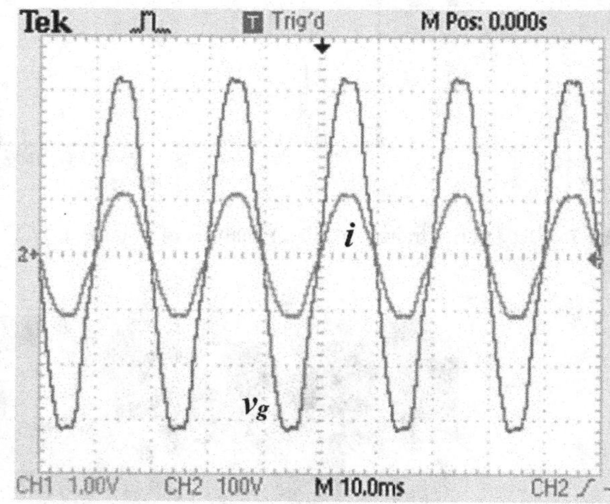

Fig. 10 Current delivered to grid and v_g

the current at the point of common coupling for various powers are observed and presented in Fig. 10 for a power delivery of 325 W. The current is observed as a voltage across a 0.5 Ω sensing resistance.

5 Conclusions

Single phase single stage grid connected PV inverter for a power rating of 500 W was designed and simulated using MATLAB. The grid connected inverter system facilitated power transfer from SPV to grid by using δ as the main parameter which controlled the power flow. The system frequency was matched to the grid frequency by updating the frequency every four cycles. The harmonics in the output voltage were filtered using a low-pass L-C filter. The simulated system was found to comply with all the standards specified by IEC 61727. The peak fundamental component of the voltage at PCC is found to be 331.7 V. The current and voltage THD at PCC were found to be 0.26 and 0.07 %, respectively, when the power was 500 W.

References

1. Kenerly AT (2013) Solar power and India's Energy future'. http://www.atkearney.com/utilities/ideas-insights/article/-/asset_publisher/LCcgOeS4t85g/content/solar-power-in-india-preparing-to-win/10192
2. Li Q, Wolfs P (2008) A review of the single phase photovoltaic module integrated converter topologies with three different DC link configurations. IEEE Trans Power Electron 23 (3):1320–1333
3. Calais M, Myrzik J, Spooner T, Agelidis VG (2002) Inverters for single-phase grid connected photovoltaic systems-an overview. In: 2002 IEEE 33rd Annual Power Electronics Specialists Conference, 2002. PESC 02, vol 4, pp 1995–2000. IEEE
4. Kjaer SB, Pedersen JK, Blaabjerg F (2005) A review of single-phase grid-connected inverters for photovoltaic modules. IEEE Trans Indus Appl 41(5):1292–1306
5. Kojabadi HM, Yu B, Gadoura IA, Chang L, Ghribi M (2006) A novel DSP-based current-controlled PWM strategy for single phase grid connected inverters. IEEE Trans Power Electr 21(4):985–993
6. Jain S, Agarwal V (2007) A single-stage grid connected inverter topology for solar PV systems with maximum power point tracking. IEEE Trans Power Electr 22(5):1928–1940
7. Jain S, Agarwal V (2007) Comparison of the performance of maximum power point tracking schemes applied to single-stage grid-connected photovoltaic systems. IET Electr Power Appl 1 (5):753–762
8. Patel H, Agarwal V (2009) MPPT scheme for a PV-fed single-phase single-stage grid-connected inverter operating in CCM with only one current sensor. IEEE Trans Energ Conv 24(1):256–263
9. Prasad BS, Jain S, Agarwal V (2008) Universal single-stage grid-connected inverter. IEEE Trans Energ Conv 23(1):128–137
10. Araújo SV, Zacharias P, Mallwitz R (2010) Highly efficient single-phase transformerless inverters for grid-connected photovoltaic systems. IEEE Trans Indus Electr 57(9):3118–3128
11. Ciobotaru M, Teodorescu R, Blaabjerg F (2005) Control of single-stage single-phase PV inverter. In: 2005 European conference on power electronics and applications, pp. 10-pp. IEEE
12. Kadri R, Gaubert JP, Champenois G (2011) An improved maximum power point tracking for photovoltaic grid-connected inverter based on voltage-oriented control. IEEE Trans Indus Electr 58(1):66–75
13. Zhou Y, Huang W, Zhao P, Zhao J (2014) A transformerless grid-connected photovoltaic system based on the coupled inductor single-stage boost three-phase inverter. IEEE Trans Power Electr 29(3):1041–1046
14. Ozdemir S, Altin N, Sefa I (2014) Single stage three level grid interactive MPPT inverter for PV systems. Energy Convers Manag 80:561–572
15. Ko SH, Lee SR, Dehbonei H, Nayar C (2005) A comparative study of the voltage controlled and current controlled voltage source inverter for the distributed generation system. AUPEC

Secure Electrocardiograph Communication Through Discrete Wavelet Transform

V. Sai Malathi Anandini, Y. Hemanth Gopalakrishna and N.R. Raajan

Abstract An ECG signal produces the electrical movements of the heart. In this paper, a secure way to communicate ECG data has been attempted by embedding it into the high frequency plane obtained by performing discrete wavelet transform using Haar wavelet on a color image. The PSNR and MSE metrics have been calculated and the ECG data has also been successfully retrieved by isolating the high frequency plane and extracting the data from it.

Keywords ECG · PSNR · MSE · OFDM · Communication · Discrete wavelet transform

1 Introduction

Digital image processing is one of the fields of major research. Having a wide range of applications which start from image compression to several security applications such as image steganography [1], image processing further includes techniques which can be used for tracking of the eyes [2], creating damage assessments under various military operations [3], various filtering techniques which are being widely used in these areas to study and identify the various properties of the tissues [4], in identifying the discrete neural pathways in the brain and their disturbances in persons with neuropsychiatric disorders [5]. Image processing can also be used along with OFDM in DWT–OFDM-based works [6]. In the current work, image processing has been used for transmitting the ECG data securely over a network by

V. Sai Malathi Anandini (✉) · Y. Hemanth Gopalakrishna · N.R. Raajan
School of EEE, SASTRA University, Thanjavur, Tamilnadu, India
e-mail: anandiniv@gmail.com

Y. Hemanth Gopalakrishna
e-mail: hemanth.krishna1@gmail.com

N.R. Raajan
e-mail: nrraajan@ece.sastra.edu

L.P. Suresh and B.K. Panigrahi (eds.), *Proceedings of the International Conference on Soft Computing Systems*, Advances in Intelligent Systems and Computing 397, DOI 10.1007/978-81-322-2671-0_44

hiding it in a high frequency plane. This helps in the prevention of misuse of ECG data and ensures the authenticity.

2 ECG and DWT

2.1 Discrete Wavelet Transform

A linear transformation that operates on a data vector whose length is an integer power of 2, transforming it into a numerically different vector of the same length is known as the discrete wavelet transform (DWT). It separates data into different frequency components, and based on the scale each component is matched. A cascade of filtering followed by a factor 2 subsampling is shown below [7] (Fig. 1).

High frequency and low frequency filtering operations are denoted by H and L, respectively, ↓ 2 denotes subsampling. Equations (1) and (2) give the results of these filters.

$$a_{j+1}[p] = \sum_{n=-\infty}^{+\infty} l[n - 2p]a_j[n] \tag{1}$$

$$d_{j+1}[p] = \sum_{n=-\infty}^{+\infty} h[n - 2p]a_j[n] \tag{2}$$

For scaling elements a_j are used and elements d_j, called wavelet coefficients, determine the output of the transform. l[n] and $h[n]$ are the coefficients of low-pass and high-pass filters. On scale $j + 1$ there is only half from number of 'a' and d elements on scale j. This causes that DWT can be done until only two a_j elements remain in the analyzed signal. These elements are called scaling function coefficients.

DWT algorithm for two-dimensional pictures is similar. The DWT is performed firstly for all image rows and then for all columns [7].

In wavelet transform the signal is first decomposed by a low-pass (LP) and a high-pass (HP) filter. Half of the frequency components have been filtered out at filter outputs and hence can be downsampled [8]. The original signal is then reconstructed by performing the reverse operation of this decomposition [9].

This DWT is used in various aspects of image processing such as improving fingerprint quality [10], optimizing gray-scale image watermarking [11], stationary gridline artifact suppression method [12], etc. In our current application, DWT is used to bring out the various frequency planes in the images (Fig. 2).

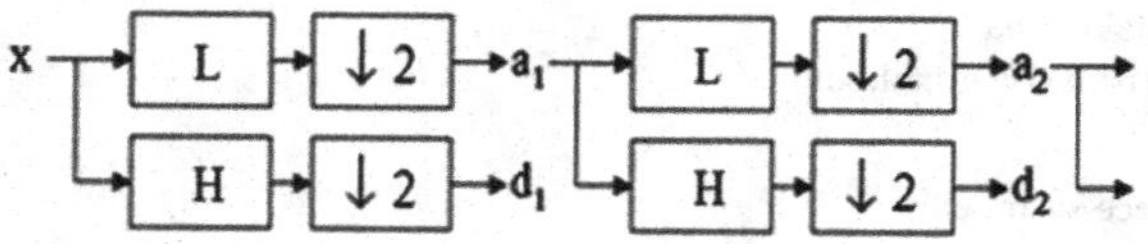

Fig. 1 DWT tree

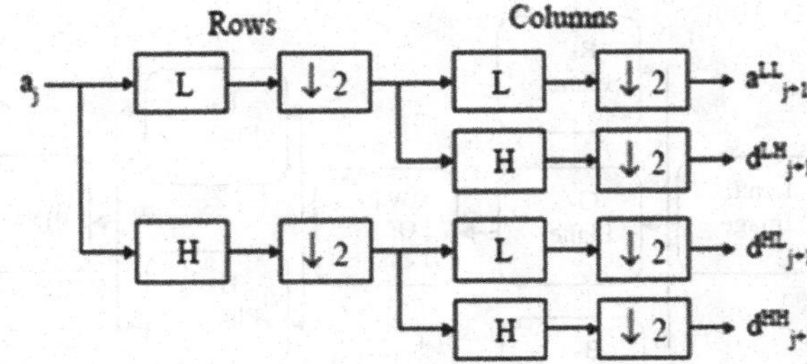

Fig. 2 DWT for two-dimensional images

2.2 *Electrocardiogram*

The most common cardiac test carried out on cardiac patients is electrocardiograph (ECG). The electrical activity of the heart (heartbeat) onto paper is taken by ECG. It gives the doctor information about the heart rate, heart rhythm, and evidence of heart disease. It is used in various applications such as designing vital sensors [13], etc.

2.3 *Securing ECG Using DWT*

In the current work, G-plane of a color image has been chosen for performing DWT. Haar wavelet is the oldest and most basic wavelet system, which is a group of square waves with magnitude of ±1 in the interval [0,1) [14]. It is used in various applications such as signal-to-noise ratio estimation for cardiovascular bio-signals [15], in various convergence theorems [16], etc. In the current work, it has been chosen to apply DWT. The high frequency h_HH plane has been separated and the ECG data has been merged into this plane by multiplying the coefficients of the plane with the ECG data. IDWT operation has been performed using this modified h_HH plane and a modified G-plane has been generated. This plane has been replaced in the original image and thus, the color image containing the ECG data has been generated.

The following is the block diagram for embedding ECG data (Fig. 3).

While recovering, the received color image has been split and the G-plane has been separated. DWT has been performed on this plane and the modified h_HH plane has been extracted. ECG data is recovered by dividing the coefficients of the plane by the initial h_HH values.

The following is the block diagram for recovering ECG data (Fig. 4).

3 Simulation Results

The following were the results obtained after simulation.

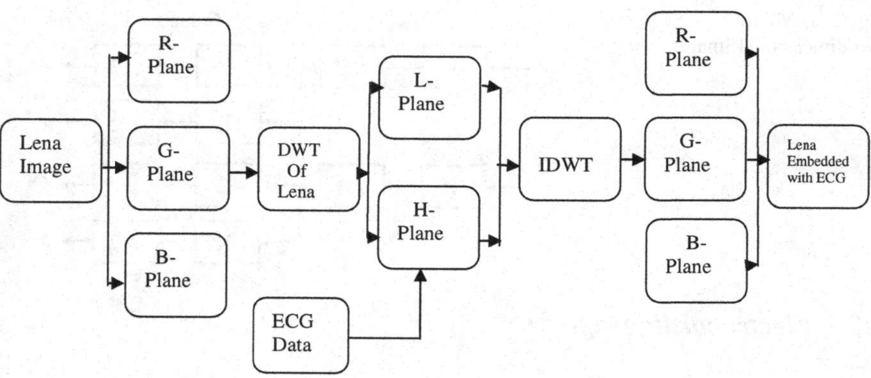

Fig. 3 Embedding ECG data

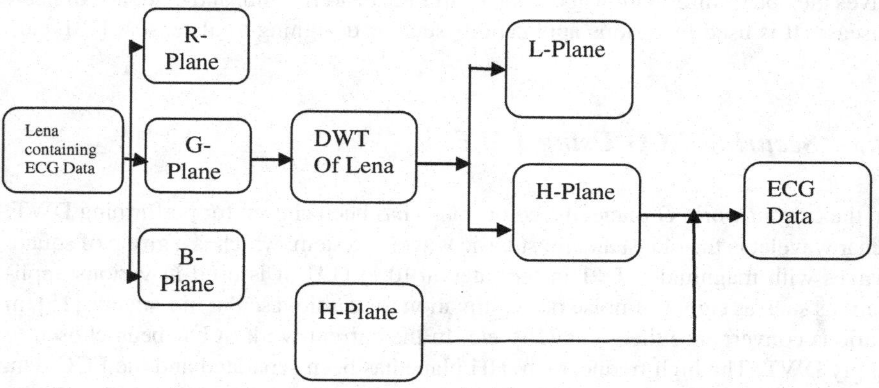

Fig. 4 Extracting ECG data

3.1 Lena

ECG data has been embedded in the h_HH plane and the following are the results obtained for various images (Fig. 5).

3.2 Baboon

ECG data has been embedded in the h_HH plane and the following are the results obtained for various images (Fig. 6).

PSNR, MSE metrics have been calculated for the above images and are tabulated as follows (Table 1).

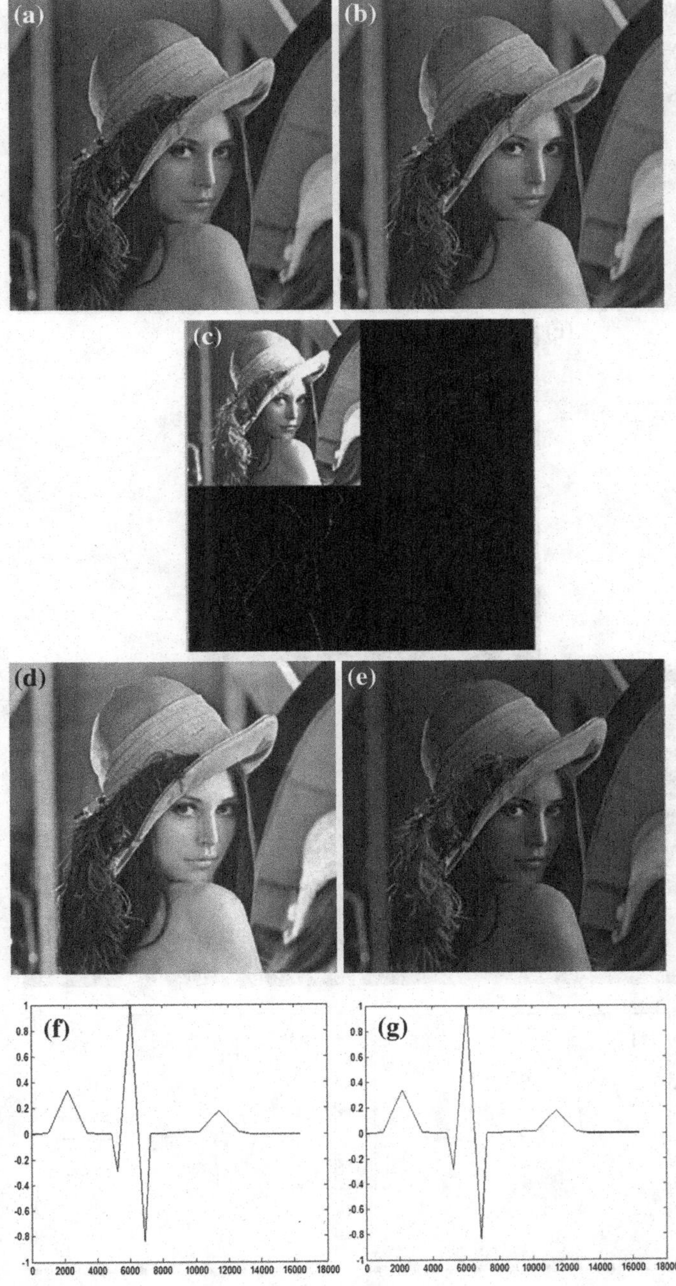

Fig. 5 **a** Original color image. **b** Color image with ECG data. **c** Original G-plane. **d** G-plane with h_HH containing ECG data. **e** DWT image. **f** Original ECG data. **g** Extracted ECG data

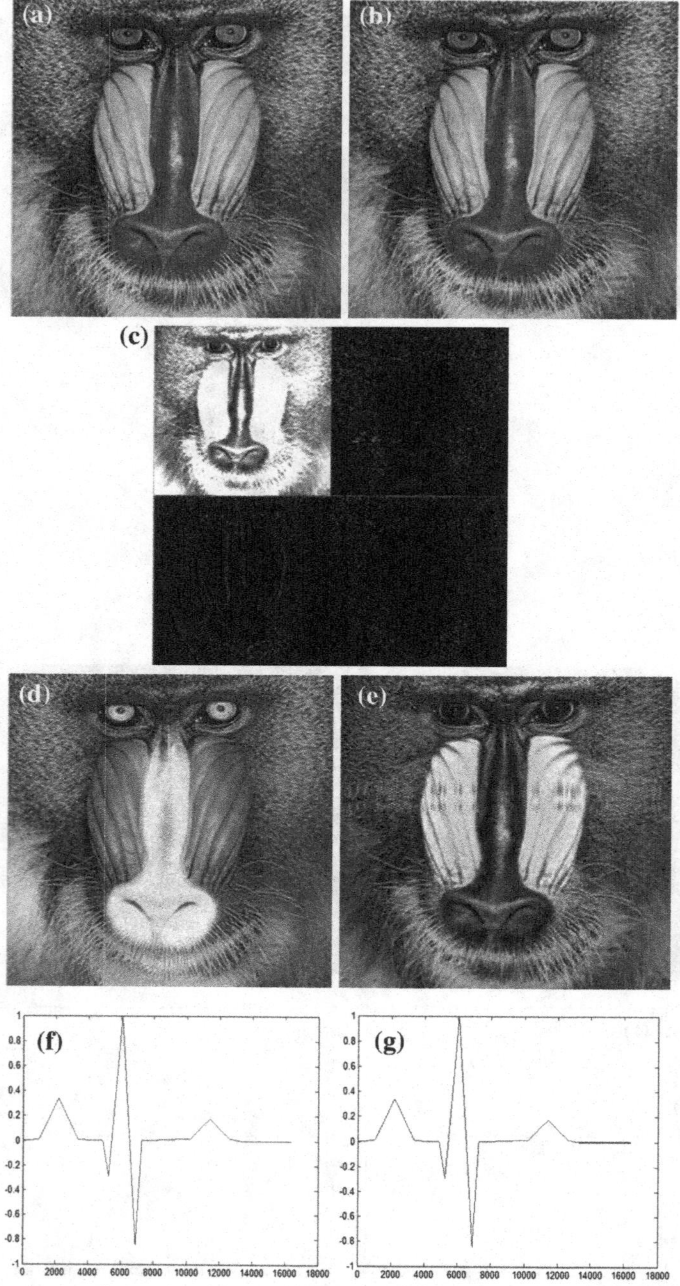

Fig. 6 **a** Original color image. **b** Color image with ECG data. **c** Original G-plane. **d** G-plane with h_HH containing ECG data. **e** DWT image. **f** Original ECG data. **g** Extracted ECG data

Table 1 PSNR and MSE

	PSNR	MSE
Lena	35.0129	20.6628
Baboon	30.8153	54.3190

Table 2 NPCR and UACI

	NPCR	UACI
Lena	0.27057393391	0.00571247175
Baboon	0.95515441894	0.05696070053

The NPCR and UACI metrics were calculated and are tabulated as follows (Table 2).

4 Conclusion

The ECG data was successfully embedded and was retrieved accurately. Thus, a secure method for the transmission of ECG data was implemented. The correlation value between the ECG data before and after embedding is found to be 1.0000 (Same data recovered).

References

1. Ibrahim R, Kuan TS (2011) PRIS: Image processing tool for dealing with criminal cases using steganography technique. In: 2011 Sixth International Conference on Digital Information Management (ICDIM), pp 193, 198, 26–28 Sept 2011
2. Liu T, Pang C (2008) Eye-gaze tracking research based on image processing. In: Congress on Image and Signal Processing, 2008, CISP '08, vol 4, pp 176, 180, 27–30 May 2008
3. Smith RWM (1998) A generic scaleable image processing architecture for real-time military applications. High Performance Architectures for Real-Time Image Processing (Ref. No. 1998/197), IEE Colloquium on, pp 3/1, 3/6, 12 Feb 1998
4. Gu Y, Qian Z, Chen J, Dana B, Ramanujam N, Britton C (2002) High-resolution three-dimensional scanning optical image system for intrinsic and extrinsic contrast agents in tissue. Rev Sci Instrum 73(1):172–178
5. (2015) Morphological covariance in anatomical MRI scans can identify discrete neural pathways in the brain and their disturbances in persons with neuropsychiatric disorders. NeuroImage 111:215–227
6. Juwono FH, Putra RS, Gunawan D (2014) A study on peak-to-average power ratio in dwt-ofdm systems. TELKOMNIKA Indonesian J Electric Eng 12(5):3955–3961
7. Ye Z, Mohamadian, H, Ye Y (2009) Quantitative effects of discrete wavelet transforms and wavelet packets on aerial digital image denoising. In: 2009 6th International Conference on Electrical Engineering, Computing Science and Automatic Control, CCE, pp 1, 5, 10–13 Jan 2009
8. Hasan MM, Singh SS PAPR analysis of FFT and wavelet based OFDM systems for wireless communications. UITS, Dhaka

9. Moholkar SV, Deshmukh A (2014) PAPR analysis of FFT and wavelet based OFDM systems. Int J Electron Commun Comput Eng 5(4), Technovision-2014. ISSN 2249–071X
10. Wang J-W, Le NT, Wang C-C, Lee J-S (2015) Enhanced ridge structure for improving fingerprint image quality based on a wavelet domain. IEEE Signal Process Lett 22(4)
11. Ali M, Wook Ahn C (2015) Comments on Optimized gray-scale image watermarking using DWT-SVD and Firefly Algorithm. Short Communication, Expert Syst Appl 42:2392–2394
12. Tang H, Tong D, Dong Bao X, Dillenseger J-L (2015) A new stationary gridline artifact suppression method based on the 2D discrete wavelet transform. Med Phys 42:1721. doi:10.1118/1.4914861
13. (2014) Design of wireless waist-mounted vital sensor node for athletes performance evaluation of microcontrollers suitable for signal processing of ECG signal at waist part. Biowireless. 978-1-4799-2316-8/14/$31.00 © 2014 IEEE
14. Ratna Kumari V, Subba Rao Garu P (2014) Haar wavelet-based ofdm system with reduced paper for different modulation techniques Bitra. IJECT 5(3)
15. Kehan Z, Mingchui D (2015) Signal-to-noise ratio estimation based on Haar wavelet for cardiovascular bio-signals in web-based e-healthcare system. Multimed Tools Appl. Springer Science+Business Media NewYork. doi:10.1007/s11042-015-2544-2
16. Majak J, Shvartsman BS, Kirs M, Pohlak M, Herranen H (2015) Convergence theorem for the Haar wavelet based discretization method. Compos Struct 126:227–232

Bandwidth Enhancement by Using Spiral Back Reflector for Wireless Communication

Rajashree Dhume and Puran Gour

Abstract A novel back-reflector antenna for bandwidth enhancement is presented. The proposed antenna is designed by using two reflectors: one is the main reflector with a U-shaped slot at four truncated shape corners and the second is a subreflector with square spiral shaped antenna. The proposed antenna is designed for 10 GHz and we achieved a bandwidth of 129 % for VSWR < 2, minimum return loss = −28.44 dB, and maximum directivity 7.061 dBi. The antenna, which is suitable for C, X, and Ku, is used.

Keywords Spiral backfire antenna (SBA) · VSWR · Bandwidth · Directivity · Coaxial/probe feeding · Flame retardant 4 (FR-4)

1 Introduction

In recent years, there has been an increasing need for a low profile antenna for wireless and mobile communications. Many types of antennas have been designed to cater to variable applications and to be suitable for their needs. The backfire antenna was initially described in 1960 by H.W. Ehernspeck and a much simplified version, called the short backfire antenna, has been subjected to extensive experimental studies [1]. Figure 1 shows the basic geometry of the backfire antenna. Antenna consists of two reflectors, small reflector $R1$ and big reflector $R2$. The antenna length is the distance between the reflectors $R1$ and $R2$, denoted by L. The antenna has a source F *and* a surface-wave structure S. The spherical wave radiated from the source F is transformed into a surface wave $SW1$ by the surface-wave structure S. It terminates by the big reflector $R2$, which reflects the surface wave

R. Dhume (✉) · P. Gour
Department of Electronics & Communication, NIIST Bhopal, Bhopal, India
e-mail: dhume.rajashree@gmail.com

P. Gour
e-mail: purangour@rediffmail.com

L.P. Suresh and B.K. Panigrahi (eds.), *Proceedings of the International Conference on Soft Computing Systems*, Advances in Intelligent Systems and Computing 397, DOI 10.1007/978-81-322-2671-0_45

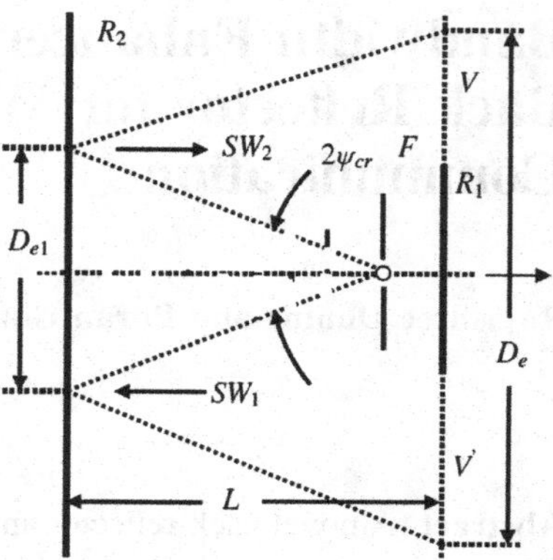

Fig. 1 Basic geometry of backfire antenna [1]

*SW*2 toward the small reflector *R*1, where it is radiated from the antenna aperture *VV'* into the space. Thus, the radiation of the antenna is directed in inverse direction in comparison with the radiation of the ordinary end-fire antenna used as a backfire antenna. Because of this reason it is called backfire antenna. The backfire antennas have attractive features such as low profile and planner surface. The reflector antennas have been widely used in various wireless systems, they may be used for long-range point-to-point applications, say WLAN, WISP, and satellite links. Although the backfire antenna suffers from narrow bandwidth different methods have been proposed to widen the bandwidth of the back reflector [1, 2, 4]. A multiplate reflector structure was designed to widen the operation bandwidth of the unidirectional antenna. The reflector structure was cut into a few sections of smaller reflectors [2]. The antenna was designed with a via-hole above the slot to create the second resonance. By adjusting the location of the via-hole, the second resonance can be moved to enhance the bandwidth of the main resonance, the antenna bandwidth can be increased up to 60 % [3]. The antenna was designed with two reflectors, one rectangular and the other circular. The antenna bandwidth can be enhanced approximately 12 times (from 4.5 to 52.8 %) [4].

2 Mathematical Analysis

Mathematical analysis is used for calculating the exact dimension of an antenna. By mathematical analysis we evaluate the width and length of the patch, ground plane, and reflectors. For mathematical purpose the main important point is how we can choose dielectric material. The bandwidth of proposed antenna is inversely proportional to the square root of substrate dielectric constant (ε_r). The substrate thickness is another important design parameter. Height of the substrate improved the fringing field at the patch edges like low dielectric constant and thus increases the radiated power. It gives lower quality factor and so higher bandwidth. The low value of dielectric constant increases the fringing field at the patch periphery and thus increases the radiated power. A small value of loss tangent is always preferable in order to reduce dielectric loss and surface-wave losses and which increases the efficiency of the antenna [4, 5].

There are three essential parameters that should be known while performing mathematical analysis:

- Frequency of operation (f_0): The resonant frequency of the antenna must be selected appropriately.
- Dielectric constant of the substrate (ε_r): The dielectric material selected for our design for microstrip patch and ground plane is FR-4, which has a dielectric constant of 4.4. Also an effective dielectric constant (ε_{eff}) must be obtained in order to account for the fringing and the wave propagation in the line.
- Height of dielectric substrate (h): The height of the dielectric substrate is selected as 1.5 mm.

Mathematical calculations for proposed antenna: [5]

$$W = \frac{c}{2f\sqrt{(\varepsilon_r + 1)/2}}$$

$$\varepsilon_{reff} = \left(\frac{\varepsilon_r + 1}{2}\right) + \left(\frac{\varepsilon_r - 1}{2}\right)\left[1 + 12\frac{h}{W}\right]^{-1/2}$$

$$\Delta L = 0.412h\frac{(\varepsilon_{reff} + 0.3)\left(W/h + 0.264\right)}{(\varepsilon_{reff} + 0.258)\left(W/h + 0.8\right)}$$

$$L = \frac{c}{2f\sqrt{\varepsilon_{reff}}} - 2\Delta L$$

$$L_0 = L + 6h$$

$$W_0 = W + 6h$$

where

f Operating frequency
ε_r Permittivity of the dielectric
ε_{reff} Effective permittivity of the dielectric
W Patch's width
L Patch's length
h Thickness of the dielectric
L_0 Length of ground plane
W_0 Width of ground plane (Table 1)

Figure 2 shows a schematic cross-sectional view of an inductor fabricated on a FR4 substrate. The line width, spacing, and outer dimensions are as shown in Table 2. The resonant frequency is about 10 GHz. The proposed spiral subreflector is used to reflect the wave to main reflector and the height between the main reflector and spiral back reflector is 0.5λ. Due to back reflector the loss of the wave is less.

Table 1 Mathematical calculation for a proposed antenna

S. no.	Parameters	Values for proposed antenna
1	Design frequency (*f*)	10 GHz
2	Dielectric constant (ε_r)	4.3
3	Height of substrate (*h*)	1.5 mm
4	Loss tangent for FR4	0.019
5	Width of excitation patch (*W*)	9.1875 mm
6	Length of excitation patch (*L*)	6.541 mm
7	Width of main reflector (W_0)	18.1875 mm
8	Length of main reflector (L_0)	15.541 mm
9	Height of subreflector	15 mm
10	Feed location:	
	X_f (along length)	1.715 mm
	Y_f (along width)	4.593 mm

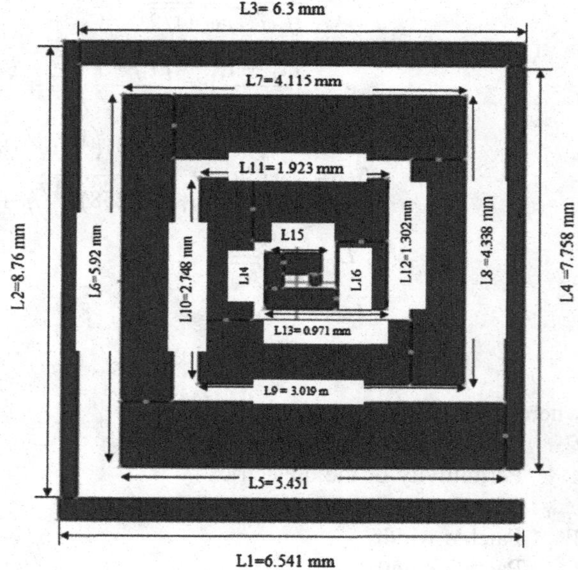

Fig. 2 Geometry of subreflector with square spiral shape

3 Antenna Design and Result Analysis

3.1 Antenna Design by Using IE3D Software

The proposed antenna is designed in flame retardant 4 (FR4) such that the different parameters can be obtained by using the EM simulator IE3D (version 9.0) software.

Table 2 Mathematical calculation for the square spiral reflector (*R*2)

S. no.	Parameters	Values for spiral subreflector (mm)
1	*L*1	6.541(0.21λ)
	*W*1	0.411(0.01λ)
2	*L*2	8.76(0.29)λ
	*W*2	0.245(8.16 × $10^{-3}\lambda$)
3	*L*3	6.3(0.21λ)
	*W*3	0.411(0.013λ)
4	*L*4	7.758(0.25λ)
	*W*4	0.245(8.16 × $10^{-3}\lambda$)
5	*L*5	5.451(0.18λ)
	*W*5	1.23(0.04λ)
6	*L*6	5.928(0.19λ)
	*W*6	0.736(0.02λ)
7	*L*7	4.115(0.13λ)
	*W*7	1.23 m(0.04λ)
8	*L*8	4.338(0.14λ)
	*W*4	0.736(0.02λ)
9	*L*9	3.019(0.10λ)
	*W*9	1.23(0.04λ)
10	*L*10	2.748(0.09λ)
	*W*10	0.736(0.02λ)
11	*L*11	1.923(0.064λ)
	*W*11	1.23(0.041λ)
12	*L*12	1.302(0.043λ)
	*W*12	0.736(0.024λ)
13	*L*13	0.971(0.032λ)
	*W*13	0.411(0.013λ)
14	*L*14	0.675(0.022λ)
	*W*14	0.245(8.66 × 10^{-3})
15	*L*15	0.51(0.017λ)
	*W*15	0.411(0.013λ)
16	*L*16	0.135(4.5 × $10^{-3}\lambda$)
	*W*16	0.245(8.66 × $10^{-3}\lambda$)

Based on the simulations and mathematical calculation we find the length and width of rectangular microstrip patch of short backfire antenna.

The proposed antenna is designed with two reflectors: small reflector *R*2 and big reflector *R*1.

3.2 Antenna with Main Reflector (R1)

The result of main reflector and exiting patch is as shown in Fig. 3. The corner of main reflector is truncated as shown in Fig. 3 (Figs. 4 and 5).

For the proposed design we have two notches, the return loss for the first notch is effective between 7.4 and 13.4 GHz and that for the second notch is effective between 11.6 and 13.4 GHz (Fig. 6).

The VSWR for first and second notches are effective between 7.4 to 13.4 GHz and 11.6 to 13.4 GHz, respectively, for this value the return loss is less than 2.

3.3 Antenna with Subreflector (R_2)

See Fig. 7.

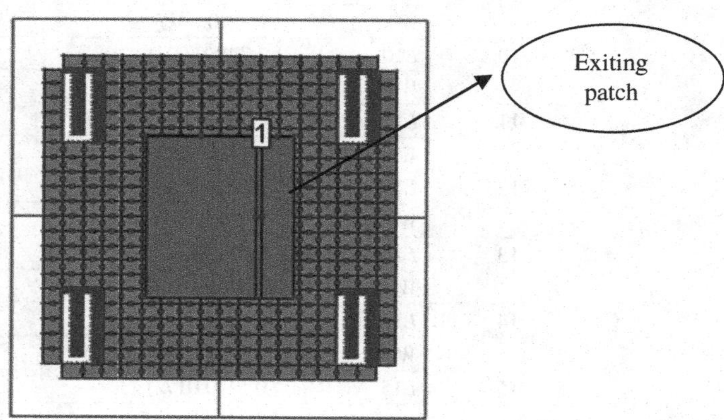

Fig. 3 Geometry of an antenna with main reflector (*R*1)

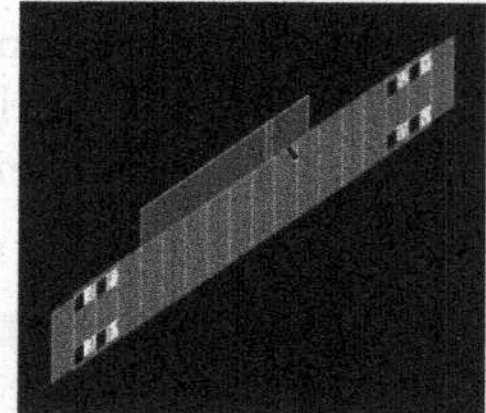

Fig. 4 3D view proposed antenna with main reflector (*R*1)

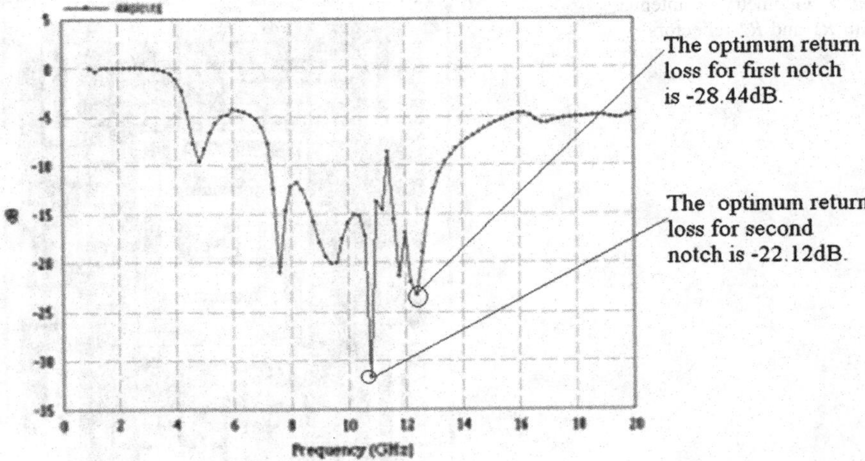

The optimum return loss for first notch is -28.44dB.
The optimum return loss for second notch is -22.12dB.
Frequency (GHz)

Fig. 5 Return loss curve for antenna with reflector (*R*1)

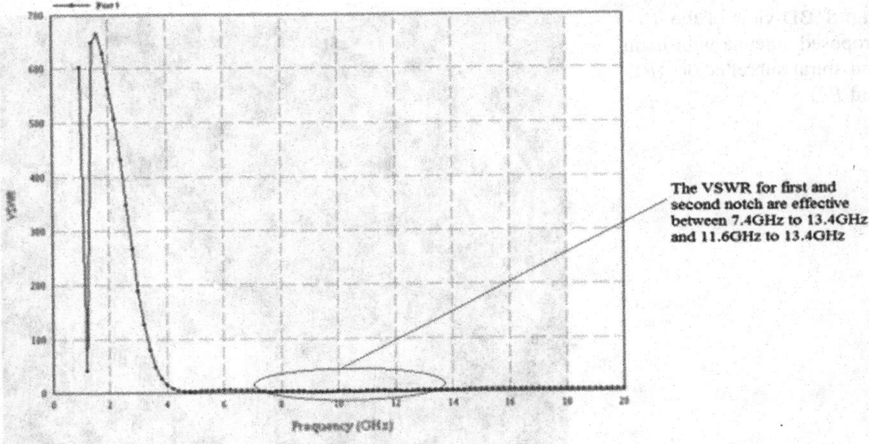

The VSWR for first and second notch are effective between 7.4GHz to 13.4GHz and 11.6GHz to 13.4GHz
Frequency (GHz)

Fig. 6 VSWR curve for antenna with reflector (*R*1)

4 Result and Discussion of Antenna with *R*1 and *R*2 Reflectors

For the proposed design the return loss value is effective between 4.2 and 19.6 GHz (Figs. 8 and 9).

The VSWR is a parameter of a transmission line or an antenna to show how well the antenna matched the cable impedance or the transmission line impedance.

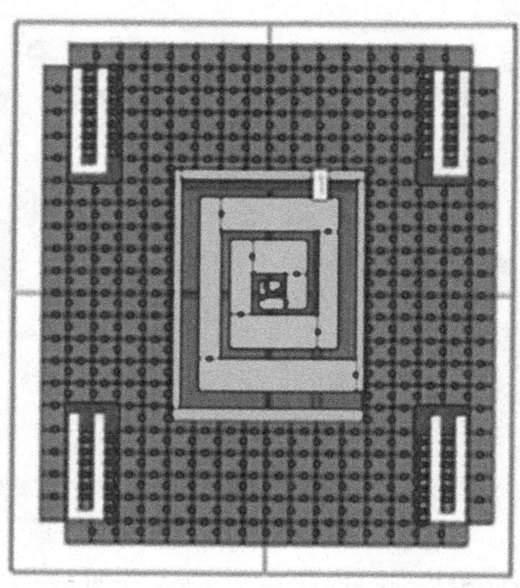

Fig. 7 Geometry of antenna with *R*1 and *R*2 reflectors

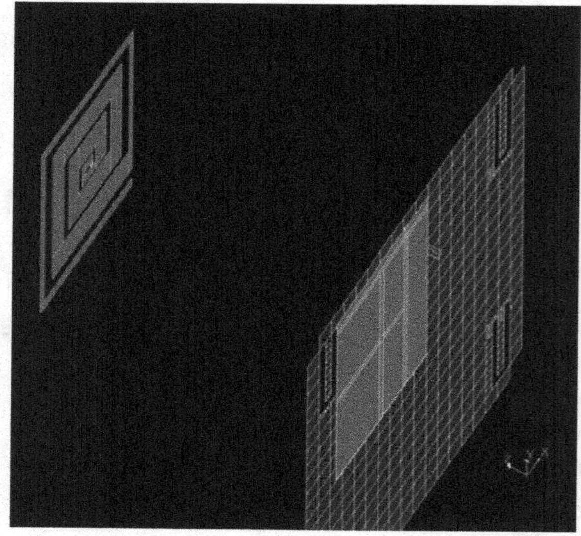

Fig. 8 3D view of the proposed antenna with main and spiral subreflectors (*R*1 and *R*2)

A proposed antenna would have a VSWR of less than 4.2–19.6 GHz. This indicates that the reflection from load is less. The VSWR obtained from the simulation is less than 2, which is approximately equal to 1. 1:1 as shown in Fig. 10. This is considered a good value as the level of mismatch is not very high because high VSWR implies that the port is not properly matched.

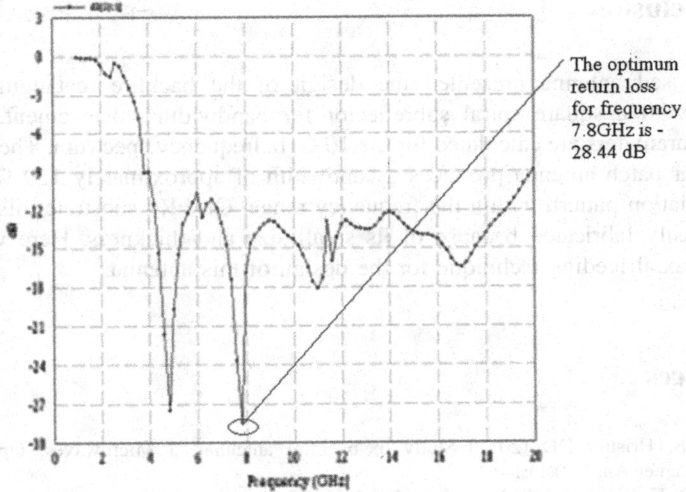

Fig. 9 Return loss curve for antenna with reflectors $R1$ and $R2$

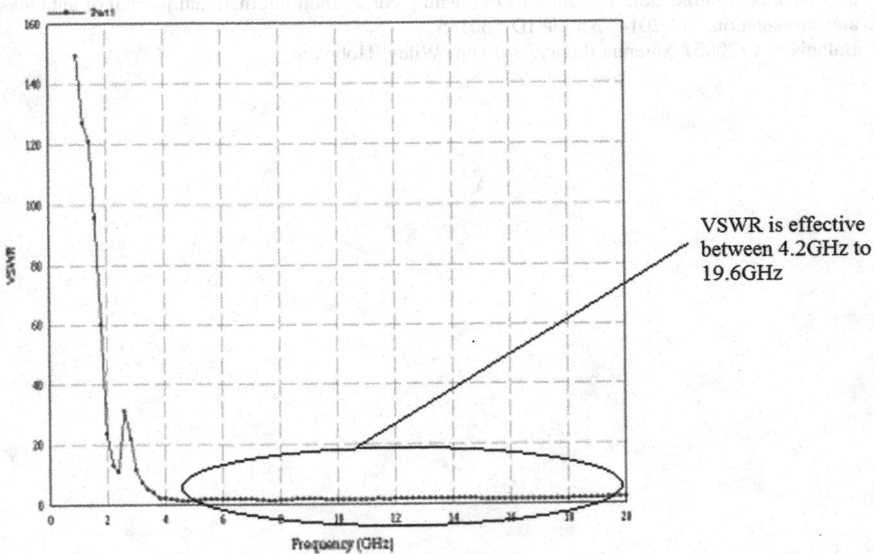

Fig. 10 VSWR curve for antenna with reflectors $R1$ and $R2$

5 Conclusion

The proposed antenna presented the design of the backfire rectangular patch antenna, with a square spiral subreflector for bandwidth enhancement. All the antenna parameters are calculated for the 10 GHz frequency spectrum. The backfire rectangular patch antenna produces a bandwidth of approximately 129 %, with a stable radiation pattern within the frequency range. On FR4 substrate this antenna can be easily fabricated because of its small size and thickness. Here we use a simple coaxial feeding technique for the design of this antenna.

References

1. Kirov GS, Hristov HD (2011) Study of backfire antennas. J Microwaves, Optoelectron Electromagnet Appl 10(1):1–12
2. Gao X, Qi Y, Jiao Y-C (2013) Design of multiplate back reflector for a wideband slot antenna. IEEE Antennas Wirel Propag Lett 12, 2013
3. Yun S, Kim D-Y, Nam S (2012) Bandwidth enhancement of cavity-backed slot antenna using a via-hole above the slot. IEEE Antennas Wirel Propag Lett 11, 2012
4. Gour P, Mishra R (2014) Bandwidth enhancement of a backfire microstrip patch antenna for pervasive communication. In: Hindawi Publishing corporation international journal of antennas and propagation, vol 2014, Article ID 560185
5. Balanis CA (2005) Antenna theory, 3rd edn. Wiley, Hoboken

Comparative Assessment of the Performances of LQR and SMC Methodologies in the Control of Longitudinal Axis of Aircraft

V. Rajeswari and L. Padma Suresh

Abstract The automatic flight control system of an aircraft is very complicated in nature with lot of logics and operational constraints. The study of the detailed automatic flight control and thrust control involves complicated logics for the control functions with closed loop control systems which can be realized in the software. The work is focused on the redesign of automatic flight control system using robust control methods in the MATLAB/Simulink environment. This paper aims at a complete study, analysis, design, and performance monitoring of new generation autopilot system using linear quadratic regulator (LQR) and sliding mode control (SMC) methods for the control of pitch angle of aircraft. A comparison of the results of simulation of the above gives an optimal performance of the autopilot system.

Keywords Autopilot · Flight control · FMS · LQR · SMC · Pitch control

1 Introduction

An aircraft can be capable of many very intensive tasks: helping the pilot focus on the overall status of the aircraft and flight. Autopilots can automate tasks such as maintaining the altitude, climbing or descending to an assigned altitude, turning to and maintaining an assigned heading, intercepting a course, guiding aircraft between the way points that make up a route programmed into a flight management system (FMS), and flying a precision or non-precision approach. The fast growth of

---

V. Rajeswari (✉)
ICE Department, G. Narayanamma Institute of Technology and Science, Hyderabad, India
e-mail: rajiviswanath28@gmail.com

L. Padma Suresh
EEE Department, Noorul Islam University, Kanyakumari, India
e-mail: suresh_lps@yahoo.co.in

L.P. Suresh and B.K. Panigrahi (eds.), *Proceedings of the International Conference on Soft Computing Systems*, Advances in Intelligent Systems and Computing 397, DOI 10.1007/978-81-322-2671-0_46

aircraft designs from less capable airplane to the present-day aircraft required the development of many technologies. The automatic control system plays an important role in monitoring and controlling of many of the aircraft's subsystems for which an autopilot is designed that controls the principal axes of the aircraft leading to the safe landing of the aircraft during adverse weather conditions [1]. In this paper, optimal and robust controllers, LQR and SMC are developed for controlling the pitch axis of the aircraft. The performances of these controllers are investigated and compared by simulating in the MATLAB environment.

2 Modeling of Pitch Control System

There are two dynamic equations of motion present for an aircraft, where lateral dynamic equations of motion represent the dynamics of aircraft with respect to the lateral axis and longitudinal dynamic equations of motion represent the dynamics of aircraft with respect to the longitudinal axis [2, 3]. Figure 1 shows the forces, moments, and velocity components in the body-fixed coordinate where X_B, Y_B, and Z_B denote the aerodynamic force components. φ and δe represent the orientation of aircraft elevator deflection angle in the earth axis system and aileron deflection angle, respectively. L, M, and N denote the aerodynamic moment components. p, q, and r represent the angular rates of roll, pitch, and yaw axes and u, v, and w represent the velocity components of roll, pitch, and yaw axes. Figure 2 shows the angular orientation and velocities of gravity vector. By applying Newton's second law to the rigid body, the equations of motion of aircraft can be established in terms of the translational and angular accelerations. During the modeling of pitch control systems, few assumptions are considered.

1. The aircraft is in steady-state cruise at constant altitude and velocity, thus the thrust and drag cancel each other out.
2. The changes in pitch angle do not change the speed of an aircraft under any circumstances.

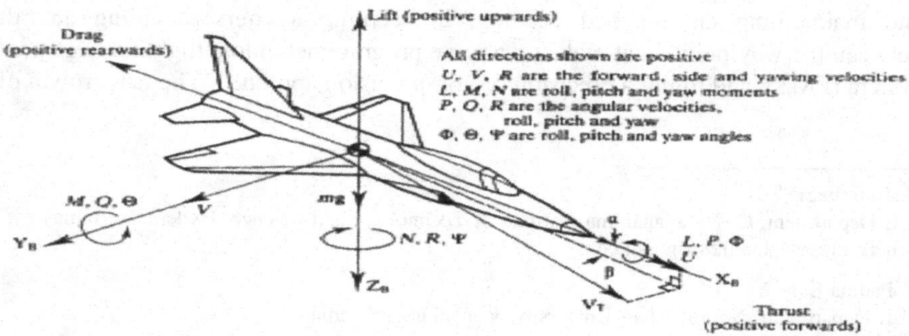

Fig. 1 Aerodynamic forces, moments, and velocity components

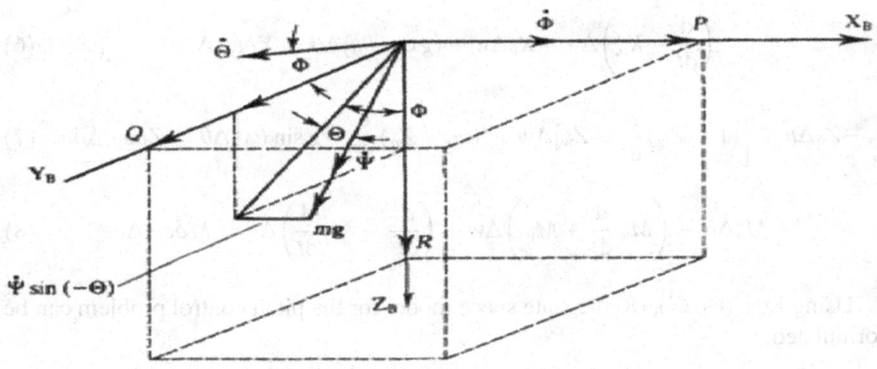

Fig. 2 Angular orientation and velocities of gravity vector

The pitch control problem is a longitudinal problem and the work is developed to control the pitch angle of an aircraft in order to stabilize the system when an aircraft performs the pitching motion. With respect to Figs. 1 and 2, the force and moment equations are given in Eqs. (1)–(3).

$$x - \mathrm{mg}\sin\theta = m(\dot{u} + qw - rv) \tag{1}$$

$$z + \mathrm{mg}\ \cos\theta\cos\varphi = m(\dot{w} + pv - qu) \tag{2}$$

$$M = I_{yy}\dot{q} + qr(I_{xx} - I_{zz}) + I_{xz}(p^2 - r^2) \tag{3}$$

The above equations are nonlinear and simplified by considering the aircraft to comprise two components: a mean motion that represents the equilibrium or trim conditions and a dynamic motion which accounts for the perturbations about the mean motion. Thus every motion variable is considered to have two components.

$$U \triangleq U_o + \Delta u,\ \triangleq Q_o + \Delta q,\ R \triangleq R_o + \Delta r,\ M \triangleq M_o + \Delta m,\ Y \triangleq Y_o + \Delta y,\ P \triangleq P_o + \Delta p,$$
$$L \triangleq L_o + \Delta l,\ V \triangleq V_o + \Delta v,\ \delta \triangleq \delta_o + \Delta\delta \tag{4}$$

The reference flight condition is assumed to be symmetric and the propulsive forces are assumed to be constant.

$$v_o = q_o = u_o = r_o = \varphi_o = \psi_o = 0 \tag{5}$$

The complete linearized equations of motion are obtained as below where the sideslip angle is used. They are obtained as below:

$$\left(\frac{\mathrm{d}}{\mathrm{d}t} - X_u\right)\Delta u - X_u \Delta w + (g \cos\theta_0)\Delta\theta = X\delta e \cdot \Delta\delta e \tag{6}$$

$$-Z_u \Delta u + \left[(1 - Z_w)\frac{d}{\mathrm{d}t} - Z_w\right]\Delta w + \left[(u_0 - \dot{Z}_q)\frac{\mathrm{d}}{\mathrm{d}t} - g \sin\theta_0\right]\Delta\theta = Z\delta e \cdot \Delta\delta e \tag{7}$$

$$-M_u \Delta u - \left(\dot{M}_w \frac{\mathrm{d}}{\mathrm{d}t} + M_w\right)\Delta w + \left(\frac{d^2}{\mathrm{d}t^2} - M_q \frac{\mathrm{d}}{\mathrm{d}t}\right)\Delta\theta = M\delta e \cdot \Delta\delta e \tag{8}$$

Using Eqs. (6), (7), (8) the state space model for the pitch control problem can be formulated.

$$\begin{bmatrix} \dot{\Delta u} \\ \dot{\Delta w} \\ \dot{\Delta q} \\ \dot{\Delta\theta} \end{bmatrix} = \begin{bmatrix} X_u & X_w & 0 & 0 \\ Z_u & Z_w & u_o & 0 \\ M_u + \dot{M}_u + Z_u & M_u + M_w + Z_w & M_q + \dot{M}_w + u_o & 0 \\ 0 & 0 & 1 & 0 \end{bmatrix} \begin{bmatrix} \Delta u \\ \Delta w \\ \Delta q \\ \Delta\theta \end{bmatrix} + \begin{bmatrix} X_{\delta e} \\ Z_{\delta e} \\ M_{\delta e} + Z_{\delta e} \\ 0 \end{bmatrix} [\Delta_{\delta e}] \tag{9}$$

The pitch control problem has the input as the elevator deflection angle with the output as the pitch angle of the aircraft. For this study, the data of the general aviation airplane Navion is considered [1]. The longitudinal stability derivatives are given in Table 1 [4] from where the state space representation can be obtained [5].

Equation (10) gives the state space representation for pitch control problem.

Table 1 Longitudinal stability derivative

Longitudinal derivatives	Components		
Yawing velocities	X-Force derivatives	Z-Force derivatives	Pitching moment
	$X_w = 0.254$	$Z_w = -2$	$M_w = -0.05$
Rolling velocities	$\dot{X}_w = 0$	$\dot{Z}_w = 0$	$\dot{M}_w = -0.051$
	$X_u = -0.04$	$Z_u = -0.37$	$M_u = 0$
Angle of attack	$X_\alpha = 0$	$Z_\alpha = -355$	$M_\alpha = -8.8$
	$\dot{X}_\alpha = 0$	$\dot{Z}_\alpha = 0$	$\dot{M}_\alpha = -0.898$
Pitching rate	$X_q = 0$	$Z_q = 0$	$M_q = -2.05$
Elevator deflection	$X_{\delta e} = 0$	$Z_{\delta e} = -28.15$	$M_{\delta e} = -11.88$

$$\begin{bmatrix} \Delta\dot{u} \\ \Delta\dot{w} \\ \Delta\dot{q} \\ \Delta\dot{\theta} \end{bmatrix} = \begin{bmatrix} -0.043 & 0.034 & 0 & 32.2 \\ -0.351 & -1.925 & 175.96 & 0 \\ 0.00193 & -0.0484 & -3.42 & 0 \\ 0 & 0 & 1 & 0 \end{bmatrix} \begin{bmatrix} \Delta u \\ \Delta w \\ \Delta q \\ \Delta\theta \end{bmatrix} + \begin{bmatrix} 0 \\ -26.82 \\ -14.182 \\ 0 \end{bmatrix} [\Delta\delta e] \quad (10)$$

3 Design of Control Methodologies

The work is emphasized to develop a pitch control scheme for controlling the pitch angle of aircraft. For this two control methodologies (i.e.) linear quadratic regulator (LQR) and sliding mode control (SMC) methodologies were proposed. The performance of both the control strategies was investigated and compared.

3.1 Design of Linear Quadratic Regulator (LQR)

Linear quadratic optimization [6, 7] is a basic method for designing controllers of dynamical systems. LQR is a powerful method for designing flight control systems. This method is based on the manipulation of the equations of motion in state space and the system can be stabilized using full state feedback system. Consider the state and output equations describing the longitudinal equations of motion.

$$\dot{X}(t) = Ax(t) + Bu(t) \quad (11)$$

$$y(t) = Cx(t) + Du(t) \quad (12)$$

In the LQR design [8], the lqr function in MATLAB can be used to determine the value of the vector '*K*', which is used to find the feedback control law.

$$u(t) = -kx(t) + \Delta\delta e \cdot N \quad (13)$$

This control law has to minimize the performance index,

$$J = \int_0^\infty (X^T QX + u^T Ru)\mathrm{d}t$$

where Q is the state cost matrix and R is the performance index matrix. Here, $R = 1$ and $Q = C^T C$.

Figure 3 shows the full state feedback controller with reference input. For the present study, the value of '*K*' is to be determined.

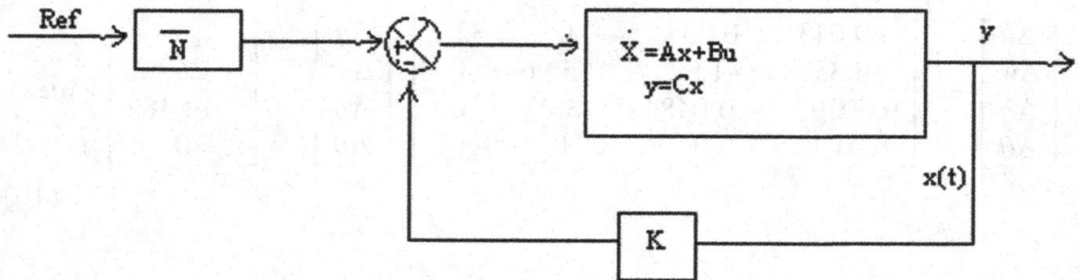

Fig. 3 Full state feedback controller

The controller is tuned by varying the elements in Q matrix which is done in the m-file code. With the below values,

$R = 1$ and $Q = [0000;\quad 0000;\quad 0000;\quad 000500]$, the values of K with $\bar{N}$ can be obtained.

3.2 *Design of Sliding Mode Control (SMC)*

Sliding mode control is a nonlinear method [9] which alters the dynamics of the nonlinear system by the application of a high frequency switching control. This is a variable structure control. The aim of this is to drive the system state from an initial condition $x(0)$ to the state space origin as $t \to \infty$. The jth component $U_j(j = 1,2,\ldots.$ $m)$ of the state feedback control vector $U(x)$ has a discontinuity on the jth switching surface which is a hyperplane 'M_j' passing through the state origin.

Defining the hyperplane by

$$M_j = \{x : C_j x = 0\}, j = 1, 2, \ldots m \tag{14}$$

Consider the system

$$\dot{x}(t) = Ax(t) + Bu(t) \tag{15}$$

From Eq. (14), the sliding mode satisfies the condition

$$S = Cx(t) = 0,\ t \geq t_s \tag{16}$$

C = mxn matrix, t_s is the time when the sliding subspace is reached. Differentiating Eq. (14) with respect to time and substituting for Eq. (16), we get

$$C\dot{x}(t) = CAx(t) + CBu(t) = 0, t \geq t_s \tag{17}$$

$$CBu(t) = -CAx(t) \tag{18}$$

C is the hyperplane matrix so that $|CB| \neq 0$.

The switching surface design is predicted based on the knowledge of the system behavior in the sliding mode and this behavior depends on the parameters of the switching surface. So achieving the switching surface design requires analytically specifying the motion of the state trajectory in a sliding mode. Therefore, the method of equivalent control is essential. Equivalent control constitutes a control input when exciting the system produces the motion of the system on the sliding surface whenever the initial state is on the surface. So Eq. (18) can be rearranged to get equivalent control as below

$$\begin{aligned} U_{\text{eqt}} &= -(CB)^{-1}CAx(t) \\ U_{\text{eqt}} &= -Kx(t) \end{aligned} \tag{19}$$

$$\dot{x}(t) = [A - Bk]x(t) \tag{20}$$

Equation (18) gives the system equation for the closed loop system dynamics during sliding. The choice of 'C' determines the matrix 'K'. The purpose of the control 'U_{eqt}' is to drive the state into the sliding subspace and thereafter maintain in it. The $(n - m)$ dimensional switching surface imposes 'm' constraints on the plant dynamics in a sliding mode. So the plant dynamics should be converted to a regular form. The nominal linear system can then be expressed as

$$\begin{aligned} \dot{y}_1(t) &= A_{11}y_1 + A_{12}y_2 \\ \dot{y}_2(t) &= A_{21}y_1 + A_{22}y_2 + B_2u(t) \end{aligned} \tag{21}$$

This plays an important role in bringing the solution of the reachability problem and the sliding condition is equivalent to

$$\begin{aligned} C_1y_1(t) + C_2y_2(t) &= 0 \\ y_2(t) &= -\frac{C_1}{C_2}y_1(t) \\ y_2(t) &= -Fy_1(t) \end{aligned} \tag{22}$$

$$\dot{y}_1(t) = [A_{11} - FA_{12}]y_1(t) \tag{23}$$

Equation (21) is the reduced order equivalent system.

(a) Design of Sliding hyperplane [10, 11]

To design the hyperplane, the quadratic performance $J = \frac{1}{2}\int_{t_s}^{\infty} x^T Qx\mathrm{d}t$ can be minimized, where $Q > 0$ is the positive definite symmetric matrix and t_s is the time of attaining the sliding mode. The matrix '*F*' is determined once the matrix Riccati equation is solved.

(b) Design of feedback control law

Once the sliding surface has been selected, then the reachability problem can be solved. This involves the selection of a state feedback control function which will drive the state '*x*' into the sliding subspace and thereafter maintain in it. This control law has two components:

1. A linear control law (U_l) to stabilize the nominal linear system.
2. Discontinuous component (U_n).

And the control law,

$$U(t) = U_l(t) + U_n(t) \tag{24}$$

where

$$U_l(t) = U_{eq}(t) = -(CB)^{-1}CAx(t) = -(CB)^{-1}[CA - \varphi C]x(t) \tag{25}$$

The nonlinear component is defined to be

$$U_n(t) = \rho\frac{S}{\|S\|} = \rho\frac{Cx(t)}{\|Cx(t)\|}, \rho > 0 \tag{26}$$

'*C*' is the symmetric positive definite matrix satisfying the Lyapunov equation, $C\varphi + \varphi^T C = -I$, φ is any stable design matrix.
The control law is

$$U(t) = -(CB)^{-1}[CA - \varphi C]x(t) + \rho\frac{Cx(t)}{\|Cx(t)\|} \tag{27}$$

4 Simulation and Results

A control system for the pitch axis is simulated using LQR and SMC and the results of simulation are analyzed and presented for comparison. To investigate the performance of the control strategy, the time domain specifications are analyzed. The values of *K* for the pitch control problems in LQR are obtained as *K* = [−0.002 0.0033 −1.5283 −22.3605] and $\bar{N} = -22.3667$ with weighting factor x = 500.

The control law in SMC has the value, *U*(*t*) = [−0.0212 −0.013 0.368].

Figures 4 and 5 show the response of the pitch control system using LQR and SMC, respectively.

Table 2 shows the performance characteristics of both LQR and SMC controllers for pitch control of aircraft. By comparing the performance characteristics in

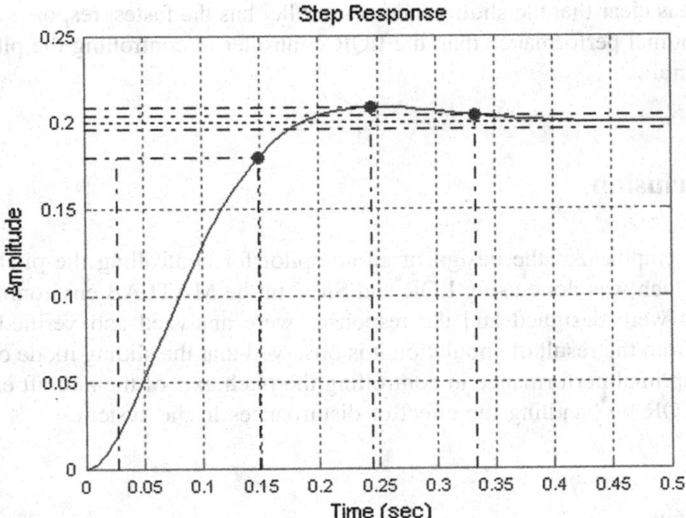

Fig. 4 Response of system for LQR

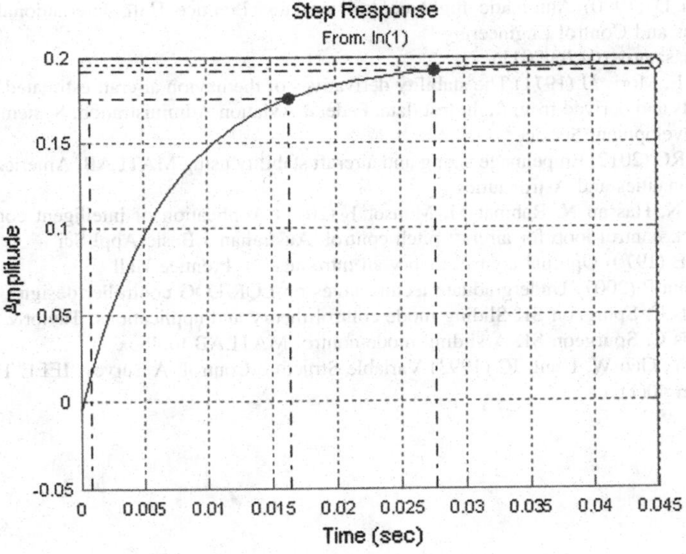

Fig. 5 Response of system for SMC

Table 2 Comparison of performance characteristics

Performance characteristics	LQR	SMC
Rise time (t_r)	0.121 s	0.0156 s
Settling time (t_s)	0.334 s	0.0276 s
% Overshoot (%Mp)	4.35 %	0.0341 %

Table 2, it is clear that the sliding mode controller has the fastest response and gave a more optimal performance than the LQR controller in controlling the pitch angle of the aircraft.

5 Conclusion

The work emphasizes the design of an autopilot for controlling the pitch axis of aircraft, which was done using LQR and SMC in the MATLAB environment. The controllers were designed and the responses were analyzed and verified in time domain. From the result of simulation it is observed that the sliding mode controller gives an optimal performance in controlling the pitch axis of the aircraft efficiently than the LQR by handling the effect of disturbances in the system.

References

1. Nelson C (1998) Flight stability and automatic control. McGraw Hill, 2nd edn
2. Mclean D (1990) Automatic flight control systems. Prentice Hall, International Series in Systems and Control Engineering
3. www.nasa.gov (01.03.2011)
4. Seckel E, Moris JJ (1971) The stability derivatives of the navion aircraft estimated by various methods and derived from flight test data. Federal Aviation Administration, Systems Research and Development Service
5. Struett RC (2012) Empennage sizing and aircraft stability using MATLAB. American Institute of Aeronautics and Astronautics
6. Wahid N, Hassan N, Rahmat M, Mansor F (2011) Application of intelligent controllers in feedback control loop for aircraft pitch control. Australian J Basic Appl Sci
7. Kirk DE (1970) Optimal control theory-an introduction. Prentice Hall
8. Hespanha P (2007) Undergraduate lecture notes on LQR/LQG controller design
9. Edwards C, Spurgeon SK Sliding mode control theory and applications. Taylor& Francis
10. Edwards C, Spurgeon SK A sliding mode control MATLAB toolbox
11. Hung JY, Gao W, Hung JC (1993) Variable Structure Control: A Survey. IEEE Trans Indus Electron 40(1)

A Hybrid Divide—16 Frequency Divider Design for Low Power Phase Locked Loop Design

Supraja Batchu, Ravi Nirlakalla, Jayachandra Prasad Talari and Venkateswarlu Surisetty

Abstract In this paper, we present a divide by 16 frequency divider (FD) for high frequency and low power phase locked loop (PLL) designs. The FDs have shown efficiency in different parameters. Divide by 16 FDs are proposed with a hybrid model which combined with a true single-phase clock (TSPC) and E-TSPC for low power PLL. Results of FDH1 have shown low power consumption.

Keywords Frequency divider · PLL · TSPC · E-TSPC · Low power

1 Introduction

Nowadays, mobile wireless localization systems are commonly operating at several GHz. Frequency synthesizer (FS) is the fundamental block of these systems. A FS is usually realized by using PLL. PLLs rely on the feedback divider chain, for implementing the division of the frequency of the voltage-controlled oscillator (VCO) output to make it equal to the reference signal frequency (Cref). A PLL is a negative feedback control system, which produces a signal that has a fixed relation to the phase and frequency of a "reference" signal. FS is usually comprehended by a PLL which is one of the crucial blocks in total power consumption, because this block is often used as a transmitter and receiver. Basically, the maximum power loss is allied to the primary layers of frequency divider (FD) which consumes half of the total power. Conventional static complementary metal oxide semiconductor (CMOS) logic cannot be used for the realization of the divider, because the input frequency in prime layer is high [1–3]. The frequency divider has to operate at the maximum output frequency of the synthesizer with low power dissipation.

S. Batchu · R. Nirlakalla (✉) · J.P. Talari · V. Surisetty
RGM College of Engineering and Technology (Autonomous), Nandyal 518501
Andhra Pradesh, India
e-mail: ravi2728@gmail.com

S. Batchu
e-mail: batchusupraja@gmail.com

L.P. Suresh and B.K. Panigrahi (eds.), *Proceedings of the International Conference on Soft Computing Systems*, Advances in Intelligent Systems and Computing 397, DOI 10.1007/978-81-322-2671-0_47

Memory elements, which read a value, save it for some time, and then write that stored value somewhere else, even if the element's input value has consequently changed, are the basic building blocks of sequential machine. A Boolean logic gate can calculate values, only when the change in the input results in an immediate change in the output. Each alternative circuit used as a memory element has its own advantages and disadvantages.

A generic memory element has an internal memory, which can be accessed by the control of the clock input. Whenever the clock instructs, memory element reads its data and stores it in its memory. This stored value will be reflected by the output after some delay.

Conventional high-speed FDs, operating at multi-gigahertz frequencies, are mainly based on current mode logic (CML), injection-locked topology, and true single-phase clock (TSPC) logic/extended true single-phase clock (E-TSPC) logic.

A conventional clock divider can be implemented by using the D-type flip flop and an inverter. The output flips, during each cycle of the input signal. Due to the wide range of digital logic families there are numerous implementations of the clock divider circuit. One such possible implementation, which allows for a higher frequency operation than would be allowed with static logic, is the TSPC divider. It minimizes power and reduces the number of transistors.

D flip-flop is a basic building block in the modern digital circuit. PLL with an excellent performance is widely reported. FD and phase frequency divider (PFD) are vital modules of PLL, which are designed by D flip-flops. Dynamic logic is used for implementing edge triggered D flip-flops used for integrated high-speed operations. This indicates that while the device is not transitioning, the digital output is deposited on parasitic device capacitance, which discharges at one or more nodes representing the reset operation. Thus, these dynamic flip flops enable reset. The dynamic power consumption is prevailing at higher frequencies. As the conventional D flip-flop implemented by E-TSPC logic operates at higher frequencies, it causes a small increase in power dissipation.

With the limitations of crystal oscillator design, the PLLs rely on FDs for dividing the output frequency in order to allow for a much lower reference frequency. Even though FDs do not pose many design challenges as other blocks of PLL, like the charge pump (CP) or the oscillator, the divider is still an essential circuit with design issues which can be addressed. Current consumption can be significant because one of the few circuits operates at the higher frequencies. Whenever the maximum frequencies are of interest, divider design frequencies are not easy, often requiring much thorny designs like current mode logic (CML) dividers or tuned injection-locked dividers. But these designs use more transistors than used at lower frequencies and require passive structures like on-chip inductors [4].

In this paper, various FD designs are discussed for low power and high-speed applications. A divider by 16 is proposed with hybrid models such as TSPC and E-TSPC designs. Divider E-TSPC is discussed in Sect. 2. Frequency divider flip flops are given in Sect. 3. Divider 2 information is given in Sect. 4. Section 5 gives the information of frequency divider 16. Simulation results and conclusion of the work are discussed in Sects. 6 and 7, respectively.

2 E-TSPC Logic Divider

In general, TSPC logic employs three transistors in each stage whereas E-TSPC uses only two transistors in every stage. E-TSPC is shown in Fig. 4. Hence, when the supply voltage is above 1-V for E-TSPC logic, the circumventing of stacked MOS structure permits higher operating frequency, serving as the best solution for low-voltage operation compared with the TSPC logic. However, E-TSPC does not operate when the supply voltage falls to 0.5 V [5, 6].

$$C_{L\text{-TSPC}} = C_{\text{dbM8}} + 2C_{\text{gdM8}} + C_{\text{dbM7}} + 2C_{\text{gdM7}} + C_{\text{gM3}} + C_{\text{gM1}} \quad (1)$$

$$C_{L\text{-ETSPC}} = C_{\text{dbM6}} + 2C_{\text{gdM6}} + C_{\text{dbPM5}} + 2C_{\text{gdM5}} + C_{\text{gM1}} \quad (2)$$

From (1) and (2) the load capacitance is high for TSPC flip-flop when compared to the E-TSPC flip-flop. As shown in (3) the switching power directly depends on the output load capacitance. Hence, TSPC flip-flop has higher switching power.

$$P_{\text{switching}} = f_{\text{clk}} C_L V_{\text{dd}}^2 \quad (3)$$

where C_L is the load capacitance
f_{clk} is the operating frequency

During a period a direct path exists between the supply voltage and ground, resulting in the short-circuit power. The transistor size, rise and fall times of the input signal will affect the short-circuit power. The analysis indicates that short-circuit power is less when the load capacitance is less. As we know that in a 2/3 prescaler, the E-TSPC flip-flop uses smaller switching power, but considerably more short-circuit power. For every fourth clock cycle when the E-TSPC flip-flop is operated as a divide-by-2 circuit the short-circuit power is present. E-TSPC circuits possess high short-circuit power which depends on the size of transistor assuming that voltage controlled oscillator (VCO) output is a full swing signal. But VCO output is not full swing signal and it has a certain DC level in many applications in the GHz range. This affects the short-circuit power of both logic types [7].

3 Frequency Divider Flip Flops

Various flip flops are proposed in the recent years in order to meet the requirements like low power high-speed performance. This section deals with some of the basic flip flops that are generally used. All the flip flops discussed in this section are single ended, i.e., single ended input and output. Figure 1 shows the conventional flip flop [8]. Later modifications are done to conventional TSPC in order to reduce the number of transistors as shown in Figs. 2, 3, 4 and 5.

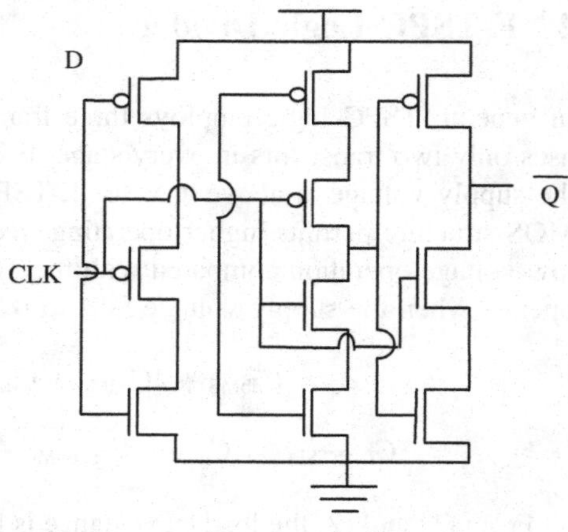

Fig. 1 Conventional TSPC (DIVCON) [8]

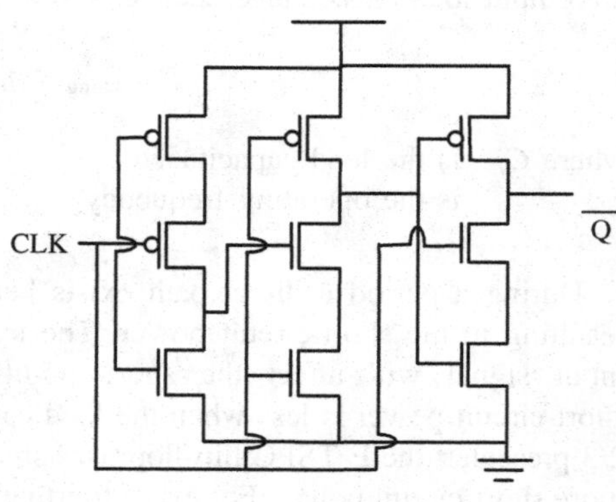

Fig. 2 Edge triggered nine-transistor D-FF of Yuan and Svensson [9] (DIV)

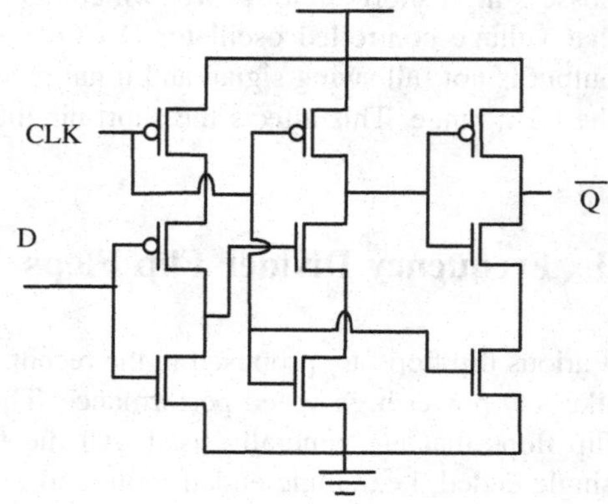

Fig. 3 Nine-transistor single-phase D-FF for higher speed [10] (DIVS)

Fig. 4 Extended TSPC (E-TSPC) [11]

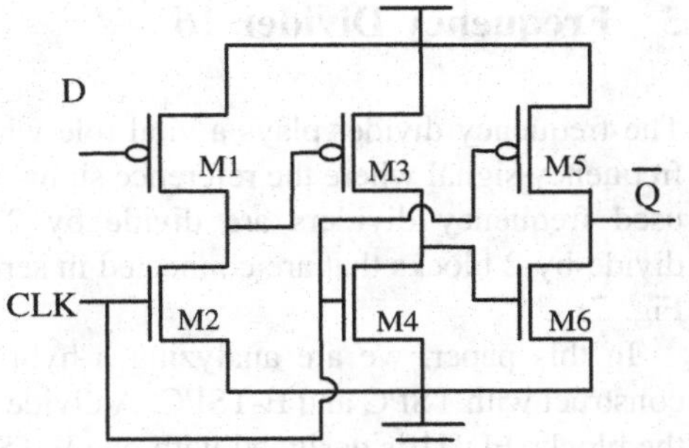

Fig. 5 A nine-transistor negative edge triggered D flip-flop (DIVN) [11]

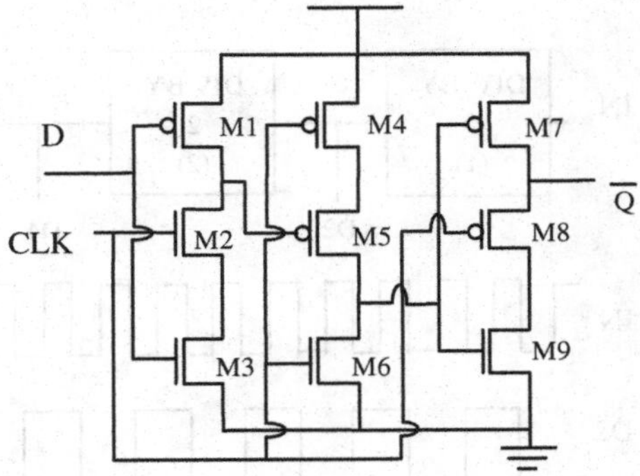

4 Divider 2

The basic divided by 2 block is obtained by driving the D input of the flip flop with the Q bar output of the flip flop as shown in Fig. 6. For every input clock cycle, the output changes its polarity resulting in the generation of signal (OUT) whose frequency is half of the frequency of CLK of the flip flop, i.e., Signal IN.

$$f_{\text{out}} = \frac{1}{2} f_{\text{in}} \tag{4}$$

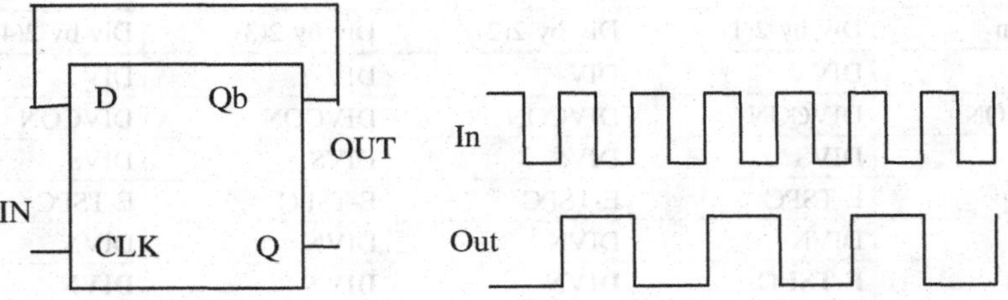

Fig. 6 Divide-by-2 design and its waveforms

5 Frequency Divider 16

The frequency divider plays a vital role whenever there is a need to generate high frequency signal where the reference signal is a low frequency signal. The generally used frequency dividers are divide by 2^n circuits, where n is the number of divide-by-2 blocks that are connected in series. A basic divide by 16 is as shown in Fig. 7.

In this paper, we are analyzing a hybrid divide by 16, which is proposed to construct with TSPC and E-TSPC. A divide by 16 comprises of four blocks. Among the blocks FDH1 is designed with one E-TSPC and three TSPC. FDH2 is designed

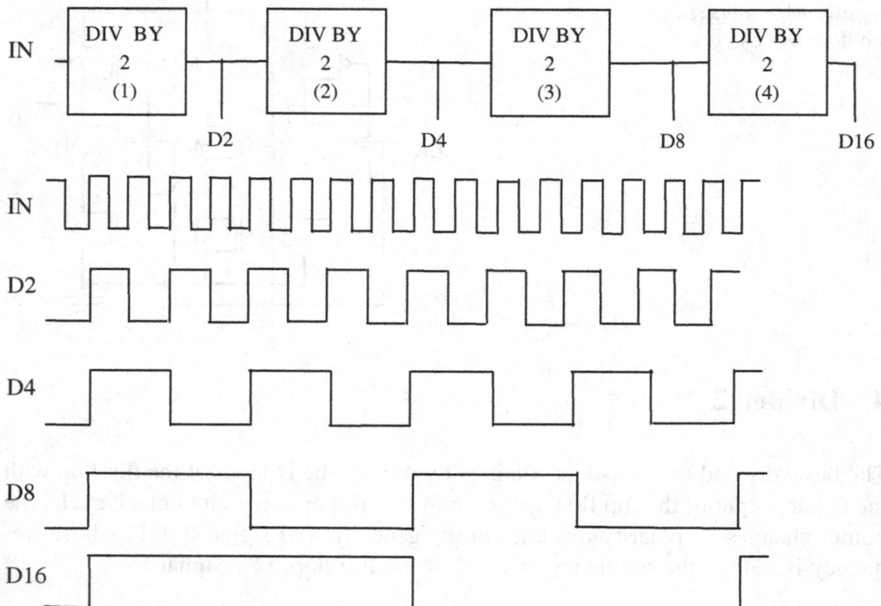

Fig. 7 Schematic diagram of divide by 16 design and its waveforms

Table 1 Divide by 16 using various flip flops

Design	Div by 2(1)	Div by 2(2)	Div by 2(3)	Div by 2(4)
FD1	DIV	DIV	DIV	DIV
DIVCON	DIVCON	DIVCON	DIVCON	DIVCON
FDs	DIVS	DIVS	DIVS	DIVS
FDetspc	E-TSPC	E-TSPC	E-TSPC	E-TSPC
FDn	DIVN	DIVN	DIVN	DIVN
FDH1	**E-TSPC**	**DIVN**	**DIVN**	**DIVN**
FDH2	**E-TSPC**	**E-TSPC**	**DIVN**	**DIVN**

with two E-TSPC and two TSPC. The waveforms of divide 2, 4, 8, and 16 are shown in Fig. 7. Table 1 shows different FDs' implementation with various flip flops.

6 Simulation Results

Different FDs are simulated and verified the functionality using Tanner EDA tool. The power consumption and propagation delay of the designs are evaluated using HSPICE at 2.0 V for 180 nm technology. Among all the FDs in terms of power consumption, FDH1 has consumed less power 0.127 mW as shown in Table 2.

Propagation delays of the FD designs are calculated for each input and output of the designs. The worst-case delays of the FDs are given in Table 2.

Powerdelay product (PDP) and energy delay product (EDP) of the FD designs are evaluated for energy efficiency of the designs.

Table 2 Power and delay analysis of different FDs

Design	Power (mW)	Delay (nS)	PDP	EDP	No. of MOS devices
FD1	0.381	66.9	2.54E-11	1.705E-18	36
DIVCON	0.536	0.878	4.70E-13	4.124E-22	40
FDs	0.407	26.4	1.074E-11	2.838E-19	36
FDetspc	7.29	58.1	4.233E-10	2.459E-17	24
FDn	0.479	25.8	1.234E-11	3.183E-19	36
FDH1	0.127	129	1.630E-11	2.100E-18	33
FDH2	3.89	11.8	4.574E-11	5.380E-19	30

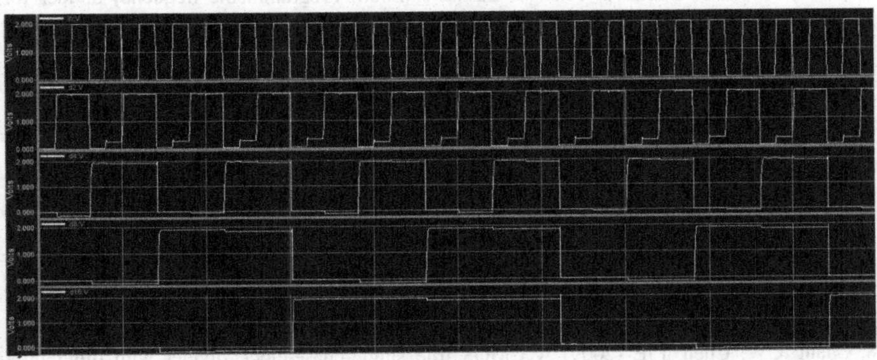

Fig. 8 Output waveforms of divide by 16 FDH1

DIVCON shows high speed among all FD designs with more area occupation. FDetspc occupies less area, but has more power consumption. But the proposed hybrid divide by 16 FD concentrates on a small area and low power applications of the PLL. Every design possesses its own advantages and disadvantages. Based on the requirement the designers may choose the suitable one. The body biasing technique is used to meet the functionality of the FD which may cause to increase the delay. Therefore, the proposed designs FDH1 and FDH2 have not shown high performance. The output waveform of FDH1 is as shown in Fig. 8.

7 Conclusion

The TSPC divider is widely used in many integrated circuits because of its simplicity in terms of implementation and clock distribution. In this paper, various divide by 16 frequency dividers which are designed by using TSPC and E-TSPC flip flops are studied. A hybrid method is also proposed to see the power and delay optimization. Among all designs, the FDH1 design with hybrid method shows lower power consumption which will be useful for high frequency PLL designs.

References

1. Rahnamaei A, Akbarimajd A, Torabi A, Vajdi B (2007) Design and optimization of ÷8/9 divider in PLL frequency synthesizer with dynamic logic (E_TSPC). In: Proceedings of the 6th WSEAS international conference on electronics, hardware, wireless and optical communications, Corfu Island, Greece, pp 46–50, 16–19 Feb 2007
2. Jung M, Fuhrmann J, Ferizi A, Fischer G, Weigel R, Ussmueller T (2012) A 10 GHz low-power multi-modulus frequency divider using extended true single-phase clock (E-TSPC) logic. In: Proceedings of the 7th European microwave integrated circuits conference, Amsterdam, The Netherlands, pp 508–511, 29–30 Oct 2012
3. Barale F, Sen P, Sarkar S, Pinel S, Laskar J (2008) Programmable frequency-divider for millimeter-wave PLL frequency synthesizers. In: Proceedings of the 38th European microwave conference, Amsterdam, The Netherlands, pp 460–463, Oct 2008
4. Lam J, Plett C (2012) Modified TSPC clock dividers for higher frequency division by 3 and lower power operation. In: IEEE conference proceedings, pp 437–440, 2012
5. Soares JN Jr, Van Noije WAM (1999) A 1.6-GHz dual modulus prescaler using the extended true-single-phase-clock CMOS circuit technique (E-TSPC). IEEE J Solid-State Circ 34(1):97–102
6. Deng W, Okada K, Matsuzawa A (2010) A 0.5-V, 0.05-to-3.2 GHz, 4.1-to-6.4 GHz LC-VCO using E-TSPC frequency divider with forward body bias for sub-picosecond-jitter clock generation. In: IEEE Asian solid-state circuits conference, Nov 8–10, 2010, Beijing, China, 3–4
7. Krishna MV, Do MA, Yeo KS, Boon CC, Lim WM (2010) Design and analysis of ultra low power true single phase clock CMOS 2/3 prescaler. In: IEEE transactions on circuits and systems I: regular papers, Vol 57, No. 1, pp 72–82, Jan 2010
8. Chang CW, Chen YJE (2009) A CMOS true single-phase-clock divider with differential outputs. IEEE Microwave Wirel Compon Lett 19(12):813–815

9. Yuan J, Svensson C (1989) High-speed CMOS circuit technique. IEEE J Solid-State Circ 24 (1):62–70
10. Huang Q, Rogenmoser R (1996) Speed optimization of edge—triggered CMOS circuits for gigahertz single-phase clocks. IEEE J Solid-State Circ 31(3):456–465
11. Guo C, Zhu S, Hu J, Diao J, Sun H, Lv X (2010) Design and optimization of dual modulus prescaler using the extended true single phase clock. In: IEEE conference ICMMT proceedings, pp 636–638, 2010

Study of Gabor Wavelet for Face Recognition Invariant to Pose and Orientation

R. Karthika and Latha Parameswaran

Abstract Gabor filters have achieved enormous success in texture analysis, feature extraction, segmentation, iris and face recognition. Face recognition is one of the most popular biometric modalities which has wide range of applications in biometric authentication. The most useful property of a Gabor filter is that it can achieve multi-resolution and multi-orientation analysis of an image. This paper presents an algorithm using Gabor wavelets in capturing discriminatory content, obtained by convolving a face image with coefficients of Gabor filter with different orientations and scales. Support vector machine (SVM) has been used to construct a robust classifier. This method has been tested with publicly available ORL dataset. This algorithm has been tested, cross-validated and the detailed results are presented. It is inferred that the proposed method offers a recognition rate (94 %) with tenfold cross-validation.

Keywords Support vector machine · Gabor wavelets

1 Introduction

Human could inherently distinguish different people based on their faces. The visual cortex processes the information to recognize the faces. Developing a model to make computer recognize human faces is challenging. The facial features were made up of different valleys and peaks. Face recognition has been commonly used

---

R. Karthika (✉)
Department of Electronics and Communication Engineering, Amrita School of Engineering, Amrita Vishwa Vidyapeetham, Coimbatore, India
e-mail: r_karthika@cb.amrita.edu

L. Parameswaran
Department of Computer Science Engineering, Amrita School of Engineering, Amrita Vishwa Vidyapeetham, Coimbatore, India
e-mail: p_latha@cb.amrita.edu

L.P. Suresh and B.K. Panigrahi (eds.), *Proceedings of the International Conference on Soft Computing Systems*, Advances in Intelligent Systems and Computing 397, DOI 10.1007/978-81-322-2671-0_48

in automatic attendance system in the classrooms, video indexing, law enforcement, airport security, and immigration.

Gabor filters [1] are a type of wavelet which provides optimized time and frequency resolution. Gabor filters are Gaussian filters modulated by a sinusoid. Its cosine part provides directional blur operator and its sine part provides directional edge operator. It is robust to changes in illumination, rotation, scale, and translation. This property helps in face recognition as human face images captured under different conditions like varying poses, illumination, and expressions. It can be tuned to operate at different scales and orientations, so that we can analyze image at different depths and angles. It acts as a tunable band-pass filter. Due to its implementation of multi-channel filter bank and tuning at varying scales and orientations, it mimics the response of human visual system.

2 Related Work

In the eigenface method [2], the face is represented using the eigenvectors instead of isolated features. Fisher faces [3] appear better in handling expressions and lighting than the eigenfaces. One-dimensional HMM [4] is used for face recognition using 2D-DCT coefficients as feature vectors. Gabor wavelets play a significant role in elastic bunch graph matching [5]. They used fiducial points and bunch graph to reduce the computational complexity. In evolutionary pursuit [6] after projecting the face image into lower dimensional space, whitening transformation is applied to reduce the mean square error. The random rotations of whitened PCA are driven by evolutionary pursuit. Gabor wavelet has been applied to face recognition using dynamic link architecture (DLA) [7]. The local feature detector, which is placed centered at one of its points, is called jet. Links are used to match the model domain within the image domain. Dynamic binding is done in the matched graph.

AdaBoosted Gabor [8] features are discriminant and not high dimensional. In the Gabor feature space, the difference between the intra-face and extra face is computed to change the multi-class problem to the two class problem. By using this technique, Gabor features are reduced. The different features [9] like best selection individual features, forward selection, floating forward search, and genetic algorithm are used to discriminate the faces. In the first three techniques, the face image is represented by face graphs and lattice. In GA, each pixel in the image is applied with convolution of Gabor wavelet. In [1] the face image is divided into small blocks and the volume local binary pattern (VLBP) is computed. Support vector machine (SVM) is investigated for different applications like face detection [10], expression recognition [11], and gender classification [12]. In [13] log-Gabor filters are analyzed for vehicle verification and Gabor filters' performance was compared with that of log-Gabor filters. They found that log-Gabor filters could represent image frequencies in a superior way. They compared [14] the different features like complex moments, Gabor energy, and grating cell operator. They claimed that the grating cell operator responds only to texture features. Fisher discriminant analysis and classification result analysis method were used to compare the features. In [15], neural network based on face recognition is presented. The neural network is based on multi-layer perception architecture with back propagation algorithm.

The organization of this paper is as follows: Sect. 2 describes the proposed algorithm. The experimental results are discussed in Sect. 3. Inferences and possible extensions are discussed in Sect. 4.

3 Proposed Algorithm

3.1 Feature Extraction Using Gabor Wavelets

Gabor wavelet [1] representation of face features shows optimal joint frequency and spatial locality. It is applied to face recognition problems because its spatial frequencies mimic the profiles of neurons in the mammalian cortical cells. Gabor wavelets are defined as follows:

$$\Phi_{u,v}(x) = \frac{\|k_{u,v}\|^2}{\sigma^2} \exp\left(\frac{-\|k_{u,v}\|^2 \|x\|^2}{2\sigma^2}\right)\left[\exp\left(ik_{u,v} \cdot x\right) - \exp\left(-\frac{\sigma^2}{2}\right)\right] \quad (1)$$

where v and u denote the scale and orientation of the Gabor filters, and the wave vector $k_{u,\ v}$ is defined as follows:

$$k_{u,v} = k_v e^{i\phi_u} \quad (2)$$

where $\Phi_u = \pi/8$, $k_v = k_{\max}/f^v$, $k_{\max}$ is the maximum frequency, and f is the Spacing factor between kernels in the frequency domain. The values for the various parameters given in Eq. (1) are as follows:

Gabor kernels are applied in eight orientations $u = \{0, 1, 2, 3, 4, 5, 6, 7\}$ and five scales $v = \{0, 1, 2, 3, 4\}$, $\sigma = 2\pi$, $k_{\max} = \pi/2$ and $f = \sqrt{2}$.

The Gabor filters defined are self-similar because they can be produced from mother wavelet by rotation and dilation. The product of complex plane wave and Gaussian envelope is called kernel.

The multiplicative factor $\frac{|k_{u,v}|^2}{\sigma^2}$ ensures the oscillatory behavior of the kernel. The factor $\exp\left(-\frac{\sigma^2}{2}\right)$ denotes that the filter is illumination insensitive.

Algorithm

- Read the input Image $I(x,y)$.
- Convolve $I(x, y)$ with Gabor kernel $\varphi(u, v)$ as defined in (1).
  $O_{u,v}(x, y) = I(x, y)^* \varphi(u, v)$. $O_{u,v}(x, y)$ is the convolution result obtained from Gabor kernel at different scales and orientations. Here * corresponds to the convolution operator.

- For $v = \{0, 1, 2, 3, 4\}$ representing the 5 scales and $u = \{0, 1, 2, 3, 4, 5, 6, 7\}$ denoting the 8 orientations, 40 sets of gabor filters are obtained.
- For each filter, mean and standard deviation is calculated, extracting totally 80 features.
- The extracted features have been fed to the SVM classifier for further analysis.

Figures 1 and 2 show the sample input image and the resultant image after convolution with Gabor kernel.

Fig. 1 Sample image

Fig. 2 Convolved output of Gabor filter with five orientations and eight scales with a sample face image shown in Fig. 1

3.2 SVM Classifier

Support vector machine (SVM) is most widely used in different disciplines because of its high accuracy and high dimensional data handling. C-SVM has been used in this work for experimentation. SVM can be implemented with different kernel functions and its parameter selection depends on the trial-and-error method.

The SVM-based kernels used in this experimentation are as follows.

Linear kernel: $K(x, y) = \langle x, y \rangle$
Polynomial kernel: $K(x, y) = [\gamma \langle x, y \rangle + b]^d$
RBF kernel: $K(x, y) = \exp(-\gamma \|x - y\|^2)$
Sigmoid kernel: $K(x, y) = \tan h(\gamma < x, y > + b)$

where γ is the smoothing parameter in the radial basis function (RBF) kernel; scaling parameter in sigmoid and polynomial kernel, b is a positive constant, and d represents the degree. The influence of kernel parameters on the decision boundary is quite significant. The resulting classifier will get altered on varying the degree of the polynomial and width of the Gaussian kernel.

4 Experimental Results

The ORL database [16] is used to evaluate the efficacy of the proposed method. The facial features extracted are implemented using MATLAB. The open source tool LibSVM is used for training and testing the SVM model.

In this experimentation, 300 images with 30 instances have been used for training. The classifier has been evaluated by cross-validation, using the number of folds.

Tables 1 and 2 give the performance of different kernels using mean, standard deviation, and combining mean and standard deviation. To carry out these analyses, 300 images and 30 instances with tenfold cross-validation have been used.

From Tables 1 and 2 it may be observed that the RBF kernel achieves more accurate results than other kernels. In the RBF kernel, the samples are nonlinearly mapped into a higher dimensional space. Moderate value of γ is preferred, since the large values of γ lead to overfitting. The decision boundary becomes linear in case of small values of γ.

Linear kernel also provides good accuracy nearer to the RBF kernel. The linear kernel is the lowest degree polynomial, which will not handle the nonlinear relationship that exists between the features. The polynomial kernel will categorize between the two classes with a noticeable margin. The minimum accuracy indicates that the sigmoid kernel is not suitable for face recognition problems.

Figure 3 shows the performance of ORL dataset with RBF kernel using different features. The good recognition rate 94 % has been achieved by combining the features mean and standard deviation.

Table 1 Performance evaluation of different kernels using mean and standard deviation

Parameter	Mean				Standard deviation			
Kernel	Linear	RBF	Polynomial	Sigmoid	Linear	RBF	Polynomial	Sigmoid
Correctly classified instances	273	278	269	10	273	276	269	10
Incorrectly classified instances	27	22	31	290	27	24	31	290
Recognition rate	91.00 %	92.66 %	89.64 %	3.33 %	91.00 %	92.00 %	90.00 %	3.33 %
Error rate	9.00 %	7.34 %	10.34 %	96.67 %	9.00 %	8.00 %	10.00 %	96.67 %

Table 2 Performance evaluation of different kernels using standard deviation and mean

Parameter	Mean and standard deviation			
Kernel	Linear	RBF	Polynomial	Sigmoid
Correctly classified instances	280	282	279	3
Incorrectly classified instances	20	18	21	87
Recognition rate	93.33 %	94.0 %	93 %	3.33 %
Error rate	6.67 %	6 %	7 %	96.67 %

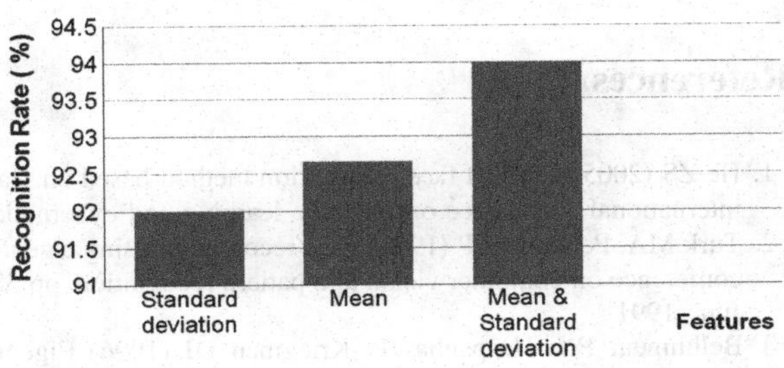

Fig. 3 Performance of ORL dataset with RBF kernel

Table 3 Face recognition performance comparisons between SVMs of different kernels

Performance	Kernel			
	Linear kernel	RBF kernel	Polynomial kernel	Sigmoid kernel
Correctly classified instances	83	84	79	3
Incorrectly classified instances	7	6	12	87
Recognition rate	92.22 %	93.33 %	87.77 %	3.33 %
Error rate	7.78 %	6.67 %	12.23 %	96.67 %

In the earlier experiment, cross-fold validation has been the strategy for classification. In the following experimentation, 210 training images and 90 testing images with 30 instances have been used. Table 3 depicts the performance of different kernels using mean and standard deviation as the features.

Table 3 shows that RBF kernel achieves more accurate results than other kernels.

5 Conclusion

In this work, an algorithm, which does face recognition based on local features, has been explored. Gabor filters exhibit fundamental invariance properties that make them applicable to face recognition invariant to pose and orientation. All facial features are obtained by convolving the input face image with different scales and orientations. The mean and standard deviation of 40 Gabor filter have been extracted as the features. Classification has been done using the SVM classifier. A good recognition rate of 94 % is achieved with RBF kernel using combined features of mean and standard deviation. This proposed technique comparatively outperforms other face recognition methods. In future the recognition rate can be improved by using more number of statistical features. A wide variety of applications prove the need for such a robust technique.

References

1. He ZS (2005) A SVM face recognition method based on Gabor-featured key points. In: 2005 international conference on machine learning and cybernetics, 2005
2. Turk MA, Pentland AP (1991) Face recognition using eigenfaces. In: Proceedings of the IEEE conference on computer vision and pattern recognition, pp 586–591, Maui, Hawaii, USA, 3–6 June 1991
3. Belhumeur PN, Hespanha JP, Kriegman DJ (1996) Eigenfaces vs. fisherfaces: recognition using class specific linear projection. In: Proceedings of the 4th European conference on computer vision, ECCV'96, pp 45–58, Cambridge, UK, 15–18 April 1996
4. Nefian AV, Hayes MH III (1998) Hidden Markov models for face recognition. In: Proceedings of the IEEE international conference on acoustics, speech, and signal processing, ICASSP'98, vol 5, pp 2721–2724, Seattle, Washington, USA, 12–15 May 1998
5. Wiskott L, Fellous J-M, Krueuger N, von der Malsburg C (1999) Face recognition by elastic bunch graph matching, Chapter 11. In: Jain LC et al (eds) Intelligent biometric techniques in fingerprint and face recognition, CRC Press, pp 355–396, 1999
6. Liu C, Wechsler H (2000) Evolutionary pursuit and its application to face recognition. IEEE Trans Pattern Anal Mach Intell 22(6):570–582
7. Lades M, Vorbruggen J, Buhmann J, Lange J, von der Malsburg C, Wurtz R, Konen W (1993) Distortion invariant object recognition in the dynamic link architecture. IEEE Trans Comput 42:300–311
8. Yang P, Shan S, Gao W, Li S, Zhang D (2004) Face recognition using Ada-boosted Gabor features. In: Proceedings of IEEE international conference on automatic face and gesture recognition, pp 356–361, 2004
9. Gokberk B, Irfanoglu M, Akarun L, Alpaydm E (2003) Optimal Gabor kernel location selection for face recognition. In: Proceedings of international conference on image processing, vol 1, pp 677–680, 2003
10. Osuna E, Freund R, Girosit F (1997) Training support vector machines: an application to face detection. In: Proceedings of IEEE computer society conference on computer vision and pattern recognition, pp 130–136, 1997
11. Ramanathan R, Nair AS, Sagar VV, Sriram N, Soman KP (2009) A support vector machines approach for efficient facial expression recognition. In: International conference on advances in recent technologies in communication and computing, 2009. ARTCom '09
12. Moghaddam B, Yang MH (2000) Gender classification with support vector machines. In: Proceedings of the fourth IEEE international conference on automatic face and gesture recognition, pp 306–311, 2000
13. Arróspide J, Salgado L (2013) Log Gabor filters for image based vehicle verification. IEEE Trans Image Process 22(6), June 2013
14. Grigorescu SE, Petkov N, Kruizinga P (2002) Comparison of texture features based on Gabor filters. IEEE Trans Image Process 11(10), Oct 2002

15. Bhuiyan AA, Liu CH (2007) On face recognition using Gabor filters. Int J Comput Inf Syst Control Eng 1(4), 2007
16. The ORL face database is available at http://www.cl.cam.ac.uk/research/dtg/attarchive/facedatabase.html
17. Bartlett MS, Movellan JR, Sejnowski TJ (2002) Face recognition by independent component analysis. IEEE Trans Neural Networks 13(6):1450–1464
18. Daugman J (1988) Complete discrete 2-D Gabor transforms by neural networks for image analysis and compression. IEEE Trans Pattern Anal Mach Intell 36:1169–1179
19. Zhao GY, Pietik¨ainen M (2007) Dynamic texture recognition using local binary patterns with an application to facial expressions. IEEE Trans Pattern Anal Mach Intell 29(6):915–928
20. Soman KP, Loganathan R, Ajay V Machine Learning with SVM and other Kernel methods, Prentice Hall of India
21. Kong A (2008) An evaluation of Gabor orientation as a feature for face recognition. In: 19th international conference on pattern recognition, 2008
22. Sharif M, Khalid A, Raza M, Mohsin S (2011) Face recognition using Gabor filters. J Appl Comput Sci Math 11

Recognizing Faces in Corrupted Images

Steven Lawrence Fernandes and G. Josemin Bala

Abstract In this paper we have developed two novel techniques. Firstly to find out if the face image is fake using fake biometric detection. Secondly recognizing images corrupted with blurring effects. Recognizing faces using fake biometric detection is based on image quality assessment technique, IIITD Look Alike face database is considered to validate the proposed system. Recognizing faces under blurring effects is based on using score level fusion algorithm, ATT database is considered to validate the proposed system.

Keywords Fake biometric detection · Image quality assessment · Face recognition

1 Introduction

Biometric details change from one person to another person; hence, they can be used in security system. Images can be corrupted by spoofing and blurring effects [1–6]. We need to identify if the input image is spoofed or if it is corrupted by blurring effects. Images can also be corrupted by blurring. In this paper, we have developed two novel techniques. Firstly to find if the face image is fake, i.e., to identify if it is spoofed using fake biometric detection. Secondly recognizing images corrupted with blurring effects. Recognizing faces using fake biometric detection is based on image quality assessment technique, IIITD Look Alike face database is considered to validate the proposed system. Recognizing faces under blurring effects is based on using score level fusion algorithm, ATT database is considered to validate the proposed system. Section 2 describes both the proposed systems; Sect. 3 presents the Result and Discussion, Sect. 4 draws the Conclusion.

S.L. Fernandes (✉) · G. Josemin Bala
Department of Electronics & Communication, Karunya University, Coimbatore, India
e-mail: steva_fernandes@yahoo.com

G. Josemin Bala
e-mail: josemin@karunya.edu

L.P. Suresh and B.K. Panigrahi (eds.), *Proceedings of the International Conference on Soft Computing Systems*, Advances in Intelligent Systems and Computing 397, DOI 10.1007/978-81-322-2671-0_49

2 Proposed System

In this paper, we have worked on two difficulties in recognizing faces through images. We have proposed two systems to work on these problems having one methodology for each problem. The proposed two methodologies are explained below.

2.1 *Fake Biometric Detection Using Image Quality Assessment*

In this paper, we are proposing a novel software-based multi-biometric using image quality assessment (IQA). It is not only capable of operating with a very good performance under different biometric systems (multi-biometric) and for diverse spoofing scenarios, but it also provides a very good level of protection against certain non-spoofing attacks (multi-attack).

2.1.1 Modules for Fake Biometric Detection Using Image Quality Assessment

- Pre-process
- Filter
- Extract the feature
- Classification

Pre-process
The pre-process step is an important one in the image processing. In that process, we remove the noise and resize the image and do some process for output.

Filter
During filtering we denoise the input image. The noise is the unwanted pixel of the image. We use Gaussian filter to remove the noise in input image. In one dimension, the Gaussian function is:

$$G(x) = \frac{1}{\sqrt{2\pi\sigma^2}} \mathrm{e}^{-\frac{x^2}{2\sigma^2}} \quad (1)$$

The Gaussian function has important properties, which are verified with respect to its integral:

$$I = \int_{-\infty}^{\infty} \exp(-x^2)\mathrm{d}x = \sqrt{\pi} \quad (2)$$

Extract the Feature
In our proposed system, we use (IQA) technique for feature extraction. Here low-pass Gaussian kernels are used to filter input grayscale image I which has a size of N × M. Gaussian kernel is applied in order to generate a smoothed version I.

Classification
We use quadratic discriminant analysis to classify if the image is original or fake.

Here Gaussian distribution is used to model the likelihood of each class. Maximum likelihood estimation is used to obtain Gaussian parameters for each class.

2.2 *Face Recognition in the Presence of Blurring Effect*

In this paper, we perform recognition of faces in the presence of blurring effect. We developed a new score level fusion algorithm as follows:

- Here we have six algorithms for identifying the blur images such as principal component analysis (PCA), fisher faces (FF), independent component analysis (ICA), fourier spectra (FS), singular value decomposition (SVD), and sparse representation (SR).
- These algorithms generate score for each train image in the database.
- From the score generated in train image, PCA, FF, and FS use Euclidean distance to find its match with test image.
- SVD uses Frobenius norm to find its match with test image.
- ICA and SR use normalized correlation to find its match with test image.
- We have six algorithms and M images in the database, hence we will have six vectors of size 1 × M.
- Next all the scores of 1 × M vectors are normalized to find their mean and standard deviation value.
- Then all the normalized scores are summed.
- At last, maximum final score is recognized as the perfect match.

The score level fusion algorithm is implemented on i.MX6 board, which is an embedded kit used in real-time applications. Features of i.MX6 are as follows:

- Implementing it on i.MX6 development board.
- CPU: ARM Cortex-A9 core, 1 GHz processor.
- 1 GB DDR3 RAM.
- Storage: 4 GB flash.
- It supports operating system (OS) such as Linux 3.0.35, Android 4.3 JB, and Windows Embedded Compact 7.
- It supports python language, OpenCV.

3 Result and Discussion

Fake biometric detection is done based on image quality assessment by considering IIITD Look Alike face database.

3.1 *Fake Biometric Detection Using Image Quality Assessment*

The steps to be followed when we run the main code are:

- A main window will open first when we press load image, we have to load the input image. After loading convert the input image to gray scale.
- Next press the resize, the grayscale image will be resized.
- Then press the filter for removing the noise and for other artifacts here we use the Gaussian filter.
- Next step is the identification process, for that press next step.
- Now the output graphical user interface (GUI) window will open; in this first we have to find the image quality by pressing the IQA button and will get 11 image quality types.
- Now press the Identify button to know whether the input image is fake or original.
- Finally, press the Analysis button to know the existing and proposed system accuracies and confusion matrix.

Step-by-step implementation of fake biometric detection using image quality assessment is shown in Fig. 1.

A comparative study of fake biometric detection using image quality assessment on IIITD Look Alike face database is given in Table 1.

Table 1 indicates the proposed recognizing faces across fake biometric detection using image quality assessment technique which gives the best recognition rate on IIITD Look Alike face database.

3.2 *Face Recognition in the Presence of Blurring Effect*

Recognizing blur faces is tested using ATT face database which consists of blur images. This system has two folders: they are—Train Images and Test Images. The Train Images folder consists of one image, which is not affected by blur and the Test Images folder consists of all blur images from the ATT face database.

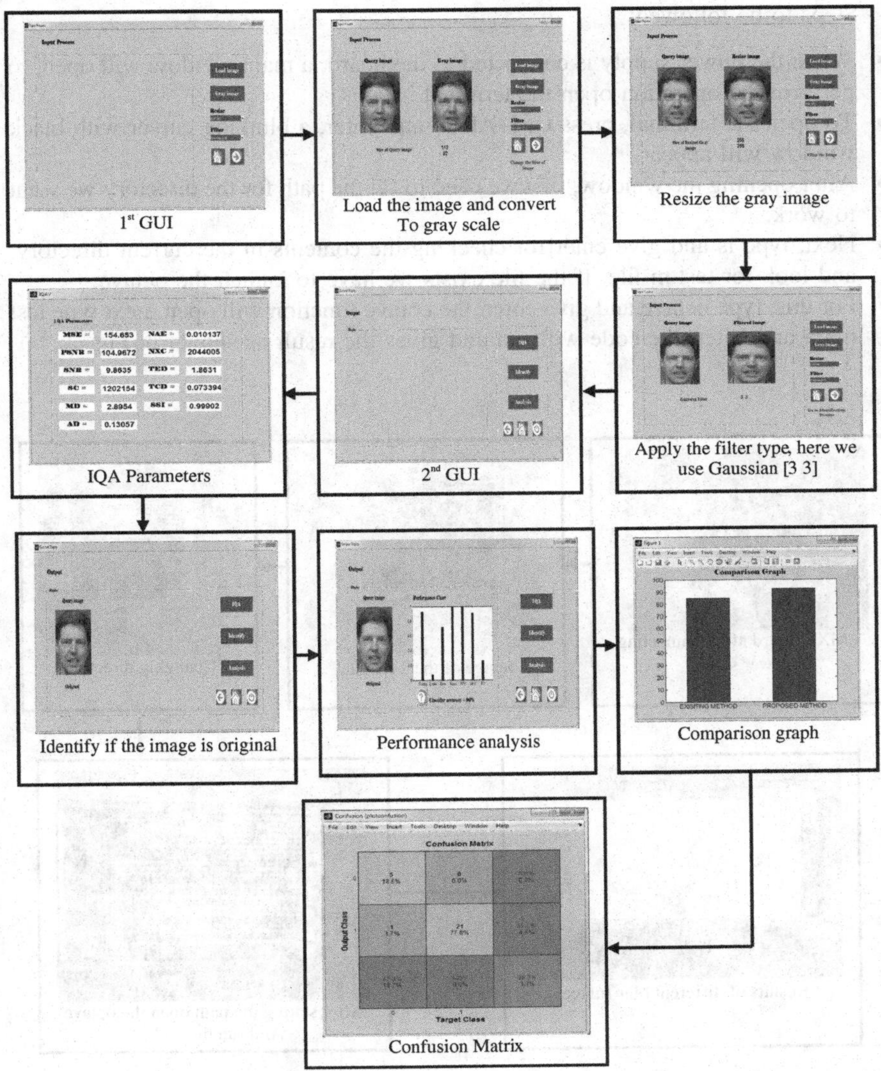

Fig. 1 Fake biometric detection using image quality assessment

Table 1 Fake biometric detection rate

Author	Method	Database	Recognition set (%)
Galbally et al. [7]	Feature level fusion	IIITD look alike	76
Gaud et al. [8]	PCA and ICA	IIITD look alike	80
Steven L. Fernandes	IQA (proposed system)	IIITD look alike	97

Steps to be followed:

- When the power supply is connected to the board, a main window will open, to perform the operation open the terminal.
- To open the terminal, press Ctrl+Alt+T and enter, a blinking cursor with black window will appear.
- After opening the window, first we need to set the path for the directory we want to work.
- Next, type is and give enter for checking the contents in the current directory, and look for test.m file. If the file exists we have to launch the octave.
- For this, type octave and give enter, the octave function will open, next type test main and enter, the code will run and gives the result as shown in Fig. 2.

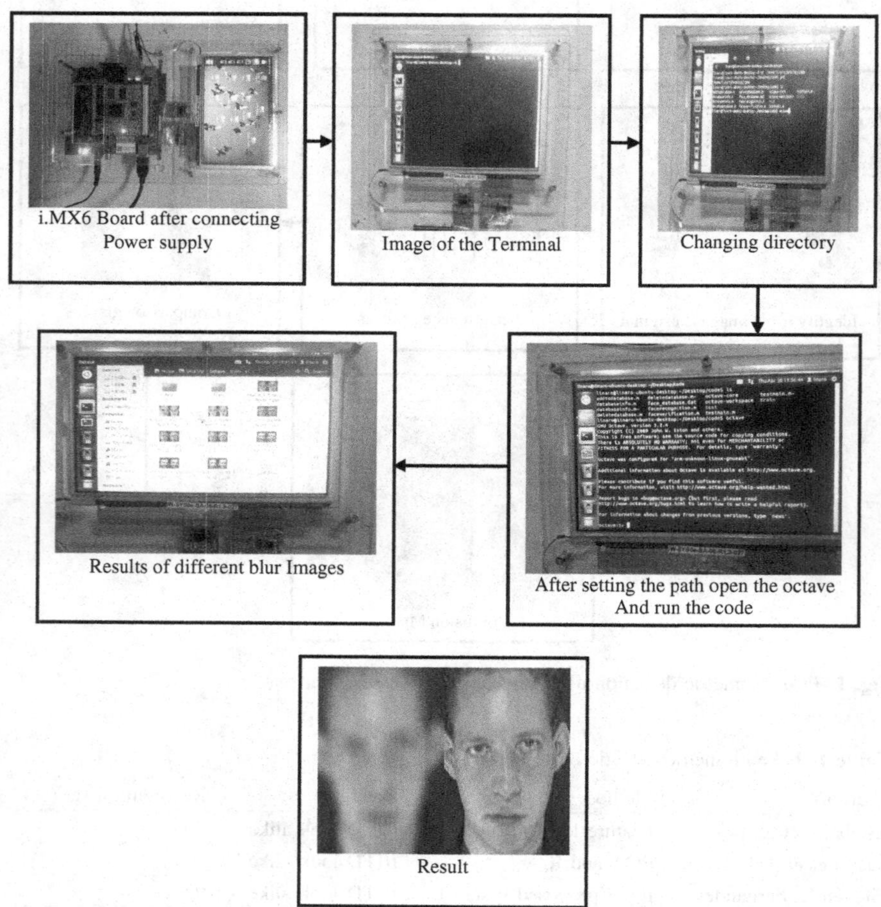

Fig. 2 Face recognition in the presence of blurring effect

Table 2 Face recognition rate of images corrupted by blurring effects

Author	Method	Database	Recognition set (%)
Bala et al. [9]	SVD and ICA	ATT	92
Amith et al. [10]	PCA	ATT	76
Steven L. Fernandes	Score level fusion (proposed system)	ATT	98

Step-by-step implementation for recognizing blurred faces on i.MX6 board is shown in Fig. 2.

A comparative study of various face recognition techniques under blurring effect on ATT database is given in Table 2. It indicates that the proposed score level fusion gives the best recognition rate on ATT face database which is corrupted by blurring effects.

4 Conclusion

In this paper, we have developed two novel techniques. Firstly to find out if the face image is fake using fake biometric detection. Secondly recognizing images corrupted with blurring effects. Recognizing faces using fake biometric detection is based on image quality assessment technique, IIITD Look Alike face database is considered to validate the proposed system. Recognizing faces under blurring effects is based on using score level fusion algorithm, ATT database is considered to validate the proposed system. From our analysis we have found that the proposed fake biometric detection using image quality assessment technique gives the best recognition rate of 97 % on IIITD Look Alike face database and recognizing faces under blur using the score level fusion algorithm gives the best recognition rate of 98 % on an ATT face database.

Acknowledgments The proposed work was made possible because of the grant provided by the vision group on science and technology (VGST), Department of Information Technology, Biotechnology and Science & Technology, Government of Karnataka, Grant No. VGST/SMYSR/GRD-402/2014-15 and the support provided by the Department of Electronics and Communication Engineering, Karunya University, Coimbatore, Tamil Nadu, India.

References

1. Li X, Da F (2009) Robust 3D face recognition based on rejection and adaptive region selection. In: ACCV'09 proceedings of the 9th asian conference on computer vision. VPart III, pp 581–590
2. Popat H (2012) Determining normal and abnormal lip shapes during movement for use as a surgical outcome measure. School of Dentistry, Cardiff University
3. Verma T, CSIT Durg D, Sahu RK (2013) PCA-LDA based face recognition system & results comparison by various classification techniques. In: IEEE international conference on green high performance computing (ICGHPC)

4. Chen H-K, Lee Y-C, Chen C-H (2013) Gabor feature based classification using enhance two-direction variation of 2DPCA discriminant analysis for face verification. In: IEEE international symposium on next-generation electronics (ISNE)
5. Fernandes SL, Bala GJ (2013) A comparative study on ICA and LPP based face recognition under varying illuminations and facial expressions. Int Conf Signal Process Image Process Pattern Recogn (ICSIPR)
6. Maodong S, Jiangtao C, Ping L (2012) Independent component analysis for face recognition based on two dimension symmetrical image matrix. In: 24th Chinese control and decision conference (CCDC)
7. Galbally J, Marcel S, Fierrez J (2013) Image quality assessment for fake biometric detection: application to iris, fingerprint, and face recognition. Image Process IEEE Trans Biometrics Compendium 23:710–724
8. Gaud JV, Bohra SU (2015) Review on fake iris detection method using image quality assessment. Int J Curr Eng Technol 5(1):352–354
9. Fernandes SL, Bala GJ, Nagabhushan P (2013) Robust face recognition in the presence of noises and blurring effects by fusing appearance based techniques and sparse representation. In: Proceedings of the 2013 2nd international conference on advanced computing, networking and security, pp 84–89
10. Amith GK, Kumar S (2015) A novel approach for blurred and noisy face image recognition. Int J Sci Res (IJSR) 4:1325–1329

A Crowdsourcing-Based Platform for Better Governance

Vishal Chandrasekaran, Shivnesh V. Rajan, Romil Kumar Vasani, Anirudh Menon, P. Bagavathi Sivakumar and C. Shunmuga Velayutham

Abstract The world's population has been increasing as every year passes by, and Governments across the world face a stupendous challenge of governing each country. These challenges include providing proper sanitation facilities, efficient disaster management techniques, effective resource allocation and management, etc. Crowdsourcing methodologies, which empower the common man to provide valuable information for better decision making, have gained prominence recently to tackle several challenges faced by several governments. In this paper, we introduce a crowdsourcing-based platform that makes use of information provided by the common man for better governance. We illustrate how this platform can be used in several instances to attend to the problems faced by people.

Keywords Crowdsourcing · Smart environments and applications · Mobile phone · Location-based services

V. Chandrasekaran (✉) · S.V. Rajan · R.K. Vasani · A. Menon ·
P. Bagavathi Sivakumar · C. Shunmuga Velayutham
Department of Computer Science and Engineering, Amrita School of Engineering,
Amrita Vishwa Vidyapeetham (University), Coimbatore, India
e-mail: c.vishal1993@gmail.com

S.V. Rajan
e-mail: shivneshr@live.com

R.K. Vasani
e-mail: romil.vasani93@gmail.com

A. Menon
e-mail: anirudhmenon763@gmail.com

P. Bagavathi Sivakumar
e-mail: pbsk@cb.amrita.edu

C. Shunmuga Velayutham
e-mail: cs_velayutham@cb.amrita.edu

L.P. Suresh and B.K. Panigrahi (eds.), *Proceedings of the International Conference on Soft Computing Systems*, Advances in Intelligent Systems and Computing 397, DOI 10.1007/978-81-322-2671-0_50

1 Introduction

Over the years, mobile phone based applications and location-based services have become a medium for users to provide valuable information for decision making, for spreading awareness, etc. OpenStreetMap [1] is one such application, wherein users spread across different geographical locations can contribute toward building an accurate online map of the world, by creating maps of their locality and incorporating it onto the world map. Airbnb [2] is another application that provides a platform for people across the world to rent their apartments to tourists. OpenStreetMap and Airbnb are just two examples out of several location-based services available in the market. The purpose of these applications is to empower the common man with a platform that helps him reach out to several other people and spread any valuable information he might possess.

In today's fast paced lifestyle, we see that almost every individual carries a mobile phone with him/her. Thus, there is immense opportunity for several organizations, including the Governments, to make use of information provided by the people to provide better service to them. The information provided by the people can be reporting a crime at a location, reporting an epidemic outbreak, reporting public infrastructure problems in a locality, etc. Ushahidi, a crisis mapping platform, was established in 2007 and deployed in Kenya [3] and Haiti [4] deploys crowdsourcing methodologies to provide an up-to-date, openly available crisis map. In January 2008, Kenya faced a massive outbreak of physical violence during the presidential elections' period. At that time, Ushahidi was set up in Kenya to continuously log the violence cases in several parts of Kenya and report it to the public, so that they could take the necessary precautions to be safe. In January 2010, a 7.0 magnitude earthquake struck Haiti and a vast population of Haiti were affected very badly. Traditional disaster response methodologies failed to succeed due to the massive destruction of property and land, rescue workers found it hard to identify the places where people were stuck and where supplies were necessary. Ushahidi provided a way to share critical information coming directly from the Haitians. Reports about trapped people, medical emergencies, and specific needs, such as food, water, etc., were received and plotted on a map, and published online. Thus, the emergency response teams were able to assist and speed up the rescue process. Ushahidi has been very effective in mapping national crisis and helping the people in need. One other way of getting information from the people is by making use of the social media. Twitter, a prominent social media platform, was used by the government officials of the State of New York when superstorm Sandy and winter blizzard Nemo struck New York in October 2012 and February 2013, respectively [5], to convey storm-related information and gather information from the public who needed help. The above platforms find their use during national emergencies. There are several other scenarios [6, 7] where crowdsourcing plays a crucial role for Governments to reach out to the public, collect information from them, and serve them better. However, with the increasing population across the world, a platform is needed for the people to report public issues to the government authorities on a daily basis. This would improve the way a government functions right from the lower-level administration to the top-level management. The government would know the issues faced by the public and can take the necessary steps to rectify the problems, thus improving the quality of life in their country.

In several countries across the world, the reported grievances are not yet digitized. Digitization of the grievances reported by the public would ease the process of solving the problems faced by the public. In this paper, we demonstrate such a system that helps people to report their day-to-day grievances, such as uncleared public trash cans, open potholes, faulty street lamps, lack of medical facilities in a locality, etc. The primary contributions of this work are:

- To provide a platform for the public to report day-to-day grievances on-the-go and not rely on phone calls or any other mode of reporting issues.
- To make sure that the right people in the administration department is notified of the issues and an efficient tracking of the reported grievances is also made possible.
- To digitize the grievances reported and improve the service offered by the government officials.

The next section mentions details about the system architecture and the modules in the system and how the proposed system serves as a platform for the public to report grievances.

2 The Grievance Reporting System

In this section, we describe the set of mandatory inputs required by the system to report a grievance and later describe how the system works with the input provided.

2.1 A Broad Classification of the Grievances to Be Reported

To ensure a formal classification of grievances reported, users have to classify their grievances into one of the following categories:

- Infrastructure Grievance

A grievance is deemed as an infrastructure grievance if it is related to the damage of public buildings, government establishments, etc.

- Medical Grievance

A grievance is deemed as a medical grievance if it involves providing information related to a spread of a disease in an area, lack of medical facilities in an area (no hospitals, lack of medicines, and lack of qualified medical personnel in the area), etc.

- Public Grievance

A grievance is deemed as a public grievance if it affects a majority of the people in the locality. For example, lack of traffic sign boards and traffic lights in a densely populated and busy area, uncleared trash cans in a locality, etc.

The above classification also serves as a method for identifying the type of problem that majorly prevails in an area. It must be noted that, the above classification can be made flexible depending on the application and that more classifiers can be included in the future. However, we demonstrate our public grievance addressal platform based on this classification.

2.2 *Reporting a Grievance*

To report a grievance on-the-go, the users can use a mobile application. Sending a report requires that the users provide the following mandatory information:

- Type of grievance

As previously explained, the users should classify the grievance according to the provided options.

- Location of the grievance

The location of a grievance can be manually provided by the user or it can be taken automatically using a global positioning system, available in majority of the mobile handsets available in the market today. In [8, 9], the various methodologies for obtaining the location from a mobile phone have been discussed. It has been proved that the accurate location information is obtained using the global positioning system and thus, the proposed crowdsourcing platform deploys the same for obtaining the location information.

- Description

Represents the observation made by the user, for example, uncleared trash cans in a locality.

2.3 *The Process of Pruning Invalid Grievance Reports*

The platform crowdsources information about various grievances in a given place. It is known that, there is a possibility of receiving erroneous reports of grievances. In such a case, tending to this grievance might prove to be costly for the authorities responding to the report. In order to validate a reported grievance, the following mechanism is used:

1. View the grievances in the user's locality.
   The users are provided a visual interface, wherein they can view the grievances reported in their locality.
2. Vote for a grievance.
   The user is given the opportunity to vote for a particular grievance. A user is expected to "upvote" a particular grievance, if he finds that the reported grievance is valid, or he can "downvote" a particular grievance, if he finds out that the reported grievance is erroneous.
3. Pruning the erroneous reports.
   The difference in the number of upvotes and downvotes for a particular reported grievance is calculated, and a threshold value is used to remove erroneous reports.

The above mechanism stresses the fact that a system built for the people requires some amount of genuine participation from their side, to make sure that the authorities are informed about genuine grievances in a locality. Apart from ensuring that the genuine grievances are sent to the authorities, this pruning mechanism also improves the confidence of several organizations and governments in this platform. This would be crucial for obtaining an increased support from governments, Non-Governmental Organizations, etc., to respond to the grievances reported.

In summary, the following steps will be used to obtain a grievance and report it to the authorities:

- Obtain grievance specifications from the user.
- Initiate a voting mechanism to validate the issue.
- The valid grievances are reported to the authorities.

2.4 The Polling Mechanism

It is known that there can be multiple ways of resolving a particular problem. For example, suppose in a region, an increase in the number of road accidents is reported. The traffic police department can either install more traffic signals in the area or deploy more policemen, etc. The public who reported the grievance might want a particular solution to be put into action. In order to facilitate this, the proposed crowdsourcing platform deploys a polling mechanism for the concerned organizations to poll the users for obtaining the most preferred solution for the reported grievance. The polling mechanism also eases the decision-making process for the concerned organizations because they would exactly know what the public wants. Further, it ensures better decision making as it ensures an increased public participation.

3 Implementation and Results

In this section, we describe the implementation of the above system implementing the features mentioned along with the pruning mechanism. We developed an application programming interface (API) using Node.js. An Android application was then developed to implement the above features of the grievance reporting system. In the testing phase, the application was deployed in a suburb of Coimbatore (India) and data was collected to understand the type of grievances in this region of Coimbatore.

Figure 1a depicts the various grievances reported by the users in the testing region and Fig. 1b depicts the percentage of each type of grievance reported. The green colored markers denote the location of the grievance and when clicked, the user can view the description and the nature of the grievance along with the upvotes and downvotes received for the particular grievance. In general, when this application is deployed over a larger region, then the regions within the state where maximum grievances are reported can be identified. Subsequently, the local authorities of the region can be notified to take the necessary steps to tackle the grievances reported and improve the quality of living in these regions. Figure 1b can be used to effectively analyze the type of issues prevailing in a region.

•

3.1 *Applications of This System*

The digitization of the public grievances makes sure that the data reported by the public is made available to the public, which includes Non-Governmental Organizations and the government officials. Thus, it enables ease of access of data which will aid in better functioning of the organizations and industries related to the grievances reported by the public. This will ensure that some credible action is taken to resolve the grievances reported. The following are some of the industries where the proposed crowdsourcing applications can be effectively used. It is to be noted that the proposed system is not restricted to the below mentioned industries, and can be used in several other industries where inputs from public can lead to better decision making.

Healthcare Industry
Recently, Governments and organizations across the world are taking measures to fight deadly diseases like Ebola and Swine Flu. The proposed system can be used as a platform by medical personnel, to report individual cases of people diagnosed with the disease. As the number of reported incidents increases, the corresponding State Health Department and National Health Department can analyze the places of origin of these incidents and the pockets where they prevail the most. Subsequently, measures can be taken to assist the people in these localities and preventive measures could also be taken to make sure that the disease does not spread to other places. Further, these organizations can analyze the healthcare infrastructure across

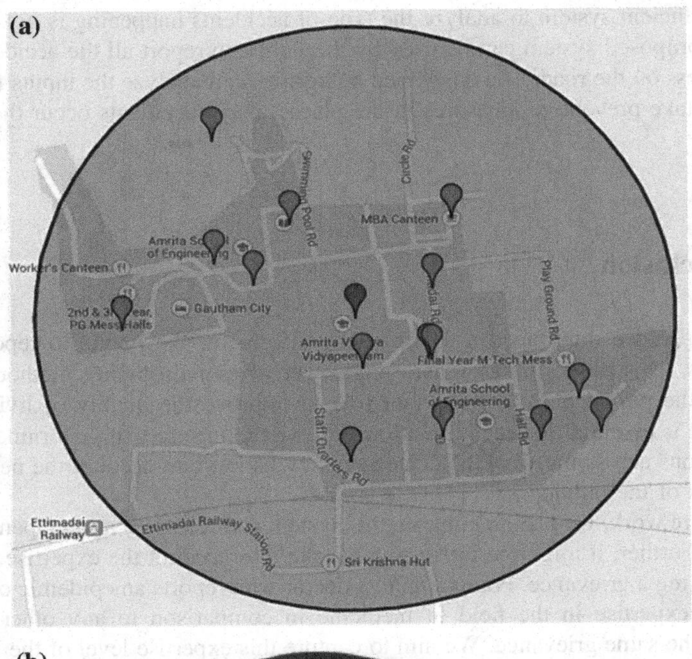

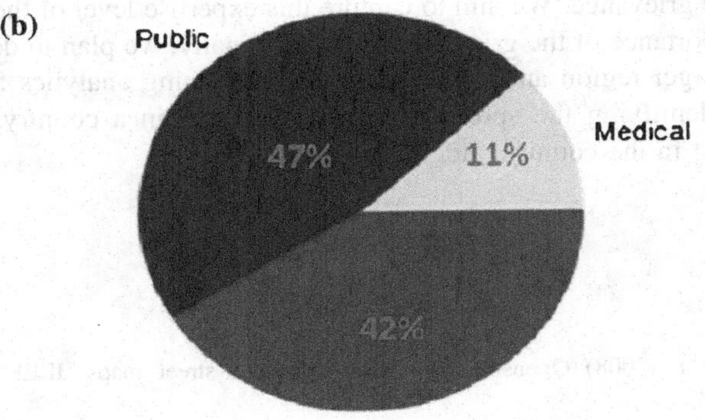

Fig. 1 **a** Grievances reported in the test region. **b** Percentage of different types of grievances reported

the country and accordingly allocate funds and set up establishments to improve the healthcare facilities in the regions which lack basic facilities to tackle such epidemic outbreak.

Road Traffic Department and Transportation Departments
Road rage on the national highways has been a matter of concern in recent years. Although, the traffic departments and transportation departments have been continuously working hard to reduce the number of road accidents occurring every

year, an efficient system to analyze the type of accidents happening is not in place yet. The proposed system can be used by the public to report all the accidents that they witness on the road. The concerned authorities can analyze the inputs from the public to take preventive measures in the places where accidents occur the most.

4 Conclusion

In this paper, we introduced a transparent platform for the public to report their grievances. This platform showcases the power of crowdsourcing methodologies and how the public play an important role to improve the quality of living of a country. It is essential for such a platform to exist to ensure that governments and organizations across the world allocate funds by taking into account the needs and grievances of the public.

As future work, we plan to enhance this system and release it as an Open Source platform. Further, it might be interesting to take into account the expertise of each user reporting a grievance. For example, a doctor who reports an epidemic outbreak has more expertise in the field of medicine in comparison to any other person reporting the same grievance. We aim to capture this expertise level of the user for improving the importance of the grievance reported. Finally, we plan to deploy the platform over a larger region and collect data for performing analytics for applications such as identifying the spread of several diseases in a country, type of accidents occurring in the country, etc.

References

1. Haklay M, Weber P (2008) Openstreetmap: user-generated street maps. IEEE Pervasive Comput 7(4):12–18
2. Airbnb, Inc. http://www.airbnb.com
3. Meier P, Brodock K (2008) Crisis mapping Kenya's election violence. Harvard humanitarian initiative (HHI)
4. Heinzelman J, Waters C (2010) Crowdsourcing crisis information in disaster-affected Haiti. US Institute of Peace
5. Barnes MD, Hanson CL, Novilla LM, Meacham AT, McIntyre E, Erickson BC (2008) Analysis of media agenda setting during and after Hurricane Katrina: implications for emergency preparedness, disaster response, and disaster policy. Am J Public Health 98(4):604
6. Kaigo M (2012) Social media usage during disasters and social capital: twitter and the Great East Japan earthquake. Keio Commun Rev 34:19–35
7. Degrossi LC, de Albuquerque JP, Fava MC, Mendiondo EM (2014) Flood citizen observatory: a crowdsourcing-based approach for flood risk management in Brazil. In: Proceedings of the 26th international conference on software engineering and knowledge, pp 1–3

8. Alt F, Shirazi AS, Schmidt A, Kramer U, Nawaz Z (2010) Location-based crowdsourcing: extending crowdsourcing to the real world. In: Proceedings of the 6th nordic conference on human-computer inter-action: extending boundaries. ACM, pp 13–22
9. Von Watzdorf S, Michahelles F (2010) Accuracy of positioning data on smartphones. In: Proceedings of the 3rd international workshop on location and the web. ACM

Design of Ripple Carry Adder Using GDI Logic

Shoba Mohan and Nakkeeran Rangaswamy

Abstract Full adders are essential building modules in applications such as digital signal processor (DSP) architectures and microprocessors. In this paper, 8 bit ripple carry adder (RCA) using gate diffusion input (GDI) logic is presented. Simulation program with integrated circuit emphasis (SPICE) simulations are carried out using 250 nm technology parameters with a power supply voltage of 5 V. Simulation results indicate that the RCA consumes 69 % lesser than complementary metal oxide semiconductor (CMOS) based, whereas the transistor count is also reduced 63 % than with CMOS. The proposed design has the best PDP in comparison with existing designs found in the literature. The simulation results also confirm that GDI-based design gives better performance than standard CMOS-based design.

Keywords Low power · GDI logic · Full adder · Digital design

1 Introduction

The increase in number of portable applications with minimal amount of power on hand demands the researchers to invent a design with small area, low power and high throughput. Addition is a basic and widely used arithmetic operation in a digital circuit design [1]. It impacts the system performance as a whole. It acts as nucleus of many other arithmetic operations such as subtraction, multiplication and division. The adder lies at critical path in most of the digital systems [2]. So enhancing the performance of an adder enhances the system as a whole.

S. Mohan (✉) · N. Rangaswamy
Department of Electronics Engineering, School of Engineering and Technology, Pondicherry University, Puducherry 605014, India
e-mail: shobamalar@gmail.com

N. Rangaswamy
e-mail: nakkeeranpu@gmail.com

L.P. Suresh and B.K. Panigrahi (eds.), *Proceedings of the International Conference on Soft Computing Systems*, Advances in Intelligent Systems and Computing 397, DOI 10.1007/978-81-322-2671-0_51

Several designs are reported to realize addition functions using different circuit techniques and approaches [3–8]. The various static logic styles reported in the literature are namely complementary metal-oxide semiconductor (CMOS), pass transistor logic (PTL) and transmission function adder (TFA). They differ in circuit topologies and arrangement of transistor to improve the circuit performance.

The standard implementation of full adder (FA) uses complementary metal-oxide semiconductor logic [9]. It uses 28 transistors and consists of both pull-up and pull-down transistor. This logic needs an inverter at the output to produce normal output from the intermediate complementary signal. The input capacitance is large due to increase in transistor count. The lesser number of transistor realization is possible using pass transistor logic as explained in [10]. This logic has a drawback of threshold voltage problem, which can be mitigated using restoring circuit or keeper transistor. The additional circuit possesses increase in transistor and the presence of inverter in the restoration circuit increases the power consumption considerably [11]. This problem is alleviated using a variant of pass transistor logic, the complementary PTL called CPL, but the minimum requirement is generation of complementary signal which increases circuit overhead [12].

The full adder is designed using transmission gate which consists of PMOS and NMOS connected in parallel and they are controlled by their gate inputs [13]. It is suitable for designing XOR/XNOR (exclusive OR/exclusive NOR) gates with less number of transistors and its suffered by lack of driving capability when connected in series. The full adder designs based on CMOS, PTL and TFA are given in Fig. 1.

To overcome the drawbacks of existing logic in terms of power consumption and transistor count, the gate diffusion input (GDI) logic has evolved. It offers the design implementation with fewer transistors but it also suffers by threshold voltage drop at the output. However, the threshold voltage problem is overcome by the use of buffer at the output. The paper is organized as follows. In Sect. 2, introduction about the GDI logic is discussed. Full adder and ripple carry adder (RCA) based on the GDI logic is discussed in Sect. 3. The simulation setup and results are described in Sect. 4. Conclusion is drawn in Sect. 5.

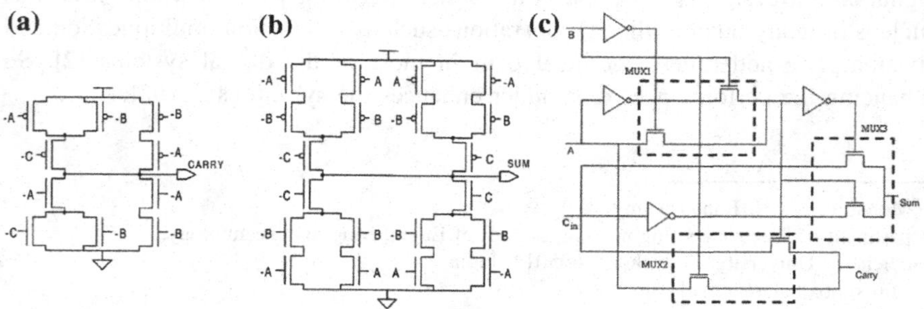

Fig. 1 **a** CMOS-based adder, **b** PTL adder and, **c** TFA

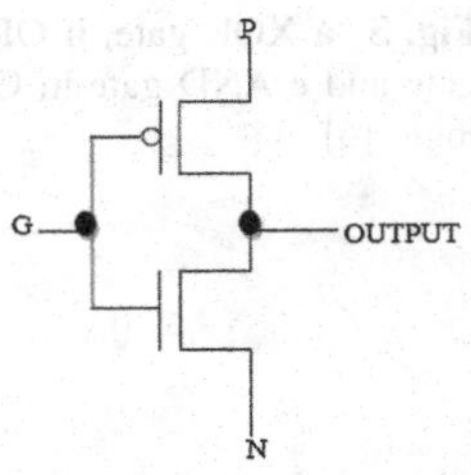

Fig. 2 Basic GDI cell [1]

Table 1 Different logic functions using GDI cell

N	P	G	Out	Logic function
0	B	A	A′B	F1
B	1	A	A′ + B	F2
1	B	A	A + B	OR
B	O	A	AB	AND
C	B	A	A′B + AC	MUX
0	1	A	A′	NOT

2 GDI Logic

GDI is a low-power design technique which facilitates the implementation of digital logic function using basic GDI cell. The basic GDI cell is shown in Fig. 2.

It looks like a CMOS inverter at first glance but it is not so. It contains three inputs G, P and N. G is a common gate input for both PMOS and NMOS transistor, P is a source/drain diffusion input of PMOS transistor and N is a source/drain diffusion input of NMOS transistor. The basic cell can be used for realizing various logic functions and is listed in Table 1.

Any complex function can be easily implemented using GDI cell with less number of transistors. The designs of AND, OR and XOR gates are shown in Fig. 3.

3 Implementation of Adder in GDI Logic

3.1 Full Adder Design

A full adder is a combinational circuit which adds three inputs *A*, *B* and *Cin* and produces two outputs *Sum* and *Carry*. The equations for *Sum* and *Carry* are given below [8]

$$SUM = A\,XOR\,B\,XOR\,Cin \quad (1)$$

$$CARRY = (\overline{A\,XOR\,B})Cin + (A\,XOR\,B)A \quad (2)$$

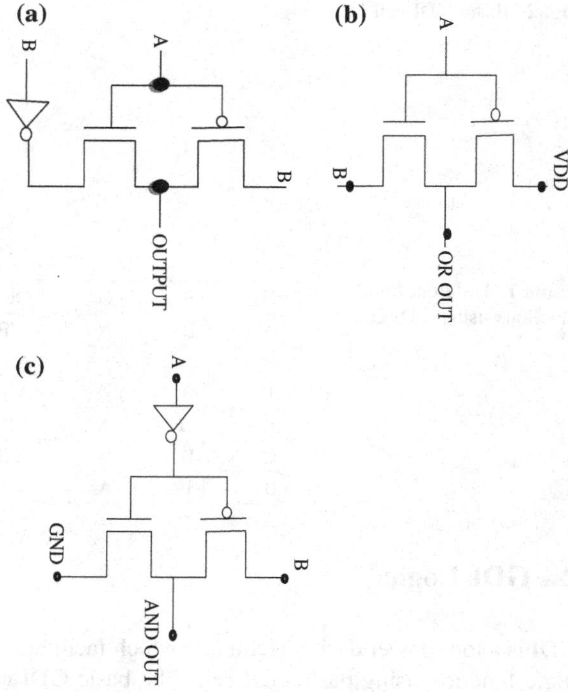

Fig. 3 **a** XOR gate, **b** OR gate and **c** AND gate in GDI logic [6]

XOR implementation can be made easily using GDI logic since it requires only four transistors compared with conventional logical types such as CMOS, PTL and TFA-based design. The full adder design can be realized using only 10T with the help of GDI logic. In the full adder design, there is no direct use of supply rail either by V_{DD} or by GND, therefore the short circuit power consumption can be minimized. The full adder design is shown in Fig. 4.

Sum output is obtained by the XOR operation of inputs *A*, *B* and *Cin*. The carry output is obtained by multiplexing the inputs *A* and *Cin* with select input as XOR output of *A* and *B*. The GDI logic allows easy implementation of multiplexer (MUX) operation with only two transistors thereby reducing the circuit complexity.

3.2 RCA Design

As an example, an 8 bit RCA is designed to validate whether the GDI logic is suitable for higher bit design. The advantage of RCA lies in its simple structure and delay increases with increase in the number of input bits. It can be found that it is a suitable candidate in the case of non-real-time application where delay is not an important issue. An 8 bit RCA design is shown in Fig. 5.

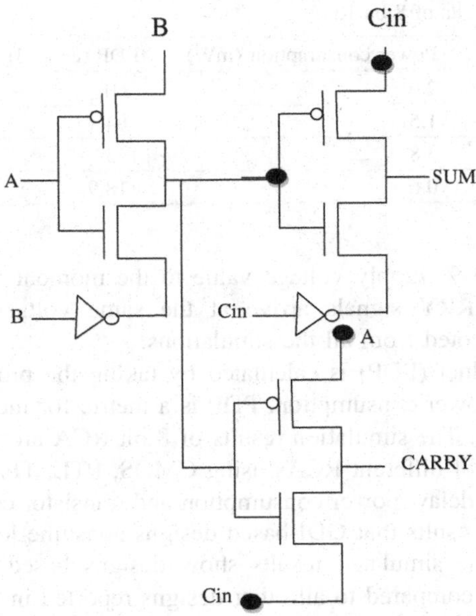

Fig. 4 FA based on GDI logic

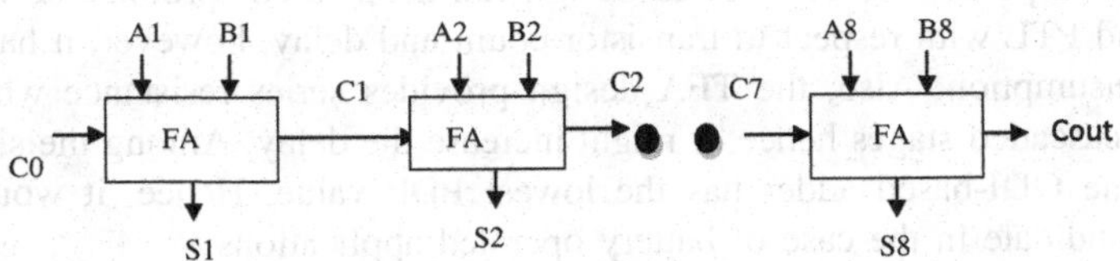

Fig. 5 RCA adder architecture

4 Simulation Results and Comparison

The simulations are performed by the SPICE tool with 0.25 μm CMOS technology. Each input is driven by buffered signals and each output is loaded with buffers to reflect the real-time simulation environment [14]. Four different RCA designs are simulated. All the circuits are simulated with a supply voltage of 5 V. The channel length of both PMOS and NMOS transistor is chosen as 0.25 μm.

Power calculation is done using detailed circuit simulation with the frequency of 125 MHz. The circuit reacts differently with different input sequences. So the circuits are simulated with all 256 input combinations. During a simulation time a single power measurement is taken by averaging the instantaneous power over total period of the time for a given input signal. The delay is measured from the moment

Table 2 Simulation results of 8 bit RCA

Logic	Delay (ns)	Power consumption (mW)	PDP (e−12 J)	Transistor count
CMOS	30	2.0	60	224
PTL	40	1.5	64	80
TFA	35	3.8	133	144
GDI	31	0.6	18.9	80

the input reaches 50 % supply voltage value to the moment the latest arrival of either SUM or CARRY signals arrive at the same voltage value [15]. The worst-case delay is noted from all the simulations.

Power delay product (PDP) is calculated by taking the product of worst-case delay and average power consumption. PDP is a metric for measuring the performance of the system. The simulation results of 8 bit RCA are given in Table 2.

The performance of different RCAs using CMOS, PTL, TFA and GDI logic is analysed in terms of delay, power consumption and transistor count. It is observed from the simulation results that GDI-based designs consume low power and have low PDP values. The simulated results show designs based on the GDI logic consume low power compared to all other designs reported in the literature.

The GDI-based adder requires 10 transistors for a single bit adder whereas in CMOS-based designs it needs 28 transistors for the same operation. Though PTL uses the same number of transistor-like GDI designs, it lacks in terms of delay and power consumption. The TFA-based design can act as a compromise between the CMOS and PTL with respect to transistor count and delay, however, it has higher power consumption. Also, the TFA design provides series resistance when connected in cascaded stages hence, it might increase the delay. Among the simulated designs, the GDI-based adder has the lowest PDP value. Hence, it would be a suitable candidate in the case of battery operated applications.

5 Conclusion

In this paper, a GDI-based adder is designed. The GDI-based adder along with other existing adder circuits are simulated in the SPICE tool. The comparison is carried out in terms of power, delay, transistor count and PDP for circuit optimization. The GDI-based adder shows lesser PDP and transistor count compared to other designs for consideration. Hence, the designs based on GDI may be suitable at low voltage arithmetic systems.

References

1. Morgenshtein A, Shwartz I, Fish A (2010) Gate diffusion input (GDI) logic in standard CMOS nanoscale process. In: Proceedings of IEEE convention of electrical and electronics engineers in Israel, pp 776–780
2. Morgenshtein A, Fish I, Wagner A (2002) Gate-diffusion input (GDI)—a power-efficient method for digital combinatorial circuits. IEEE Trans VLSI Syst 10(5):566–581
3. Goel S, Kumar A, Bayoumi MA (2006) Design of robust, energy-efficient full adders for deep-sub micrometer design using hybrid-CMOS logic style. IEEE Trans Very Large Scale Integr (VLSI) Syst 14(12)
4. Morgenshtein A, Shwartz I, Fish A (2014) Full swing gate diffusion input (GDI) logic—case study for low power CLA adder design. Integr VLSI J 47(1):62–70
5. Foroutan V, Reza M, Taher I, Navi K, Azizi Mazreah A (2014) Design of two low-power full adder cells using GDI structure and hybrid CMOS logic style. Integr VLSI J 4(1):48–61
6. Uma R, Dhavachelvan P (2012) Modified gate diffusion input technique: a new technique for enhancing performance in full adder circuits. Proc ICCCS 6:74–81
7. Ramana Murthy G, Senthil Pari C, Velraj Kumar P, Lim TS (2012) A new 6-T multiplexer based full adder for low power and leakage current optimisation. IEICE Electron Express 9 (17):1434–1441
8. Lee PM, Hsu CH, Hung YH (2007) Novel 10-T full adders realized by GDI structure. Proc. IEEE Int Symp Integr Circ
9. Chaddha KK, Chandel R (2010) Design and analysis of a modified low power CMOS full adder using gate-diffusion input technique. J Low Power Electron 6(4):482–490
10. Rabaey JM, Chandrakasan A, Nikolic B (2003) Digital integrated circuits: a design perspective, 2nd edn. Prentice-Hall
11. Weste NHE, Harris DM (2010) Integrated circuit design, 4th edn. Pearson Education
12. Shams AM, Darwish TK, Bayoumi MA (2002) Performance analysis of low-power 1-bit CMOS full adder cells. IEEE Trans Very Large Scale Integr (VLSI) Syst 10(1):20–29
13. Uming KO, Poras T, Balsara WL (1995) Low power design techniques of high performance CMOS adders. IEEE Trans Very Large Scale Integr (VLSI) Syst 3(2):327–333
14. Zimmermann R, Fichtner W (1997) Low-power logic styles: CMOS versus pass-transistor logic. IEEE J Solid-State Circuits 32(7):1079–1090
15. Vesterbacka M (1999) 14-transistor CMOS full-adder with full voltage swing nodes. In: Proceedings of the IEEE workshop on signal processing systems, pp 713–722

Brain-Actuated Wireless Mobile Robot Control Through an Adaptive Human–Machine Interface

L. Ramya Stephygraph and N. Arunkumar

Abstract Electroencephalogram (EEG) signal generated by the brain's spontaneous recording of electrical activity is often utilized in diagnosis for brain disorders and also employed in rehabilitation devices compared to other biosignals. This work proposes an EEG-based wireless mobile robot control using brain–computer interface (BCI) for people with motor disabilities can interact with robotic systems. An experimental model of mobile robot is explored and it can be controlled by human eye blink strength and attention level. A closed neuro-feedback loop is used to control different mental fatigue. Here, the EEG signals are acquired from neurosky mindwave sensor (single channel prototype) and features of these signals are extracted by adopting discrete wavelet transform (DWT) to enhance signal resolution. Preprocessed signals are impart to robot module, different movements are detected such as right, left, forward, backward, stop positioned on eye blink strength. The proposed adaptive human–machine interface system provides better accuracy and navigates the mobile robot based on user command, so it can be adaptable for disabled people.

Keywords Brain–computer interface (BCI) · Electroencephalography (EEG) · Neurosky mindwave sensor · Wavelet transform · Eye blink detection · Brain-actuated mobile robot

1 Introduction

Progress in the field of medical science empowers for disabled people to exercise their will power. Most of these paralyzed people have great straggle to communicate with physical devices. Preferment in cognitive neuroscience and brain

L.R. Stephygraph (✉) · N. Arunkumar
Department of Electronics and Instrumentation, Sastra University, Thanjavur, India
e-mail: stephyloganathan92@gmail.com

N. Arunkumar
e-mail: arunnura@gmail.com

L.P. Suresh and B.K. Panigrahi (eds.), *Proceedings of the International Conference on Soft Computing Systems*, Advances in Intelligent Systems and Computing 397, DOI 10.1007/978-81-322-2671-0_52

computer technologies has evoked to provide ability to directly interact with the user brain. This can be done using BCI technique which facilitates direct communication between human brain and physical device. Use of sensors can monitor the physical processes corresponds to various mental fatigue [1]. By employing BCI, the users absolutely betray their mental activity in place of using motor movements to accomplish brain signals that can be used to control robot and also for other communication devices.

At present, a brain–machine interface registers bioelectrical activity of brain with electrodes. Many people in the world suffer from mobility impairments are incapable of doing activities for their daily life [2]. Especially, people getting motor neuron disease (MND) or paralyzed by amyotrophic lateral sclerosis (ALS) can use advanced brain–robot interface with computer being an effectual communication device. There are various techniques and paradigms in the implementation of brain–computer interface. It is a system based on mental activity that allows controlling external devices which are independent of peripheral nerves and muscle [3–5]. It plays a crucial role and provides a new form of direct communication and control that assist interaction tool for people with severe motor disorders, to increase the integration into the society. There are two types of BCI: Invasive—implanted directly into the brain cortex and have the highest quality signals (requiring surgery) and Noninvasive—medical scanning devices or sensors are mounted on scalp which read brain signals. This method is less intrusive and safest when compared to others [6]. Hence, automatic recognition systems exist; it is although assured to develop an alternative interface which can be used for communicating with autonomous systems. Brain–computer interfaces (BCI) have been initiated to endeavor many real-time challenges. It is a natural way of providing interaction with paralyzed person with the outside world [7]. Figure 1 shows the overview of BCI system. Furthermore, people need advanced technology to implement their task even they lose all their muscle contraction [8].

The objective of this work is to implement a brain-controlled mobile robot in a noninvasive manner. Here, two mental activities are exploited, sensing of eye blink strength and attention using wireless sensor (single channel). These activities are analogized with five movements of robot, respectively: forward, backward, right,

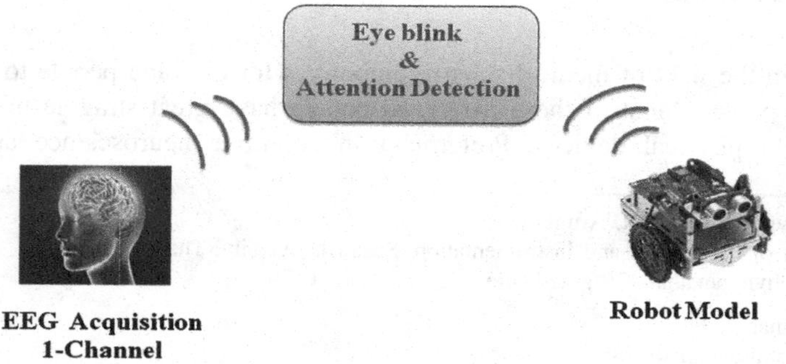

Fig. 1 Overview of BCI

left, and stop. Wavelet transform is enforced to elevate the signal strength, features, and also for denoising to get better accuracy. The main intention is to enable paralyzed people to impart with their environment [9].

2 Methods and Materials

2.1 EEG Signal Acquisitions

Electroencephalogram (EEG) has been shown to reflect the state of electrical activity of brain and can be viewed in either the time domain or the frequency domain. The EEG signal is the most predictable and reliable physiological indicators to measure the level of alertness. EEG-based technology has become more popular in various control applications. It is generally described in terms of its frequency band. The amplitude of acquired EEG is in the range of 50–100μv. Data are collected from neuroSky mindwave sensor (single channel). Figure 3 shows the user's different mental fatigue.

2.2 NeuroSky Mindwave Sensor

In this paper, we use dry EEG mindset called "NeuroSky Mindwave" sensor (see Fig. 2). Numerous circumstances of user's relaxing state, meditative state, and attentive state and during eye blink artifacts are identified from recording of EEG signal. ThinkGear is the technology which present inside in every neuroSky product that enables a device to interact with user brainwave from the serial port to an open network socket [10, 11]. This module contains the onboard chip that process the data and it is used to filter the electrical noise. Both raw brainwaves and the esense (attention, meditation, and eye blink) values are calculated on the TGAM chip.

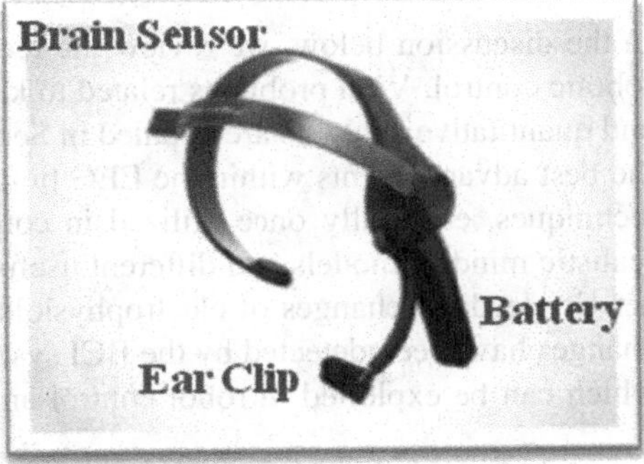

Fig. 2 NeuroSky mindwave Sensor

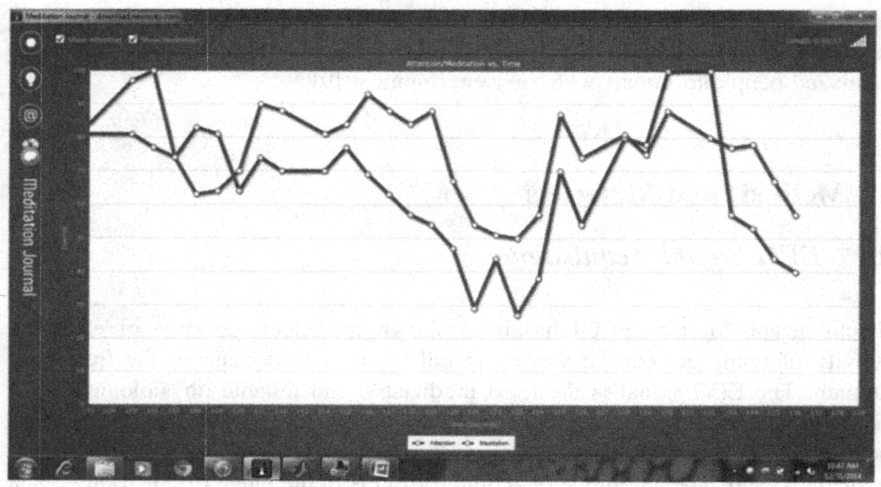

Fig. 3 Sensor status indication (*red* indicates attention and *blue* eye blink signal)

This unsigned one byte value represents the intensity of the user's most recent eye blinks which is used to control the robot. Its value ranges from 0 to 256 and operating frequency 0.05–0.5 Hz. The function of mindset, (Fig. 2), wireless headset extracts the raw EEG data from the scalp in a noninvasive manner (without the need of gel) and transmits it via RF transmitter to the processing unit. This device composed of headset, an ear clip, and a sensor arm. The specification of the device is given as follows. The weight of the sensor is 90 g, optimized frequency ranges from 2.42 to 2.47 GHz. It operates at maximum power of 50 mw and the signals are sampled at a rate of 512 Hz. esense rate (NeuroSky proprietary algorithm for representing the mental states) is 1 Hz and 576000 baud is the baud rate of UART [12].

2.3 BCI Implementation

In the discussion below, we review the recent advances in the field of EEG-based robotic control. Vital problems related to knowledge acquisition, signal processing, and quantitative analyses are detailed in Sect. 3. The discussion made by us, may be the best advancements within the EEG field within the development of localization techniques, especially once utilized in concert with high-density EEG recording, realistic mindset models and different useful neuroimaging techniques [13]. Mental activity leads to changes of electrophysiological signals like EEG and EOG. Such changes have been detected by the BCI system and transform it into a control signal which can be exploited in robot control and other applications [9] (Fig. 3).

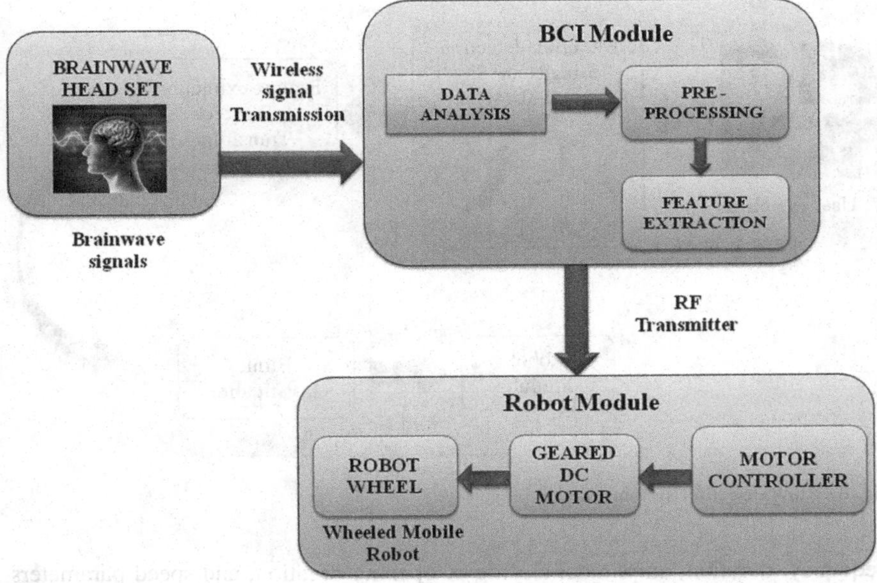

Fig. 4 Block diagram of brain-controlled system

The general idea of BCI is to convert human brain patterns into real commands (thought, motor imagery, eye blink, attention, and meditation. Figure 4 shows the system architecture of the overall process which is carried out in three steps: EEG acquisition, signal processing, and design of robot module. Initially, raw EEG signals are extracted from the mindset which debugging the information and send to the processing unit wirelessly. In the PC, the features are extracted using wavelet, an algorithm is created and set of commands also implemented to direct and control the robot module. These commands will be transmit to robot module through motor controller and geared DC motor. The main function of motor controller is to accelerate the speed of robot, when the processor Intel core i7 receives signal, the DC motor will start toward the specific direction.

3 Preprocessing of Acquired EEG

3.1 Eye Blinks Detection

Eye blink is considered as an eminent component in most of the applications such as human–machine interaction, mobile interface, and health care safety. Human eye blink detection is an important technique for disable to interact with robotic systems. It is a well-known indicator of arousal and cognitive state [14]. Indeed, their

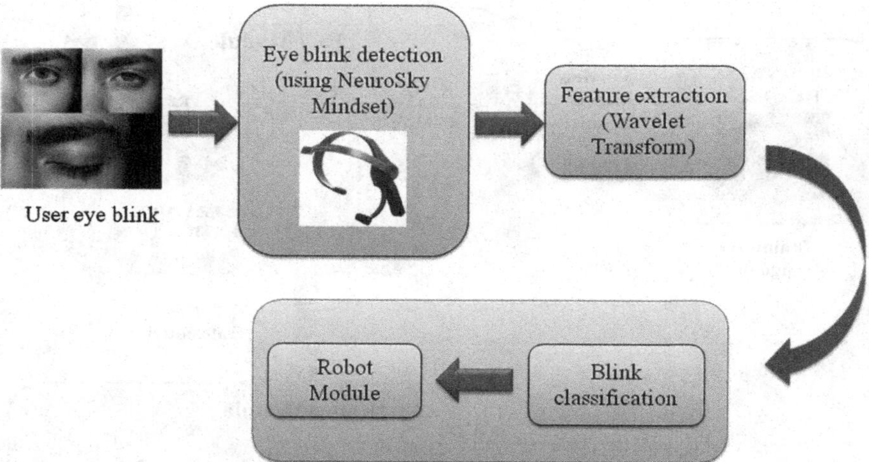

Fig. 5 Blink detection system

frequency, duration, amplitude, closing or opening duration, and speed parameters are subject to fluctuations depending on the operator's mental fatigue. All those parameters are the measures of eye blink characteristics [15] which can be performed using EEG signal. In order to detect and characterize the eye blinks using EEG signal, several processing steps are need to be performed. In this paper, a new method to extract and characterize blink activity from EEG signals usually used to monitor cerebral activity which can control the direction of the mobile robot. The mindset sensor can detect the blink count, blink strength and it can send information wirelessly. Figure 5 shows the blink detection system in which the user' blink status is detected by neurosky headset; it will sense the signal and send it wirelessly to the processing unit. Feature extraction, denoising, and signal reconstruction have been done in the BCI module with the help of wavelet packet decomposition.

3.2 Feature Extraction Using Wavelet Transform

To make the developed BCI system, most effective and have better enforcement, signal features can accordingly recognize different activities of brain. Since we are interested in detecting eye blink rate and attention level of user based on EEG. The raw signal is collected from the neurosky mindset. These signals have some electrical distortion; so we use wavelet transform to extract the features in order to elevate the accuracy [16]. The international standard 10–20 electrode system consists of 16 channels such as FP2, T4, T6, FP1, F7, T3, F3, F4, C3, C4, P3, P4, O2, F8, O1, T5 placed on its corresponding location. Instead of using these electrodes system, here we propose a new technology of using mindset (NeuroSky Technology) to obtain the

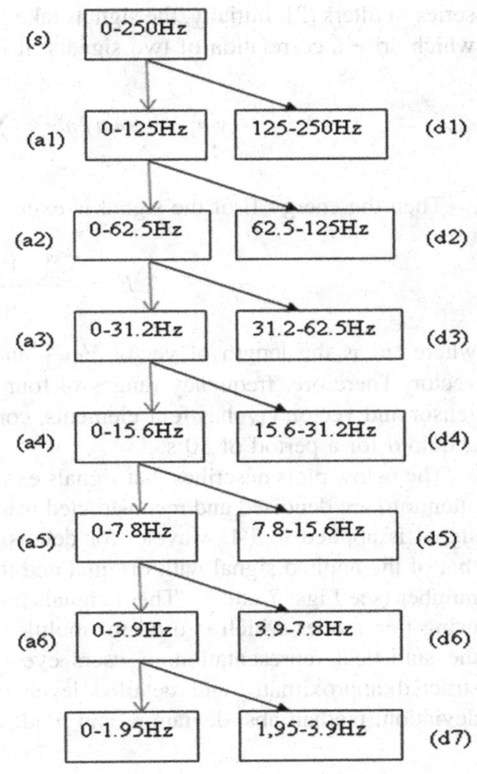

Fig. 6 Tree decomposition of frequency ranges using DWT; (*d5*) alpha wave (8–13 Hz); (*d4*) beta wave (13–30 Hz); (*a6*) delta wave (0–4 Hz); (*d6*) theta wave (4–8 Hz)

EEG signal. This single channel prototype is associated with 10–20 electrodes system.

In 10–20 electrode system, the function of channel A1 correlated with sensor ear clip, FP1 is associated with brain sensor and T4 with headset reference point as shown in (Fig. 6). In this work, tree decomposition of 7 level decibel (db) wavelet is chosen for preprocessing. Wavelet transform is suitable for both time and frequency domains. It decomposes a signal into a set of basis function called *wavelets*. These are obtained from a single channel prototype called *mother Wavelet.* Wavelet transform can splits up the signal from bunch of signal, and can represent the same signal correlate with different frequencies. Signal acquired from the single channel sensor are decomposed and reconstruct using discrete wavelet transform (DWT). This preprocessing can be performed in order to attain the various brain signal frequencies. Here, tree decomposition is depleted by adopting six levels of db to proportionately indicate the frequency ranges of four brain signals [16]. Figure 6 shows the wavelet tree of different bands. In tree decomposition, each level of output node assigns half the frequency range of corresponding input node; hence intensity of frequency can be doubly increased [2]. where '*a*' and '*d*' symbolize approximation and detailed coefficients. The DWT of signal x is computed using a series of filters [2]. Initially, the signals take place in low-pass filter with impulse 'h' which arise a correlation of two signals. It is expressed as

$$y[n] = (z * h)[n] = \sum_{k=-\infty}^{\infty} z[k]h[n-k] \tag{1}$$

Then the energy E of the signal is expressed as

$$E = \frac{\sum_{i=1}^{n} V_{\mathrm{DWT}} .j^2}{n} \tag{2}$$

where 'n' is the length of vector V_{DWT} and 'i' is the element of corresponding vector. Therefore, frequency ranges of four bands correlate to the single channel sensor and vector V_{FC} has four elements, concord with different eye blink strength acquired for a period of 10 s.

The below plots describes that signals extracted from mind sensor (eye blink and attention) are denoised and reconstructed using wavelet toolbox. The acquired EEG signal is applied to 1-D wavelet for denoising which produces effective result as that of the applied signal with original and threshold coefficients based on its level number (see Figs. 7 and 8). These signals are processed with db wavelet of 7 level using tree modes which is used for multilevel decomposition. Figure 9a, b shows the statistical representation of users eye blink which includes original, reconstructed approximate, and detailed levels with mean, median, and mode, std. deviation, median abs. deviation, and mode abs. deviation.

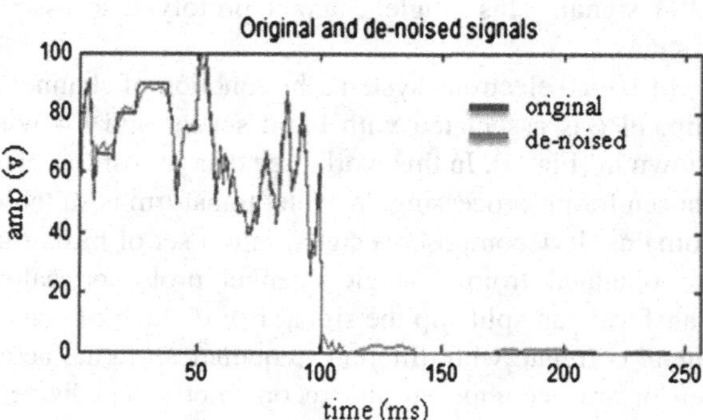

Fig. 7 Original and denoised signal

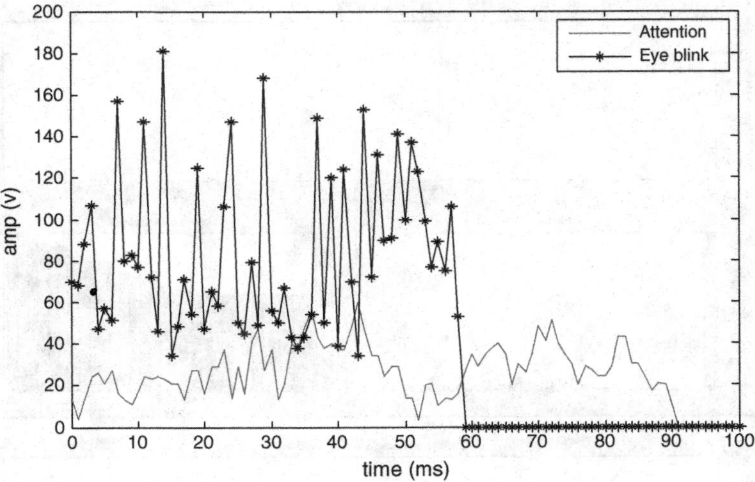

Fig. 8 Preprocessed eye blink signal

4 Experimental View of Mobile Robot

4.1 Evaluation of Mobile Robot

To evaluate the prospective approach, the advanced BCI is used to execute the wireless mobile robot (3-wheeled robot) named "Phidget21" (see Fig. 10). This mobile robot has been chosen for our work which is available in the robotics laboratory. It works under MATLAB platform to follow commands based on user's different states of mental activities. Phidget21 robot module consists of Phidget Accelerometer, Encoder, Interface kit and Motor control. The translational velocity, V for each moment can be expressed as

$$V = V_{\max} + \sum_{i=1}^{N} \omega_i \theta_i \tag{3}$$

where θ_i denotes the angle position of robot, choosing of ω_i values corresponding to the dynamics of mobile robot, and fidelity of sensors.

4.2 Controlling a Mobile Robot Using EEG-Based BCI

The EEG signal acquired from the user is processed in the processing unit. Preprocessing is executed under MATLAB environment which encompass acquisition of EEG and conditioning, preprocessing and sending the commands using serial communication port. The controlled signals are send to wheeled mobile robot

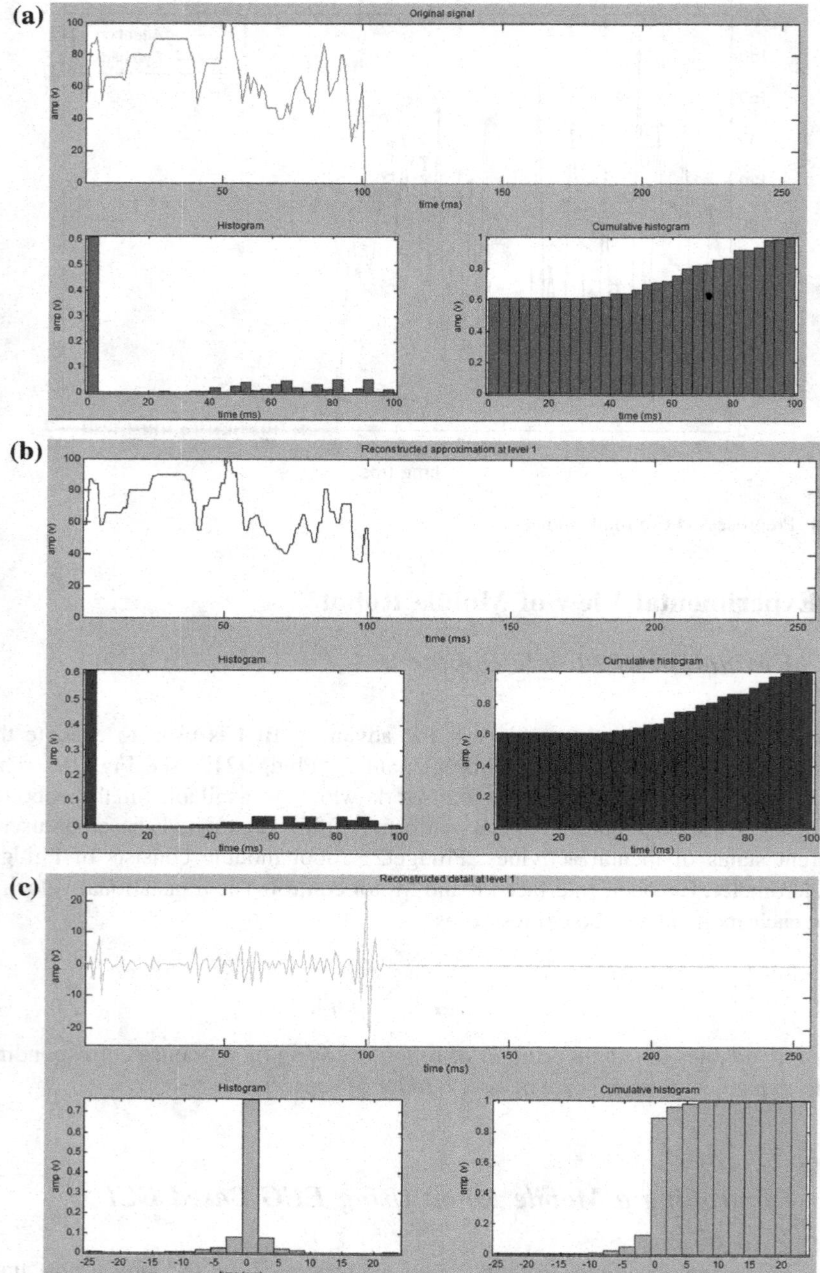

Fig. 9 Histogram of original signal. **a** Histogram of reconstructed approximate signal. **b** Histogram of reconstructed detailed signal

Fig. 10 Experimental setup

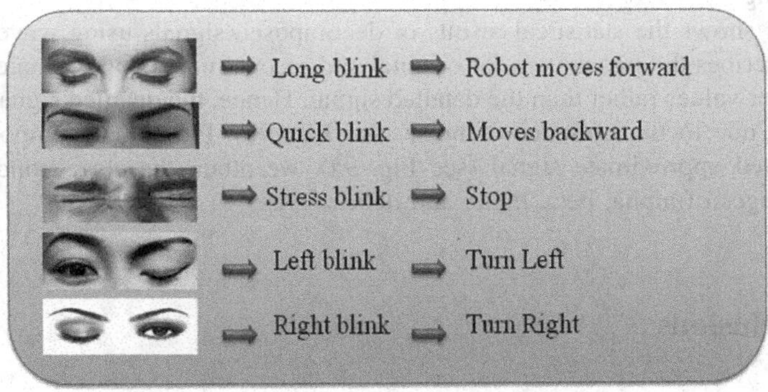

Fig. 11 Commands for mobile robot

(WMR) through geared DC motor. The calibration is carried out in two steps. First, the user need to wear the mindset in which the sensor arm should placed his/her forehead in the left side. Thus, the user can make sure that his/her brain waves are interfaced with processing unit can able to activate the robot. In this section, we can read the value of raw EEG signal with the optimized frequency of 512 Hz.

The value of signal and time are written to array data. Data stored in array will be compared with threshold points given by the user [17–19]. Here, the MATLAB section waits for three consecutive blink with threshold value greater than 50 and less than 100 in order to send the activation signal for robot (the motor will start and move forward). After three consecutive blink, the BCI system will scan for a left blink and right blink to turn the robot left and right, respectively (See Fig. 11). If the threshold value is less than 50, the robot will stop. Note that the movement of robot depends on users blink rate, as well as the attention level can maintain the position of robot in the suitable direction until it receives the next command given by user. This process is repeated for different movements of robot based on the algorithm and loop function will be performed for 30 s. Both threshold and statistical

Table 1 Statistical values

Parameters	Original signal	Reconstructed approximation level	Reconstructed detail level
Mean	26.8	26.5	−0.0003261
Mode	1.667	1.654	0.4598
Mean abs. deviation	32.67	32.54	1.104
Std. deviation	35.15	35.01	2.991
Range	100	100	49.33

implementations (see Figs. 8 and 9) are evaluated based on the different mental activities performed by the user. Successful commands are detected and wrong eye blinks are eliminated based on the threshold value set by the user and signal conditioning.

Table 1 shows the statistical results of decomposed signals using wavelet toolbox. It describes the parameters for original and reconstructed approximated signal have similar values rather than the detailed signal. Hence, the detailed signal will be eliminated due to the presence of noise (see Fig. 9b). Further, decomposing the reconstructed approximate signal (see Fig. 9a), we attain the corresponding frequency ranges of alpha, beta, theta, and delta waves.

5 Conclusions

This work describes a new method for detecting and characterizing eye blinks recorded on the scalp through the use of neurosky technology (mindset) based on EEG. The aim of this work is to implement a robot module that can assist people with disabilities, especially people affected by motor neuron disease (MND) can able to withstand of their own. The result demonstrated that mental activities records for the specific users generated by the mindset can be used to interact with robots and other applications. This can greatly improve the BCI recognition rate and well control the direction of robot module to reach the destination based on the user's eye blink strength and attention level. This method allows for a practical mental state monitoring with the use of single channel prototype as a recording modality of user mental fatigue, since blink activity can be monitored along with cerebral activity, without need for other devices and feature extraction is performed using wavelet transform to elevate the accuracy. Future scope is to navigate the humanoid robot system using eye blink and meditation signals with great accuracy.

Acknowledgments We like to express our heartiest thanks to Dr. S. Jayalalitha, Associate Dean, Dr. K. Ram Kumar Associate Prof., S. Rakesh Kumar, Asst. Prof. of EIE Dept., and V. Venkatraman, Dept. of mathematics for their valuable advices and support.

References

1. Gandhi V, Prasad G, Coyle D, Behera L, McGinnity TM (2014) EEG based mobile robot control through an adaptive brain-robot interface. IEEE Trans Syst Cyberne 44(9)
2. Barbosa AOG, Achanccaray DR, Meggiolaro MA (2010) Activation of a mobile robot through a brain computer interface. IEEE international conference on robotics and automation (2010) 3–8
3. Ubeda A, Ianez E, Jose, Azorin M, Jose, Sabater M, Fernandez E (2013) Classification method for BCIs based on the correlation of EEG maps. Neurocomputing 114:98–106
4. Bi L, Fan X-A, Liu Y (2013) EEG Based brain-controlled mobile robots: a survey. IEEE Trans Hum-Mach Syst 43(2)
5. Rebsamen B, Guan C, Zhang H, Wang C, Teo C, MH Ang, Burdet E (2010) A brain controlled wheelchair to navigate in familiar environments. IEEE Trans Neural Syst Rehabil Eng 18(6):590–598
6. Millan JR, Renkens F, Mourino J (2004) Noninvasive brain-actuated control of a mobile robot by human EEG. IEEE Trans Biomed Eng 51(6)
7. Diez PF, Torres Muller SM, Mut VA, Laciar E, Avila E, Bastos-Filho TF, Sarcinelli-Filho M (2013) Commanding a robotic wheelchair with a high-frequency steady-state visual evoked potential based brain computer interface. Med Eng Phys 35:1155–1164
8. Lopes AC, Pires G, Nunes U (2013) Assisted navigation for a brain-actuated intelligent wheelchair. Rob Auton Syst 61:245–258
9. Lee P-L, Chang H-C, Hsieh T-Y, Deng H-T, Sun C-W (2012) A brain-wave actuated small robot car using ensemble empirical mode decomposition based approach. IEEE Trans Syst Man, Cybern 42(5)
10. Vourvopoulos A, Liarokapis F (2014) Evaluation of commercial brain-computer interfaces in real and virtual world environment: a pilot study. Comput Electr Eng 40:714–729
11. Ting W, Guo-zheng Y, Bang-hua Y, Hong S (2008) EEG feature extraction based on wavelet packet decomposition for brain computer interface. Measurements 41:615–625
12. NeuroSky related information's. www.neurosky.com
13. Salami KMJE (2011) EEG signal classification for real-time brain-computer interface applications: a review. In: Proceedings of the 4th international conference mechatronics, pp 1–7, 2011
14. Raphaelle N (2014) Roy, Sylvie Charbonnier, Stephane Bonnet: eye blink characterization from frontal EEG electrodes using source separation and pattern recognition algorithms. Biomed Signal Process Control 14:256–264
15. Borghetti D, Bruni A, Fabbrini M, Murri L, Sartucci F (2007) A low-cost interface for control of computer function by means of eye movements. Comput Biol Med 37:1765–1770
16. Jahankhani P, Kodogiannis V, Revett K (2006) EEG signal classification using wavelet feature extraction and neural networks. In: IEEE international symposium on modern computing, pp 120–124, 2006
17. Norani NAM, Mansor W, Khuan LY (2010) A review of signal processing in brain computer interface system. Proc Conf Rec IEEE/EMBS Conf Biomed Sci 2010:443–449
18. Chae Y, Jeong J, Jo S (2012) Toward brain-actuated humanoid robots: asynchronous direct control using an EEG-based BCI. In: IEEE Transactions on Robotics (2012)
19. Sarac M, Koyas E, Erdogan A, Cetin M, Patoglu V (2013) Brain computer interface based robotic rehabilitation with online modification of task speed. IEEE international conference on rehabilitation robotics (2013)
20. Ubeda A, Ianez E, Azorin JM (2013) Shared control architecture based on RFID to control a robot arm using a spontaneous brain machine interface. Rob Aut Syst 61:768–774

Analysis of Speed Estimation for Sensorless Induction Motor Using Model-Based Artificial Intelligent Estimation Techniques

D. Rajalakshmi, V. Gomathi, K. Ramkumar and G. Balasubramanian

Abstract In this paper, speed tracking capability of model reference adaptive system (MRAS) with model-based flux/speed observers and artificial neural network (ANN)-based adaptive speed estimators for sensorless induction motor (IM) drives has been analyzed. In model-based technique, mathematical model of IM is used to estimate the rotor speed. The current and flux observers are used as the reference model to estimate the rotor flux. The estimated rotor flux signals are used as the input signal for the adaptive observer to estimate the speed. In ANN-based method, adaptive model is constructed with a feedforward neural network to estimate the rotor speed. Feedforward ANN algorithm is used to train the network. The training algorithm decides the learning speed, stability, and dynamic performance of the system. Both methods have good speed tracking capability. Simulation results are presented to know the accuracy of the proposed methods. The proposed speed estimation techniques have great potential in industrial applications.

Keywords Induction motor · Flux/current observer · Model-based method · ANN-based method · Speed estimation · Sensorless control

1 Introduction

Induction machines are widely employed in industrial applications owing to their simplicity, cost, and operation on wider speed ranges. In order to provide the load requirements, speed control of IM is very important. Scalar control of IM provides satisfactory performance under steady-state conditions. Vector control provides better dynamic performance. To improve the performance of induction motor drives field-oriented control (FOC) of IM is widely used. Various speed control methods are presented [1]. FOC method requires the IM state variables, i.e., electromagnetic torque, rotor speed, or stator currents. But in some IM drive applications like

D. Rajalakshmi (✉) · V. Gomathi · K. Ramkumar · G. Balasubramanian
Department of Electronics and Instrumentation, Sastra University, Thanjavur, India
e-mail: drajalakshmi92@gmail.com

L.P. Suresh and B.K. Panigrahi (eds.), *Proceedings of the International Conference on Soft Computing Systems*, Advances in Intelligent Systems and Computing 397, DOI 10.1007/978-81-322-2671-0_53

undermining the speed sensors in IM drives reduces the robustness of the whole drive. In order to provide highly robust IM drives, many researchers are going to the sensorless control of IM drives [8]. Sensorless IM drives provide reduced hardware complexity, lower cost, better noise immunity, and less maintenance. Operation in hostile environment also possible.

Various techniques like signal injection, fundamental excitation, etc., are used for sensorless control [2–12]. In signal injection methods [2, 3], high-frequency signals are injected into the machine. The injected signals produce harmonic current, electrical losses, and torque oscillations. In fundamental excitation method, machine model is used to estimate the state variables. Kalman filter, sliding mode observer, MRAS are used to estimate these state variables. State estimation based on sliding mode techniques is reviewed in [4–7]. Sliding mode techniques are widely used due to their order reduction, robustness against parameter variations, and disturbance rejection. In sliding mode techniques, designing the sliding surface is very difficult. MRAS-based speed estimation schemes are discussed in [9–12]. An adaptive rotor speed observer based on artificial neural network has been presented for sensorless IM drives [9]. The main challenge in this method is to construct an effective approximation base.

In this paper, model reference adaptive system (MRAS) with different adjustable mechanisms has been established to estimate the rotor speed and flux for sensorless induction motor drives. The first method discuss about model-based speed estimation technique. The model-based method contains a flux observer and a current observer. They act as the reference model for speed estimation. Measured stator terminal current and estimated rotor flux linkages are used to estimate the rotor speed. Model-based adaptive observer is used to estimate the rotor speed. Estimated speed is fed back to the system to provide closed-loop speed control of induction motor [13]. The estimated speed feedback signals improve the operating range and robustness of the whole drive. In the second method, feedforward ANN (FFANN) has been employed as the adjustable mechanism to estimate the rotor speed. ANN-based estimation scheme does not require any precise analytical expression of the motor model. Simulation results are presented to know the speed tracking capability of the proposed methods.

2 Model-Based Speed Estimation

The proposed MRAS-based speed estimator consists of two models: Reference model—which estimates the rotor flux in stationary reference frame by employing the motor stator voltage and current as input signal. Adjustable model—which estimates the rotor flux in stationary reference frame by employing the motor stator current and the estimated speed as input signal. The output of the adjustable model and reference model is compared and the error signal is calculated. The calculated rotor flux error signal is given to the propotional integral (PI)-based controllers. PI controller tuning is done by Ziegler–Nichols tuning method. The estimated rotor speed signal is employed as the adaptive signal for adjustable model. Accuracy of this method highly depends on the construction of the reference model Fig. 1 shows the block diagram of MRAS-based speed estimation scheme.

The model-based flux and speed observer is addressed in this section. The proposed closed-loop observer model accurately tracks the rotor and stator flux linkages. The measured stator terminal current and voltage are taken as input signal to the observer. Rotor flux in *xy* coordinate frame is calculated as follows [1]:

$$\hat{\psi}_{rtx} = \frac{\hat{\psi}_{stx} - \sigma L_{st}\hat{i}_{stx}}{K_r} \quad (1)$$

$$\hat{\psi}_{rty} = \frac{\hat{\psi}_{sty} - \sigma L_{st}\hat{i}_{sty}}{K_r} \quad (2)$$

The current observing system is designed by using the following equations

$$\hat{i}_{stx} = \frac{\hat{\psi}_{stx} - K_r\hat{\psi}_{rtx}}{\sigma L_{st}} \quad (3)$$

$$\hat{i}_{sty} = \frac{\hat{\psi}_{sty} - K_r\hat{\psi}_{rty}}{\sigma L_{st}} \quad (4)$$

where K is the observer gain

$$K_r = L_m/L_{rt},\ \sigma = 1 - \frac{L_m^2}{L_{st}L_{rt}},\ \tau_{st}' = \frac{\sigma L_{st}}{R_{st}}$$

Estimation of rotor speed is done using the adaptive observer [1]. The magnitude and angle position of estimated rotor flux components are obtained as

$$\hat{\omega}_{sp} = R_{rt}\frac{L_m}{L_{rt}}\frac{\left(\hat{\psi}_{rt\alpha}\hat{i}_{st\beta} - \hat{\psi}_{rt\beta}\hat{i}_{st\alpha}\right)}{|\hat{\psi}_{rt}|^2} \quad (5)$$

$$|\hat{\psi}_{rt}| = \sqrt{\hat{\psi}_{rt\alpha}^2 + \hat{\psi}_{rt\beta}^2} \quad (6)$$

$$\hat{\rho}_{\psi rt} = \mathrm{arc}\, tg\frac{\hat{\psi}_{rt\beta}}{\hat{\psi}_{rt\alpha}} \quad (7)$$

Subtract the rotor flux speed and slip speed the motor mechanical speed is obtained as

$$\hat{\omega} rt = \hat{\omega}\psi rt - \hat{\omega} sp \quad (8)$$

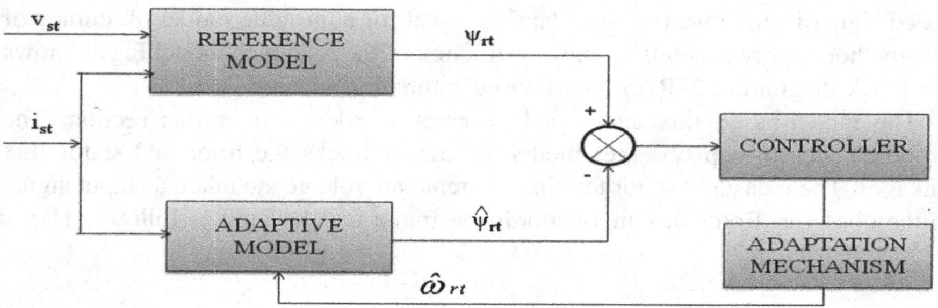

Fig. 1 Block diagram of MRAS

Rotor flux pulsation is obtained as

$$\hat{\omega}_{\psi rt} = \frac{d\hat{\rho}_{\psi rt}}{d\tau} \tag{9}$$

Rotor slip pulsation is obtained as

$$\hat{\omega}_{sp} = R_{rt}\frac{L_m}{L_{rt}}\frac{\left(\hat{\psi}_{rt\alpha}\hat{i}_{st\beta} - \hat{\psi}_{rt\beta}\hat{i}_{st\alpha}\right)}{|\hat{\psi}_{rt}|^2} \tag{10}$$

Here $\hat{\psi}_{rtx}, \hat{\psi}_{rty}$ is the estimated rotor flux linkages in *xy* (rotating reference) coordinate frame, $\hat{i}_{stx}, \hat{i}_{sty}$—Estimated stator current in *xy* coordinate frame, $\hat{\psi}_{stx}, \hat{\psi}_{sty}$—Estimated stator flux linkages in *xy* coordinate frame. Calculated stator flux components and measured stator current components in the stationary reference frame are given as input signal to the rotor flux estimator. To improve the rotor flux estimation, the difference between estimated and measured current signal is given as input to the flux estimator.

3 Ann-Based Speed Estimation

In this section, we discuss about the adaptive model which utilizes ANN. Feedforward ANN (FFANN) algorithm is used to derive the adaptation rule. Error in current estimation is minimized using a FFANN learning algorithm. The weights are proportional to the estimated speed, motor parameters, and torque. The motor model itself acts as the reference model. Stator voltages and currents in the *xy* reference frame are given as input signal to the ANN. To train the network, previously estimated rotor speed is given to the approximation base. The main disadvantage of this method is to provide proper knowledge base. The simplified block diagram of ANN-based MRAS is shown in Fig. 2. In stationary reference frame, rotor flux is estimated by utilizing the following equations [9],

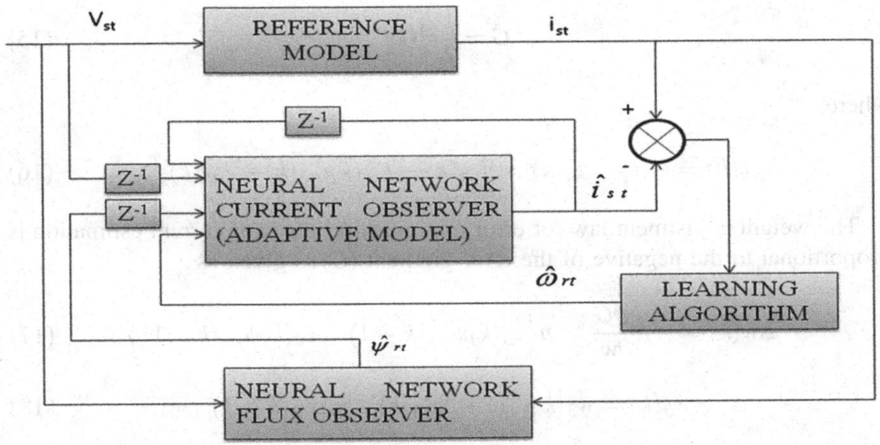

Fig. 2 Block diagram of ANN-based MRAS

$$p\hat{\psi}_{rt\alpha} = \frac{L_m}{T_{rt}} i_{st\alpha} - \frac{1}{T_{rt}} \hat{\psi}_{rt\alpha} - \hat{\omega}_{rt} \hat{\psi}_{rt\beta} \tag{11}$$

$$p\hat{\psi}_{rt\beta} = \frac{L_m}{T_{rt}} i_{st\beta} - \frac{1}{T_{rt}} \hat{\psi}_{rt\beta} - \hat{\omega}_{rt} \hat{\psi}_{rt\alpha} \tag{12}$$

Here, T_{rt}—sampling time for rotor current observer

The discretized stator current observer system is designed using the following equations:

$$\hat{i}_{st\alpha}(k) = w_1 \hat{i}_{st\alpha}(k-1) + w_2 \hat{\psi}_{rt\alpha}(k-1) + w_3 \hat{\psi}_{rt\beta}(k-1) + w_4 v_{st\alpha}(k-1) \tag{13}$$

$$\hat{i}_{st\beta}(k) = w_1 \hat{i}_{st\beta}(k-1) + w_2 \hat{\psi}_{rt\beta}(k-1) + w_3 \hat{\psi}_{rt\alpha}(k-1) + w_4 v_{st\beta}(k-1) \tag{14}$$

Here

$$w_1 = 1 - \frac{T_{st} R_{st}}{\sigma L_{st}} - \frac{T_{st} L_m^2}{\sigma L_{st} L_{rt} T_{rt}}, w_2 = \frac{T_{st} L_m}{\sigma L_{st} L_{rt} T_{rt}}, w_3 = \frac{T_{st} L_m}{\sigma L_{st} L_{rt}} \hat{\omega}_{rt}, w_4 = \frac{T_{st}}{\sigma L_{st}}$$

Where T_{st}—stator current observer sampling time.

The energy function G to minimize the error in estimated stator current is given as

$$G = \frac{1}{2}\varepsilon^2(k) \tag{15}$$

where

$$\varepsilon(k) = i_{st}(k) - \hat{i}_{st}(k) = [i_{st\alpha}(k) - \hat{i}_{st\alpha}(k) i_{st\beta}(k) - \hat{i}_{st\beta}(k)]^T \tag{16}$$

The weight adjustment law for error minimization of stator current estimation is proportional to the negative of the error gradient (G) is given as

$$\Delta w(k) = -\mu \frac{\partial G}{\partial w_3} = \mu \left\{ \varepsilon_\alpha(k)\hat{\psi}_{rt\beta}(k-1) - \varepsilon_\beta(k)\hat{\psi}_{rt\alpha}(k-1) \right\} \tag{17}$$

$$w_3(k) = w_3(k-1) + \Delta w_3(k) + \eta \Delta w_3(k-1) \tag{18}$$

where μ—Learning rate, η—Momentum constant

The estimated rotor speed is given as

$$\hat{\omega}_{rt}(k) = \frac{\sigma L_{st} L_{rt}}{T_{st} L_m} w_3(k) \tag{19}$$

From the above equations, the adaptive speed estimator is designed. Figure 2 shows the block diagram of MRAS technique utilizing ANN for rotor flux estimation. By training the network for various motor parameters, we can improve the speed tracking capability of this method.

4 Sensorless Foc of IM

See Fig. 3.

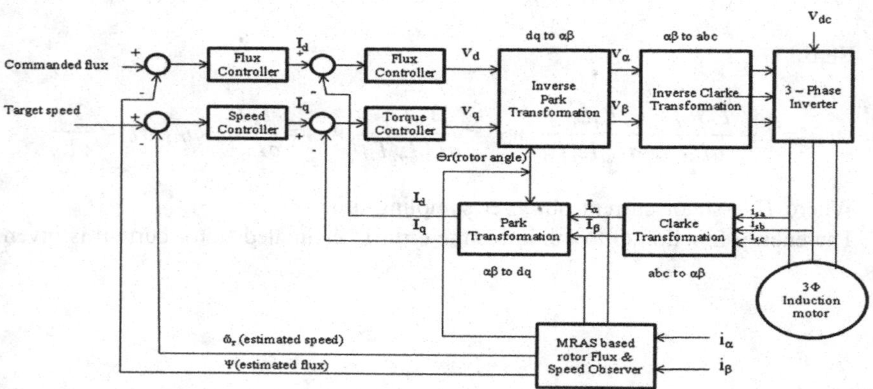

Fig. 3 Overall schematic of sensorless FOC of IM

5 Simulation Results

The proposed sensorless schemes are simulated using MATLAB/Simulink and SimPower system library. The simplified block diagram is presented in Fig. 3. Simulated induction motor parameters are listed in Table 1. These motor parameters are taken from [4].

Figure 4 shows actual and estimated rotor speed for square pulse demand. Figure 5 shows the estimated rotor flux when the rotor flux command is set at 1 pu. The estimated direct and squadrature axis stator current components are shown in Figs. 6 and 7. Figure 8 shows the actual and estimated speed of ANN-based speed estimators. The estimated stator current in *xy* rotating reference frame is shown in Fig. 9. The input signals to the ANN are shown in Fig. 10.

Table 1 Parameters of induction motor

Parameters	Values
S_{rated}	1.5 HP
Stator resistance (R_{st})	3.24 Ω
Rotor resistance (R_{rt})	4.96 Ω
Stator leakage inductance (L_{st})	0.4024 H
Rotor leakage inductance (L_{rt})	0.4047 H
Mutual inductance (L_m)	0.3886 H
Moment of inertia (J)	59 kg m^2

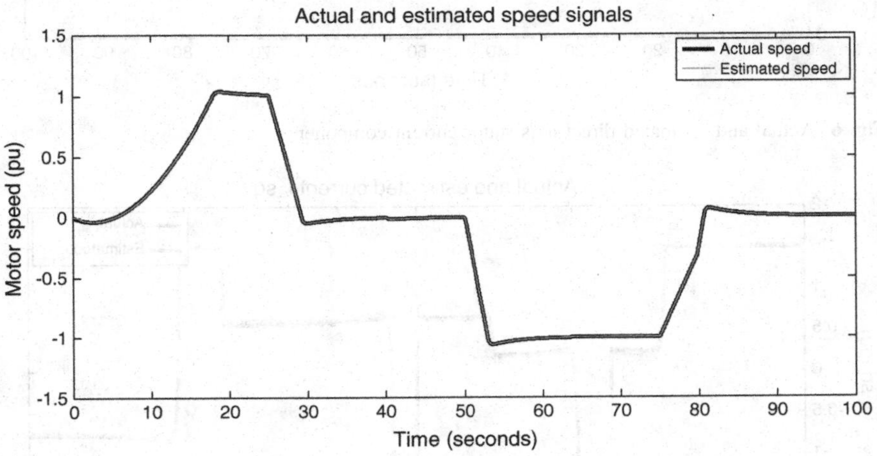

Fig. 4 Actual and estimated rotor speed response of model-based estimator

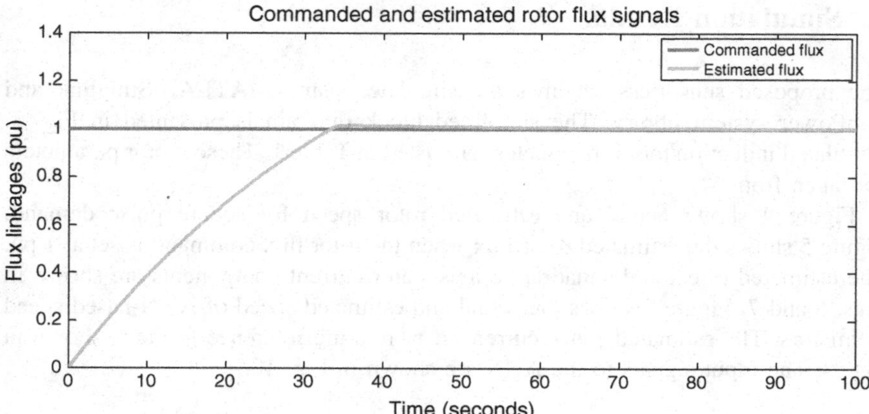

Fig. 5 Commanded and estimated rotor flux signals

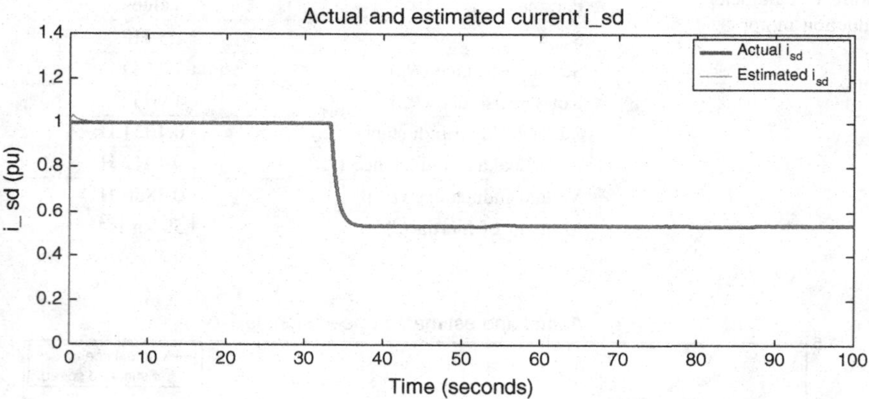

Fig. 6 Actual and estimated direct axis stator current components

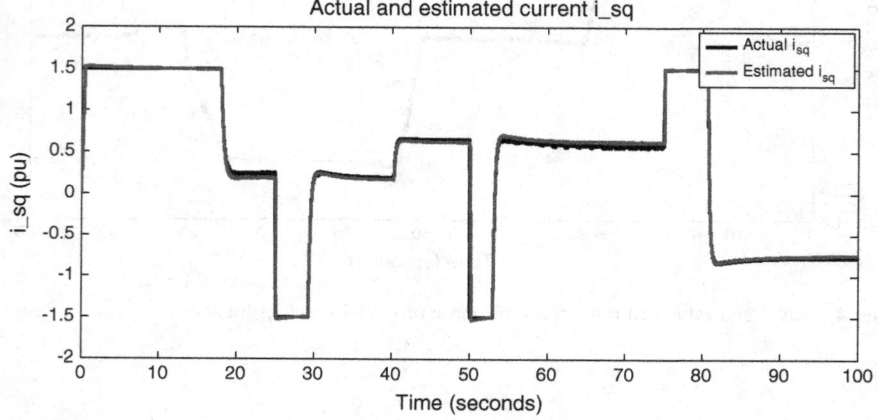

Fig. 7 Actual and estimated quadrature axis stator current components

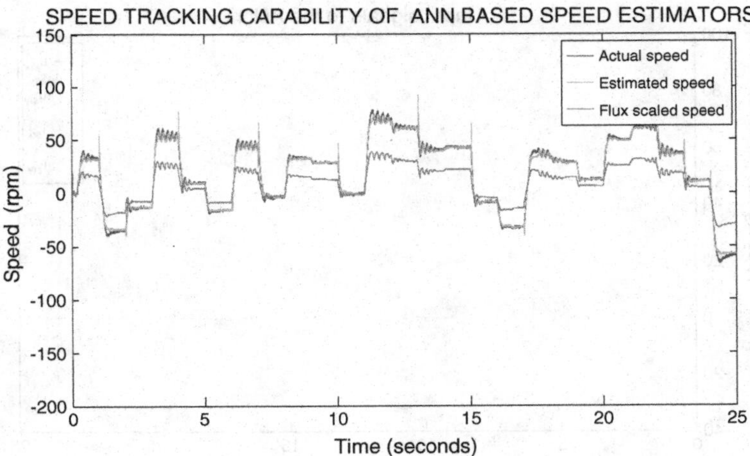

Fig. 8 Actual and estimated speed of ANN-based estimator

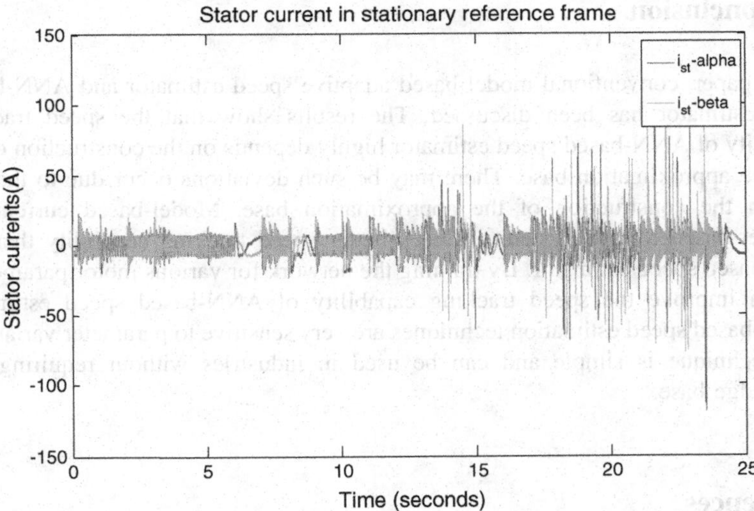

Fig. 9 Response of stator current in $\alpha\beta$ stationary reference frame for ANN-based speed estimation

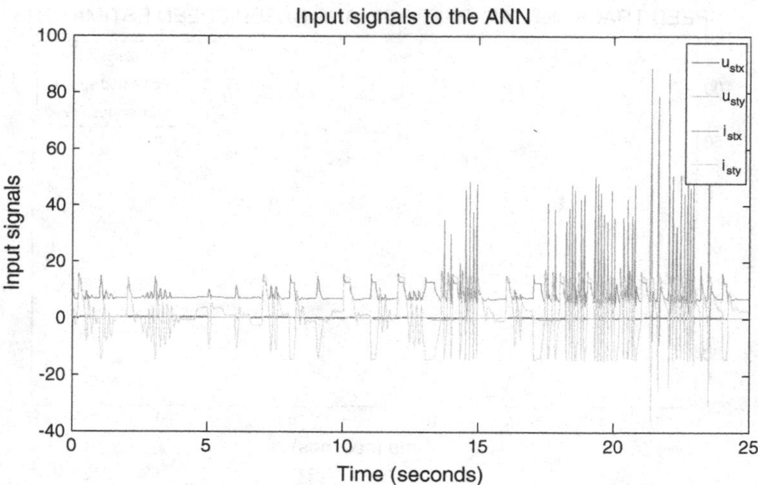

Fig. 10 Input signals to the ANN (u_{stx}, u_{sty}, i_{stx}, i_{sty})

6 Conclusion

In this paper, conventional model-based adaptive speed estimator and ANN-based speed estimator has been discussed. The results show that the speed tracking capability of ANN-based speed estimator highly depends on the construction of the effective approximation base. There may be such deviations occur due to the lagging in the construction of the approximation base. Model-based current/flux observer-based MRAS system provides better speed tracking capability than the ANN-based speed estimator. By training the network for various motor parameters, we can improve the speed tracking capability of ANN-based speed estimator. Model-based speed estimation techniques are very sensitive to parameter variations. This technique is simple and can be used in industries without requiring any knowledge base.

References

1. Abu-Rub H, Iqbal A, Guzinski J (2012) High performance control of AC drives. Wiley, West Sussex, United Kingdom
2. Holtz J (2006) Sensorless control of induction machines—with or without signal injection. IEEE Trans Ind Electron 53(1):7–30
3. Wang G, Hofmann HF, El-Antably A (2006) Speed-sensorless torque control of induction machine based on carrier signal injection and smooth-air-gap induction machine model. IEEE Trans Energy Convers 21(3):699

4. Vieira RP, Gastaldini CC, Azzolin RZ, Grundling HA (2014) Sensorless sliding-mode rotor speed observer of induction machines based on magnetizing current estimation. IEEE Trans Ind Electron 61(9):4573–4582
5. Traore d, Plestan F, Glumineau A, de Leon J (2008) Sensorless induction motor: high-order sliding-mode controller and adaptive interconnected observer. IEEE Trans Ind Electron 55 (11):3818–3827
6. Derdiyok A (2005) Speed-sensorless control of induction motor using a continuous control approach of sliding mode and flux observer. IEEE Trans Ind Electron 52(4):1170–1176
7. Zhao Y, Qiao W, Wu L (2014) Improved rotor position and speed estimators for sensorless control of interior permanent-magnet synchronous machine. IEEE J Emerg Sel Top Power Electron 2(3):627–639
8. Alonge F, D'Ippolito F, Sferlazza A (2014) Sensorless control of induction-motor drive based on robust Kalman filter and adaptive speed estimation. IEEE Trans Ind Electron 61(3):1444–1453
9. Gadoue SM, Giaouris D, Finch JW (2013) Stator current model reference adaptive systems speed estimator for regenerating—mode low-speed operation of sensorless induction motor drives. IET Electr Power Appl 7(7):597–606
10. Abdelsalam AK, Masoud MI, Hamad MS, Williams BW (2013) Improved sensorless operation of a CSI-based induction motor drive—long feeder case. IEEE Trans Power Electron 28(8):4001–4012
11. Peng FZ, Fukao T (1994) Robust speed identification for speed-sensorless vector control of induction motors. IEEE Trans Ind Appl 30(5):1234–1240
12. Cirrincione M, Accetta A, Pucci M, Vitale G (2013) MRAS speed observer for high-performance linear induction motor drives based on linear neural networks. IEEE Trans Power Electron 28(1):123–134
13. Maiti S, Verma V, Chakraborty C, Hori Y (2012) An adaptive speed sensorless induction motor drive with artificial neural network for stability enhancement. IEEE Trans Ind Inf 8 (4):757–766

Optimal Multilevel Image Thresholding to Improve the Visibility of *Plasmodium* sp. in Blood Smear Images

N. Siva Balan, A. Sadeesh Kumar, N. Sri Madhava Raja and V. Rajinikanth

Abstract Malaria is one of the mosquito-borne communicable diseases for humans caused due to *Plasmodium* sp. During the treatment process, it is necessary to identify the exact *Plasmodium* sp. in order to give the specific antimalarial drug. Hence, in this paper, an image segmentation procedure is attempted to enhance the visibility of the *Plasmodium* sp. in microscopic blood smear images. In this paper, two RGB blood smear images of *Plasmodium ovale* (300 × 300) are considered and segmented using Otsu and heuristic algorithms, such as PSO, DPSO, and FODPSO available in the literature. During the segmentation procedure, maximization of a multiple objective function is adopted to guide the heuristic algorithm-based exploration. The performances of considered algorithms are analyzed using the popular image parameters, such as Otsu's function, SSIM, RMSE, and PSNR. This study shows that FODPSO offers improved segmentation result compared to PSO and DPSO algorithms. The similar procedure can be used to identify other *Plasmodium* sp. using the microscopic blood smear images.

Keywords Plasmodium species · RGB image · PSO algorithm · Otsu · Performance measure

---

N. Siva Balan (✉) · A. Sadeesh Kumar · N. Sri Madhava Raja · V. Rajinikanth
Department of Electronics and Instrumentation Engineering, St. Joseph's College of Engineering, Chennai 600119, Tamil Nadu, India
e-mail: balansiv@gmail.com

A. Sadeesh Kumar
e-mail: sat22ish@gmail.com

N. Sri Madhava Raja
e-mail: nsrimadhavaraja@stjosephs.ac.in

V. Rajinikanth
e-mail: rajinikanthv@stjosephs.ac.in

L.P. Suresh and B.K. Panigrahi (eds.), *Proceedings of the International Conference on Soft Computing Systems*, Advances in Intelligent Systems and Computing 397, DOI 10.1007/978-81-322-2671-0_54

1 Introduction

Image segmentation is an important process used to extract the valuable information from a photo or video frame. An extensive number of classical and heuristic algorithm-assisted image segmentation methods have been proposed and implemented by most of the researchers in the literature [1–5]. In recent years, biomedical image processing attracted the research community because of its significance. Recently, heuristic algorithm-based segmentation process is applied to extract the meaningful information from fundus retinal image [6], angiogram [7], and microscopic images [8–12].

In this paper, an attempt is made to segment the malaria-infected blood smear images. Malaria is a most common and most dangerous mosquito-borne infectious disease. Recent paper by Manickavasagam et al., presents the detailed segmentation procedure and the available antimalarial drug based on the *Plasmodium* sp. [9]. Literature also presents the image processing methodology for Plasmodium infected blood smear images [13, 14]. In most of the cases, the RGB images are segmented using the grayscale images and the outcome of this process also in gray form. In this paper, the RGB images are segmented using Otsu-based RGB segmentation procedure proposed by Ghamisi et al. [4, 5].

In this work, thick and thin blood smear images of *Plasmodium ovale* (*P. ovale*) was acquired from the Parasite Image Library (DPDx) of the Centers for Disease Control and Prevention [15] and is then considered for segmentation process.

In this work, PSO, DPSO, and FODPSO algorithms are employed to solve bilevel and multilevel RGB image segmentation problem. Initially, the attempted method is tested on mandrill (512 × 512) image and later it is implemented on blood smear images. The image performance measures, such as mean-squared error (MSE), root mean-squared error (RMSE), peak-to-signal Ratio (PSNR), and the objective functions are considered to evaluate the segmentation accuracy.

2 Otsu-Based Segmentation

A considerable number of classical and optimization algorithm-based thresholding methods are existing in the literature to segment grayscale and RGB images [1]. In this work, Otsu-based image thresholding method proposed in 1979 [16] is adopted. Compared to other existing methods, this technique is simple and presents the optimal thresholds by maximizing the between class variance function. In this work, the RGB image thresholding technique elaborately discussed in [4, 5, 10–12] is considered.

In order to validate the performance of the segmented image quality, the common performance measures such as root mean square error (RMSE), the peak signal-to-noise ratio (PSNR), and the structural similarity index matrix (SSIM) are considered.

- PSNR is used to find the similarity of the segmented image against the original image based on the root mean square error (RMSE) of each pixel [2, 3]:

$$\mathrm{PSNR}(o,s) = 20\log_{10}\left(\frac{255}{\sqrt{\mathrm{MSE}(o,s)}}\right); \mathrm{dB} \tag{1}$$

$$\mathrm{RMSE}_{(o,s)} = \sqrt{\mathrm{MSE}_{(o,s)}} = \sqrt{\frac{1}{\mathrm{MN}}\sum_{i=1}^{H}\sum_{j=1}^{W}[o(i,j)-s(i,j)]^2} \tag{2}$$

where o and s are original and segmented images of size H × W.

- SSIM is generally used to estimate the image superiority and interdependencies between the original and the processed image [17].

$$\mathrm{SSIM}_{(o,s)} = \frac{(2\mu_o\mu_s + C_1)(2\sigma_{os} + C_2)}{(\mu_o^2 + \mu_s^2 - C_1)(\sigma_{o^2} + \sigma_{s^2} + C_2)} \tag{3}$$

The segmentation accuracy mainly relies on the objective function. In this paper, weighted sum of multiple objective function (OF) recently proposed by Rajinikanth and Couceiro [18] is considered.

This objective function is presented below:

$$J_{\max} = \text{Otsu's function} + W^{*}\mathrm{SSIM} \tag{4}$$

where $J_{\max}$ is the OF to be maximized and W is the weighting parameter (assigned with a value of 100).

3 Overview of Heuristic Algorithms in This Study

In this paper, the classical particle swarm optimization (PSO) and its recent improved forms, such as Darwinian PSO (DPSO), and fractional order DPSO (FODPSO) are considered [4, 5].

Classical PSO algorithm was initially proposed by Kennedy and Eberhart with the inspiration of social activities within a flock of birds and a school of fish [19]. Based on the concepts inherit to the PSO algorithm, recent improvements, such as DPSO [20] and FODPSO [4, 5] have been developed. In the FODPSO algorithm, a group of swarms battles using Darwin's theory and the fractional calculus to regulate the convergence rate. FODPSO enhances the performance of the traditional PSO to escape from local optima by running several simultaneous and parallel, PSO algorithms. Table 1 presents the algorithm parameters considered in this work.

Table 1 Initial parameters of heuristic algorithms

Parameter	PSO	DPSO	FODPSO
Number of iterations	200	200	200
Population	50	50	50
Cognitive coefficient (ρ_1)	2.0	2.0	2.0
Social coefficient (ρ_2)	1.5	1.5	1.5
Inertial coefficient (W)	0.75	–	–
Maximum particles' position (X_{max})	255	255	255
Minimum particles' position (X_{min})	0	0	0
Min population	–	10	20
Max population	–	50	60
Number of swarms	–	8	8
Min swarms	–	2	2
Max swarms	–	6	6
Stagnancy	–	15	15
Fractional coefficient (α)	–	–	0.75
Stopping criteria	J_{max}	J_{max}	J_{max}

4 Result and Discussions

The multilevel thresholding problem deals with finding optimal thresholds within the threshold range [0, L − 1] that maximize a fitness criterion J_{max}. In this work, search dimension of the optimization problem is allocated based on the chosen threshold (m) values. For each image and each threshold value, the segmentation procedure is repeated ten times and mean value of the trial is chosen as the optimal value.

Otsu-based multilevel thresholding method have been tested on the RGB mandril image (512 × 512) and *P. ovale* RGB image (300 × 300) available at [15]. Generally, the data space in gray level histogram is simple [0, 255], whereas the RGB histogram holds a larger pixel value with a data space of $[0, 255]^3$ [21, 22].

Figure 1 shows the RGB mandrill image and the corresponding histograms. The RGB histogram is more complicated and has three different color patters namely red (R), green (G), and blue (B), which may increase the computational time of segmentation process. From Fig. 1b, c, it can also be noted that the RGB histogram of the mandrill is similar to its grayscale histogram. Hence, in this RGB segmentation work, gray histogram-based threshold is considered.

Initially, the attempted technique is implemented on the mandrill image with $m = 2$ using PSO algorithm. The dimension of the problem is two and the algorithm explores the pixel distribution which has a data space of $[[0, 255]^3]^2$. Later, the similar segmentation procedure is repeated with m = 3–5 and the corresponding threshold values are recorded for m = 3–5. Similar procedure is repeated with DPSO and FODPSO. From Table 2, it is observed that the FODPSO algorithm

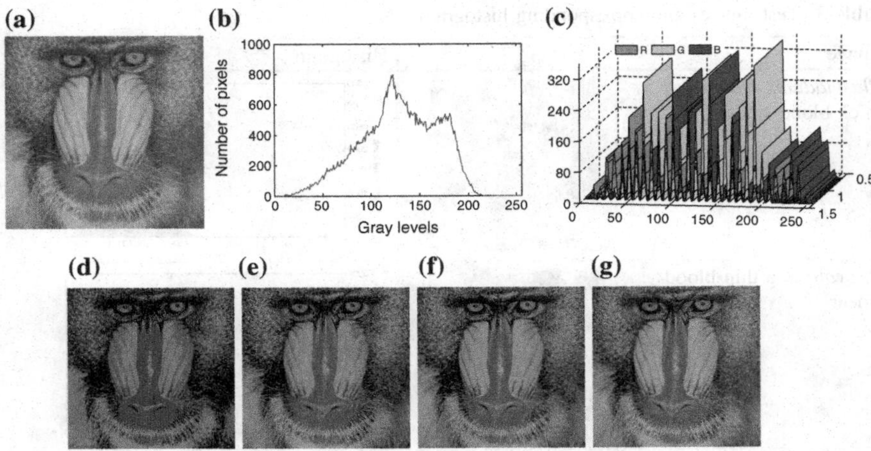

Fig. 1 Multilevel thresholding for 512 × 512 sized mandrill image. **a** Mandrill. **b** Gray histogram. **c** RGB histogram. **d** $m = 2$. **e** $m = 3$. **4** $m = 4$. **5** $m = 5$

Table 2 Performance measure values for the test images (mean value of 10 trials)

m	Otsu's function			SSIM			PSNR (dB)		
	PSO	DPSO	FODPSO	PSO	DPSO	FODPSO	PSO	DPSO	FODPSO
2	1247.55	**1249.18**	1248.91	0.7488	**0.7604**	0.7621	10.4902	**11.1720**	11.1627
3	1250.69	1251.02	**1252.71**	0.7933	0.7871	**0.7955**	14.3104	14.3312	**14.3505**
4	1251.71	1254.07	**1254.47**	0.8288	0.8259	**0.8315**	16.6803	16.7006	**16.9592**
5	1254.86	1255.98	**1257.02**	0.8529	0.8694	**0.8711**	18.9141	**18.9967**	18.9569

The bold numbers in the table represent optimal values obtained in this study.

offers better performance measure values compared to PSO and DPSO for $m = 3$–5 and DPSO offers better result for $m = 2$.

The proposed method is then tested on the *P. ovale* blood smear images are presented in Table 3.

Initially, the multilevel segmentation procedure is proposed on thick blood smear image with 2, 3, 4, and 5 threshold values with PSO, DPSO, and FODPSO. Later, similar procedure is implemented on the thin blood smear image and the results are depicted in Tables 4, 5, 6 and 7.

Table 4 shows the segmented *P. ovale* images obtained by Otsu and heuristic algorithms for various threshold levels. When comparing the segmented image (Table 4) with the original image (Table 3), it is clearly noticeable that the segmented image has improved visibility of *Plasmodium* sp. (for $m = 2$ in the case of thick blood smear image and for $m = 3$, 4 in the case of thin blood smear) compared to original image. In the segmented image, the infected blood cells are clearly noticeable compared to the unsegmented blood smear images.

Table 3 Test images and corresponding histogram

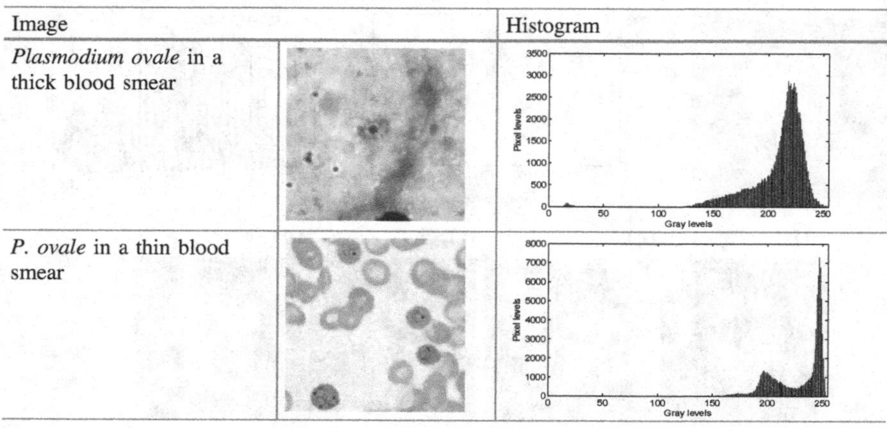

Image		Histogram
Plasmodium ovale in a thick blood smear		
P. ovale in a thin blood smear		

Table 4 Segmented *P. ovale* images for m = 2–5

Image	Method	$m = 2$	$m = 3$	$m = 4$	$m = 5$
P. ovale in a thick blood smear	PSO				
	DPSO				
	FODPSO				
P. ovale in a thin blood smear	PSO				
	DPSO				
	FODPSO				

Table 5 Maximized objective function value

Image	*m*	Otsu's function			SSIM		
		PSO	DPSO	FODPSO	PSO	DPSO	FODPSO
P. ovale in a thick blood smear	2	**456.58**	448.37	447.91	**0.6499**	0.6382	0.6377
	3	467.15	472.18	**480.09**	0.6824	0.6913	**0.6941**
	4	483.06	483.28	**488.12**	0.7102	0.7207	**0.7232**
	5	492.14	497.10	**499.36**	0.7286	0.7306	**0.7319**
P. ovale in a thin blood smear	2	424.84	**426.77**	424.12	0.6618	0.6583	**0.6636**
	3	432.93	441.04	**443.67**	0.6924	0.6935	**0.6988**
	4	451.00	458.36	**460.32**	0.7192	0.7205	**0.7219**
	5	463.26	472.52	**475.13**	0.7216	0.7286	**0.7292**

The bold numbers in the table represent optimal values obtained in this study.

Table 6 Optimal thresholds for m = 2–5

Image	*m*	Optimal threshold values		
		PSO	DPSO	FODPSO
P. ovale in a thick blood smear	2	173, 216	171, 217	171, 218
	3	161, 194, 224	164, 190, 225	165, 190, 223
	4	158, 182, 206, 232	156, 180, 207, 234	157, 180, 209, 232
	5	151, 179, 204, 216, 247	150, 175, 202, 217, 248	151, 174, 204, 218, 250
P. ovale in a thin blood smear	2	203, 244	201, 240	200, 242
	3	197, 231, 246	195, 230, 245	194, 233, 246
	4	192, 205, 238, 249	187, 203, 234, 247	184, 201, 232, 248
	5	174, 191, 203, 240, 251	172, 188, 206, 241, 251	170, 189, 203, 241, 252

Table 7 Performance measure values of *P. ovale* images for m = 2–5

Image	*m*	RMSE			PSNR (dB)		
		PSO	DPSO	FODPSO	PSO	DPSO	FODPSO
P. ovale in a thick blood smear	2	66.3001	86.7760	84.9530	**11.7005**	9.3628	9.5472
	3	63.7313	37.0936	37.0740	12.0437	16.7448	**16.7494**
	4	37.0870	26.2219	26.2883	16.7464	**19.7575**	19.7355
	5	26.2166	20.7947	21.3625	19.7593	**21.7717**	21.5378
P. ovale in a thin blood smear	2	119.908	120.288	119.268	6.5538	6.5263	**6.6003**
	3	59.8280	61.1443	49.0467	12.5927	12.4037	**14.3186**
	4	57.2397	45.6597	43.7435	12.9769	14.9402	**15.3125**
	5	43.7435	31.2706	18.9451	15.3125	18.2281	**22.5809**

The bold numbers in the table represent optimal values obtained in this study.

From these results, the conclusion is that, the overall performance (Otsu's function, SSIM, RMSE, and PSNR) offered by FODPSO better than PSO and DPSO algorithms for m = 3, 4, 5. Similar segmentation procedure can also considered to segment other *Plasmodium* sp., such as *p. Falciparum*, *p. Vivax*, *p. Malariae,* and *p. Knowlesi* existing in the microscopic blood smear database.

5 Conclusion

In this paper, an attempt is made to segment the RGB images using Otsu and heuristic algorithms, such as PSO, DPSO, and FODPSO. Maximization of Otsu's function and SSIM is chosen as the objective function. The performance of heuristic algorithm-based segmentation is initially tested on mandrill image. Similar segmentation procedure is then implemented on RGB blood smear images of *P. ovale*. From this study, it is noted that the FODPSO algorithm offers improved performance measure values compared with PSO and DPSO for higher threshold values on the considered RGB images.

References

1. Tuba M (2014) Multilevel image thresholding by nature-inspired algorithms: A short review. Comput Sci J Moldova 22(3):318–338
2. Rajinikanth V, Sri Madhava Raja N, Latha K (2014) Optimal multilevel image thresholding: an analysis with PSO and BFO algorithms. Aust J Basic Appl Sci 8(9):443–454
3. Sri Madhava Raja N, Rajinikanth V, Latha K (2014) Otsu based optimal multilevel image thresholding using firefly algorithm. Model Simul Eng 2014:17 p, Article ID 794574
4. Ghamisi P, Couceiro MS, Benediktsson JA, Ferreira NMF (2012) An efficient method for segmentation of images based on fractional calculus and natural selection. Expert Syst Appl 39(16):12407–12417
5. Ghamisi P, Couceiro MS, Martins FML, Benediktsson J (2014) A: Multilevel image segmentation based on fractional-order Darwinian particle swarm optimization. IEEE Trans Geosci Remote Sens 52(5):2382–2394
6. Raja NSM, Kavitha N, Ramakrishnan S (2012) Analysis of vasculature in human retinal images using particle swarm optimization based Tsallis multi-level thresholding and similarity measures. In: Panigrahi BK et al (ed) SEMCCO 2012, LNCS 7677, pp 380–387
7. Kamalanand K, Ramakrishnan S (2015) Effect of gadolinium concentration on segmentation of vasculature in cardiopulmonary magnetic resonance angiograms. J Med Imaging Health Inf 5(1):147–151
8. Manickavasagam K, Sutha S, Kamalanand K (2014) An automated system based on 2d empirical mode decomposition and k-means clustering for classification of Plasmodium species in thin blood smear images. BMC Infect Dis 14(Suppl 3):P13. doi:10.1186/1471-2334-14-S3-P13
9. Manickavasagam K, Sutha S, Kamalanand K (2014) Development of systems for classification of different plasmodium species in thin blood smear microscopic images. J Adv Microsc Res 9(2):86–92
10. Kamalanand K, Jawahar PM (2012) Coupled jumping frogs/particle swarm optimization for estimating the parameters of three dimensional HIV model. BMC Infect Dis 12(Suppl 1):82
11. Atchaya A, Aashiha JP, Vijayarajan R (2015) Optimal image segmentation of cancer cell images using heuristic algorithms, information systems design and intelligent applications. Adv Intell Syst Comput 339:269–278
12. Joyce Preethi B, Rajinikanth V (2014) Improving segmentation accuracy in biopsy cancer cell images using otsu and firefly algorithm. Int J Appl Eng Res 9(25):8502–8506
13. Le MT, Bretschneider TR, Kuss C, Preiser PR (2008) A novel semi-automatic image processing approach to determine Plasmodium falciparum parasitaemia in Giemsa-stained thin blood smears. BMC Cell Biology 9:15. doi:10.1186/1471-2121-9-15

14. Frean JA (2009) Reliable enumeration of malaria parasites in thick blood films using digital image analysis. Malaria J 8:218. doi:10.1186/1475-2875-8-218
15. www.dpd.cdc.gov/dpdx/HTML/ImageLibrary/Malaria_il.htm
16. Otsu N (1979) A threshold selection method from gray-level histograms. IEEE Trans Syst Man Cybern 9(1):62–66
17. Wang Z, Bovik AC, Sheikh HR, Simoncelli EP (2004) Image quality assessment: from error measurement to structural similarity. IEEE Trans Image Process 13(1):1–14
18. Rajinikanth V, Couceiro MS (2015) Optimal multilevel image threshold selection using a novel objective function, information systems design and intelligent applications. Adv Intell Syst Comput 340:177–186. doi:10.1007/978-81-322-2247-7_19
19. Kennedy J, Eberhart RC (1995) Particle swarm optimization. In: Proceedings of IEEE international conference on neural networks, pp 1942–1948. doi:10.1109/ICNN.1995.488968
20. Tillett T, Rao TM, Sahin F, Rao R (2005) Darwinian particle swarm optimization. In: Proceedings of the 2nd Indian international conference on artificial intelligence, Pune, India, pp 1474–1487
21. Sarkar S, Das S (2013) Multilevel image thresholding based on 2D histogram and maximum Tsallis entropy—a differential evolution approach. IEEE Trans Image Process 22(12):4788–4797
22. Su Q, Hu Z (2013) Color image quantization algorithm based on self-adaptive differential evolution. Comput Intell Neurosci 2013:8 p, Article ID 231916

A Study of Household Object Recognition Using SIFT-Based Bag-of-Words Dictionary and SVMs

Aadarsh Sampath, Aravind Sivaramakrishnan, Keshav Narayan and R. Aarthi

Abstract In the era of computational intelligence, computer vision-based techniques for robotic cognition have gained prominence. One of the important problems in computer vision is the recognition of objects in real-time environments. In this paper, we construct a SIFT-based SVM classifier and analyze its performance for real-time object recognition. Ten household objects from the CALTECH-101 dataset are chosen, and the optimal train-test ratio is identified by keeping other SVM parameters constant. The system achieves an overall accuracy of 85 % by maintaining the ratio as 3:2. The difficulties faced in adapting such a classifier for real-time recognition are discussed.

Keywords SIFT · Bag of visual words · SVM · Real time · Object recognition

1 Introduction

Computer vision is the field that attempts to mimic human visual perception in a computer. An important application of computer vision is in the area of object recognition. Object recognition requires the use of object representation supported by learning and inference algorithms in order to achieve good results. Most object representations refer to the high-level description of an object and its attributes, such as edges, texture, etc. There are a variety of machine learning algorithms that learn the representation of an object, and assign labels to the object based on the inference obtained. However, most of these methods involve a trade-off between accuracy and speed. This paper considers a real-time environment where time taken for recognition of the object must be low.

A. Sampath (✉) · A. Sivaramakrishnan · K. Narayan · R. Aarthi
Department of CSE, Amrita Vishwa Vidyapeetham, Coimbatore, India
e-mail: aadarsh.sampath@gmail.com

R. Aarthi
e-mail: aarthi.r4@gmail.com

L.P. Suresh and B.K. Panigrahi (eds.), *Proceedings of the International Conference on Soft Computing Systems*, Advances in Intelligent Systems and Computing 397, DOI 10.1007/978-81-322-2671-0_55

Over the past five decades, many kinds of object representations have been proposed to aid object recognition, such as volumetric parts and context-based image retrieval [1]. These representations are formally called as features or feature descriptors. Local feature-based methods have gained prominence due to their robustness in occluded environments, as well as their invariance to primitive geometric transforms. One such descriptor is the scale invariant feature transform (SIFT), proposed by Lowe [2] and has been used in a variety of applications, ranging from robot localization [3] to panorama stitching [4]. A survey by Mikolajczyk et al. [5] reveals that the scale invariant feature transform (SIFT) show significant recall and 1-precision values among other descriptors surveyed, like GLOH (gradient location and orientation histogram) and PCA-SIFT (principal component analysis SIFT) [6]. Thus, dense SIFT, which is a version of SIFT has been selected for recognition purposes as it is both fast to compute, and does not compromise on accuracy.

However, SIFT features alone cannot be used to assign a label to an object, as there might be many outliers in the extracted features, and the number of such features may vary from one object to another, and even within different instances of the same object. Hence, a robust model must be used for classification. The bag-of-words approach was proposed by Joachims [7] for text classification purposes. In this model, documents are represented as a multiset of the words present in the document, often referred to as a bag. Sivic et al. [8] applied the bag-of-words model to computer vision problems such as image classification by treating image features as words. In recent work [9], the SIFT keypoints are clustered into visual "words" using k-means. A support vector machine (SVM) is used to aid in the process of object recognition by performing linear classification on the BoVW dictionary.

In this paper, we analyze the classifier built using VLFEAT library [10] for objects in real-time environments. Further, an optimal train-to-test ratio is obtained as a result of experiments performed using the CALTECH-101 [11] dataset. We also study the class-wise accuracy for a subset of the dataset, and discuss the reasons for misclassification in the same. Methods for improving the results of classification have also been suggested.

2 Method Overview

A high-level representation of the implemented system has been provided in Fig. 1. Initially, dense multiscale SIFT descriptors are extracted from the training images. This results in the identification of distinct keypoints in the training images that are robust to affine transformations and change in illuminations. For each image, a collection of keypoint descriptors is obtained. Each keypoint descriptor is a 128-dimensional vector.

Elkan k-means [12] is then used for fast visual word dictionary construction where several SIFT vectors are grouped together to form a single visual word. Thus,

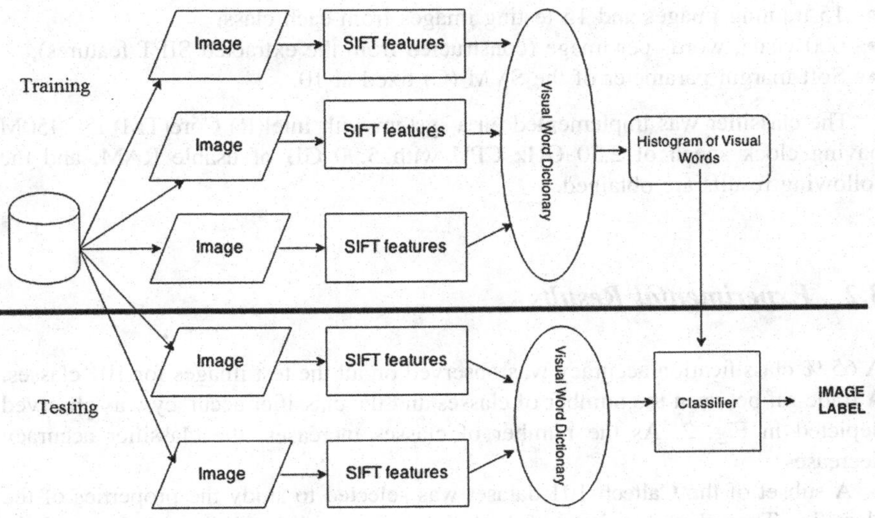

Fig. 1 System architecture of the classifier

each patch in the image is mapped to a particular visual word. The spatial histograms of the visual words obtained are then used as the image descriptors. Each histogram is in the form of a linear vector, where each element corresponds to the occurrence of a particular visual word. Hence, a linear classifier can be trained for the purpose of image classification. SVMs are chosen for this purpose over other tools such as Naive–Bayes classifiers as they are able to achieve high accuracy on new test samples. As multiple image labels (classes) are involved with the problem, a one-vs-all classifier [13] is used on the dataset.

3 Experimental Analysis

3.1 Experimental Setup

The Caltech-101 dataset [11] is a collection of pictures of objects belonging to 101 categories. This dataset was chosen over other benchmark datasets such as the PASCAL VOC 2009 due to the well-annotated category names, as well as the incorporation of everyday objects, such as cups and chairs. The number of images in each category varies from 40 to 800, with each image size roughly 300 × 200. Most categories have about 50 images.

The VLFeat open source library [10] is used for implementation. As it is written in C, the library is highly efficient and compatible. A MATLAB interface was used in obtaining the experimental results. The classifier was initially trained using the default VLFeat configurations:

- 15 training images and 15 testing images from each class,
- 600 visual words per image (constructed from the extracted SIFT features),
- Soft margin parameter of the SVM (C) fixed at 10.

The classifier was implemented on a system with Intel(R) Core(TM) i5-2450M having clock speed of 2.50 GHz CPU with 5.90 GB of usable RAM, and the following results are obtained.

3.2 Experimental Results

A 65 % classification accuracy was observed on all the test images for 101 classes. A trade-off between the number of classes and the classifier accuracy was observed depicted in Fig. 2. As the number of classes increases, the classifier accuracy decreases.

A subset of the Caltech-101 dataset was selected to study the properties of the classifier. The subset consisted of the following everyday objects: cellphone, chair, cup, headphones, lamp, laptop, scissors, stapler, and umbrella. An overall accuracy of 87.50 % was obtained with the configuration mentioned above. Keeping the SVM configurations constant, the number of training and test images was varied to identify its influence on the classifier accuracy. In order to do so, 40 images from each class were chosen at random. To estimate the optimal ratio between the training and testing data, different ratios of training and testing images were taken, and the classifier was trained for each case and the classifier accuracy on the test images was measured. Cross-validation sets of 10 and 20 training images, respectively, were taken, and the classifier accuracy on those sets (denoted by CV-10 and CV-20) were also measured. The results are shown in Table 1.

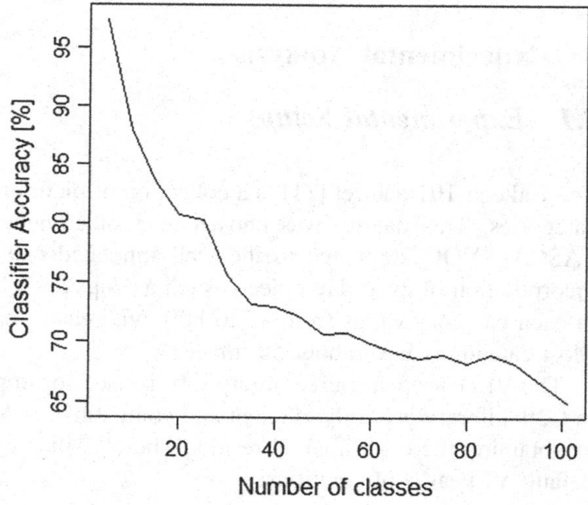

Fig. 2 Classifier accuracy versus number of classes

Table 1 Classifier accuracy percentages for different training and test image ratios

Percentage of images used for training	Test accuracy (%)	CV-10 accuracy (%)	CV-20 accuracy (%)
50	87.50	98.0	97.5
55	88.69	98.0	96.5
60	86.88	98.0	97.0
65	87.14	99.0	97.5
70	85.00	97.0	96.5
75	85.00	97.0	97.0
80	85.00	98.0	97.0
85	90.00	97.0	96.5
90	97.50	97.0	97.0

Although the classifier accuracy is very high, if 90 % of the images are chosen for training, the SVM should not be trained under such a configuration as the classifier would tend to overfit the data, and it might misclassify new images that are not present in the training or the test set. Also, as the number of training images grows, the time taken to train the SVM classifier would be very high. Hence, it is advisable to not use such a large number of training samples while training the classifier, especially for a real-world scenario. Underfitting occurs when few number of images (less than 50 % of the available data) are used for training, as the trained classifier might generalize more on the given data. Hence, a few misclassifications may occur when new examples are presented to it. After verifying the performance of the system from Table 1, we can infer that the classifier shows high accuracy on both the test and cross-validation sets when the percentage of images used for training is about 60–65 %. Hence, the optimal train-to-test ratio for the classifier is fixed at approximately 3:2, yielding an overall accuracy of 86–87 % on the test set.

However, in real-time applications, an accuracy of greater than 90 % is desirable. Hence, the class-wise accuracy of the classifier has been investigated to gain further insights into the reason for lower accuracy scores, so that the scores may be improved in the future (Table 2). From Table 2, it is easy to divide the classes into

Table 2 Class-wise accuracy of the classifier

Class	Classifier accuracy (%)
Cellphone	98.31
Chair	90.32
Cup	98.25
Headphones	100
Lamp	86.69
Laptop	96.30
Scissors	100
Stapler	95.56
Umbrella	89.33
Watch	93.72

two subgroups: Subgroup-I, consisting of the categories that have a classifier accuracy of >95 % (cellphone, cup, headphones, laptop, scissors, and stapler); and Subgroup-II, consisting of those categories which have a classifier accuracy of <95 % (chair, lamp, umbrella, and watch). An analysis on the discrepancies of the class-wise accuracy are presented in the next section.

4 Analysis of Results

Although the classifier achieves an accuracy of 86–87 % on the training set, its class-wise accuracy ranges from 86 to100 %, as shown in Table 2. This might be attributed to a variety of reasons, which are discussed in detail below.

4.1 Variety of Training Images Within a Class

In the case of subgroup-I objects, the training images tend to more or less look the same, as in real life. For example, images of two different pairs of scissors will look very similar, even if procured from different sources (Fig. 3a). As the training images look very similar, SIFT features extracted from these images, and the subsequent visual word dictionary constructed for these classes will not display high values of variance. Thus, the classification error on such classes will be very low.

On the other hand, in the case of subgroup-II objects, one training image may differ vastly from another training image. For example, in a real-world scenario, two chairs may not look the same. One may be a chair with handles, one may be a rocking chair, etc. (Fig. 3b). Thus the SIFT features, as well as the visual word dictionaries associated with these classes would display high values of variance, and this would lead to a classification error when new samples are presented to the classifier. Lower values of classifier accuracy, relative to Subgroup-I objects would be displayed.

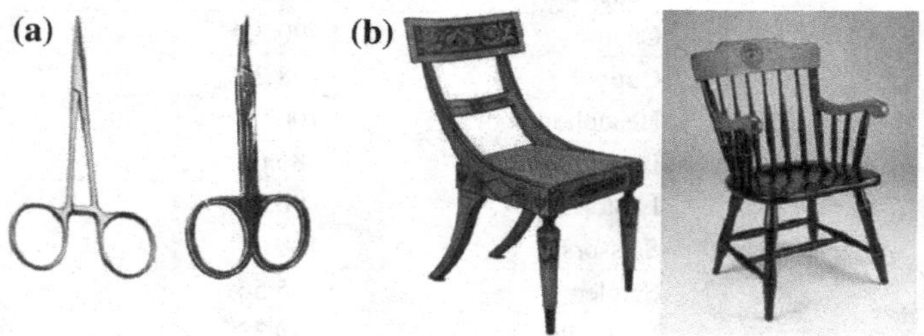

Fig. 3 Examples of images within categories. **a** Scissors and **b** Chair

Since the selection of training and test images from the various object categories is randomized, it is difficult to ascertain which subset of training images is suitable for the classifier to achieve high accuracy values. Manual selection of optimal training images for the classifier might cause the classifier to overfit. Thus, a more robust feature representation must be developed for the real-world scenario, especially for such object categories, or significant changes in the algorithm must be made to achieve low classification error.

4.2 Features Extracted from Each Object Class

In the case of Subgroup-I objects, the objects are made of well-defined geometric lines. Hence, clustering of SIFT features into visual words will result in more features per visual word, leading to an overall better representation of the object. With such a strong feature descriptor, the classifier will be able to achieve high classification rate for these classes.

However, this is not the case with some objects in Subgroup-II, such as a watch, which have rounded edges and no well-defined geometric shape, in which case, the number of features may be very less, leading to a poor representation of the object involved. Thus, generally, with such weak feature representations of the object, the classifier will not be able to achieve the same level of classification rates as Subgroup-I.

5 Conclusion and Future Work

In this paper, a bag of SIFT representations and linear SVM classifier was constructed, and the accuracy of the classifier was studied. The optimal train–test ratio for the classifier was identified as 3:2. The class-wise accuracy for various household objects was observed, and the reasons for misclassification were elucidated. The challenges involved in household object recognition in real-time environments are during the image acquisition phase, where there are constraints such as illumination and background. We envision a system that can perform preprocessing to locate the object of interest for accurate classifications in a real-time environment.

References

1. Andreopoulos A, Tsotsos JK (2013) 50 Years of object recognition: directions forward. Comput Vis Image Underst 117(8):827–891
2. Lowe DG (2004) Distinctive image features from scale-invariant keypoints. Int J Comput Vis 60(2):91–110

3. Se S, Lowe D, Little J (2001) Vision-based mobile robot localization and mapping using scale-invariant features. In: Proceedings of the IEEE international conference on robotics and automation, 2001 ICRA, vol 2, pp 2051–2058. IEEE, 2001
4. Brown M, Lowe DG (2007) Automatic panoramic image stitching using invariant features. Int J Comput Vis 74(1):59–73
5. Mikolajczyk K, Schmid C (2005) A performance evaluation of local descriptors. IEEE Trans Pattern Anal Mach Intell 27(10):1615–1630
6. Ke Y, Sukthankar R (2004) PCA-SIFT: a more distinctive representation for local image descriptors. In: Proceedings of the 2004 IEEE computer society conference on computer vision and pattern recognition, 2004. CVPR 2004, vol 2, pp II-506. IEEE, 2004
7. Joachims T (1998) Text categorization with support vector machines: Learning with many relevant features. Springer, Berlin Heidelberg
8. Sivic J, Zisserman A (2009) Efficient visual search of videos cast as text retrieval. IEEE Trans Pattern Anal Mach Intell 31(4):591–606
9. Issolah M, Lingrand D, Precioso F (2013) Sift, bow architecture and one-against-all support vector machines. In: Working notes of CLEF 2013 conference, 2013
10. Vedaldi A, Fulkerson B (2010) VLFeat: an open and portable library of computer vision algorithms. In: Proceedings of the international conference on Multimedia, pp 1469–1472. ACM, 2010
11. Fei-Fei L, Fergus R, Perona P (2007) Learning generative visual models from few training examples: an incremental bayesian approach tested on 101 object categories. Comput Vis Image Underst 106(1):59–70
12. Elkan C (2003) Using the triangle inequality to accelerate k-means. ICML 3:147–153
13. Duan KB, Sathiya Keerthi S (2005) Which is the best multiclass SVM method? An empirical study. In: Multiple classifier systems, pp 278–285. Springer Berlin Heidelberg, 2005

Optimization of Boiler Efficiency at Mettur Thermal Power Station

Muhamed Muzaiyen S.M. Bijli, T. Paneerselvam, K. Alahiyanambi and M.A. Jafar Basha

Abstract Electricity generation and environment protection are two important aspects of a coal-fired power plant. In this framework, it is required to optimize the boiler efficiency which in turn helps to generate maximum amount of power in an optimized environment. Boiler efficiency can be increased by reducing losses inside the boilers which are done using good quality fuels (Coal). This could provide environmental protection by minimizing the amount of green house gases that exit the boilers which also indirectly increase the efficiency of the boilers. Particle swarm optimization (PSO) is a nontraditional approach, applied to the observed and collected data from mettur thermal power station (MTPS). MATLAB tool is used for programming and simulating the results. The optimal result obtained from PSO is compared to the existing parameters which control the boiler efficiency and PSO technique is found to give better result to improve the boiler efficiency.

Keywords Boiler efficiency · Particle swarm optimization · Indirect method · Heat loss method · Thermal power plant

M.M.S.M. Bijli (✉)
Department of Instrumentation and Control, SASTRA University, Tirumalaisamudram
Thanjavur 613401, India
e-mail: muzaiyan30@gmail.com

T. Paneerselvam
Department of Mechanical Engineering, SASTRA University, Tirumalaisamudram,
Thanjavur 613401, India
e-mail: tpansel@mech.sastra.edu

K. Alahiyanambi · M.A. Jafar Basha
Control and Instrumentation Department, Mettur Thermal Power Station, Mettur Dam, Salem
636406, India
e-mail: nambhi77@gmail.com

M.A. Jafar Basha
e-mail: jafar67@gmail.com

L.P. Suresh and B.K. Panigrahi (eds.), *Proceedings of the International Conference on Soft Computing Systems*, Advances in Intelligent Systems and Computing 397, DOI 10.1007/978-81-322-2671-0_56

1 Introduction

A thermal power station is a type of power station that burns chemical combustibles to produce electricity. A thermal power station uses energy conversion in three sequential steps. At first, the chemical energy of a combustible is converted into heat. Second, the heat is converted into mechanical energy that is finally converted into electricity. The heat is generated during burning processes in a boiler, burning chamber or fuel cell; the mechanical energy is converted from heat during gas expansion in some type of rotating machine, which finally operates an electrical generator to produce the electricity. The rotating machine can be a steam turbine, a gas turbine, or a piston engine. A basic technological scheme of a thermal power station is given in Fig. 1.

Boiler is a kind of vessel that is used to provide a combustion heat which is transferred to water until it converts into heated water or steam. A combustible (coal) is burned with air supply in the boiler; the chemical energy of the combustible is converted into a heat and transferred into a working media (water). Pressurized feed-water is warmed at constant pressure up to boiling temperature; this is converted into to a saturated steam and superheated to required higher temperature. Water is a cheap and easily available medium which is used to transfer heat to a process. Volume of the water increases about 1600 times when it is converted into steam, resulting in production of a force that may result in an explosion like gunpowder [1] due to this the boiler becomes dangerous equipment to handle.

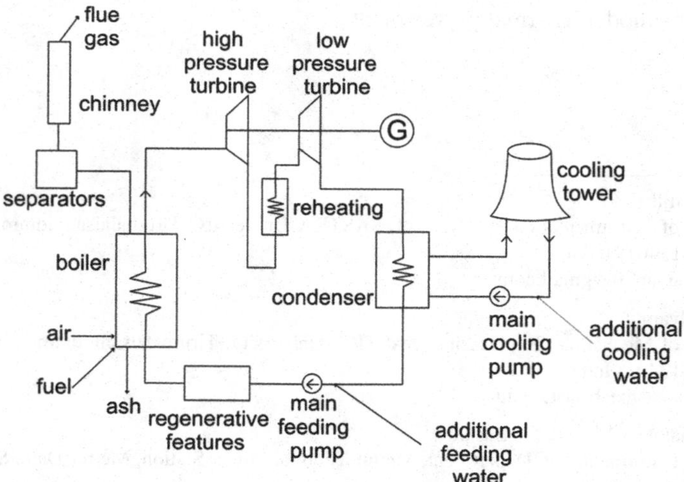

Fig. 1 Basic technological scheme of thermal power plant

Maximization of boiler efficiency will lead to obtaining maximized output. This maximization is done by minimizing various losses. To minimize these losses, we are using particle swarm optimization (PSO). This optimization is done by taking into account constraints such as ambient temperature, flue gas temperature, air temperature, etc.

Boiler efficiency is optimized by PSO Technique. PSO is based on population where stochastic optimization is done. The system is initialized with population of solutions randomly which helps in finding out the optimum value by updating generations [2–4]. In PSO, the likely solutions known as particles search through the problem space by following the optima particles at that stage. In each and every iteration, the updating of particle is done by two "best" values. Best result (fitness), attained so far called p-best. Best values attained into that point by any particle in the population is known as g-best [5–7].

1.1 Operation of Boilers

The boiler is used to convert water into steam. For a boiler to run more efficiently, the deaerator and economizer processes are important. Water tube boilers are used in MTPS. The boiler consists of steam drum and water tubes. In these kinds of boilers, the tubes contain the water and hot gases are contained outside the tubes and drums. Feed-water enters the boiler and passes through the tubes which are connected external to drums. Water in tubes is converted into steam by the hot gases that surround the tubes. This steam in the tube is deposited at the top of the drum because of its light weight which is later collected for other processes. Rate of heat transfer becomes high because of the movement of water in water tubes is high, resulting in greater efficiency. These types of boilers produce high pressure, easy accessibility and during steam demand they can respond quickly. Basic boiler operation diagram is given in Fig. 2 [1].

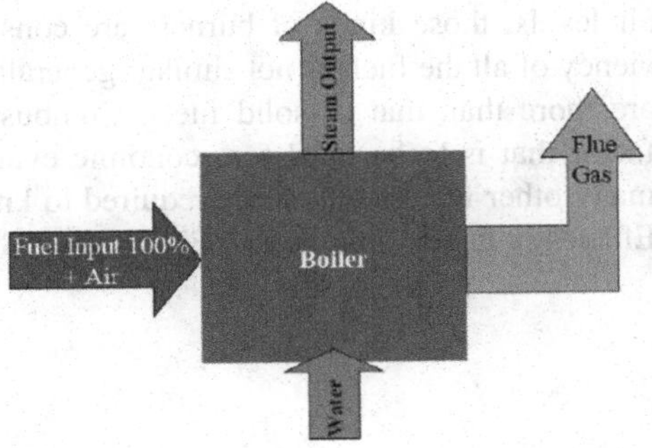

Fig. 2 Basic block diagram of boiler operation

2 Boiler Efficiency

Efficiency is described as the degree to which cost, effort or time, is well-utilized for the desired job. It is the capability through which precise appliance of effort is used to obtain a intended result with a smallest amount or measure of expense, unnecessary exertion, or waste [1]. In general, efficiency is given by the ratio of output to input. The efficiency of boilers is depended on thermal efficiency, combustion efficiency, and overall efficiency.

2.1 Thermal Efficiency

The measure of heat exchanged in the boiler is known as thermal efficiency. It is used to show how good a heat exchanger is utilized for the purpose of generating heat that is given to water from combustion process. The convection and conduction losses from the boilers are omitted [8, 9].

2.2 Combustion Efficiency

The combustion process of boilers in general is described as burning of carbon contained fuels (coal, oil, and gas) with oxygen to produce heat [10]. The necessary oxygen is supplied to the burner of the boiler as air. The type of fuel used determines the quantity of air needed. The complete combustion of fuel is ensured by providing excess air (Secondary air). This excess air leads to lowering of boiler efficiency [1] (when this excess air passes through the boiler, the heat is removed from the boiler as it is a cold air), so the quantity of excess air that is to be supplied should be optimum. Combustion efficiency indicates the burner's capability of burning the fuel and also shows the capability of boiler to soak up the generated heat. When the unburned fuel levels are extremely low with respect to low excess air levels, those kinds of burners are considered very efficient. Combustion efficiency of all the fuels is not similar; generally, liquids and gaseous fuels efficiencies are more than that of solid fuels. Combustion efficiency should not be the only factor that is to be used for economic evaluations as it does not give account of many other factors which are required to know the boilers fuel usage. Combustion Efficiency is also known as "Flue loss or Stack loss Efficiency" [11].

2.3 Overall Efficiency

The overall boiler efficiency gives the quantity of how well the boiler can transfer the heat obtained from the combustion process to the steam. Boiler's overall efficiency can be measured using direct method and indirect method. Direct method is also called as "input-output method" [1]. The determination of efficiency using input–output method would require precise measurement of all inputs as well as all outputs. In the direct method, the output (steam) and input (coal) are measured and the ratio output/input is given as efficiency. In the indirect method, the input is assumed to be 100 % and various losses encountered are calculated and subtracted from 100. Ash is formed from flame-resistant materials in the field for solid fuels like wood and coal [12], this ash could also affect efficiency. Various Losses considered for calculations are as follows:

- Loss due to unburned carbon in refuse
- Loss due to heat in dry gas
- Loss due to moisture in fuel
- Loss due to moisture from burning of hydrogen
- Loss due to moisture in air
- Loss due to heat in atomizing steam
- Loss due to radiation and convection.

2.4 Boiler Efficiency-Significance

Abundant amount of electricity would be required in India in fast approaching years. In India, gas, coal, and oil are used to produce 80 % of electricity. Boilers are one of the important equipment for generating electricity. Operating boiler efficiently is a critical aspect in power plants. Understanding boiler's performance characteristics would help us in improving overall efficiency by optimizing boiler operation in power plant besides this it also helps us to decrease emission. Boiler performance, like evaporation ratio and efficiency decreases with time because of operation and maintenance and deprived combustion. Worsening of fuel and water class may also lead to deprived performance of boiler. Testing efficiency online aid us to identify what is the boiler efficiency at present and how much it moves away from the best efficiency. It is mandatory to know the exact point of efficiency for evaluation of performance, which is required for energy conservation action in industries [13, 14].

3 Indirect Method

Boiler Efficiency is calculated from Eq. (1). Maximization of boiler efficiency is obtained minimizing the losses.

$$\text{EFBL} = 100 - \text{LUBC} - \text{LHDG} - \text{LFM} - \text{LMBH} - \text{LMA} - \text{LHAS} - \text{LRC} \quad (1)$$

where

EFBL boiler efficiency
LUBC loss due to unburned carbon in refuse
LHDG loss due to heat in dry gas
LFM loss due to moisture in fuel
LMBH loss due to moisture from burning of hydrogen
LMA loss due to moisture in air
LHAS loss due to heat in atomizing steam
LRC loss due to radiation and convection

3.1 *Varying Parameters*

The parameters which affect the efficiency of the boiler as per the calculations done are:

- Ambient temperature
- Flue gas (FG) oxygen % at air preheater (APH) outlet
- Gas temp at APH inlet
- Gas temp at APH outlet
- Air temp at APH inlet (Ambient temperature)
- Total coal flow

Ambient temperature, air temperature at APH inlet, and Gas temp at APH inlet are taken as constant for calculations.

Average FG oxygen % at APH outlet, gas temp at APH outlet, and total coal flow are taken as varying parameters in PSO algorithm.

3.2 *PSO Algorithm*

- Each particle is initialized.
- Fitness value for every particle is calculated. Fitness values are compared and the best fitness value is updated. This is the new p-best.

- The best fitness value of all the particles is known as g-best.
- For each particle, calculate particle velocity according to equation.

$$\text{(a) } v[] = c1 * \text{rand}() * (\text{pbest}[] - \text{present}[]) + c2 * \text{rand}() * (\text{gbest}[] - \text{present}[]) \quad (2)$$

Update the particle position according to equation

$$\text{(b) New } v[] = \text{present}[] + v[] \quad (3)$$

3.3 Technical Data of Power Plant

Power generated per unit: 210 MW
Boiler type: single drum hanging type boiler
Ambient temperature = 40 °C
Air temperature at APH inlet = 40 °C
Gas temp at APH inlet = 350 °C
Average FG oxygen % at APH outlet = 2–3 %
Gas temp at APH outlet = 150–170 °C
Total coal flow = 120–150 T/h

3.4 Efficiency Calculation Formulae

Loss L1 due to heat in dry gas Loss due to Latent heat in dry gas

$$\text{LHDG} = (\text{LHDG1} * 100)/\text{HHOA} \quad (4)$$

$$\text{LHDG1} = \text{FDG} * \text{CDG} * (\text{TGA} - \text{TAA}) \quad (5)$$

where

FDG Dry flue gas flow at AH outlet
CDG Specific heat of dry flue gas
TGA Avg FG temp at AH outlet
TAA Ambient temperature
HHOA High heat value of fuel gas fired

Loss L2 by hydrogen moisture in the fuel Loss caused by evaporation of the moisture produced from hydrogen in the fuel and the contained moisture during combustion of the fuel, moreover the loss caused by heating up to the temperature of exhaust gas and discharged:

$$LMBH = (4.053 * FH * (HE - HRW) * 100)/HHOA \quad (6)$$

where

FH	Hydrogen content of fuel
HE	Enthalpy of vapor in flue gas
HRW	Enthalpy of saturated liquid at draft fan outlet
HHOA	High heat value of fuel gas fired

Loss L3 due moisture in the air The loss caused by latent heat of moisture contained in the air for combustion is

$$LMA = (FMA * (HE - HRV) * 100)/HHOA \quad (7)$$

where

FMA	Moisture from air
HE	Enthalpy of vapor in flue gas
HRV	Enthalpy of saturated vapor at draft fan outlet
HHOA	High heat value of fuel gas fired

Loss L4 by radiation heat It is difficult to accurately obtain the heat loss radiated into the atmosphere from the peripheral walls of the boiler. This loss becomes proportionally smaller with large capacity boilers because their surface area becomes relatively smaller and also because the radiation heat amount is roughly constant irrespective of the load; the proportion of loss becomes smaller as the load becomes larger.

This loss is determined using the curve which is shown in Fig. 3.

X-axis	Main Steam flow—T/H
Y-axis	Radiation and Convection Loss—%

Loss L5 due to moisture in fuel This is the heat loss due to the combustible gas remaining such as CO in the fuel gas because of incomplete combustion.

$$LFM = (LFM1 * 100)/HHOA \quad (8)$$

$$LFM1 = H2OT * (HE - HRW) \quad (9)$$

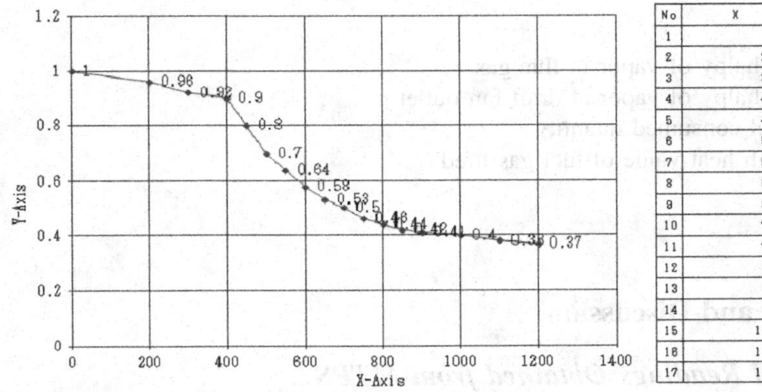

No	X	Y
1	0	1
2	200	0.96
3	300	0.92
4	400	0.9
5	450	0.8
6	500	0.7
7	550	0.64
8	600	0.58
9	650	0.53
10	700	0.5
11	750	0.46
12	800	0.44
13	850	0.42
14	900	0.41
15	1000	0.4
16	1100	0.38
17	1200	0.37

X-axis – Main Steam flow – T/H
Y-axis – Radiation and Convection Loss - %

Fig. 3 Radiation and convection loss *curve*

where

H2OT Total H2O in Fuel
HE Enthalpy of vapor in flue gas
HRW Enthalpy of saturated liquid at draft fan outlet
HHOA High heat value of fuel gas fired

Loss L6 due to unburned CO in refuse This is heat loss mainly by unburned carbon in the combustion residue by combustion of solid fuel.

$$\text{LUBC} = (\text{HHC} * \text{KKF} * 100)/\text{HHOA} \quad (10)$$

where

HHC Heat value of carbon
KKF Unburned carbon in fuel per 1 kg
HHOA High heat value of fuel gas fired

Loss L7 due to heat in atomizing steam

$$\text{LHAS} = (\text{LHAS1} * 100)/\text{LHAS2} \quad (11)$$

$$\text{LHAS1} = 17 * (\text{HE} - \text{HVFA}) \quad (12)$$

$$\text{LHAS2} = \text{MLTFH} * \text{HHOA} \quad (13)$$

where;

HE	Enthalpy of vapor in flue gas
HVFA	Enthalpy of vapor at draft fan outlet
MLTFH	Fuel consumed quantity
HHOA	High heat value of fuel gas fired

4 Results and Discussion

4.1 Actual Readings Obtained from MTPS

- Ambient Temperature = 40.1 °C
- Flue Gas (FG) Oxygen % APH Outlet = 2.659 %
- Gas Temp at APH inlet = 356.01 °C
- Gas Temp at APH outlet = 161.06 °C
- Air Temp at APH inlet = 40.1 °C
- Total Coal Flow = 125.9

Efficiency obtained = 89.5 %

4.2 Results Obtained Using PSO

Ambient temperature, gas temp at APH inlet, and air temp at APH inlet are kept as constant for theoretical purpose as controlling these parameters would require a costly setup.

- Ambient Temperature = 40 °C
- Air Temp at AH inlet = 40 °C
- Gas temp at AH inlet = 350 °C
- Total Coal Flow = 150 T/h
- Flue Gas (FG) Oxygen % AH Outlet = 2 %
- Gas temp at AH outlet = 150 °C

Boiler efficiency obtained using PSO in simulation = 90.23 %.

Figure 4 shows how three particles converge on to maximize the efficiency of boiler. Blue line denotes particle 1 which reaches the maximum efficiency at second iteration. Red line denotes particle 2 which attains maximum efficiency at 6th

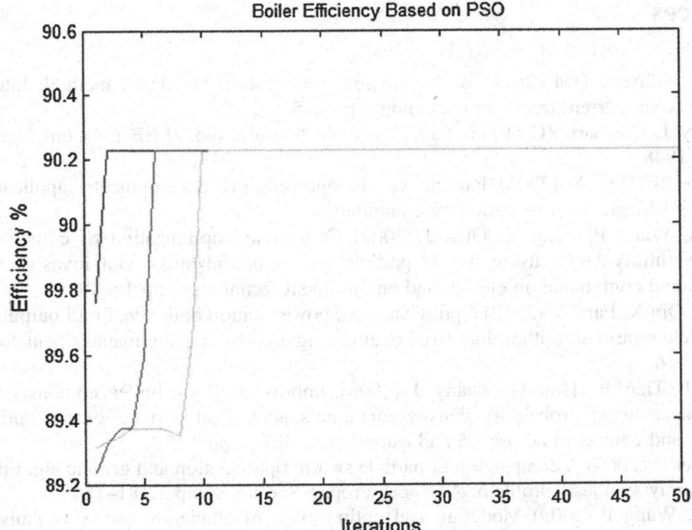

Fig. 4 Efficiency attained by different particles

iteration. Green line denotes 3rd particle which attains maximum efficiency at 11th iteration. For simple calculations, easy understanding and reducing the time only three particles are shown in the graph.

5 Conclusion

In this paper, indirect method (i.e., heat loss method) is used to calculate boiler efficiency. Boiler Efficiency at MTPS is taken as reference and PSO algorithm is employed to increase the efficiency of the boiler. The simulation results show that there is an increase of 0.7 % (from 89.5 to 90.23 %) in efficiency of the boiler if all the parameters are maintained to the above-mentioned optimized values. Using PSO algorithm in MATLAB, the results are obtained in very quick time. The above outcomes illustrate that the proposed technique is very easy and gives the required improvement in output. The results obtained in this paper were obtained using industrial data and concepts were demonstrated in industry.

Acknowledgments The authors like to thank Chief Engineer K. Siva Prakasam and all the employees of MTPS for providing valuable guidance. The authors would also like to thank Dr. S. Jayalalitha, Associate Dean, Dr. K. Ramkumar, Senior Associate Professor and other staffs of Department of Electronics and Instrumentation Engineering, SASTRA University for the encouragement provided to take up the work and for offering valuable inputs as and when required.

References

1. Shah S, Adhyaru DM (2011) Boiler efficiency analysis using direct method. International conference on current trends in technology, pp 1–5
2. Kennedy J, Eberhart RC (1995) Particle swarm optimization. IEEE Int Conf Neural Netw 4:1942–1948
3. Eberhart RC, Shi Y (2001) Particle swarm optimization: developments, applications and resources. Congress on evolutionary computation
4. Zhao H, Wang P, Peng X, Qian J (2009) Constrained optimization of combustion at a coal-fired utility boiler using hybrid particle swarm optimization with invasive weed. In: International conference on energy and environment technology, pp 16–18
5. Fang Y, Qin X, Fang Y (2012) Optimization of power station boiler coal mill output based on the particle swarm algorithm. Industrial engineering and engineering management conference, pp 612–616
6. Zhang J, Tian B, Hou G, Zhang J (2008) Improvement of boiler combustion control performance using probability density function shaping and particle swarm optimization. Systems and control in aerospace and astronautics, IEEE, pp 1–5
7. Boeringer D (2003) A comparison of particle swarm optimization and genetic algorithms for a phased array synthesis problem. Antennas Propag Soc Int Symp 1:181–184
8. Zhao H, Wang P (2009) Modeling and optimization of efficiency and NOx emission at a coal-fired utility boiler. Power and energy engineering conference, pp 1–4
9. Li Y, Jilin C, Gao H (2010) On-line calculation for thermal efficiency of boiler. Power and energy engineering conference, pp 1–4
10. Zhang S, Zhou H, Peng M et al (2002) A study on combustion control of a coal-fired power generation unit based on furnace radiant energy signal. In: Proceedings of the CSEE, pp 660–668
11. Combustion analysis basics: an overview of measurements, methods and calculations used in combustion analysis. TSI Incorporated
12. Turner WC, Doty S (2007) Energy management handbook, 6th edn. The Fairmont Press Inc
13. 'Boilers' bureau of energy efficiency
14. Cleaver Brooks. Boiler efficiency guide

A Comparative Study of Inverted Pendulum

S. Nemitha, B. Vijaya Bhaskar and S. Rakesh Kumar

Abstract The aim of this paper is to design a controller for Inverted Pendulum. With the help of PID Controller, it develops for tuning Fuzzy Logic Controller, Adaptive Neuro Fuzzy Inference System (ANFIS) Controller. PID Controller has been certain limitations. These limitations can be taken to acumen methods. This paper presents the acumen methods based on Fuzzy Logic, ANFIS for tuning a PID Controller used for the control of Inverted Pendulum. It reveals Fuzzy Logic Controller and ANFIS are acumen methods provide better performance than the PID Controller.

Keywords PID Controller · Fuzzy Logic Controller · ANFIS · Inverted Pendulum

1 Introduction

The Inverted Pendulum is one of the most ancient problems of control engineering. Inverted Pendulum is highly nonlinear system and unstable (upwards) in open loop. It consists of a moving cart and oscillating pendulum. The aim of the work is to move the cart to its commanded position without causing the pendulum to tip over. PID Controller is a commonly used in controlling industrial loops owing to its simple structure, easy implementation, robust nature, and less numbers of

S. Nemitha (✉)
Department of Instrumentation and Control, SASTRA University, Thirumalaisamudram, Thanjavur 613401, India
e-mail: nemithasugumar@gmail.com

B. Vijaya Bhaskar · S. Rakesh Kumar
Department of Electronics and Instrumentation, SASTRA University, Thirumalaisamudram, Thanjavur 613401, India
e-mail: vijayabhaskar@eie.sastra.edu

S. Rakesh Kumar
e-mail: srakesh@eie.sastra.edu

L.P. Suresh and B.K. Panigrahi (eds.), *Proceedings of the International Conference on Soft Computing Systems*, Advances in Intelligent Systems and Computing 397, DOI 10.1007/978-81-322-2671-0_57

parameters [1, 2]. Proportional, integral, and derivative are the controllers output, a control variable (CV) is a measure of error (e) between set point (SP) and process variable (PV) [3]. However, the method is not suitable for higher order systems. Fuzzy control techniques can provide a good solution for those above problems by introducing linguistic information. Fuzzy systems are knowledge-based or rule-based systems that contain descriptive IF-THEN rules that are created from human knowledge and experience. ANFIS is a combination of both Neural Network and Fuzzy system. The ANFIS control is developed using supervised learning (i.e., the input–output data sets are known). ANFIS refers to a set of adaptive networks, having a similar function corresponds to fuzzy inference systems.

The acumen method based on Fuzzy Logic and ANFIS for tuning a PID Controller has been compared. The controller tuned by various methods has been used for the control of Inverted Pendulum. The acumen methods provide better performance in terms of various performance specifications than the PID Controller.

This paper is organized in four sections. Section 1 gives the general introduction of the paper. Section 2 represents the problem formulation. Section 3 the tuning of PID Controller using various acumen methods such as Fuzzy Logic and ANFIS has been discussed. The results, comparison, and discussion are given in Sect. 4. At the end conclusion and brief list of references are given.

2 Problem Formulations

The aim of this paper is to design a PID, Fuzzy Logic and ANFIS controller. The Inverted Pendulum is unstable in an open loop system (i.e., without any controller). To control this system, we have to design a controller. Finally, these controllers which will give the effective output are compared. The dynamic equation of the cart and pendulum in Laplace transform form is as follows:

$$\left(I + ml^2\right)\Phi(s)s^2 - mgl\Phi(s) = mlX(s)s^2 \tag{1}$$

$$(M + m)X(s)s^2 + bX(s)s - ml\Phi(s)s^2 = U(s) \tag{2}$$

The plant has two transfer function model. The first is the Cart's position and the second is the pendulum of the angle given below [4]:

$$T \cdot F_{\text{Position}} = \frac{X(s)}{U(s)} = \frac{5.841}{s^2}$$

$$T \cdot F_{\text{Angle}} = \frac{\Phi(s)}{U(s)} = \frac{3.957}{s^2 + 6.807}$$

3 Conventional and Acumen Methods

3.1 *PID Controller*

A control loop feedback device for a Proportional-Integral-Derivative (PID) controller is extensively used in industrial control loops [5]. It calculates an error value as the deviation between PV and SP. The output of the PID Controller in the time domain is as follows [6]:

$$u(t) = k_p e(t) + k_i \int e(t)\mathrm{d}t + k_d \frac{\mathrm{d}e}{\mathrm{d}t} \quad (3)$$

The control signal (u) to the plant is the super-positioned output of the proportional (k_p), derivative (k_d) and integral term (k_i) of the error with respect to time.

3.2 *Fuzzy Logic Controller*

Nonlinear, unstable, process dynamics system proves that the conventional PID Controller will not produce better performance. To get better response, it is essential to tune automatically the PID Controller. Fuzzy Logic controller will do the automatic tuning of PID Controller. Fuzzy Logic system translates the linguistic control scheme into robotic control scheme. A fuzzy is characterized by a membership function which provides a grade between 0 and 1 for all the objects defined in the set. Fuzzy sets allow partial membership which means that an element may partially belong to more than one set [7]. The inputs are error and the rate of change of error (Δe) while the output is controller output (y). For finding the universe of discourse for the input and output membership functions, the PID Controller has been tuned using Ziegler–Nicholas method, and hence the input output membership functions will be found out [8].

The position FIS has nine rules with the three membership functions triangular type. The input1 (error) membership function ranges are from −1 to 1 and the input2 (derivative) membership function ranges are from −1 to 1 respectively. The output membership function ranges are from −1 to 1 respectively. The angle FIS structure has twenty five rules with the five membership functions triangular type. The input1 (error) membership function ranges are from −5 to 5 and the input2 (derivative) membership function ranges are from −7 to 7 respectively. The output membership function ranges are from −35 to 35 respectively.

3.3 ANFIS Controller

ANFIS is a combination of neural network and Fuzzy Logic features [9]. ANFIS develops a Takagi Sugeno fuzzy inference system (FIS) with the help of the input output dataset. Using the back propagation algorithm, error tolerance, and epochs the membership functions are acquired. The input–output dataset will be taken from the PID Controller tuning using conventional method for both position and angle [10–13]. The inputs are error and the rate of change of error, while the output is controller output (y). The input–output dataset will be loaded in the ANFIS toolbox for both position and angle.

Initially, the position data set will be loaded and then by assigning the above terms will acquire the membership function. We have to reduce the average error into minimum range and then it will be saved as a FIS structure. Figure 1 shows that the loading of training data in ANFIS and Fig. 2 shows that the training of FIS in ANFIS for position control.

A total of 736 data points are considered which are collected from PID position control and stored in workspace of MATLAB ANFIS toolbox. FIS are generated using subtractive clustering with two membership function and two rules.

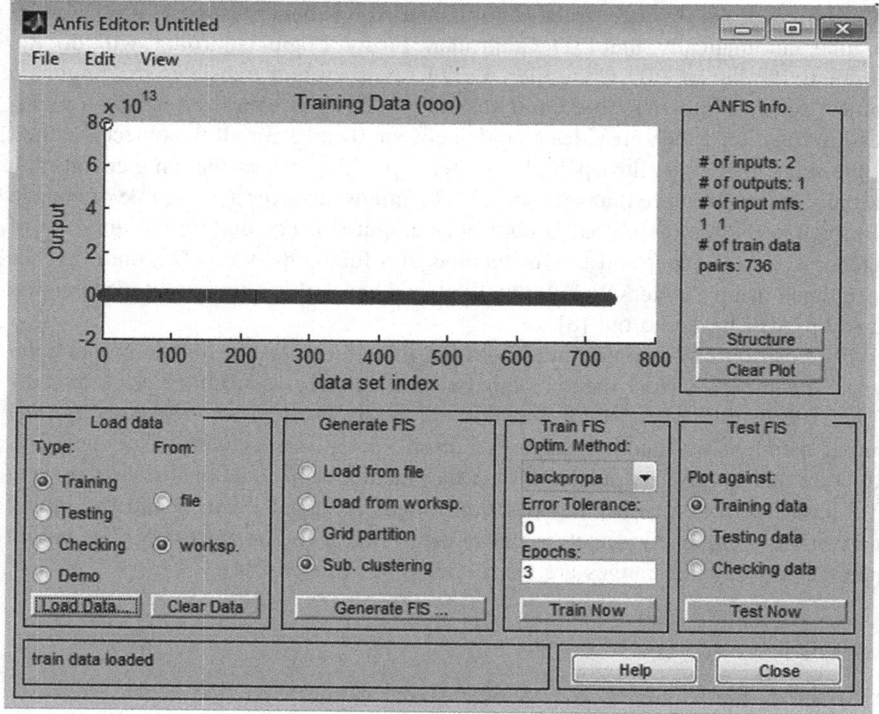

Fig. 1 Loading of training data in ANFIS

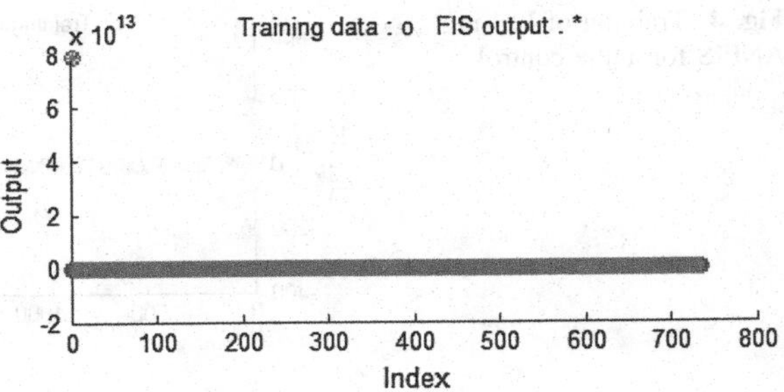

Fig. 2 Training of FIS in ANFIS for position control

The average testing error of position control is as 0.23624. Figure 3 shows that the loading of training data in ANFIS and Fig. 4 shows that the training of FIS in ANFIS for angle control.

A total of 2740 data points are considered which are collected from PID angle control and stored in workspace of MATLAB ANFIS toolbox. FIS are generated using grid partition with five membership function and twenty five rules. The error of angle control is as 0.57216. That FIS structure should be placed in the control

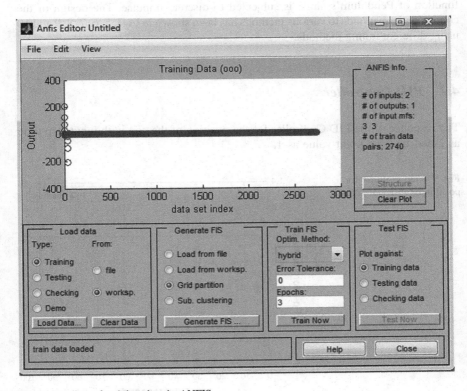

Fig. 3 Loading of training data in ANFIS

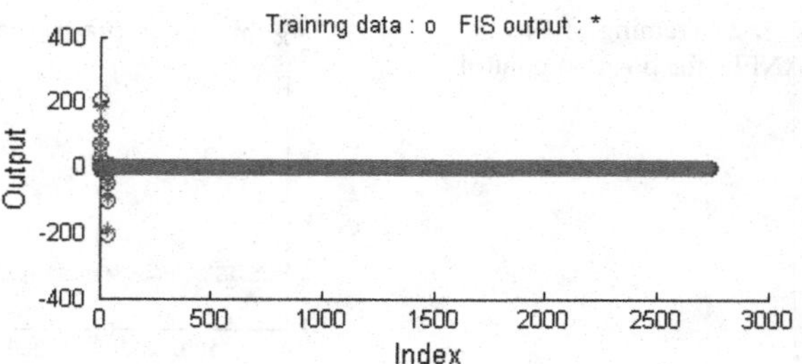

Fig. 4 Training of FIS in ANFIS for angle control

system block instead of controller. The input can be any test signal, usually we use step signal as input for the position control and we use discrete impulse as input for the angle control. Hence, the output should track the input which we are giving.

4 Results, Comparison, and Discussion

The transfer function of Cart's Position is subjected to step input and the transfer function of Pendulum's angle is subjected to discrete impulse. The design of the controller corresponds to various methods has been shown in Figs. 5, 6, 7, 8, 9 and 10 using different tuning methods.

4.1 PID Controller

The response of the PID Controller for the position of the cart with input step time as 2 and final set point value as 1.

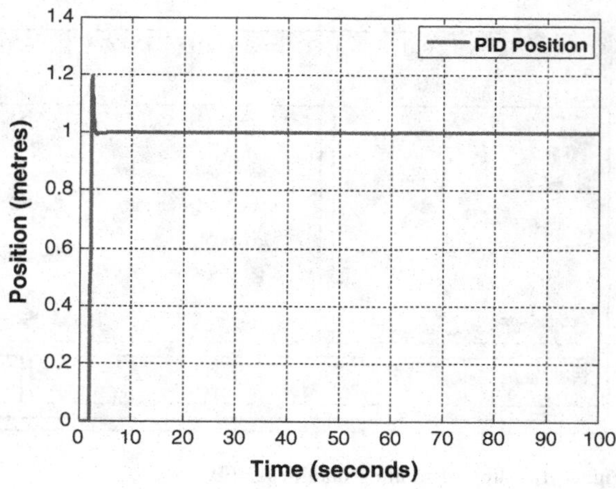

Fig. 5 Step response for position of PID Controller

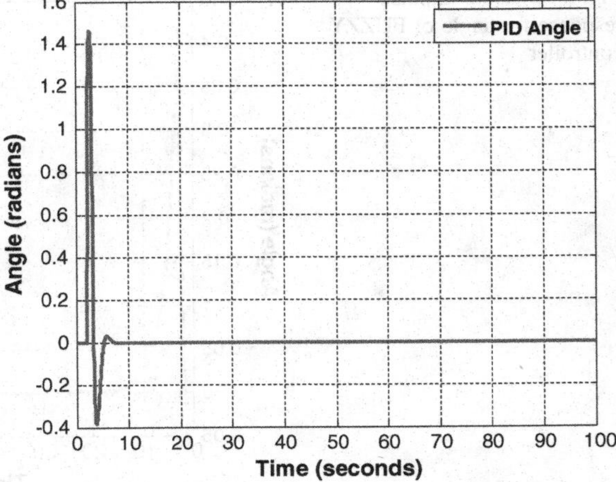

Fig. 6 Discrete impulse response for angle of PID Controller

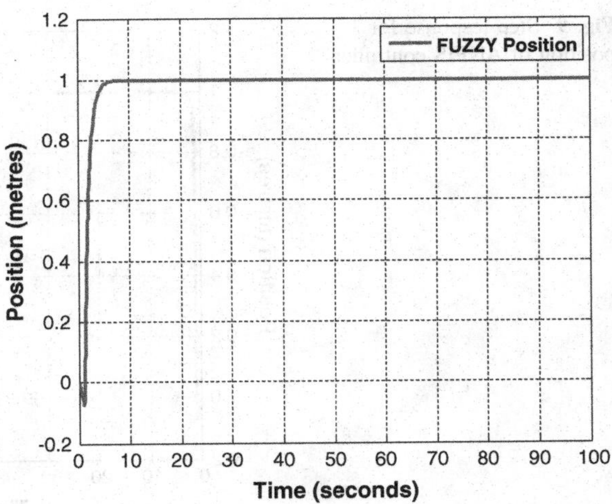

Fig. 7 Step response for position of FUZZY controller

The response of the PID Controller for the Pendulum's angle with discrete impulse input delay as 2 and final set point value as 0.

4.2 *Fuzzy Logic Controller*

The response of the Fuzzy Logic controller for the position of the cart with input step time as 1 and final set point value as 1.

The response of the Fuzzy Logic controller for the angle of the pendulum with discrete impulse input delay as 2 and final set point value as 0.

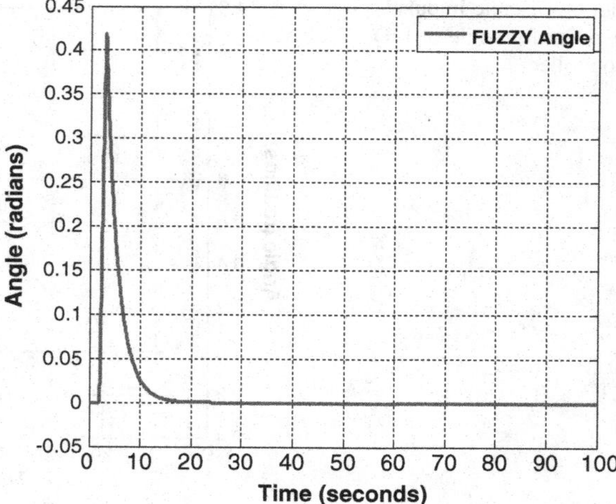

Fig. 8 Discrete impulse response for angle of FUZZY controller

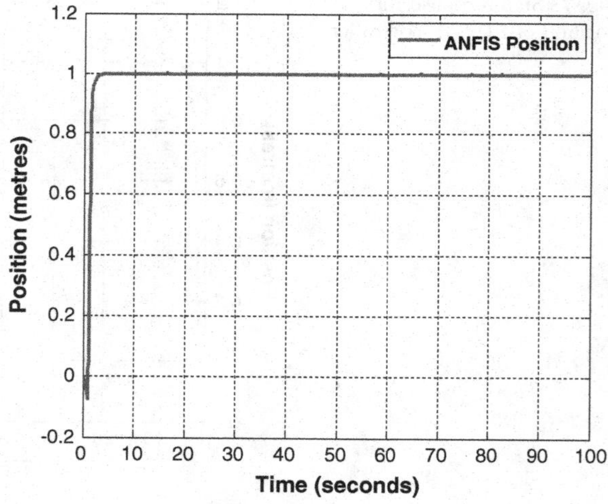

Fig. 9 Step response for position of ANFIS controller

4.3 ANFIS Controller

The response of the ANFIS controller for the position of the cart with input step time as 1 and final set point value as 1.

The response of the ANFIS controller for the angle of the pendulum with discrete impulse input delay as 2 and final set point value as 0.

The comparison among different acumen methods in terms of various performance specifications using acumen and conventional methods has been shown in Table 1.

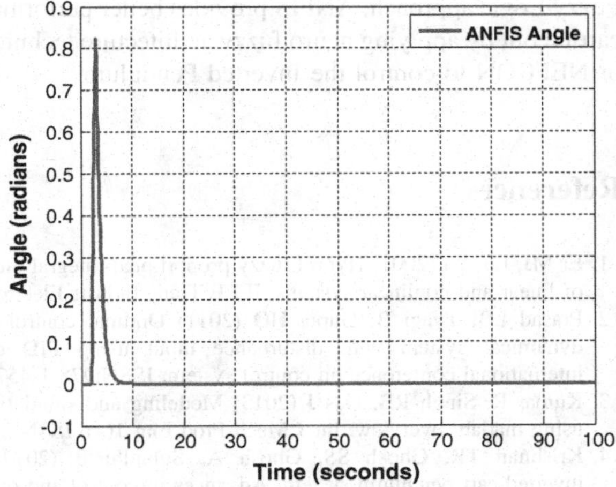

Fig. 10 Discrete impulse response for angle of ANFIS controller

Table 1 Comparison among two acumen methods for tuning a PID Controller

Controllers	Methods	Rise time	Overshoot	Settling time
PID	Position	2.14	0.2	8.25
	Angle	2.19	0.46	9.41
FLC	Position	5.66	0	7.27
	Angle	0.45	0	16
ANFIS	Position	3.09	0	4
	Angle	0.82	0	8.91

From comparison among two acumen methods for tuning a PID Controller can be concluded that

- The rise time is less in conventional PID Controller than the acumen methods.
- Settling time and overshoot has given best performance in ANFIS and PID.
- Hence, the acumen methods give the better performance when compared with conventional PID Controller.

5 Conclusion

The controller for Inverted Pendulum has been designed. The tuning of a PID Controller is compared with Fuzzy Logic and Adaptive Neuro Fuzzy Inference System (ANFIS). The controller tuned by the acumen methods is used to control the Inverted Pendulum. This investigation result shows that by using both PID and

Fuzzy Logic approach, ANFIS provides better performance. The future work can be carried out by applying neuro fuzzy architecture techniques like FALCON, GARIC, or NEFCON to control the Inverted Pendulum.

References

1. Er MJ, Lei Ya (2001) Hybrid fuzzy proportional-integral plus conventional derivative control of linear and nonlinear systems. IEEE Trans Industr Electron 48(6):1109–1117
2. Prasad LB, Tyagi B, Gupta HO (2011) Optimal control of nonlinear inverted pendulum dynamical system with disturbance input using PID controller and LQR. In: IEEE international conference on control system. ISSN:978-1-4577-1642-3
3. Kumar R, Singh RB, Das J (2013) Modeling and simulation of inverted pendulum system using matlab: overview. Int J Mech Prod Eng 1(4). ISSN:2320-2092
4. Krishnan TR, Ghosh SS, Ghosh A, Subudhi B (2012) Periodic compensation of an inverted-cart pendulum system. Advances in control and optimization of dynamical systems (ACODS), IISC Bangalore
5. Astrom KJ, Hagglund T (1995) PID controllers: theory, design and tuning, 2nd edn. Instrument Society of America
6. Coughanowr DR (1991) Process system analysis and control, 2nd edn
7. Ji CW, Lei F, Kin LK (1997) Fuzzy logic controller for an inverted pendulum system. IEEE Int Conf Intell Process Syst. ISSN:0-7803-4253-4
8. Prasad LB, Tyagi B, Gupta HO (2011) Intelligent control of nonlinear inverted pendulum dynamical system with disturbance input using fuzzy logic systems. Int Conf Recent Adv Electr. ISSN:978-1-4577-2149-6
9. Shing J, Jang R (1993) ANFIS: adaptive network based fuzzy inference system. IEEE Trans Syst Man Cybern 23(3)
10. Kharola A, Gupta P (2014) Stabilization of inverted pendulum using hybrid adaptive neuro fuzzy (ANFIS) controller. Eng Sci Lett. ISSN:2052-5257
11. Goswami A (2013) The analysis of inverted pendulum control and its other applications. J Appl Math Bioinform 3(2). ISSN:1792-6602 (print), 1792-6939 (online)
12. Gite AV, Bodade RM, Raut BM (2013) ANFIS controller and its application. Int J Eng Res Technol (IJERT) 2(2). ISSN:2278-0181
13. Somwanshi K, Srivastava M, Panchariya R (2012) Analysis of control of inverted pendulum using adaptive neuro fuzzy system. Int J Adv Res Comput Eng Technol (IJARCET) 1(6). ISSN:2278-1323

Speech Signal Enhancement Using Stochastic Resonance

R. Prakash Kumar, V.S. Balaji and N. Raju

Abstract The quality of speech signal needs to be enhanced for applications like teleconferencing, hearing aids, speech recognition etc. Basically, speech-enhancement techniques are responsible for the removal of additive noise, to recover the noise free speech signal. This paper deals with an effective method, stochastic resonance (SR), in which the weak speech signal can be improved by increasing its signal-to-noise ratio (SNR) through addition of white Gaussian noise. Noise, though considered obnoxious in signal processing, can also be used constructively. Pertinence of stochastic resonance in signal processing is aimed at examining the detection of a weak speech signal buried in strong noise. Realizing the effect of stochastic resonance depends on the effective nonlinear system which is mathematically optimized through fourth-order Runge–Kutta algorithm and the system parameters were adjusted adaptively to examine the stochastic resonance (SR) effect on output signal-to-noise ratio and peak of the output signal spectrum.

Keywords Stochastic resonance · Signal-to-noise ratio (SNR) · White gaussian noise · Fourth-order Runge-Kutta algorithm

---

R.P. Kumar (✉)
Department of Instrumentation and Control, SASTRA University, Thirumalaisamudram, Thanjavur 613401, India
e-mail: prakashthambhu@gmail.com

V.S. Balaji
Department of Electronics and Instrumentation, SASTRA University, Thirumalaisamudram, Thanjavur 613401, India
e-mail: biobala_mtech@eie.sastra.edu

N. Raju
Department of Electronics and Communication, SASTRA University, Thirumalaisamudram, Thanjavur 613401, India
e-mail: raju@ece.sastra.edu

L.P. Suresh and B.K. Panigrahi (eds.), *Proceedings of the International Conference on Soft Computing Systems*, Advances in Intelligent Systems and Computing 397, DOI 10.1007/978-81-322-2671-0_58

1 Introduction

Users of modern communication devices, like cell phones are rapidly increasing day-by-day, are annoyed usually by any source of background hiss leading to lack in signal communication. In recent years, the development in digital speech communication applications, e.g., speech coding and speech recognition, are focused at eliminating noise for the perceptual quality of speech. The degradation in speech is the main concern in communication caused by the environment and acoustical noise leading to uncertainty. Speech-enhancement methods like wavelet transform, spectral subtraction originated for improving the speech intelligibility and quality by eliminating noise. Noise is always considered as a deteriorating component in signal processing applications. In 1980s, Italian scholar Roberto Benzie researched the positive side of noise through the phenomena stochastic resonance in periodic recurrence of earth's ice ages [1]. The stochastic resonance, an extra dose of additive noise can in fact enhance the weak signal rather than affect the signal information under certain circumstances [2]. It has been considerably used in the numerous areas of biology, communication, electronics, and so on. The output of SR contains rapid increase in the signal-to-noise ratio and the presence of peak in the output speech spectrum.

Signals and Systems—signals often refer to a function of independent variables, changing with time and nature that represents some information. A system refers to a set of physical components which takes and produces signals. Robustness to the source of noise can be an important design consideration for certain systems. In general, noise is a kind of degradation to the signals and systems, arising from limitations, unmodeled fluctuations, and environmental factors. It is randomly distributed with zero mean, having wide band of frequencies. Signal-to-noise ratio (SNR) is the measure of signal power to the noise power, which describes the quality of the signal in communications.

$$\mathrm{SNR}_{\mathrm{dB}} = 10\log_{10}\left(\frac{P_{\mathrm{signal}}}{P_{\mathrm{noise}}}\right) \tag{1}$$

Basically, speech signals are easily degraded by noise leading to low SNR value. In order to increase the output SNR, the optimal noise is added to the speech to realize SR [3]. The output SNR is examined for the quality of the speech signal and some adaptive filtering is carried out to enhance the speech intelligibility.

2 Speech Enhancement

Speech enhancement is a way of improving the speech quality and intelligibility. The way of approaching the nonstationary signal improvement is quite difficult in the field of communication. Speech signal is easily attenuated by the background noise but, can be enhanced through various speech-enhancement algorithms. A method called

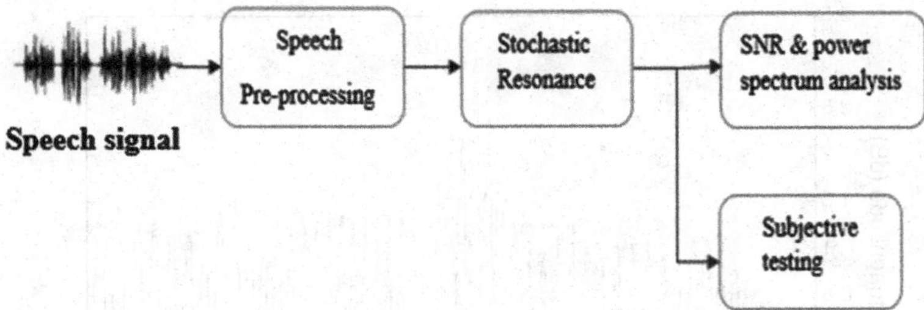

Fig. 1 Functional block diagram of speech enhancement

stochastic resonance reveals the better side of noise [4] by processing the weak speech signal through nonlinear system as already explained above. The main objectives of speech enhancement technique are noise reduction and subjective or objective quality measures. The functional block diagram is given in Fig. 1.

2.1 Speech Signal Acquisition

An input speech signal has to be acquired from the environment which contains several harmonics, region of silence, and low SNR. The input sample speech, e.g., we.wav at sample frequency of 8000 Hz is given below Fig. 2.

2.2 Speech Preprocessing

The above input speech has to be preprocessed to remove the silence region. The waveform corresponds to after preprocessing as shown in Fig. 3.

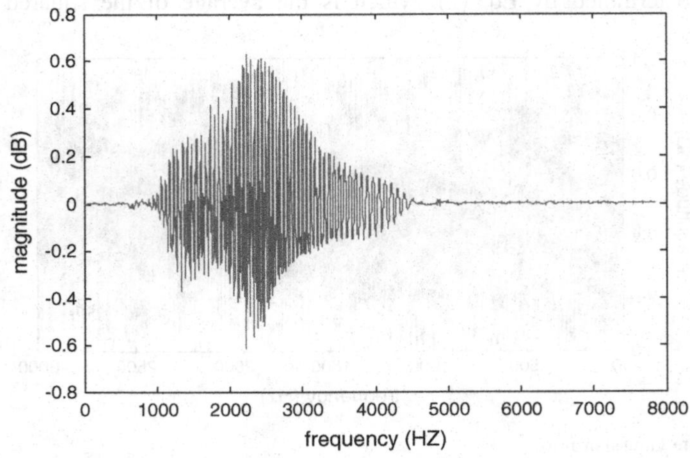

Fig. 2 A input speech signal, e.g., we.wav

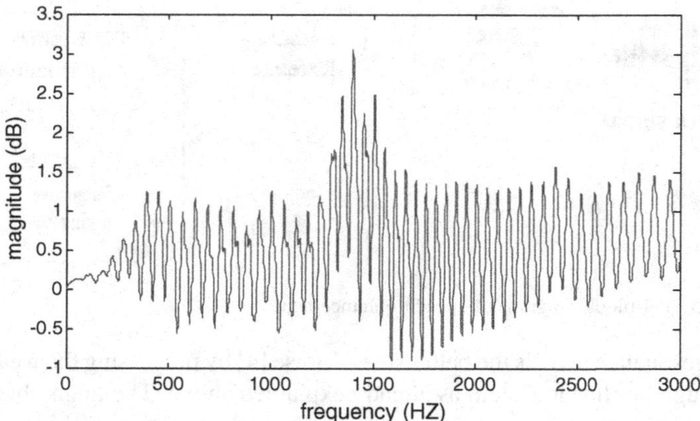

Fig. 3 Removal of silence region in the input speech

2.3 Stochastic Resonance

The effect of stochastic resonance depends on the effective nonlinear system [5] and the intensity of noise as shown in Fig. 4. The white noise is added with the above processed input speech signal, which is mathematically optimized through fourth-order Runge–Kutta algorithm and the system parameters were adjusted adaptively to examine the stochastic resonance effect.

2.4 SNR and Power Spectrum Analysis

The output signal-to-noise ratio (SNR) and power spectrum of the optimized speech signal is determined by Eq. (2), which is the average of the squared Fourier

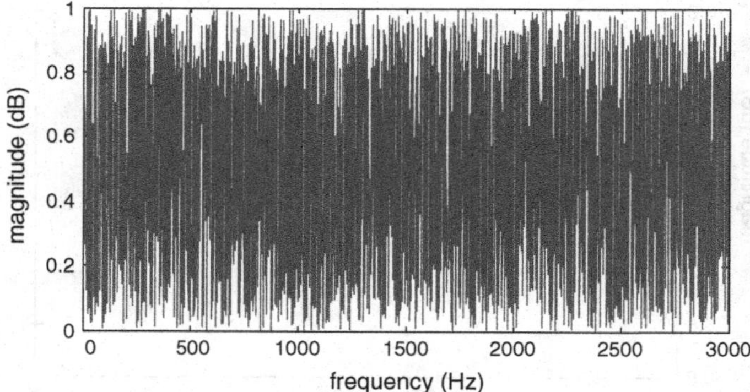

Fig. 4 White Gaussian noise

transform magnitude, over a large time interval and the output signal is validated through subjective testing.

$$S_x(f) = \lim_{T\to\infty} E\left\{ \frac{1}{2T} \left| \int_{-T}^{T} x(t) e^{-j2\pi ft} dt \right|^2 \right\} \quad (2)$$

3 Principle of Stochastic Resonance

Stochastic resonance is realized by three factors such as bistable or multistable system, input weak speech signal, and an additive white Gaussian noise [6]. The nonlinear multistable system is described by nonlinear Langevin equation (LE) which is expressed as follows:

$$\dot{x}(t) = Ax(t) - Bx^3(t) + S(t) + n(t) \quad (3)$$

where $S(t)$ is the input speech signal, $n(t)$ is the additive white Gaussian noise whose mean is zero, $E(n(t)) = 0$ and autocorrelation is given by $E(n(t)n(t-\tau)) = \sigma^2\delta(t)$. The output signal $x(t)$ is the response of stochastic resonance system which represents the enhanced speech signal.

The corresponding potential function of Eq. (3) is given as follows:

$$U(x) = \frac{1}{2}ax^2 - \frac{1}{4}bx^4 + [s(t) + n(t)]x \quad (4)$$

The above function can be expressed in a double-well potential diagram [7] in Fig. 5 which represents a motion of the Brownian particle [5], where $S(t)$ is given by a sinusoidal signal generally. The potential diagram describes an over damped Brownian particle movement, consisting of two strap points $(-x_m, x_m)$ and the coordinate origin. The potential well point $x_m = \pm\frac{a}{b}$ occurs, when the input signal

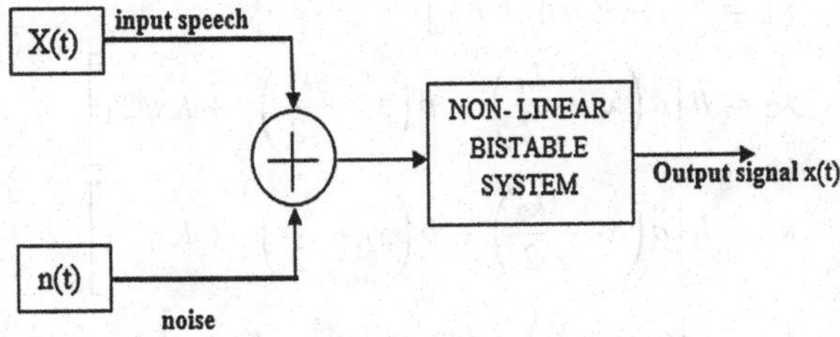

Fig. 5 System model of stochastic resonance (SR)

amplitude and the intensity of noise are zero and the barrier height corresponding to this condition is given by $\Delta U = \frac{a^2}{4b}$.

Suppose, the amplitude A is greater than zero, the trap point changes according to the noise intensity to establish SR. From the relationship, when the parameter of nonlinear system remains the same [8], the amplitude and noise intensity of the speech signal must meet a certain scope for SR to occur. The signal amplitude is inversely proportional to the noise intensity basically. When the speech signal is getting smaller with amplitude for smaller noise intensity [2], the system does not exhibit SR. When the input speech extracts energy from the noise resonation takes place, increasing the output SNR [9]. Consequently, the output SNR is very high when compared to input SNR and also the amplitude of the power spectrum reaches maximum peak [10]. The SNR could be expressed as follows:

$$\mathrm{SNR} = \frac{S}{N} \approx \frac{\sqrt{2}a^2A^2}{bD^2}\exp\left(-\frac{2\Delta U}{D}\right) \tag{5}$$

It is clear that the SNR of the output signal can be increased by adjusting the potential barrier ΔU and the noise intensity. Where ΔU is determined by the system parameters a and b.

4 Runge–Kutta Algorithm

The nonlinear Langevin equation (LE) is numerically solved by fourth-order Runge–Kutta algorithm [2], exhibits SR when the system reaches optimal state. It is an iterative algorithm in which each parameter is corresponding for the system linearity. The algorithm is described as follows:

$$x_{l+1} = x_l + \frac{1}{6}(k_1 + 2k_2 + 2k_3 + k_4) \quad l = 0, 1\ldots L-1 \tag{6}$$

where k_1, k_2, k_3, k_4, could be expressed as

$$\begin{aligned}
k_1 &= h\left[ax_l - bx_l^3 + Ks_l\right] \\
k_2 &= h\left[a\left(x_l + \frac{k_1}{2}\right) - b\left(x_l + \frac{k_1}{2}\right)^3 + Ks_{l+1}\right] \\
k_3 &= h\left[a\left(x_l + \frac{k_2}{2}\right) - b\left(x_l + \frac{k_2}{2}\right)^3 + Ks_{l+1}\right] \\
k_4 &= h\left[a(x_l + k_3) - b(x_l + k_3)^3 + Ks_{l+1}\right]
\end{aligned}$$

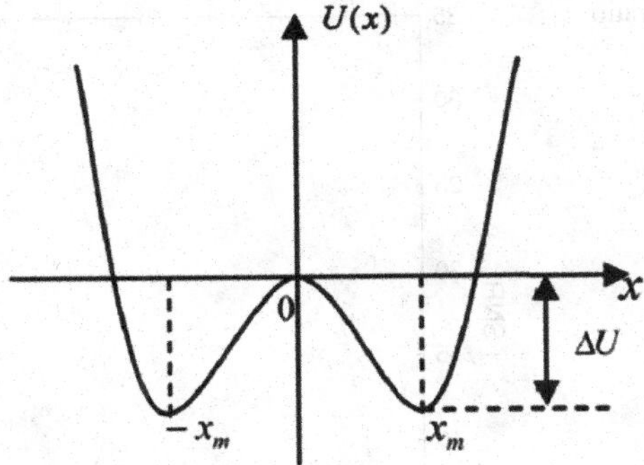

Fig. 6 Double-well potential diagram

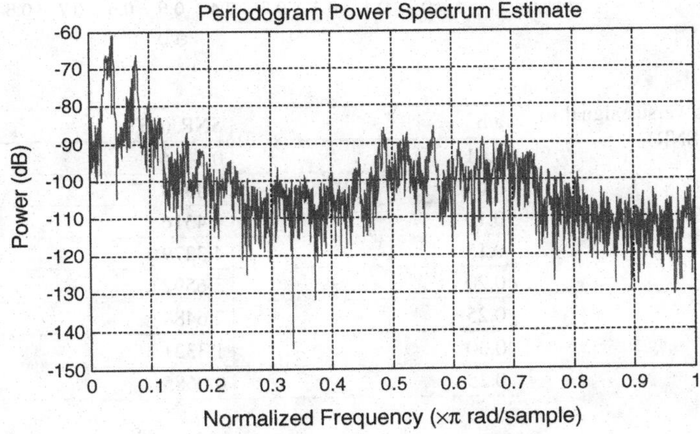

Fig. 7 Power spectral density of the output signal

In the above equations, x_P, s_P are the first sample of input $s(t)$ and output $x(t)$, where h. an integral step size, can be fixed according to the frequency of the enhanced speech signal (Figs. 6 and 7).

5 Procedure to Produce SR

1. Fix the system parameters for the input speech signal to be examined with noise, also fix the step size initially [11], and set the noise intensity value to d and then it is gradually increment it.
2. Adjust the system parameter a, with the step size gradually incremented and also the selection of appropriate parameter value of the nonlinear system [12].

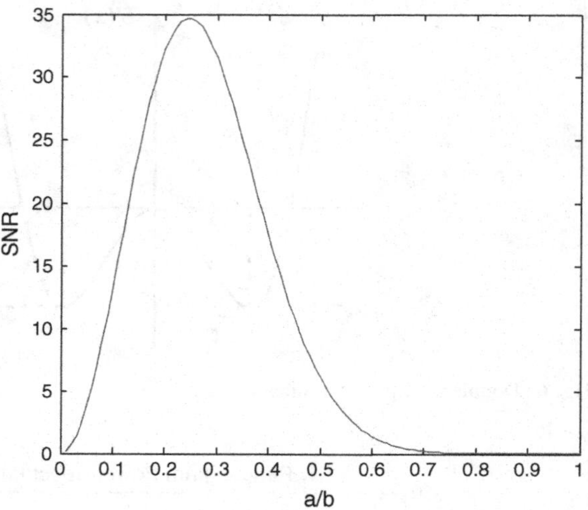

Fig. 8 Signal-to-noise ratio (SNR) versus a/b

Table 1 a/b versus signal to noise ratio (SNR)

a/b	SNR (dB)
0.01	0.3414
0.05	3.4403
0.10	4.4516
0.15	4.3379
0.20	3.6592
0.25	2.6488
0.30	1.3321
0.35	0.0685

3. A Runge–Kutta-IV algorithm is mathematically solved to get the optimized solution of the output speech signal [13] for every value of the parameter a.
4. Based on the optimal parameters of the system, the speech signal has been enhanced when the system exhibits stochastic resonance [15].
5. Then output SNR has to be calculated according to the best value of a and the noise intensity as shown in table and the relationship between output SNR and a/b is shown in Fig. 8.
6. The power spectrum of the output speech signal [14] is to be estimated to determine the spectrum peak corresponding to the frequency of the speech signal which is shown in Fig. 7 and Table 1.

6 Simulation and Result

The output speech spectrum obtained from the SR system is optimized using fourth-order Runge–Kutta algorithm. Figure 9 shows the optimized signal using the stated algorithm. Table 2 shows the output signal-to-noise ratio (SNR) corresponding to the noise intensity. The frequency domain spectrum of the optimized signal is given in Figs. 9 and 10.

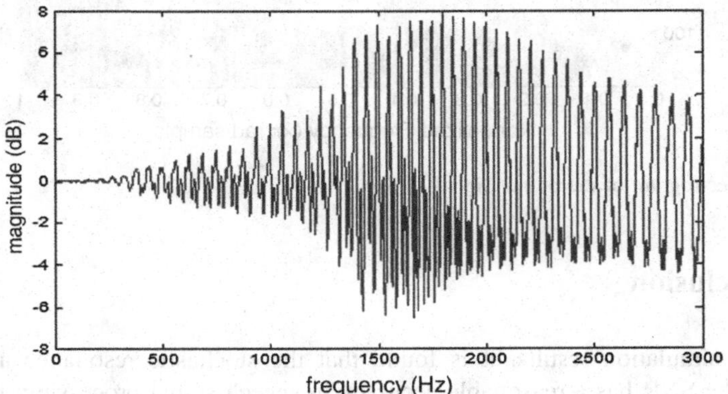

Fig. 9 Frequnecy spectrum of the optimized signal

Table 2 SNR difference corresponding to noise intensity

Noise intensity	Output SNR	SNR difference
0.01	24.9563	20.8191
0.02	22.7483	18.5925
0.03	21.4048	17.2929
0.04	20.2986	16.1869
0.05	19.4521	15.2535
0.06	18.7808	14.6474
0.07	18.1469	14.0242
0.08	17.6449	13.4874
0.09	17.2639	13.1635
0.1	16.7773	12.6386

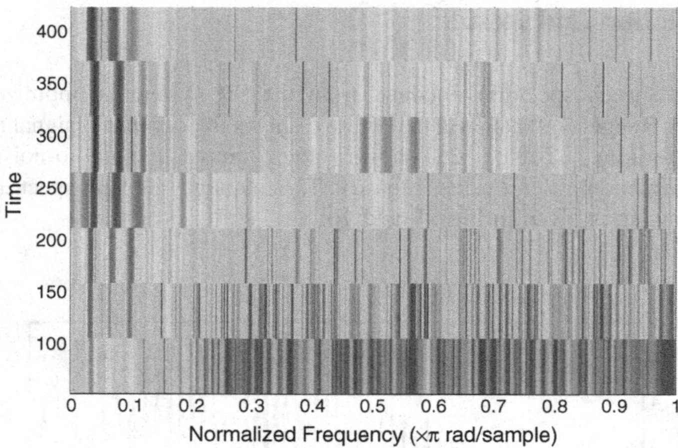

Fig. 10 Spectrogram of the output signal

7 Conclusion

From the simulation results, it is found that the stochastic resonance aimed to improve the SNR has a reasonable accuracy in speech signal processing. It is also proved stochastic resonance that has been used to detect the presence of weak sinusoidal signals can also be used to improve SNR of speech signals.

The future work would be extended to examine a high-frequency speech signal to enhance the intelligibility using stochastic resonance with the aid of wavelet packet decomposition.

Acknowledgments The authors like to thank Dr. S. Jayalalitha, Associate Dean, Dr. K. Ramkumar, Senior Associate Professor, and other staffs of Department of Instrumentation and Control, SASTRA University for the encouragement provided to take up the work and offering valuable inputs as and when required.

References

1. Benzi R, Sutera A, Vulpiani A (1981) The mechanism of stochastic resonance. J Phys A: Math Gen 14:453–457
2. Hou Z, Yang J, Wang Y, Wang K (2008) Weak signal detection based on stochastic resonance combining with genetic algorithm. IEEE Singapore international conference on communication systems
3. Ando B, Graziani S (2001) Adding noise to improve measurement. IEEE Instrum Meas Mag
4. Ando B, Graziani S (2000) Stochastic resonance-theory and its applications. Kluwer Academic Publishers, London

5. Zou H, Zheng L, Liu C (2013) Detecting parameters of high frequency signals with frequency modulation stochastic resonance. IEEE Int Conf Image Signal Process
6. Mu F, Zhang J, Du J (2011) A weak signal detection technology based on stochastic resonance system. IEEE Int Congr Comput Sci Serv Syst
7. Chen M, Hu NQ, Qin GJ, Yang YM (2008) A study on additional-signal-enhanced stochastic resonance in detecting weak signals. IEEE Int Conf Netw Sens Control
8. Yan G, Liying X (2010) Simulation of weak signal detection based on stochastic resonance. In: Proceedings of the third international symposium on electronic commerce and security workshops (ISECS '10). Guangzhou, P. R. China, pp 329–331
9. Wang J, Xiao Q, Li X (2009) The high-frequency weak signal detection based on stochastic resonance. Int Conf Test Meas
10. Yang D, Hu Z (2011) Extraction of weak pulse in crack inspection based on stochastic resonance. Fourth international conference on intelligent computation technology and automation
11. Chen H, Varshney PK (2008) Theory of the stochastic resonance effect in signal detection Part II: variable detectors. IEEE Trans Signal Process 56(10):5031–5041
12. Lu Z, Yang T, Zhu M (2013) Study of the method of multi-frequency signal detection based on the adaptive stochastic resonance. Hindawi Publishing Corporation. Abstr Appl Anal
13. Papadopoulos HC, Wornell GW A class of stochastic resonance systems for signal processing applications. Massachusetts Institute of Technology, Cambridge, p 02139
14. Moskowitz MT, Diction BW (2001) Stochastic resonance in natural speech recognition: a role for environmental noise" master's project. Princeton University, Princeton
15. Fan B, Hu N (2011) Method of weak signals detection based on array of stochastic resonance. IEEE conference on prognostics & system health management

Multiloop IMC-Based PID Controller for CSTR Process

M. Manimozhi and R. Meenakshi

Abstract In this paper, we have designed a multiloop proportional-integral-derivative (PID) controller for a nonlinear plant CSTR system. CSTR exhibits extremely nonlinear behaviors and habitually have broad operating ranges. The tuning of controller for each operating points of CSTR (continuous stirred tank reactor) is based on internal model control (IMC) tuning method. The main objective of this paper is to design a multiloop PID controller for the control of variable specifically concentration and temperature of multivariable nonlinear system CSTR. A multiple input multiple output (MIMO) process that merges the output of several linear PID controllers, each describing process dynamics at a precise level of operation. The global output is an interruption of the individual multiloop PID controller outputs weighted based on the current value of the deliberated process variable. A common approach to crack the nonlinear control problem such as CSTR is using gain scheduling with linear multiple PID controllers.

Keywords Multiloop gain-scheduled IMC-PID controller · MIMO process · CSTR process · Local linear model

1 Introduction

Continuous stirred tank reactors (CSTRs) are general chemical devices and also significant industrial divisions of the chemical process industry, which reveal extremely nonlinear behaviors and habitually have wide operating ranges. Proportional-integral-derivative (PID) controllers are the most widely used controller in the chemical process industries because of their ease and robustness. For control application, multiloop PID controllers are often preferred to the multivariable approach at regulatory level. Multiloop PID controller for CSTR process

M. Manimozhi (✉) · R. Meenakshi
School of Electrical Engineering, VIT University, Vellore 632014, India
e-mail: manimozhim@gmail.com

L.P. Suresh and B.K. Panigrahi (eds.), *Proceedings of the International Conference on Soft Computing Systems*, Advances in Intelligent Systems and Computing 397, DOI 10.1007/978-81-322-2671-0_59

merges the output of several linear PID controllers, each relating process dynamics at a precise level of operation. The controller tuned based on IMC tuning method. In paper [1], they have used numerous linear DMC controllers. It will give fine response over linear controllers. It will not useful, if the plant has nonlinear uniqueness. In paper [2, 3], they have designed a CSTR model and obtained the steady-state values for different operating levels. In paper [4], they have used neural network for controlling the nonlinear dynamics of CSTR process.

2 Methodology

In this study, we have used the operating point data (Table 1) and the standard model of CSTR as given in the Pottman and Seborg paper [2]. CSTR process is extremely nonlinear and it is very familiar in chemical and petrochemical industries. Here we have considered CSTR process for model study (as shown in Fig. 1), is an irreversible, exothermic reaction *Product A* → *Product B* occurs in constant volume reactor that is frozen by a solitary coolant stream. The CSTR system has two state variables, specifically the reactor temperature and the reactor concentration. The process is modeled by the following equations:

$$\frac{\mathrm{d}(M_\mathrm{C})}{\mathrm{d}t} = f_1(M_\mathrm{C}, R_\mathrm{T}) = \frac{F}{V} \times (C_\mathrm{F} - M_\mathrm{C}) - \left[k_0 \times \exp\left(\frac{-E}{\mathrm{RT}}\right) \times M_\mathrm{C}\right]$$

$$\frac{\mathrm{d}R_T}{\mathrm{d}t} = f_2(M_\mathrm{C}, R_\mathrm{T}) = \frac{F}{V}(T_\mathrm{F} - R_\mathrm{T}) - \left[\frac{(-\Delta \mathrm{H})}{\rho C_p} \times k_0 \times \exp\left(\frac{-E}{\mathrm{RT}}\right) \times M_\mathrm{C}\right]$$
$$+ \left[\frac{\rho_\mathrm{C} C_\mathrm{pc}}{\rho V C_\mathrm{p}} \times F_\mathrm{C} \times \left(1 - \exp\frac{-hA}{\rho C_\mathrm{pc} F_\mathrm{C}}\right)\right](T_\mathrm{cin} - R_\mathrm{T})$$

Table 1 Operating data at steady state

Process variable	Nominal operating state
Measured product concentration (M_c)	0.09869 mol/l
Reactor temperature (R_T)	438.8 K
Coolant flow rate (F_c)	103 l/min
Process flow rate (F)	100 l/min
Feed concentration (C_F)	1 mol/l
Feed temperature (T_F)	350 K
Inlet coolant temperature (T_{Cin})	350 K
CSTR volume (V)	100 l
Heat transfer term (hA)	7×10^5 cal/(min·k)
Reaction rate constant (k_0)	7.2×10^{10} min^{-1}
Activation energy term (E/R)	1×10^{10} K
Heat of reaction ($-\Delta H$)	-2×10^5 cal/mol
Liquid density (ρ, ρ_c)	$1 \times 10_3$ g/l
Specific heats (C_p, C_{pc})	1 cal/(g·k)

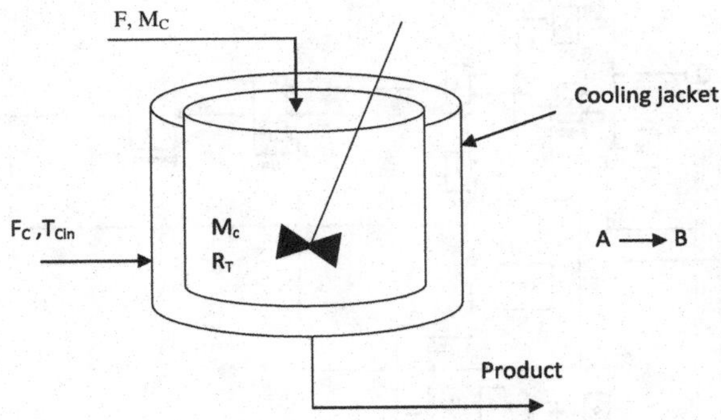

Fig. 1 Continuous stirred tank reactor

Via Taylor series expansion, a linear model will be developed around the steady-state operating point. The linearization will be respect to M_C and R_T (the output variables) and F and F_c (the input variables).

Using MATLAB/Simulink, it is possible to implement a CSTR plant model and further designing Multiloop PID for CSTR. For the completion of the proposed process simulation in MATLAB/Simulink, first the CSTR have to be modeled with the help of above differential equations. Figure 2 shows a Simulink model of a CSTR.

From this simulink model of CSTR plant, we can obtain concentration and temperature value at different operating levels. That is shown in Table 2.

2.1 *Transfer Function Matrix at Different Operating Level*

To design a Multiloop IMC-based PID controller for CSTR process, first, we have to linearize the model at different operating and find out transfer function matrix TF_1 and TF_2. From TF_1 to TF_2, get to identify about the Eigenvalues, damping frequency, and undamped frequency values. These Eigenvalues and damping frequency are used to design an IMC-based PID controller. In this work, we have proposed to interpolate five multiloop PID controllers. In this work, reactor concentration is controlled based on coolant flow rate and temperature is controlled by feed flow rate. The manipulated variables are coolant flow rate and the feed flow rate. The outputs are the concentration and reactor temperature. Using Matlab, the Eigenvalues and damping factor obtained at the individual operating points are presented in Table 3, from which it can be inferred that the process is stable at all the operating points because the Eigenvalues have negative real parts.

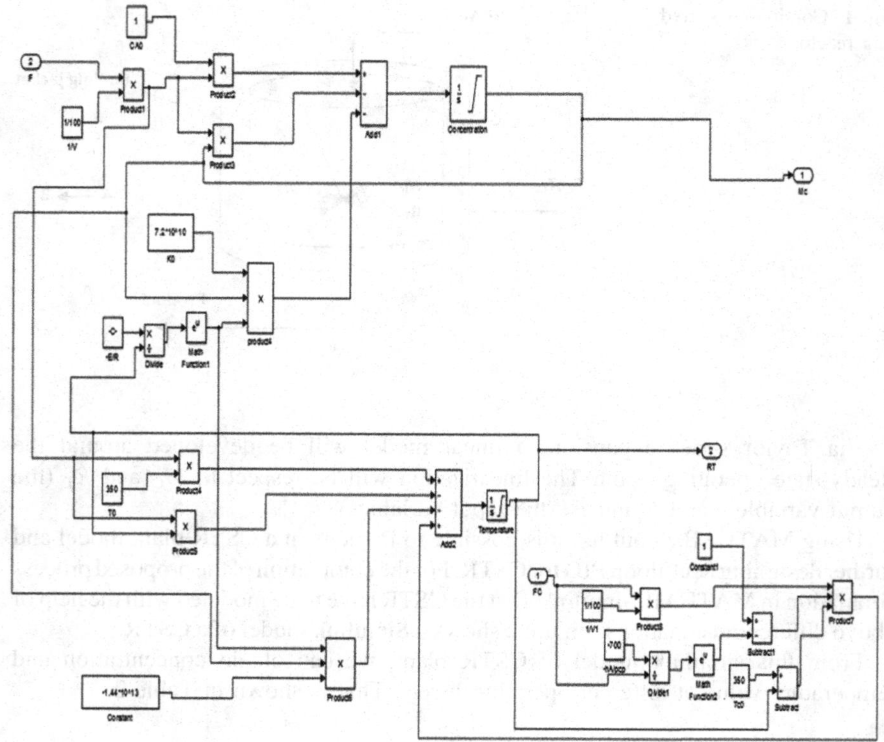

Fig. 2 CSTR plant model

Table 2 Concentration and temperature values for different operating points

Operating point	Flow rate, F (lpm)	Coolant flow rate, F_c (lpm)	Concentration, M_c (mol/l)	Temperature, R_T (K)
1	102	97	0.07624	444.7
2	101	100	0.08638	441.9
3	100	103	0.09869	438.8
4	99	106	0.1131	435.7
5	98	109	0.1317	432.3

Table 3 Eigenvalues at the five operating points of the CSTR

Operating point	Eigen value	Damping factor (ζ)	Frequency (ω)
1	$-2.92 \pm 2.89i$	0.710	4.11
2	$-2.13 \pm 3.07i$	0.569	3.74
3	$-1.04 \pm 3.04i$	0.419	3.34
4	$-0.76 \pm 2.86i$	0.262	2.96
5	$-0.199 \pm 2.54i$	0.0781	2.54

2.2 Design of PID Controller Based on IMC Tuning Method

In this project, reactor concentration is controlled based on flow rate of coolant stream and temperature is controlled by feed flow rate.

The concentration of reactor against coolant flow rate is in the form

$$\mathrm{TF}_1, j(s) = \frac{K_{p1,j}}{\tau^2 s^2 + 2\zeta\tau s + 1},$$

where $j = 1$–5 for the diverse operating points.

The temperature against feed flow rate is in the form

$$\mathrm{TF}_2, j(s) = \frac{K_{p2,j}(-\beta s + 1)}{\tau^2 s^2 + 2\zeta\tau s + 1},$$

where $j = 1$–5 for the diverse operating points.

Solving these formulas at different operating level, we can obtain values for PID controller parameters K_{c1}, K_{c2}, τ_I, and τ_D. Table 4 gives the value for these parameters.

Table 4 PID controller's parameters at five different operating points

Operating points	ζ	ω	K_{c1}	K_{c2}	τ_I	τ_D
Region 1 at $F = 102$, $F_c = 97$, $M_c = 0.07624$, $R_T = 444.7$	0.710	4.11	$129.97/\lambda_1$	$0.5738/0.0933+\lambda_2$	0.345	0.171
Region 2 at $F = 101$, $F_c = 100$, $M_c = 0.08638$, $R_T = 441.9$	0.569	3.74	$98.452/\lambda_1$	$0.4836/0.1045+\lambda_2$	0.304	0.235
Region 3 at $F = 100$, $F_c = 103$, $M_c = 0.09869$, $R_T = 438.8$	0.419	3.34	$68.178/\lambda_1$	$0.37621/0.119+\lambda_2$	0.251	0.357
Region 4 at $F = 99$, $F_c = 106$, $M_c = 0.1131$, $R_T = 435.7$	0.262	2.96	$39.574/\lambda_1$	$0.2477/0.1368+\lambda_2$	0.177	0.644
Region 5 at $F = 98$, $F_c = 109$, $M_c = 0.1317$, $R_T = 432.3$	0.078	2.54	$14.685/\lambda_1$	$0.10448/0.158+\lambda_2$	0.081	1.835

The IMC-based PID controller's parameters at five different operating points have been accounted in Table 4. It should be noted that the controller gain has been found to be function of the filter time constant lambda (λ_1 and λ_2). The selection of the best value of Lambda must be based on performance and robustness deliberations. As far as the tuning of the controller is concerned we have an optimum filter tuning factor λ (lambda) value which concessions the effects of inconsistencies entering into the system to attain the best performance. Thus, what we mean by the best filter structure is the filter that gives the best controller performance for the optimum λ value.

2.3 *Design of Weight Scheduler*

The planned control technique uses a linear controller to control a nonlinear system using gain scheduling approach. Gain scheduling means that the tuned parameters of the controller at each operating point are composed using gain scheduler algorithm; the global controller monitors the state of the process and decides suitable parameters from the algorithm [3]. At each sampling instant, the gain scheduler will allocate weight for each controllers and the weighted sum of the outputs will be applied as an input to the plant. On the basis of a number of different variables such as state variables and process inputs, the scheduler will make its choice. In this work, the design of multiloop IMC-based PID controllers on the basis of linear models developed at different operating points and we combined the multiloop IMC-based PID controller outputs to acquiesce a global controller output.

2.4 *Design of Multiloop IMC-Based PID Controller for CSTR*

Overall Design of multiloop IMC-based PID control scheme for CSTR process is given in Fig. 3.

3 Results

In all the simulation runs, by solving the nonlinear differential equations using differential equation solver in Matlab 10.0, the whole simulation has been achieved with the subsequent initial conditions:

$M_c = 0.09869$ mol/l; $F_c = 103$ l/min; $F = 100$ l/min; $R_T = 438.8$ K.

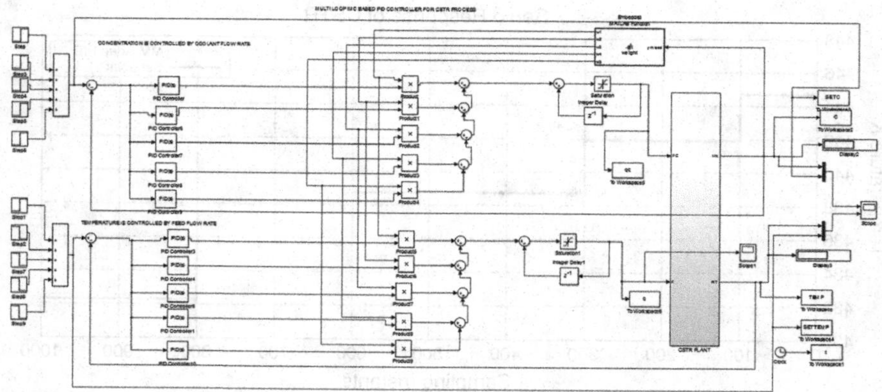

Fig. 3 Multiloop PID controller for CSTR process

3.1 Servo Response

In order to evaluate the tracking ability of considered controllers, setpoint deviations in concentration as given in Fig. 4 and setpoint deviations in temperature as given in Fig. 5 have been commenced. From the responses it can be inferred that the controller designed for the CSTR process is capable to preserve the variables concentration and temperature at the preferred setpoints.

As the setpoint concentration specifically distorted from 0.07624 to 0.08638 mol/l, 0.08638 to 0.09868 mol/l, 0.09868 to 0.1131 mol/l, and 0.1131 to 0.1317 mol/l multiloop IMC-based PID follows the setpoint deviations. Likewise as

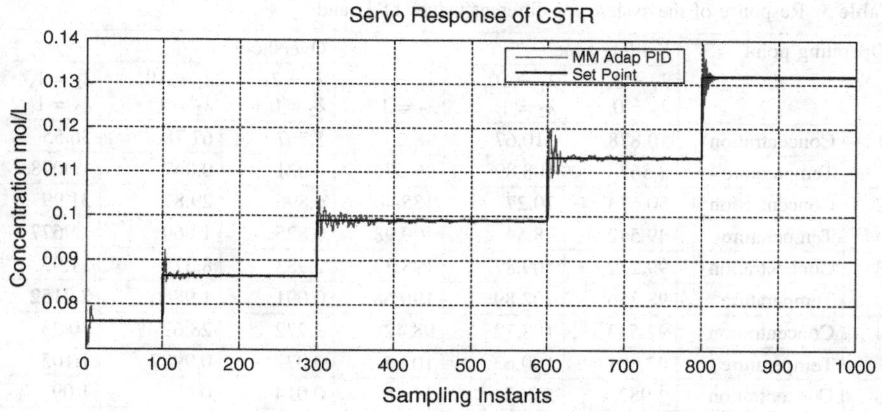

Fig. 4 Servo response of CSTR process concentration

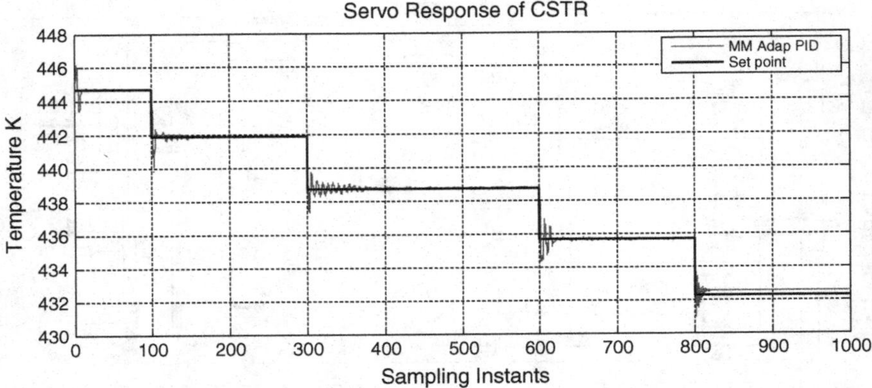

Fig. 5 Servo response for CSTR process temperature

the setpoint specifically temperature is distorted from 444.7 to 441.9 K, 441.9 to 438.8 K, 438.8 to 435.7 K, and 435.7 to 432.3 K. Multiloop IMC-based PID follows the setpoint deviations.

4 Discussion

As evident from above, value of gains K_{c1} and K_{c2} depends on λ_1 and λ_2 values for concentration and temperature, respectively. For different values of λ_1 and λ_2, the system response is argued below in Table 5.

Table 5 Response of the system for different values of λ_1 and λ_2

Operating point		Settling time			Overshoot		
		$\lambda_1 = 3$, $\lambda_2 = 0.1$	$\lambda_1 = 10$, $\lambda_2 = 1$	$\lambda_1 = 1$, $\lambda_2 = 1$	$\lambda_1 = 3$, $\lambda_2 = 0.1$	$\lambda_1 = 10$, $\lambda_2 = 1$	$\lambda_1 = 1$, $\lambda_2 = 1$
1	Concentration	10.878	110.67	98.25	5.767	67.78	56.55
	Temperature	8.567	108.89	99.67	0.021	0.067	1.0078
2	Concentration	50.673	70.27	185.46	6.896	29.87	31.99
	Temperature	49.542	78.54	199.98	0.825	1.869	1.9677
3	Concentration	99.272	109.87	198.37	2.983	8.378	11.93
	Temperature	98.356	102.89	198.68	0.091	1.986	2.9152
4	Concentration	92.543	108.72	98.47	2.772	28.67	30.25
	Temperature	92.001	100.63	102.5	0.077	0.9987	1.105
5	Concentration	4.982	5.267	11.45	0.014	0	1.09
	Temperature	4.536	4.378	10.26	0.009	0.01	0.561

At $\lambda_1 = 3$ for concentration and $\lambda_2 = 0.1$ for temperature, a smooth curve for both concentration and temperature is obtained as shown in Figs. 4 and 5. On increasing the value of λ, at $\lambda_1 = 10$ for concentration and $\lambda_2 = 1$ for temperature, oscillations in the curves have increased and it takes large time to settle as shown in Figs. 6 and 7. On decreasing the value of λ, at $\lambda_1 = 1$ for concentration and $\lambda_2 = 1$ for temperature, oscillations in the curves have increased further, settling time has increased, and there is large overshoot as shown in Figs. 8 and 9.

4.1 Simulation Results for Various Lambda Values

See Figs. 6, 7, 8 and 9.

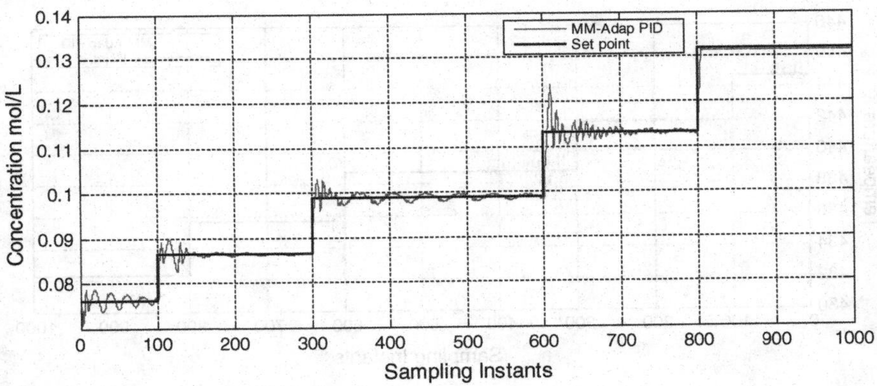

Fig. 6 Servo response of the CSTR concentration for $\lambda_1 = 10$ and $\lambda_2 = 1$

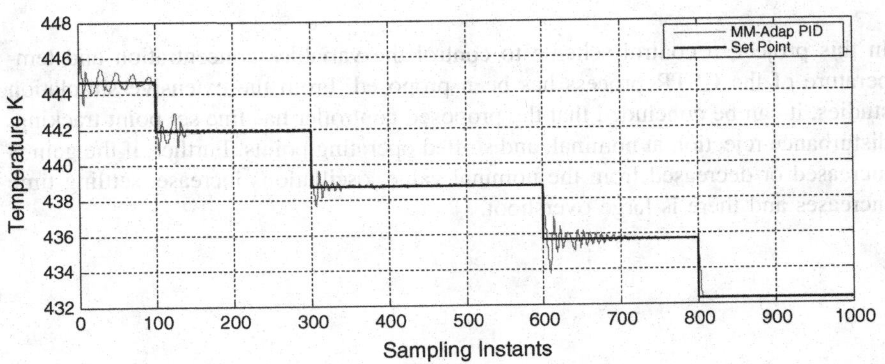

Fig. 7 Servo response of the CSTR temperature for $\lambda_1 = 10$ and $\lambda_2 = 1$

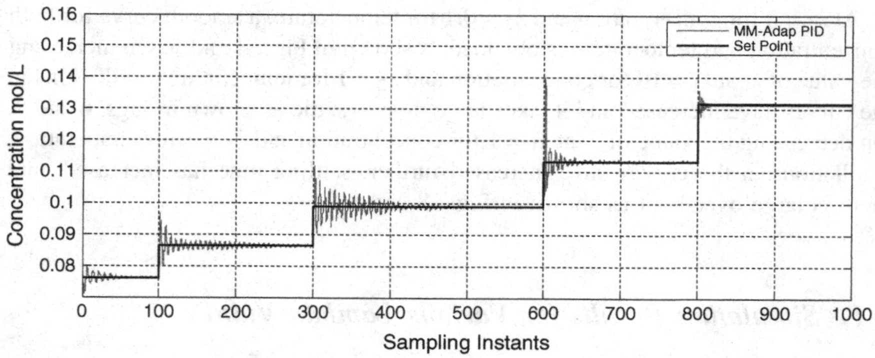

Fig. 8 Servo response of the CSTR concentration for $\lambda_1 = 1$ and $\lambda_2 = 1$

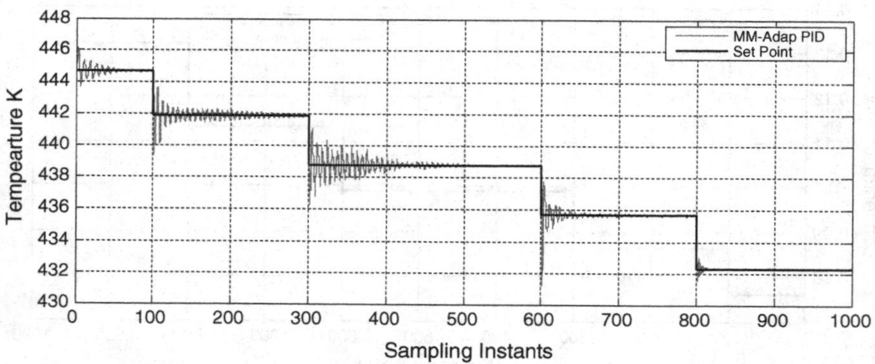

Fig. 9 Servo response of the CSTR temperature for $\lambda_1 = 1$ and $\lambda_2 = 1$

5 Conclusion

In this project, a control scheme to control the variables concentration and temperature of the CSTR process has been proposed. From the extensive simulation studies, it can be concluded that the proposed controller has fine set point tracking, disturbance rejection at nominal, and shifted operating points. Further, if the gain is increased or decreased from the nominal value, oscillations increase, settling time increases and there is large overshoot.

References

1. Danielle D, Cooper D (2003) A practical multiple model adaptive strategy for multivariable model predictive control. Control Eng Pract 11:649–664
2. Pottmann M, Seborg DE (1992) Identification of nonlinear process using reciprocal multi quadratic functions. J Process Control 2:189–203
3. Vinodha R, Abraham Lincoln S, Prakash J (2010) Multiple model and neural based adaptive multi-loop pid controller for a CSTR process. World Acad Sci Eng Technol 68
4. Jalili-Kharaajoo M (2003) Predictive control of a continuous stirred tank reactor based on neuro-fuzzy model of the process. SICE Ann Conf Fukui 57:3005–3011
5. Senthil R, Janarthanan K, Prakash J (2006) Nonlinear state estimation using fuzzy kalman filter. Ind Eng Chem Res 45(25):8678–8688

Dynamic Energy Management on a Hydro-Powered Smart Microgrid

Prasanna Vadana, Rajinikandh and Sasi K. Kottayil

Abstract Penetration of renewable energy-based microgrids onto the legacy grid is in demand to solve the global energy problems and the environmental issues. This paper attempts to employ dynamic energy management on a Grid-Connected Smart Microgrid (GCSMG) energized by a Micro Hydro Power Plant (MHPP) sans governor control. Frequency control of such SMGs poses a challenge as the latter is distributed. The concept of Dynamic Energy Management (DEM) plays a significant role in accomplishing the frequency control without perturbing the controlling facility in the conventional grid. DEM is a concept of controlling the charge–discharge transactions on the energy storage modules to oppose the frequency excursions on the grid. Support Vector Machine (SVM) algorithm is employed to automate DEM operation. The Dynamic Energy Management System (DEMS) is implemented on a Field Programmable Gate Array (FPGA) as the response time is critical for this application. The DEM scheme is validated on the SMG simulator in the Renewable Energy Laboratory of Amrita Vishwa Vidyapeetham University, Coimbatore.

Keywords DEM · SVM · FPGA · SMG simulator

1 Introduction

Penetration of microgrids onto the conventional grid is the present phase of evolution on electric power utility both nationally and globally [1]. Microgrid is the miniature of the legacy grid energized by Renewable Energy (RE) sources capable of meeting the local demands connected to it, partially or fully, in a synchronized

P. Vadana (✉) · Rajinikandh · S.K. Kottayil
Department of EEE, Amrita Vishwa Vidyapeetham University, Coimbatore, India
e-mail: d_prasanna@cb.amrita.edu

Rajinikandh
e-mail: c_rajinikandh@cb.amrita.edu

S.K. Kottayil
e-mail: kk_sasi@cb.amrita.edu

L.P. Suresh and B.K. Panigrahi (eds.), *Proceedings of the International Conference on Soft Computing Systems*, Advances in Intelligent Systems and Computing 397, DOI 10.1007/978-81-322-2671-0_60

manner [2]. Also, introduction of Demand Response Programs (DRPs) among the end-users is becoming an essential requirement to solve the global energy crisis from the demand side [3]. Integration of distributed RE-based generation along with facilitation of DRPs on the demand side in a large power grid needs to happen at a micro scale and in a distributed manner as the legacy grid is a centralized complex network which cannot be bothered to a very large extent [4, 5]. The need for an Energy Management System (EMS) is to monitor and control energy production and consumption in generation, transmission and consumer facilities [6].

Performing energy management by introducing new schemes both on the generation and demand sides become a responsibility on the microgrid, as the governor control on the conventional grid is centralized and the former is distributed [7]. Whenever there is frequency variation on the grid, microgrid should be capable of taking necessary action to help the grid in balancing the total generation with the total demand. To facilitate this, Energy Storage System (ESS) is generally recommended to preserve power quality based on the frequency diagnosis [8]. The microgrid should have the ability to monitor its operating parameters, viz., frequency, voltage, current, etc. and then communicate these to a controller for necessary corrective action.

The microgrid with real-time measurement and communication facilities becomes a Smart Microgrid (SMG). A SMG can be made to operate automatically through proper EMS besides interacting with the main grid as there is a requirement to sustain synchronized operation of all Distributed Generation (DG) schemes and various DRP schemes to maintain the system stability [9]. Several proposals of EMS for energy storage systems and microgrid have been made by researchers worldwide. EMS for energy storage takes decisions (a) to handle the rate of charging/discharging ESS based on the load and generation schedule [10], (b) to coordinate Hybrid Energy Storage Systems [11, 12], and (c) to regulate the storage system parameters [13]. The operation of an EMS on a SMG is verified by creating a constrained/unconstrained optimization model and solved using an appropriate optimization technique.

Most of the optimization models use soft computing techniques-based load forecasting [14, 15]. Microgrid Central Controllers (MGCCs) developed in [16–18] is used to handle power flow variations to minimize energy cost and optimize power exchanges with conventional grid. Optimization techniques are used to tune parameters of MGCC which handles frequency excursions on the grid [19]. In [20], an EMS control strategy proposed as "Dynamic Energy Management (DEM)" is developed using SVM to take necessary actions to handle supply–demand mismatch. DEM is stated as "the charge-discharge transactions in the energy storage systems to oppose frequency excursions on the grid in real time environment." The authors further proposed implementation of DEMS realized using SVM on a FPGA platform [21]; also concluded that SVM is better than neural networks in realizing DEM when implemented on FPGA [22]. This paper deals with the utilization of DEMS for a smart microgrid energized with a Micro Hydro Power Plant (MHPP) and emphasizes that DEM scheme works effectively irrespective of the structure and specification of the power system under consideration by validating on a SMG simulator in Renewable Energy Laboratory of Amrita Vishwa Vidyapeetham University.

2 DEM Scheme

DEM, an intelligent and self-decisive scheme houses a DEMS that is capable of monitoring the status of SMG at regular intervals and take appropriate actions with respect to the storage modules installed in the SMG. Figure 1 shows the DEM scheme for the power system under consideration. Pumped Hydro (PH)—a slow responding large capacity storage module—aids the implementation of DEM in the SMG. DEMS activates the Variable Speed Drive (VSD) to increase or decrease the speed of the PH unit.

Based on a qualitative analysis, the status parameters of the SMG are chosen and are listed in Table 1.

The DEM decisions can be listed as

- Maintain status quo
- Increase the speed of PH
- Decrease the speed of PH.

These decisions are decided based on the storage modules installed in the SMG. Though there is a slight variation in the status parameters and DEM decisions due to the absence of a storage module as compared to [22], this paper is an attempt to insist on the fact that DEM is independent of the power system chosen for implementation.

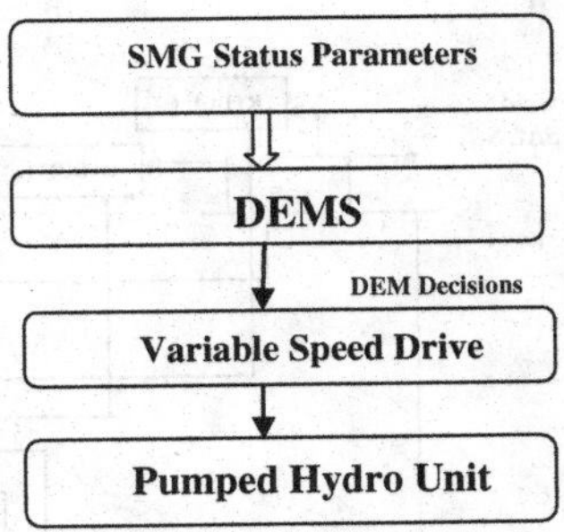

Fig. 1 DEM scheme

Table 1 SMG parameters

Parameter	Definition	Conditions
S_{W1}	Status of power exchange with the main grid	1: Imported from the grid
		−1: Exported to the grid
		0: No power exchange
S_{W2}	Status of local load	1: Power is consumed
		0: Power is not consumed
S_{W3}	Status of PH	1: Pumping
		0: Idling
$S_{\Delta f}$	Status of frequency	1: f < 50 Hz
		−1: f > 50 Hz
		0: f = 50 Hz

3 Smart Microgrid Simulator (SMGS)—Test Bed for Realization of DEM Scheme

DEM Scheme demands the microgrid to be 'smart,' i.e., having the capability of monitoring and communicating the real-time operational data of all the systems installed in it. Figure 2 shows the bus diagram of a SMGS employed to realize the DEM scheme. The SMGS is energized by MHPP. PH unit with VSD are connected to the SMG which act as the Energy Storage System (ESS). The MHPP meets the power requirement of the consumer load center, sometimes fully and otherwise partially based on the generation schedule.

The storage capacity of PH is larger but it responds slowly. The MHPP is assumed to have no governor control and it works on a preplanned schedule based on the load forecasting performed by the available soft computing techniques. The purpose of PH is to lift the water from the lower reservoir to the upper reservoir. Two separate machines are used to model the MHPP, one for generation and the other for pumping operation. DEM helps frequency regulation by controlling the increase or decrease of speed of PH in real time.

The SMGS, as shown in Fig. 3, is facilitated with a Real-Time Data Collection Unit (RTDCU) at every bus to measure the currents and voltages and transmit the data to FPGA through RS232 serial communication. Every RTDCU and the FPGA is installed with a serial communication module in it. To measure the grid frequency

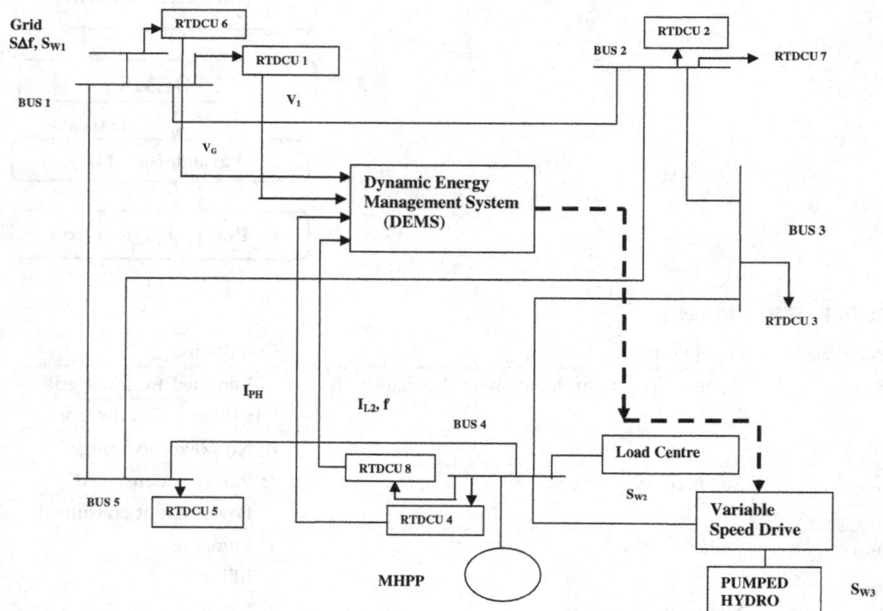

Fig. 2 Bus diagram of grid-connected smart microgrid simulator employed with DEMS

Fig. 3 Smart microgrid simulator available in the renewable energy laboratory of Amrita Vishwa Vidyapeetham University, Coimbatore

and grid voltage, a remote RTDCU (RTDCU 6 as shown in Fig. 2) is installed. A specific data format is followed to transmit the real-time operational data to the FPGA.

4 Simulation and FPGA Implementation

SVM is used to automate DEM in a more intelligent way. SVM is better than any other machine learning technique as the solution given by the algorithm is a unique one [23] supported by the strong statistical learning theory and optimization techniques.

Table 2 Simulation results of SVM

Model	γ (gamma)	C (cost)	Cross validation accuracy (%)	Testing accuracy of unknown patterns (%)	
ABCD	1	2	91.8239	80 (32/40)	8
BCDE	0.0625	64	95.625	94.8718 (37/39)	2
CDEA	1	8	86.1635	80 (32/40)	8
DEAB	1	16	89.3082	85 (34/40)	6
EABC	0.0625	64	93.0818	100 (40/40)	0

Table 2 is listed for reference and it varies from [22] in the classification accuracy being tuned to 100 %, which is necessary for such applications. Since the SMG is distributed, it should be capable of handling itself to the maximum extent possible and get aligned with the conventional grid in all aspects without perturbing the centralized control in it. Classification accuracy tuned to 100 % is necessary to accomplish distributed frequency control in the SMG effectively.

Figure 4 shows DEMS implemented on a Xilinx Spartan FPGA housed on an Altium NanoBoard 3000. The status word to the DEMS as per [22] should have six status parameters (two extra parameters for the additional battery storage which is not present in this power system under consideration). This status word represents the expected status word as shown in Fig. 4.

The parameters S_{Ic} and S_{SOC} represent the status of the battery storage module installed in the SMG. Here, due to the absence of battery modules, these two parameters can be considered as '0' which indicates the applicable status word. DEMS implemented on FPGA can be used to predict the necessary actions to be performed on the storage modules to oppose the frequency excursions on the grid.

In [22], seven decisions are identified with respect to the storage modules. The decisions 1, 2, and 3 being the same as listed above. The decisions 4–7 are: 4—charge the battery; 5—discharge the battery; 6—charge the battery and increase the speed of PH; 7—discharge the battery and decrease the speed of PH. As shown in Fig. 3, decision 4 or 6 can be considered as decision 2, i.e., to increase the speed of PH.

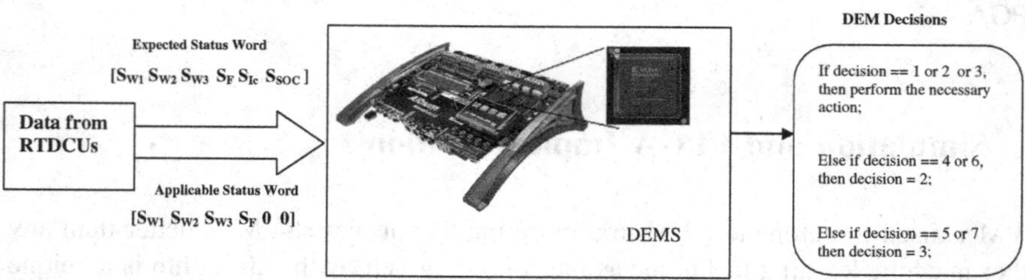

Fig. 4 DEMS on FPGA

Table 3 FPGA implementation report (generated by altium designer 10)

Logic utilization	SVM
Number of slice flipflops	1651/22528 (7 %)
Number of four input LUTs	4037/22528 (17 %)
Logic distribution	
Number of occupied slices	2604/11264 (23 %)
Number used as logic	3781
Number used as route-thru	205
Total number of four input LUTs	4242/22528 (18 %)
Peak memory usage (MB)	239
Total REAL time to MAP (s)	16

On the contrary, decision 5 or 7 can be considered as decision 3, i.e., to decrease the speed of PH. DEM scheme implemented on FPGA is validated on the SMGS in the Renewable Energy Laboratory in Amrita Vishwa Vidyapeetham University, Coimbatore and the results turned out to be successful. Appropriate decisions are displayed on the mini touch screen embedded on the Altium NB3000 board. The implementation report of DEMS as generated by the Altium Designer IDE is presented in Table 3.

5 Conclusion

Renewable energy has penetrated to the tune of 12 % in Indian power grid and the present power policy and planning of the power sector is to enhance the penetration further. It is therefore certain that an advanced power management technique like DEM is essential for the future power grid operation in the country. Presently, the carbon foot print of Indian power grid is very high. The country has recently risen to a vulnerable position as the third largest nation in the world in GHG emission. The only remedial measure in this regard is adoption of more of RE and this is impossible in future without DEM which urges the need for a smart environment at the microgrid level.

The versatility of employing DEM scheme irrespective of the structure of the power system is verified in the SMGS in Renewable Energy Lab, Amrita Vishwa Vidyapeetham University. Quick response of an EMS is vital for an SMG as compared to the legacy grid which is accomplished with the FPGA implementation. The presence of DEMS in an SMG would be appreciated when many such RE energized SMGs penetrate the public power grid and operate in a synchronized manner which forms the urgent need for solving the global energy crisis.

References

1. Farhangi H (2010) The path of the smart grid. Power Energy Mag IEEE 8(1):18–28
2. Lasseter RH (2002) Microgrids. In: Power engineering society winter meeting, 2002. IEEE, vol 1, pp 305–308. IEEE, 2002
3. Rahimi F, Ipakchi A (2010) Overview of demand response under the smart grid and market paradigms. In: Innovative Smart Grid Technologies (ISGT), pp 1–7. IEEE, 2010
4. Balijepalli VSKM, Pradhan V, Khaparde SA, Shereef RM (2011) Review of demand response under smart grid paradigm. In: Innovative smart grid technologies-India (ISGT India), 2011 IEEE PES, pp 236–243, IEEE, 2011
5. Siano P (2014) Demand response and smart grids—a survey. Renew Sustain Energy Rev 30:461–478
6. Kondolen D, Ten-Hope L, Surls T, Therkelsen RL (2003) Microgrid energy management system. In: California energy commission consortium for electric reliability technology solutions (CERTS) consultant report, Oct 2003
7. Chen C, Duan SX, Cai T, Liu B, Hu G (2011) Smart energy management system for optimal microgrid economic operation. IET Trans Renew Power Gener 5(3):258–267
8. Vasquez JC, Guerrero JM, Miret J, Castilla M, De Vicuna LG (2010) Hierarchical control of intelligent microgrids. IEEE Ind Mag 22–29
9. Kanchev H, Lu D, Colas F, Lazarov V, Francois B (2011) Energy management and operational planning of a microgrid with a PV-based active generator for smart grid applications. IEEE Trans Ind Electron 58(10):4583–4592
10. Manjili YS, Rajaee A, Jamshidi M, Kelley BT, Intelligent decision making for energy management in microgrids with air reduction policy. In: 7th international conference on systems of systems engineering (SoSE), pp 13–18, 2012
11. Etxeberria A, Vechiu I, Camblong H, Vinassa J-M (2010) Hybrid energy storage systems for renewable energy sources integration in microgrids: a review. In: Conference proceedings of IPEC, pp 532–537, IEEE, 2010
12. Zhou H, Bhattacharya T, Tran D, Siew TST, Khambadkone AM (2011) Composite energy storage system involving battery and ultracapacitor with dynamic energy management in microgrid applications. IEEE Trans Power Electron 26(3):923–930
13. Zhang Y, Jia HJ, Guo L (2012) Energy management strategy of islanded microgrid based on power flow control. In: IEEE PES innovative smart grid technologies (ISGT), pp 1–8, January 2012
14. Palma-Behnke R, Benavides C, Aranda E, Llanos J, Saez D (2011) Energy management system for a renewable based microgrid with a demand side management mechanism. In: 2011 IEEE symposium on computational intelligence applications in smart grid (CIASG), pp 1–8, 2011
15. Chakraborty S, Weiss MD, Simoes MG (2007) Distributed intelligent energy management system for a single-phase high-frequency AC microgrid. IEEE Trans Ind Electron 54(1):97–109
16. Chamorro HR, Ramos G (2011) Microgrid central fuzzy controller for active and reactive power flow using instantaneous power measurements. In: IEEE power and energy conference at illinois (PECI), pp 1–6. IEEE, 2011
17. Hooshmand A, Asghari B, Sharma R (2013) A novel cost-aware multi-objective energy management method for microgrids. In: IEEE PES innovative smart grid technologies (ISGT), pp 1–6. IEEE, 2013
18. Tsikalakis AG, Hatziargyriou ND (2011) Centralized control for optimizing microgrids operation. In: 2011 IEEE power and energy society general meeting, pp 1–8, 2011
19. Mishra S, Mallesham G, Jha AN (2012) Design of controller and communication for frequency regulation of a smart microgrid. IET Renew Power Gener 6(4):248–258
20. Prasanna Vadana D, Kottayil SK (2012) Support vector machine based dynamic energy management on smart grid. In: IEEE ICIIS 2012, IIT Chennai, Aug 2012

21. Prasanna Vadana D, Kottayil SK (2015) Dynamic energy management on smart micro grid. In: ISGW 2015, Mar 2015
22. Prasanna Vadana D, Kottayil SK (2014) Energy aware controller for dynamic energy management on smart micro grid. In: Proceedings of IEEE PSETSE 2014, Mar 2014
23. Soman KP, Loganathan R, Ajay V (2009) Kernel methods and evolution of SVM. In: Machine learning With SVM and other Kernel methods. PHI Learning Private Limited, Delhi, pp 116–178

Performance Evaluation of a Speech Enhancement Technique Using Wavelets

R. Dhivya and Judith Justin

Abstract In this work, a novel speech enhancement algorithm is proposed based on multiband spectral subtraction speech enhancement technique and wavelet thresholding. The algorithm is tested with noisy speech signal produced by a prosthetic device for laryngectomy patients. The performance of the proposed algorithm is compared with the multiband spectral subtraction algorithm in terms of perceptual evaluation of speech quality (PESQ). The objective measures such as signal-to-noise-ratio (SNR), log-likelihood ratio (LLR), segmental signal-to-noise-ratio (SegSNR), weighted spectral slope (WSS), itakura-saito distance (IS), Cepstral Distance, and frequency-weighted segmental signal-to-noise-ratio (fwSNR) are used to test the effectiveness of the algorithm.

Keywords Speech enhancement · Wavelet thresholding · Pearson's correlation · Speech quality measures

1 Introduction

Advanced stage of laryngeal cancer leads to total laryngectomy, which is the surgical process of removal of larynx. A person who has undergone laryngectomy (laryngectomee) cannot produce voice in a conventional manner, because the vocal folds are removed. An alternative speaking method is required to produce voice, using sound sources which produce voice without vibrating the vocal folds, and the voice thus produced is called alaryngeal speech. The voice of a laryngectomee

R. Dhivya (✉) · J. Justin
Department of Biomedical Instrumentation Engineering, Faculty of Engineering, Avinashilingam Institute for Home Science and Higher Education for Women, Coimbatore, Tamil Nadu, India
e-mail: dhivyaramasamy21@gmail.com

J. Justin
e-mail: judithvjn@yahoo.co.in

L.P. Suresh and B.K. Panigrahi (eds.), *Proceedings of the International Conference on Soft Computing Systems*, Advances in Intelligent Systems and Computing 397, DOI 10.1007/978-81-322-2671-0_61

implanted with a prosthetic device is recorded through a microphone in an anechoic room and is stored on a computer. This alaryngeal voice is utilized for the study.

Speech Enhancement algorithms are proposed in order to improve the performance of devices working in a noisy environment. Numerous algorithms have been proposed to evaluate the quality of speech enhancement. Although two different algorithms may produce equal word intelligibility scores, listeners may perceive the speech of one of the algorithms as being more natural, pleasant, and acceptable. So, there is a need to measure the attributes of a speech signal. Reliable rating of speech quality is a challenging task because quality assessment is highly subjective and reliability of subjective measurements becomes an issue. Accurate assessment is extremely important in selecting an algorithm for a suitable application. Objective measures for assessment of quality involve the computation of the numerical distance or the distortion between the original and processed speech signals.

The performance of an objective measure is established from its correlation with a subjective measure. Objective measure having high correlation with mean opinion score is considered as an effective measure of perceived quality. Regression analysis helps to analyse the contribution of the enhanced signal in three different dimensions—the signal alone, the background intrusiveness, and overall quality.

A new technique of speech enhancement using wavelets is proposed and the performance of the enhanced speech is evaluated using validated metrics. The speech signal is enhanced with two different algorithms namely, multiband (mband) spectral subtraction and a combination of mband and wavelet thresholding (proposed method). The quality of the enhanced signal is assessed through subjective and objective measures. Statistical analysis analyses the correlation between subjective and objective scores. Two figures of Merit—the Pearson's correlation coefficient and standard deviation of the estimated error are computed.

2 Materials and Methods

Speech sounds are produced with the help of an implanted prosthetic device, created by the surgeon, placed using a tracheo esophageal puncture (TEP). The speaker pushes air into the esophagus and then pushes it back up to produce articulating sounds. A male speaker implanted with the Blom-singer voice prosthesis is considered for the study. The valve used is 18 mm long with a 16Fr. puncture and voice generation practice helped him generate an alaryngeal voice, which is clear with distinct pronunciations. A sentence is presented from the IEEE sentence database. These sentences are phonetically balanced. The sentence, "Kick the ball straight and follow through" is taken for the study. The voice generated by the speaker after implantation with Blom-Singer Duckbill Voice Prosthesis is recorded using a unidirectional microphone in an anechoic room with the help of a computer. In this study, a combination of mband spectral subtraction speech enhancement algorithm with wavelet thresholding is explored. The basic principle of the spectral subtraction method is to subtract the magnitude spectrum of noise from that of the noisy speech.

The noise is assumed to be uncorrelated and additive to the speech signal. Multiband spectral subtraction algorithm is proved to be better among speech enhancement algorithms.

2.1 *Proposed Method*

The proposed algorithm takes the advantage of both mband spectral subtraction algorithm and wavelet thresholding technique [1]. The mband filter is applied to the noisy signal to enhance the additive colored noise. Here, seven different wavelet filters such as Haar, Daubechies, Symlet, Demy, Coiflet, Biorthogonal, and Reverse Biorthogonal are tested [2] and Symlet7 is found to give best results based on SNR. The enhanced signal is decomposed by discrete wavelet transform into approximation (LL) and detail (LH, HL, and HH) coefficients using the symlet7 wavelet filter. Approximation coefficient (LL) contains less noise, and the detail wavelet coefficients are thresholded using soft thresholding technique. The inverse wavelet transform is applied to the modified wavelet coefficients to reconstruct the speech signal.

2.1.1 Multiband Spectral Subtraction Algorithm

The magnitude spectrum of noise is subtracted from that of the noisy speech. A general expression for the spectral subtraction algorithms [3, 4] is given as

$$\left|\widehat{X}\right|p = |Y(\omega|p - \left|\widehat{D}(\omega)^{\rho}\right| \tag{1}$$

where, p is the power exponent; when $p = 1$ we get the original magnitude spectral subtraction and $p = 2$ yields the power spectral subtraction algorithm. $|\widehat{D}(\omega)|$ is the estimate of the magnitude noise spectrum made during non-speech activity. $|Y(\omega)|$ is the magnitude spectrum of the corrupted noisy signal. $\left|\widehat{D}(\omega)^{\rho}\right|$ is the magnitude spectrum of the clean signal.

2.1.2 Wavelet Thresholding

Wavelet transform is applied to the noisy signal to decompose it into approximation and detail coefficients. Soft thresholding is applied to the detail coefficients to denoise the noisy signal [5].

Assume that $y(n)$ is a noisy signal and is given as

$$y(n) = x(n) + d(n) \tag{2}$$

where $x(n)$ is original signal and $d(n)$ is noise signal. High-frequency features present in original signal are well-preserved by wavelet transform. The detail coefficient of the noisy signal is shrinked by soft thresholding as

$$\hat{x} = y - \operatorname{sgn}(y)T \quad \text{if}|y| > T$$

$$0 \quad \text{if}|y| < T \tag{3}$$

2.2 Quality Metrics

Quality is known to possess many dimensions, encompassing many attributes of the processed signal such as "naturalness," "clarity," "pleasantness," etc. Speech quality and speech intelligibility are not synonymous terms, hence different methods need to be used to assess the quality and intelligibility of processed speech. The performance of the algorithm is assessed using the subjective measures such as signal distortion, background intrusiveness, overall quality and objective measures such as signal-to-noise ratio (SNR), segmental signal-to-noise ratio (SNRseg), frequency-weighted segmental signal-to-noise ratio (fwSNRseg), weighted spectral slope (WSS), itakura-saito distance (IS), Cepstral Distance, and log-likelihood ratio (LLR).

Subjective Quality Measures Quality is highly subjective in nature and it is difficult to evaluate reliably. The subjective listening tests are designed according to ITU-T recommendation P.835 [6, 7].

The P.835 methodology is designed to reduce the listener's uncertainty in a subjective test, from the basis of their rating of overall quality based on the component of a noisy speech signal, the signal distortion or the background noise. 20 subjects with normal hearing activity are recruited for measuring subjective quality.

Scale of signal distortion Five-point scale of signal distortion of speech signal [8], i.e., a score of rating of 1-2-3-4-5 is given for very unnatural, very degraded-fairly unnatural, fairly degraded-somewhat natural, somewhat degraded-fairly natural, little degradation-very natural, and no degradation.

Scale of background intrusiveness Five-point scale of background intrusiveness [8], i.e., a score of rating of 1-2-3-4-5 is given for very conspicuous, very intrusive-fairly conspicuous, somewhat intrusive-noticeable, somewhat intrusive-somewhat noticeable, not noticeable.

Overall Quality The overall effect on a scale of the mean opinion score [8], i.e., a score of rating of 1-2-3-4-5 is given for bad-poor-fair-good-excellent.

Objective Quality Measures Objective quality measure typically defies interpretation and also involves an impartial measure without bias or prejudice.

Perceptual Evaluation of Speech Quality (PESQ) [9, 10] gives a mean opinion score (MOS), i.e., a score of rating of 1-2-3-4-5 is given for inadequate-deprived-reasonable-worthy-outstanding on a listening quality scale as per international telecommunication union (ITU) for assessing speech quality.

$$\text{PESQ} = a0 + a1 \times \text{dsym} + a2 \times \text{dasym} \tag{4}$$

where, $a0 = 4.5$, $a1 = -0.1$, and $a2 = -0.0309$.

Signal-to-Noise Ratio (SNR) [6]

$$\text{SNR} = 10 \log_{10} \frac{\sum_{i=1}^{N} x^2(i)}{\sum_{i=1}^{N} (x(i) - \hat{x}(i)} \tag{5}$$

where $x(i)$ and $\hat{x}(i)$ are the original and processed speech samples which are indexed by i and N is the total number of samples.

Segmental Signal-to-Noise Ratio (SNRseg) [6]

$$\text{SNRseg} = \frac{10}{M} \sum_{m=0}^{M-1} \log_{10} \frac{\sum_{n=Nm}^{Nm+N-1} x^2(n)}{\sum_{n=Nm}^{Nm+N-1} (x(n) - (\hat{x}(n))^2} \tag{6}$$

where, $x(n)$ is the original signal, $\hat{x}(n)(n)$ is the enhanced signal, N is the frame length, and M is the number of frames in the signal.

Frequency-weighted segmental SNR (fwSNRseg) [6]

$$\text{fwSNRseg} = \frac{10}{M} \sum_{m=0}^{M-1} \frac{\sum_{j=1}^{k} \log_{10} \left[\frac{F^2(m,j)}{(F(m,j) - \widehat{F}(m,j))^2} \right]}{\sum_{j=1}^{k} B_j} \tag{7}$$

where B_j is the weight placed on the jth frequency band, K contains 25 number of bands, M is the total number of frames, (m, j) is the filter bank amplitude of the clear signal in the jth frequency band at the mth frame.

Log-Likelihood ratio (LLR) [6]

$$d_{\text{LLR}}(a_x, \bar{a}_{\hat{x}}) = \log \frac{\bar{a}_{\hat{x}}^T R_x \bar{a}_{\hat{x}}}{a_x^T R_x a_x} \tag{8}$$

where a_x^T are the LPC coefficients of the clean signal, $\bar{a}_x^T$ are the coefficients of the enhanced signal, and R_x is the $(p + 1)*(p + 1)$ autocorrelation matrix of the clean signal.

Itakura-Saito (IS) [6]

$$d_{\text{IS}}(a_x, \bar{a}_{\hat{x}}) = \frac{G_x}{\bar{G}_{\hat{x}}} \frac{\bar{a}_{\hat{x}}^T R_x \bar{a}_{\hat{x}}}{a_x^t R_x a_x} + \log \left\{ \frac{\bar{G}_{\hat{x}}}{G_x} \right\} - 1 \tag{9}$$

where G_x and $\overline{G}_{\hat{x}}$ are the all-pole gains of the clean and enhanced signals, respectively.

Cepstral Coefficients [6]

$$d_{\text{cep}}(c_x, \bar{c}_{\hat{x}}) = \frac{10}{\log_e 10}\sqrt{2\sum_{k=1}^{p}\left[c_x(k) - c_{\hat{x}}(k)\right]^2} \tag{10}$$

where $C_x(k)$ and $C_{\hat{x}}(k)$ are the cepstral coefficients of the clean and enhanced signals, respectively.

Weighted Spectral Slope (WSS) [6]

$$d_{\text{wss}}(j) = K_{\text{spl}}(K - \widehat{K}) + \sum_{k=1}^{36} w_a(k)S(k) - \widehat{S}(k))^2 \tag{11}$$

where K, $\widehat{K}$ are related to overall sound pressure level of the original and enhanced utterances, and K_{spl} is a parameter which can be varied to increase overall performance.

Evaluation of Objective Quality Measures Objective measures are optimized for a particular type of distortion and may not be significant for another type of distortion. The task of evaluating the weight of objective measures over a wide range of distortions is enormous. The distorted database needs to be evaluated by human listeners using one of the subjective listening tests as shown in Fig. 1. For the objective measure to be reliable, it needs to correlate well with subjective listening tests [6, 7]. Regression analysis of objective measures is carried out using the following equations

$$\text{SIG} = 0.567 + \sum_{i=1}^{N} \text{B\_Sig.} * \text{WSS}) \tag{12}$$

$$\text{BAK} = 1.013 + \sum_{i=1}^{N} \text{B\_Bak.} * \text{WSS}) \tag{13}$$

$$\text{OVRL} = 0.446 + \sum_{i=1}^{N} \text{B\_Ovrl.} * \text{WSS}) \tag{14}$$

where B_Sig, B_Bak, and B_Ovrl are coefficients of multiple linear regression analysis and WSS is the mean value of Weighted spectral slope.

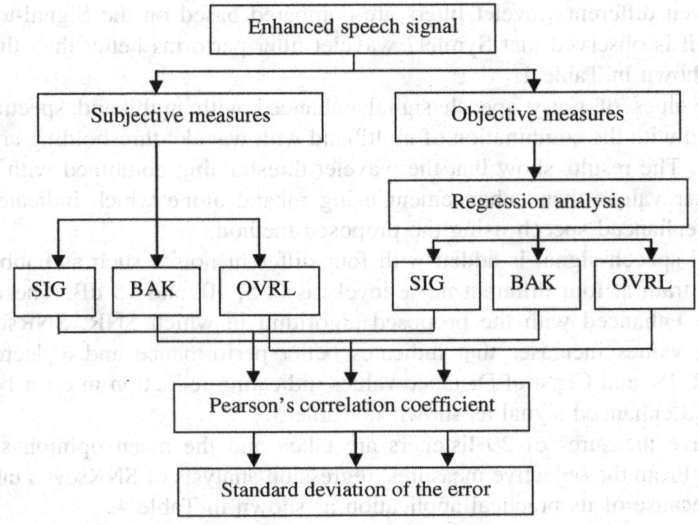

Fig. 1 Performance evaluation of the enhanced speech signal

2.2.1 Figure of Merit

Pearson's correlation coefficient [6, 7] measures the relationship between two items can range from 0 to 1. Here, the correlation is between subjective listening scores and objective measures

$$\rho = \frac{\sum_{\mathrm{d}} (S_{\mathrm{d}} - \overline{S}_{\mathrm{d}})(O_{\mathrm{d}} - \overline{O}_{\mathrm{d}})}{\sqrt{\left[\sum_{\mathrm{d}} (S_{\mathrm{d}} - \overline{S}_{\mathrm{d}})^2\right]}\sqrt{\left[\sum_{\mathrm{d}} (O_{\mathrm{d}} - \overline{O}_{\mathrm{d}})^2\right]}} \quad (15)$$

Standard Deviation [6, 7] is the prediction error obtained by using the objective measures to predict the subjective listening scores.

$$\sigma_{\mathrm{e}} = \sigma_p \sqrt{1 - \rho} \quad (16)$$

where σ_{e} is the standard error of estimation.

3 Results and Discussion

The proposed method is tested on the voice recorded from laryngectomee speakers implanted with voice prosthesis corrupted with additive noise.

The seven different wavelet filters are compared based on the Signal-to-Noise Ratio and it is observed that Symlet7 wavelet filter performs better than the other filters as shown in Table 1.

PESQ values of noisy speech signal enhanced with multiband spectral subtraction and with the combination of multiband with wavelet thresholding are listed in Table 2. The results show that the wavelet thresholding combined with mband gives higher values than enhancement using mband alone which indicate better quality of enhanced speech using the proposed method.

Original speech signal is added with four different noises such as babble, car, street, and train at four different noise levels as 0, 5, 10, and 15 dB. These noisy signals are enhanced with the proposed algorithm in which SNR, SNRseg, and fwSNRseg values increases that indicates better performance and a decrease in WSS, LLR, IS, and Cepstral Distance values indicating reduction in error between original and enhanced signal as shown in Table 3.

Subjective measures of 20 listeners are taken and the mean opinion score is calculated. From the objective measures, regression analysis of SNRseg is taken for analysis because of its practical application as shown in Table 4.

Correlation coefficients are obtained by computing the Pearson's correlation coefficient between the subjective and objective scores. A value closer to one indicates better correlation between subjective and objective measures. Standard deviation of error closer to 0 indicates reduced error between the two measures (shown in Table 5).

Table 1 Selection of wavelet filters using SNR values

Wavelt filters	SNR
Haar	9.5631
db9	9.9256
Coif4	9.9200
Sym7	9.9258
Demy	9.9183
Bior3.7	9.9220
rbio1.3	9.9252

Table 2 Comparison of MBAND and wavelet thresholding with MBAND using PESQ scores tested on various noises at 10 dB

Noise at 10 dB	PESQ scores	
	MBAND	Wavelet thresholding with MBAND
Babble	2.8397	2.8455
Car	3.7126	3.7138
Street	3.1769	3.1875
Train	3.5035	3.5834

Table 3 Signal quality metrics obtained for the algorithm wavelet thresholding combined with MBAND tested on various noises at different dB levels

Noise	SNR	SNR seg	LLR	WSS	Cepstral distance	fwSNRseg	IS
Babble 0 dB	4.0925	−1.3428	0.8125	57.4311	4.9930	5.46140	2.2685
Babble 5 dB	6.3228	1.2830	0.7576	52.3814	4.9888	7.05070	2.1988
Babble10 dB	7.3676	3.4342	0.6266	41.3963	4.5691	9.38200	1.7787
Babble15 dB	8.6346	5.3956	0.5728	34.3946	4.3892	11.1853	1.5242
Car 0 dB	3.7582	−0.4516	0.6515	55.0498	5.0232	8.35340	3.6873
Car 5 dB	6.5145	2.4727	0.5795	40.3130	4.6062	10.1532	2.8392
Car 10 dB	7.6460	4.0547	0.5163	31.5988	4.1976	11.8948	2.7985
Car 15 dB	8.9243	6.1360	0.4770	23.9343	3.9534	13.0638	2.3943
Street 0 dB	4.4992	−1.2560	0.8044	57.8679	5.4690	6.64470	2.2451
Street 5 dB	5.7472	0.9594	0.6666	48.3415	4.9063	8.64440	2.4665
Street 10 dB	7.8377	3.7856	0.6309	43.9714	4.7119	10.4093	1.5666
Street 15 dB	8.6264	5.4133	0.5407	33.6243	4.3318	12.1411	1.4890
Train 0 dB	1.3679	−0.4385	0.6787	51.4693	5.0898	8.27300	3.5824
Train 5 dB	4.5026	2.5119	0.6132	38.6560	4.7761	9.87090	3.3079
Train 10 dB	7.7156	4.6307	0.5233	26.5049	4.2382	11.9333	2.2064
Train 15 dB	8.6090	5.7071	0.4908	23.9793	4.0384	12.5115	1.7776

Table 4 Mean opinion score of SIG, BAK, and OVRL of the objective measure and the subjective measures obtained from the listeners

Noise	Subjective measures			Regression analysis of SNRseg		
	SIG	BAK	OVRL	SIG	BAK	OVRL
Babble 0 dB	1.7	2.9	1.1	1.6621	2.6535	1.1217
Babble 5 dB	2.2	2.1	1.6	2.2535	2.2482	1.8408
Babble 10 dB	3.1	2.5	1.9	3.3430	2.8220	2.0467
Babble 15 dB	2.6	2.6	2.8	3.5705	2.7650	2.8238
Car 0 dB	3.3	2.5	2.5	3.4250	2.9809	3.0332
Car 5 dB	3.4	2.9	2.9	3.6165	3.1375	3.0927
Car 10 dB	2.4	3.1	2.7	3.4125	3.2286	2.5558
Car 15 dB	2.3	3.1	1.9	2.9122	3.5453	2.1032
Street 0 dB	2.3	1.4	1.5	2.4795	1.6202	1.6689
Street 5 dB	3.1	2.2	2.1	3.2806	1.8079	2.1972
Street 10 dB	3.2	2.5	2.1	3.6114	2.8356	2.6738
Street 15 dB	3.4	3.6	2.5	3.5751	3.3503	2.6466
Train 0 dB	3.3	2.7	2.6	3.2772	2.8283	2.9103
Train 5 dB	3.7	2.9	2.4	3.6031	3.0062	3.1971
Train 10 dB	3.1	3.1	1.9	3.1015	3.2894	2.4503
Train 15 dB	2.9	3.2	2.3	3.0906	3.3086	2.5226

Table 5 Evaluation of objective measures of SNRseg using the subjective measures obtained from the listeners

	Pearson's correlation coefficient	Standard deviation of error
SIG	0.89097	0.24520
BAK	0.89642	0.23141
OVRL	0.90336	0.21671

4 Conclusions

The higher PESQ value for the proposed algorithm wavelet thresholding technique along with mband proves to be marginally better than the existing algorithm mband spectral subtraction algorithm taken alone. The higher values of objective measures such as SNR, SNRseg, fwSNRseg, and lower values of IS, WSS, Cepstral Distance, and LLR indicates that the speech quality is improved by the proposed method and it is suitable for higher noise levels and for real-world situations. From the subjective measures, it can be realized that the denoised signal is closer to the original clean speech signal.

References

1. Boutaleb R, Ykhlef F, Boucetta Y, Bendaouia L (2013) Comparative performance study between spectral subtraction and discreet wavelet transform for speech enhancement. Comput Syst Appl. ISSN:2161-5322
2. Jain R, Parveen S (2013) Analysis of different wavelets by correlation. Int J Eng Adv Technol. 2(4). ISSN:2249-8958
3. Verteletskaya E, Simak B (2010) Spectral subtractive type speech enhancement methods. Inf Commun Technol Serv 8(3)
4. Kamath S, Loizou P (2002) A multi-band spectral subtraction method for enhancing speech corrupted by colored noise. IEEE ICASSP
5. Kumari VSR, Devarakonda DK (2013) A wavelet based denoising of speech signal. Int J Eng Trends Technol 5(2)
6. Hu Y, Loizou PC (2008) Evaluation of objective quality measures for speech enhancement. IEEE Trans Audio Speech Lang Process 16(1)
7. Philipos C (2011) Loizou: speech quality assessment. Multimedia analysis, processing & communications. Studies Comput Intell 346:623–654
8. Hu Y, Loizou PC (2007) Subjective comparison and evaluation of speech enhancement algorithms. Speech Commun 588–601
9. Yen T-Y, Chen J-H, Chi T-S (2009) Perception-based objective speech quality assessment. IEEE ICASSP
10. Perceptual Evaluation of Speech Quality (PESQ) (2000) An objective method for end-to-end speech quality assessment of narrowband telephone networks and speech codecs. ITU-T Recommendation P.862

D-Mine: Accurate Discovery of Large Pattern Sequences from Biological Datasets

Prasanna Kottapalle, Seetha Maddala and Vinit Kumar Gunjan

Abstract Exploring interesting associations on gene variables help to assess the accuracy of the pattern sequence mining. Exploration of genetic structures like DNA, RNA, and protein sequences from biological datasets will boost up new innovations in Pathology diagnosis. For this mission, very large genetic pattern sequences are to be discovered. To do this, doubleton pattern mining (DPM) is considered as very constructive for analyzing these datasets. In this paper, D-Mine, a new approach for discovering very large gene pattern sequences from Biological datasets is discussed. D-Mine effectively discovers doubleton patterns which are further enriched to generate gene pattern sequences with vector intersection operator and Markov probabilistic grammars. D-Mine is described as a solution to diminish the set of discovered patterns. D-Mine makes use of a new integrated data structure called 'D-struct,' as combination of a virtual data matrix and 1D array pair set to dynamically discover doubleton patterns from biological datasets. D-struct has a diverse feature to facilitate which is that it has extremely limited and accurately predictable main memory and runs very quickly in memory-based constraints. The algorithm is designed in such a way that it takes only one scan over the database to discover large gene pattern sequences by iteratively enumerating D-struct matrix. The empirical analysis on D-Mine shows that the proposed approach attains a better mining efficiency on various biological datasets and outperforms with CARPENTER in different settings. The performance of D-Mine on biological data set is also assessed with accuracy and F-measure.

P. Kottapalle (✉)
JNIAS-JNTUH, AITS, Rajampet, Hyderabad, AP, India
e-mail: prasanna.k642@gmail.com

S. Maddala
JNIAS, GNITS, Hyderabad, AP, India
e-mail: smaddala2000@yahoo.com

V.K. Gunjan
Department of CSE, AITS, Rajampet, Hyderabad, AP, India
e-mail: vinitkumargunjan@gmail.com

L.P. Suresh and B.K. Panigrahi (eds.), *Proceedings of the International Conference on Soft Computing Systems*, Advances in Intelligent Systems and Computing 397, DOI 10.1007/978-81-322-2671-0_62

Keywords Gene pattern sequences · Doubleton pattern mining · Gene association analysis · Biological data · CARPENTER

1 Introduction

Modern Computational Biology known as Bioinformatics exploration is gaining much importance in the extraction of knowledge from biological data sets. The best part is its strong relationship with medicine. The bioinformatics has developed various important algorithms for biological data analysis. The development in medical technology in past decade has introduced new form of datasets called biological datasets well known as gene expression datasets and microarray datasets. Unlike transactional datasets, these high-dimensional databases usually have few rows (samples) and a huge number of columns (genes). In fact, from the genome sequences or system biology, the principal challenge is to identify functional genes for effective analysis. In bioinformatics, the biologists can make use of the advances in computational biology to analyze large and complex datasets. Knowledge discovery and data mining has concerned as an imminent need to extract useful information and knowledge from these datasets.

It is broadly believed that in an accustomed living organism, the accumulation of genes and their substances like DNA, RNA, and protein sequences are usually active in a complicated and orchestrated way. The traditional molecular biology analysis works on the basis of 'one gene in one experiment' and it infers an extremely constrained throughput. So, it is difficult to assess the gene functionality. With the progression of DNA microarray data, it has presented a mixture of information and data analysis issues which are not discussed under conventional molecular biology. The data extracted from different microarray studies is represented normally in matrix form $N \times M$ of articulation levels, where N rows relate to different experimental conditions and M columns relate to genes under study. Due to this high dimensionality, it needs efficient data mining methods to discover interesting knowledge from datasets including the exploration of DNA sequences for scientific or medical process.

After its introduction in data mining, frequent pattern mining (FPM) gained a prominent data mining paradigm that assists to extract patterns that conceptually symbolize associations among discrete attributes and performs an imperative role in information mining and data exploration tasks as well as applications. Based on the intricacy of those relations, different types of patterns can occur. The most common kind of patterns tend to be mining association rules [1, 2], episodic [3], correlations [4], sequential patterns [5, 6], maximal patterns and frequent closed patterns [7–9], classification [10, 11], and clustering [12].

There are numerous algorithms developed for fast and efficient mining of frequent patterns, which are classified into three categories. The very first category candidate generation approach, such as Apriori [2] and its subsequent studies are in view of Apriori property [2]: if a pattern seriously is not frequent, then its super pattern cannot be frequent. The Apriori-based algorithm achieved good diminution around the sized candidate sets. Nevertheless, when there are quite a few frequent patterns or even long patterns, it will take multiple scans over large databases to build candidate sets. The second category, *pattern growth approach*, including *FP-growth* [2] also uses the Apriori property. However, it recursively partitions the database into sub databases to generating candidate sets. It makes limited scans over the database. The third category is vertical data approach. It is an enumeration based approach. It gives better performance and greatly reduces the possibility to prune the search space.

In the literature, several algorithms were developed under pattern growth approach for discovering frequent patterns and closed patterns [8, 13, 14]. It uses enumeration-based approaches [8, 14, 15] in which item combinations are searched for frequent closed patterns. In view of this, their running time increases exponentially with increase in the average length of the records and makes minimum of two scans over the database. These will consume large extent of memory usage and predictably takes enough time when memory based constrains are present. These algorithms are rendering to be impractical on high-dimensional microarray datasets. The complete set of frequent closed patterns are obtained using row enumeration space was first shown in [15], which was also observed in [16].

Nevertheless, the existing frequent pattern mining approaches still encounter the following difficulties.

- All item enumeration-based mining methods are based on singleton patterns and take much time to compute these patterns.
- *Huge main memory is required for effective mining*. When memory constraints are present, an Apriori-like algorithm will not be effective since it produces enormous candidates for long patterns. Enough memory space is required to store candidate sets for discovering frequent patterns of different size. *FP-growth* [6] evades candidate generation by condensing into an *FP-tree.*
- *Real-time databases hold all the cases.* Most of the datasets in real-time applications are either sparse or dense. It is difficult to choose a proper mining method on the fly which suits for all cases.
- *Real-time applications require to be high dimensional and scalable.* Several existing approaches are efficient for smaller size data sets. However, as the dataset size increases, the existing methods show fit falls on core data structures and requires enough memory.
- *Multiple scans over the database.* Most of the existing Apriori and FP-growth approaches make several scans over the databases. Efficient data storage structures are needed to store intermediate results.

During the past decade, the biologists have explored information concerning cell characteristics of many genes. The data can be extrapolated to different species utilizing evolutionary principles. To do this, gene pattern sequences are very useful. These sequences are useful to infer human behaviors from other species and also disclose the biological relevant information between gene associations and tend to discover gene networks [16] and biclustering of expressions [17].

Row enumeration search can be explored by constructing projected database recursively [15]. However, there is a need to consider column enumeration algorithms specified in many algorithms which are proposed to mine frequent closed patterns. However, for high-dimensional datasets, the pattern mining problem consumes more time and space. If a dataset is with 100 rows and 1000 columns, the existing enumeration algorithms work well if threshold is set to low while discovering closed patterns and often generates huge number of discovered patterns with no suitable information. However, traditional FPM methods are having fit falls in dealing with high-dimensional datasets because of its dimensionality, size and main memory utilization. These pretenses a novel challenge on design and developing a new method which is *efficient in pattern mining on large databases where space requirement is limited.* For this reason, DPM is considered for analyzing biological datasets.

In this paper, we study an efficient new algorithm D-Mine that is specially designed to discover very large pattern sequences over biological datasets is described. D-Mine makes use of a new data structure called doubleton data matrix D-struct which can be used to discover pattern sequences by performing attribute enumeration as depth first row-wise enumeration, and efficiently reduces the searching time over the dataset. D-Mine has the following stages; first, a doubleton pattern discovery algorithm is proposed for the reduced datasets using D-struct that can fit into the memory. Second, D-Mine uses a new attribute enumeration column vector based intersection operator to discover pattern sequences efficiently by reducing the search time and database scans. The experimental results show that this approach produces better results when mining biological datasets and outperforms CARPENTER on different settings.

2 Basic Preliminaries

In this chapter, first we present the basic concepts of doubleton pattern mining and a formal problem statement and in the second one we describe the related work of the mining task by using an example.

2.1 Basic Concepts and Problem Formulation

Doubleton patterns have been extensively used to analyzing massive datasets by means of discovering interesting relationship between the attributes in high-dimensional databases.

Basic definitions

Let G = $\{g_1, g_2....g_m\}$ be a set of m gene attributes, also called gene variables. An attribute X is a subset of attributes such that $X \subseteq G$. in short, an attribute $G = \{g_1, g_2....g_m\}$ is also denoted as $G = g_1, g_2....g_m$. Let $C = \{C_1, C_2, ... C_n\}$ be a set rows representing experimental conditions defied over biological dataset, where each C_i is a set of n subsets called genes. Each row in C identifies a subset of items. $C = (Rid, X)$ is a 2-tuple, where *Rid* is a row-id and X an attribute. A row $C = (Rid, X)$ is said to contain attribute Y if and only if $Y \subseteq X$. Table 1 shows an example of the dataset in which the genes are represented from $g1$ to $g11$. Let the first two columns of Table 1 be our sample data set. Table 1 shows a dataset *DB* is the set of experimental conditions. Each C_i contains a subset of genes represented in lexicographic order.

Definition 1 A support (s) is defined as the number of transactions in DB, which contains both X and Y, represented as its frequency.

$$\text{Support } (X \to Y) = P(XUY) \tag{1}$$

Definition 2 Confidence (c) of the rule $X \to Y$ is true in the database DB, if it contains the number of transactions containing X that also contains Y, represented as

$$\text{Confidence } (X \to Y) = P(Y|X) = P(XUY)|P(X) \tag{2}$$

Definition 3 The *Relative frequency* of an attributes, X, Y is contained in database, the relative frequency is defined as

$$\text{Relative Frequency (RF)} = \frac{\text{Frequency}(X, Y)}{\text{Frequency}(X)} \tag{3}$$

Definition 4 An *Association Rule* is an inference of the form $X \to Y$ between two attributes X and Y where $X, Y \subseteq I$ and $X \cap Y = \phi$, which satisfies user supplied *Support S* and *Confidence C*.

Table 1 Sample transaction database DB

Rid	Gene attributes
c1	g1, g2, g3, g5, g7, g8, g9
c2	g1, g3, g4, g5, g6, g8, g10
c3	g2, g5, g6, g7, g8
c4	g1, g2, g3, g4, g5, g6, g7, g11
c5	g1, g2, g4, g6, g7
c6	g2, g5, g7, g8, g9, g10, g11

Definition 5 *Pattern sequence* An attribute set $X \subseteq I$, is a pattern sequence if and only if sup(X) ≥ *minsup* and is a doubleton pattern

Definition 6 *Doubleton pattern* A doubleton pattern can be frequent if both items in the set are frequent by themselves. A doubleton pattern set (X, Y) is frequent if both X and Y in the set are also frequent and it is true in database DB, if it is having its *minsup* above two.

2.2 Problem Statement

The problem of doubleton pattern mining is to *find the complete set of pattern sequences in a given biological data set*. The main objective is to discover all gene pattern sequences in a given biological dataset D with regard to user minimum support threshold.

3 Related Work

For a given set features in biological dataset, we define a gene expression matrix (M) with m rows and n columns in such a way that experimental conditions on rows and genes on columns. Table 2 represents a bit matrix M, which is the equivalent gene expression matrix of the database DB, where 1-means 'overexpressed' and 0-means 'underexpressed.' A transaction in gene expression data is associated with 'overexpressed' data.

Column-wise pruning will be performed on gene matrix M based on minsup and eliminate the columns whose total occurrences are less than minsup. The pruned gene expression matrix is shown in Table 3 with the minsupp is 3.

The support is given as the frequency of the rows in the dataset that contain a set of features G'. By definition, **Support** is defined as the maximal frequency set of

Table 2 Gene expression matrix (M) of the sample database (DB)

Rid	g1	g2	g3	g4	g5	g6	g7	g8	g9	g10	g11
c1	1	1	1	0	1	0	1	1	1	0	0
c2	1	0	1	1	1	1	0	1	0	1	0
c3	0	1	0	0	1	1	1	1	0	0	0
c4	1	1	1	1	1	1	1	0	0	0	1
c5	1	1	0	1	0	1	1	0	0	0	0
c6	0	1	0	0	1	0	1	1	1	1	1
Support count	**4**	**5**	**3**	**3**	**5**	**4**	**5**	**4**	**2**	**2**	**2**

Table 3 Pruned gene matrix with support as three

Rid	g1	g2	g3	g4	g5	g6	g7	g8
c1	1	1	1	0	1	0	1	1
c2	1	0	1	1	1	1	0	1
c3	0	1	0	0	1	1	1	1
c4	1	1	1	1	1	1	1	0
c5	1	1	0	1	0	1	1	0
c6	0	1	0	0	1	0	1	1
Support count	**4**	**5**	**3**	**3**	**5**	**4**	**5**	**4**

the transaction that contains *X*. The relative frequency of rows in the dataset that contain *X* is called its support of *X*, for a given set of items $X \subseteq I$. For set patterns, there exists a pattern sequences with a maximum length. The pattern sequences are discovered using doubleton pattern mining.

4 D-Mine

In this section, we study efficient mining of pattern sequences from biological dataset. We first illustrate the mining process of D-Mine with an example. Then, we present the D-Mine algorithm.

4.1 Discovering Pattern Sequences with Vector Database

In this section, we describe the process of discovering pattern sequences using an example dataset described above.

In the literature there are different ways to analyze the biological datasets. High-dimensional databases characterized as experimental conditions as rows and large gene variables as columns. This distinctive characteristic will minimize the number of experimental conditions in pattern mining process by constructing a doubleton data matrix with vertical search strategies. Row enumeration algorithms work well when the dataset size is in low dimensions. Horizontal search strategy cannot do efficient mining of patterns since the possibility of discovering exponential order of items. In this paper, we used vertical search strategies along with vector intersection operator to generate pattern sequences from biological datasets. D-struct matrix is constructed using vertical search strategy as shown in Fig. 1.

For the same gene expression data in Table 1 with minimum support = 3, we introduce a doubleton pattern mining method for mining pattern sequences. This method explores the concept of vector databases as shown in Fig. 1.

(a)

	g1	g2	g3	g4	g5	g6	g7	g8
g1	--	1,4,5	1,2,4	2,4,5	1,2,4	2,4,5	1,4,5	1,2
g2	-	-	1,4	1,5	1,3,4,6	3,4,5	1,3,4,5,6	1,3,6
g3	-	-	-	2,4	1,2,4	2,4	1,4	1,2
g4	-	-	-	-	2,4	2,4,5	4,5	2
g5	-	-	-	-	-	2,3,4	1,3,4,6	1,2,3,6
g6	-	-	-	-	-	-	3,4,5	2,3
g7	-	-	-	-	-	-	-	1,3,6
g8	-	-	-	-	-	-	-	-

(b)

g1.g2
1
0
0
1
1
0

g5.g8
1
1
1
0
0
1

(c)

Pattern pair set		'K'
{g7,g8}	3	28
	.	
{g5, g8}	4	25
{g5,g7}	4	24
{g5,g6}	3	23
	.	
{g3,g5}	3	13
{g2,g7}	5	12
	.	
{g2,g5}	4	10
	.	
{g1, g7}	3	6
{g1, g6}	3	5
{g1, g5}	3	4
{g1, g4}	3	3
{g1, g3}	3	2
{g1, g2}	3	1

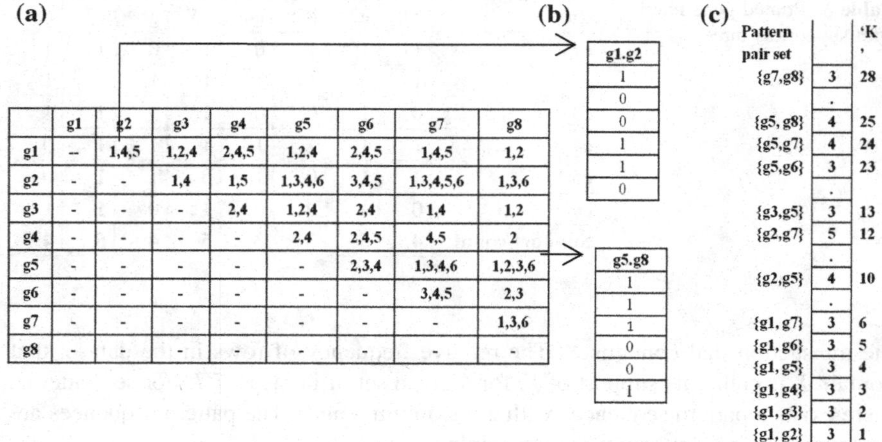

Fig. 1 Doubleton data matrix with column vector database and triple count arrays. **a** 2-level Doubleton Pattern sequences. **b** Column vector. **c** One dimensional triple

4.1.1 Finding Doubleton Frequent Patterns

"A doubleton pattern can be frequent if the items in the set are also frequent by themselves."

Using vertical search strategies, construct a doubleton data matrix such that each attributes is a bitwise column vector and their corresponding genes are those which there are in the all rows of this column vector. Now, scan the dataset and mark the row number corresponding to each row and column. Each entry in the matrix is a column vector that contains set of bit fields and storing with binary values. Its support values are stored in triple count array as shown in Fig. 1.

4.1.2 One Dimensional Triple Array Pair Set

In general, the Association Rule mining algorithms maintain different item count frequency values throughout a scan over database. For instance, it is essential to have adequate main memory to hoard each pattern count that the number of times a pattern pair sets occurs in the transaction database. It is hard to update a 1 to a count set where the counting sequences are stored in different memory locations and difficult in loading the page to main memory. In such cases, these algorithms will be slow in finding that pattern pair count in main memory as it takes extra overhead on processing time and increases the time to discover frequent pattern set. So it is difficult to count a value that requires enough main memory. When it comes to high-dimensional datasets, it is difficult to maintain all in one memory. So a new 1D triple array set is used to count all the pattern occurrences in the given database.

To optimize main memory, a pattern pair (i, j) occurrence in the dataset should be counted in one place. If the pattern sequence order is $i < j$, and uses only one

Table 4 Two-level pattern pairs discovered using doubleton matrix

1	{g1, g2}, {g1, g3}, {g1, g4}, {g1, g5}, {g1, g6}, {g1, g7}
2	{g2, g5}, {g2, g6}, {g2, g7}, {g2, g8}
3	{g3, g5}
4	{g4, g6}
5	{g5, g6}, {g5, g7}, {g5, g8}
6	{g6, g7}
7	{g7, g8}

entry, a [*i*, *j*] in two dimensional array a. This approach makes half of the array as useless. Count array (CA) is a more efficient way to store pattern sequences in memory. A count array is defined as a 1D triple array set which will store a count as CA[*index*] for the pair (*i*, *j*), with $1 \leq i < j \leq n$, where

$$index = K + (i - 1)\left(n - \frac{i}{2}\right) + (j - i) \tag{4}$$

and k is position at (k-1) frequent subset in count array. To discover pattern sequences, D-Mine performs a iterative depth first search (DFS) on column enumeration strategy. By imposing backtracking search order on column sets, we are able to perform a systematic search over pattern sequences. Two-level doubleton pattern pairs, using vector databases and triple array, are discovered as shown in Table 4.

4.1.3 Pruning the Search Space and Creating a Doubleton Database

Each pattern sequence corresponds to a unique set of genes. By enumerating all possible combinations on genes, we discovered all patterns sequences in the dataset. However, pruning the search space must be introduced to minimize unnecessary exploring on set of genes.

Let *R* be the gene sequence discovered from doubleton matrix, gene pattern sequences are identified as *R*-gene doubleton database which exclusively contains a particular gene and its Rid count must be above the min support threshold as shown in Table 5. All the discovered doubleton pattern pair sets can be divided into seven nonoverlap subsets based on the doubleton data matrix

Table 5 A gene doubleton database

S. no.	Gene	Conditioned on	Rid numbers
1	g1	{g2, g3, g4, g5, g6, g7}	1, 2, 4, 5
2	g2	{g5, g6, g7, g8}	1, 3, 4, 5, 6
3	g3	{g5}	1, 2, 4
4	g4	{g6}	2, 4, 5
5	g5	{g6, g7, g8}	1, 2, 3, 4, 6
6	g6	{g7}	3, 4, 5
7	g7	{g8}	1, 3, 6

- The ones containing gene g1,
- The ones containing g2 but not g1,
- The ones containing item g3 but no g1 nor g2,
- The ones containing g4 but no g1, g2 nor g3,
- The one containing g5 but no g1, g2, g3, nor g4,
- The one containing only g6, g7,
- The one containing only g7 and g8.

Once all the pattern pair sets are found, the complete set of vector database is done.

4.1.4 Discovering Pattern Sequences

From the discovered gene doubleton databases, we can expand each *i*-level pattern pair sets to form a new bitwise column vector to determine the equivalent pattern sequences which is frequent or not. In column enumeration search strategy, each pattern sequence is a column set and its adjunct gene is those which there are all rows of this column set. A column bit vector, which is the result of performing intersection operation on column vectors to determine the corresponding pattern sequence. The discovered doubleton pattern pair sets can be mined to discover pattern sequences by construing a 1D triple array pair set and mine each pattern sequences recursively. Clearly, it takes one scan over the dataset to build D-struct along with triple array pair set.

- Finding pattern sequences containing g1
- Finding pattern sequences containing g2
- Finding pattern sequences containing g3
- Finding pattern sequences containing g4
- Finding pattern sequences containing g5 and so on

The left over mining process can be performed on D-struct, only without referring the original database. Pattern sequences are discovered one by one using triple array pair set values. For every doubleton pattern there is a k value. Using this k values recursively over the doubleton patterns and vector intersection operator collectively discovers pattern sequences as shown in Fig. 2.

g1.g2	g3
1	1
0	1
0	0
1	1
1	0
0	0

→

g1g2ANDg3
1
0
0
1
0
0

Fig. 2 Column vector intersection operator between g1, g2, and g3

Table 6 Pattern sequences generated using column vector intersection operator

S. no.	Pattern sequence
1	{g1, g2, g3, g5, g7}
2	{g1, g2, g7}
3	{g1, g2, g4, g6, g7}
4	{g1, g3, g4, g5, g6}
5	{g1, g4, g6}
6	{g2, g5, g6, g7}
7	{g2, g5, g7}
8	{g2, g6, g7}
9	{g2, g7}
10	{g5, g8}
11	{g2, g5, g7, g8}
12	{g1, g3, g5, g8}
13	{g5, g6, g8}

In the above example, the pair set g1·g2 can be explored on g3 to create a new pattern pair set as g1·g2 AND g3. It performs bitwise vertical intersection on g1g2 and g3 and discovers a new doubleton pattern pair set. Its corresponding count value is stored on a 3-level triple count array. g1g2 AND g3 is not equal to either g1g2 or g3. Hence, it is also called as closed doubleton patterns and its count is two, which is stored in triple array. By performing the above process recursively, we can discover long gene pattern sequences which are doubleton or closed patterns as in Table 6

4.2 *Finding the Level of Accuracy of the Discovered Pattern Sequences*

After discovering the pattern sequences form doubleton pattern mining, the level of correctness of our algorithm is measured using the "Accuracy" and "F-measure" to evaluate the overall performance; these are characterized by using the formulas.

Let $X \rightarrow Y$ be the discovered doubleton pattern sequence, then

$$F\text{-measure (FM) of } (X, Y) = \frac{2 * P(\frac{Y}{X}) * P(\frac{X}{Y})}{P(\frac{Y}{X}) + P(\frac{X}{Y})} \quad (5)$$

$$\text{Accuracy (ACC) of } (X, Y) = P(XY) + P(\neg X \neg Y) \quad (6)$$

4.3 D-Mine Algorithm

In a given gene database, a relevance frequency (RF), the problem of mining the set of doubleton patterns can be considered as partitioning into n-sub problems. The jth problem is to find the complete set of doubleton patterns containing in $+1 - j$ but not ik. The problem of partitioning can be performed recursively that is each subset of DPM can be further divided when necessary. This forms a divide and prune framework. To mine the subsets of DPM, we construct corresponding doubleton databases.

Given a DB, let i be the frequent attribute or gene constituted as a column vector in DB. The *i-level* doubleton DB denoted as $DT|_i$ is the subset of DB containing i, and all occurrences are stored in count array (CA). All the genes that precedes i can be formed as pattern sequence.

Let j be a frequent attribute in i-doubleton database and i is a doubleton pattern set. The *ji*-DB is the set of transactions containing i and j denoted as $DB|_{ij}$ by performing bitwise intersection operation on column vector DB and all the occurrences are stored in its *i-level* count array crated dynamically. Thus, pattern sequences are discovered by repeating the process for all attributes.

Algorithm
Input: Gene Database and relevance frequency as min support threshold.
Output: the complete set of pattern sequences.

1. Initialize PS = 0, let PS as set of pattern sequences
2. Scan the database and compute doubleton data matrix and discover all doubleton patterns pair set and create a doubleton database $DT|_i$.
3. Call Dmine(*0*, $DT|_i$, A_i, *PS*)

Procedure Dmine(*iX*, $DT|_i$, A_i, *PS*)

Let iX : the doubleton patterns if DB is *x-doubleton database*,
$DT|i$: Doubleton Database
A_i: Attribute list.

1. Set $DT|_i \leftarrow 0$,
2. Let j be the set of attributes in DB, such that they appear in every transaction of DB, then PS $\leftarrow$ $i \otimes j$ and create its column vector DB's, count array and verify $i \otimes j$ count should be above 3.
3. Set $PS = PS \cup \{i \otimes j\}$
4. For each remaining attribute i in *Ai,* recursively call Dmine(*iX*, $Dt|_i$, A_i, *PS*) to build its *i-level* doubleton Database $DT|i$ and discover all its patterns using dynamically created count array.

5 Experimental Analysis

In this section we will revise the performance of our algorithm with CARPENTER. Experiments were performed on a 2.8 GHz dual core CPU, with 1 GB RAM, and running on Windows7. All programs are written in Java. The run time is measured as elapsed time and IO seek time. CARPENTER is an enumeration-based algorithm that has shown its better performance on discovering pattern sequences. We implemented this algorithm and compare our method with them.

In our performance study, we used the variant size of the datasets; it is difficult to assess the minsup threshold as an absolute number. Instead, minsup threshold is determined by relative frequency (RF). Experiments are performed on four real datasets from UCI [18] to compare the algorithm. Table 7 shows the characteristic information about the datasets.

Table 8 show the result of running two algorithms D-Mine and CARPENTER on a real standard dataset lung cancer (LC). There are 12,533 genes and 181 samples. It is notices that with increasing minimum support (RF) all the algorithm performance in data set will be decreased.

Often with frequent pattern mining algorithms, it is observed that the performance is poor when minsup is small than the large because with smaller minsup the algorithms will find more frequent items and the searching time will be increased dramatically. However, in DPM, when the minsup decrease, the level of data matrix will be decreased and searching time also minimized. Therefore, when the minsup is small, D-Mine has a good mining efficiency. From Fig. 3, it is observed that the difference of efficiency of D-Mine with CARPENTER is very much when the minsup is small. Tables 9 and 10, shows the correctness of D-Mine on various datasets is presented.

Table 7 characteristics of test datasets

S. no	Dataset name	Size
1	Lung cancer	12533 genes × 181 samples
2	Breast cancer	25 genes × 699 samples
3	Heart	28 genes × 303 samples
4	Diabetes	17 genes × 768 samples

Table 8 Performance on LC in (sec)

Support	Carpenter	D-Mine
0.03	60	49
0.05	48	35
0.06	41	28
0.08	31	21
0.09	26	16

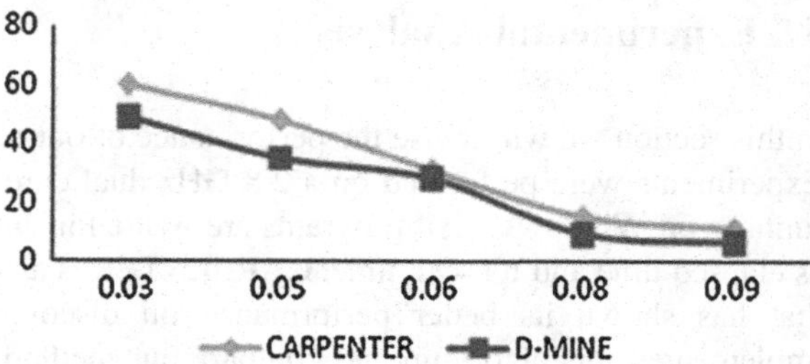

Fig. 3 Performance on LC (s)

Table 9 Pattern sequences discovered using D-Mine (uniform relative requency)

S. no.	Dataset	No. of pattern sequences	Maximum possible accuracy	Average		Highest	
				F-measure (FM)	Accuracy (ACC)	F-measure (FM)	Accuracy (ACC)
1	Breast cancer	6,936	100	94.55	95.12	96.08	96.42
2	Heart	41,096	100.00	66.05	70.27	80.37	80.85
3	Diabetes	923	97.79	66.87	73.83	68.54	74.09

Table 10 Pattern sequences discovered using D-Mine (varying relative frequency)

S. no	dataset	No. of pattern sequences	Maximum possible accuracy (MA)	Average		Highest	
				F-measure (FM)	Accuracy (ACC)	F-measure (FM)	Accuracy (ACC)
1	Breast cancer	11,338	100.00	94.22	94.84	95.74	96.13
2	Heart	62,833	100.00	64.74	69.94	79.40	79.87
3	Diabetes	1,133	97.79	67.20	73.70	68.26	73.70

6 Conclusion

According to doubleton pattern mining "A doubleton pattern is frequent only it all items in the set are also frequent." This property leads to a generation large number of redundant patterns. However, frequent pattern mining over biological datasets, applied on small and midsized pattern discovery. Doubleton pattern mining is a relatively new approach applied on high-dimensional datasets which enhance the process time than Frequent Pattern Mining.

In this paper, a new algorithm D-Mine is presented to mine long biological datasets. In our algorithm we iteratively constructed a data matrix D-struct, a bit-wise representation of the dataset for effective discovery of doubleton patterns. We used a vector column intersection bitwise operation to facilitate the algorithm to extract pattern sequences. We also used triple count array along with D-struct to improve the efficiency of the mining process when memory constraints are present. The empirical study shows that our algorithm has attained good mining efficiencies under different settings. Furthermore, our performance analysis demonstrate that this algorithm also attains highest accuracy and F-measure in discovering pattern sequences and significantly the best compared to formerly developed algorithms.

References

1. Agrawal R, Srikant R (1994) Fast algorithms for mining association rules. In: VLDB'94, pp 487–499
2. Mannila H, Toivonen H, Verkamo AI (1997) Efficient algorithms for discovering association rules. In: KDD'94, pp 181–192
3. Manila H, Toivonen H, Verkamo AI (1997) Discovery of frequent episodes in event sequences. Data Min Knowl Discovery 259–289
4. Brin S, Motwani R, Silverstein C (1997) Beyond market basket: generalizing association rules to correlations. In: Proceedings of the ACM-SIGMOD international conference on management of data, pp 265–276
5. Srikant R, Agrawal R (1996) Mining sequential patterns: generalizations and performance improvements. In: EDBT'96, pp 3–17
6. Pei J, Han J, Mortazavi-Asl B, Pinto H, Chen Q, Dayal U, Hsu M-C (2001) PrefixSpan: mining sequential patterns efficiently by prefix-projected pattern growth. In: ICDE'01, pp 215–224
7. Bayardo RJ (1998) Efficiently mining long patterns from databases. In: SIGMOD'98, pp 85–93
8. Pei J, Han J, Mao R (2000) CLOSET: an efficient algorithm for mining frequent closed itemsets. In: Proceedings of the 2000 ACM-SIGMOD international workshop data mining and knowledge discovery (DMKD'00), pp 11–20
9. Zaki M (2000) Generating non-redundant association rules. In: KDD'00, pp 34–43
10. Cheng Y, Church GM (2000) Biclustering of expression data. In: Proceedings of the 8th international conference on intelligent systems for mocular biology
11. Cong G, Tung AKH, Xu X, Pan F, Yang J (2004) FARMER: finding interesting rule groups in microarray datasets. In: Proceedings of the 23rd ACM international conference on management of data
12. Yang J, Wang H, Wang W, Yu PS (2003) Enhanced biclustering on gene expression data. In: Proceedings of the 3rd IEEE symposium on bioinformatics and bioengineering (BIBE), Washington DC
13. Pasquier N, Bastide Y, Taouil R, Lakhal L (1999) Discovering frequent closed itemsets for association rules. In: Proceedings of the 7th international conferrence on database theory (ICDT)
14. Zaki MJ, Hsiao C (2002) CHARM: an efficient algorithm for closed association rule mining. In: Proceedings of the SIAM international conference on data mining (SDM)
15. Pan F, Cong G, Tung AKH, Yang J, Zaki MJ (2003) CARPENTER: finding closed patterns in long biological datasets. In: Proceedings of the ACM SIGKDD international conference on knowledge discovery and data mining (KDD), 2003
16. Creighton C, Hanash S (2003) Mining gene expression databases for association rules. Bioinformatics 19
17. Zhang Z, Teo A, Ooi B, Tan K-L (2004) Mining deterministic biclusters in gene expression data. In: 4th symposium on bioinformatics and bioengineering
18. UCI machine learning data sets. http://archive.ics.uci.edu/ml/datasets/

Comparison of Multipliers to Reduce Area and Speed

V.L.V. Pratyusha, P.Y. Narendra Babu, S. Nivetha and P. Jagadeesh

Abstract In an average processor, multiplication is one of the fundamental number-crunching operations and it requires considerably more equipment assets and preparing time than expansion and subtraction. Actually, 8.72 % of all the direction in regular transforming units are multipliers (Asadi and Navi (2007) A new lower power 32*32-bit multiplier) [1]. Increase is one of the fundamental number-crunching operations and it requires considerably more equipment assets and handling time than expansion and subtraction. In this task, we think about the working of three multipliers by executing each of them independently.

Keywords Dadda multiplier · Wallace multiplier · Booth multiplier

1 Introduction

The multiplier is imperative and wide spreading building squares of advanced outlines. At the point when multiplier postponement is a fundamental layout parameter, composed tend to use unprecedented logarithmic multipliers like Dadda and Wallace. Multipliers are taking into account lessening tree in which diverse plans of pressure of fractional items can be executed. Multipliers assume a vital part in today's computerized sign handling and different applications [2].

V.L.V. Pratyusha (✉) · P.Y. Narendra Babu · S. Nivetha · P. Jagadeesh
Department of Electronics and Communication Engineering, Saveetha School of Engineering, Saveetha University, Chennai 602105, India
e-mail: v.l.v.pratyusha@gmail.com

P.Y. Narendra Babu
e-mail: pynarendra225@gmail.com

S. Nivetha
e-mail: nivethasundar27@gmail.com

P. Jagadeesh
e-mail: pjagadeesh89@gmail.com

L.P. Suresh and B.K. Panigrahi (eds.), *Proceedings of the International Conference on Soft Computing Systems*, Advances in Intelligent Systems and Computing 397, DOI 10.1007/978-81-322-2671-0_63

Force dissemination emerges from the charging and releasing of the circuit hub capacitances found on the yield of every method of reasoning entryway. Power organization is the watchful masterminding of force plan for each subsystem of a VLSI chip. This is particularly an essential issue throughout today's complex frameworks. The most essential and fruitful utilization of force administration is to deactivate a segment of circuit when its calculation is not needed. Each low-to-high rationale move in a digital circuit brings about a change of voltage, drawing vitality from the force supply.

A creator at the innovative and structural level can attempt to minimize the variables in these comparisons to minimize the general vitality utilization. On the other hand, power minimization is often a complex methodology of exchange offs between velocity, zone, and force utilization. The current work proposes decrease of element exchanging force, to diminish exchanging action and diminishment of gate counts in 16 * 16 complex multipliers by utilizing higher request compressors. Multipliers oblige high amount of force and deferral amid the fractional items development and expansion of the numbers. Time to market, quick innovation progression, increment in chip size, and many-sided quality of digital frameworks are a portion of the variables that prompted the improvement of equipment portrayal languages (VHDL and Verilog) alongside the mechanized rationale union devices. The VHDL program describing the computerized framework is innovation autonomous; however, the blend instrument that creates the circuit requires an innovation document joined to it. The nonappearance of such record causes the issue of non-plausibility of time delay extraction. In this work, the coding of distinctive multipliers is created in VHDL. In the wake of the coding, they are centered on the FPGA and by utilizing Xilinx synthesis report, the execution of these multipliers has been watched and power utilization is watched utilizing force analyzer in Xilinx 10.1 bundle.

2 Dadda Multiplier

Dadda multipliers are the refinement of parallel multipliers at first shown by Wallace in 1964. Dadda multiplier executes diminish at every point major. The best roof of every stage is controlled by last stage extension which embodies two lines of defective things. These two sections are then consolidated using a brisk pass on causing snake Carry Propagation Adder. In the Wallace framework, the midway things are lessened at the most punctual open door. Alternately, Dadda method does the base diminishment imperative at each level to perform the diminishing in the same number of levels as required by the Wallace methodology achieving a layout with less full adders, additionally half adders. The downside of Dadda philosophy is that it obliges a fairly broader, fast Carry proliferation adder and has a less steady structure than Wallace. Figure 1 explains about around 8 × 8 Dadda multipliers [3].

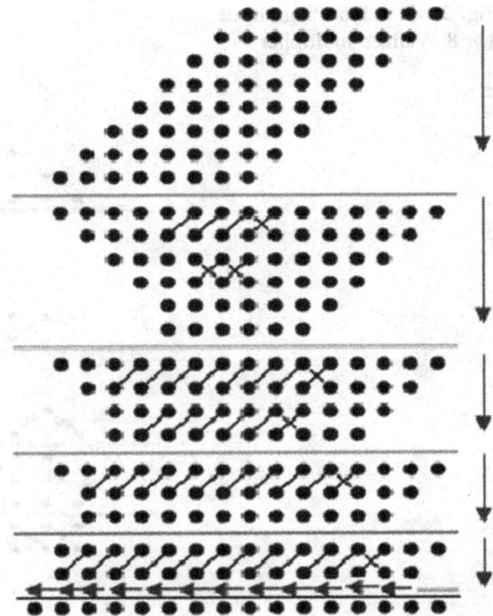

Fig. 1 [10] Dot diagram of 8 × 8 Dadda multiplier

Dadda summed up and opened up Wallace's results by observing that a full serpent can be like half adders. The disadvantage of Dadda philosophy is that and yields the number in 2-bit twofold structure. Using such a counter, Dadda prescribed that at every stage, basically scarcest measure of diminishment ought to be finished recollecting the completed goal to reducing the deficient structure by a section of 1.5. Dadda system requires the same number of levels as that of Wallace methodology. However, Dadda framework does the base backslide fundamental at each level. This results in an outline with less full adders and half adders [4].

3 Wallace Multiplier

A changed Wallace multiplier is a profitable gear execution of cutting edge circuit imitating two numbers. Generally, in common, in Wallace multipliers various full adders and half adders are used as a piece of their diminishment stage. Half adders do not diminish the amount of midway it emanates. Thus, minimizing the amount of half adders used as a piece of a multiplier diminishment will diminish the diverse quality. Therefore, a change to the Wallace abatement is done in which the deferment is the same concerning the customary Wallace dropping. The balanced abatement methodology immensely decreases the amount of half adders with a slight augmentation in the amount of full adders.

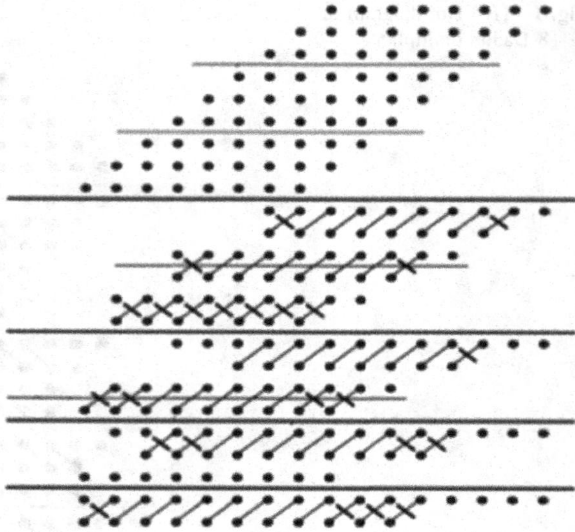

Fig. 2 [11] Dot diagram of 8 × 8 Wallace multiplier

Reduced unconventionality Wallace multiplier diminishment includes three stages. To begin with stage, the $N \times N$ structure is organized and before proceeding ahead to the second stage, the system is enhanced to take the condition of modified pyramid [5]. In the midst of the second stage, the overhauled system is assembled into non-covering social occasion of three as showed in Fig. 2, single bit and two bits in the get-together will be gone ahead to the accompanying stage and three bits are given to a full snake. The amount of segments in every period of the reduction stage is figured by the formula.

$$r_i + 1 = 2[r_i/3] + r_i \bmod 3$$

$$\text{If } r_i \bmod 3 = 0, \text{ then } r_i + 1 = 2r_i/3$$

In case the value figured from the above correlation for number of sections in every stage in the second stage and the amount of segment that are confined in every period of the second stage does not organize, at precisely that point, the half snake will be used [6]. The last consequence of the second stage will be in the stature of two bits and went ahead to the third stage. In the midst of the third stage, the yield of the second stage is given to the pass on spread snake to deliver the last yield.

4 Array Multiplier

There is a coordinated topological correspondence between this equipment structure and the manual augmentation. The era of n fractional items obliges $N * M$ good for nothing AND entryway. A large portion of the region of the multiplier is given to the including of incomplete items, which obliges $N - 1$ and M-bit adders. The moving of the midway things for their fitting plans performed by direct controlling and does not require any clarification. The general structure can easily be compacted into rectangle, realizing particularly powerful plan (Fig. 3).

5 Booth Multiplier

Corner increase calculation gives a methodology for reproducing paired numbers in marked—2's supplement representation. Taking after steps are utilized for actualizing the stall calculation: Let X and Y are two double numbers and having m and n quantities of bits (m and n are meet) individually.

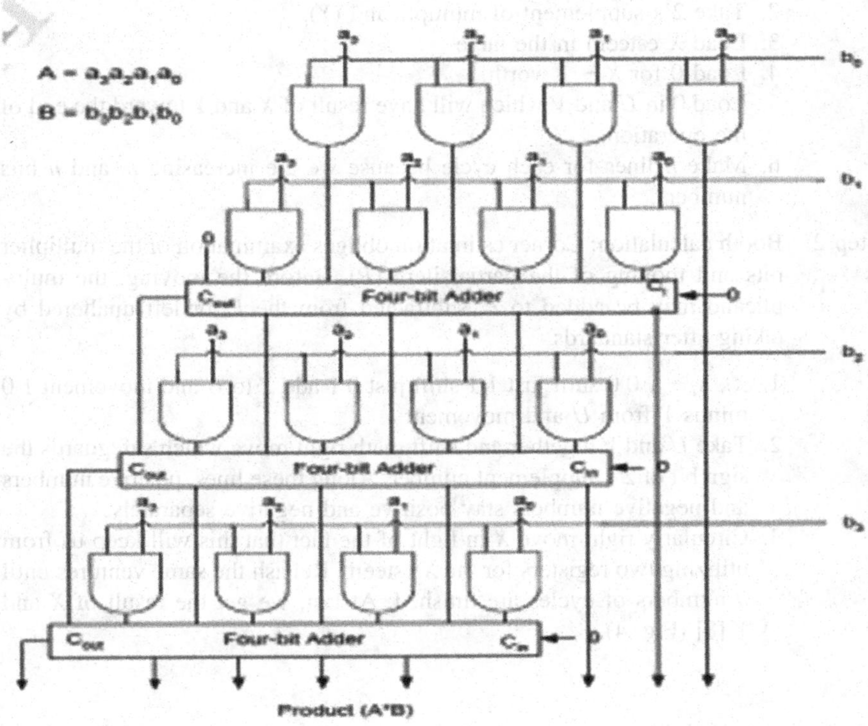

Fig. 3 [12] Array multiplier

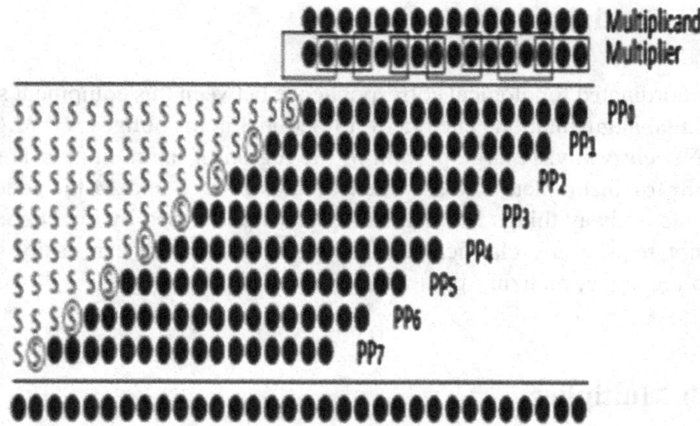

Fig. 4 [12] Booth multiplier

Step 1: Making stall table: In the corner table, we will take four sections; one segment for multiplier, second for past first LSB of multiplier, and other two (*U* and *V*) for incomplete item collectors [7].

1. From two numbers, pick the multiplier (*X*) and multiplicand (*Y*).
2. Take 2's supplement of multiplicand (*Y*).
3. Load *X* esteem in the table
4. Load 0 for $X - 1$ worth.
5. Load 0 in *U* and *V* which will have result of *X* and *Y* toward the end of the operation.
6. Make *n* lines for each cycle because we are increasing *m* and *n* bits numbers.

Step 2: Booth calculation: Corner estimation obliges examination of the multiplier bits and moving of the partial item (*P*). Before the moving, the multiplicand may be added to *P*, subtracted from the *P*, or left unaltered by taking after standards

1. X_i, $X_i - 1$ 0 0 shift just 1 1 shift just 0 1 add *Y* to *U* and movement 1 0 minus *Y* from *U* and movement
2. Take *U* and *V* together and shift math right move which safeguards the sign bit of 2's supplement number. Along these lines, positive numbers and negative numbers stay positive and negative separately.
3. Circularly right move *X* in light of the fact that this will keep us from utilizing two registers for the *X* esteem. Rehash the same ventures until *n* numbers of cycles are finished. At last, we get the result of *X* and *Y* [8] (Fig. 4).

6 Comparison of 32 Bit Designs

6.1 Area Comparison

See Table 1.

6.2 Area Optimized

See Table 2.

7 Comparision and Discussion

FPGA usage results demonstrates that multiplier Nikhilam Sutra in light of Vedic arithmetic for augmentation of twofold numbers is quicker than multipliers taking into account Array and Booth multiplier [9]. It additionally demonstrates that as the quantity of bits increments to *N*, where *N* can be any number, the deferral time is enormously diminished in Vedic Multiplier when contrasted with different multipliers. Vedic Multiplier has the points of interest as over different multipliers additionally for force and normality of structures (Table 3).

Table 1 [13] Gate counts for various modes

Area optimized				
Size	Array	Wallace	Dadda	RA
4 × 4	75	81	78	82
8 × 8	401	432	405	413
16 × 16	1,744	1,905	1,782	1,802
32 × 32	7,231	7,598	7,114	7,198
Speed optimized				
Size	Array	Wallace	Dadda	RA
4 × 4	113	118	115	114
8 × 3	628	609	676	633
16 × 16	2,855	2,732	3,014	2,780
32 × 32	11,961	11,661	12,503	11,808
Auto optimized				
Size	Array	Wallace	Dadda	RA
4 × 4	74	76	75	81
8 × 3	391	419	397	400
16 × 16	1,738	1,840	1,738	1,739
32 × 32	7,193	7,692	7,241	7,289

Table 2 [13] Delay data counts for various methods

Area optimized				
Size	Array	Wallace	Dadda	RA
4 × 4	3.26	2.77	2.64	2.77
8 × 8	9.17	6.08	5.16	6.14
16 × 1(3	20.64	11.6	10.97	12.03
32 × 32	44.33	22.77	22.06	22.63
Speed optimized				
Size	Array	Wallace	Dadda	RA
4 × 4	3.10	2.40	2.54	2.59
8 × 8	8.36	5.19	5.44	5.36
16 × 16	19.14	9.82	10.47	10.54
32 × 32	40.88	18.69	19.63	19.74
Auto optimized				
Size	Array	Wallace	Dadda	RA
4 × 4	3.30	3.30	2.87	2.91
8 × 8	9.54	5.86	6.15	6.31
16 × 16	22.55	11.67	11.50	11.80
32 × 32	48.56	23.19	22.63	22.31

Table 3 [11] Comparison of multipliers w.r.t. delay (ns)

Names of multiplier	Array multiplier		Booth multiplier		Vedic multiplier	
	8 × 8 bit	16 × 16 bit	8 × 8 bit	16 × 16 bit	8 × 8 bit	16 × 16 bit
Delay (ns)	47	92	117	232	27	39

8 Conclusion

From the results and discussions, both the Wallace Multiplier and Dadda Multiplier is similar in design and working process but there are some differences between these two multipliers in certain parameters like speed, delay, and power consumption. This paper shows the comparison of parameters all the multipliers with different bits. The Carry Propagation Adder is used in Dadda Multiplier to perform the addition operation of partial products after the reduction process. This CPA takes more area when compared to the Carry Select Adder (CSA). Hence, to reduce the area Carry Propagation Adder is replaced with Carry Select Adder in Dadda Multiplier.

References

1. Asadi P, Navi K (2007) A new lower power 32*32-bit multiplier. ISSN 818-4952, IDOSI publications, 2007
2. Eriksson H, Larsson-Edefors P, Sheeran M, Sjalander M, Johansson D, Scholin M (2006) Multiplier reduction tree with logarithmic logic depth and regular connectivity. In: Proceedings of the 2006 IEEE international symposium on circuits and systems, ISCAS 2006, pp 4–8
3. Priyanka Brahmaiah V, Dharma Raja L, Padma Sai Y Dr (2013) Study on comparison of various multipliers, IJECT, Oct 2013, pp 133
4. Dadda L (1965) Some schemes for parallel multipliers. Alta Frequenza 34(5):349–356
5. Waters RS, Swartzlander EE Jr (2010) A reduced complexity Wallace multiplier reduction. IEEE Trans Comput 59(8), Aug 2010
6. Wallace CS (1964) A suggestion for a fast multiplier. IEEE Trans Electron Comput EC-13 (1):14–17, Feb 1964
7. Getahun A Booth multiplication algorithm, Fall 2003 CSCI 401
8. Soniya, Suresh Kumar (2013) A review of different type of multipliers and multiplier-accumulator unit, pp 1–2, July–August 2013
9. Thapliyal H, Rarbania H A novel parallel multiply and accumulate (V-MAC) architecture based on ancient Indian vedic mathematics
10. Praveenkumar Reddy S (2014) Efficient hybrid method for binary floating point multiplication, April 2014, Ijera publications, vol 4, pp 37–42
11. Dadda L (1965) Some schemes for parallel multipliers. Alta Frequenza 34:349–356
12. Seo YH, Kim DW (2010) New VLSI Architecture of parallel multiplier–accumulator based on Radix-2 modified booth algorithm. IEEE Trans Very Large Scale Integr (VLSI) Syst 18(2), Feb 2010
13. Lee CYH, Hiung LH, Lee SWF, Hamid NH A performance comparison study on multipliers design. IEEE publications, pp 3–4
14. Lee CYH, Hiung LH, Lee SWF, Hamid NH A performance comparison study on multiplier designs, pp 3–6
15. Vaidya S, Dandekar D (2010) Delay power performance comparison of multipliers invlsi circuit design, 2(4):8, July 2010, IJCNC publications
16. Bhombe DL, Deshmukh SR (2014) Performance comparison of different multipliers using booth algorithm, (IJERT) 3(2), Feb 2014
17. Chen SK, Liu CW, Member, IEEE, Wu TY, Tsai AC (2013) Design and implementation of high-speed and energy-efficient variable-latency speculating booth multiplier (VLSBM). IEEE Trans Circ Syst—I: Regular Papers, 60(10):2631, Oct 2013
18. Qi H, Kim YB, Choi M (2012) A high speed low power modulo 2n + 1 multiplier design using carbon-nanotube technology, '2012 IEEE
19. Prasad BKV, Satish Kumar P, Stephen Charls B, Prasad T (2012) Low power design of Wallace tree multiplier. Int J Electron Commun Eng Technol (IJECET) 3(3):258–264. ISSN print: 0976-6464, ISSN Online: 0976-6472
20. Kavitha, Umesh Goyal (2013) FPGA implementation of vedic multiplier. Int J Adv Res Eng Technol (IJARET) 4(4):150–158
21. Abdul Sattar S Dr, Yousuf Khan M Dr, Qadeer S (2013) A new Radix-4 FFT algorithm. Int J Adv Res Eng Technol (IJARET) 4(3):251–256. ISSN print: 0976-6480. ISSN Online: 0976-6499
22. Taralabenchi J, Hegde K, Hegde S (2012) Implementation of binary multiplication using booth and systolic algorithm on FPGA using VHDL. In: International conference & workshop on recent trends in technology, (TCET) 2012, Proceedings published in International Journal of Computer Applications® (IJCA)

Gesture Controlled Automation for Physically Impaired

Vishnu Reghu and T. Senthil Kumar

Abstract Hand gesture recognition system is widely used for interfacing between computer and human using hand gesture. This work presents a room for automation system using hand gestures, which is meant for physically impaired. The objective of project is to develop an algorithm for recognition of hand gestures with reasonable accuracy. Most of the gesture recognition system fails when the background is complex, here in our method we use hand detection in complex background. Hu moments are used as feature which is used to classify the gestures. Kinect sensor is used to get the input video which will give the depth map from which we will get the location of the people performing gesture. Two gestures are used to switching on and off function for an Electrical appliance. Arduino Board is used to interface between the computer and the appliances.

Keywords K-means · Gesture recognition · Kinect · Arduino

1 Introduction

Nowadays, gesture recognition technology has drawn attention as a promising man–machine communication. Gesture technology is being implemented in many areas and has applications like sign language, interactive game technology, automated homes, remote control, etc. The advantage is ability of communication in the distance without any contact with the system that we need to operate. Performance is highly depended on the accuracy of hand segmentation from the background. Many methods for hand detection and tracking are using various sensors which are directly attached to the hand and special feature gloves.

V. Reghu (✉) · T. Senthil Kumar
Department of CSE, Amrita Vishwa Vidhyapeetham, Coimbatore, India
e-mail: vishnureghu007@gmail.com

T. Senthil Kumar
e-mail: t_senthilkumar@cb.amrita.edu

L.P. Suresh and B.K. Panigrahi (eds.), *Proceedings of the International Conference on Soft Computing Systems*, Advances in Intelligent Systems and Computing 397, DOI 10.1007/978-81-322-2671-0_64

The importance of human–machine interaction using computer vision is increasing. Many video processing techniques have been developed in the recent years for recognizing the gestures. Many gesture recognition systems work under restricted environment because of complex background and illumination conditions. The gesture recognition techniques are also limited in the distance between camera and the person.

In the proposed method, we considered a complex background for gesture recognition and Hu moments are used as features for hand gesture recognition in order to control the electrical appliances.

2 Related Work

As the project is based on gesture recognition concept, many papers have been analyzed in order to understand the existing framework. Two common methodologies are static hand gesture classification and dynamic hand gesture classification. Static gestures utilize only spatial information. Structure of different gesture is identified by extraction of the features from the images and then using machine learning technique the gesture is identified. As the number of predefined gestures is increased, the differences between gestures become harder to distinguish. Features which can be extracted for static hand gesture classification are radial signature [1], histogram of gradients (HOG) [2], and local binary patterns [3]. Motion-based recognition is done by identifying the motion of an object in the sequence of frames in the video. One of the widely used gesture recognition is using optical flow [4, 5].

Most of the works related to hand gesture techniques [6] can be categorized into two: glove-based [7, 8] and vision-based. In glove-based method, if user have to wear special gloves in the hand, and generally any sensor. Examples for vision-based techniques are model-based [9] and state-based [10]. Recently, there has been an increasing number of gesture recognition researches using vision-based methods. 3D neural network is used by Huang [11] for sign language recognition. David and Shah [12] used a finite state machine model to find different generic gestures. Hand gesture recognition using HMM [13, 14] have also been developed. In our system, gesture recognition is done using vision-based method without using any sensors or gloves in hand which is scale, rotation, and translation invariant.

3 Proposed System

In the proposed method, we have four main steps

- Face detection
- Hand segmentation from complex background
- Gesture recognition
- Switching the electrical appliances

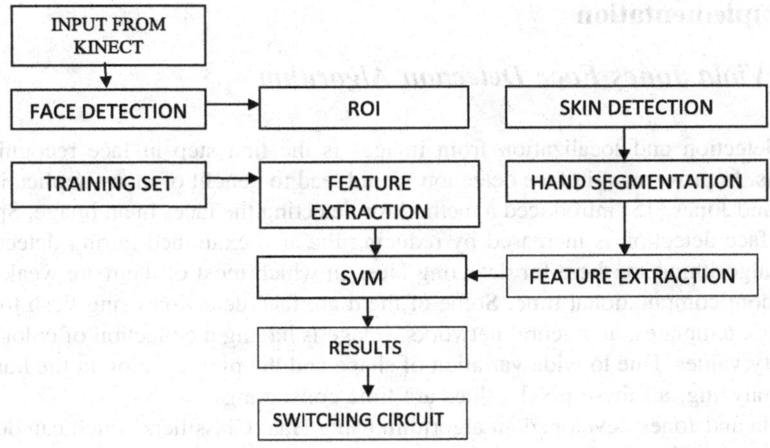

Fig. 1 Block diagram

Initially, face detection is done in the input video; this is for two main reasons one is to know whether a person is present or not and the other one is to check if multiple people are present or not. After face detection, a ROI is found near the face; this is the region in the frame where the gesture is performed. ROI is a region to right side of a face having a resolution of 150 * 150. If multiple people are present, we will get different ROI for each face and the regions can be separated by using the 3D information of the face in the frame. Finding the ROI will reduce the search area and increase the accuracy of gesture recognition.

In most of the gesture recognition system, hand is segmented based on skin color but if the background is complex this system will fail and give very bad result, this is because of the presence of skin color in the background. In our method we are working in a complex background, so we have to segment the hand from a background having a skin color. This is done by an additional step for background elimination after skin color detection.

After the segmentation of hand the features are extracted from the shape. Hu moments are used as features to classify the gesture. Seven Hu moments are extracted from each image of the training set and trained using SVM. The moments of the segmented hand are extracted and the gesture recognition is done. These results are used to switch on/off the electrical appliances which are done using an Arduino board and a relay circuit (Fig. 1).

4 Implementation

4.1 Viola Jones Face Detection Algorithm

Face detection and localization from images is the first step in face recognition systems. Improvement in face detection would lead to benefit of many applications. Viola and Jones [15] introduced a method for detecting the faces in an image. Speed of the face detection is increased by reducing the area examined during detection. Many algorithms are there for detecting faces in which most of them are weak and need more computational time. Some of them are face detection using flesh tones, contours, templates, and neural networks. Image is having a collection of color and intensity values. Due to wide variation of shape and the pigmentation in the human face analyzing, all these pixel values are time consuming.

Viola and Jones developed an algorithm using Haar Classifiers which can detect an object rapidly. AdaBoost classifier cascades use Haar like features for classification. While training two sets of samples are used, positive and negative samples. Positive samples contain the images which have similar properties of object to be detected. Negative samples contain the images which does not contain the object. Once the training is completed, it can be used to detect the face.

Face detection using Haar features may also give many false detection so that one more condition is checked for the detected region to conclude that the detected region is face or not. This is done by using percentage of skin color in the detected region. Face is a region containing skin color so that we took a detected region with more than 60 % of skin color. This will increase the accuracy of face detection (Fig. 2).

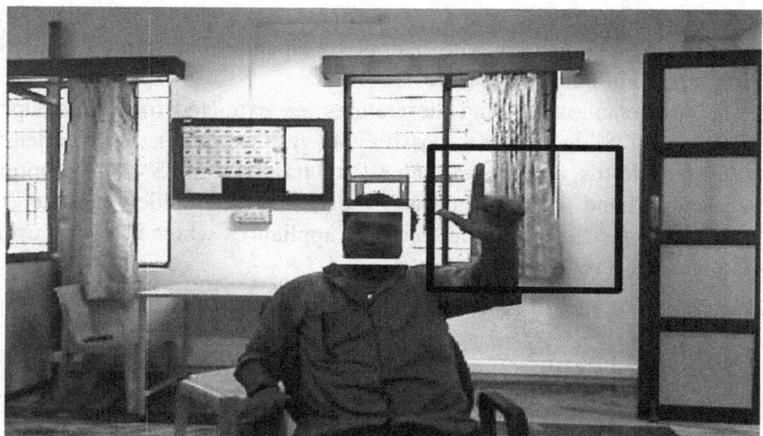

Fig. 2 Face detection and ROI w.r.t face

4.2 Hand Segmentation

In our method, we are using the Kinect kit to get the input video which gives the depth information for each pixel, thus foreground skin pixels can be found using this depth information. We use K-means clustering by using depth values of each pixel as sample points which can separate the background and foreground pixels.

If (ROI (xi) = skin pixel && ROI (xi) = foreground)
Handimage (xi) ==1
Else
Handimage (xi) ==0
Where x_i is the 2D position of the pixel

4.2.1 Skin Color Detection

YCbCr color space The YCbCr color space also known as family of color spaces because the Chroma components Cr and Cb can be easily calculated. Luminance is denoted by Y. Blue difference and red difference Chroma components are denoted by Cb and Cr. The three components of YCbCr can be easily calculated by linear combinations of R, G, and B components of image.

$$[\mathrm{Y} \quad \mathrm{Cb} \quad \mathrm{Cr}] = [\mathrm{R} \quad \mathrm{G} \quad \mathrm{B}] \begin{bmatrix} 0.299 & -0.168935 & 0.499813 \\ 0.587 & -0.331665 & -0.418531 \\ 0.114 & 0.50059 & -0.081282 \end{bmatrix}$$

Following conditions are applied to detect skin region

140 <= Cr <= 165 AND
140 <= Cb <= 195

After skin color segmentation, there remains some small noise. Those are reduced by using image erosion through morphological structure. Image erosion shrinks the object. The binary erosion of A by B, denoted A Θ B, is defined as the set operation. $A \Theta B = \{z | Bz \subseteq A\}$.

4.2.2 Background Subtraction Using Depth

After the skin color detection in the ROI, there will be many pixels having the skin color so that this will affect the accuracy of the system and also the hand cannot be segmented efficiently. Due to this reason we have used the depth information of each pixel in the image to separate the skin pixels in the foreground from the background so that the hand can be segmented accurately. This is done by a

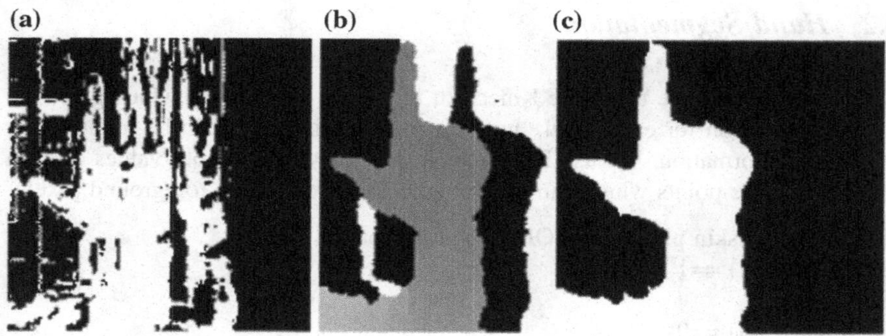

Fig. 3 **a** Skin color detection. **b** Depth image. **c** Segmented hand after clustering

K-means clustering algorithm which will take the depth data of the pixels as the data samples and cluster them into two classes, i.e., foreground and background.

K-Means Clustering

K-means [16] is an unsupervised learning algorithm which is used for clustering the given samples. The input samples will be clustered into different clusters (2 in our problem). Initially we have to define the k centers for each cluster. Next step is to group the given data points based on the nearest Centre. Whenever no points are pending then the first step is completed. Now the k new centroids are recalculated using the clusters obtained from the previous step. Now a loop is generated so that centers are updated using the clusters from the previous step each and every time. Whenever two successive iterations give the same clusters, the final clusters are obtained (Fig. 3). Thus, the algorithm will minimize an objective function given as

$$J(V) = \sum_{i=1}^{C} \sum_{j=1}^{C_i} \left(\left\| x_i - v_j \right\| \right)^2$$

'$\|x_i - v_j\|$' gives Euclidean distance between x_i and v_j
'c_i' is the total number of data points
'c' is the number of cluster centers considered

Algorithmic Steps for K-Means Clustering

Let $X = \{x_1, x_2, x_3, \ldots\ldots\ldots, x_n\}$ be given set of data points and $V = \{v_1, v_2, \ldots\ldots\ldots, v_c\}$ be the centers.

1. Select *'c'* cluster centers randomly.
2. Calculate the distance between the data points and the current centers.
3. Group the data point to the cluster center based on the minimum distance to the cluster centers.
4. Update cluster center by using

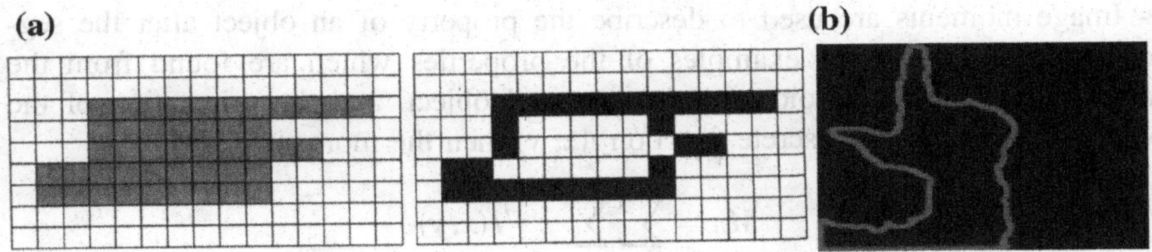

Fig. 4 **a** Binary image and the boundary. **b** Contour of segmented hand

$$v_i = \left(\frac{1}{c_i}\right)\sum_{j=1}^{c_i} x_j$$

where, 'c_i' represents number of data points in the ith cluster.

5. Calculate the distance between the data points and current cluster centers.
6. If cluster remains same as before then stop, otherwise goto step 3.

4.3 Gesture Recognition

Gesture recognition is done by using machine learning algorithm. In our method we extracted the feature from the shape of the segmented hand. The contour of the segmented hand is found and the Hu moments are calculated from the contour which used as the feature for the classification. SVM is used for classification of gesture (Fig. 4).

4.3.1 Contour Extraction

Scan the image from top to bottom until a 1 pixel is found

(1) Stop if it is starting pixel.
(2) If value is 1, add the current pixel to boundary.
(3) Go to 0 in the 4-neighbor on left.
(4) Select the first 1 in the clockwise direction by checking the 8-neighbors of the current pixel.
(5) Go to step 2

4.3.2 Feature Extraction

In the image processing-related fields, image moments are weighted average of the intensities of the pixels or such function are mostly chosen to have some attractive property and interpretation.

Image moments are used to describe the property of an object after the segmentation. Some of the examples of the properties which are found from the moments are area, centroid of the segmented object, and the orientation of the object. Consider a 2D discrete function *f*(*x*, *y*) then the moment is given as

$$M_{ij} = \sum\sum x^i y^j I(x, y)$$

The value of *Mpq* is determined by *f*(*x*,*y*) where *f*(*x*,*y*) is a pixel position of a digital image. The moments can be found of different orders by changing the value of *p* and *q*; the central moment of a digital image *f*(*x*,*y*) is given as

$$\mu_{pq} = \sum_x \sum_y (x - \bar{x})^p (y - \bar{y})^q f(x, y)$$

where,

$\bar{x} = \frac{M_{10}}{M_{00}}\ \bar{y} = \frac{M_{01}}{M_{00}}$ are components of centroid.

If we divide the above central moment by scaled (00)th moment, we can construct the moments which will be both scale- and translation-invariant, which is given by η_{ij} where $i + j \geq 2$.

$$\eta_{ij} = \frac{\mu_{ij}}{\mu_{00}^{1+\frac{i+j}{2}}}$$

Rotation Invariant Moments

Now the moments need to be rotation invariant along with invariant to scale and translation, thus we use Set of Hu moments [17]. It is proved that these moments are invariant to scale, translation and rotation except the seventh one. This invariance is proved with for infinite image resolution.

These are seven Hu moments that are the features which are extracted from the motion energy image for gesture recognition. SVM is used to classify the gestures based on these seven Hu moments extracted from the contour of segmented hand.

$$I_1 = \eta_{20} + \eta_{01}$$

$$I_2 = (\eta_{20} + \eta_{02})^2 + 4\eta_{11}^2$$

$$I_3 = (\eta_{30} + 3\eta_{12})^2 + (3\eta_{21} + \eta_{03})^2$$

$$I_4 = (\eta_{30} + \eta_{12})^2 + (\eta_{21} + \eta_{03})^2$$

$$I_5 = (\eta_{30} + 3\eta_{12})(\eta_{30} + \eta_{12})\left[(\eta_{30} + \eta_{12})^2 - 3(\eta_{21} + \eta_{03})^2\right] + ((3\eta_{21} - \eta_{03})(\eta_{21} + \eta_{03})[3(\eta_{30} + \eta_{12})^2 - (\eta_{21} + \eta_{03})^2]$$

$$I_6 = (\eta_{20} + \eta_{02})\left[(\eta_{30} + \eta_{12})^2 - (\eta_{21} + \eta_{03})^2\right] + 4\eta_{11}(\eta_{30} + \eta_{12})(\eta_{21} + \eta_{03})$$

$$I_7 = (3\eta_{21} + \eta_{03})(\eta_{30} + \eta_{12})\left[(\eta_{30} + \eta_{12})^2 - 3(\eta_{21} + \eta_{03})^2\right] - (\eta_{30} - 3\eta_{12})(\eta_{21} + \eta_{03})[3(\eta_{30} + \eta_{12})^2 - (\eta_{21} + \eta_{03})^2]$$

4.4 Switching the Electrical Appliances

The switching is done by using an Arduino board and a relay circuit. Based on the classification of gesture, a serial communication is done in between the system and the Arduino-Uno which is connected with a relay circuit which can switch on and switch off the electrical appliances. Arduino is ATmega328-based microcontroller which has 14 digital input or output pins, it has six analog inputs along with 16 MHz Crystal Oscillator. Based on the result after the classification the digital values are sent serially to the Arduino board through usb. Here, it is a binary classifier so 0 or 1 will be sent to the board. The Arduino is connected to a relay circuit. Relays are electrically controlled mechanical switches. We need to add a transistor and a fly back diode in between Arduino and the relay circuit rather than connecting Arduino directly with the relay.

5 Results and Discussions

5.1 System Specifications

We have implemented our proposed system with Microsoft Kinect as the acquisition system. Our algorithm runs on Microsoft Windows 8 with 4 GB RAM powered by Intel's Core i5 of 2.4 GHz. Since we exploited single core of CPU, we have not achieved the expected frames per second. And also exploiting the GPU would have increased the frames per second.

5.2 Training Phase

After finding the contour of the segmented hand seven Hu moments are extracted as features during training. These Features are scale and rotation invariant. 200 positive and negative samples are considered while training. The features are saved in a text file. SVM is used in the training stage for the classification of gestures.

5.3 Result Analysis

We have compared the results of our method with convex hull method and other static gesture recognition techniques and found that convex hull method is more sensitive to noise and illumination changes than our method. The accuracy of the gesture classification reduces in case of features like radial signature if the distance from the camera is increased.

In our method, we are using the feature which is also less sensitive to illumination changes; the features are invariant to scaling, rotation, and translation. The gestures are classified correctly to a maximum distance of 2.5 m.

6 Conclusion

An algorithm for gesture recognition system for physically impaired based on feature from shape in complex background has been proposed. In this method, face detection is done initially and an ROI is calculated near the face. Gesture can be performed in this ROI. Hu moments are extracted as features from these images of segmented hand from the complex background. SVM is used to train the images for further classification.

The proposed method can be used for gesture recognition when the distance of the person from the camera is at a distance of 2.5 m maximum. The proposed method is invariant to illumination changes. As of now, recognition for only two classes is done and the features are static. This can be extended to many classes and the features can be extracted from motion or the gesture can be recognized by any hand tracking method so that the system will give good results even if a person is at a far distance.

References

1. Choudhari WM, Mishra P, Rajankar R, Sawarkar M (2014) Hand gesture recognition using radial length metric. Int J Sci Res (IJSR). ISSN-2319-7064, April 2014
2. Misra A, Takashi A, Okatani T, Deguchi K (2011) Hand gesture recognition using histogram of oriented gradients and partial least squares regression. In: MVA2011 IAPR conference on machine vision applications, June 13–15, 2011
3. Trigueiros P, Ribeiro F, Reis LP (2013) A comparative study of different image features for hand gesture machine learning. In: 5th international conference on agents and artificial intelligence, 2013
4. Saikia P, Das K (2013) Head gesture recognition using optical flow based classification with reinforcement of GMM based background subtraction. Int J Comput Appl 0975–8887, March 2013
5. Park SY, Le EJ (2011) Hand gesture recognition using optical flow field segmentation and boundary complexity comparison based on hidden Markov model. J Korea Multimedia Soc 14 (4), April 2011
6. Pavlovic VI, Sharma R, Huang TS (1997) Visual interpretation of hand gestures for human-computer interaction: a review. IEEE Trans Pattern Anal Mach Intell 19(7):677–695
7. Baudel T, Baudouin-Lafon M (1993) Charade: remote control of objects using free-hand gestures. Comm. ACM 36(7):28–35
8. Sturman DJ, Zeltzer D (1994) A survey of glove-based input. IEEE Comput Graph Appl 14:30–39
9. Takahashi T, Kishino F (1992) A hand gesture recognition method and its application. Syst Comput Jpn 23(3):38–48

10. Bobick AF, Wilson AD (1995) A state-based technique for the summarization and recognition of gesture. In: Proceedings of the fifth international conference on computer vision, 1995, pp 382–388
11. Huang CL, Huang WY (1998) Sign language recognition using model based tracking and a 3D Hopfield neural network. Mach Vis Appl 10:292–307
12. Davis J, Shah M (1994) Visual gesture recognition. IEE Proc Vis Image Signal Process 141(2)
13. Campbell LW, Becker DA, Azarbayejani A, Bobick AF, Plentland A (1996) Invariant features for 3-D gesture recognition. In: Proceedings IEEE second international workshop on automatic face and gesture recognition, 1996
14. Schlenzig J, Hunter E, Jain R (1994) Recursive identification of gesture inputers using hidden Markov models. In: Proceedings of the second annual conference on applications. of computer vision, 1994, pp 187–194
15. Viola P, Jones MJ (2004) Robust real-time face detection. Int J Comput Vis 57(2):137–154
16. Yadav J, Sharma M (2013) A review of K-mean algorithm. Int J Eng Trends Technol (IJETT) 4(7), July 2013
17. Hu MK (1962) Visual pattern recognition by moment invariants. IRE Trans Inf Theory 8:179–187

GIS-Based Ground Water Quality Monitoring in Thiruvannamalai District, Tamil Nadu, India

M. Kaviarasan, P. Geetha and K.P. Soman

Abstract Ground water is a vital resource for drinking water around the world. The economic and ecological stability of many countries heavily relay upon groundwater availability. With rapid developments in industrial and agricultural sectors, the need for ground water is greater than ever before. Consequently, the quality of ground water is affected by fertilizers, effluents run off from industries, chemical dumping sites, domestic sewage, etc. Hence, it is necessary to constantly monitor ground water quality as it has a serious impact on human health. In this paper, we have analyzed ground water quality of Thiruvannamalai district of Tamil Nadu, India. The ground water samples are taken from 13 locations per area. Water Quality Index (WQI) is estimated for each area to ascertain for the potability of water. The physicochemical parameters like pH, Electrical Conductivity (EC), nitrates, fluorides, and chlorides sample data are compared against World Health Organization (WHO) standards. Geographical information system (GIS), an efficient tool for estimating water quality is used both in spatial and temporal domain. The results are useful in efficient monitoring and assessment of ground water and thus, for taking relevant measures to curb unrestrained exploitation.

Keywords GIS · Ground water · Thiruvannamalai · Water quality index

M. Kaviarasan (✉) · P. Geetha · K.P. Soman
Center for Excellence in Computational Engineering and Networking, Amrita Vishwa Vidyapeetham, Coimbatore 641112, India
e-mail: kaviarasan.mano@gmail.com

P. Geetha
e-mail: p_geetha@cb.amrita.edu

K.P. Soman
e-mail: kp_soman@cb.amrita.edu

L.P. Suresh and B.K. Panigrahi (eds.), *Proceedings of the International Conference on Soft Computing Systems*, Advances in Intelligent Systems and Computing 397, DOI 10.1007/978-81-322-2671-0_65

1 Introduction

Ground water is a replenishable resource that is heading toward severe exploitation owing to human activities. Groundwater has a number of advantages over surface water in terms of higher quality, better protected from possible pollution, less subject to seasonal and perennial fluctuations, and a wide area of uniform spread than surface water [1]. The agriculture and industrial development heavily rely on ground water availability thus, having a direct impact on the economic growth and prosperity of the nation. According to the survey by World Bank [2], India is the world's largest user of ground water with more than 60 % of it utilized for agriculture and almost 85 % of ground water is utilized for drinking purposes. The pollution of land and air has a direct effect upon the ground water in a given area. The usage of bore well technology has also lead to unabated exploitation of ground water giving less time for its recharge which in turn leads to severe water depletion.

The study area Thiruvannamalai is an agrarian district with farming zone in almost 40 % of its geographical area. According to the central ground water board, the ground water in the district is potable and suitable for irrigation and industrial applications except for localized areas. However, it is observed that ground water in the shallow zone of the major part of district is likely to cause high or medium level of alkalinity hazard. Hence, proper soil management strategies are required when it is used for irrigation [3]. Periodic analysis of water quality is essential in order to find the changes in ground water for its efficient management. The quality of ground water is essentially determined by its physicochemical parameters like pH, chloride, nitrates, permanent hardness, Electrical Conductivity (EC), fluorides which are subject to change based on the pollution level, season, and ground water exploitation pertaining to an area. The overall quality of water at certain area can be evaluated by means of WQI. It is an effective tool that works by taking into consideration certain water quality parameters and provides grades for the study areas based on its suitability for drinking.

In addition to this, Geographical Information System (GIS) is useful for a clear understanding of environmental changes especially, the analysis of ground water including a site suitability, managing site inventory data, estimating vulnerability of groundwater to pollution, modeling groundwater movement, modeling solute transport, etc. [4]. GIS software also saves time taken for collecting geographical data during both preprocessing and postprocessing stages that is required for ground water modeling [5]. It serves in creating a comprehensive ground water model by comparative study of water quality in an area based on the factors such as recharge zones, canal network, etc. Statistical analysis is also a mathematical method widely used for assessing the water quality. The study of correlation between the physicochemical parameters is useful in deciding the suitable water management techniques.

2 Study Area

Thiruvannamalai district is famous for its world renowned temples and ashrams. The district lies between 11°55′ and 13°15′ North latitudes and 78°20′ and 79°50′ East longitudes. It is bounded on the North by Vellore Region, on the west and South by the district of Dharmapuri and Villupuram, Kancheepuram district on the east, respectively. Geographical area of Thiruvannamalai is 6191 km^2..There are no perennial rivers. The major source of irrigation in the district are wells and tanks with an area of 2,19,150 ha under irrigation. Agriculture is the chief occupation and the major crops cultivated in this district are paddy, sugarcane, and groundnut. The climate is tropical and it enjoys a rainfall around 439.80 mm during North East Monsoon and 465.80 mm during the South West Monsoon. The study area of Thiruvannamalai district is shown in Fig. 1.

3 Methodology

The study is based on data collected from Central Ground Water Board and Geographical society of India pertaining to the district of Thiruvannamalai. The maps which have a scale of 1:25,000 are digitized using on-screen digitization to

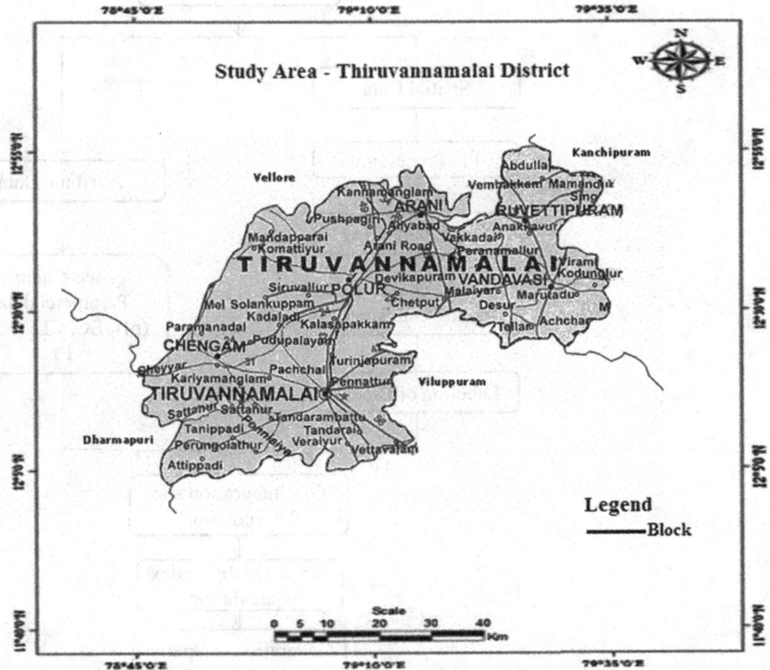

Fig. 1 Topographic map of Thiruvannamalai district

UTM coordinate system. Following the process of digitization, the maps need to be traced out for creating the spatial database of the taluks, blocks and labeling them. GIS software package ArcGIS 9.2 and ArcGIS Geostationary analyst extension are used for georeferencing and assessing the vulnerable zones based on five physicochemical parameters namely pH, electrical conductivity, chloride, nitrates, and fluorides. The study of the thematic maps output aids in the identification of the vulnerable zones as well as exploitation levels of Thiruvannamalai district. The flowchart of the proposed work is described in Fig. 2. The World Health Organization (WHO) standard is adopted to estimate the water quality [6].

3.1 Water Quality Index

Water Quality Index (WQI) is an effective tool to assess the state of an ecosystem, and this method is based on a group of physicochemical and biological characteristics of water samples [7–9]. From the rating given to the quality of water in a

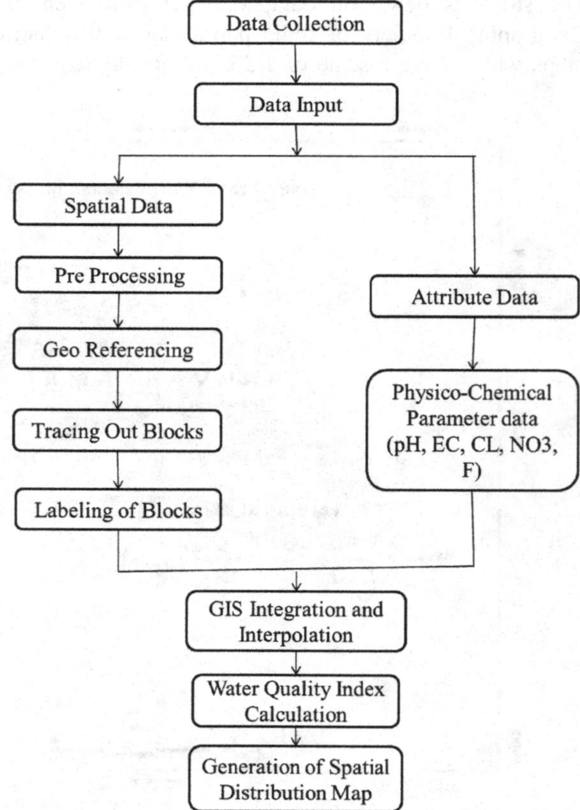

Fig. 2 Flowchart of the proposed work

given area, the authorities can decide upon the suitable water treatment method that is required.

The calculation of WQI is followed from the work by [10].

$$W_i \alpha \frac{1}{P_i} \tag{1}$$

Therefore,

$$W_i = \frac{K}{P_i} \tag{2}$$

where K is a constant of proportionality. The value of K is calculated as,

$$K = \frac{1}{\sum_{i=1}^{5} \frac{1}{P_i}} \tag{3}$$

The quality ratings for each of the physicochemical parameters considered is calculated from,

$$Q = O_i * \frac{100}{P_i} \tag{4}$$

Thus, WQI is calculated as,

$$\text{WQI} = \frac{\sum_{i=1}^{n} (Q_i W_i)}{\sum_{i=1}^{n} (W_i)} \tag{5}$$

The parameters used are explained as follows,

i Number of parameter
W_i Unit weightage given to the ith parameter
P_i Highest permitted value for the ith parameter
Q_i Subindex of the ith parameter
O_i Observed value seen from the ith parameter.

3.2 Statistical Analysis

The excel spreadsheet is used to calculate the standard deviation, maximum, minimum, average and correlation coefficient. For this, each pair of physicochemical parameters are taken and the calculations are performed.

The formulas used are as follows:
Mean:

$$\mu = \frac{\sum x}{N} \tag{6}$$

where,
x values of observation
N total number of observation.

Standard Deviation:

$$\sigma = \frac{\sqrt{n\sum x^2 - (\sum x)^2}}{n(n-1)} \tag{7}$$

where,
x values of parameter
N number of observation.

The correlation between the parameters x and y is given by the Karl Pearson Coefficient, given by the formula [11]

$$r = \frac{n\sum xy - \sum x \sum y}{\sqrt{n\sum x^2 - (\sum x)^2}\sqrt{n\sum y^2 - (\sum y)^2}} \tag{8}$$

where,
x value of array 1
y value at array 2
n number of observation.

4 Results and Discussion

The study areas considered are Arni, Chetpet, Cheyyar, Jamunamarattur, Melmakootroad, Pachel, Polur, Thandarampattu, Thanipadi, Thiruvannamalai, Turinjapuram, Vandavasi, and Sathanur. WHO standards of water quality parameter and its assigned unit weights are given in Table 1. The physicochemical parameters in the water samples collected from these areas are discussed in Table 2 and WQI rating is given in Table 3 which is estimated for each area.

Table 1 WHO [6] standards of water quality parameter assigned unit weights

Parameter	Standards (P_i)	$1/P_i$	K	Unit weightage W_i
pH	8.5	0.11765	1.236051	0.14542
EC	1400	0.00071	1.236051	0.00088
Cl	250	0.004	1.236051	0.00494
N	50	0.02	1.236051	0.02472
F	1.5	0.66667	1.236051	0.82403
		0.80903		1

Table 2 Water quality parameters of study areas

Code no.	Location	pH	EC μS/cm	Cl Mg/l	No3	F	WQI	Quality of water
L1	Arni	8.03	1985	220	56	0.42	40.13968	Good
L2	Chetpet	7.65	1702	386	34	0.59	48.05136	Good
L3	Cheyyar	7.66	532	64	11	0.20	24.79581	Excellent
L4	Jamunamarattur	7.99	684	35	29	1.00	70.15105	Moderate
L5	Melmakootroad	7.41	1524	266	147	0.32	38.14656	Good
L6	Pachel	7.84	1366	241	63	0.80	61.03875	Moderate
L7	Polur	8.15	812	35	27	0.52	43.96488	Good
L8	Thandarampattu	7.55	915	106	45	0.51	43.42591	Good
L9	Thanipadi	8.01	1440	177	177	0.43	46.5179	Good
L10	Thiruvannamalai	8.20	1212	135	28	0.40	37.73058	Good
L11	Turinjapuram	7.90	1420	192	129	0.53	49.47846	Good
L12	Vandavasi	7.81	927	162	2	0.77	56.13947	Moderate
L13	Sathanur	7.58	1768	319	79	0.40	39.59039	Good

Table 3 Water quality index rating

Value of WQI	Quality of water
0–25	Excellent
26–50	Good
51–75	Moderate
76–100	Very poor
Above 100	Unsuitable for drinking water

4.1 pH

pH is a parameter that is used to determine the corrosivity of water. The optimum pH as specified by WHO is in the range of 6.5–8.5. It is considered to be a crucial operational parameter. pH exceeding 11 can result in irritation of eye and skin whereas lower pH gives a metallic and sour taste to water. On analysis, we find that Thiruvannamalai has the highest value of 8.20. The minimum value of 7.41 is seen at Melmakootroad. It is observed that all areas have pH values well below the specified limit. The pH of ground water in study area is shown in Fig. 3.

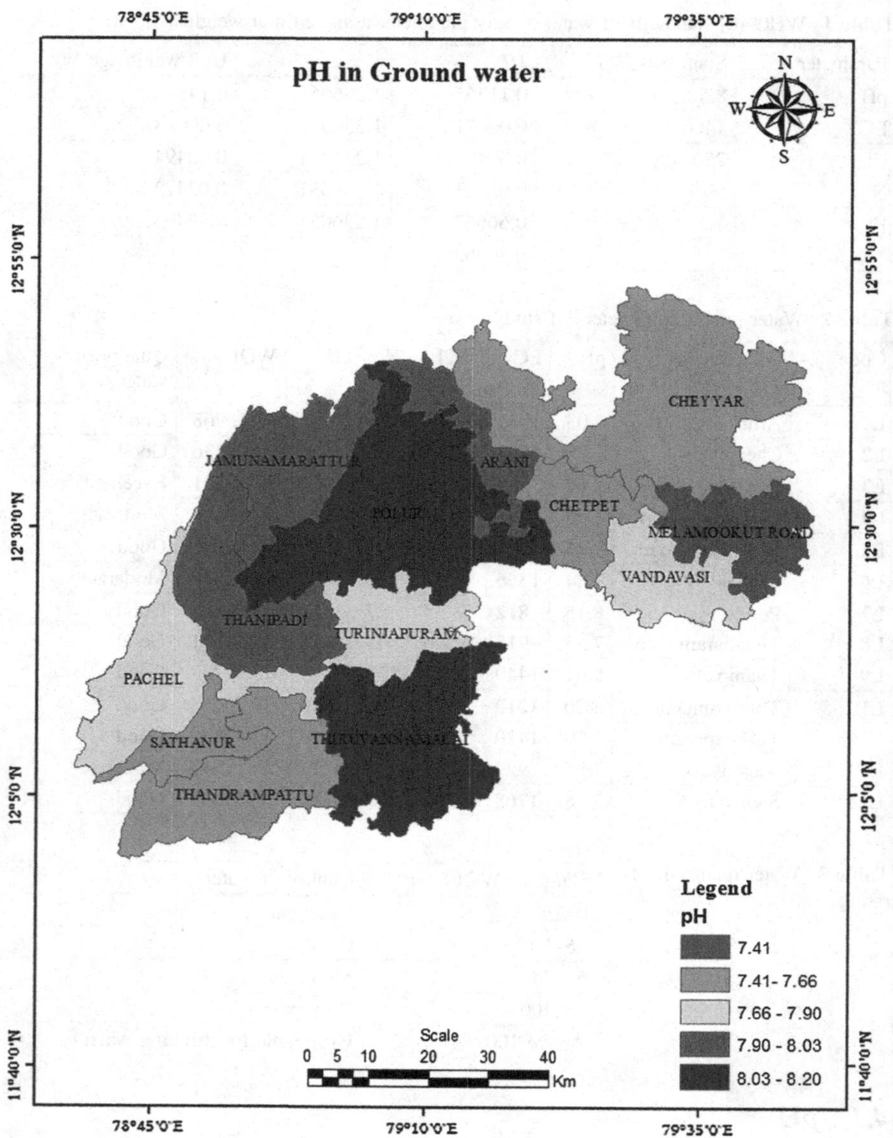

Fig. 3 pH in ground water

4.2 *EC*

Electrical Conductivity is the ability of an aqueous solution to conduct electric current. EC is considered to be a rapid and good measure of dissolved solid and it is a major criterion in determining the suitability of water for irrigation [12].

The recommended limit given by WHO is around 1400 µS/cm. Below this limit, water is suitable for drinking and irrigation purpose. When EC values lies between 800 and 2500 µS/cm, the lower range is suitable for drinking whereas special care must be taken for higher range during irrigation like soil suitability, salt tolerance level of plants. Cheyyar shows minimum value of 532 µS/cm. Arni shows a maximum value of 1985 µS/cm. Figure 4 shows the EC of ground water for various locations.

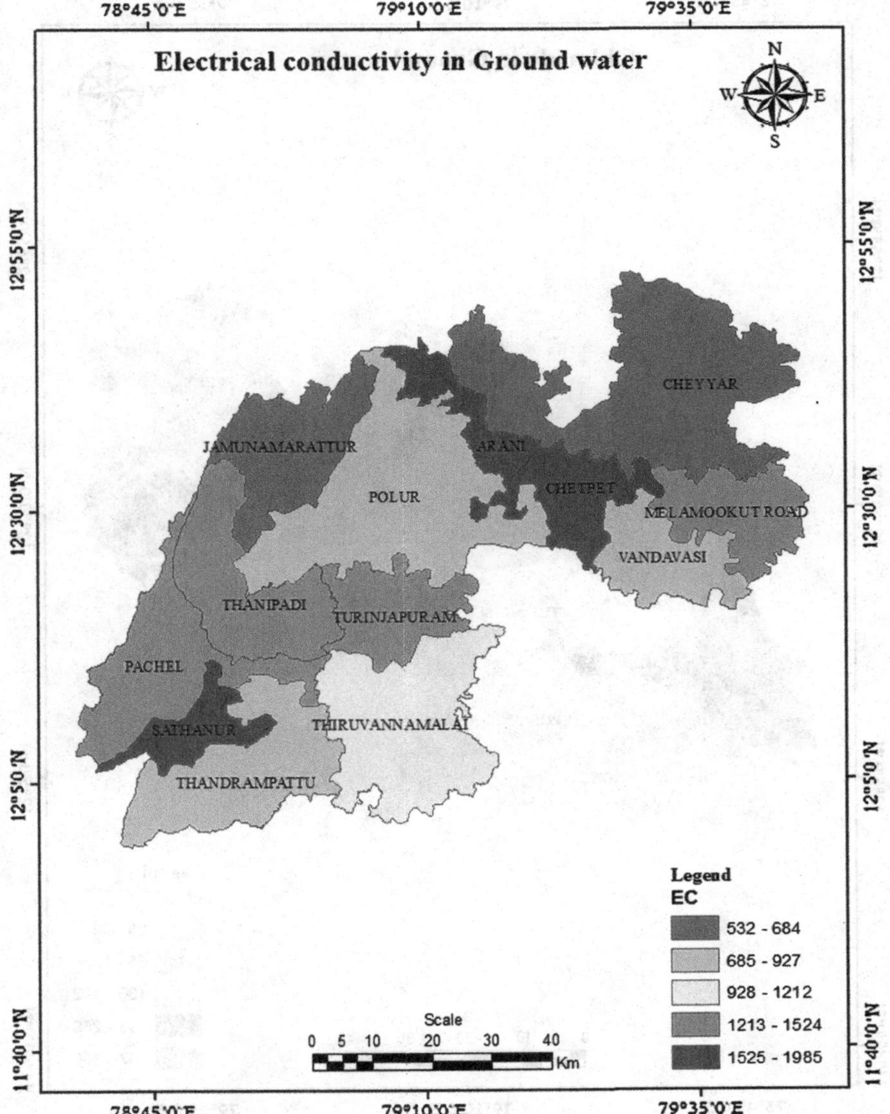

Fig. 4 Electrical conductivity in ground water

4.3 Chloride

The permissible limit for chloride is given as 250 mg/L. The major source of chloride is the rocks through which water passes, sea water intrusions and also anthropogenic factors like landfill leachates, industrial effluents, etc. Chloride

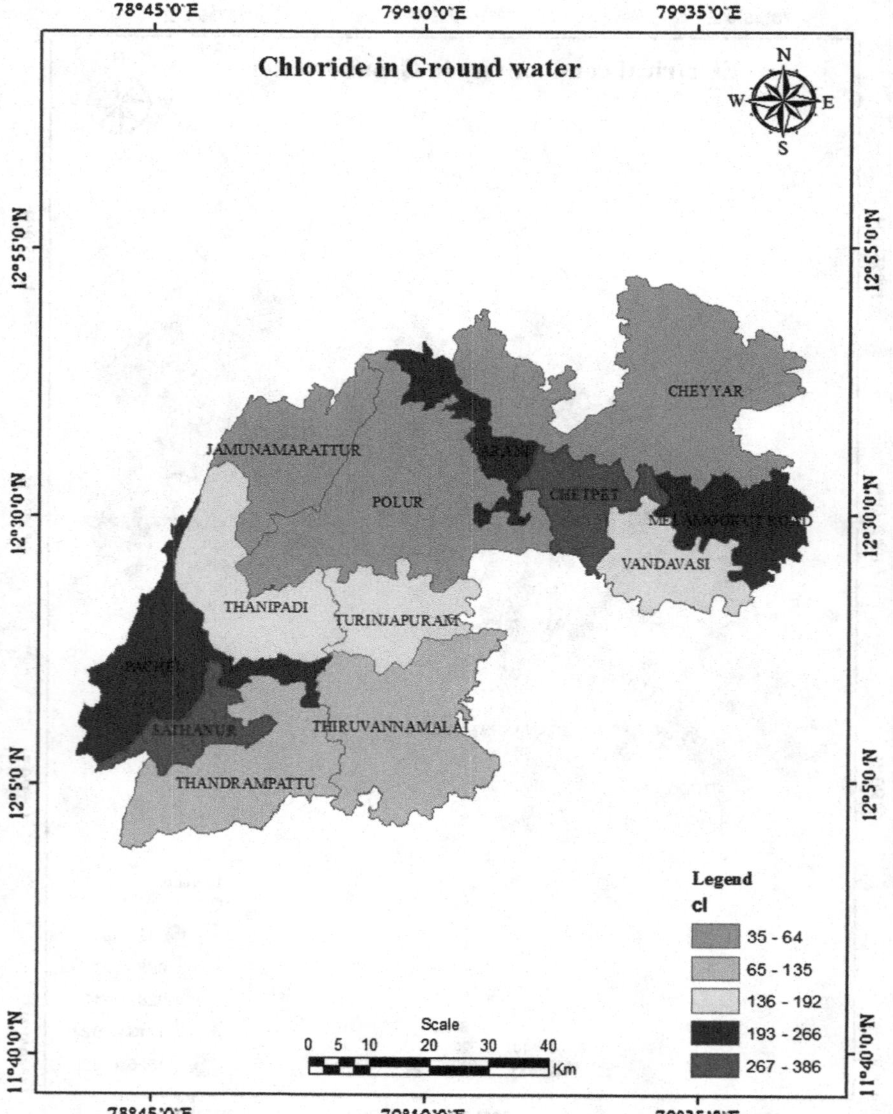

Fig. 5 Chloride in ground water

increases the EC of water and thus increases its corrosivity [13] and also induces a change in taste when it exceeds the given limit. Chetpet shows the maximum value of 386 mg/L. The minimum value is seen at Jamunamarattur and polur (35 mg/L). The presence of chloride in ground water for various locations in study area is depicted in Fig. 5.

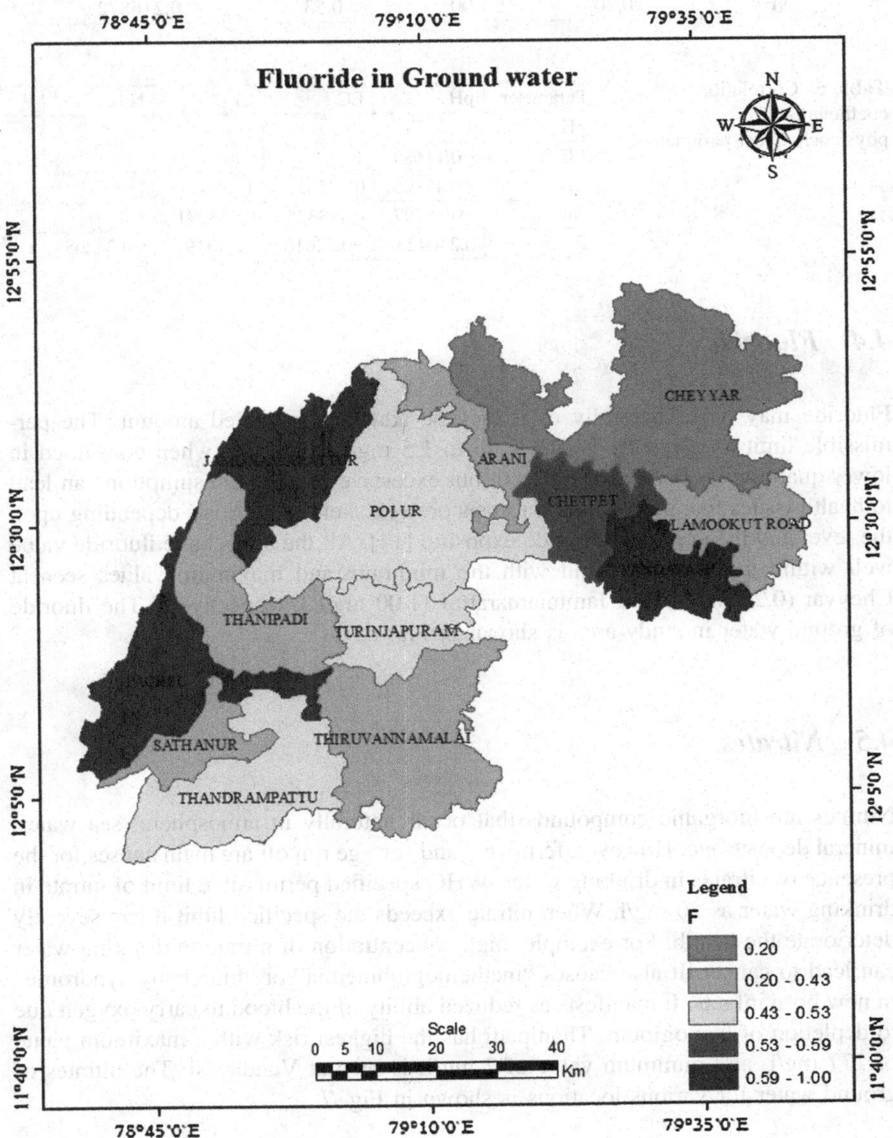

Fig. 6 Fluoride in ground water

Table 4 Statistical analysis of physicochemical parameters

S. no	Parameter	Minimum	Maximum	Average	Standard deviation
1	pH	7.41	8.20	7.83	0.245169
2	EC	532	1985	1252.846	448.3304
3	Cl	35	386	180	107.2216
4	NO_3	2	177	64	54.75327
5	F	0.20	1.00	0.53	0.216872

Table 5 Correlation coefficients of physicochemical parameters

Parameter	pH	EC	Cl	No_3	F
pH	1				
EC	−0.11684	1			
Cl	−0.45805	0.84565	1		
No_3	−0.15597	0.493325	0.338291	1	
F	0.240423	−0.22016	−0.13192	−0.28268	1

4.4 *Fluoride*

Fluoride may occur naturally or it may be added in specified amount. The permissible limit as given by WHO is up to 1.5 mg/L. Fluoride, when consumed in lower quantities is beneficial for teeth but excessive fluoride consumption can lead to health issues like mild dental fluorosis or even skeletal fluorosis depending upon the level and the period of fluoride exposure [14]. All the areas have fluoride value well within the specified limit with the minimum and maximum values seen at Cheyyar (0.20 mg/L) and Jamunamarattur (1.00 mg/L), respectively. The fluoride of ground water in study area is shown in Fig. 6.

4.5 *Nitrates*

Nitrates are inorganic compounds that occur naturally in atmosphere, sea water, mineral deposits etc. However, fertilizers and sewage run off are main causes for the presence of nitrates in drinking water. WHO specified permissible limit of nitrate in drinking water as 50 mg/l. When nitrate exceeds the specified limit it can severely deteriorate the health. For example, high concentration of nitrate in drinking water can lead to cancer. It also causes "methemoglobinemia" or "blue baby syndrome" in new born infants. It manifests as reduced ability of the blood to carry oxygen due to depletion of hemoglobin. Thanipadi has the highest risk with a maximum value of 177 mg/L and minimum value of 2 mg/L is seen at Vandavasi. The nitrates of ground water for various locations is shown in Fig. 7.

4.6 Statistical Analysis

An analysis of the correlation coefficient table reveals the following results. A significant positive correlation (>0.8) is found only between the Cl–EC pair. A weak positive correlation ($0.5 > x > 0.8$) exists between three values (NO_3–EC, NO_3–Cl,

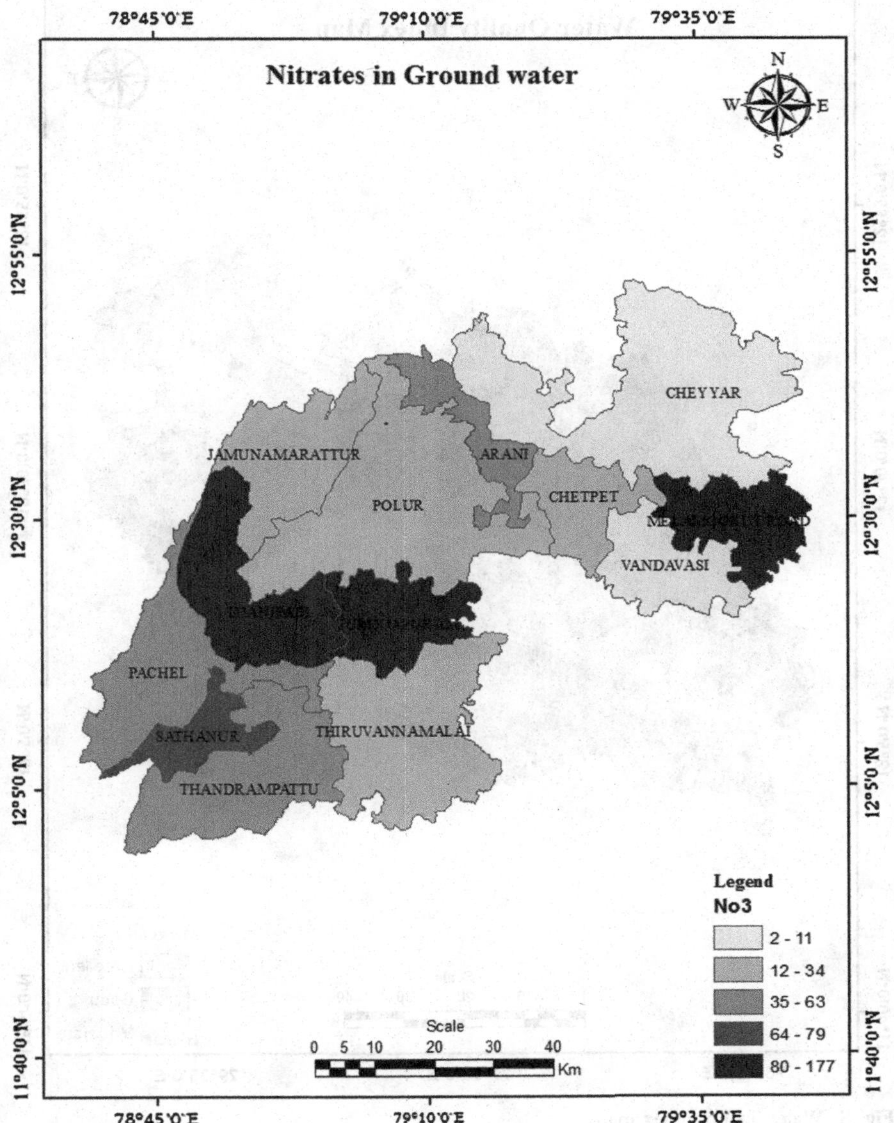

Fig. 7 Nitrate in ground water

F–pH). A weak negative correlation is found to exist between the remaining values. The statistical analysis and correlation coefficients of physicochemical parameters are given in Tables 4 and 5 respectively. The water quality index map is shown in Fig. 8 and also the variation of WQI values is shown in Fig. 9.

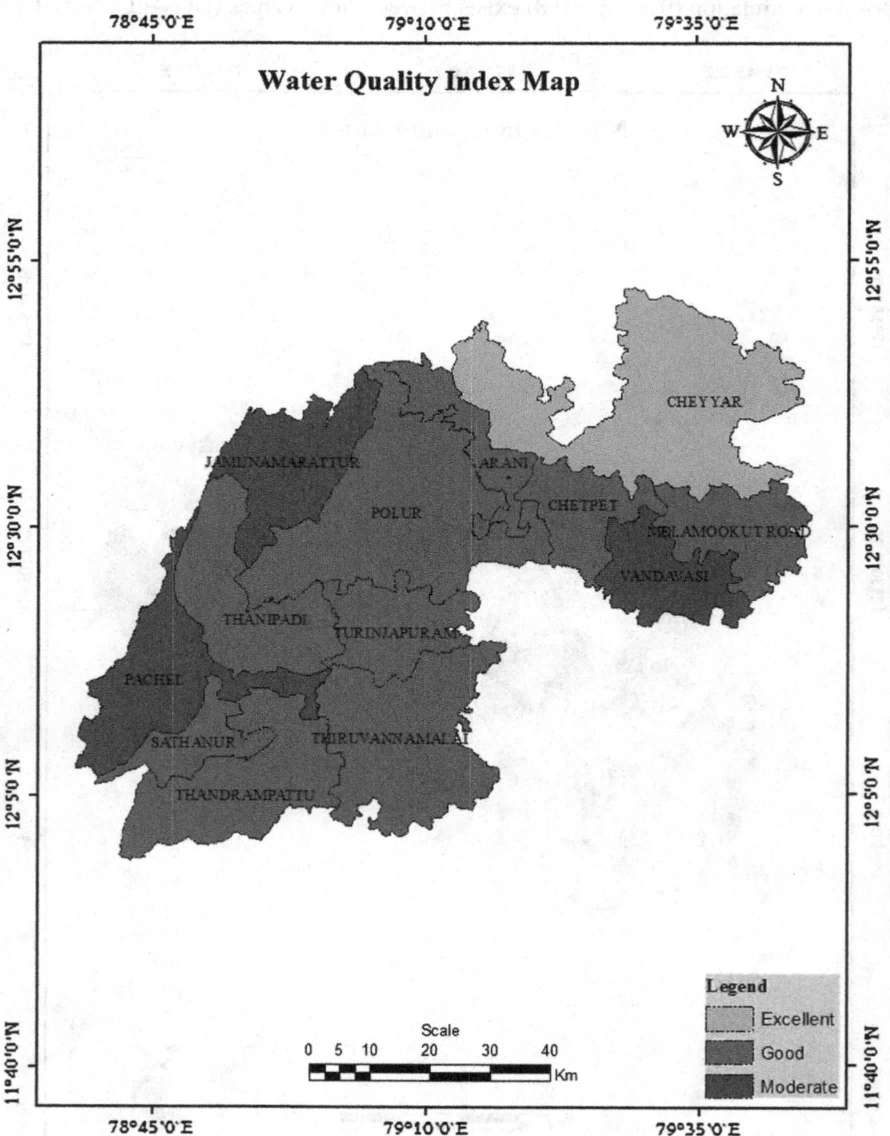

Fig. 8 Water quality index map

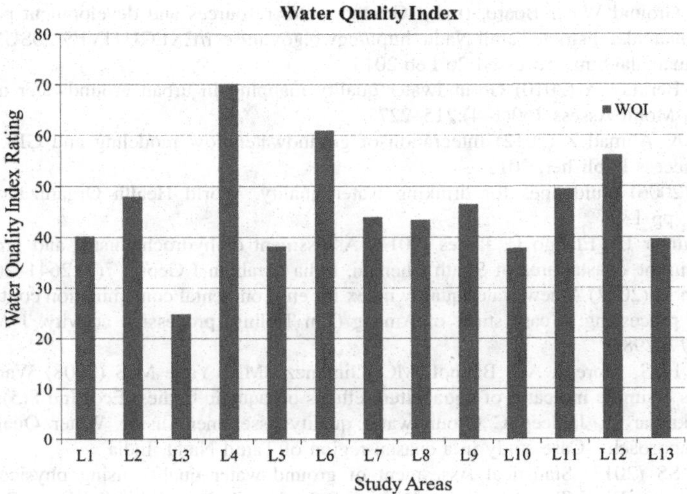

Fig. 9 The variation of WQI values for various ground water sampling stations

5 Conclusion

This paper presents map creation and evaluation of ground water quality. From the experimental analysis, it is observed that pH and EC values are well below the limit. In Chetpet, the chloride value is high, which may affect aquatic life as well as change in water taste. Thanipadi is found to have high value for nitrate content which requires immediate measures to be taken. From the WQI values, it is evident that all areas have acceptable ratings of water quality with the exception of Jamunamarattur, Pachel, and Vandavasi for which immediate attention is required. From the study, it is clear that GIS and WQI can be effectively used to ascertain for water quality and alert the policy makers to take suitable measures.

Acknowledgments We are grateful to Central Ground Water Board and Geological Survey of India for providing us with data for this project. We thank Mrs.V. Sowmya Assistant Professor of our center for her valuable support.

References

1. Zektser IS, Lorne E (2004) Groundwater resources of the world: and their use. In: IHP Series on groundwater, no. 6. Unesco, 2004
2. The World Bank, India Groundwater: a valuable but diminishing resource. http://www.worldbank.org/en/news/feature/2012/03/06/india-groundwater-critical-diminishing. Accessed 20 Feb 2015

3. Central Ground Water Board. (n.d.). Ground water resources and development potential of thiruvannamalai district, Tamil Nadu. http://cgwb.gov.in/secr/EXECUTIVE%20SUMMARY-Tiruvannamalai.htm. Accessed 26 Feb 2015
4. Nas B, Berktay A (2010) Groundwater quality mapping in urban groundwater using GIS. Environ Monit Assess 160(1–4):215–227
5. Ashraf A, Ahmad Z (2012) Integration of groundwater flow modeling and GIS. INTECH Open Access Publisher, 2012
6. WHO (2006) Guidelines for drinking water quality. World Health Organization, vol 1, Geneva, pp 139
7. Sajil Kumar PJ, Elango L, James (2014) Assessment of hydrochemistry and groundwater quality in the coastal area of South Chennai, India. Arabian J Geosci 7(7):2641–2653
8. Nasirian M (2007) A new water quality index for environmental contamination contributed by mineral processing: a case study of Amang (Tin Tailing) processing activity. J Appl Sci 7 (20):2977–2987
9. Simoes FDS, Moreira AB, Bisinoti MC, Gimenez SMN, Yabe MJS (2008) Water quality index as a simple indicator of aquaculture effects on aquatic bodies. Ecol Ind 8(5):476–484
10. Ganeshkumar B, Jaideep C Groundwater quality assessment using Water Quality Index (WQI) approach—Case study in a coastal region of Tamil Nadu, India
11. Saxena SS (2013) Statistical Assesment of ground water quality using physico-chemical parameters in Bassi Tehsil of Jaipur District, Rajasthan, India. Glob J Sci Front Res 13(3)
12. Choudhary R, Rawtani P, Vishwakarma M (2011) Comparative study of drinking water quality parameters of three manmade reservoirs ie Kolar, Kaliasote and Kerwa Dam. Curr World Environ 6(1):145–149
13. Kumar M, Puri A (2012) A review of permissible limits of drinking water. Indian J Occup Environ Med 16(1):40
14. Fawell JK, Bailey K (2006) Fluoride in drinking-water. World Health Organization

Random Forest Ensemble Classifier to Predict the Coronary Heart Disease Using Risk Factors

R. Ani, Aneesh Augustine, N.C. Akhil and O.S. Deepa

Abstract Heart diseases are the major cause of death in today's modern age. Coronary heart disease is one among them. This disease attacks the normal person instantly. Proper diagnosis and timely attention to the patients reduce mortality rate. Proper diagnosis has become a challenging task for the medical practitioners. The cost involved in the immediate treatment or intervention methods are also very expensive. Early diagnosis of the disease using mining of medical data prevents the inattention of occurrence of sudden CHD events. Today, almost all hospitals are using hospital information system and it has huge volume of patient records. This study results in the development of a decision support system using machine intelligence techniques applied on the medical records stored in hospital databases. Classification algorithms are used to evaluate the accuracy of the early prediction of coronary heart events. The classification techniques analyzed are K-nearest neighbor, decision tree-C4.5, Naive Bayes, and the random forest. The accuracy of each technique is found to be 77, 81, 84, and 89 %, respectively. In this study 10-fold cross-validation method is used to measure the unbiased estimate of these prediction models.

Keywords Coronary heart disease · Prediction · Classification · Random forest · Cleveland heart data · Risk factors

1 Introduction

Coronary heart disease is one of the major causes of sudden death among the people. This disease occurs when the coronary arteries become narrowed by a fat deposits or plaque inside the artery walls. Early detection of heart disease has become one

R. Ani (✉) · A. Augustine · N.C. Akhil
Department of Computer Science and Applications, Amrita Vishwa Vidyapeetham, Amritapuri, India
e-mail: anir@am.amrita.edu

O.S. Deepa
Department of Computer Science and Applications, Amrita Vishwa Vidyapeetham, Coimbatore, India

L.P. Suresh and B.K. Panigrahi (eds.), *Proceedings of the International Conference on Soft Computing Systems*, Advances in Intelligent Systems and Computing 397, DOI 10.1007/978-81-322-2671-0_66

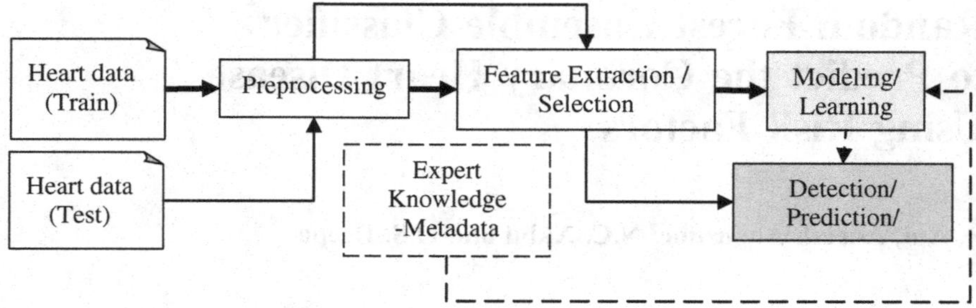

Fig. 1 Architecture of the system

among the important research areas in medical field [1]. The presence of more number of risk factors increases the probability of getting a disease. The risk prediction models are very helpful in the prevention of cardiac diseases. Accurate diagnosis depends on the interpretation and analysis of large data collected from the medical data base and from medical test results. This study compares the performances of classification techniques which are used to predict the presence of coronary heart disease. A decision support system based on the best classification algorithm analyzed is used to perform the diagnosis. This type of systems will help the medical practitioners to have a better judgment in diagnosis [2].

Supervised learning is a technique of deducing a function y = f(x) from a labeled training dataset. Each training example consists of a pair of input vector and desired output of the form (x, f(x)) [3]. This deduced function is used for mapping new examples. The decision support system which was developed uses the medical test results and predicts the patient to any of the two classes. This paper analyses four classification algorithms such as K-nearest neighbor, Naive Bayes, C4.5, and random forest. The study begins with KNN and finally got better results from random forest. It is an ensemble method of decision tree. The paper proposes an effective clinical decision support system using random forest, in which 200 trees are used for decision making. In preprocessing, the missing values of attributes are filled using measures of central tendency [4]. It uses gain ratio as the splitting criteria for selecting the attribute for each node in the tree. After creating trees the extraction of IF. THEN rules are derived from these trees. The rules generated are used to classify the given test data to any of the two class (Fig. 1).

2 Related Work

Various studies have been done relating to heart disease prediction using data mining techniques worldwide [5]. The main goal of these studies is to improve the accuracy of prediction using data mining and machine learning techniques.

2.1 Classification of Heart Disease Using K-Nearest Neighbor and Genetic Algorithm [6]

This paper has been proposed for classification of heart disease culminating the algorithms K-nearest neighbor and genetic algorithms. Genetic algorithms have been used to perform global search in complex multimodal landscape and provide precise solution. Experimental results conclude that this algorithm intensify the diagnosis of heart disease.

2.2 A Data Mining Technique for Prediction of Coronary Heart Disease Using Neuro-Fuzzy Integrated Approach Two Levels [7]

A two-layer approach has been used for identifying the disease probability. The risk factors that are compulsory for a coronary heart disease to occur are at two levels, the mandatory ones as first and the not so necessary ones as second. Disease chances can be accurately predicted with the aid of this two level approach.

2.3 Assessment of the Risk Factors of Coronary Heart Events Based on Data Mining With Decision Trees [8]

The aim of this data mining project is to analyze the different risk circumstances in which myocardial infarction, percutaneous coronary intervention, and coronary artery bypass graft surgery can occur.

2.4 Using Methods from the Data-Mining and Machine-Learning Literature for Disease Classification and Prediction: A Case Study Examining Classification of Heart Failure Subtypes [9]

The modern flexible tree-based methods from the data mining literature is of considerable importance in improving prediction and classification of HF subtype compared with conventional classification and regression trees. Logistic regression had a superior performance for predicting the probability of the presence of HFPEF when compared with other methodologies proposed.

3 Proposed Work

3.1 K-Nearest Neighbor

In K-nearest neighbor (KNN), the distances between the input and the training data are calculated and from the least values, select the class which comes most frequent [10]. Simple K-nearest neighbor algorithm is shown below

- Input: D = {(X1, Y1), . . . , (Xn, Yn)}
- **x** = (x1, . . . , xn) new patient data to be classified
- FOR each labeled data (Xi , Yi) calculate distance d(Xi , **x**)
- Sort d(Xi , **x**) ascending
- Select the K nearest instances to **x**: H
- Assign to x the most frequent class in H

Euclidean distance is a good measure, because it is the shortest distance between two points.

$$D(X,Y) = \sqrt{\sum_{K=1}^{n} (X_k - Y_k)^2}.$$

3.2 Naïve Bayes

The Bayesian classifier is capable of calculating the most probable output depending on the input [11]. The input is being classified on the basis of the posteriori probability of the class with high value.

- A priori probability of event X= P[X]
    - Probability of event before evidence has been seen
- A posteriori probability of X= P[X|E]
    - Probability of event after evidence has been seen
- Probability of event H given evidence E:
    - $P[X \mid E] = \frac{P[E \mid X] * P[X]}{P[E]}$
- Classification learning
    - E = instance
    - X = class value for instance
- Naive Bayes assumption: evidence can be split into independent parts (i.e. attributes of instance

$$P[X \mid E] = \frac{P[E1 \mid X] * P[E2 \mid X] * \ldots * P[En \mid X]\, P[X]}{P[E]}$$

3.3 C4.5

C4.5, which is an enhanced algorithm of ID3 supports the classification of continues as well as discreet attributes. In each stage of iteration, the attribute which has the highest gain ratio is assigned to the nodes of the tree after splitting. All the branches will end up in the class label which is the leaf of the tree. Algorithm to generate decision tree

```
Input: Dataset D
Tree = Null
GenerateTree (D) ---------- (1)
If D is pure then
        exit
else
        for all attributes
                find out gain ratio
        end for
        Tree = the best attribute as root node
        Finding induced sub-data set corresponding to the root node
        for all induced sub data sets
                Branch = Apply C4.5
                Attach branches to the Tree
        end for
end if
return Tree.
```

3.4 Random Forest

The random forests [6] is an effective prediction tool in data mining. This consists of collection of decision trees which are called as forest. The bagging technique is used to generate random set of training datasets for each tree. Let the total number of attributes be P. Different subsets of attributes are selected for bagging. For each of the above datasets, it creates a tree based on the gain ratio for splitting. Finally there will be N number of trees. Random forest will classify a new input by combining the results from all N trees and assign it into the most frequent class. The number of randomly selected attributes is taken as square root (P). In this study, 200 decision trees are generated using random selection of attributes. The class for the test data is predicted using the majority voting technique on the generated 200 decision trees.

```
n = sqrt (P)
N = required number of classifiers
Forest = { }
For 1 to N
        Forest = Random Forest (D)
End for
Random Forest (D):
        Select n random attributes from D
        Create new Di using sampling with replace from D
        Return GenerateTree (Di) --------- (1)
Prediction (input):
For all trees in Forest
        Traverse through each tree and store the class
End for
Class= majority of stored classes
```

4 Dataset Details

The dataset used in this study is taken from UCI repository. The preprocessed 13 attributes and corresponding class values are considered for the study [12]. These 14 attributes are as follows [13] (Fig. 2).

- Age- Age of the patient- numeric data
- Sex– gender– nominal data - Male represented by 1 and female by 0
- Chest pain- type of chain pain- nominal data
  - Typical angina (angina)
  - Atypical angina (abnang)
  - Non-anginal pain (notang)
  - Asymptomatic (asympt)
- Trestbps- Resting blood pressure - numeric data
- Chol- Serum cholesterol in mg/dl- numeric data
- Fbs- fasting blood sugar- nominal data
  - True, if >= 120 mg/dl; False, if < 120 mg/dl
- Restecg- Electrocardiographic result when patient at rest- nominal data
  - normal (norm)
  - ST–T wave abnormality
  - ventricular hypertrophy (hyp)
- Thalach- Maximum heart rate- numeric data
- Exang- Exercise induced angina- nominal data
- Oldpeak- ST depression by exercise relative to rest- numeric data
- Slope- Slope characteristics of the peak exerciseST segment- nominal data-
  - up sloping, flat, down sloping
- Ca- number of fluoroscopy colored major vessels-values (0–3) numeric data
- Thal- heart status – nominal data
  - normal, fixed defect, reversible defect
- Class- Patient with/with-out heart related disease.

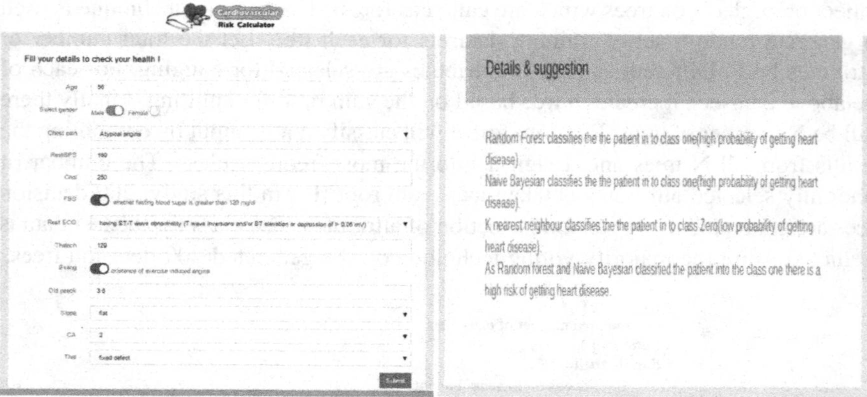

Fig. 2 The interface of the developed system used for the prediction

5 Results and Discussions

The four different classification algorithms are being implemented and the performance measures are calculated. In this study, 10-fold cross-validation technique is used to validate the data. This technique divides the dataset into 10 equal partitions. One among the 10 is selected as the test sample and the remaining 9 partitions as the training sample. This process is repeated 10 times and the average of all results obtained is calculated. The accuracy of the model is evaluated using the confusion matrix. The *Mij*th entry in this square matrix represents the number of tuples in class i that are classified by the model as class j. This matrix provides the percentage of test set tuples that are correctly classified (TP + TN)/(P + N), where TP and TN are the positive tuples that are correctly labeled by the classifier and the negative tuples that are correctly labeled by the classifier, respectively. The model's ability to predict whether the patient is diseased or not is calculated as sensitivity (TP/P) and specificity (TN/N), respectively.

5.1 *Evaluation of KNN*

See Table 1.

5.2 *Evaluation of Naïve Bayes*

See Table 2.

Table 1 Out of 138 patients without heart disease and 165 with heart disease KNN classifies 129 and 105 patients correct in each class, respectively

Confusion matrix prediction			
		Zero	One
Model	Zero	129	36
	One	33	105

Table 2 Out of 138 patients without heart disease and 165 with heart disease Naïve Bayes classifies 147 and 109 patients correct in each class, respectively

Confusion matrix prediction			
		Zero	One
Model	Zero	147	18
	One	29	109

5.3 Evaluation of C4.5

See Table 3.

5.4 Evaluation of Random Forest

See Figs. 3, 4, 5, 6 and Table 4.

5.5 Accuracy Assessment

See Table 5.

Table 3 Out of 138 patients without heart disease and 165 with heart disease C4.5 classifies 140 and 106 patients correct in each class, respectively

Confusion matrix prediction			
		Zero	One
Model	Zero	140	25
	One	32	106

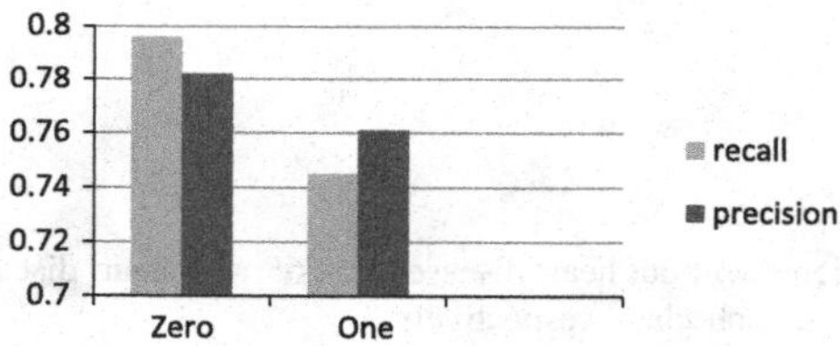

Fig. 3 78.2, 76.1 % of randomly retrieved document is relevant and 0.796, 0.745 probability of relevant document retrieved in a search in each class, respectively

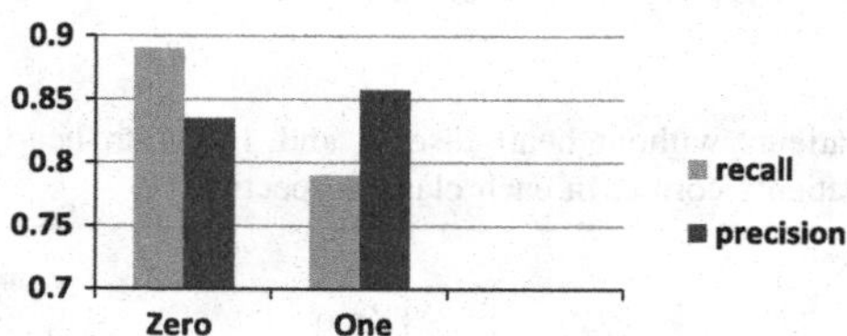

Fig. 4 83.5, 85.8 % of randomly retrieved document is relevant and 0.891, 0.79 probability of relevant document retrieved in a search in each class, respectively

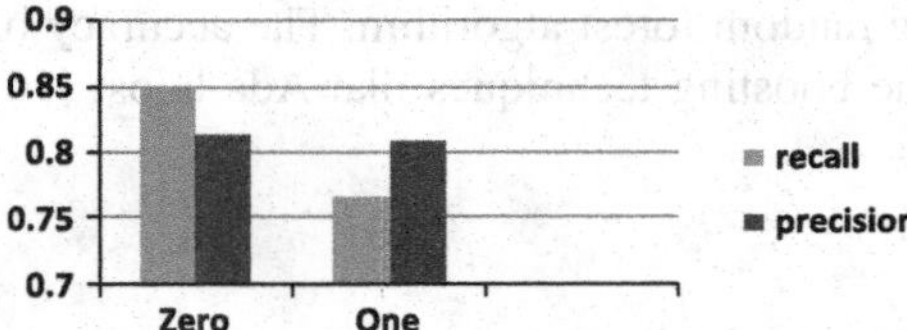

Fig. 5 81.4, 80.9 % of randomly retrieved document is relevant and 0.848, 0.768 probability of relevant document retrieved in a search in each class, respectively

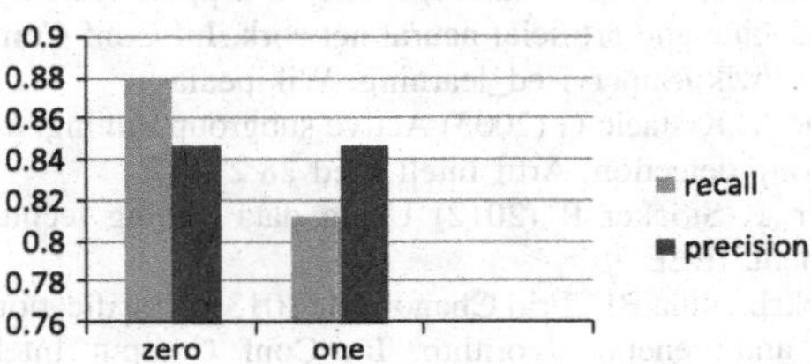

Fig. 6 84.8, 84.8 % of randomly retrieved document is relevant and 0.879, 0.812 probability of relevant document retrieved in a search in each class, respectively

Table 4 Out of 138 patients without heart disease and 165 with heart disease random forest classifies 145 and 124 patients correct in each class, respectively

Confusion matrix prediction			
		Zero	One
Model	Zero	145	20
	One	26	124

Table 5 The result obtained from 10-fold cross-validation testing on KNN, C4.5, Naïve Bayes and random forest are 77, 81, 84 and 89 %, respectively

	Correctly classified instances	Incorrectly classified instances
C4.5	246	57
KNN	234	69
Naïve Bayes	256	47
Random forest	269	34

6 Conclusions

Computer-based decision support systems can reduce poor clinical decisions and it also minimizes the cost involved in unnecessary clinical tests. The accuracy obtained in the random forest algorithm is higher when compared with the accuracy of KNN, Naïve Bayesian, and C4.5 algorithms. The decision support system which was designed uses the random forest algorithm. The accuracy of the system can be improved by using the boosting techniques like Ada-boost [16], which needs further study.

References

1. Framingham heart study. https://www.framinghamheartstudy.org/
2. Gudadhe M, Wankhade K, Dongre S (2010) Decision support system for heart disease based on support vector machine and artificial neural network. Int Conf Comput Commun Technol
3. http://en.wikipedia.org/wiki/Supervised_learning. Wikipedia
4. Gamberger D, Lavrac N, Krstacic G (2003) Active subgroup mining: a case study in coronary heart disease risk group detection. Artif Intell Med 28:27–57
5. Shouman M, Turner T, Stocker R (2012) Using data mining techniques in heart disease diagnosis and treatment. IEEE
6. Akhil Jabbar M, Deekshatulua BL, Priti Chandra B (2013) Classification of heart disease using K-nearest neighbor and genetic algorithm. Int Conf Comput Intell: Model Tech Appl (CIMTA)
7. Sen AK, Patel SB, Shukla DP (2013) A data mining technique for prediction of coronary heart disease using neuro-fuzzy integrated approach two level. Int J Eng Comput Sci 2(9):2663–2671. ISSN:2319-7242
8. Karaolis MA, Moutiris JA, Hadjipanayi D, Pattichis CS (2010) Assessment of the risk factors of coronary heart events based on data mining with decision trees. IEEE Trans Inf Technol Biomed 14(3)
9. Austin PC, Tu JV, Ho JE, Levy D, Lee DS (2013) Using methods from the data-mining and machine-learning literature for disease classification and prediction: a case study examining classification of heart failure subtypes. J Clin Epidemiol 66:398e407
10. Krishnaiah V, Srinivas M, Narsimha G, Subhash Chandra N (2012) Diagnosis of heart disease patients using fuzzy classification technique. IEEE
11. Srinivas K, Raghavendra Rao G, Govardhan A (2010) Analysis of coronary heart disease and prediction of heart attack in coal mining regions using data mining techniques. Int Conf Comput Sci Educ
12. Muthukaruppan S, Er MJ (2012) A hybrid particle swarm optimization based fuzzy expert system for the diagnosis of coronary artery disease. Expert Syst Appl 39:11657–11665
13. UCI heart disease data set. http://archive.ics.uci.edu/ml/
14. Bhatla N, Jothi K (2012) An Analysis of heart disease prediction using different data mining techniques. Int J Eng Res Technol
15. Han J, Kamber M, Pei J (2010) Data mining concepts and techniques
16. Zhang Z, Xie X (2010) Research on AdaBoost.Ml with random forest. Int Conf Comput Eng Technol

Design and Implementation of Reconfigurable VLSI Architecture for Optimized Performance Cognitive Radio Wideband Spectrum Sensing

S. Gayathri and K.S. Sujatha

Abstract Spectrum sensing performs major function in cognitive radio to efficiently use the underutilized spectrum. It has to detect the presence of primary user signal in a channel and has to utilize in primary user's absence. In wideband spectrum sensing, a wide frequency has to be sensed. In this paper, a reconfigurable VLSI architecture is designed to perform cooperative spectrum sensing for wideband, this needs a local detection, which is fundamental for cooperative sensing. Each and every individual secondary user has to perform energy detection. In this paper, energy detection technique used is based on the Neyman Pearson criterion. Then the cooperative decision is taken which increases the sensing performance. The designed architecture is then implemented in Xilinx Virtex-4 Field programmable Gate array.

Keywords Neyman pearson criterion · Xilinx Virtex-4 field programmable gate array · MATLAB · Cognitive radio network

1 Introduction

Radio spectrum is a natural resource which is essential for wireless systems. Progress in wireless communication system has increased the demand of radio spectrum. Cognitive radio has become a reassuring approach to mitigate the spectrum scarcity and to increase the efficiency of the spectrum utilization. In cognitive radio systems, the secondary user (unlicensed user) is allowed by primary user to follow the dynamic spectrum access policy. Spectrum sensing is the basic function of cognitive radio. In order to make use of the radio spectrum, CR has to

S. Gayathri (✉) · K.S. Sujatha
Department of ECE, Easwari Engineering College, Chennai, India
e-mail: gayathrisubramanian90@gmail.com

K.S. Sujatha
e-mail: sujatha.ks@srmeaswari.ac.in

L.P. Suresh and B.K. Panigrahi (eds.), *Proceedings of the International Conference on Soft Computing Systems*, Advances in Intelligent Systems and Computing 397, DOI 10.1007/978-81-322-2671-0_67

continuously monitor the spectrum to identify whether spectrum is idle or busy. This mechanism is known as spectrum sensing. Spectrum sensing symbolizes a major role in cognitive radio systems because the other functions of CR depend on spectrum sensing. Spectrum sensing has to execute two functions as follows; first, it has to detect the white space or unoccupied bands; and second, it has to continuously monitor the spectrum to sense the arrival of primary user during transmission. Many spectrum sensing schemes were introduced which are as follows: energy detection technique, coherent detection method, and cyclostationary feature detection. These approaches are performed for local sensing. From the above approaches, energy detection technique is known to be the easiest approach in existence. The channel degradation caused by the shadowing and multipath effects will arise the hidden terminal problem in local sensing. Henceforth, multinode sensing (or) cooperative sensing helps us to manipulate this problem. It is of two types namely centralized and distributed. In centralized cooperative sensing approach, multiple secondary users will send either the one bit hard decision control channel or energy statistics to the fusion centre which in turn decides whether the band is occupied or not. In distributed approach, the cognitive radios will take their own decision based on the sensing report from the neighbouring cognitive radios. Both the hidden terminal problem and designing a hardware to perform wide band sensing are the major design challenges in cognitive radio. Therefore, a reconfigurable architecture is proposed in this work to perform collaborative wideband sensing to overcome this problem.

2 Related Works

The existing literature for wideband spectrum sensing for CR system is scarce or little. A wavelet transform based spectrum sensing was introduced [1] which uses wavelet transform to estimate power spectral density; however, it is not feasible in real-time sensing. Thomson multitaper spectral estimation was proposed in [2] to estimate PSD. Whereas, its dependency of eigenvalue decomposition, hardware realization of this method is impractical. In [3], spectral estimator based on FFT is proposed where filter banks are used to convert the wideband into narrowband and then PSD is estimated for each narrowband. This method turns out to be impractical because of improper filtering.

In [4], a sensing processor for wideband is designed using multitaper windowed frequency domain power detector. It has a drawback of partial realization in FPGA.

In [5], a cooperative spectrum sensing design is implemented in FPGA but it is restricted for narrowband sensing. In this paper, a wideband has to be sensed henceforth a reconfigurable architecture with multiple direct down converter channelized cognitive radios for multinode detection is designed. The proposed architecture is then implemented in FPGA and its area requirements are reported.

3 System Architecture

This section proposes an architecture for cooperative spectrum sensing cognitive radio for wideband sensing. As shown in Fig. 1, the cooperative sensing architecture of a cognitive radio consists of two units. The two units are as follows: Local detection unit and hard decision logic unit. Local detection unit performs local sensing using energy detection method. Whereas in the hard decision logic unit along with the local decision of the cognitive radio, local decision form multiple cognitive radios is compared with the threshold and gives out the final sensing output.

3.1 Local Detection Unit

The proposed work is for wideband cooperative sensing; therefore, multiple CR undergoes local sensing and computes a one bit decision control output. The local detection unit in the proposed architecture consists of IF sampling wideband receiver architecture with digital IQ baseband processing unit. The receiver architecture consists of RF front end and ADC block and IQ Digital down converter-based channelizer. In the RF front end, bandpass sampling is done in digital domain by a digital bandpass filter (antialising filter) and mixed with an LO to produce the IF signal which is given to the ADC. The ADC performs sampling and gives out time domain samples, $x_k(nT_s)$. Consider the received wideband signal $x(t)$, is given as a input to RF front end and ADC block, where bandpass sampling is done on the received signal. The bandpass sampled output is given as input to IO digital down converted channelizer block [6], which gives out K subbands with discrete time domain samples $X_k = x_k(1), x_k(2), \ldots x_k(N)$ where $k = 1, 2, \ldots, k$. The time domain samples are given as input to the local detection unit as shown in Fig. 2. ADC cannot be realizable in FPGA and the time domain samples are

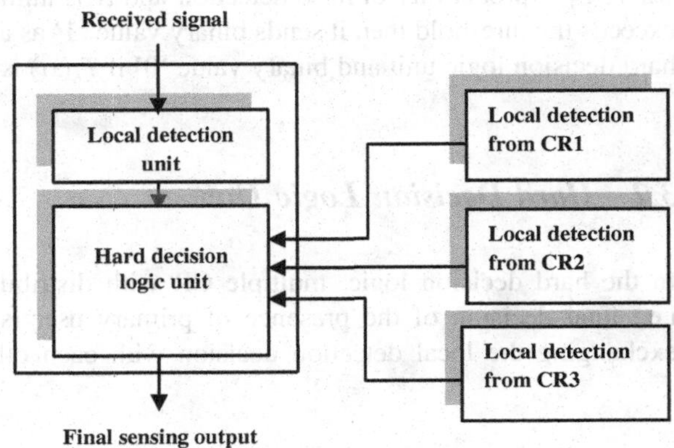

Fig. 1 Proposed architecture of cooperative spectrum sensing cognitive radio

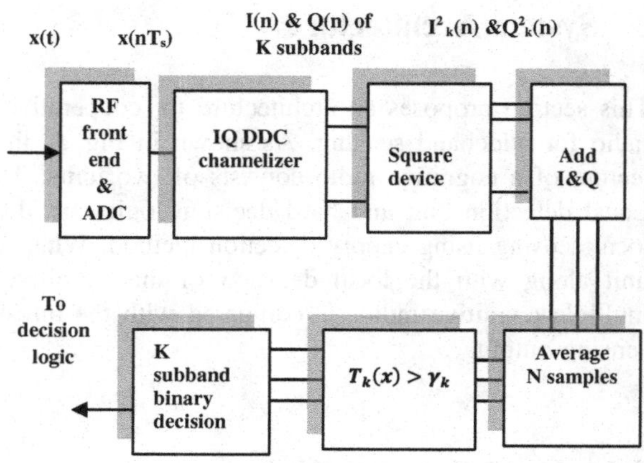

Fig. 2 Local detection unit for a cognitive radio

demultiplexed to K subbands. Further operation takes place as narrowband channelization process. The samples are given to K IQ digital down converters where samples are divided into inphase components, $I_k(n)$ and quadrature components, $Q_k(n)$ using a demultiplexer and followed by multiplying it with monobit multiplier [7]. The computed samples undergo digital lowpass FIR filtering and digital downsampling to decimate the sampling rate. The filtered samples are then squared using a multipler. The squared inphase and quadrature samples are then added by sum block and computes $|x_k(n)|^2$. Finally, the average of N samples is taken using the average FIR filter to take the mean of the sample which gives out the energy statics of the signal in K subband, $\sum_{n=1}^{N} |x_k(n)|^2$. The test static, $T_k(x_k)$ is compared with the derived threshold value, r_k for $p_f = 0.1$. The threshold r_k for local detection is obtained from the formula [8] as follows:

$$r_k = Q^{-1}\left(\frac{p_f\left(\sigma_v^2\sqrt{2N}\right)}{N\sigma_v^2}\right)$$

where P_f is probability of false detection and N is number of samples. If the $T_k(x_k)$ exceeds the threshold then it sends binary value '1' as a binary decision input to the hard decision logic unit and binary value '0' if $T_k(x_k)$ is less than threshold (Fig. 3).

3.2 Hard Decision Logic Unit

In the hard decision logic, multiple CR with distributed approach is performed. i.e., final decision of the presence of primary user is taken after collaboratively exchanging the local detection decision with each other. Multiple CR gives out

Fig. 3 IQ digital down converted channelization

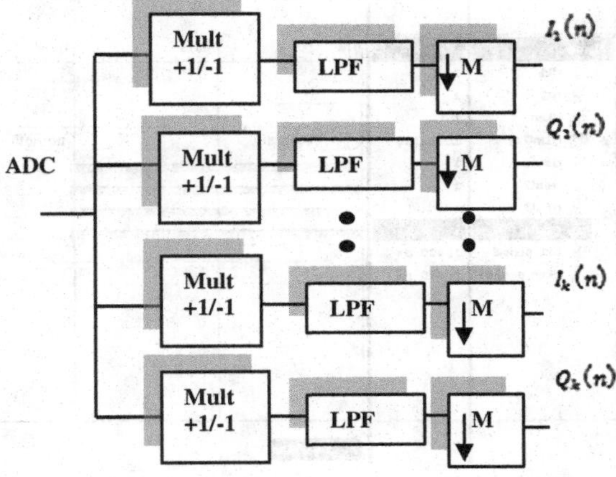

K subband local detection sensing results in the form of vector which is represented as $Y_k = T_{k,1}(1), T_{k,2}(2), \ldots T_{k,m}(N)$. The 1 bit binary decision output for K subands from local detection of M number of CRs are weighted and AND operation is performed, i.e., $Z_k = \prod_{i=1}^{M} W_{k,i}^{T} Y_{k,i}$. Z_k is then compared with threshold, r^k. The threshold, r^k value is '1' because of using hard decision AND logic. Finally, the hard decision logic gives out '1' if the primary user is present and if '0' is obtained then the band is identified as free, i.e., the primary user is absent at the particular band. The binary decision for K subband from M CRs should be '0' for further data transmission by cognitive radio.

4 Results and Discussions

The RF front end of the receiver architecture is simulated in MATLAB. A 10 GHz Wideband signal is modeled and bandpass filtering is done. The filtered signal is then modulated with a carrier wave and given to ADC. The ADC computes time domain samples. The discrete time domain samples computed from ADC sampling is given as the input to the VHDL module of cooperative spectrum sensing. The VHDL module has four local detection submodules meant for four Cognitive radios, which gives it binary decision output after comparing it with the average submodule output with the threshold. The threshold value used is based on $P_f = 0.1$. The binary decision is given to hard decision unit where the cooperative decision is taken and a final binary decision out is obtained. The simulation result of the module of cooperative spectrum sensing is given in Fig. 4.

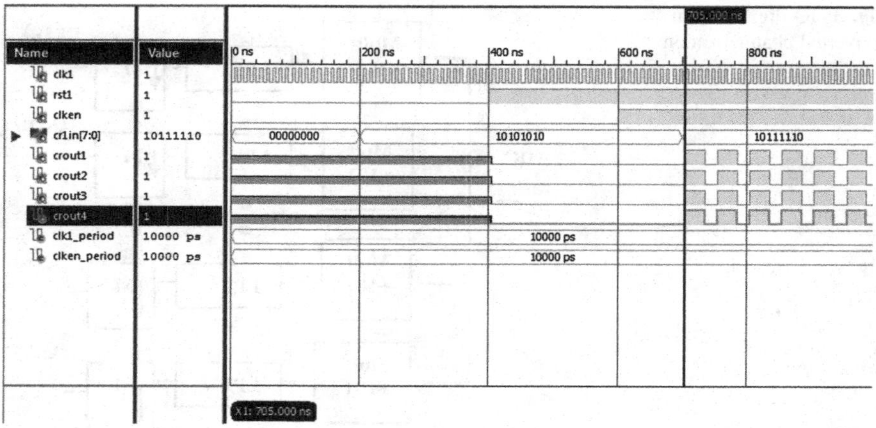

Fig. 4 Simulation result of wideband cooperative spectrum sensing cognitive radio

Logic Utilization	Used	Available	Utilization
Number of Slices	494	5472	9%
Number of Slice Flip Flops	480	10944	4%
Number of 4 input LUTs	921	10944	8%
Number of bonded IOBs	15	240	6%
Number of GCLKs	1	32	3%
Number of DSP48s	32	32	100%

Fig. 5 Design summary report

4.1 Implementation Results in Virtex Board

The proposed architecture is implemented in virtex-4 FPGA after simulating and synthesizing it with Xilinx Isim 13.2 suite. The net list can then be generated and downloaded in Virtex-4 FPGA kit. The design summary report of the implementation are given in Fig. 5 shows number of multipliers, look up table and slices has been used. This summary report clearly points out that the proposed architecture outperforms the other implementations by consuming less area and completely realizable in FPGA.

5 Conclusion and Future Work

In this paper, a cooperative spectrum sensing scheme has been proposed for wideband sensing cognitive radio. Thus the cooperative sensing method improves the spectrum sensing performance over wide frequency range. Simultaneously, it solves the hidden terminal problem in the cognitive radio network and it can be easily implemented in FPGA.

In future, the work will be focused on optimizing the power consumed by the architecture by applying low power techniques like clock gating, etc. and furthermore, replacing filters with multiplierless filters.

Acknowledgments The authors would like to thank Dr. K. Kathiravan for his valuable suggestions to improve the quality of work.

References

1. Haykin S (2005) Cognitive radio: brain—empowered wireless communications. IEEE JSAC 23(2):201–20
2. Tian Z, Giannakis GB (2006) A wavelet approach to wideband spectrum sensing for cognitive radios. In: Proceedings international conference on cognitive radio oriented wireless networks communication (CROWNCOM), pp 1–5, June 2006
3. Sheikh F, Bing B (2008) Cognitive spectrum sensing and detection using polyphase DFT filter banks. In: Proceedings consumer communication and networking conference (CCNC), pp 973–977, Jan 2008
4. Srinu S, Sabat L (2010) FPGA implementation of spectrum sensing based on energy detection for cognitive radio. In: 2010 IEEE international conference on communication control and computing technologies (ICCCCT), Oct 2010
5. Yu T-H, Sekkat O, Rodriguez-Parera S, Marković D, Cabric D (2011) A wideband spectrum-sensing processor with adaptive detection threshold and sensing time. IEEE Trans Circuits Syst—I: Regular Papers 58(11)
6. Pucker L (2003) Channelization techniques for software defined radio. In: Proceedings of spectrum signal processing Inc., Burnaby, B.C, Canada 17–19, Nov 2003
7. Ghasemi A, Sousa ES (2008) Spectrum sensing is cognitive radio networks, challenges and design trade–offs. IEEE Commun 46:32 –39
8. Quan Z, Cui S, Poor H, Sayed A (2008) Collaborative wideband sensing for cognitive radios. IEEE Signal Process Mag 25(6):60–73
9. Mitola J, Maguire CQ (1999) Cognitive radio: making software radio more personal. IEEE Commun Mag 6:13–18 (1999)
10. Ziomek C, Corredoura P Digital I/Q demodulator. Stanford Linear Accelerator Center, Stanford
11. Farhang-Boroujeny B (2008) Filter bank spectrum sensing for cognitive radios. IEEE Trans Signal Process 56(5):1801–1811
12. Chiang T-W, Lin J-M, Ma H-P (2009) A wavelet approach to wideband spectrum sensing for cognitive radios. In: Proceedings IEEE global telecommunication conference (GLOBECOM), pp 1–6, Nov 2009
13. Srinu S, Sambat L (2013) Cooperative wideband sensing based on entropy and cyclic features under noise uncertainty. Signal Process IET 7(8)
14. Lohmiller P, Elsokary A, Schumacher H, Chartier S (2014) Towards a broadband front-end for cooperative spectrum sensing networks. In: Europaen microwave conference, Oct 2014

State Estimation of Interior Permanent Magnet Synchronous Motor Drives Using EKF

Lakshmi Mathianantham, V. Gomathi, K. Ramkumar and G. Balasubramanian

Abstract Interior permanent magnet synchronous motor (IPMSM) is more widely used and attractive in modern hybrid electric vehicles (HEV) and industrial applications due to its relatively simple motor structure and reduced processing cost. This paper proposes the field-oriented controlled (FOC) IPMSM drive using extended Kalman filter (EKF) algorithm for state estimation is established with Matlab/Simulink. Using this algorithm, the phase currents (i_a and i_c), rotor position (θ_{re}), and angular speed (ω_{re}) are estimated. Moreover, simulations results are presented to show that the control system works smoothly with load and that indicate the good functionality of proposed algorithm.

Keywords Interior permanent magnet synchronous motor (IPMSM) · Field-oriented vector control (FOC) · State estimation · Extended Kalman filter (EKF) · Matlab/Simulink

1 Introduction

IPMSM possesses good dynamic performance, smaller size, reduced processing cost, and higher efficiency, when compared to other form of motors. With the development of the permanent magnets, the application area of IPMSM has been extended in recent years. Meanwhile, elimination of brushes will improve the reliability of the IPMSM.

Therefore, IPMSM has many good features such as small size, modest structure, high power factor, high competence and moment of inertia is low, when compared with conventional electrical motor. With the improvement of permanent magnet, the field-oriented control system of IPMSM has good position and speed control

L. Mathianantham (✉) · V. Gomathi · K. Ramkumar · G. Balasubramanian
Department of Electronics and Instrumentation Engineering,
SASTRA University, Thanjavur, India
e-mail: lakshmiash91@gmail.com

L.P. Suresh and B.K. Panigrahi (eds.), *Proceedings of the International Conference on Soft Computing Systems*, Advances in Intelligent Systems and Computing 397, DOI 10.1007/978-81-322-2671-0_68

with high precision. Further, IPMSM can be controlled by two ways, one method is controlling with sensor and other method is sensorless controlling.

In IPM motor drive, system performance is declared by function of sensors. Thus, the fault diagnosis in system is necessary for good control of the system. Now-a-days, a controller modification technique is used in induction motor drives in case of sensor fault [1]. An observer-based fault tolerant control that focuses induction machine sensor fault [2]. The current sensor fault diagnosis in doubly fed induction generator is presented [3] and it is very insensitive to noise. The stator currents and resistance in rotor of induction motor are estimated using adaptive observer [4] and this observer is implemented in $\alpha\beta$ stationary frame of IPMSM. The fault diagnosis method derived from parity space approach for IPMSM [5].

The faults in current sensor and voltage sensor are detected in DTC controlled IPMSM [6]. The rotor position and speed are estimated signal rejection algorithm [7]. An offline fault detection for current sensor in IPMSM drives is presented [8]. Using adaptive observer, the resistance in rotor and voltage obtained from DC link of induction motor is estimated [9]. In order to get better performance of AC drives, [10] uses various type of square root EKF and compared its performance.

The observability of state matrix of IPMSM is get from [11]. The real-time implementation of fault diagnosis of IPMSM with EKF algorithm is given [12]. The EKF tuning in sensorless PMSM drives is presented with a direct matrices choice [13]. The carrier frequency injection method is used and the rotor magnet's polarity gives the magnetic saturation effect [14].

The EKF usage in sensorless drives is abundant, but it is virtually absent in sensor fault diagnosis drives. The EKF will efficiently overcome the error occurred in system and measurement on state estimation. In this present paper, IPMSM model is linearized using FOC with Matlab modeling capabilities and the EKF simultaneously estimates the rotor speed and the phase current of the IPMSM drives.

This paper is ordered as follows. In Sect. 2, IPMSM mathematical model is given. In Sect. 3, the algorithm for EKF is given. In Sect. 4, results are analyzed. In Sect. 5, the final conclusion statements are given.

2 IPMSM Mathematical Model

2.1 Basic Model

In this paper, IPMSM motor model is represented in *dq*-rotating axis is given as follows:

$$\frac{\mathrm{d}i_d}{\mathrm{d}t} = -\frac{R_s}{L_d} i_d + \frac{L_q}{L_d} \omega_{\mathrm{re}} i_q + \frac{1}{L_d} v_d \tag{1}$$

$$\frac{di_q}{dt} = -\frac{R_s}{L_q} i_q - \frac{L_d}{L_q} \omega_{re} i_d - \frac{1}{L_q} \omega_{re} \lambda_f + \frac{1}{L_q} v_q \tag{2}$$

$$\frac{d\omega_{re}}{dt} = 0 \tag{3}$$

$$\frac{d\theta_{re}}{dt} = \omega_{re} \tag{4}$$

λ_d and λq are expressed as follows:

$$\lambda_d = L_d i_d + \lambda_f \tag{5}$$

$$\lambda_q = L_d i_q \tag{6}$$

where,

R_s	Stator resistance
L_d, L_q	Inductance of *dq*-axes
λ_f	Flux linkage in permanent magnet
λ_d, λ_q	Stator flux linkage—*dq* axis
i_a, i_b, i_c	Current in *abc* reference frame
i_d, i_q	Current in d*q* rotating axis
i_α, i_β	Current in $\alpha\beta$ stationary axis
v_d, v_q	Voltage in *dq* rotating axis
v_α, v_β	Voltage—$\alpha\beta$ stationary axis
ω_{re}	Angular speed of rotor
θ_{re}	Angular position of rotor
T_M	Mechanical time constant
t_l	Load torque

2.2 *Coordinate Transformation and Field-Oriented Vector Control*

Figure 1 gives the coordinate of IPMSM rotor and stator. In order to concern field-oriented control, first decouples the three-phase time variant system into the two-phase *dq* coordinates system. FOC has torque part (associated with quadrature axis) and flux part (associated with direct axis). This work is similar to DC machine control.

The Clarke and Park transformation is given as follows:

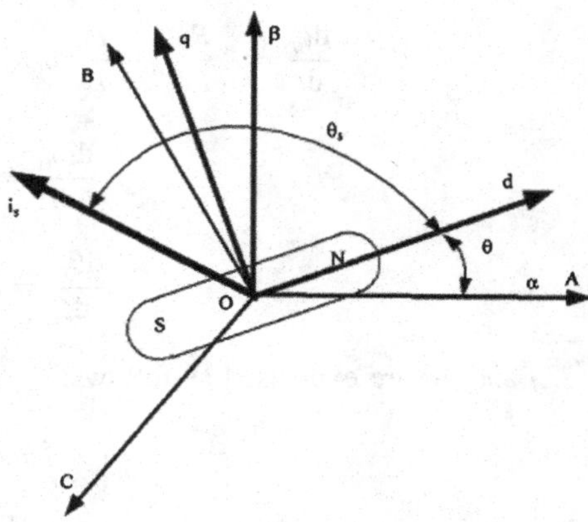

Fig. 1 Coordinate system of IPMSM

$$\begin{pmatrix} i_\alpha \\ i_\beta \end{pmatrix} = \begin{pmatrix} \frac{2}{3} & -\frac{1}{3} & -\frac{1}{3} \\ 0 & \frac{1}{\sqrt{3}} & -\frac{1}{\sqrt{3}} \end{pmatrix} \begin{pmatrix} i_a \\ i_b \\ i_c \end{pmatrix} \tag{7}$$

where

$$i_a + i_b + i_c = 0 \tag{8}$$

$$\begin{pmatrix} i_d \\ i_q \end{pmatrix} = \begin{pmatrix} \cos\theta_{\text{re}} & \sin\theta_{\text{re}} \\ -\sin\theta_{\text{re}} & \cos\theta_{\text{re}} \end{pmatrix} \begin{pmatrix} i_\alpha \\ i_\beta \end{pmatrix} \tag{9}$$

The inverse park transformation is also used in this control. By adjusting the torque current component (i_q) and the flux current component (i_d), we get linearized IPMSM control and high performance of motor is ensured by doing this control.

3 Extended Kalman Filter (EKF)

3.1 EKF State Estimation

The extended Kalman filter (EKF) is used for nonlinear model. It is an optimal estimator and it provides the information about parameter variations, model inaccuracy, and measurement noise. This nonlinear filter is used to linearize the current mean and covariance. The EKF is a group of mathematical expressions which produces the optimal estimation of the state system and it estimates the process using feedback control. The EKF provides significant reliable tolerance for the mathematical model error. The IPMSM with EKF state estimation is given in Fig. 2. The mathematical model of the IPMSM given by (1) and (2) is the nonlinear

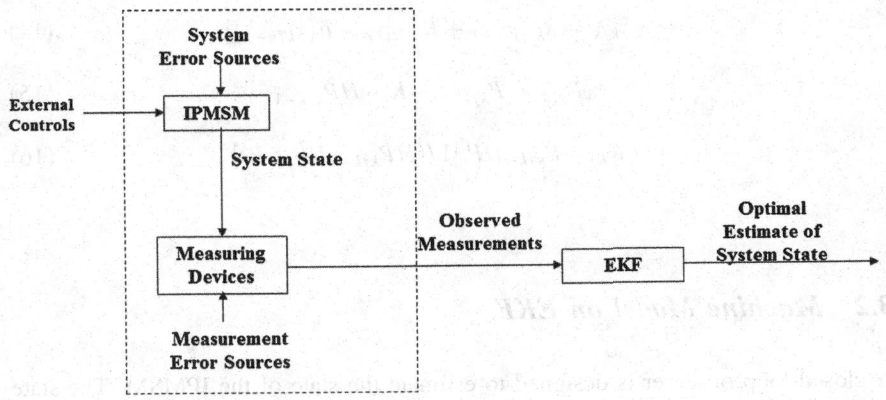

Fig. 2 IPMSM with EKF estimation

system. To implement the EKF algorithm, the system needs to be discretized and linearized. The discrete state model for nonlinear system is given as follows:

$$x_{k+1} = f(x_k, u_k) + W_k. \tag{10}$$

$$y_k = h(x_k) + V_k. \tag{11}$$

where,

- x_k State vector of the system
- u_k Input voltage vector to the system
- y_k Measured output current vector of the system
- W_k System noise vector
- V_k Measurement noise vector

The EKF algorithm contains two important steps. In first step, it performs the state prediction $\hat{X}_{k/k-1}$ and its covariance matrix prediction $\hat{P}_{k/k-1}$ is given in (12). Here, T_s is the sampling period and F is the system gradient matrix.

(i) Prediction

$$\hat{x}_{k/k-1} = \hat{x}_{k-1/k-1} + T_s f(x_{k-1/x-1}, u_{k-1}) \tag{12}$$

$$\hat{P}_{k/k-1} = \hat{P}_{k-1/k-1} + F\hat{P}_{k-1/k-1}F^T + Q \tag{13}$$

In second step, it performs correction with measurement matrix H is given in (13). Here, the Kalman gain K_k is considered by (14) as follows:

(ii) Correction

$$\hat{x}_{k/k} = \hat{x}_{k-1/k-1} + K_k \cdot [y_k - h(\hat{x}_{k/k-1})] \tag{14}$$

$$\hat{P}_{k/k} = \hat{P}_{k/k-1} + K_k \cdot H\hat{P}_{k/k-1} \tag{15}$$

$$K_k = \hat{P}_{k/k-1}H^T[1/(H\hat{P}_{k/k-1}H^T + R)] \tag{16}$$

3.2 Machine Model on EKF

A closed-loop observer is designed to estimate the state of the IPMSM. The state vector is taken to be $x = (i_a, i_c, \omega_{\mathrm{re}}, \theta_{\mathrm{re}})^T$, the input voltage vector to be $u = (v_\alpha, v_\beta)^T$, and the output current vector to be $y = (i_a, i_c)^T$, since these measures can be get easily from the available measurements. The EKF is carried out in IPMSM machine model given in (1) and (2). The system observability is given in (17).

$$\frac{\mathrm{d}}{\mathrm{d}t}\begin{pmatrix} i_a \\ i_c \\ \omega_{\mathrm{re}} \\ \theta_{\mathrm{re}} \end{pmatrix} = \begin{pmatrix} \frac{\lambda_d + \lambda_f}{L_d}\cos\theta_{\mathrm{re}} - \frac{\lambda_q}{L_q}\sin\theta_{\mathrm{re}} \\ \frac{\lambda_d - \lambda_f}{L_d}\left[\frac{\sqrt{3}}{2}\sin\theta_{\mathrm{re}} + \frac{1}{2}\cos\theta_{\mathrm{re}}\right] + \frac{\lambda_q}{L_q}\left[\frac{1}{2}\sin\theta_{\mathrm{re}} - \frac{\sqrt{3}}{2}\cos\theta_{\mathrm{re}}\right] \\ 0 \\ \omega_{\mathrm{re}} \end{pmatrix} \tag{17}$$

3.3 Computation of F and H Matrices

System gradient matrix F (Jacobian matrix) defined as follows:

$$F = f'(\hat{x}_k, u_k) = \begin{pmatrix} f_{11} & f_{12} & f_{13} & f_{14} \\ f_{21} & f_{22} & f_{23} & f_{24} \\ f_{31} & f_{32} & f_{33} & f_{34} \\ f_{41} & f_{42} & f_{43} & f_{44} \end{pmatrix} \tag{18}$$

The Jacobian matrix elements can be calculated as follows:

$$
\begin{aligned}
f_{11} &= \frac{-R_s}{L_d} \\
f_{12} &= \omega_{\text{re}} \\
f_{13} &= \lambda_q \\
f_{14} &= -v_\alpha \sin\theta_{\text{re}} + v_\beta \cos\theta_{\text{re}} \\
f_{21} &= -\omega_{\text{re}} \\
f_{22} &= \frac{-R_s}{L_q} \\
f_{23} &= -\lambda_d \\
f_{14} &= -v_\alpha \cos\theta_{\text{re}} - v_\beta \sin\theta_{\text{re}} \\
f_{31} &= f_{32} = f_{33} = f_{34} = f_{41} = f_{42} = f_{44} = 0 \\
f_{43} &= 1
\end{aligned}
$$

Measurement matrix H is given as

$$
\begin{aligned}
H &= h'(\hat{x}_k) \\
&= \begin{pmatrix} h_{11} & h_{12} & h_{13}\ h_{14} \\ h_{21} & h_{22} & h_{23}\ h_{24} \end{pmatrix}
\end{aligned} \tag{19}
$$

$$
\begin{aligned}
h_{11} &= \frac{1}{L_d}\cos\theta_{\text{re}} \\
h_{12} &= \frac{1}{L_q}\sin\theta_{\text{re}} \\
h_{13} &= 0 \\
h_{14} &= -\frac{\lambda_d - \lambda_f}{L_d}\sin\theta_{\text{re}} - \frac{\lambda_q}{L_q}\cos\theta_{\text{re}} \\
h_{21} &= \frac{1}{L_d}\left[\frac{\sqrt{3}}{2}\sin\theta_{\text{re}} + \frac{1}{2}\cos\theta_{\text{re}}\right] \\
h_{22} &= \frac{1}{L_q}\left[\frac{1}{2}\sin\theta_{\text{re}} - \frac{\sqrt{3}}{2}\cos\theta_{\text{re}}\right] \\
h_{23} &= 0 \\
h_{24} &= -\frac{\lambda_d - \lambda_f}{L_d}\left[\frac{\sqrt{3}}{2}\cos\theta_{\text{re}} - \frac{1}{2}\sin\theta_{\text{re}}\right] - \frac{\lambda_q}{L_q}\left[\frac{1}{2}\cos\theta_{\text{re}} + \frac{\sqrt{3}}{2}\sin\theta_{\text{re}}\right]
\end{aligned}
$$

The general block diagram of field-oriented vector controlled IPMSM with EKF state estimation is shown in Fig. 3. The block diagram is shown in Fig. 3, including speed controller, flux controller, torque controller, space vector modulation

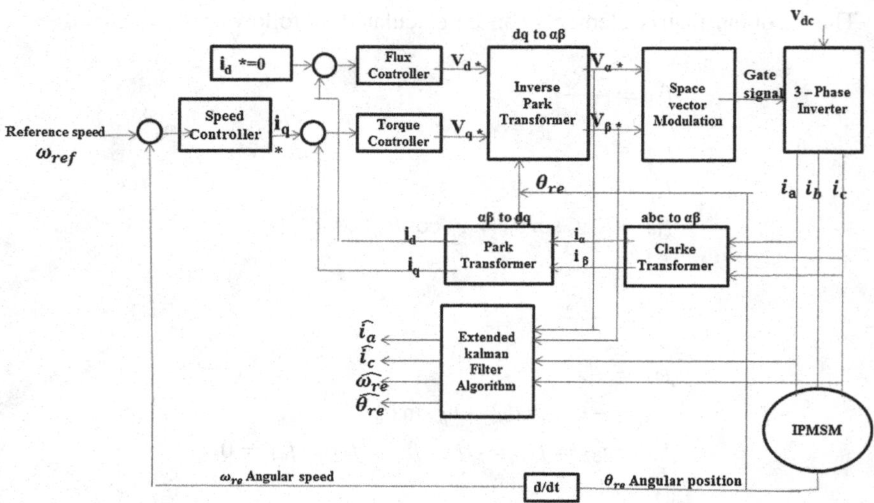

Fig. 3 Block diagram of IPMSM drive under field-oriented control with EKF state estimation

(SVM) module, Clarke–Park transformation, three-phase inverter block, IPMSM motor, and EKF state estimation block.

The block diagram gives EKF estimation, which is implemented in field-oriented controlled IPMSM drive. EKF gives estimates of stator phase currents (i_a and i_c), rotor position (θ_{re}), and angular speed (ω_{re}). In this block, ω_{re} is the comment variable given to the speed control loop.

4 Simulation Analysis of IPMSM

In Matlab/Simulink, the simulation model of IPMSM is created EKF estimation. By using this algorithm, the estimates of stator currents, rotor speed and rotor position of IPMSM is obtained. The simulation model diagram is shown in Fig. 4. The IPMSM motor parameter is given in below Table 1.

In the Matlab function block for extended Kalman filter, the initial state x, the initial P, Q, and R matrices are configured. To compute the Kalman filter gain K, the Q, R, and P are needed. The accuracy of the EKF estimation is not concluded by Q and R. By repeating the simulation, initial values of the system is constructed.

From Figs. 5, 6, 7 and 8, the straight line shows the true values and the dotted line indicates the EKF estimated values. The standard deviations of estimation error are obtained in command window for the estimated $i_a = 0.024165$, $i_c = 0.0093664$, $\omega_{\mathrm{re}} = 0.024898$, and $\theta_{\mathrm{re}} = 0.0011157$, respectively.

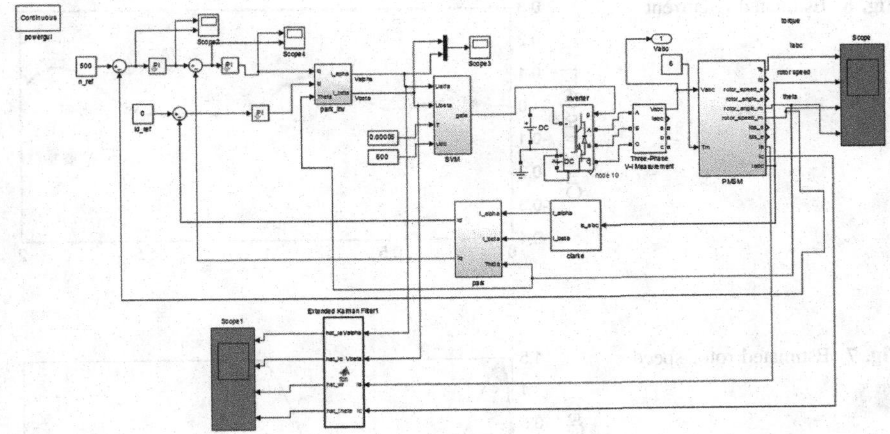

Fig. 4 Simulation model of IPMSM with EKF state estimation

Table 1 IPMSM motor parameter

Parameter		Values
Number of poles	P	4
Stator resistance	R_s	4.7 Ω
Magnet flux linkage	λ_f	0.667 Wb
Inductance in d axis	L_d	0.0235 H
Inductance in q axis	L_q	0.0325 H
Supply voltage	V	135 V
Supply current	I	2.5 A
Rated torque	T	6 Nm

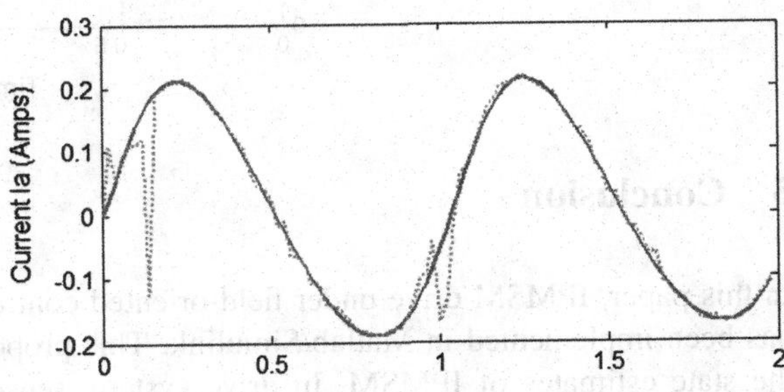

Fig. 5 Estimated i_a current

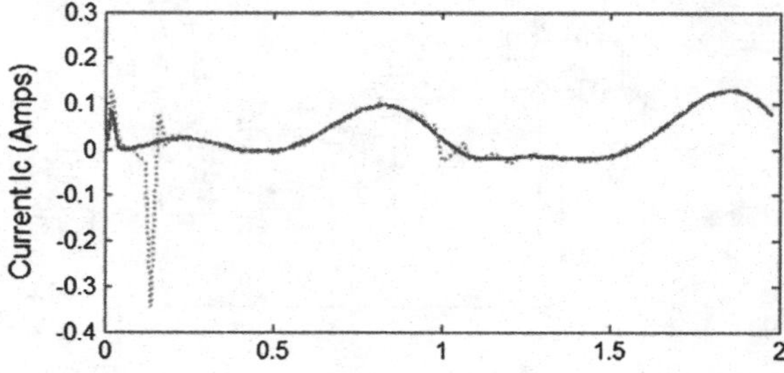

Fig. 6 Estimated i_a current

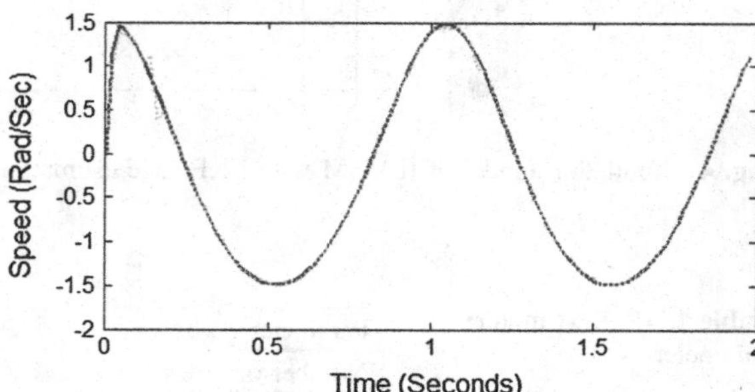

Fig. 7 Estimated rotor speed ω_{re}

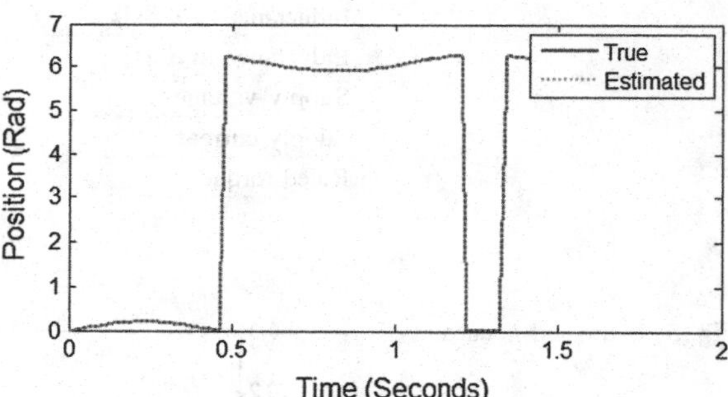

Fig. 8 Estimated rotor position θ_{re}

5 Conclusion

In this paper, IPMSM drive under field-oriented control with EKF state estimation has been implemented in Matlab/Simulink. This proposed method is used to offer the state estimates of IPMSM. In drive system, sensors are used to measure the current and DC-link voltages and encoder are used to measure the rotor position. Due to external noise in the sensor and system, the control performances get worsen. Using this proposed methodology, the estimated value from EKF gives correct value and this will overcome the noises in system and sensors. In this paper, simulation results show that EKF algorithm exactly follows stator current, rotor position, and rotor speed of IPMSM when the perfect IPMSM machine equation is offered.

References

1. Benbouzid MEH, Diallo D, Zeraoulia M (2007) Advanced faulttolerant control of induction-motor drives for EV/HEV traction applications: from conventional to modern and intelligent control techniques. IEEE Trans Veh Technol 56(2):519–528
2. Romero ME, Seron MM, De Dona JA (2010) Sensor fault-tolerant vector control of induction motors. IET Control Theory Appl 4(9):1707–1724
3. Rothenhagen K, Fuchs FW (2009) Current sensor fault detection, isolation, and reconfiguration for doubly fed induction generators. IEEE Trans Ind Electron 56(10):4239–4245
4. Najafabadi TA, Salmasi FR, Jabehdar-Maralani P (2011) Detection and isolation of speed, DC-link voltage and current sensors faults based on an adaptive observer in induction motor drives. IEEE Trans Ind Electron 58(5):1662–1672
5. Berriri H, Naouar MW, Slama-Belkhodja I (2012) Easy and fast sensor fault detection and isolation algorithm for electrical drives. IEEE Trans Power Electron 27(2):490–499
6. Foo G, Rahman MF (2010) Sensorless direct torque and flux controlled IPM synchronous motor drive at very low speed without signal injection. IEEE Trans Ind Electron 57(1):395–403
7. Sayeef S, Foo G, Rahman MF (2010) Rotor position and speed estimation of a variable structure direct torque controlled IPM synchronous motor drive at very low speeds including standstill. IEEE Trans Ind Electron 57(11):3715–3723
8. Jeong Y-S, Sul S-K, Schulz SE, Patel NR (2005) Fault detection and fault-tolerant control of interior permanent-magnet motor drive system for electric vehicle. IEEE Trans Ind Appl 41 (1):46–51
9. Salmasi FR, Najafabadi TA, Maralani PJ (2010) An adaptive flux observer with online estimation of DC-link voltage and rotor resistance for VSI-based induction motors. IEEE Trans Power Electron 25(5):1310–1319
10. Smidl V, Peroutka Z (2012) Advantages of square-root extended Kalman filter for sensorless control of AC drives. IEEE Trans Ind Electron 59(11):4189–4196
11. Vaclavek P, Blaha P, Herman I (2013) AC drives observability analysis. IEEE Trans Ind Electron 60(8):3047–3059
12. Foo G, Zhang X, Vilathgamuwa DM (2013) A sensor fault detection and isolation method in interior-permanent-magnet synchronous motor drives based on an extended Kalman Filter. IEEE Trans Ind Electron 60(8):3485–3495
13. Bolognani S, Tubiana L, Zigliotto M (2003) Extended Kalman Filter tuning in sensorless PMSM drives. IEEE Trans Ind Appl 6(39):1741–1747
14. Jeong Y, Lorenz RD, Jahns TM et al (2005) Initial rotor position estimation of an interior permanent magnet synchronous machine using carrier frequency injection methods. IEEE Trans Ind Appl 41(1):38–45

Texture and Correlation-Based 3D Image Analysis

J. Shankar, M. Lenin, R. Siva Raman, V. Arjun and N.R. Raajan

Abstract In the recent days, stereoimage processing is used for the real-time applications like estimation of velocity, distance estimation of an object from the camera, and security purposes. From the stereoimage, one can infer the depth information. Using depth information, height from ground point can be inferred. In this work, few metrics have been computed using the stereoimage. Texture map and depth discontinuous regions have been obtained from the stereoimage. Then based on correlation metrics various forms of disparity map has been computed using SAD and SSD functions. Computational time has been also computed for the above and the difference is noted. These metrics define the quality for a disparity map. Further, using depth information one can easily split the forth ground and background which provides us easy means of segmentation. Further scaling can be done to enhance the stereoimage.

Keywords Texture · Disparity map · Depth · Discontinuous regions · SSD · SAD

J. Shankar (✉) · M. Lenin · R. Siva Raman · V. Arjun · N.R. Raajan
Department of Electronics and Communication Engineering, School of Electrical and Electronics Engineering, SASTRA University, Thanjavur, Tamilnadu, India
e-mail: shankece123@gmail.com

M. Lenin
e-mail: leninmkarthi@gmail.com

R. Siva Raman
e-mail: sivaece26492@gmail.com

V. Arjun
e-mail: vechhamarjun@gmail.com

N.R. Raajan
e-mail: nrraajan@ece.sastra.edu

L.P. Suresh and B.K. Panigrahi (eds.), *Proceedings of the International Conference on Soft Computing Systems*, Advances in Intelligent Systems and Computing 397, DOI 10.1007/978-81-322-2671-0_69

1 Introduction

Stereoscopy is also referred 3D imaging. The name has been its origin from Greek word stereo and skopeo, in which former is a solid and latter is 'to see'. Capturing an object or scene using stereo camera gives us two pictures known as left and right images. From the below figure, the basic setup for the stereo camera system is given. Point p is the pixel where the two cameras capture it. Stereo is preferred to overcome the difficulties found in single camera systems, which is to obtain depth information [1, 2] (Fig. 1).

Left Camera gives out the equation

$$X_l = \frac{fX}{Z}; Y_l = \frac{fY}{Z} \tag{1}$$

Right camera gives the equation as

$$X_r = \frac{f(X - S)}{Z}; Y_r = \frac{fY}{Z} \tag{2}$$

where

X_l and Y_l Coordinate points of left camera.
X_r and Y_r Coordinate points for the right camera.
f Focal Length.
Z Depth.

2 Flow Chart

See Fig. 2.

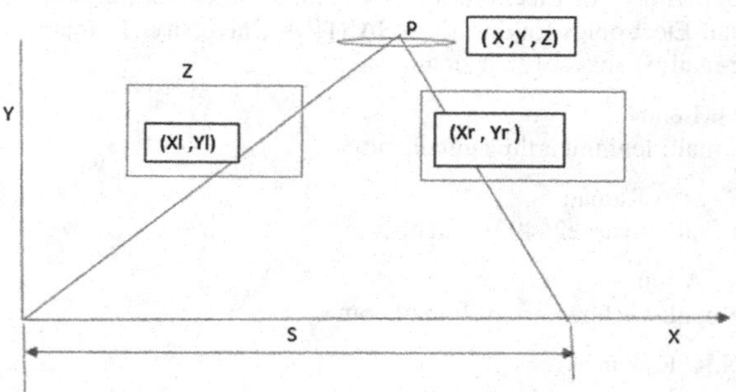

Fig. 1 Stereo vision

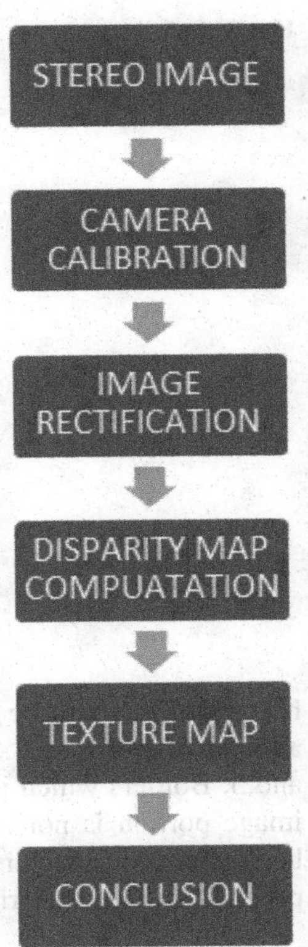

Fig. 2 Process implementation

3 Left and Right Image

Stereoimage consists of two images in it namely left and right images which is obtained from a stereo camera. Stereogram is another term used for a stereoimage. Human visual system is an example for stereo system. In the initial step, camera calibration is done with help of default checker board pattern. The whole pattern must be covered in stereo camera. After this process, real-time scene or object is captured. The captured image is stored only in lossless format like PNG which provides calibration [3, 4] (Fig. 3).

4 Image Rectification

Rectification process is one of the vital steps in Stereo vision. It gives out two or more images into a common plane. This process helps to refine our search and matching of points by narrowing it to 1D search. Therefore, pixels in same row or

Fig. 3 Left image

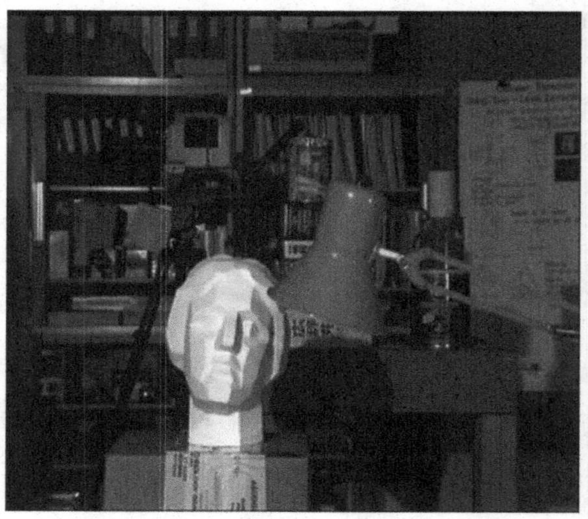

horizontal line will be used for matching. Hence, correspondence problem will be avoided after this step. It is also to be noted that there is a difference in both Figs. 4 and 5. Borders which is black in color marks out that there is no pixel value or the image portion is not captured by left and right images. After rectification, it has been removed (which is termed to be bad pixel) to avoid confusions while matching process. Borders of original image are discarded [4, 5] (Fig. 6).

Fig. 4 Right image

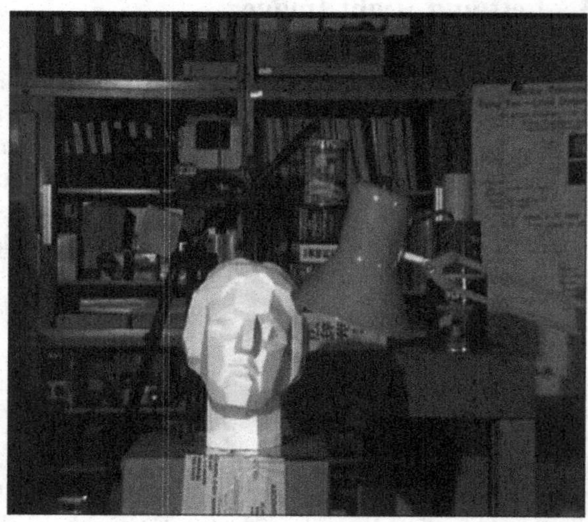

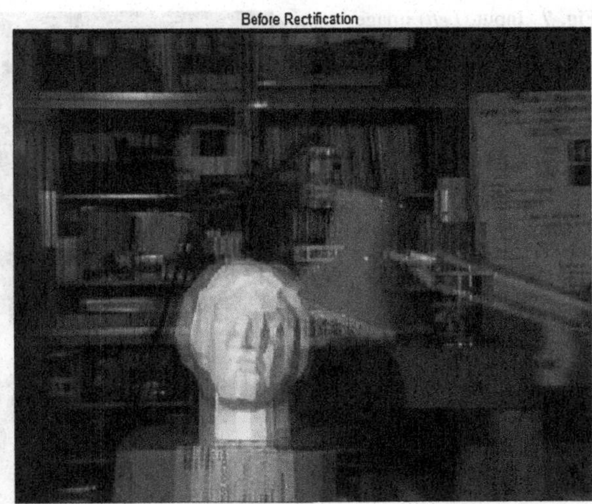

Fig. 5 Before rectification

Fig. 6 After rectification

5 Texture Map

Stereo vision algorithms typically compute erroneous results in regions where there is a little or no texture in the scene. They are defined as regions where the squared horizontal intensity gradient averaged over a square window of a given size is below a given threshold. From the below results, the inference obtained is the black region denotes no texture or texture less regions and the rest is texture region [1, 3] (Figs. 7 and 8).

Fig. 7 Input (*Left*) image

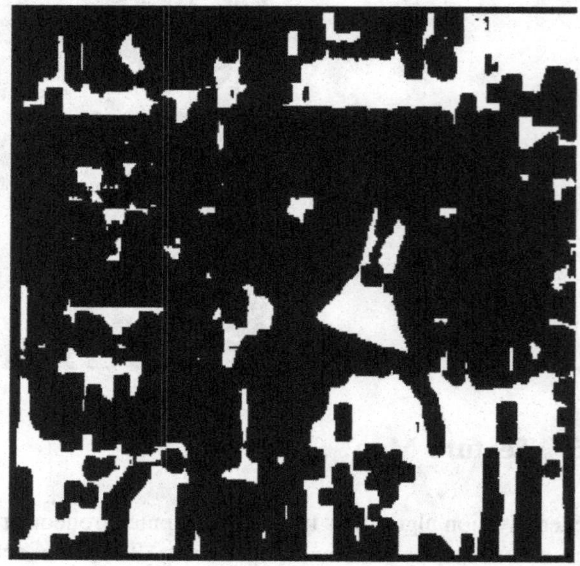

Fig. 8 Texture map

6 Depth Discontinuous Regions

Stereo vision algorithms typically compute accurate results in regions where there is a sudden change in the depth between objects in the scene. They are defined as regions where neighboring disparities differ by more than a certain gap, dilated by a window of a given width. From the below figure, the pixels marked in white are called as the regions of depth discontinuous [2, 6] (Figs. 9 and 10).

7 Disparity Map Computation

Correlation-based matching gives out dense depth maps by calculating the disparity at each pixel within a neighborhood. This work helped by taking a square window of certain size. Then we have to choose the region (pixel of interest) in the reference image and finding the similar kind of pixel within the window in the target image. The goal is to find the corresponding (correlated) pixel within a certain disparity range d (dE [0... d_{max}]). Ultimate aim is to minimize the error and the similarity must be high. The images used for matching in two ways. First, using left image as reference, left to right matching is done. This work is known as direct matching. Second, Using right image as reference right to left matching is done. This is known as reverse matching [5, 6].

Fig. 9 Input image

Fig. 10 Depth discontinuous regions

8 SAD

Sum of absolute differences (SAD) is one of the techniques used for the similarity measures between two images. It is calculated by subtracting pixels within a square neighborhood between the two images.Within the square window, the pixels will be correlated in the left and right images. Exact match of pixel gives out the resultant will be zero.

SAD is given by the following equation

$$\sum_{(i,j)\in W} |I_1(i,j) - I_2(x+i, y+j)|$$

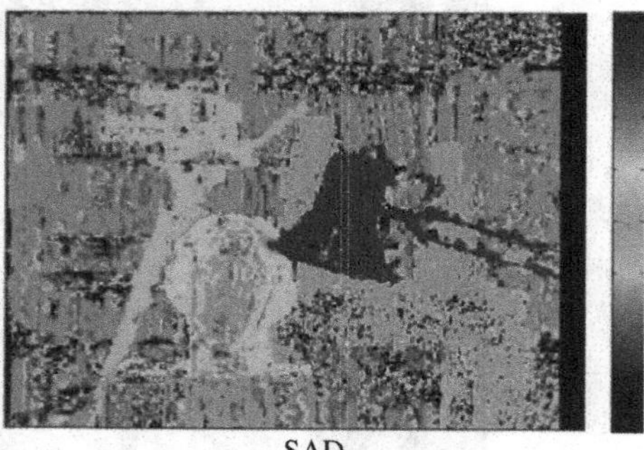

SAD

ZSAD is given by the equation—Zero denotes zero mean

$$\sum_{(i,j)\in W} |I_1(i,j) - \bar{I}_1(i,j) - I_2(x+i,y+j) + \bar{I}_2(x+i,y+j)|$$

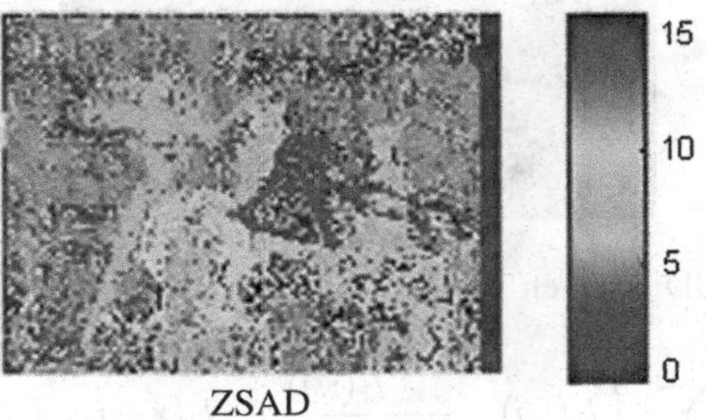

ZSAD

LSAD

$$\sum_{(i,j)\in W} \left(I_1(i,j) - \frac{I_1(i,j)}{I_2(x+i,y+j)} I_2(x+i,y+j) \right)$$

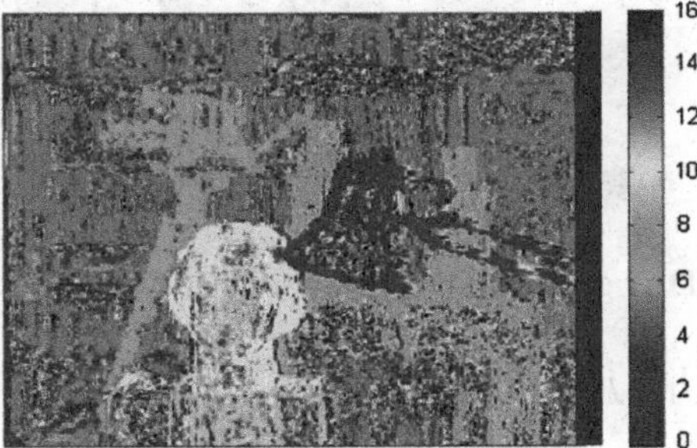

SSD

In sum of squared differences (SSD) technique, the differences between the two images namely left and right are squared and aggregated. Square window has been selected for process. Optimization is done and by Winner Take All strategy the best pixel is chosen and output has been obtained. Computational complexity is high compared to SAD system since it uses very high number of the multiplication operations.

$$\sum_{(i,j)\in W} (I_1(i,j) - I_2(x+i,y+j))^2$$

SSD

LSSD: The LSSD is given by the following equation as

$$\sum_{(i,j)\in W}\left(I_1(i,j)-\frac{\bar{I}_1(i,j)}{\bar{I}_2(x+i,y+j)}I_2(x+i,y+j)\right)^2$$

LSSD

ZSSD: The ZSSD is given by the below disparity map

$$\sum_{(I,J)\in W}(I_1(i,j)-\bar{I}_1(i,j)-I_2(x+i,y+j)+\bar{I}_2(x+i,y+j))^2$$

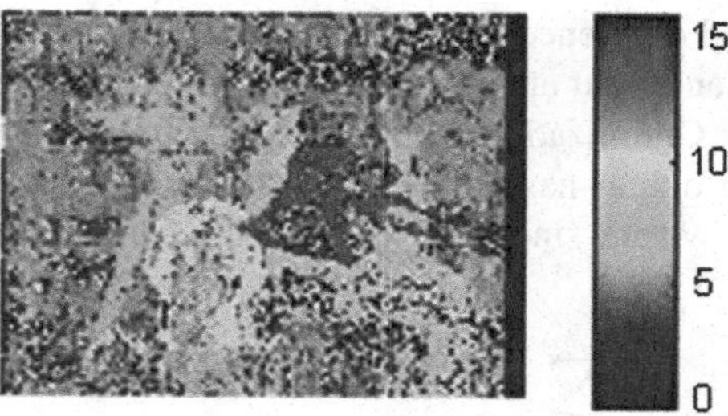

9 Conclusion

Stereoimages have been preferred mainly due to the depth information inferred from it. Disparity map and the depth are the most important key words in the stereo literature. Computing the effective disparity map is very much essential in the process. From the above work, texture map has been calculated to keep a note on texture region to focus that region more. Further using correlation metrics, the disparity map has been calculated with the run time also and comparative study is done. Effective one can be used for further processing to infer the depth information and analysis of the 3D image.

References

1. Raajan NR, Ram Kumar M, Monisha B, Jaiseeli C, Venkatesan SP (2012) Disparity estimation from stereo images. In: International conference on modelling optimization and computing, code 105169, April 2012
2. Howard J, Morse B, Cohen S, Price B (2014) Depth based patch scaling for content aware stereo image completion. In: IEEE winter conference on applications of computer vision, March 2014
3. Raajan NR, Ramkumar M, Monisha B, Rengarajan M, Ravi Shankar LV, Prasanna Venkatesan S, Neha VC (2012) Stereopsis based information identification on real world. In: International conference on modeling optimization and computing
4. https://www.youtube.com/watch?v=hab07nMeUzA
5. Xiaoyan and Harbin Normal (2008) Pseudo disparity based stereo image coding. In: 15th IEEE conference on ICIP 2008, 12–15 Oct 2008
6. Hwa Lee S, Sharma S (2011) Real time disoarity estimation algorithm for stereo camera systems. IEEE Trans Consum Electron 57(3)
7. Kaaniche M, Miled W, Benazza-Benyahia B (2009) Dense disparity map representations for stereo image coding. In: 16th IEEE International Conference, Nov 2009
8. Kwak J-H, Kim K-T, Bong-Hyun K (2008) Intermediate view image generation based on disparity path search of disparity space image. In: 2008 Second international conference on future generation communication and networking symposia
9. Zamarin M, Forchhammer S (2012) Lossless compressiom of stereo disparity maps for 3D. In: IEEE international conference on multimedia and expo workshops 2012
10. Zhang Z, Ai X, Dahnoun N (2013) Efficient disparity calculation based on stereo vision with ground obstacle assumption. In: Signal processing conference (EUSIPCO), 9–13 Sep 2013

A Novel Approach to Compress an Image Using Cascaded Transform and Compressive Sensing

S. Nirmalraj and T. Vigneswaran

Abstract Transmission of high-frequency signals require a channel with very high bandwidth. But effective use of bandwidth always remains as an important criterion in communication. So, compression of the message signal is important before transmission. Many transforms are used such as discrete Fourier transform, discrete cosine transform, etc. The objective of this paper is to compress the given image using a cascaded transform and a novel technique known as compressive sensing.

Keywords Sparsity · Compressive sensing · Cascaded image transform

1 Introduction

Communicating a digital signal is more advantageous than an analog signal. Before converting an analog signal into a digital signal, it is better to compress the signal in order to save the bandwidth required to transmit the signal. A new technique known as compressive sensing shows how to do both tasks. Initially, the analog signal of interest is converted into sparse and then it is sampled. This helps one to sample the analog signal in information rate rather than the Nyquist rate. In order to decompose a signal many transforms can be used. For one-dimensional signals DFT and for two-dimensional signals DWT can be used. When DFT and DWT were cascaded together, it provides good results for the image signals. After decomposing the image, the larger coefficients are retained because they are the information bearing coefficients. These larger coefficients when applied with the inverse transform the original image is reconstructed without much loss which can be measured with some performance measures like PSNR, MSE, SSI, etc. This is explained by

S. Nirmalraj (✉)
Department of E.E.E, Sathyabama University, Chennai, India
e-mail: snirmal4u@yahoo.co.in

T. Vigneswaran
Department of E.C.E, VIT University, Chennai, India

L.P. Suresh and B.K. Panigrahi (eds.), *Proceedings of the International Conference on Soft Computing Systems*, Advances in Intelligent Systems and Computing 397, DOI 10.1007/978-81-322-2671-0_70

sparsity. All the available natural signals can be converted into sparse if they are represented with a proper transform. The energy density can be varied after applying the transform to an image, which can produce better results. On reducing the energy density, the performance measures of the reconstructed image improves. In order to compress the image, apply proper transform to the image and also reduce the energy density in order to obtain a better reconstructed image. Then the compressed signal is sampled which is explained by compressive sensing [1, 2]. This method of compressing the image signal is compared with SPHIT algorithm to prove that the proposed algorithm is better than the SPIHT algorithm.

2 Cascaded Transform

To decompose a signal a proper transform is required, for one-dimensional signals DFT can be used and for two-dimensional signals DWT can used. Various wavelets are SYM2, COIF1, DB2, DB10, and DMDEY. When DFT and DWT are cascaded, then it becomes a perfect transform to decompose an image signal [3–6]. When various wavelets are cascaded with DFT, it was found that COIF1 cascaded with DFT provides better results. This paper utilizes this cascaded transform to decompose the given image signal to a number of coefficients. For an $N \times N$ image, the two-dimensional DFT is given by Eq. 1,

$$F(k,l) = \sum_{i=0}^{N-1} \sum_{j=0}^{N-1} f(i,j) e^{-i2\Pi\left(\frac{ki}{N} + \frac{lj}{N}\right)} \tag{1}$$

Coif wavelet can be mathematically represented as follows:

$$A_k = (-1)^k B_{N-1-k} \tag{2}$$

where, k is the coefficient index, A is the wavelet coefficient, and B is the coefficient of the scaling function.

3 Compressive Sensing and SPIHT

Compressive sensing is a recent algorithm which explains how to compress and sample a signal. It deals with two properties that are sparsity and incoherence. First, the given signal is converted into sparse and then sampled [7–9]. This technique samples even a high frequency signal at a low rate. The given signal which when represented with a proper transform gets decomposed into a number of coefficients, in which only the higher order coefficients are retained whereas all the lower order coefficients are made zero. These information bearing larger coefficients are then

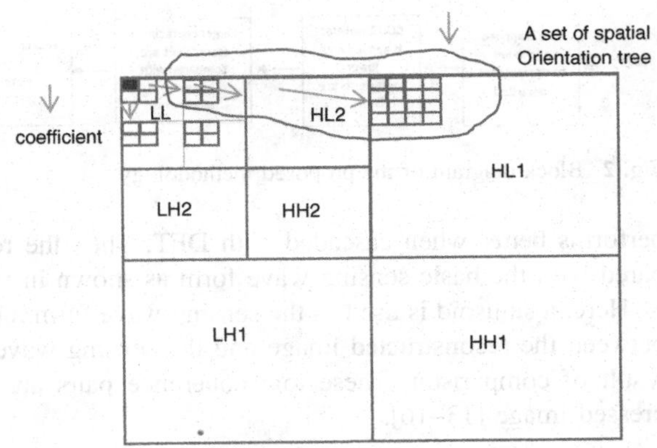

Fig. 1 Illustrates the four subbands of decomposition due to SPIHT

correlated with basic sensing waveforms to find only the low coherence pairs from them. These low coherence pairs when applied with the inverse transform, gives the compressed signal [10, 11].

Set partitioning in hierarchical trees algorithm (SPIHT) is a wavelet-based technique which is used to compress an image [12]. Initially, an image is decomposed using wavelet transform into four subbands. Out of these four subbands one will be a low frequency subband and the remaining three will be high frequency subbands. Likewise decomposition process will be repeated until a final scale is reached. SPIHT decomposes the image signal using the following equation.

$$A_i(T) = \{1, \max_{(m,n) \in T}\{|C_{m,n}|\} \geq 2^i\} \tag{3}$$

where, $A_i(T)$ is the importance of a set of coordinate T and $C_{m,n}$ is the coefficient value at each coordinate (m, n). Figure 1 shows the four subbands of image decomposition using wavelet transform.

4 Proposed Algorithm

An image is taken as input which is applied initially with a discrete Fourier transform so that the image gets decomposed into a number of coefficients. A threshold value is set based on the energy density of the given image, above the threshold value the coefficients are considered to be the larger coefficients and below which the coefficients are said to be lower coefficients. All the lower coefficients are discarded and only the larger coefficients which are said to have the information content of the image are taken and applied with inverse DFT. The reconstructed image is once again applied with second order wavelet transform and the process is repeated to pick only the larger coefficients. These larger coefficients are applied with inverse DWT. Out of the various wavelets, COIF1 wavelet

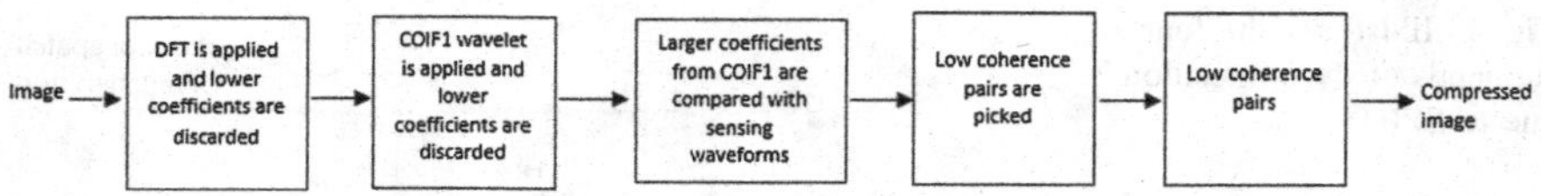

Fig. 2 Block diagram of the proposed methodology

performs better when cascaded with DFT. Then the reconstructed image is compared with the basic sensing wave form as shown in Fig. 2.

Here, a sinusoid is used as the sensing wave form. Only the low coherence pairs between the reconstructed image and the sensing wave form are picked up as the result of comparison. These low coherence pairs are integrated to form a compressed image [13–16].

5 PSNR and MSE

The quality of the reconstructed image is estimated using performance measures like PSNR and MSE. PSNR is defined as the ratio of maximum power of the signal to a noise corrupted signal that affects the reliability of the signal.

$$\mathrm{PSNR} = 10 * \log_{10}\left(\max_i^2/\mathrm{MSE}\right) \quad (4)$$

where, $\max_i$ represents the maximum pixel value of the image. Higher the PSNR value, better the quality of the reconstructed image.

MSE is a performance measure which is used to find the difference between the original and the reconstructed image in terms of squared error value.

$$\mathrm{MSE} = \frac{1}{mn}\sum_{i=o}^{m-1}\sum_{j=0}^{n-1}[I(i,j) - K(i,j)]^2 \quad (5)$$

where, I is the monochrome image and K is noisy approximation of the image.

6 Results and Discussions

In this paper, Lena image is taken as the reference and initially SPIHT algorithm is applied and compressed. In order to analyse the quality of the compressed image, PSNR and MSE values are calculated. The original Lena image and the compressed image are shown in Fig. 3. The PSNR and MSE values of the compressed Lena image are shown in Table 1. Then, the Lena image is compressed using cascaded image transform and compressive sensing [17–20]. First, the Lena image is applied with the cascaded transform DFT and COIF1 where only the larger coefficients are picked up

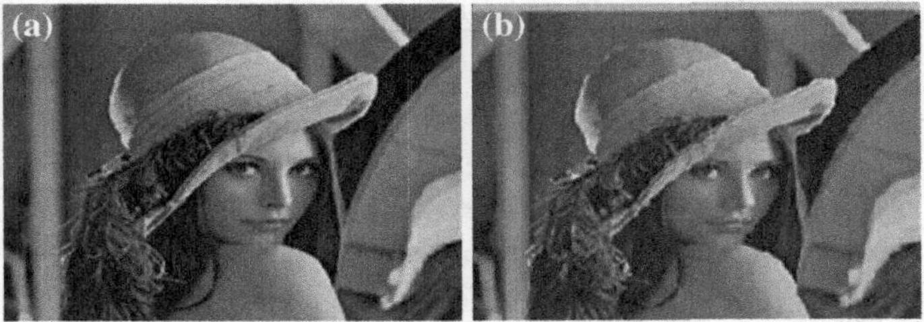

Fig. 3 **a** Original Lena image and **b** compressed Lena image using the SPIHT Algorithm

Table 1 PSNR and MSE values of the compressed image using SPIHT algorithm

PSNR	MSE
24.76	10.9826

and applied with the inverse transform. The reconstructed Lena image is compared with the basic sensing waveform; here, sinusoid is used as the sensing waveform. A sinusoid of 0.5 kHz frequency, amplitude of 2 V, and a sampling frequency of 6 kHz is generated and compared with the reconstructed Lena image. Figure 4 shows the generated sinusoid. Coherence estimation is done between the reconstructed image and the sinusoid using Welch estimator which is shown in Fig. 5. The PSNR and MSE values for the compressed Lena image are calculated, which is shown in Table 2. The PSNR value is found to be high for the proposed algorithm than the SPIHT algorithm. Also, the MSE value for the proposed algorithm is found to be less when compared with the SPIHT algorithm. Figure 6 shows the comparison of original image and the reconstructed image using proposed algorithm.

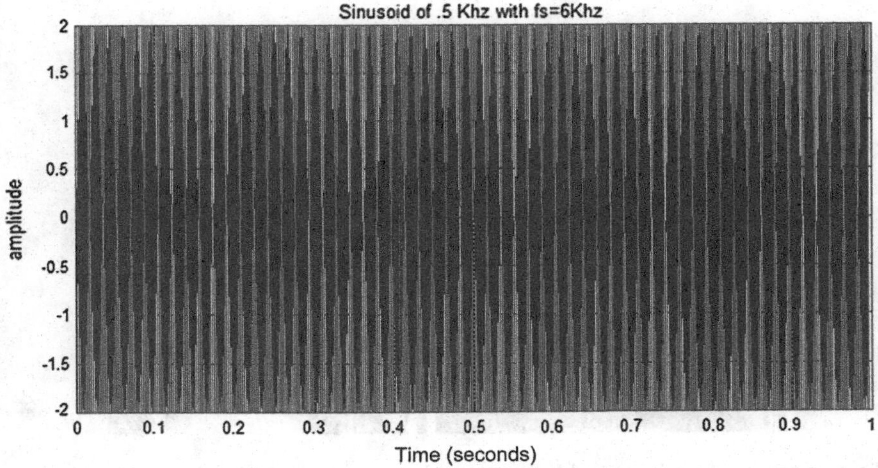

Fig. 4 Sinusoids used as the basic sensing waveform

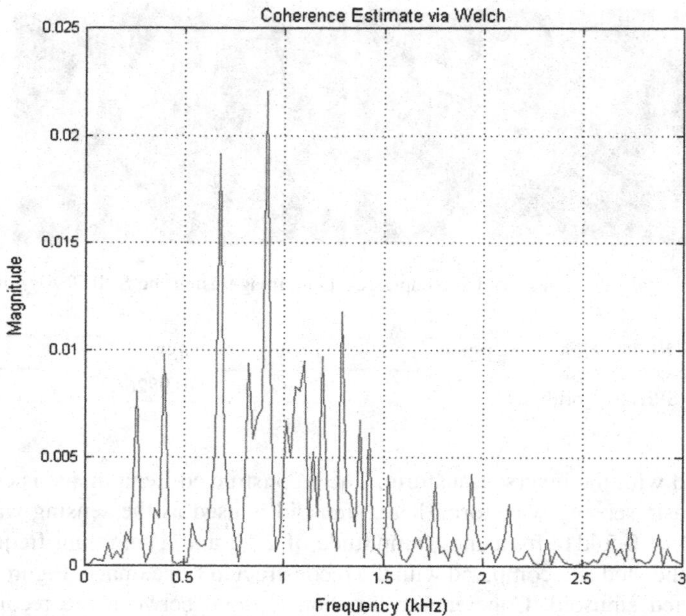

Fig. 5 Coherence estimation between the sensing wave sinusoid and the reconstructed image by applying the inverse COIF1

Table 2 PSNR and MSE values of the compressed image using the proposed algorithm

PSNR	MSE
263.7616	2.7563e−022

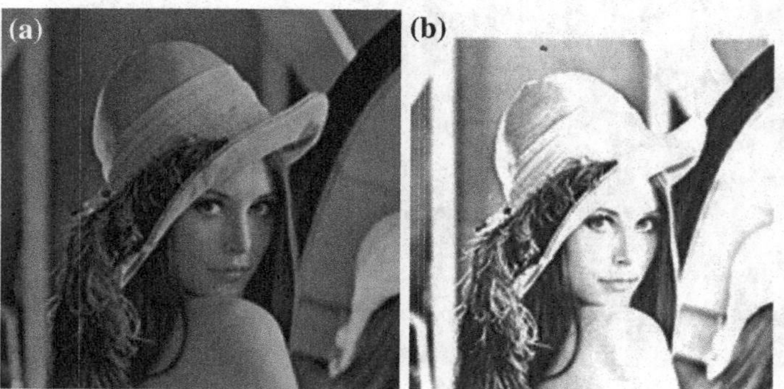

Fig. 6 **a** Original Lena image and **b** compressed Lena image using the proposed algorithm

7 Conclusion

In this paper, the Lena image is compared using SPHIT and the proposed algorithm using compressive sensing. With the performance measures obtained for the compressed Lena image from the two algorithms, the paper concludes that the proposed algorithm is better than the SPIHT algorithm.

References

1. Candèsand EJ, Wakin MB An introduction to compressive sensing. IEEE Sig Process Mag Digital Object Ident. doi:10.1109/MSP.2007.914731, March 2008
2. Baraniuk RG (2007) A lecture on compressive sensing. IEEE Signal Process Mag 24(4):118–121
3. Malioutov D, Cetin M, Willsky AS (2005) A sparse signal reconstruction perspective for source localization with sensor arrays. IEEE Trans Signal Process 53(8):3010–3022
4. Model D, Zibulevsky M (2006) Signal reconstruction in sensor arrays using sparse representations. Sig Process 86:624–638
5. Donoho DL, Stark PB (1989) Uncertainty principles and signal recovery. SIAM J Appl Math 49(3):906–931
6. Vikalo H, Parvaresh F, Hassibi B (2007) On recovery of sparse signals in compressed DNA microarrays. In: Proceedings of the asilomar confernce on signals, systems and computers (ACSSC 2007), Pacific Grove, CA, Nov 2007, pp 693–697
7. Marvasti F, Amini A, Haddadi F, Soltanolkotabi M (2012) A unified approach to sparse signal processing. In: EURASIP journal on advances in signal processing, 2012:44 (SPRINGER)
8. Tang G, Nehorai A (2011) Performance analysis of sparse recovery based on constrained minimal singular values. arXiv:1004.4222v2 [cs.IT] 24 Feb 2011
9. Marvasti F (2001) Nonuniform sampling: theory and practice. Springer, New York
10. Candès E, Romberg J, Tao T (2006) Robust uncertainty principles: exact signal reconstruction from highly incomplete frequency information. IEEE Trans Inform Theory 52(2):489–509
11. Candès E, Tao T (2006) Near optimal signal recovery from random projections: universal encoding strategies? IEEE Trans Inform Theory 52(12):5406–5425
12. Nirmalraj S (2015) SPIHT: a set partitioning in hierarchical trees algorithm for image compression. Contemp Eng Sci 8(6):263–270 HIKARI Ltd
13. Taubman DS, Marcellin MW (2001) JPEG 2000: image compression fundamentals, standards and practice. Kluwer, Norwell, MA
14. Donoho DL, Huo X (2001) Uncertainty principles and ideal atomic decomposition. IEEE Trans Inform Theory 47(7):2845–2862
15. Candès E, Tao T (2005) Decoding by linear programming. IEEE Trans Inform Theory 51(12):4203–4215
16. Tropp, Gilbert AC (2007) Signal recovery from partial information via orthogonal matching pursuit. IEEE Trans Inform Theory 53(12):4655–4666
17. Donoho D (2006) Compressed sensing. IEEE Trans Inform Theory 52(4):1289–1306
18. Nirmalraj S, Vigneswaran T (2014) Analysis of image transforms for sparsity evaluation in compressive sensing. Int J of Appl Eng Res 9(24):30309–30322 ISSN: 0973-4562
19. Nirmalraj S, Vigneswaran T, A novel method to compress voice signal using compressive sensing. In: International conference on circuit, power and computing technologies 2015, Noorul Islam University
20. Nirmalraj S, Vigneswaran T, A novel cascaded image transform by varying energy density to convert an image into sparse. In: Indian journal of science and technology, to be published

Portable Text to Speech Converter for the Visually Impaired

K. Ragavi, Priyanka Radja and S. Chithra

Abstract The portable text to speech converter is designed to help the visually impaired listen to an audio read-back of any scanned text. The system consists of a handheld page scanner, android phone to which the scanned image is sent over Bluetooth, an application to extract the text from the scanned image and to convert the extracted text to speech. The additional advantage of this system is that it employs a page scanner which scans the entire page containing the text. Therefore, the visually impaired need not take photos focusing on the region of text to be read, and then crop it to remove background pictures, which is the case in existing systems. The scanned image may contain text with background pictures which are simply ignored and only the text in the scanned image is extracted by the optical character recognition application. The text may also contain special characters and equations.

Keywords Optical character recognition · Text to speech · Text extraction · Tesseract

1 Introduction

Visually impaired people depend on a third person to read text since fewer than 10 % of the blind people can read Braille. Also, Braille books are hard to find and expensive. Therefore, the need arises for a cheaper device that will serve as the perfect solution and companion for the visually impaired. Optical character recognition (OCR) and text to speech (TTS) can be employed to aid the visually impaired. These two technologies are used to enable the visually impaired listen to

K. Ragavi (✉) · P. Radja · S. Chithra
SSN College of Engineering, Chennai, India
e-mail: ragavik@outlook.com

P. Radja
e-mail: radja.priyanka@gmail.com

S. Chithra
e-mail: chithras@ssn.edu.in

L.P. Suresh and B.K. Panigrahi (eds.), *Proceedings of the International Conference on Soft Computing Systems*, Advances in Intelligent Systems and Computing 397, DOI 10.1007/978-81-322-2671-0_71

an audio read-back of textual information from books, newspapers, or any documents. The first step consists of recognizing the text to be read. The second step involves converting the recognized text to speech.

Tesseract is an open-source OCR engine by HP. It was first developed between 1984 and 1994. Processing is done step by step. The input (color or gray image) to the Tesseract engine is assumed as a binary image by adaptive thresholding. On connected component analysis, outlines of component are stored. Blobs, which are formed by gathering the nested outlines together, are organized into text lines. The text lines which are analyzed for fixed pitch and proportional text are broken into words by analyzing the character spacing. Two passes are done to recognize the words accurately.

Text to speech (TTS) is a speech synthesis application that converts text into spoken output. Text to Speech API is offered by the Android Operating System for mobile application development. However, Apple offered the Text to Speech API only in the recent release of its Operating System, iOS7. The Text to Speech Synthesizer creates a completely synthetic voice output using a model of the vocal tract and human voice characteristics. The input text is detected and analyzed; following which normalization and linearization is done. The output speech is obtained on phonetic analysis and acoustic processing.

The organization of this paper is as follows. In Sect. 2, a detailed literature survey regarding the existing systems in this domain has been presented, followed by Sect. 3 that describes the system architecture, followed by Sect. 4 that tells about the proposed system further followed by Sect. 5 that depicts the implementation and results. Finally, Sect. 6 concludes the work.

2 Literature Survey

Mithe [1] Smart mobile phones of android platform with an application to perform image to speech conversion are used in various fields like banking, legal industry, office automation, etc. to help people read any type of text document. This application requires the use of android phones with higher quality camera.

Gaudissart [2] SYPOLE project is a text reading assistant for the visually impaired. It uses a personal digital assistant (PDA) and innovative algorithms. Text detection, optical character recognition, and speech synthesis are also used.

Pazio [3] The algorithm recognizes text embedded in natural scene images to help the visually impaired. Text-like image regions are localized and color image segmentation is used followed by segment shape analysis. This preprocessing makes OCR work more efficiently.

On the whole, the existing systems are capable of reading aloud text from documents using android phone with high quality camera, PDA with camera and character recognition in natural scene images. The Portable TTS Converter is capable of optical character recognition and speech synthesis of images from a handheld scanner.

3 System Architecture

The portable text to speech converter is based on the architecture shown in Fig. 1. The system architecture consists of a page scanner, a Bluetooth module and an android phone.

The page scanner sends the scanned image file to the android phone with the help of either the in-built Bluetooth module or in the absence of one, the external Bluetooth module. The android phone on reception of the scanned image file uses the OCR application to extract the text from the image file using the Tesseract OCR library in. Once the text has been extracted, the text can be converted to speech using the existing text to speech library present in Java.

3.1 Page Scanner

The page scanner or the document scanner is a handheld A4 size scanner which can scan documents, books, or sheets of paper containing pictures, graphs, or plain text. The scanner must be kept horizontal to the document to be scanned. Then, the scanner can scan these documents or books when it is slid over their surface from top to bottom. The scanned file is either an image file or a pdf. The scanner used in this project is shown in Fig. 2.

The in-built Bluetooth module or the external Bluetooth module interfaced with the scanner can then be employed to transfer the file to the android phone. The scanner specification is given in Table 1.

3.2 Bluetooth Module

This external Bluetooth module is used when the scanner does not have any in-built Bluetooth settings to transfer the scanned image file or pdf file to the android phone.

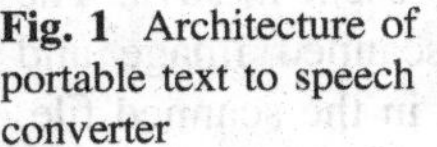

Fig. 1 Architecture of portable text to speech converter

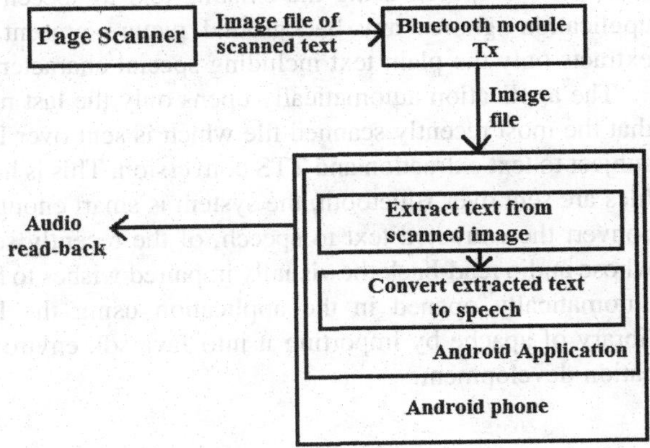

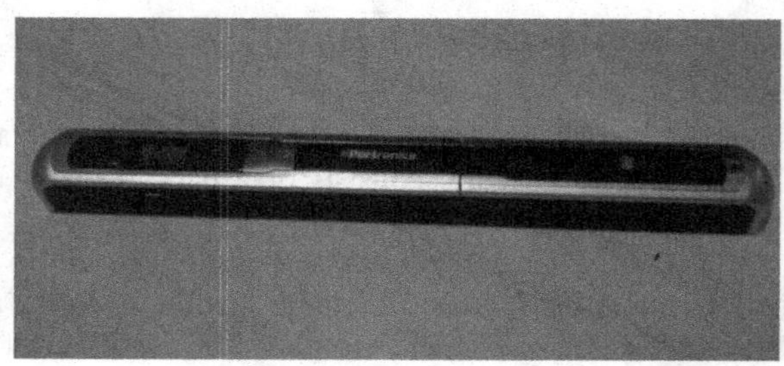

Fig. 2 Page scanner

Table 1 Page scanner specification

Model name	Portronics scanny 3 bluetooth
Image sensor	Color contact image sensor
Number of sensor	5136 dots
Resolution	Low resolution: 300 × 300 dp
	High resolution: 600 × 600 dpi
Scan width	Approx. 8.5″
Scan length	300 dpi: 53″(Max), 600 dpi: 26″(Max)
File format	JPEG

The input to this module is the scanned image file/pdf file from the scanner which is then transferred as such to the android phone connected to it.

The scanner moreover has a unique key used for pairing it with the android phone. The key is used to authenticate and verify the Bluetooth connection with the android device and to send the files to the correct android phone.

3.3 The Android Application

An android application is developed which uses the Tesseract OCR library, the TessBaseAPI which is built separately and imported to the Java sdk environment to extract the text from the image file sent to the android phone over Bluetooth by the scanner or to convert the pdf file to doc format so that the extracted text can be converted to speech using the existing text to speech library present in Java. The application ignores any background picture present in the scanned image and extracts only the plain text including special characters present in the scanned file.

The application automatically opens only the last modified file for extraction so that the most recently scanned file which is sent over Bluetooth is the one which is subject to text extraction and TTS conversion. This is helpful because when multiple files are sent over Bluetooth, the system is smart enough to extract text from and to convert the extracted text to speech, of the recently scanned file, which is the file whose audio read-back the visually impaired wishes to hear. This last modified file is automatically opened in the application using the LastModifiedFileComparator library of apache by importing it into Java sdk environment for the android application development.

4 System Model

In this system, the handheld scanner is used to scan the document. The visually impaired can scan the entire document and need not focus the region of text to scan. Once the scan is complete, the scanned image is transferred to the android phone paired with the scanner over Bluetooth. The android phone must be within the Bluetooth range of the scanner. As soon as the scanned image is received in the android phone, the OCR application developed automatically opens the scanned image received and extracts the text from the image. Furthermore, the application converts the extracted text to speech which is conveyed to the visually impaired. This process is depicted with a flow diagram in Fig. 3.

Algorithm

Step 1 Scan the document using the page scanner.

Step 2 Transfer the scanned image file/pdf file to the paired android phone using the in-built Bluetooth module or the external Bluetooth module.

Step 3 On the reception of the scanned file on the android phone, automatically open the last modified or the most recently received scanned file in the application using LastModifiedFileComparator library of apache which is imported into the Java sdk environment.

Step 4 Extract the text from scanned file in the application making use of Tesseract (TessBaseAPI) library by building it separately and importing the built library files into Java sdk environment.

Step 5 Convert the extracted text to speech using Text to speech library present in Java.

Step 6 Play the speech form of the scanned text to the visually impaired.

The application to perform OCR and speech synthesis is developed and is loaded into the android phone which is paired with the scanner. When a scanned file is sent over Bluetooth to the android phone with the application preloaded, the optical character recognition, and speech synthesis of the scanned image file takes place. The scanner and android application are shown in Fig. 4.

5 Implementation and Result

The portable text to speech converter was implemented and screenshot of the Android application is shown in Fig. 5. On clicking the button, character recognition is performed and a sample of the scanned image is displayed along with the extracted text at the bottom. Also, speech synthesis is performed immediately after extracting the text without any intimation from the user.

Fig. 3 Flow diagram of the system

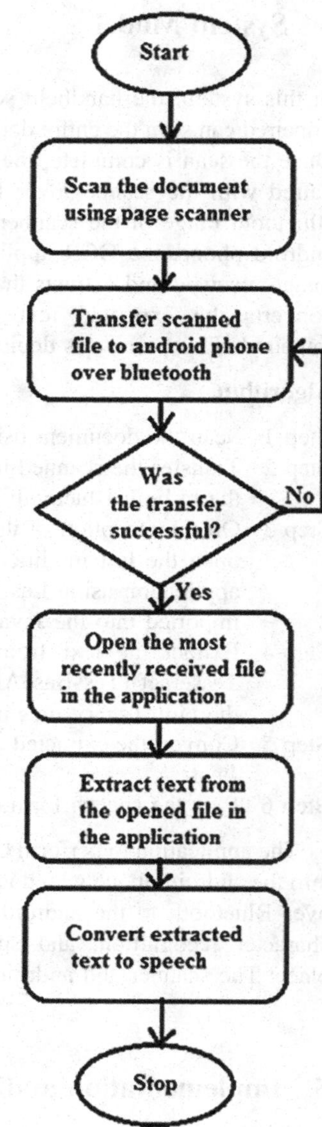

Advantages of this system over the existing systems which use mobile phones with high resolution cameras can be observed from Table 2. Size of the scanned image is lesser than the image from camera. This reduces the processing speed, text detection is done faster and speech output is obtained quicker. This system is about 92.3 % more accurate than the existing system.

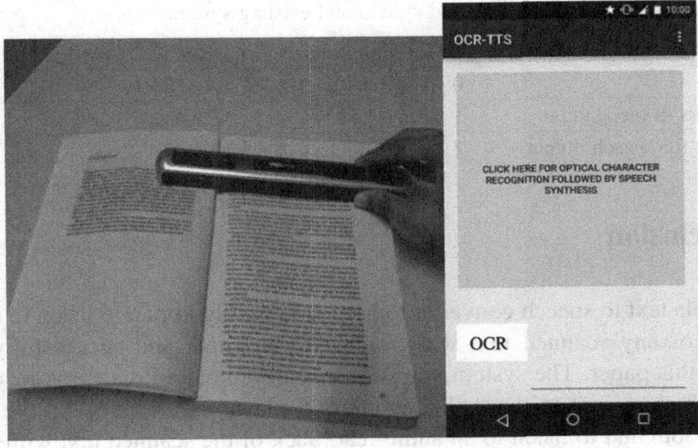

Fig. 4 System model of portable text to speech converter

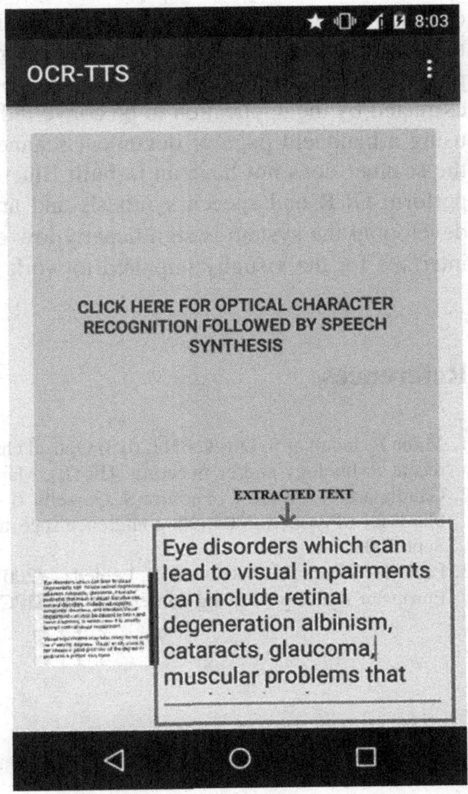

Fig. 5 Screenshot of the android application

Table 2 Comparison between proposed system and existing system

Sample image (with 90 words)	Taken with scanner	Taken with 5 MP phone camera
Size	0.45 MB	1.4 MB
No. of errors in recognition	1 word	13 words
Time delay for speech output	7 s	12 s

6 Conclusion

The portable text to speech converter to help the visually impaired listen to an audio read-back of any scanned text was proposed, designed, and successfully implemented in this paper. The system helps the visually impaired scan any document, a copy of which is sent to the authentic android phone connected with the scanner over Bluetooth and to listen to an audio read-back of the scanned text with the help of an android application. The authenticity of the android device is checked with the help of a unique key used for pairing over Bluetooth. The major advantage of this system is that it employs a page or document scanner which scans the entire page containing the text, and therefore, the visually impaired need not focus the region of text to be scanned. Moreover, the scanned image may contain text with background pictures which are simply ignored and only the text in the scanned image is extracted by the application to be converted to speech. This project is implemented using a handheld page or document scanner, an external Bluetooth module when the scanner does not have an in-built Bluetooth module, an android application to perform OCR and speech synthesis and an android phone. The cost involved in developing the system is significantly low and the system provides a friendly user interface for the visually impaired to work with.

References

1. Mithe R, Indalkar S, Divekar N (2013) Optical character recognition. In: International journal of recent technology and engineering (IJRTE), March 2013
2. Gaudissart V, Thillou C, Ferreira S, Gosselin B (2004) SYPOLE: mobile reading assistant for blind. In: SPECOM'2004: 9th conference speech and computer, St. Petersburg, Russia, 20–22 Sept 2004
3. Pazio M, Wiecki MN, Kowalik R, Lebiedz J (2007) Text detection system for the blind. In: 15th European signal processing conference (EUSIPCO 2007), Poznan, Poland, 3–7 Sept 2007

Outage Analysis of PLNC-Based Multicast Half Duplex Cognitive Radio Relay System in Nakagami-*m* Fading Channels

S. Anjanaa, Gayathri, Velmurugan and Thiruvengadam

Abstract In this paper, the concept of physical layer network coding (PLNC) is applied in decode and forward (DF) protocol-based cognitive radio relay (CRR) system with two source nodes and two destination nodes to operate in multicast environment. DF relay node enables the two source nodes to send data to both the destination in two time slots compared to four time slots in conventional system. The performance of the proposed PLNC-based CRR system is analyzed in terms of outage probability in Nakagami-*m* fading channel environment for different configurations. The performance of the proposed system is compared with non-PLNC-based CRR system.

Keywords Physical layer network coding (PLNC) · Decode and forward (DF) protocol · Cognitive radio relay (CRR) · Nakagami-m fading channel

1 Introduction

Cognitive radio (CR) has emerged as a promising solution to poor spectrum efficiency problem by fully exploiting the underutilized spectrum where in unlicensed (secondary) users are allowed to opportunistically access the un-used licensed spectrum without interfering with primary users [1]. As CR system deals with problem of spectrum scarcity and spectrum underutilization efficiently, it can play a vital role in high data rate wireless communications [2]. Two basic modes of operation of CR system are spectrum overlay and spectrum underlay. In spectrum overlay mode, the secondary user uses a channel from a primary user only when it is not occupied. The secondary user senses the frequency spectrum and detects the

S. Anjanaa (✉) · Gayathri · Velmurugan · Thiruvengadam
Department of ECE, Thiagarajar College of Engineering, Madurai, India
e-mail: anjanasri14@gmail.com

L.P. Suresh and B.K. Panigrahi (eds.), *Proceedings of the International Conference on Soft Computing Systems*, Advances in Intelligent Systems and Computing 397, DOI 10.1007/978-81-322-2671-0_72

presence of primary user then the spectrum band will not be used. In spectrum underlay mode, signals with a very low spectral power density can coexist as a secondary with the primary users of the frequency band [3]. The major challenges of CR system are spectrum sensing and opportunity exploitation. This paper focuses on opportunity exploiting after spectrum holes are identified to maximize the data rate of secondary communication without interfering with primary transmission. Cooperative relays are used to improve the throughput of wireless communication [4]. In bidirectional relay networks, two source nodes exchange their information with a help of single-relay node in two time slots using PLNC concept [5]. However, the concept of PLNC has not been used for multicast system. In this paper, multicast CRR system is proposed in which two secondary source node transmit their information to two secondary destination with the help of PLNC concept at relay node. Outage probability is one of the major performance metrics for analyzing the performance of the wireless systems in fading environments. In [6], outage probability for CRR system in underlay spectrum sharing method is derived. System model consists of a source, relays, and a destination. Out of relays, best relay is used for transmission. Based on the dependence of null decoding set, outage probability is derived.

In [7], a three node CRR system consists of multiple relays is proposed and the exact outage performance of the CR networks with maximum power transmit limit for secondary nodes is derived. The outage probability and diversity-multiplexing tradeoff for a three node CR relay networks with incremental DF and AF protocols over Nakagami-*m* fading channels are investigated in [8]. The end-to-end performance for single-relay employing DF protocol is analyzed in [9]. Outage expressions are investigated in two-hop amplify and forward systems in the presence of Rayleigh faded multiple interferers [10]. However, in all the previous works, outage analysis is carried out for three node half duplex (HD) CRR systems. Moreover, from the literature survey, it is observed that most of the outage analysis has been carried out in Rayleigh fading environment for various configurations of CRR systems. But, Nakagami-*m* fading distribution is the most general fading distribution in which parameters are adjusted to fit a variety of empirical measurements [11].

The major contribution of this paper is to analyze the outage performance of the proposed PLNC-based multicast CRR system in Nakagami-*m* fading channel environment. End-to-end outage probability expression for the proposed PLNC-based CRR system is derived for underlay and overlay spectrum sharing schemes.

The rest of the paper is organized as follows: Section 2 describes the system model for two source and destination nodes. In Section 3, analytical expressions of outage probability are derived for the proposed PLNC-based half duplex CRR system and non-PLNC system in both spectrum sharing schemes. In Section 4, numerical result in terms of outage and BER for the proposed PLNC-based CRR system is presented. Section 5 concludes the paper.

2 System Model

Consider a multicast HD-CRR system shown in Fig. 1. It consists of two secondary source nodes S_1 and S_2, one DF relay node R, and two secondary destination nodes D_1 and D_2. Both secondary source nodes have to transmit $h_{s2}^{d_2}$ symbols to both secondary destination nodes, while there is no direct link between S_1 (or S_2) and D_2 (or D_1) due to path loss and large-scale fading. PLNC concept is employed at the relay node R, which enables both the secondary nodes transmit their information to two secondary destination nodes.

Time Slot I

In time slot I, the receive data symbol at half duplex DF relay node R is given by

$$y_R = \sqrt{P_{s_1}} h_{s_1}^r x_1 + \sqrt{P_{s_2}} h_{s_2}^r x_2 + \sqrt{P_{p_t}} h_{p_t}^r x_p + w_r. \quad (1)$$

P_{s_1} and P_{s_2} are the transmit powers of source nodes S_1 and S_2, respectively. $h_{s_j}^r = \left|h_{s_j}^r\right| e^{i\theta_j}$ are the fading channel coefficients between source node S_j, j = 1, 2 and relay node R. The magnitude of fading channel coefficients $\left|h_{s_j}^r\right|$ are modeled as Nakagami-m distributed random variables with shape parameters $m_{s_j r}, j = 1, 2$ and scale parameters $\Omega_{s_j r}$, j = 1, 2.

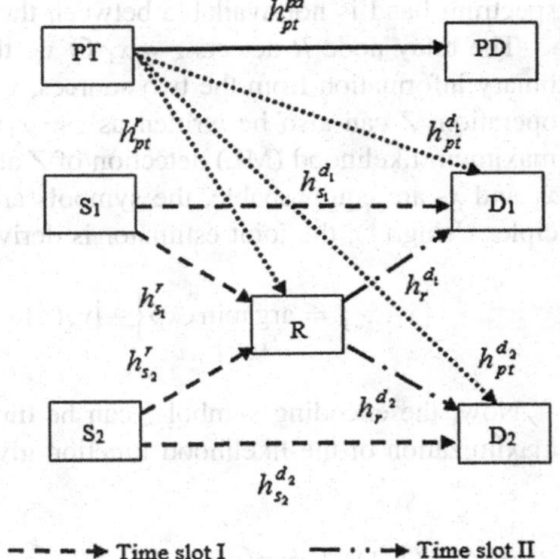

Fig. 1 PLNC-based half duplex cognitive radio relay system

Probability density function (PDF) of $\left|h_{s_j}^r\right|$ is

$$P_X(x) = \frac{2}{\Gamma(m_{s_jr})}\left(\frac{m_{s_jr}}{\Omega_{s_jr}}\right)^{m_{s_jr}} x^{2m_{s_jr}-1}\exp\left(-\frac{m_{s_jr}}{\Omega_{s_jr}}x^2\right), x \geq 0, j = 1, 2 \text{ and } m_{s_jr} \geq 0.5\,. \tag{2}$$

where $\Gamma(\cdot)$ is the gamma function. The shape parameter m_{s_jr} signifies the fading severity and smaller values of m_{s_jr} represent more fading in the channel. θ is assumed to be uniformly distributed over the interval (0, 2π). X_1 and x_2 are the binary phase-shift keying (BPSK) modulated data symbols of secondary source nodes S_1 and S_2, respectively, satisfying $E[|x_1|^2] = 1$ and $E[|x_2|^2] = 1$. P_{p_t} is the transmit power of primary node, and the magnitude of $h_{p_t r}$ is modeled as Rayleigh fading channel coefficient between primary node P_{Tx} and relay node R with shape parameter $m_{p_t r}$ and scale parameter $\Omega_{p_t r}$ w_r represents the complex additive white Gaussian noise (AWGN) with zero mean and unit variance at relay node R. The interference from the primary to relay node is considered to be Gaussian with mean zero and unit variance. Hence the received signal at relay is also considered to be Gaussian.

The symbols x_1 and x_2 from the secondary source nodes S_1 and S_2, respectively, are jointly received by the relay node R. The symbols x_1 and x_2 are considered to be interfering with each other. The interfering signals are jointly decoded at the relay node R. Based on the decoded information of the interfering signals, the relay node R sends a coded signal to the destinations. This concept is referred to as PLNC. This enables the data transmission between S_1 and D_2, S_2 and D_1, though a common spectrum band is not available between them.

The relay node R detects $z = x_1 \oplus x_2$, the XORed version of the two received binary information from the two sources, where the symbol $\oplus$ is the bitwise XOR operation. Z can also be written as $z = x_1x_2$ for BPSK symbol. Let $\tilde{z}$ denote the maximum likelihood (ML) detection of Z at relay node R. When the input symbols x_1 and x_2 are equiprobable, the symbols are jointly estimated using the ML principle. Using (1), the joint estimator is derived as

$$\{\hat{x}_1, \hat{x}_2\} = \underset{\{x_1, x_2\}}{\arg\min}\exp\left\{-\left|y_R(n) - \sqrt{P_{s_1}}h_{s_1}^r x_1 - \sqrt{P_{s_2}}h_{s_2}^r x_2\right|^2\right\}. \tag{3}$$

Now, the encoding symbol $\tilde{z}$ can be directly estimated by modifying (3). The maximization of the likelihood function given by [8],

$$\exp(-L(1,1)) + \exp(-L(-1,-1)) \underset{\tilde{z}=-1}{\overset{\tilde{z}=1}{\gtrless}} \exp(-L(1,-1)) + \exp(-L(-1,1))\,. \tag{4}$$

where $L(x_1, x_2) = |y_R - \sqrt{P_{s_1}} h_{s_1}^r x_1 - \sqrt{P_{s_2}} h_{s_2}^r x_2|^2$. In time slot I, the receive data symbols y_{D1_1} and y_{D2_1} at secondary destination nodes D_1 and D_2, respectively, are given by

$$y_{D_1\_1} = \sqrt{P_{s_1}} h_{s_1}^{d_1} x_1 + \sqrt{P_{p_t}} h_{pt}^{d_1} x_p + w_{d_1\_1}. \tag{5}$$

$$y_{D_2\_1} = \sqrt{P_{s_2}} h_{s_2}^{d_2} x_2 + \sqrt{P_{pt}} h_{p_t}^{d_2} x_p + w_{d_2\_1}. \tag{6}$$

$h_{s_1}^{d_1}$ and $h_{s_2}^{d_2}$ are the fading channel coefficients between source nodes S_j, $j = 1, 2$ and destination nodes D_j, $j = 1, 2$. The magnitude of $h_{s_1}^{d_1}$ and $h_{s_2}^{d_2}$ are modeled as Nakagami-m distributed with parameters $m_{s_jd_j}, j = 1, 2$ and $\Omega_{s_jd_j}, j = 1, 2$, $w_{d_1\_1}$ and $w_{d_2\_1}$ are AWGN with zero mean and unit variance at destination nodes D_1 and D_2, respectively. $h_{p_t}^{d_j}, j = 1, 2$ are fading channel coefficients between primary nodes P_{Tx} to destination nodes D_j, $j = 1, 2$. The magnitude of $h_{p_t}^{d_j}$, $j = 1, 2$ are modeled as Nakagami-m fading channel coefficients with shape parameter $m_{p_td_j}$ and scale parameter $\Omega_{p_td_j}$. It is assumed that CSI is known at the secondary destination nodes D_j, $j = 1, 2$. Applying ML criterion, symbols x_1 and x_2 are detected from the received symbols $y_{D_1\_1}$ and $y_{D_2\_1}$, respectively. They are expressed as,

$$\tilde{x}_{1\_1} = \underset{x_1 \in \{1,-1\}}{\arg\min} \left|y_{D_1\_1} - \sqrt{P_{s_1}} h_{s_1}^{d_1} x_1\right|^2. \tag{7}$$

$$\tilde{x}_{2\_1} = \underset{x_2 \in \{1,-1\}}{\arg\min} \left|y_{D_2\_1} - \sqrt{P_{s_2}} h_{s_2}^{d_2} x_2\right|^2. \tag{8}$$

Time Slot II

In time slot II, the source nodes S_1 and S_2 remain silent; the relay node R forwards the detected symbol $\tilde{z}$ to the destination nodes D_1 and D_2. The receive data symbol at destination nodes D_1 and D_2 are given by,

$$\tilde{y}_{D_1\_2} = \sqrt{P_r} h_r^{d_1} \tilde{z} + \sqrt{P_{p_t}} h_{p_t}^{d_1} x_p + w_{d_1\_2}. \tag{9}$$

$$\tilde{y}_{D_2\_2} = \sqrt{P_r} h_r^{d_2} \tilde{z} + \sqrt{P_{p_t}} h_{p_t}^{d_2} x_p + w_{d_2\_2}. \tag{10}$$

P_r is the transmit power at the relay node R. $h_r^{d_1}$ and $h_r^{d_2}$ are the fading channel coefficients between the relay node R and destination nodes D_1 and D_2, respectively. The magnitude of $h_r^{d_1}$ and $h_r^{d_2}$ are also modeled as Nakagami-m fading channels with parameters m_{rd_j} and $\Omega_{rd_j}, j = 1, 2$ $w_{d_1\_2}$, and $w_{d_2\_2}$ are the AWGN with zero mean and unit variance at D_1 and D_2. The symbols at D_1 and D_2 are detected using the ML criterion and they are given by,

$$\tilde{x}_{1\_2} = \arg \min_{\tilde{z} \in \{1,-1\}} \left| \tilde{y}_{d_1} - \sqrt{P_r} h_r^{d_1} \tilde{z} \right|^2 . \tag{11}$$

$$\tilde{x}_{2\_2} = \arg \min_{\tilde{z} \in \{1,-1\}} \left| \tilde{y}_{d_2} - \sqrt{P_r} h_r^{d_2} \tilde{z} \right|^2 . \tag{12}$$

Now, the symbols at x_1 and x_2 are detected at destination nodes D_2 and D_1 using the expressions

$$\hat{x}_{D_2} = \tilde{x}_{2\_1} \tilde{x}_{2\_2} . \tag{13}$$

$$\hat{x}_{D_1} = \tilde{x}_{1\_1} \tilde{x}_{1\_2} . \tag{14}$$

3 Outage Analysis

In this section, analytical expressions are derived for outage probability of the proposed PLNC-based CRR system and non-PLNC-based CRR system with two source nodes and two destination nodes in Nakagami-*m* fading environment.

3.1 Outage Analysis of PLNC-Based CRR System

In the first time slot, both source nodes S_1 and S_2 send data to the relay node R, and destination nodes D_1 and D_2. Hence, outage probability of the PLNC-based CRR system for the data rate of R_d b/s/Hz is defined as

$$P_\mathrm{out}^\mathrm{I}(R_\mathrm{d}) = \left[\Pr\left(\log_2\left(1 + \min\left(\gamma_{s_1 r}, \gamma_{s_2 r}, \gamma_{s_1 d_1}, \gamma_{s_1 d_2}\right)\right)\right) < R_d \right] . \tag{15}$$

where $\gamma_{s_1 r}$ and $\gamma_{s_2 r}$ are the SINR at relay node R. The signal from source nodes S_1 and S_2 are the desired signals and $\gamma_{s_1 r}$, $\gamma_{s_2 r}$ are the SINR at destination nodes D_1 and D_2. The threshold SINR for the data rate R_d b/s/Hz is $\gamma_\mathrm{th} = 2^{R_\mathrm{d}} - 1$. The CDF of $\gamma_{s_j r}$, $j = 1, 2$ and $\gamma_{s_j d_j}$, $j = 1, 2$ are calculated as

$$F_{\gamma s_j r} = \Pr\left[\frac{\mathrm{SNR}_{s_j}^r \left| h_{s_j}^r \right|^2}{\mathrm{INR} \left| h_{p_t}^r \right|^2 + 1} < \gamma_\mathrm{th} \right] ; j = 1, 2 \tag{16}$$

$$F_{\gamma s_j d_j} = \Pr\left[\frac{\mathrm{SNR}_{s_j}^{d_j}\left|h_{s_j}^{d_j}\right|^2}{\mathrm{INR}\left|h_{p_t}^{d_j}\right|^2 + 1} < \gamma_{\mathrm{th}}\right] \; ; j = 1,\, 2 \tag{17}$$

where $\mathrm{SNR}_{s_j}^{r}$ and $\mathrm{SNR}_{s_j}^{d_j}$ are the SNR at relay node R and destination node D_j, respectively.

Assume INR is very high, (16) and (17) become,

$$F_{\gamma s_j r} = \Pr\left[\frac{\mathrm{SNR}_{s_j}^{r}\left|h_{s_j}^{r}\right|^2}{\mathrm{INR}\left|h_{p_t}^{r}\right|^2} < \gamma_{\mathrm{th}}\right]; j = 1,2 \tag{18}$$

$$F_{\gamma s_j d_j} = \Pr\left[\frac{\mathrm{SNR}_{s_j}^{d_j}\left|h_{s_j}^{d_j}\right|^2}{\mathrm{INR}\left|h_{pt}^{d_j}\right|^2} < \gamma_{\mathrm{th}}\right] \; ; j = 1,2 \tag{19}$$

The random variables $z_j = \left|h_{s_j}^{r}\right|^2$ and $y = |h_{pt}^{r}|^2$ are exponentially distributed with variance $\beta_{s_j r}$ and $\beta_{p_j r}^2$, respectively.

The PDF of Nakagami-m fading channel is defined as

$$P_{z_j}(z) = \frac{z^{m-1} e^{\frac{-z}{\beta_{s_j r}^{m}}}}{\Gamma(m)\beta_{s_j r}^{m}}. \tag{20}$$

The PDF of Rayleigh fading channel is written as

$$f_y(y) = \frac{1}{\beta_{ptr}^2}\exp\left(\frac{-y}{\beta_{ptr}^2}\right). \tag{21}$$

Using the PDFs of z_j and y, the CDF of $\gamma_{s_j r}$ is written as

$$F_{\gamma_{s_j r}}\left(\gamma_{s_j r}\right) = \int_0^{\infty}\int_0^{\varepsilon} P_{z_j}(z) f_y(y)\,\mathrm{d}z\mathrm{d}y. \tag{22}$$

Let $\varepsilon = \frac{y\gamma_{s_j r}\mathrm{INR}}{\mathrm{SNR}_{s_j}^{r}}$. By integrating with respect to the PDF of z_j and y, the CDF of $\gamma_{s_j r}$ is calculated as

$$F_{\gamma_{s_jr}}\left(\gamma_{s_jr}\right) = \frac{1}{\Gamma(m)\beta_{ptr}^2}\left[\frac{\left(\frac{1}{\beta_{s_jr}}\frac{Y\gamma_{s_jr}\mathrm{INR}}{\mathrm{SNR}_{s_j}^{r}}\right)^m \Gamma(1+m)}{m\left(\frac{1}{\beta_{s_jr}}\frac{Y\gamma_{s_jr}\mathrm{INR}}{\mathrm{SNR}_{s_j}^{r}} + \frac{1}{\beta_{ptr}^2}\right)^{1+m}}\right] \times {}_2F_1\left(1, 1+m; 1+m; \frac{\frac{1}{\beta_{s_jr}}\frac{Y\gamma_{s_jr}\mathrm{INR}}{\mathrm{SNR}_{s_j}^{r}}}{\left(\frac{1}{\beta_{s_jr}}\frac{Y\gamma_{s_jr}\mathrm{INR}}{\mathrm{SNR}_{s_j}^{r}} + \frac{1}{\beta_{ptr}^2}\right)}\right). \quad (23)$$

where ${}_2F_1(a, b, c; z)$ is the confluent hypergeometric function of 4 variables which is a solution to a second-order linear ordinary differential equation (ODE)

Similarly, the CDF of $\gamma_{s_jd_j}$ is determined as

$$F_{\gamma_{s_jd_j}}\left(\gamma_{s_jd_j}\right) = \int_0^{\infty}\int_0^{\mu} \frac{z^{m-1}e^{-z/\beta_{s_jd_j}}}{\Gamma(m)\beta_{s_jd_j}^m}\frac{1}{\beta_{ptd_j}^2}\exp\left(\frac{-y}{\beta_{ptd_j}^2}\right)\mathrm{d}z_j\mathrm{d}y. \quad (24)$$

Let $\mu = \frac{Y\gamma_{s_jd_j}\mathrm{INR}}{\mathrm{SNR}_{s_j}^{d_j}}$. CDF of $\gamma_{s_jd_j}$ is determined as

$$F_{\gamma_{s_jd_j}}\left(\gamma_{s_jd_j}\right) = \frac{1}{\Gamma(m)\beta_{ptd_j}^2}\frac{\left(\frac{1}{\beta_{s_jd_j}}\frac{Y\gamma_{s_jd_j}\mathrm{INR}}{\mathrm{SNR}_{s_j}^{d_j}}\right)^m \Gamma(1+m)}{m\left(\frac{1}{\beta_{s_jd_j}}\frac{Y\gamma_{s_jd_j}\mathrm{INR}}{\mathrm{SNR}_{s_j}^{d_j}} + \frac{1}{\beta_{ptd_j}^2}\right)^{1+m}} \times {}_2F_1\left(1, 1+m; 1+m; \frac{\frac{1}{\beta_{s_jd_j}}\frac{Y\gamma_{s_jd_j}\mathrm{INR}}{\mathrm{SNR}_{s_j}^{d_j}}}{\left(\frac{1}{\beta_{s_jd_j}}\frac{Y\gamma_{s_jd_j}\mathrm{INR}}{\mathrm{SNR}_{s_j}^{d_j}} + \frac{1}{\beta_{ptd_j}^2}\right)}\right). \quad (25)$$

CDF of first time slot transmission is given by

$$F_{\gamma^{(1)}}(\gamma_{\mathrm{th}}) = 1 - \prod_{j=1}^{2}\left(1 - F_{\gamma_{s_jr}}\left(\gamma_{s_jr}\right)\right)\left(1 - F_{\gamma_{s_jd_j}}\left(\gamma_{s_jd_j}\right)\right). \quad (26)$$

In the second time slot, relay R sends data to destination nodes D_1 and D_2. Hence, outage probability of the PLNC-based CRR system for the data rate of R_{d} b/s/Hz is defined as

$$P_{\text{out}}^{\text{II}} = \left[\Pr\left(\log_2\left(1+\min\left(\gamma_{rd_1},\gamma_{rd_2}\right)\right) < R_{\text{d}}\right)\right]. \tag{27}$$

$\gamma_{rd_j}, j = 1,2$ is the SINR at destination node D_j. The CDF of γ_{rd_j} is defined as

$$F_{\gamma_{\text{rd}_j}} = \Pr\left[\frac{\text{SNR}_r^{d_j}\left|h_r^{d_j}\right|^2}{\text{INR}\left|h_{p_t}^{d_j}\right|^2+1} < \gamma_{\text{th}}\right]\ ; j = 1,2 \tag{28}$$

Assume INR is very high, (28) becomes,

$$F_{\gamma_{rd_j}} = \Pr\left[\frac{\text{SNR}_r^{d_j}\left|h_r^{d_j}\right|^2}{\text{INR}\left|h_{p_t}^{d_j}\right|^2} < \gamma_{\text{th}}\right]\ ; j = 1,2 \tag{29}$$

Using the PDFs of z_j and y, the CDF of γ_{rd_j} is written as

$$F_{\gamma_{rd_j}}\left(\gamma_{rd_j}\right) = \int_0^\infty \int_0^\rho P_{z_m}(z) f_y(y)\,\text{d}z\text{d}y\,. \tag{30}$$

Let $\rho = \frac{y\gamma_{rd_j}\text{INR}}{\text{SNR}_r^{d_j}}$.

Substituting Eqs. (20) and (21) in (30),

$$F_{\gamma_{rd_j}}\left(\gamma_{rd_j}\right) = \int_0^\infty \int_0^\rho \frac{z^{m-1}\text{e}^{-z/\beta_{rd_j}}}{\Gamma(m)\beta_{rd_j}^m} \frac{1}{\beta_{ptd_j}^2} \exp\left(\frac{-y}{\beta_{ptd_j}^2}\right) dzdy\,. \tag{31}$$

By integrating with respect to the PDF of z_j and y, the CDF of γ_{rd_j} is

$$\begin{aligned} F_{\gamma_{rd_j}}\left(\gamma_{rd_j}\right) = {} & \frac{1}{\Gamma(m)\beta_{ptd_j}^2} \frac{\left(\frac{1}{\beta_{rd_j}} \frac{y\gamma_{\text{rd}_j}\text{INR}}{\text{SNR}_r^{d_j}}\right)^m \Gamma(1+m)}{m\left(\frac{1}{\beta_{rd_j}} \frac{y\gamma_{\text{rd}_j}\text{INR}}{\text{SNR}_r^{d_j}} + \frac{1}{\beta_{ptd_j}^2}\right)^{1+m}} \\ & \times {}_2F_1\left(1, 1+m; 1+m; \frac{\frac{1}{\beta_{rd_j}} \frac{y\gamma_{rd_j}\text{INR}}{\text{SNR}_r^{d_j}}}{\left(\frac{1}{\beta_{rd_j}} \frac{y\gamma_{rd_j}\text{INR}}{\text{SNR}_r^{d_j}} + \frac{1}{\beta_{ptd_j}^2}\right)}\right). \end{aligned} \tag{32}$$

CDF of second time slot transmission is given by

$$F_{\gamma^{(2)}}(\gamma_{\text{th}}) = 1 - \prod_{j=1}^{2}\left(1 - F_{\gamma_{rd_j}}\left(\gamma_{rd_j}\right)\right). \tag{33}$$

The end-to-end outage probability of the proposed PLNC-based CRR system is determined by combining the outage probabilities at first and second time slots. It is defined as

$$P_{\text{out\_PLNC}}^{\text{End}}(\gamma_{\text{th}}) = F_{\gamma^{(1)}}(\gamma_{\text{th}}) + \left(1 - F_{\gamma^{(1)}}(\gamma_{\text{th}})\right)F_{\gamma^{(2)}}(\gamma_{\text{th}}). \tag{34}$$

3.2 *Outage Analysis of Non-PLNC-Based CRR System*

When the concept of PLNC is not applied to the proposed system, it would require four time slots. In time slot I, source node S_1 sends data to the relay node R and destination node D_1. In time slot II, relay forwards the data to the destination nodes D_1 and D_2. In time slot III, source node S_2 sends data to the relay node R and destination node D_2. In time slot IV, relay node R forwards the data to the destination nodes D_1 and D_2. In the non-PLNC-based CRR system, analytical expression is derived for the end-to-end outage probability considering the transmission of data from S_1 to D_2.

In the first time slot, both source nodes S_1 send data to the relay node R, and destination node D_1. Hence, outage probability of the non-PLNC-based CRR system for the data rate of R_{d} b/s/Hz is defined as

$$P_{\text{out}}^{\text{I\_NP}}(R_{\text{d}}) = \left[\Pr\left(\left(\frac{1}{2}\log_2\left(1 + \min\left(\chi_{s_1r}, \chi_{s_1d_1}\right)\right) < R_{\text{d}}\right)\right)\right]. \tag{35}$$

where χ_{s1r} is the signal-to-interference plus noise ratio (SINR) at relay node R considering the signal from source nodes S_1 as the desired signal and $\chi_{s_1d_1}$ is the SINR at destination node D_1. They are defined as

$$\chi_{s_1r} = \frac{\text{SNR}_{s_1}^{r}\left|h_{s_1}^{r}\right|^2}{\text{INR}\left|h_{p_t}^{r}\right|^2 + 1}. \tag{36}$$

$$\chi_{s_1d_1} = \frac{\text{SNR}_{s_1}^{d_1}\left|h_{s_1}^{d_1}\right|^2}{\text{INR}\left|h_{p_t}^{d_1}\right|^2 + 1}. \tag{37}$$

As non-PLNC-based CRR system requires four time slots, it is two times slower than the PLNC-based CRR system. The threshold SINR for the data rate R_d b/s/Hz is $\chi_{th} = 2^{2R_d} - 1$

Similar to (23), the CDF of $\chi_{s_1 r}$ is determined as

$$F_{\chi_{s_1 r}}(\chi_{s_1 r}) = \frac{1}{\Gamma(m)\beta_{ptr}^2}\left[\frac{\left(\frac{1}{\beta_{s_1 r}}\frac{y\chi_{s_1 r}\mathrm{INR}}{\mathrm{SNR}_{s_1}^{r}}\right)^m \Gamma(1+m)}{m\left(\frac{1}{\beta_{s_1 r}}\frac{y\chi_{s_1 r}\mathrm{INR}}{\mathrm{SNR}_{s_1}^{r}} + \frac{1}{\beta_{ptr}^2}\right)^{1+m}}\right] \times {}_2F_1\left(1, 1+m; 1+m; \frac{\frac{1}{\beta_{s_1 r}}\frac{y\chi_{s_1 r}\mathrm{INR}}{\mathrm{SNR}_{s_1}^{r}}}{\left(\frac{1}{\beta_{s_1 r}}\frac{y\chi_{s_1 r}\mathrm{INR}}{\mathrm{SNR}_{s_1}^{r}} + \frac{1}{\beta_{ptr}^2}\right)}\right). \quad (38)$$

Similarly, the CDF of $\chi_{s_1 d_1}$ is determined as

$$F_{\chi_{s_1 d_1}}(\chi_{s_1 d_1}) = \frac{1}{\Gamma(m)\beta_{ptd_1}^2}\frac{\left(\frac{1}{\beta_{s_1 d_1}}\frac{y\chi_{s_1 d_1}\mathrm{INR}}{\mathrm{SNR}_{s_1}^{d_1}}\right)^m \Gamma(1+m)}{m\left(\frac{1}{\beta_{s_1 d_1}}\frac{y\chi_{s_1 d_1}\mathrm{INR}}{\mathrm{SNR}_{s_1}^{d_1}} + \frac{1}{\beta_{ptd_1}^2}\right)^{1+m}} \times {}_2F_1\left(1, 1+m; 1+m; \frac{\frac{1}{\beta_{s_1 d_1}}\frac{y\chi_{s_1 d_1}\mathrm{INR}}{\mathrm{SNR}_{s_1}^{d_1}}}{\left(\frac{1}{\beta_{s_1 d_1}}\frac{y\chi_{s_1 d_1}\mathrm{INR}}{\mathrm{SNR}_{s_1}^{d_1}} + \frac{1}{\beta_{ptd_1}^2}\right)}\right). \quad (39)$$

CDF of $\min(\chi_{s_1 r}, \chi_{s_1 d_1})$ is defined as

$$F_{\chi^{(1)}}(\gamma_{th}) = 1 - \left(1 - F_{\chi_{s_1 r}}(\chi_{s_1 r})\right)\left(1 - F_{\chi_{s_1 d_1}}(\chi_{s_1 d_1})\right). \quad (40)$$

In the second time slot, relay R sends data to destination nodes D_1 and D_2. Hence, outage probability of the PLNC-based CRR system for the data rate of R_d b/s/Hz is defined as

$$P_{out}^{II} = \left[\Pr\left(\log_2\left(1 + \min\left(\chi_{rd_1}, \chi_{rd_2}\right)\right) < R_d\right)\right]. \quad (41)$$

χ_{rd_j} is the SINR at destination node D_j where $j = 1, 2$. Similar to Eq. (32), the CDF of χ_{rd_j} is defined as

$$F_{\chi_{rd_j}}\left(\chi_{rd_j}\right) = \frac{1}{\Gamma(m)\beta_{ptd_j}^2} \frac{\left(\frac{1}{\beta_{rd_j}} \frac{y\chi_{rd_j}\mathrm{INR}}{\mathrm{SNR}_r^{d_j}}\right)^m \Gamma(1+m)}{m\left(\frac{1}{\beta_{rd_j}} \frac{y\chi_{rd_j}\mathrm{INR}}{\mathrm{SNR}_r^{d_j}} + \frac{1}{\beta_{ptd_j}^2}\right)^{1+m}} \times {}_2F_1\left(1, 1+m; 1+m; \frac{\frac{1}{\beta_{rd_j}} \frac{y\chi_{rd_j}\mathrm{INR}}{\mathrm{SNR}_r^{d_j}}}{\left(\frac{1}{\beta_{rd_j}} \frac{y\chi_{rd_j}\mathrm{INR}}{\mathrm{SNR}_r^{d_j}} + \frac{1}{\beta_{ptd_j}^2}\right)}\right). \quad (42)$$

CDF of $\min\left(\chi_{rd_1}, \chi_{rd_2}\right)$ is defined as

$$F_{\chi^{(2)}}(\chi_{\mathrm{th}}) = 1 - \prod_{j=1}^{2}\left(1 - F_{rd_j}\left(\chi_{rd_j}\right)\right). \quad (43)$$

The end-to-end outage probability of the proposed non-PLNC-based CRR system is determined by combining the outage probabilities at first and second timeslots. It is defined as

$$P_{\mathrm{out\_NP}}^{\mathrm{End}}(\chi_{\mathrm{th}}) = F_{\chi^{(1)}}(\chi_{\mathrm{th}}) + \left(1 - F_{\chi^{(1)}}(\chi_{\mathrm{th}})\right)F_{\chi^{(2)}}(\chi_{\mathrm{th}}). \quad (44)$$

4 Results and Discussions

In this section, the numerical results for outage performance of HDR-based CRR system are presented. The results are compared with HDR-based CRR system with PLNC and non-PLNC environment in both overlay and underlay mode. The simulation parameters for the HDR-based CRR system are listed in Table 1.

In Fig. 2, for an outage probability of 10^{-2}, PLNC-based overlay networks require 13 dB, whereas underlay networks require 21 dB. This is due to the interference from primary in underlay networks. Thus overlay networks have better performance than underlay networks. For the same outage probability of 10^{-2}, PLNC-based overlay networks require 21 dB, whereas non-PLNC-based underlay networks require 23 dB. Thus PLNC networks have better performance.

Table 1 List of simulation parameters

Parameters	Values
$\mathrm{SNR}_{s_j}^{d_j}$	30 dB
$\mathrm{SNR}_r^{d_j}$	30 dB
$\beta_{p_t r}$	−4 dB
$\beta_{s_j r}$	2 dB
Data rate	0.5 b/s/Hz

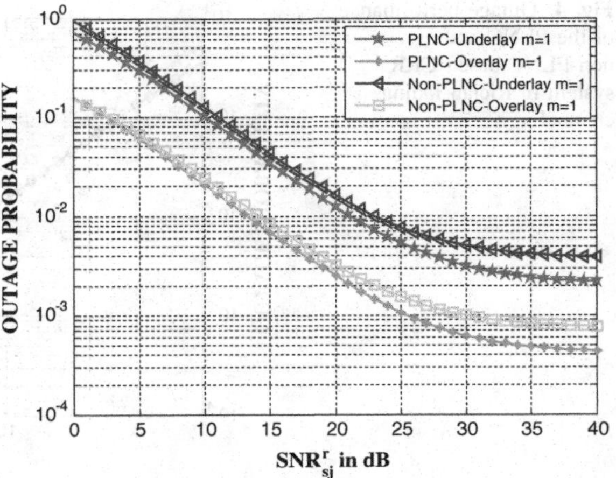

Fig. 2 Outage performance of the PLNC and non-PLNC-based CRR system in overlay and underlay mode by varying $SNR^{r}_{s_j}$

In Fig. 3, for a data rate of 1.5 b/s/Hz, PLNC-based overlay networks has an outage probability of $10^{-2.8}$, whereas underlay networks has $10^{-1.8}$. This is due to the interference from primary in underlay networks. Thus overlay networks have better performance than underlay networks. For the same data rate of 1.5 b/s/Hz, PLNC-based overlay networks has an outage probability of $10^{-2.6}$, whereas non-PLNC-based underlay networks has $10^{-1.9}$. Thus PLNC networks have better performance.

In Fig. 4, when k = 1 the channel becomes Rayleigh fading channel, for an outage probability of 10^{-2}, PLNC-based overlay networks require 8 dB, whereas underlay networks require 17 dB. This is due to the interference from primary in underlay networks. Thus overlay networks have better performance than underlay

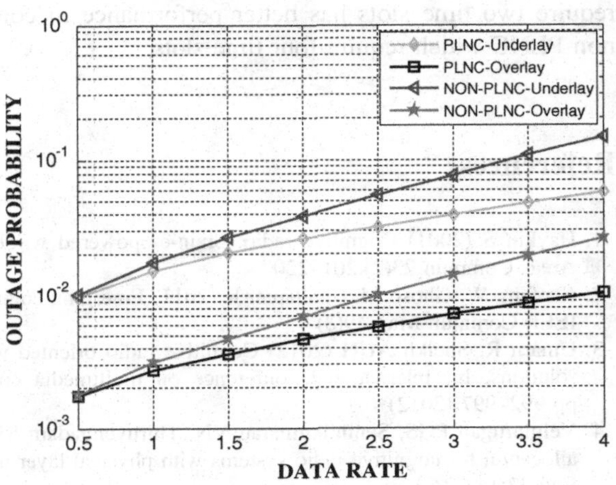

Fig. 3 Outage performance of the PLNC and non-PLNC-based CRR system in overlay and underlay mode by varying data rate

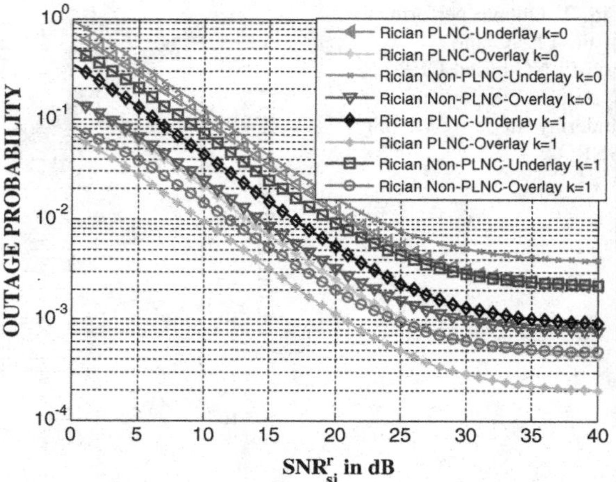

Fig. 4 Outage performance of the PLNC and non-PLNC-based CRR system in Rician fading channel

networks. For the same outage probability of 10^{-2}, PLNC-based overlay networks require 12 dB, whereas non-PLNC-based underlay networks require 20 dB. Thus PLNC networks have better performance.

5 Conclusion

In this paper, the performance of PLNC and non-PLNC-based half duplex CRR system is analyzed in Nakagami-*m* fading channels. Closed form analytical expressions for outage probabilities in first and second time slots are derived for both spectrum sharing schemes. The proposed PLNC-based CRR system is compared with non-PLNC system. It is proved that the proposed system with PLNC which require two time slots has better performance in comparison with the traditional non-PLNC which require four time slots.

References

1. Haykin S (2001) Cognitive radio: brain-empowered wireless communications. IEEE J Sel Areas Commun 23(2):201–220
2. Jianfeng W, Ghosh M, Challapali K (2011) Emerging cognitive radio applications: a survey. IEEE Commun Mag 49(3)
3. Umarn R, Sheikh AUH (2012) Cognitive radio oriented wireless networks: challenges and solutions. In: International conference on multimedia computing and systems (ICMCS), pp 992–997 (2012)
4. Velmurugan PGS, Senthilkumaran VN, Thiruvengadam SJ (2013) Joint channel and power allocation for cognitive radio systems with physical layer network coding. IJST Trans Electr Eng 37:147–159

5. Lee J, Wang H, Andrews JG, Hong D (2011) Outage probability of cognitive relay networks with interference constraints. IEEE Trans Wirel Commun 10(2):390–395
6. Chu SI (2014) Outage probability and DMT Performance of underlay cognitive networks with incremental DF and AF relaying over nakagami-m fading channels. IEEE Commun Lett 18(1): 62–65
7. Ahlswede R, Cai N, Li SYR, Yeung RW (2000) Network information flow. IEEE Trans Inform Theory 46(4):1204–1216
8. Ju M, Kim IM (2010) Error performance analysis of BPSK moduation in physical-layer network-coded bidirectional relay networks. IEEE Trans Commun 58(10):2770–2775
9. Alouini MS, Abdi A, Kaveh M (2001) Sum of gamma variates and performance of wireless communication systems over Nakagami-fading channels. IEEE Trans Veh Technol 50: 1471–1479
10. Asghari V, Assia S (2011) End-to-end performance of cooperative relaying in spectrum sharing systems with quality of service requirements. IEEE Trans Veh Technol 60(6): 2656–2668
11. Chang CW, Lin PH and Su SL (2011) A low-interference relay selection for decode-and-forward cooperative network in underlay cognitive radio. In: IEEE conference on cognitive radio oriented wireless networks and communication, pp 306–310, 2011

Estimation of Paralinguistic Features and Quality Analysis of Alaryngeal Voice

Judith Justin and Ila Vennila

Abstract This research focuses on the analysis of speech produced by a prosthetic device implanted in laryngectomees and the comparison of its paralinguistic features with that of a normal voice. Acoustic analysis was done using fundamental frequency, jitter, shimmer, and intensity. The study included eight males and the results indicated that the alaryngeal speech has values closer to natural voice in features like jitter, shimmer, pitch, and intensity. The harmonics-to-noise ratio of alaryngeal voice was slightly lower than that of normal voice. The formants and bandwidth were higher than that of normal voice. The study implies that though the alaryngeal voice is a pseudo voice produced by a speech aid (Blom-singer), it can produce a voice as close to the natural voice as possible and observations indicate that the pronunciations produced are as natural as the voice produced by the vocal cords.

Keywords Paralinguistic features · Alaryngeal speech · Vocal fold · HNR

1 Introduction

Most of the speech research carried out, use data from databases which contained information below 5 kHz (used by telephone communication channels), since intelligibility does not get seriously affected by filtering out the highest octaves of the speech spectrum. The whole voice spectrum is considered in order to identify

J. Justin (✉)
Department of Biomedical Instrumentation Engineering, Avinashilingam University, Coimbatore, India
e-mail: judithvjn@yahoo.co.in

I. Vennila
Department of Electrical and Electronics Engineering, PSG College of Technology, Coimbatore, India

L.P. Suresh and B.K. Panigrahi (eds.), *Proceedings of the International Conference on Soft Computing Systems*, Advances in Intelligent Systems and Computing 397, DOI 10.1007/978-81-322-2671-0_73

discriminating features to differentiate an alaryngeal speaker and to understand the health of a voice production mechanism.

Speech results from a complex interaction between the brain, the ear, the lungs, larynx, vocal tract, and the tongue. They work together to produce sounds of a language. The production of human voice occurs in the larynx. The vocal folds are located in the larynx and that are essential parts of the human speech mechanism. During speech, the vibrations of the vocal folds convert the air flow from the lungs into a sequence of short-flow pulses. To oscillate, the vocal folds come near together, so that sub-glottal pressure builds up beneath the larynx. The folds are pushed apart by this increased pressure, with the inferior part of each fold leading to the superior part. The rush of air vibrates the folds and generates a sound that is shaped by the resonating cavities, the mouth, the nose, and the sinuses.

The vocal positions indicate the mechanism of voice production as shown in Fig. 1 [1]. Position 1 shows the closed position, in which the column of upward air pressure moving toward vocal folds. In Fig. 1, positions 2 and 3 indicate the column of air pressure opening the bottom of the vibrating layer of vocal fold, while the body of the vocal folds stays in place. The column of air moving upward and opening the top is illustrated using positions 4 and 5. Positions 6–10 indicates the low pressure created behind the fast moving air column produces a Bernoulli's effect which makes the bottom to close followed by the top. Position 10 shows the closure of the vocal cords cutting off the air column and releasing a pulse of air.

Sounds are classified as either voiced or unvoiced. Vibrations of muscles in the larynx produce voiced sounds. Voice production, voiced sound resonance, and articulation together result in the spoken word. Pitch, loudness, resonance, quality, and flexibility are known as paralinguistic features. These features help to identify a

Fig. 1 Vocal fold positions

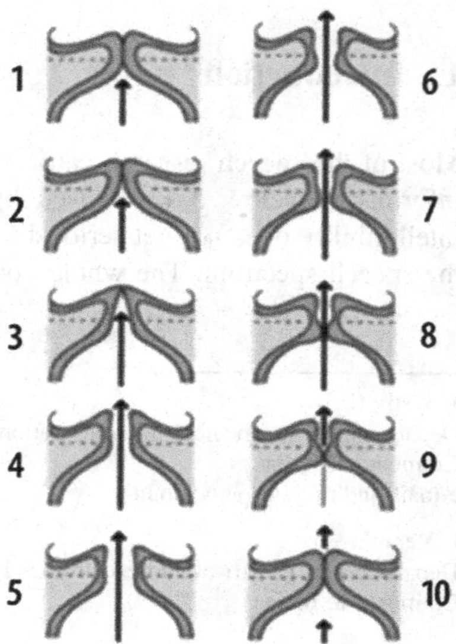

speaker and also work together with verbal language to express a meaning. Speakers vary these features to infuse their talk with emotion.

The vocal fold vibrations cannot be measured directly by noninvasive methods, because the larynx is inaccessible during phonation. So, speech signals captured by microphone yields information about vocal fold vibrations, which helps identify vocal cord pathology.

The larynx is an enlargement in the airway located superior to the trachea as shown in Fig. 2a [2]. It is a passage for air moving in and out and prevents foreign objects from entering the trachea. The larynx also houses the vocal cords. Cancer can develop in any part of the larynx as in Fig. 2b [3], and the cure rate is influenced by the location of the tumor. Consequences of laryngeal cancer can be severe and can cause difficulty in breathing and swallowing or can lead to loss of voice. Advanced stage of laryngeal carcinoma leads to total laryngectomy which is the surgical removal of the larynx.

The patient who has undergone laryngectomy cannot produce speech sounds in a conventional manner because the vocal folds have been removed in the process [4]. Therefore, they require an alternative method to produce speech, using sound sources which generate voice without vibrating the vocal folds. The top of the trachea is attached to a permanent opening, when the larynx is removed. This opening is called the stoma. The laryngectomee breathes through this opening. The laryngectomee loses his voice and learns to speak again with the help of training given by the speech therapist. Rapid and effective rehabilitation restores the confidence in the patient.

There are three ways to restore voice; the esophageal speech involves breathing in a manner that injects air into the pharynx, and then expelling it in a controlled way to form a voice. The second method is to use a speaking aid device (artificial larynx) that emits a vibrating noise when held against the throat. By making word like movements using mouth, the device converts vibrations into speech. The last option is the implantation of voice prosthesis through a tracheoesophageal puncture (TEP) [4]. The speech thus produced is called alaryngeal voice. Tracheoesophageal speech is the most popular type of alaryngeal voice.

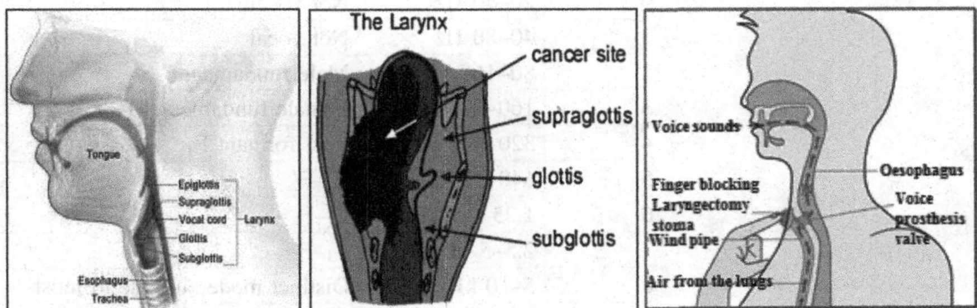

Fig. 2 **a** Anatomy of larynx; **b** larynx cancer; **c** voice prosthesis

The TEP is a method in which voice restoration occurs within 2 weeks of surgery. In this procedure, a small passage is surgically created inside the stoma, from the back wall of the trachea into the esophageal wall. The prosthesis is placed in this passage, which allows air from the lungs to pass into the esophagus as shown in Fig. 2c [2]. The air from the lungs induces the vibrations which help sounds to be produced. The valve allows air to pass into the esophagus, but prevents food and liquids from entering into the trachea. The Blom-Singer valve is an external valve, which can be taken out and cleaned by the patients themselves. The silicone voice prosthesis has silver oxide (an antifungal substance) coated on the inside surface which acts as a material preservative. The valves are available in two diameters 16 and 20 Fr. and with lengths varying from 6 to 22 mm. The size and type of TEP voice prosthesis are selected by considering four factors: the phonatory effort, candidacy for independent insertion, durability, and cost. However, it is difficult to learn the skills to produce tracheoesophageal speech. The quality and intelligibility of sound produced by a laryngectomee is severely degraded compared with that of normal sound. Moreover, alaryngeal speech sounds are of similar quality regardless of the speaker (male/female). This is called degradation of speaker individuality. Tracheoesophageal sounds appear more natural than other types of alaryngeal speech.

2 Materials and Method

The aim of the study is to determine the acoustic parameters of alaryngeal voice and compare it with that of normal speakers. The basic requirement of the study is a high-fidelity recording. The features of the raw voice signal are captured exactly as it comes from the mouth. The human audible range is between 20 Hz and 20 kHz which is spread over 9 octaves as given in Table 1 [5].

Table 1 Octave bands of voice spectrum

Octave	Frequencies	Vocal significance (general)
0	20–40 Hz	Not vocal
1	40–80 Hz	Not vocal
2	80–160 Hz	Male fundamental F_0
3	160–320 Hz	Female fundamental F_0
4	320–640 Hz	First formant F_1
5	640–1250 Hz	F_1–F_2
6	1.25–2.5 kHz	F_2–F_3
7	2.5–5 kHz	F_3–F_5
8	5–10 kHz	Distinct mode, audible to most
9	10–20 kHz	Lumped mode, audible to some

2.1 Recording Alaryngeal Voice

To analyze the entire spectrum, at least twice the sampling frequency is needed. To follow adopted standards, 44,100 Hz sampling rate is chosen [6]. Recordings were made in an anechoic room to prevent room acoustics from interfering with the features of the raw voice signals with 16-bit resolution. A continuous speech sentence is chosen from IEEE sentence database (phonetically balanced sentences). The sentence 'READ VERSE OUT LOUD FOR PLEASURE' is presented to the speaker as reading text and the speech produced is recorded through a microphone attached to a computer. The same sentence spoken by a normal person is also recorded for comparison.

Eight speakers implanted with the Blom-Singer voice prosthesis are included in the study. All the eight speakers are men, in the mean age of 69.5 years (55–84 years). They used 18–22 mm long valves with a 16 Fr. puncture, and the voice produced was clear and produced near perfect pronunciations.

The paralinguistic features considered for the study are pitch (Hz), intensity (dB), jitter (%), shimmer (dB), formants, bandwidth, and harmonics-to-noise ratio (HNR). The values are compared with that of the same sentence spoken by a normal speaker, to assess the extent to which this alaryngeal voice resembles the characteristics of a natural voice. Breathiness is observed in speakers who paused between the voiced sounds.

The duration of the recordings varied from 4 to 7 s and the duration is longer when the voice is produced with a lot of effort. All the eight speakers were able to bring forth a near natural pronunciation of the words of the sentence.

The acoustic analysis is done to assess the quality of the alaryngeal voice using PRAAT software [7]. The paralinguistic features are computed and analyzed. The values obtained are compared with that of a normal speaker. This helps to assess the extent to which the features of the alaryngeal voice resemble that of the normal voice.

3 Paralinguistic Features

3.1 Pitch

The pitch (fundamental frequency F_0) of the voice refers to the note produced by the vibrating vocal folds. The faster the vocal folds vibrate the higher the pitch. The pitch of the note is measured in terms of frequency (Hz). The factors which influence pitch variations are length of the vocal cords, the tension, the mass, and the pressure of the forced expiration [8]. The average pitch varies from person to person but, typically men will speak with a lower pitch (80–160 Hz) than women (160–320 Hz).

3.2 Intensity (Loudness)

The speed of the vibrations determines the pitch of the vocal folds and the strength of their vibration determines loudness. This is controlled by the force with which the air from the lungs passes through the larynx. The pitch of the voice can remain constant while the loudness of that particular pitch can be varied. It is possible to keep the vibration frequency the same and increase the strength of the vibration by forcing through more air. Loudness is measured in decibels (dB) [9].

3.3 Voice Perturbations (Jitter and Shimmer)

Voiced speech signals exhibit repeating deterministic pattern as well as random perturbations [10]. To be deterministic, some symmetry in the vocal folds can cause the periodicity to be achieved every second or third cycle and results in a sub-harmonic frequency.

The stochastic component consists of random variations in timing and amplitude of the signal, which make the speech signal non-periodic and characterizes rough voices. These random variations in frequency and amplitude are known as jitter and shimmer. Jitter and shimmer are given as a percentage that represents the maximum deviation from a nominal frequency or amplitude [6]. Speech pathology can be classified as mild or severe with the help of jitter [10].

3.4 Harmonics-to-Noise Ratio (HNR)

HNR is used to assess vocal fold behavior [6]. A correlation has been observed between the HNR and vocal fold disorders, and HNR is closely related to the efficacy of the vocal fold closure due to its independence from the vocal tract configuration. HNR is a good indicator of the amount of noise in speech and is efficient for voice quality analysis. HNR is defined as the log ratio of the energies of the periodic and aperiodic components. Voice pathologies can be classified by estimating the aperiodicity component in speech [11]. Two main sources of aperiodicity in glottal flow components are additive noise like aspiration noise and jitter and shimmer which are structural noises [12]. Constriction in the vocal system causes additive noise, leading to a turbulent flow. Additive noise is quasi-stationary and Gaussian and it characterizes breathy voices [13].

HNR is computed using Boersma's method [14] in PRAAT software. It uses a short-term autocorrelation function of speech to determine pitch and does not include frequency domain processing.

The autocorrelation of a signal is defined as

$$r_x(\tau) = \int x(t)\,x(t+\tau)\mathrm{d}t \tag{1}$$

The fundamental period T_0 is defined as the value of τ corresponding to the highest maximum (index zero excluded) of the short-term auto correlation function $r_x(\tau)$. The energy of the windowed speech signal is the value of the short-term autocorrelation function at its index zero:

$$r_x(0) = r_p(0) + r_{ap}(0) \tag{2}$$

where $r_p(0)$ and $r_{ap}(0)$ are the respective energies of the periodic and aperiodic components. The normalized autocorrelation is defined as:

$$\dot{r}_x(\tau) = r_x(\tau)/r_x(0) \tag{3}$$

Given the periodicity of the periodic component autocorrelation function and assuming an additive white noise, the energy of the periodic component is given by:

$$\dot{r}_p(0) = \dot{r}_p(T_0) = \dot{r}_x(T_0) \tag{4}$$

Then the energy of the aperiodic component:

$$\dot{r}_{ap}(0) = 1 - \dot{r}_p(0) = 1 - \dot{r}_x(T_0) \tag{5}$$

The HNR is defined as:

$$\mathrm{HNR} = \dot{r}_p(0)/\dot{r}_{ap}(0) \tag{6}$$

Although this algorithm is based on an additive white noise aperiodic component, it is sensitive to jitter also.

3.5 Formants

Formants are meaningful frequency components of human speech. Acoustic resonance of the human vocal tract is known as formant. It is measured as an amplitude peak in the frequency spectrum of the sound. The information that humans require to differentiate between vowels is represented quantitatively by the frequency content of the vowel sounds. The f_1 is the formant with the lowest frequency, the second f_2, and the third f_3. The quality of vowels in terms of the open/close and front/back dimensions which are associated with the position of the tongue is determined by the formants f_1 and f_2. Thus f_1 has a higher frequency for an open vowel such as [a] and a lower frequency for a close vowel such as [i] or

[u]; and f_2 has a higher frequency for a front vowel such as [i] and a lower frequency for a back vowel such as [u]. The vowels always have four or more distinguishable formants. All the parameters explained above results in high discrimination between normal and pathological voices. Changes in formants bandwidth also results in changes in voice quality [11].

4 Results

The phonatory effort brought forth by a laryngectomee had produced the sentence with strain. The perceptual quality of the voice decides on the type and size of the voice prosthesis being chosen for the implantation. The TEP has found widespread application and the results obtained from this study further confirms the superiority of the tracheoesophageal speech over the other two methods, using electro-acoustical aids and speech produced by esophageal aids. Table 2 shows the values of jitter (%), shimmer (dB), pitch (Hz), intensity (dB), HNR, and formants and their bandwidth obtained from alaryngeal speakers. In Table 3, the minimum and maximum values are identified to indicate the range of these parameters and the mean is computed. The values of these parameters for the speakers implanted with voice prosthesis are compared with the value obtained from a normal person. The same sentence 'READ VERSE OUT LOUD FOR PLEASURE' is used for the comparison of the paralinguistic features.

4.1 *Jitter and Shimmer*

The normal jitter value is 1 % [15]. Table 2 indicates the jitter (%) obtained from the eight alaryngeal speakers. The values vary from 0.47 to 2.11 % with a mean of 1.16 % as shown in Table 3. Shimmer is an average perturbation of the amplitude in dB. In Table 3, the shimmer value is 1.22 dB for the normal voice. The shimmer values for alaryngeal voice range from 1.15 to 2.29 dB with a mean value of 1.64 dB. Slightly higher jitter and shimmer values are observed in the severe cases of pathological voices.

4.2 *Pitch and Intensity*

The normal value of the fundamental frequency for men is between 80 and 160 Hz (Table 1). Mean pitch (fundamental frequency F_0) computed for the eight alaryngeal speakers is 188.74 Hz which is given in Table 3. The mean intensity of alaryngeal speakers' voice is 66.8 dB and the value observed for the normal voice is 59 dB which lie close to each other.

Table 2 Acoustic parameters of tracheoesophageal speakers Sentence: Read verse out loud for pleasure

S. no.	Jitter (%)	Shimmer (dB)	Pitch (Hz)	Intensity (dB)	HNR	F_1	B_1	F_2	B_2	F_3	B_3	F_4	B_4
1	0.99	1.99	295.61	63.34	2.11	1152.69	1196.04	2166.00	723.98	3296.41	648.15	4263.79	727.12
2	2.11	1.48	217.63	71.65	2.26	1171.85	161.94	2658.33	792.06	3569.95	561.77	4428.49	612.49
3	0.86	1.84	215.64	58.85	4.48	968.89	1284.63	1979.92	238.92	3147.61	83.04	4128.84	259.45
4	1.48	1.15	204.43	67.61	5.36	873.19	352.69	2008.26	452.52	3228.44	464.02	3951.22	359.94
5	0.47	1.19	143.69	70.85	3.52	1104.14	544.20	2173.49	788.07	3232.04	647.11	4297.47	509.70
6	0.74	2.29	145.48	62.84	3.52	1203.49	187.26	2289.53	372.4	3430.59	110.07	4202.66	184.14
7	1.63	1.79	191.35	68.53	2.43	1104.79	510.94	2112.76	745.15	3263.06	535.70	4226.81	454.62
8	1.03	1.36	96.10	70.69	3.05	1310.40	1824.87	2494.95	56.98	3456.67	287.17	4240.86	285.51

Table 3 Mean values of tracheoesophageal speakers compared with normophonic values

Features	Normophonic voice	Alaryngeal voice (min–max)	Mean
Jitter (%)	0.68	0.47–2.11	1.16
Shimmer (dB)	1.22	1.15–2.29	1.64
Pitch (Hz)	127.33	96.10–295.61	188.74
Intensity (dB)	59.09	58.85–71.65	66.80
Harmonics-to-noise ratio (HNR)	8.33	2.11–5.36	3.34
F_1	565.3	873.19–1171.85	1111.18
F_2	1396.17	1979.92-2658.33	2235.41
F_3	2033.54	3147.61–3569.95	3328.10
F_4	2730.68	3951.22–4428.49	4217.52
B_1	52.58	161.94–1824.87	757.80
B_2	47.74	56.98–792.06	521.26
B_3	630.48	110.07–648.15	417.13
B_4	64.11	184.14–727.12	424.12

From Table 2, it is observed that the pitch of the alaryngeal voice lies closer to the range specified in the voice spectrum. The estimated jitter and fundamental frequencies were found to be inversely related as shown in studies related to short-term jitter [15]. An observation is also made that a decrease in jitter is accompanied by an increase in fundamental frequency.

4.3 *Harmonics-to-Noise Ratio*

The mean value of HNR computed for the eight alaryngeal speakers is 3.34, which is lower when compared to the HNR value for a normophonic speaker (Table 3). This indicates the higher noise (aperiodic) component when compared to a lower harmonic signal (periodic). This is observed as a general phenomenon in the HNR values of all the eight tracheoesophageal speakers.

4.4 *Formants and Bandwidth*

Table 1 gives the first formant F_1 value between the ranges 320–640 Hz in the voice spectrum. In Table 3, it is observed that the value of the normal speaker is 565.3 Hz. The mean value of the eight alaryngeal speakers is observed to be 1111.18 Hz which is higher compared to normophonic value.

Table 1 shows F_1–F_2 values between the ranges 640–1250 Hz in the voice spectrum. In Table 3, it is observed that the values of the normal speaker lie between 565.3 and 1396.17 Hz. The mean value observed for alaryngeal voice is 1111.18–2235.41 Hz, which is higher compared to the normophonic value.

Table 1 shows F_2–F_3 ranges between 1.25 and 2.5 kHz in the normal voice spectrum. In Table 3, it is observed that the values for the normal speaker lie in the range 1396.17–2033.54 Hz. The mean value observed for alaryngeal voice is 2235.41–3328.10 Hz which is higher compared to the normophonic value.

Table 1 shows the F_3–F_5 values lie in the range 2.5–5 kHz in the normal voice and it is observed that the values for the normal speaker lie between 2033.54 and 2730.68 Hz (F_3–F_4). In Table 3, the mean value observed for alaryngeal voice is 3328.10–4217.52 Hz (F_3–F_4) which is higher compared to normophonic value.

5 Discussion

The best of the alaryngeal speaker also suffered from variably hoarse and harsh voice quality, periodic wet gurgly overtones, a soft volume, and a very low pitch [13]. It is sometimes accompanied by a noise from the stoma and a limited phonation time. The duration between the voiced sounds increased with the severity of the pathologic condition. This study does not identify the difficulty with which a voice produces 'f' and 's' in the sentence 'READ VERSE OUT LOUD FOR PLEASURE,' From Table 3, it is observed that all the values of paralinguistic features of alaryngeal voice lie near the values of normophonic voice.

All the mean values jitter (%), shimmer (dB), fundamental frequency, and mean intensity are found to lie very close to the normophonic voice [13]. As the severity of the pathology increases, these values increase. Harmonic-to-noise ratio is found to be lower for all tracheoesophageal speakers. The formant values indicate a higher range for the alaryngeal speakers. This implies that the vowels are pronounced with difficulty which is taken as an indicator of the dysphonic voice. Pronunciation of vowels and voiced sounds was found to be the same as that of a normophonic voice.

6 Conclusion

It is finally concluded that the alaryngeal speech produced with the help of an implanted Blom-Singer Voice prosthesis, produced sounds which is very natural and with similar characteristics as that of a natural voice. It is observed that it took longer for the alaryngeal speaker to read a sentence than a normal speaker. Also the voice is produced with a lot of effort and is breathy in nature. The same study can be extended for other prosthetic aids available that are being used by surgeons for implantation. A comparison of performance can be done and this will help identify the aids which produce a machine-like voice.

Acknowledgments The authors would like to thank Dr. V. Anand, ENT, Head and Neck Surgeon and Director of M.C. Velusamy ENT Trust Hospital, Pollachi, Coimbatore, India, for the help rendered in collecting data from laryngectomees implanted with the Blom-Singer prosthesis.

References

1. http://www.phon.ox.ac.uk/~jcoleman/phonation.htm
2. http://www.cancerresearchuk.org/about-cancer/type/larynx
3. http://www.cancer.gov/cancertopics/pdq/treatment/laryngeal/Patient/page1
4. United States of America: Speech-Language Pathology Medical Review Guidelines. American Speech Hearing Association. (2011)
5. Ternstrom SO (2008) Hi-Fi voice: observations on the distribution of energy in the singing voice spectrum above 5 kHz. In: Proceedings of ASA, presented at Acoustics'08, Paris, France, pp 3171–3176
6. Lucas Leon Oller (2008) Analysis of Voice Signals for the Harmonics-to-Noise Crossover Frequency. TMH Publications, UPC, Barcelona
7. Boersma P (1993) Accurate short-term analysis of the fundamental frequency and the harmonics-to-noise ratio of a sampled sound. In: Proceedings of IFA, pp 97–110
8. Globek D, Simunjak B, Ivkic M, Hedjever M (2003) Speech and Voice analysis after near-total laryngectomy and tracheo esophageal puncture with implantation of provox 2 prosthesis. Logoped Phoniatr Vocol 29(2):84–86
9. Kreiman J, Bruce Gerratt R (2003) Jitter, shimmer and noise in pathological voice quality perception. presented at ISCA archive, Voqual' 03, Geneva, Switzerland, pp 57–62
10. Severin F, Bozkurt B, Dutoit T (2005) HNR extraction in voiced speech, oriented towards voice quality analysis. In: Proceedings eusipco
11. Li T, Jo C, Wang SG, Yang BG, Kim HS (2004) Classification of pathological voice including severely noisy cases. Proceedings of ICSLP, Jeju, Korea, pp 77–80
12. Jody Kreiman R, Gerratt Bruce (2005) Perception of aperiodicity in pathological voice. J Acoust Soc. Am 117:2201–2211
13. Globlek D, Stajner- Katusic S, Musura M, Horga D, Liker M (2004) Comparison of alaryngeal voice and speech. Logoped Phoniatr Vocol 29(2):87–91
14. Richard G, Alessandro CD (1996) Analysis/synthesis and modification of speech aperiodic component. Speech Commun 19:221–244
15. Vasilakis M, Stylianou Y (2009) Voice pathology detection based on short-term jitter estimations in running speech. Folia Phoniatrica et Logopaedica 61(3):153–170

Control and Operation of 4 DOF Industrial Pick and Place Robot Using HMI

Akshay P. Dubey, Santosh Mohan Pattnaik and R. Saravanakumar

Abstract Human–machine interface commonly known as HMI is deployed for control and visualization interface between a human and a process, machine, application and appliance. It is a graphics-based visualization of an industrial control and monitoring system. In this paper we intend to present the controlling of an industrial pick and place robot using HMI. Implementation of logic is done for pick and place robot in HMI and design a GUI for controlling and monitoring the working of the robot. The robot is interfaced with PLC and the PLC in turn is interfaced with the HMI. The HMI can be used to control the robot by giving control signals from the HMI panel and also it simultaneously receives the signals from the system to show the status of the robot operation. The HMI panels shows which arm of the robot is active during the operation. Advantage of this HMI controlled system is that the system is automatic and supervises continuously as a result of which faults can be monitored and acknowledged to the user.

Keywords HMI · PLC · Automation · Pick and place robot · Communication

1 Introduction

Automation was a general need from the earlier days of industrial revolution where every industry and its subsystems needed to automate its various processes. The automation was done mainly to reduce operating costs, ensure safety and to reduce human work. As days passed several researches and inventions were made to

A.P. Dubey (✉) · S.M. Pattnaik · R. Saravanakumar
VIT University, Vellore, India
e-mail: akshayprasaddubey@gmail.com

S.M. Pattnaik
e-mail: pattnaik.santosh123@gmail.com

R. Saravanakumar
e-mail: rsarvanakumar@vit.ac.in

L.P. Suresh and B.K. Panigrahi (eds.), *Proceedings of the International Conference on Soft Computing Systems*, Advances in Intelligent Systems and Computing 397, DOI 10.1007/978-81-322-2671-0_74

develop efficient automation tools and softwares. The first logic used was relay logic, which comprised of large coils for switching, looked big bulky and consumed a lot of space. Then the PLCs came which were robust, portable, efficient, which we are using in our industries till date. But using PLC people were not able to monitor the process taking place, i.e. they were not able to see the status of the process, fault detection and the output at a given instant. So these problems led to the development of graphics-enabled visualization systems and of course supervisory systems like HMI and SCADA (supervisory control and data acquisition). Here we incorporate the use of HMI, which is a graphic based visualization and control of an industrial process.

The HMI is a panel comprising of a screen and several push buttons. A user can easily control a given process by visualizing the simple GUI. The HMI also acts as a supervisory unit where status of the process at every interval can be monitored. The main feature of HMI panel is that it is small, robust, compact and can be used almost anywhere in industries or home automation applications. Also touch screen panels have been developed nowadays which make the HMIs even more user friendly. The HMIs are very simple to operate upon. In this paper, we a PLC program is developed for the industrial robot and interfaced it with HMI using the PROFINET communication module.

The HMI screen was designed using the SIEMENS HMI software for acting as a simple GUI. According to the mechanism, a definite position and orientation of the end-effector for a certain serial manipulator can result in variable configurations, even in infinite configurations [1]. Here the designed HMI panel has many screens which the user can navigate for performing certain functions as well as monitor the status of the process. Also there are screens which give information about the system components. The robot is now controlled, and its status is monitored, which are explained in the later sections. The block diagram of the system is shown in Fig. 1.

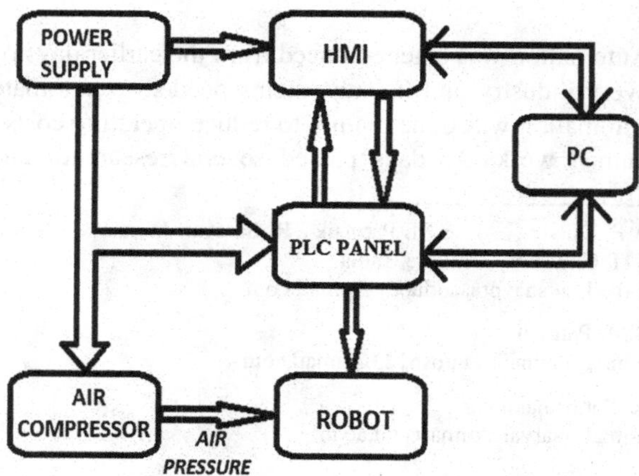

Fig. 1 Block diagram of the system

2 System Architechture

2.1 Hardware Configuration

The panel designed for controlling the industrial pick and place robot uses 230 V power source from the main supply, which is distributed to the S7-1200 PLC, STIOP PSU100L AC–DC converter, relay cards, and communication module. 24 V DC from the converter is fed to a 10 V DC–DC converter. Toggle switches and indicators are used for input and output operations of the PLC in the manual mode. Ethernet communication module is used for communication between PC and PLC. Relay cards are used to actuate the pneumatic solenoid valves for operation of the robot's arms and wrist. The PLC gives the signals to the relay cards, which in turn actuate the robot thorough the valves.

The HMI panel uses 24 V DC for its operation. The HMI panel consists of several push buttons and a screen for display. Ethernet is used for communicating between the PC and the HMI panel.

The industrial pick and place robot consists of three arms and a wrist or gripper. The body of the arms and the wrist comprises of pneumatic pistons. The pistons are controlled by solenoid valves, which get the actuating signals from the PLC.

Figures 2, 3 and 4 shows the electrical panel, robot and HMI for pick and place robot.

Hardware specifications

PLC: SIEMENS SIMATIC S7-1200, CPU 1214CACDCRLY.
HMI: SIEMENS SIMATIC HMI, KP700 COMFORT.

Fig. 2 Panel for pick and place robot

Fig. 3 Pick and place robot

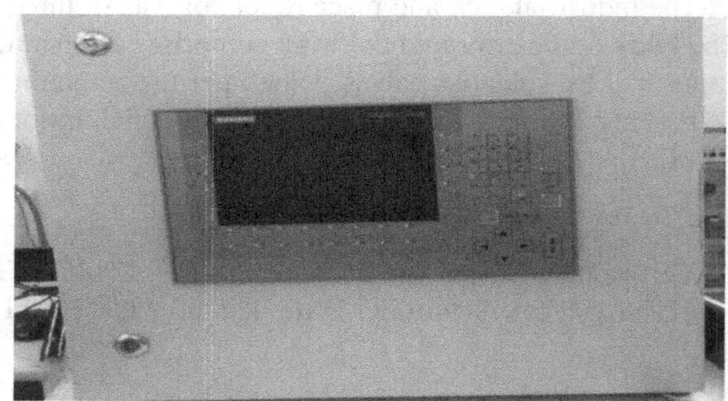

Fig. 4 HMI panel

2.2 Software Configuration

PLC Software The software used for developing the PLC logic is SIEMENS TIA (Totally Integrated Automation) version V13. The ability to communicate not only allows information and control to be communicated across the network of nodes, but also allows nodes to cooperate in performing more complex tasks like sampling, data aggregation, system health, and status monitoring in different formats [2]. This software communicates with the PLC and the PC through PROFINET. This is a very flexible and user-friendly software which allows various modes of programming like ladder logic, functional blocks, etc. But we use ladder logic because it is the simplest and easy to implement than all other methods. Figure 5 shows PLC software screen [11].

HMI Software The software for developing the HMI screen is SIEMENS TIA version V12. This software is efficiently used for programming [3] the HMI panel. HMI software divides by two types which is Visual Basic (VB) software and

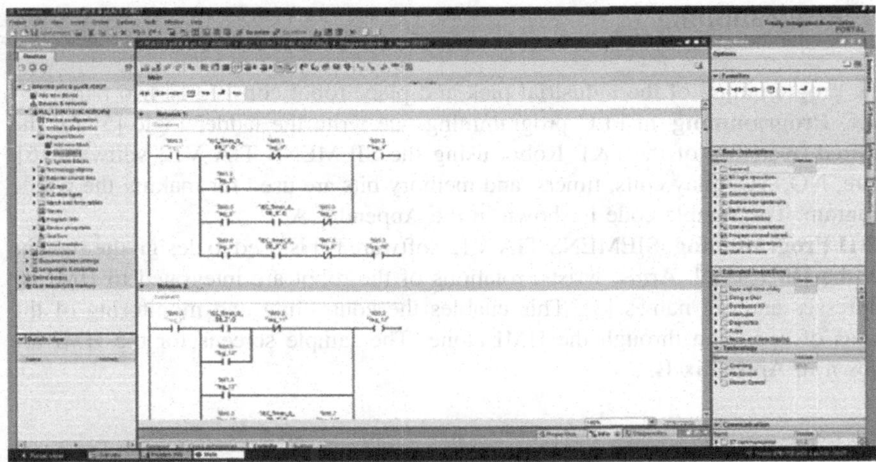

Fig. 5 PLC software screen

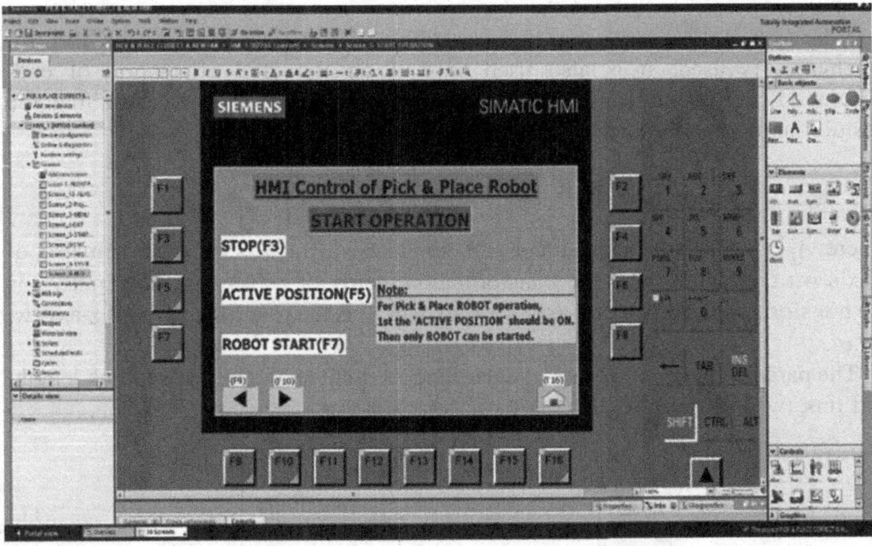

Fig. 6 HMI software screen

SCADA packages software [4]. Here we are using the COMFORT KP700 HMI panel [8]. The communication between PC and HMI is done by using PROFINET. Figure 6 shows HMI software screen [9].

3 Programming

The programming of the industrial pick and place robot consists of two parts.
PLC Programming In PLC programming, we write the ladder logic [5] for the desired operation of the P&P Robot using the SIEMENS TIA V12 software [6]. Here, NO, NC, relay coils, timers, and memory bits are used for making the whole program. The sample code is shown in the Appendix A.
HMI Programming SIEMENS TIA V12 software [7] is used to design the screens for the HMI panel. Arms, wrists, rotations of the robot are integrated to HMI by addresses and tag names [3]. This enables the controlling and monitoring of the status of operation through the HMI alone. The sample screens for the HMI are shown in Appendix B.

4 D–H Convention

In order to find the relationship between different joints of a manipulator, forward kinematics is used. It is also used to find position and orientation of end-effector [12]. Denavit–Hartenberg is the conventional method used to find the position and orientation of the end-effector [13].

$$A_i = \mathrm{Rot}_{z,\theta i}\mathrm{Trans}_{z,di}\mathrm{Trans}_{x,ai}\mathrm{Rot}_{x,\alpha i}$$

where, A_i is the homogeneous transformation matrix, $\mathrm{Rot}_{z,\theta i}$ is the rotation matrix of z-axis w.r.t. θi, $\mathrm{Trans}_{z,di}$ is the transformation matrix of z-axis w.r.t. di, $\mathrm{Trans}_{x,\alpha i}$ is the transformation matrix of z-axis w.r.t. αi, $\mathrm{Rot}_{x,\alpha i}$ is the rotation matrix of z-axis w. r.t. αi.

The parameters θi, di, αi, and ai determine the joint angle, link offset, link length, and link twist, respectively.

$$A_i = \begin{bmatrix} C_{\theta i} & -S_{\theta i}C_{\alpha i} & S_{\theta i}S_{\alpha i} & a_iC_{\theta i} \\ S_{\theta i} & C_{\theta i}C_{\alpha i} & -C_{\theta i}S_{\alpha i} & a_iS_{\theta i} \\ 0 & S_{\alpha i} & C_{\alpha i} & d_i \\ 0 & 0 & 0 & 1 \end{bmatrix} \tag{1}$$

From Fig. 7, we get the link parameters for the 4 DOF manipulator as shown in Table 1.

The homogeneous transformation matrix A_i for four respective links are given as Therefore,

Fig. 7 4 DOF manipulator

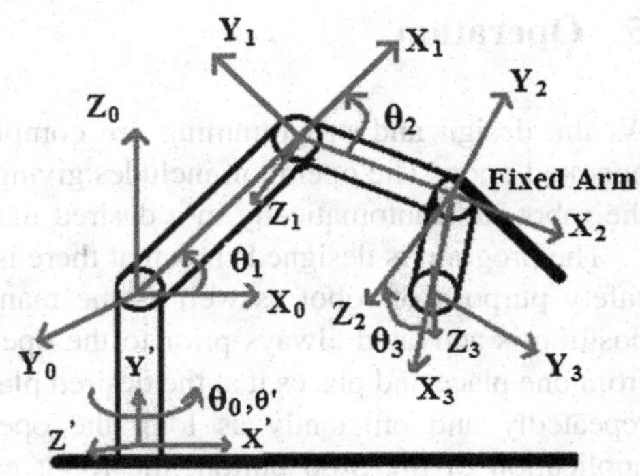

Table 1 Link parameters for 4 DOF manipulator

Link	Joint angle 'θ_i'	Link offset, 'd_i'	Link length, 'a_i'	Link twist, 'α_i'
1	θ'	d_1	0	0
2	θ_1	0	a_1	+90°
3	θ_2	0	a_2	0
4	θ_3	0	a_3	+90°

$$A_1 = \begin{bmatrix} C_{\theta'} & -S_{\theta'} & 0 & 0 \\ S_{\theta'} & C_{\theta'} & 0 & 0 \\ 0 & 0 & 1 & d' \\ 0 & 0 & 0 & 1 \end{bmatrix}, \tag{2}$$

$$A_2 = \begin{bmatrix} C_{\theta 0} & 0 & S_{\theta 0} & a_0 C_{\theta 0} \\ S_{\theta 0} & 0 & -C_{\theta 0} & a_0 S_{\theta 0} \\ 0 & 1 & 0 & 0 \\ 0 & 0 & 0 & 1 \end{bmatrix} \tag{3}$$

$$A_3 = \begin{bmatrix} C_{\theta 1} & -S_{\theta 1} & 0 & a_1 C_{\theta 1} \\ S_{\theta 1} & C_{\theta 1} & 0 & a_1 S_{\theta 1} \\ 0 & 0 & 1 & 0 \\ 0 & 0 & 0 & 1 \end{bmatrix}, \tag{4}$$

$$A_4 = \begin{bmatrix} C_{\theta 2} & 0 & S_{\theta 2} & a_2 C_{\theta 2} \\ S_{\theta 2} & 0 & -C_{\theta 2} & a_2 S_{\theta 2} \\ 0 & 1 & 0 & 0 \\ 0 & 0 & 0 & 1 \end{bmatrix} \tag{5}$$

So, final homogeneous transformation matrix will be

$$T_2^0 = A_1 * A_2 * A_3 * A_4 \tag{6}$$

5 Operation

As the design and programming are completed, the system is now ready to be operated upon. The operation includes giving the PLC a set of commands such that the robot runs automatically in a desired manner.

The program is designed such that there is always an active position of robot for safety purpose of robot as well as the manpower working around it. The safety position is activated always prior to the operation of the robot. It picks the object from one place and places it at the desired place. This process goes on continuously, repeatedly and efficiently as long the operator applies the Stop button. After application of the Stop button, the robot comes back to its home position. The user/operator can check the status of operation of robot also through the HMI screen which tells about each and every movement of all arms, wrist, and rotation. Also, by Alarm Screen in HMI, the user gets the information about the faults occurring in the system [10].

6 Results

This Robot coding is developed for the auto-operation of robot which will take the object and place it. This process will be in repeated behavior. Also, if any fault occurs in the system, the alarm window shows that fault or alarm on the Alarm Screen of HMI. The status of operation of robot arms, wrist, and rotation is also displayed on the HMI Screen. The HMI Screen shows this result by the indication of colors red or green. Some sample results of the movements of robot arms, wrist, and rotations are shown in Table 2. The respective HMI Screen status of operation is shown in Table 3 having G for green color and R for red color.

Table 2 Robot movement

Rotation	First arm	Second arm	Wrist
Right	Up	Down	Open
Right	Up	Up	Open
Left	Up	Up	Close
Left	Down	Down	Open
Left	Up	Up	Open

Table 3 HMI screen status of operation of robot movement

Rotation	First arm	Second arm	Wrist
R	G	R	R
R	G	G	R
G	G	G	G
G	R	R	R
G	G	G	R

7 Conclusion

In this paper, the HMI system is studied and designed appropriately for the control and monitoring of the industrial pick and place robot. The TIA Portal V13 for HMI Programming and TIA Portal V12 for PLC Programming are successfully used in implementing the robot system. It is found that HMI is very appropriate, robust and user-friendly tool for controlling and supervising the system. The monitoring and control of the system is very efficiently done using the HMI panel. Thus, HMI is found to be a reliable, efficient, and user-friendly automation tool which can be used for many other industrial and home automation systems.

Acknowledgments We would like to thank the management of VIT University, Vellore for giving us the opportunity to undertake this work in the Industrial Automation Lab in School of Electrical Engineering (SELECT).

Appendix A

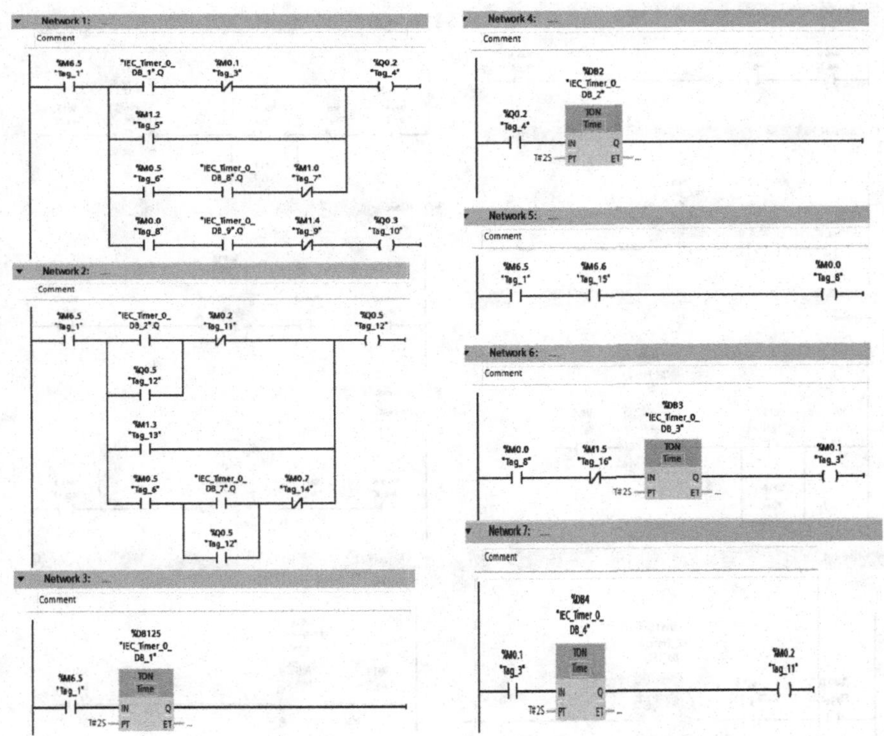

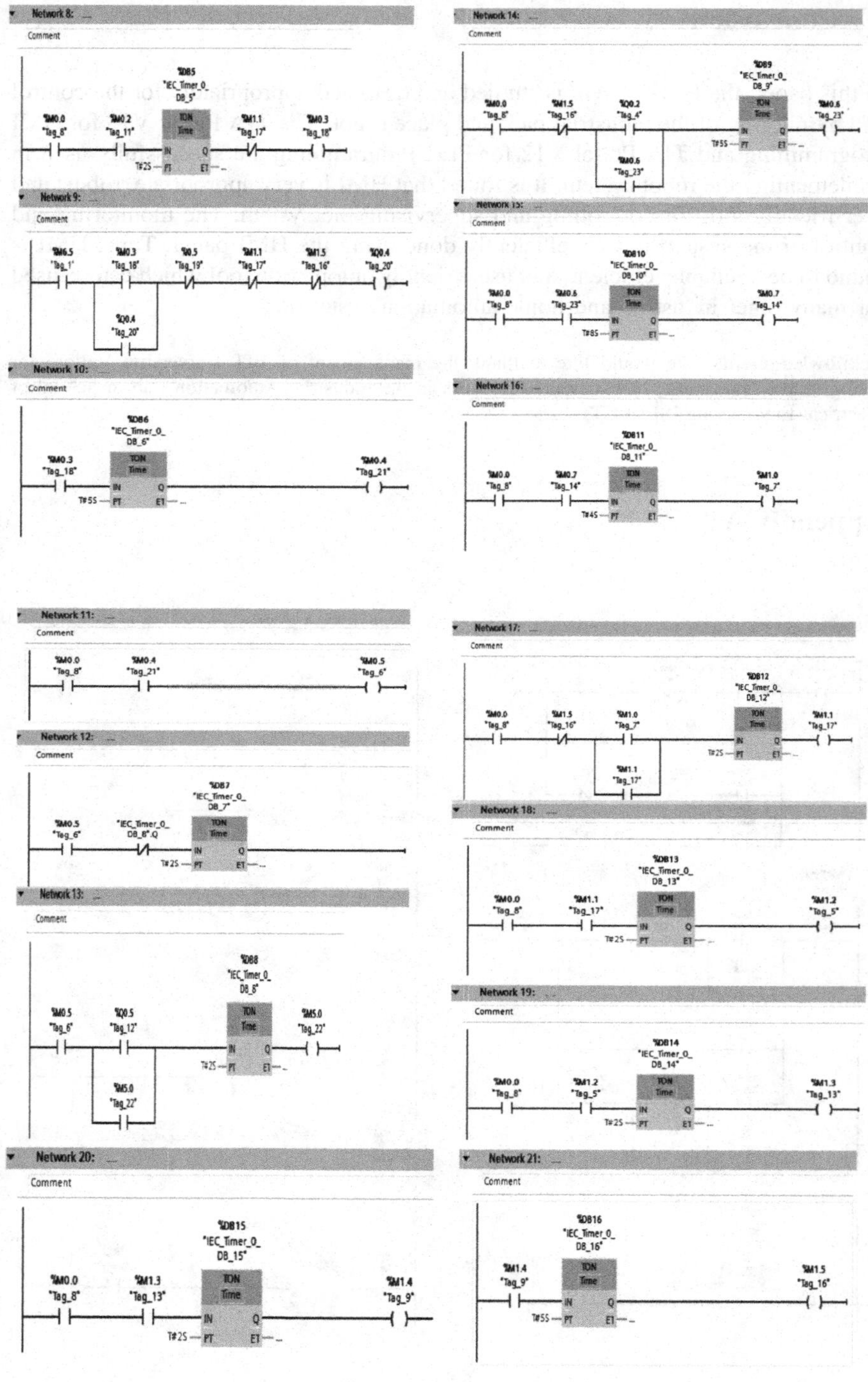

Network 8: ...
Comment
%DB5
"IEC_Timer_0_DB_5"
TON
Time
%M0.0
"Tag_8"
%M0.2
"Tag_11"
%M1.1
"Tag_17"
%M0.3
"Tag_18"
IN
Q
T#2S
PT
ET
Network 9: ...
Comment
%M6.5
"Tag_1"
%M0.3
"Tag_18"
%I0.5
"Tag_19"
%M1.1
"Tag_17"
%M1.5
"Tag_16"
%Q0.4
"Tag_20"
%Q0.4
"Tag_20"
Network 10: ...
Comment
%DB6
"IEC_Timer_0_DB_6"
TON
Time
%M0.3
"Tag_18"
%M0.4
"Tag_21"
T#5S
Network 11: ...
Comment
%M0.0
"Tag_8"
%M0.4
"Tag_21"
%M0.5
"Tag_6"
Network 12: ...
Comment
%DB7
"IEC_Timer_0_DB_7"
TON
Time
%M0.5
"Tag_6"
"IEC_Timer_0_DB_8".Q
T#2S
Network 13: ...
Comment
%DB8
"IEC_Timer_0_DB_8"
TON
Time
%M0.5
"Tag_6"
%Q0.5
"Tag_12"
%M5.0
"Tag_22"
%M5.0
"Tag_22"
T#2S
Network 14: ...
Comment
%DB9
"IEC_Timer_0_DB_9"
TON
Time
%M0.0
"Tag_8"
%M1.5
"Tag_16"
%Q0.2
"Tag_4"
%M0.6
"Tag_23"
%M0.6
"Tag_23"
T#5S
Network 15: ...
Comment
%DB10
"IEC_Timer_0_DB_10"
TON
Time
%M0.0
"Tag_8"
%M0.6
"Tag_23"
%M0.7
"Tag_14"
T#8S
Network 16: ...
Comment
%DB11
"IEC_Timer_0_DB_11"
TON
Time
%M0.0
"Tag_8"
%M0.7
"Tag_14"
%M1.0
"Tag_7"
T#4S
Network 17: ...
Comment
%DB12
"IEC_Timer_0_DB_12"
TON
Time
%M0.0
"Tag_8"
%M1.5
"Tag_16"
%M1.0
"Tag_7"
%M1.1
"Tag_17"
%M1.1
"Tag_17"
T#2S
Network 18: ...
Comment
%DB13
"IEC_Timer_0_DB_13"
TON
Time
%M0.0
"Tag_8"
%M1.1
"Tag_17"
%M1.2
"Tag_5"
T#2S
Network 19: ...
Comment
%DB14
"IEC_Timer_0_DB_14"
TON
Time
%M0.0
"Tag_8"
%M1.2
"Tag_5"
%M1.3
"Tag_13"
T#2S
Network 20: ...
Comment
%DB15
"IEC_Timer_0_DB_15"
TON
Time
%M0.0
"Tag_8"
%M1.3
"Tag_13"
%M1.4
"Tag_9"
T#2S
Network 21: ...
Comment
%DB16
"IEC_Timer_0_DB_16"
TON
Time
%M1.4
"Tag_9"
%M1.5
"Tag_16"
T#5S

Appendix B

References

1. Si Y, Jia Q, Chen G, Sun H (2013) A complete solution to the inverse kinematics problem for 4 DOF manipulator robot, 2013, IEEE
2. Abed AA, Ali AAA, Aslam N (2010) Building an HMI and demo application of WSN based industrial control system. In: 2010 first international conference on energy, power & control (EPC-IQ), 30 Nov–2 Dec 2010
3. Falkman P, Helandery E, Anderssonz M Automatic generation: a way of ensuring PLC and HMI standards
4. Rahman SH, Hanafiah MAM, Ab Ghani MR, Wan Jusoh WNSE (2014) A human machine interface(HMI) framework for smart grid system. 2014 IEEE Innovative Smart Grid Technologies-Asia

5. I.E.C. (2002) IEC 61131-3. programmable controllers—Part 3: programming languages, 2nd edn. Final draft international standard (FDIS), 2002
6. [pdf]s7-1200 easy book. https://cache.automation-siemens.com/dnl/zQ/ZQxMTMOMWAA_39710145_HB/s71200_easy_book_en_US_en_US.pdf
7. [pdf]comfort panels-SIEMENS. https://cache.automation-siemens.com/dnl/ju/juyNz10OAA_493132HB/hmi_comfort_panels_operating_instructions_en_US.pdf
8. Implementation of SCADA system for DC motor control (ICCCE 2010), Kuala Lumpur, Malaysia, 11–13 May 2010
9. KUKA Roboter GmbH (2005) KR C2 PC based robot controller: technical data. www.kuka.com, March 2005
10. Asada H, Youcef-Toumi K (1983) Analysis and design of semidirect drive robot arms. In: Proceedings of American control conference, San Francisco, CA, June 1983, pp 757–764
11. Amditis A, Andreon L, Pagle K, Markkula G, Deregibus E, Rue MR, Bellotti F, Engelsberg A, Brouwer R, Peters B, De Gloria A (2010) Towards the automotive HMI of the future: overview of the AIDE-integrated project results. IEEE Trans Intell Transport Syst 11(3), Sept 2010
12. Shukui H, Junlan C (2011) Uncompensatable error of four DOF parallel robot, 2011-IEEE
13. Cai W, Xiong T, Yin Z (2012) Vision-based kinematic calibration of a 4-DOF pick-and-place robot. In: Proceedings of 2012 IEEE international conference on mechatronics and automation, 5–8 Aug, Chengdu, China

Battery Storage Photovoltaic Grid-Interconnected System: Part-IV

Ritesh Dash and S.C. Swain

Abstract This paper basically describes about the phenomena of off-grid and on-grid electrification of renewable energy sources like photovoltaic and wind system. The main objective of the paper is to find some grid standard for interconnection of these renewable sources to the existing grid. To study the behavior of the system a matlab based model has been designed by taking solar PV, wind and a conventional grid. The simulation of the system focuses on maximum power extraction from the grid, grid code prediction, real and reactive control of power.

Keywords Grid interconnection · Solar PV system · Storage system · Maximum power point tracking · Real and reactive power control · Wind turbine

1 Introduction

A clean, green, and zero-pollution emission concept drives us to think for harvesting renewable energy. As the people demand for more and continuous supply of energy we cannot simply rely on thermal power plant. Therefore, both the wind and solar can be used together for a better performance and also can increase the system reliability. The research-based development of solar PV and wind make it possible to design a stand-alone system or a grid-interconnected system. If some other sources like biogas, biodiesel, or a storage device are interconnected then it may be treated as a hybrid system. However, the control of hybrid system is more difficult than the control of simple wind or solar system.

The integration of renewable sources requires a full bridge inverter which always creates a problem in real-time implementation. The inverter must be designed in such a way that it should not affect the voltage magnitude at the grid side or inject any reactive power of more than 5 % and this value is not fixed generally and vary

R. Dash (✉) · S.C. Swain
School of Electrical Engineering, KIIT University, Bhubaneshwar, India
e-mail: riteshfel@gmail.com

L.P. Suresh and B.K. Panigrahi (eds.), *Proceedings of the International Conference on Soft Computing Systems*, Advances in Intelligent Systems and Computing 397, DOI 10.1007/978-81-322-2671-0_75

from country to country. Under transient conditions, the inverter may sense a low voltage at its output terminal and isolates itself from the grid. So inverters must be designed to withstand these small voltage fluctuations. Under voltage sag conditions, if islanding occurs then the whole performance is affected and collapse of grid may occur. So it is very important to predict a control strategy which can operate the system safely under such adverse condition.

In a renewable-based grid-interconnected system, it is always assumed that the grid voltage is constant. However, in a practical system it is always fluctuating. So this paper tries to find the problems associated with a dynamic system where the parameters are variables.

This also allows large penetration of solar PV system into the existing grid. Here a battery storage system is also provided for increasing the performance under the absence of PV system.

This paper investigated the performance of a hybrid system under different power quality condition. The solar photovoltaic system is connected near the load end which is again located in a remote area. The load is connected through a doubly fed transmission line with the grid. The coupling capacitor which is used for the interconnection purpose may affect the grid voltage. Therefore series compensation to the system is applied to increase the performance.

2 Hybrid System

2.1 *Annual Energy Consumption*

According to Indian electricity rule, each consumer in an urban area is supplied with a 2 kW of load. However, the average energy consumption by a customer is 1200 W. So in this regard the total monthly average may be 36,000 W or 36 kW. So here an average of 40 kh is considered for operation.

2.2 *Solar Panel Calculation*

Harvesting the solar energy is a very difficult task, because of its weather-dependent property. The average efficiency of a solar photovoltaic system is only 19 %. So it requires a lot of photovoltaic panel to meet the load demand. From estimation the average solar energy that a panel receives is about 2.2 MWh. So if the entire earth surface is covered with the solar panel then it can supply the entire world demand. But it is not possible because of its low efficiency. So, considering a polycrystalline solar cell of annual producing capacity of 316 kwh of energy, the number of panels required is 68 (Fig. 1).

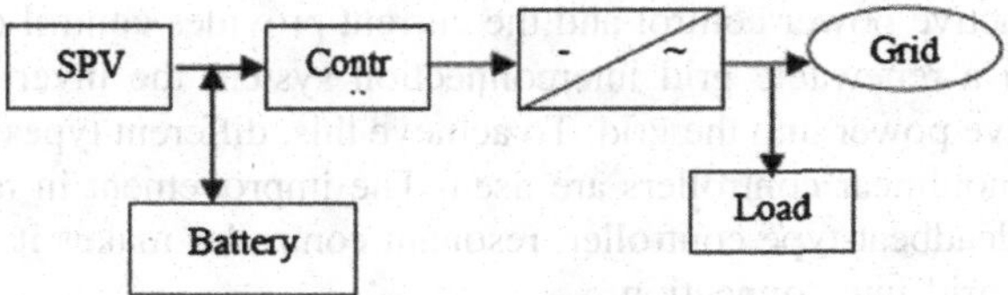

Fig. 1 Overall line diagram

2.3 Solar Panel Operating Condition

Photovoltaic system produces energy from 8 a.m. to 4 p.m. The total energy produced by the system throughout the whole day may be more than the required energy. So after utilizing the energy the surplus can be stored for future. Another advantage of storing the energy is in sudden cloudy condition the photovoltaic system may not produce the energy to withstand the peak load and it will affect the system stability. So to avoid unintentional islanding the help of battery can be taken. However, it increases the overall cost (Fig. 2).

2.4 Controller

One of the most important problems in the renewable grid interconnection is maintaining the DC the bus. There are basically two topologies like droop control method and master control method. Droop control method does not require any grid converter communication. Therefore it is the simplest method of controlling the DC voltage over the DC bus than the master control method, which requires a continuous communication between the converter and grid. Therefore, it requires a lot of control action which is complex and also it generates error during online monitoring system.

Master control can be achieved by two methods like inner loop control and outer loop control method. Inner control method provides the current control where as outer loop provides the voltage control. In a general sense, the voltage control

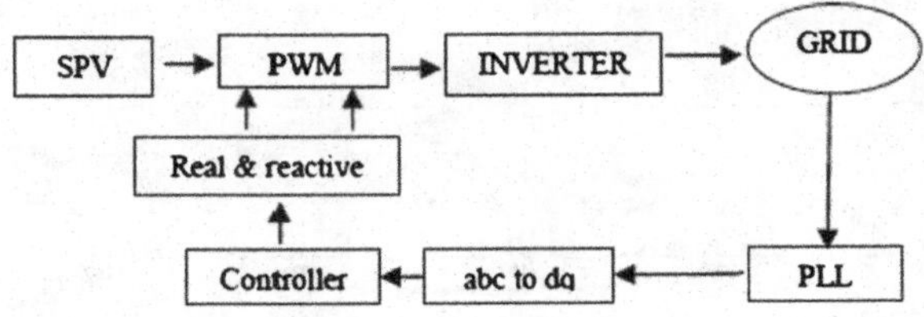

Fig. 2 Controller diagram

provides the reactive power control and the current provides control of real power. It is because in a renewable grid interconnection system the inverter should not inject any reactive power into the grid. To achieve this, different types of controllers like linear and nonlinear controllers are used. The improvement in nonlinear controller such as deadbeat-type controller, resonant controller makes it possible for a variable source grid interconnection.

3 Results and Analysis

In our earlier paper, we have simulated the grid-connected system to use it to supply a load which is located in a remote area at a distance of 15 km. Here the same system is again modeled with a storage device. The proposed storage device is a nickel–cadmium battery. The number of batteries exactly required to withstand fluctuation can be calculated through a proper simulation. The below-shown Fig. 3 shows the state of charge of the battery. Each battery in the system has a full load voltage of 12 V and has a maximum capacity of 14.856 V. The typical capacity of the entire battery system is 2000 A. The simulation for the proposed system has been tested under different solar irradiance condition like 1000, 800 and 400 W/m^2. The system consist of a radial bus system having four number of alternator each of 2500 MVA. The grid is supposed to supply a load of 1 MW located at a remote location. Figure 1 shows the daily load profile of the load (Fig. 4).

Table 1 shows the different parameters used in the simulation. Based on the different simulation results, the maximum penetration label can be determined. Figure 5 shows the charging and discharging conditions of the battery.

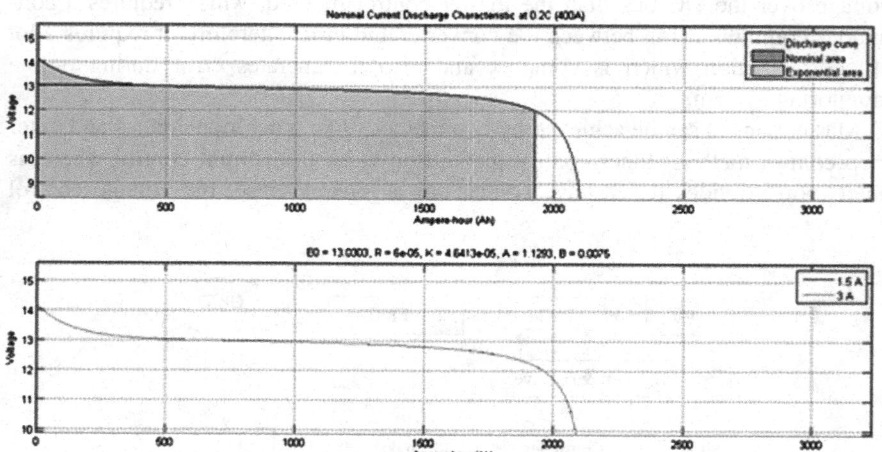

Fig. 3 Battery discharging characteristics

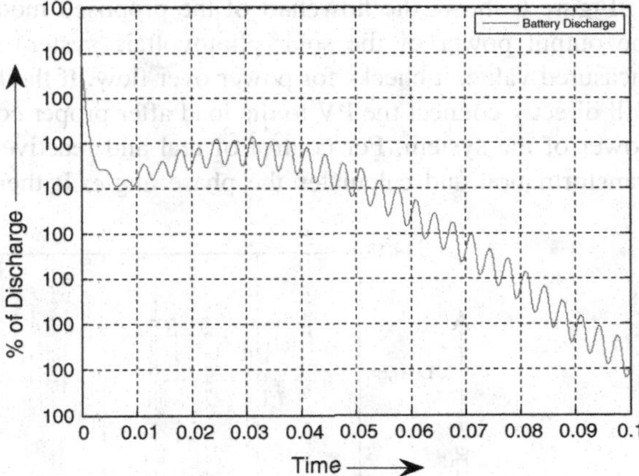

Fig. 4 State of discharge of battery

Table 1 System parameters under consideration

Bus no.	Generator (MW)	Voltage (kV)	Load (MW)	Transformer
Solar bus	1	0.4	1	400 V/33 kV, 55 MVA
1	250	11	–	11 kV/33 kV 47 MVA
2	250	11	–	11 kV/33 kV 47 MVA
3	160	11	–	11 kV/33 kV 25 MVA
4	85	13.8	–	11 kV/33 kV 25 MVA
5	–	33	45	–
6	–	33	40	–

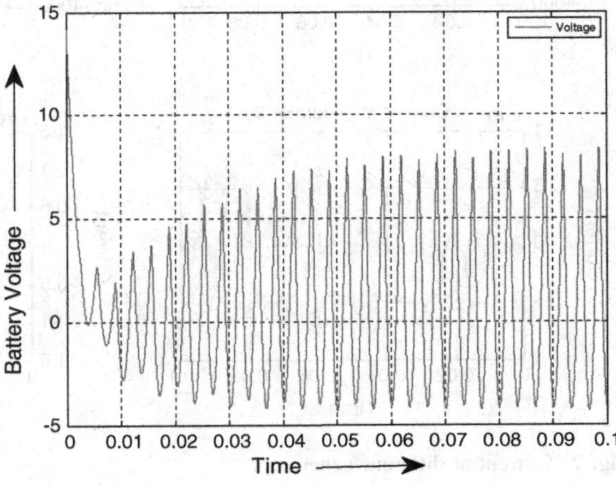

Fig. 5 Battery charging and discharging condition

Figure 6 shows the flowchart of the proposed model. The algorithm first reads the output power of the solar photovoltaic system and the grid. Based on the measured value, it checks for power over flow. If the PV power is sufficient then it will directly connect the PV to the load after proper control of the real and reactive power of the system. For control of real and reactive power, it applies the park's transformation and calculates the phase angle. It then fed the phase angle to the

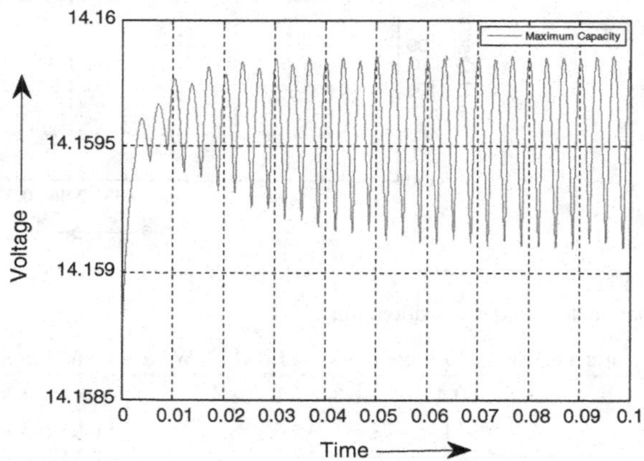

Fig. 6 Battery maximum charging and discharging condition

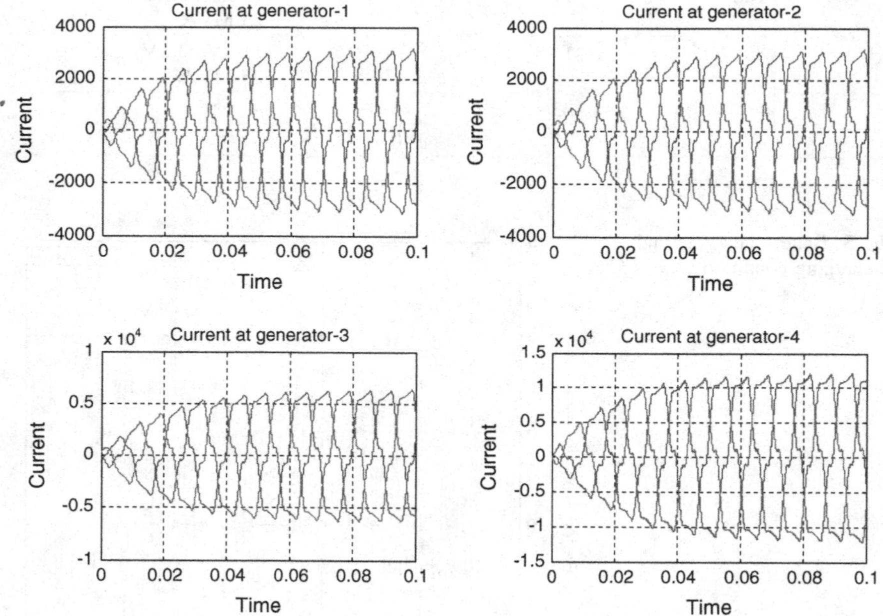

Fig. 7 Current at different generators

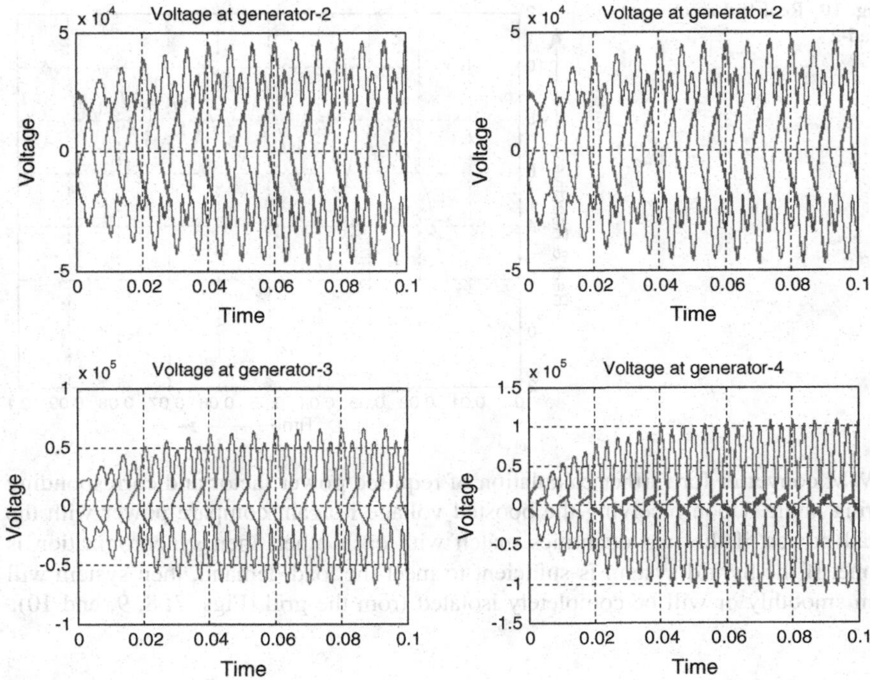

Fig. 8 Voltage at different generators

Fig. 9 Real and reactive at load-1

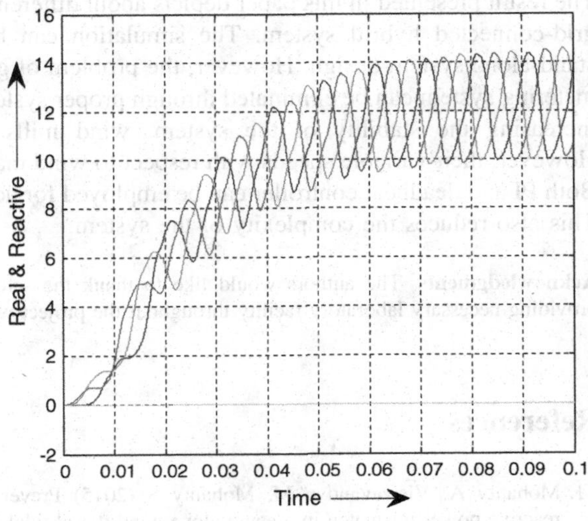

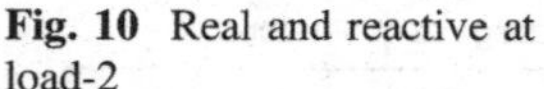

Fig. 10 Real and reactive at load-2

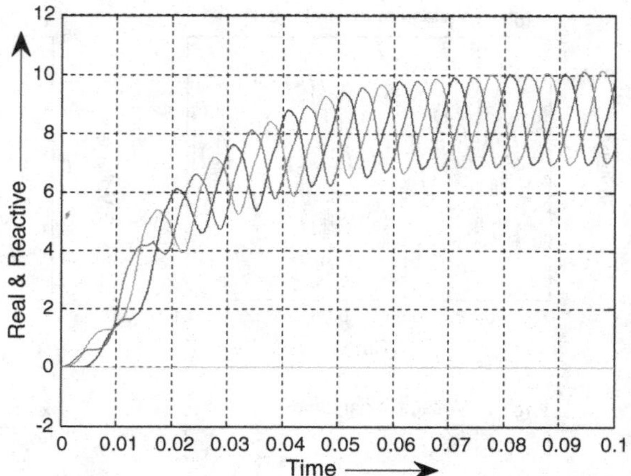

PWM converter for further calculation of required power factor and corresponding firing angle. Upon receiving the boosted voltage it again compare power with the load power, if the two quantities match with each other, then synchronization is done. If the hybrid system is sufficient to meet the load demand, then system will run smoothly or will be completely isolated from the grid (Figs. 7, 8, 9, and 10).

4 Conclusion

The result presented in this paper depicts about different problems associated with a grid-connected hybrid system. The simulation can be taken as a reference for stand-alone system design. However, the problem of grid integration and two-way metering system can be eliminated through proper system modeling and design. For increasing the stability of the system, wind mills can also be incorporated. However, their variable speeds with respect to wind may create instability problem. Both PI and deadbeat controller can be employed for achieving the system stability. This also reduces the complexity of the system.

Acknowledgments The authors would like to thank the school of electrical engineering for providing necessary laboratory facility throughout the project work.

References

1. Mohanty A, Viswavandya M, Mohanty S (2015) Prevention of transient instability and reactive power mismatch in a stand-alone wind-diesel-tidal hybrid system by an ANN based SVC. Aquatic Procedia, 4:1529–1536. ISSN 2214-241X
2. Maouedj R, Mammeri A, Draou MD, Benyoucef B (2014) Performance evaluation of hybrid photovoltaic-wind power systems. Energy Procedia 50:797–807. ISSN 1876-6102

3. Salih T, Wang Y, Awad M, Adam A (2014) Renewable micro hybrid system of solar panel and wind turbine for telecommunication equipment in remote areas in Sudan. Energy Procedia 61:80–83. ISSN 1876-6102
4. Sun H, Luo X, Wang J (2015) Feasibility study of a hybrid wind turbine system—integration with compressed air energy storage. Appl Energy 137:617–628, 1 Jan 2015. ISSN 0306-2619
5. Bocklisch T, Böttiger M, Paulitschke M (2014) Multi-storage hybrid system approach and experimental investigations. Energy Procedia 46:186–193. ISSN 1876-6102
6. Müller M, Bründlinger R, Arz O, Miller W, Schulz J, Lauss G (2014) PV-off-grid hybrid systems and MPPT charge controllers, a state of the art analyses. Energy Procedia 57:1421–1430. ISSN 1876-6102
7. Chaudhary A, Huggett A, Yap WK, Karri V (2014) Remote area hybrid solar-diesel power systems in tropical Australia. Energy Procedia 57:1485–1491. ISSN 1876-6102
8. Notton G, Diaf S, Stoyanov L (2011) Hybrid photovoltaic/wind energy systems for remote locations. Energy Procedia 6:666–677. ISSN 1876-6102
9. Prompinit K, Plangklang B, Hiranvarodom S (2011) Design and construction of a mobile PV hybrid system prototype for isolated electrification. Procedia Eng 8:138–145. ISSN 1877-7058
10. Saheb-Koussa D, Koussa M, Haddadi M, Belhamel M (2011) Hybrid options analysis for power systems for rural electrification in Algeria. Energy Procedia 6:750–758. ISSN 1876-6102
11. Li X, Li Y, Han X, Hui D (2011) Application of fuzzy wavelet transform to smooth wind/PV hybrid power system output with battery energy storage system. Energy Procedia 12:994–1001. ISSN 1876-6102
12. Wang C, Chen W, Shao S, Chen Z, Zhu B, Li H (2011) Energy management of stand-alone hybrid PV system. Energy Procedia 12:471–479. ISSN 1876-6102

Detection of Abrupt Transitions in Night Time Video for Illumination Enhancement

S. Padmavathi and G. Abirami

Abstract A large number of security-related problems could be addressed by night time video surveillance. This has additional challenges of handling nonuniformly illuminated areas in the frame as compared to day time videos. The presence of vehicle lights degrades the existing background extraction process. An intensity-based shot boundary detection technique is applied for the night time videos in this paper. This algorithm is used to identify the abrupt transition frames that are caused by the vehicle lights or moving lights. The frames occuring between such detected frames could be reliably used for background extraction and night time video enhancement. The algorithm is tested on few real-time videos and their results are provided.

Keywords Night time video enhancement · Video surveillance · Shot boundary detection

1 Introduction

Video surveillance is an emerging field where researches are carried out for higher-level analysis such as detection, identification, and recognition of persons and moving vehicles. A night time video surveillance is more challenging due to the high illumination variation that is caused due to lighting. The most of the video surveillance systems rely on the background extraction and foreground detection. The background is extracted from a sequence of frames which are then used as reference frame for further processing. The foreground and analysis relies on the

S. Padmavathi (✉) · G. Abirami
Department of Computer Science Engineering, Amrita School of Engineering, Amrita Vishwa Vidyapeetham, Coimbatore 641112, India
e-mail: s_padmavathi@cb.amrita.edu

G. Abirami
e-mail: abiramigovindarajan@ymail.com

L.P. Suresh and B.K. Panigrahi (eds.), *Proceedings of the International Conference on Soft Computing Systems*, Advances in Intelligent Systems and Computing 397, DOI 10.1007/978-81-322-2671-0_76

extracted reference frame. The background extraction process will be useful for video analysis [1] when there is no abrupt change in the video. The day time video has uniform or gradually changing illumination taken in a shot. Hence the background extraction can be done in one shot.

Shot boundary detection techniques are important in the video analysis. Shot boundary detection (SBD) is the process of detecting the transitions between shots in the video. A shot in the video may be defined as a sequence of frames from a single camera at a time. A shot boundary is the gap between two shots.

In the night time video, abrupt changes in the intensity occur due to vehicle or street lights present in the real-time environment. The abrupt changes in the video are caused due to sudden illumination variations such as immediate switch on or off of any street lamps, vehicles lights, and indoor lighting. The background extraction for a single-shot night time video will not give appropriate results when the entire video is considered. Hence a shot boundary detection technique is applied for the frames of night time video to segregate the frames that have similar illumination. These frames can then be used for background extraction and further enhancement.

The real-time application environment for this system would be monitoring the movement of the persons at night time and for home security purpose. The paper is structured as follows Sect. 2 deals with the literature survey, Sect. 3 explains about the proposed work, Sect. 4 depicts experimental results, and Sect. 5 has the concluding remarks.

2 Literature Survey

Illumination variations in the video scenes could be classified as abrupt and gradual changes. Fading and dissolving are two types of gradual changes. The abrupt transition is named as cut transition [2, 3]. Shot boundary detection techniques assume the illumination variation in the scenes as constant or as a gradually varying one. The abrupt changes in the scene are used to identify the shot boundaries. Types of transitions that occur [4]:

- A fade-in starts with a black frame; gradually the image of the next shot appears, brightening to full strength.
- A fade-out is the opposite of a fade-in.
- A dissolve consists in the superimposition of a fade-out over a fade-in.

Several shot boundary methods provides a better result on abrupt and gradual cuts [9]. The various shot boundary techniques are briefly listed in Table 1 with their pros and cons.

Table 1 Comparison between shot transition detection techniques

Title	Description	Pros	Cons
Pixel-based difference [5, 8, 9]	On comparing the pixel difference between two successive video frames	Detects discontinuity	Highly sensitive to camera motion
Statistical-based difference [5, 8]	Compares few properties of every pixel from several regions between successive frames using statistical computation parameters	A small region tends to reduce detection which is invariant with respect to motion	Detects only fade transition
Transform-based difference [6, 8]	Performs compression difference computation using several transformation methods. For example Discrete Cosine Transformation (DCT) coefficients	A global threshold is set for scene changes, and if there is an uncertainty, a few neighboring frames are selected for further decompression	It does not detect gradual transition like fade and dissolving
Histogram-based difference [7, 8]	Computes color histogram of each frame and compares it to detect shot boundaries	Pixel bases inter-frame difference based on brightness information to detect the presence of abrupt scene changes	The gradual scene changes are not taken into account
Edge-based difference [9]	The edges of successive frames are detected first and then the edge pixels are paired with nearby edge pixels in the other image to find out if any new edges have entered the image	Effective for flash detection	They cannot handle more general illumination variation

3 Proposed Work

To recognize pedestrians in the night time video, the visibility of dark regions has to be improved thereby providing enough visual appearance to the night time video. This requires a different illumination adjustment for regions within a frame. The similar regions across the frames have to be identified for such an enhancement. Due to the presence of vehicles the similar regions differ across the frames in the entire video. Hence the similar frames of the nighttime video are first grouped using an intensity-based shot boundary technique. This technique identifies the abrupt changes in the intensity occurring in the night time video due to vehicles or street lights.

3.1 Methodology

Shot Boundary Detection (SBD)

Intensity-Based Change Detection An intensity-based SBD used in this paper utilizes the histogram of intensity frames. Since the abrupt illumination change is identified as a shot boundary for a night time video, such changes are evident in the intensity frames. Hence the histogram of intensity images is used by the algorithm. The frames that are detected by the algorithm are named as highly varying frames, and the rest are named as low-varying frames.

Algorithm to obtain the highly varying and low-varying intensity frames in the night time video is as follows:

1. Extract night time frames from the night time video.
2. Convert RGB frames into HSV frames.
3. Compute histogram difference between every consecutive intensity frames, say D.
4. Calculate mean μi for each histogram difference frame, di ε D.
5. Compare the mean value μi + 1 with μi.
6. Initialize the counter for extracted frames to zero.
7. If μi + 1 > T, where T = 2 μi (experimentally varied), then extract the i + 1th frame as a shot boundary.
8. Increment the counter.

The frames obtained depict those frames that have abrupt change in the intensity. The counter value indicates the total number of highly varying frames that have occurred in the input video. The frames between the highly varying frames are considered for the background extraction process. This method gives a background that is more reliable than the one constructed from the entire video.

4 Experimental Results

The experimental dataset includes an indoor video which is collected from the website: http://web.media.mit.edu/~raskar/NPAR04/. The video has a person walking from one place to another in indoor environment. Two outdoor surveillance videos collected from the Amrita Vishwa Vidyapeetham University campus are also considered for experimentation. Both these videos are taken from same view point, same location but at different times such as day and night without altering camera parameters. The camera specification is as follows: common CCD outdoor camera, vari focal lens (2.8–12 mm) (2 mega pix), horizontal resolution 700 TV Lines, 36 IR LED, IR distance-20 m. The outdoor videos have huge number of pedestrians and vehicles crossing the security post. These videos are cut to one hour videos for experimentation which are listed in Table 2. Among the videos the night time videos are used for testing the algorithm. The implementation is performed on MATLAB R2013a platform.

Indoor Environment

Shot boundary detection as explained in Sect. 3.1 is applied to the video nite_indoor.avi. The screen shot of the input video is shown in Fig. 1 and the plot of its histogram difference frames is shown in Fig. 2. The peaks in Fig. 2 indicate the frames with large variation. These frames are extracted and their frame numbers are stored. The

Table 2 List of several night videos among the dataset

Name of the video	Day/night	Category	Time in min	No. of frames	Frame rate (fps)
Nite_outdoor.avi	Night	Outdoor	60	89,982	25
Nite_indoor.avi	Night	Indoor	5 s	154	29
Nite_outdoor1.avi	Night	Outdoor	30	43,293	25
Nite_outdoor2.avi	Night	Outdoor	40	58,372	25

Fig. 1 Screen shot of input night time video

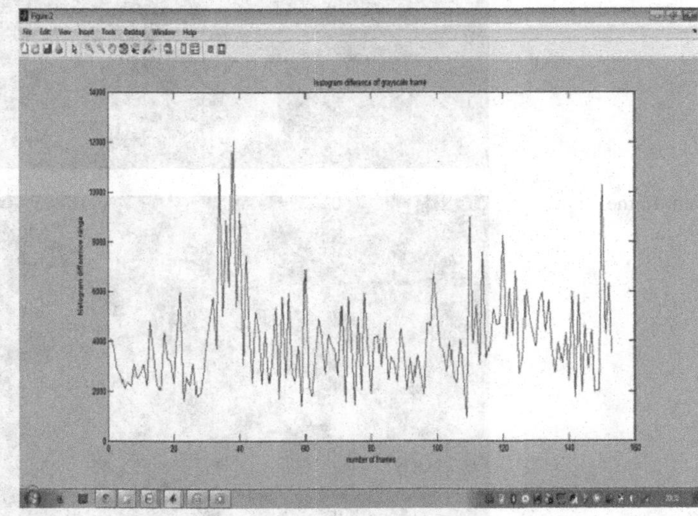

Fig. 2 Plot for histogram difference between video frames

```
Command Window
>> shot_boundary_abrupt
number of changes occured

ans =

    20

current frame number
    13
    17
    34
    42
    45
    53
    55
    60
    71
    73
    76
    78
    81
    97
   110
   114
   141
   143
   145
   150
fx
```

Fig. 3 Change detected frames

Fig. 4 Current frame number: 97

Fig. 5 Current frame number: 110

Fig. 6 Screen shot of night time video

Fig. 7 Current frame number: 8303

frame numbers stored for the indoor video is shown in Fig. 3. Two such frames that are extracted are shown in Figs. 4 and 5. It could be seen that the abrupt change in the indoor environment mainly occurred due to switching of the lights and lights hidden by person.

Outdoor Environment

Similarly the output of the shot boundary algorithm, when applied to the outdoor night time video nite_outdoor.avi with 100 frames, the screen shot of the input

Fig. 8 Current frame number: 8330

Table 3 Change detected frames

Count	1	2
Current frame number	8303	8330

video is shown in Fig. 6. The frames with number 8303 and 8330 are extracted as highly varying frames are shown in Figs. 7 and 8, respectively. Table 3 shows the list of frames with the changes.

5 Conclusion and Future Work

Shot Boundary Detection method is usually applied for the day time videos as a preliminary process for background extraction. In this paper, an intensity-based SBD method is applied for the night time videos. The frames which shows abrupt illumination changes are identified using this method. This process helps in the treatment of the frames in a better way for background extraction and video enhancement. The method in this paper uses the histogram for detecting the shot boundary. The highly illumination variant frames are detected as shot boundary. This enables the automatic grouping of similarly lighted regions. These groups can be utilized for more reliable background extraction followed by the video enhancement. The enhanced videos can be used for detecting people/animals in the darker regions.

References

1. Sowmya R, Shettar R (2013) Analysis and verification of video summarization using shot boundary detection. Am Int J Res Sci Technol Eng Math 3(1):82–86
2. Boreczky JS, Rowe LA (1996) Comparison of video shot boundary detection techniques. In: Storage and retrieval for image and video databases (SPIE), 1996, pp 170–179
3. Qi J. Video classification based on shot detection and motion activity. Beijing University of Posts and Telecommunications, Master level thesis
4. Abdeljaoued Y, Ebrahimi T, Christopoulos C, Mas Ivars I. A new algorithm for shot boundary detection. EPFL signal processing laboratory CH-1015 Lausanne, Switzerland Media Lab, Ericsson Research, Ericsson radio Systems AB, S-164 80 Stockholm, Sweden
5. Mittalkod PS, Srinivasan GN Dr (2011) Shot boundary detection algorithms and techniques: a review. Res J Comput Syst Eng-Int J. ISSN: 2230-8563; e-ISSN-2230-8571, 02(02), June 2011
6. Mohanta PP, Saha SK Member, IEEE, Chanda B, Member IEEE (2012) A model-based shot boundary detection technique using frame transition parameters. IEEE Trans Multimedia 14(1), Feb 2012
7. Sao N, Mishra R (2014) A survey based on video shot boundary detection techniques. Int J Adv Res Comput Commun Eng 3(4), April 2014
8. Lienhart R. Comparison of automatic shot boundary detection algorithms. Microcomputer Research Labs, Intel Corporation, Santa Clara, CA 95052-8819
9. Adhikari P, Gargote N, Digge J, Member IEEE (2009) Video shot boundary detection. IEEE Trans Consum Electron 1
10. Gonzalez RC, Woods RE (2011) Digital image processing. Person Prentice Hall, New Jersey, USA

PSO Algorithm in Quadcopter Cluster for Unified Trajectory Planning Using Wireless Ad Hoc Network

Deepika Rani Sona, Rashmi Rannjan Das, Sarthak Dubey and Yogesh Kaswan

Abstract This paper is aimed at developing a swarm unmanned aerial vehicle (UAV) network for motion in free space controlled by an algorithm depending on axis-aligned minimum bounding space, and a collision is avoided by the use of a collision-free trajectory planning algorithm depending on the position of the UAVs by implementing the optimization technique named particle swarm optimization (PSO) algorithm for intelligent features. PSO is an algorithm where each quad has its own information and also combined information of the environment due to their interaction. Wireless network was used in swarm UAVs implementation to connect to a server processor that analyses the PSO equation and controls the quadcopters. This experiment includes 3 UAVs and one object as target (Stationary and Moving).

Keywords MPU6050 · ETA · ESP8266 · PWM · PSO · UAV · Ad hoc · Swarm · Quadcopters

D.R. Sona (✉) · S. Dubey · Y. Kaswan
School of Electronics Engineering, VIT University, Vellore, India
e-mail: deepika.rs@vit.ac.in

S. Dubey
e-mail: sarthak.dubey@outlook.com

Y. Kaswan
e-mail: yogeshkaswan@gmail.com

R.R. Das
School of Electrical Engineering, VIT University, Vellore, India
e-mail: rashmiranjandas@vit.ac.in

L.P. Suresh and B.K. Panigrahi (eds.), *Proceedings of the International Conference on Soft Computing Systems*, Advances in Intelligent Systems and Computing 397, DOI 10.1007/978-81-322-2671-0_77

1 Introduction

Modern machines produce a marvel example of small parts/individual machines working together to serve a bigger prospect. A similar case is observed in a swarm of robots where each individual machine is exactly identical to another one. The use of quadcopter as UAV to perform tasks that require multiple hands (individual unit support) at the same time is fuelled by the observed maneuverability and stability in quadcopters [1, 2]. An UAV or drone is an aircraft capable of flight using onboard computer or remote control. In swarm of quadcopters both the control mechanisms are used for flight. Each unit contains an onboard computing system to perform calculations for flight stability, orientation, direction, and vectors. Further all the UAV's are connected to a wireless network behaving as clients to a common server under an ad hoc network [3]. The server here may be an RTOS equipped with wireless networks or a standard computer over the network. Each quad is assigned an IP and is kept on a local network. The individual UAV's send out their location vectors based on the PSO algorithm to a central server that calculates the new positions based on the governance of a geometric equation (2-D/3-D/4-D). The equation may be 2D, 3D, or 4D depending upon the position of the quadcopter in a 2D plane or within a 3D space or 4D—moving as a cluster in this particular case.

The problem with movement of multiple quadcopter as a swarm is collision avoidance among individual units and with the environment at the same time with a constraint to maintain positions [4]. Furthermore, planning optimal collision-free trajectories for multiple UAV leads to optimization problems with multiple local minimums in most cases and, thus, local optimization methods as gradient-based techniques are not well suited to solve it. The application of evolutionary techniques or PSO is an efficient and effective alternative for this problem, since they are global optimization methods [5].

The quadcopter used here is equipped with a Arduino pro mini as a local client computational controller with MPU6050 IMU as the accelerometer-gyroscope sensor for stability sensor. Further the quadcopter also contains an ESP8266 serial to wireless transreceiver module for communication over IP on ad hoc [5, 6]. Additionally the UAV is also equipped with 2 ultrasonic-based altimeters for altitude and ceiling detection in closed spaces. To drive the motors of the quad on a fast changing PWM, a Darlington transistor ULN2003A is used with a 300 mAh Li–Po battery to power the whole quadcopter assembly [7].

2 Problem Formulation

2.1 Quadcopter Design

The first problem comes with quadcopter to simple, light weighted, ergonomic, and use up the smallest possible 3D cubic space. To achieve this, a 5.5 × 5.5 cm Vero board is soldered with an Arduino Pro Mini (5 V, 16 MHz) with motor output connections from pin 3, 5, 6, and 9 PWM ports to the Darlington transistor. MPU6050 has pins SDA and SCL as serial data and serial clock connected to the A4 and A5 analog input pins. The wireless module is connected to the RX and TX pins of the controller through a logic level converter [8]. The motors are connected using carbon fiber rods of 6.0 cm in length. This makes the UAV take up not more than 14 × 14 × 4 cm in 3D space.

2.2 *Individual Flight Stability*

Kalman filter with multiple dimensions using matrix transformation is used to obtain the yaw, pitch, and roll angles in inertial frame with respect to earth. This greatly helps in improving the flight behavior through IMU using PID while changing motor rpm to maintain a stable flight and position. MPU 6050 uses a combination of a gyroscope and an accelerometer that provides the yaw, pitch and roll movements.

The altitude sensor at the bottom and the one at the top are used to maintain position and a particular height in 3D space. In closed rooms, the ratio of both sensors is used as a stimulus for altitude, while in open areas with upper sensor reading infinite values the quadcopter used only ground as reference for height [6, 9].

3 Wireless Network

ESP8266 wireless module is used to create and connect to an ad hoc network and uses AT command set. Each individual quadcopter obtains it's own IP address in the form of 192.168.X.X that is used to transmit data individually to each unit or UAV. This helps in using a lot of UAVs on a small bandwidth hence improving the data rate. The server is a program on a computer or android phone with processing IDE as the medium. All the PSO calculation is done on the processing server, and data are then sent to each IP hence the quadcopters.

4 Particle Algorithm: Basic PSO Algorithm

The implemented algorithm is based on [1]. Let S be the number of particles in the swarm, each particle is defined by a state vector x_i in the search space and a velocity vector v_i. This state vector contains the information about the location of the intermediate waypoint and the velocity in the first sector of the trajectory of each

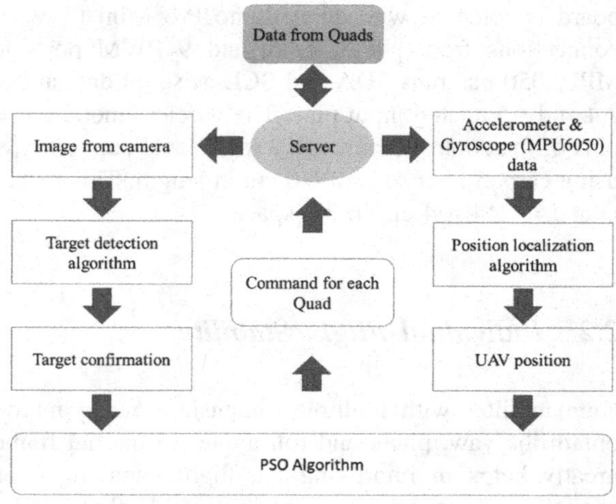

Fig. 1 The working of network through the server

UAV. Note that the speed in the second sector is calculated so the ETA to the final waypoint is the same as in the original trajectory.

In the first place, the swarm is initialized by randomly assigning initial locations and velocities with a uniform distribution. Then a special particle containing the initial solution is added to ensure the existence of one conflict-free solution at any time. Let p_i be the best-known state vector of particle *i*. Let *g* is the best-known state vector of the entire swarm. These are recalculated whenever a new iteration is obtained. Then the exploration loop is executed. At every iteration, both the state vector and the velocity of each particle are updated by applying the expressions indicated in steps 10 and 11 (see Algorithm 1) (Fig. 1).

The most important parameters in this formula are the social weight, φ_g, and the local weight, φ_p. ω is the inertia weight. r_g and r_p are vectors where each component is generated at randomly with an uniform U (0, 1) distribution. Local and global best state vectors are also updated if necessary (steps 13–15).

Using many different termination criteria can finish the exploration loop; among these criteria, a timeout condition and a convergence condition (most of the individuals lay into a tight region of the search space) are the common approaches. In this paper, the algorithm concludes when the available computation time is reached [10].

The parameters φ_g and φ_p have been tuned by performing several tests with the same conditions and only changing one parameter at a time. These parameters are usually selected in the interval [0,1]. In our case, the best values found were $\varphi_g = 0.9$ and $\varphi_p = 0.1$ [11] (Fig. 2).

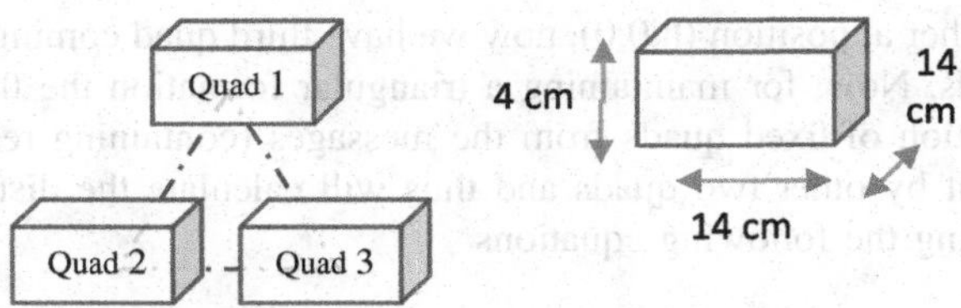

Fig. 2 displays the quadcopters in an equilateral alignment in 3D space with cuboids takes for collision avoidance [13]

Algorithm 1: [12, 13]

for Each particle **do**
  Initialize each particle's state vector x_i with the desired probability function
  Initialize particle best state vector $p_i \leftarrow x_i$
  If $f(p_i) < f(g)$ update the swarm best state vector $g \leftarrow x_i$
  Initialize each particle's velocity vector v_i.
  An uniform distribution is usually used.
end for
repeat
  for each particle do
  Pick random numbers r_g r_p with U(0,1)
  Update the particle's velocity: $v_i \leftarrow \omega v_i + \varphi_p r_p (p_i - x_i) + \varphi_g r_g (g - x_i)$
  Update the particle's state vector: $x_i \leftarrow x_i + v_i$
  if $f(x_i) < f(p_i)$ **then**
  Update the particle's best known state vector
    if $f(x_i) < f(g)$ **then**
    Update the swarm's best known state vector $g \leftarrow x_i$
    end if
  end if
for end
until A termination criterion is met.

4.1 Geometric Formations

- Quadcopters can obey formations in circle and spheres defined by $-a^2 + b^2 = r^2$
- Sphere: $a^2 + b^2 + c^2 = r^2$

The quads can be programmed to maintain triangle formation just using the simple concept of maintaining a particular distance between each quad [14, 15]. For simplicity, let us assume that we have two fixed quads with one being placed at origin and another at position (h,0,0), now we have third quad coming toward these two fixed quads. Now, for maintaining a triangular formation the third quad will check the position of fixed quads from the messages (containing respective position) being sent by other two quads and thus will calculate the distance between them by applying the following equations

$$d_1^2 = x^2 + y^2 + z^2 \tag{1}$$

and

$$d_2^2 = (h - x)^2 + y^2 + z^2 \tag{2}$$

Here, d_1 and d_2 are the distances of fixed quads from the third dynamic quad and (*x*, *y*, *z*) is the position of third quad at a given instant. Now, this is just a simple scenario but in reality this above stated quad satisfies condition and thus a perfect triangular form is maintained [14, 16, 17].

5 Result

According to the mathematical calculations and algorithms the output of the networked program could be observed as a cluster of quadcopters in swarm behaving as a single entity maintaining a particular geometric shape and following a trajectory as one entity (Figs. 3 and 4).

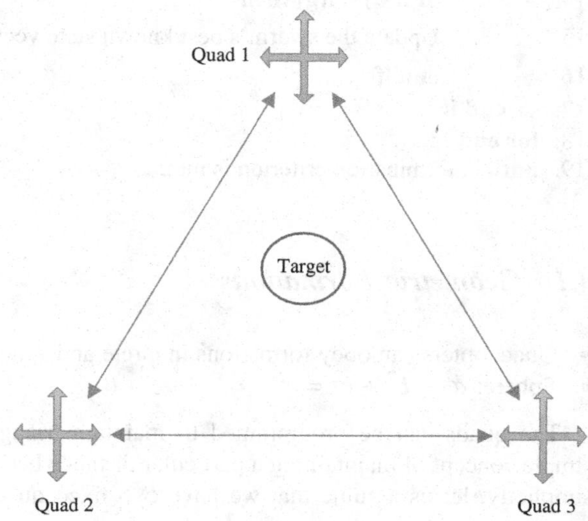

Fig. 3 Quads approaching target

Fig. 4 Fully assembled quadcopter

Acknowledgments This work was supported by School of Electronic engineering, VIT University. The authors would like to thank the Director, SENSE, and the management of VIT University, Vellore for providing the facilities to carry out this study.

References

1. Pedersen MEH (2010) Good parameters for particle swarm optimization. In: Hvass Laboratories, Technical Report no. HL1001, 2010
2. Lupashin S, Schöllig A, Sherback M, D'Andrea R (2010) A simple learning strategy for high-speed quadrocopter multi-flips. In: Proceedings of the IEEE international conference on robotics and automation (ICRA), 2010
3. Müller M, Lupashin S, D'Andrea R (2011) Quadrocopter ball juggling. In: Proceedings of the IEEE international conference on intelligent robots and systems (IROS), 2011
4. Lindsey Q, Mellinger D, Kumar V (2011) Construction of cubic structures with quadrotor teams. In: Proceedings of the robotics: science and systems (RSS), Los Angeles, CA, USA, 2011
5. Grzonka S, Grisetti G, Burgard W (2009) Towards a navigation system for autonomous indoor flying. In: Proceedings of the IEEE international conference on robotics and automation (ICRA), 2009
6. Achtelik M, Bachrach A, He R, Prentice S, Roy N (2009) Stereo vision and laser odometry for autonomous helicopters in GPS-denied indoor environments. In: Proceedings of the SPIE unmanned systems technology XI, 2009
7. Blösch M, Weiss S, Scaramuzza D, Siegwart R (2010) Vision based MAV navigation in unknown and unstructured environments. In: Proceedings of the IEEE international conference on robotics and automation (ICRA), 2010
8. Achtelik M, Achtelik M, Weiss S, Siegwart R (2011) Onboard IMU and monocular vision based control for MAVs in unknown in- and outdoor environments. In: Proceedings of the IEEE international conference on robotics and automation (ICRA), 2011
9. Huang AS, Bachrach A, Henry P, Krainin M, Maturana D, Fox D, Roy N (2011) Visual odometry and mapping for autonomous flight using an RGB-D camera. In: Proceedings of the IEEE international symposium of robotics research (ISRR), 2011

10. Engel J, Sturm J, Cremers D (2012) Camera-based navigation of a low-cost quadrocopter. In: Proceedings of the international conference on intelligent robot systems (IROS), Oct 2012
11. Kuchar JK, Yang LC (2000) A review of conflict detection and resolution modeling methods. IEEE Trans Intell Transp Syst 1:179–189
12. Alejo D, Cobano JA, Heredia G, Ollero A (2013) Particle swarm optimization for collision-free 4D trajectory planning in unmanned aerial vehicles. In: IEEE unmanned aircraft systems (ICUAS) conference, 2013
13. Engel J, Sturm J, Cremers D (2012) Accurate figure flying with a quadrocopter using onboard visual and inertial sensing. In: IEEE/RJS international conference on intelligent robot systems (IROS), 2012
14. Gilmore JF (1991) Autonomous vehicle planning analysis methodology. In: AIAAA guidance navigation control conference, 1991, pp 2000–4370
15. Szczerba RJ (1999) Threat netting for real-time, intelligent route planners. In: IEEE symposium information, decision control, 1999, pp 377–382
16. Klein G, Murray D (2007) Parallel tracking and mapping for small AR workspaces. In: Proceedings of the IEEE international symposium on mixed and augmented reality (ISMAR), 2007
17. Mellinger D, Michael N, Kumar V (2010) Trajectory generation and control for precise aggressive maneuvers with quadrotors. In: Proceedings of the international symposium on experimental robotics, Dec 2010

Stabilization of UAV Quadcopter

P. Eswaran, Mahendar Guda, Mukunda Priya and Zeeshan Khan

Abstract This paper presents the stabilization controller design and implementation of cost-effective UAV (unmanned aerial vehicle) quadcopter which can handle small disturbances and remain stable. PID controller provides stability by controlling the speed of quadcopter. Flight controller is used to control the speed of the propeller and also to monitor the orientation of the craft using accelerometer and gyroscope, and ESC continually adjusts the motor speeds to keep the airframe stable. Algorithm is also developed to control the quadcopter in a fraction of seconds. If the quadcopter deviates from the set position, flight controller automatically controls the quadcopter and maintains stability. Test cases for different tilt angle and manoeuvre along pitch, yaw, and roll were simulated using LabVIEW v11.0.1 environment and PID controller parameters are optimized to have quicker response time. Finally, the developed PID controller algorithm was implemented in the designed quadcopter and tested, which yields the desired performance.

Keywords Quadcopter design · Stability controller · UAV flight controller · H-airframe design · PID controller for quadcopter

P. Eswaran (✉) · M. Guda · M. Priya · Z. Khan
Department of Electronics and Communication Engineering, SRM University, Kattankulathur, Chennai 603203, India
e-mail: eswaran.p@ktr.srmuniv.ac.in

M. Guda
e-mail: gmr.mahendar142@gmail.com

M. Priya
e-mail: mukundapriyapmp@gmail.com

Z. Khan
e-mail: zshn25@gmail.com

L.P. Suresh and B.K. Panigrahi (eds.), *Proceedings of the International Conference on Soft Computing Systems*, Advances in Intelligent Systems and Computing 397, DOI 10.1007/978-81-322-2671-0_78

1 Introduction

Quadcopters are gaining more importance due to easy implementation in many applications such as building inspection after a disaster, search and rescue missions, transportation, agriculture/remote farming, and aerial imaging. Quadcopter is popular due to its simplicity of mechanical structure and low cost. This work presents the design and implementation of a stable quadcopter controller, which can handle small disturbances and remain stable. Quadcoptor is provided with single axis tilting mechanism for individual arm which will increase stability and augment advanced manoeuvers.

Nemati Kumar derived single-axis stabilization using dynamic modeling from the relationship between tilting-rotor angles and the quadcopter orientation in one direction [1]. Tanveer et al. developed efficient control strategy by combining PID controller with NMPC controller for altitude and attitude controlling of quad-rotor UAV under disturbance and noisy conditions [2]. Sangyam et al. compared conventional PID and fuzzy based auto-tuning PID algorithms for no disturbance force and sinusoidal force under payload variation [3]. Salhi et al. developed a PID controller algorithm to control the quad rotor system modified by simplification of the existing controller design [4]. Joyo et al. present the altitude and horizontal motion control technique of quad rotor [5]. Goodarzi et al. developed control systems that work directly on the special Euclidean group to avoid singularities of minimal attitude representations or ambiguity of quaternions [6]. Sufendi et al. proposed a PID control system with Ziegler-Nichols tuning method for the longitudinal and lateral directional dimensions, which include angular rate, attitude, altitude, and navigation control which are able to stabilize an unstable system or to improve the system response [7].

This paper is organized as follows; Sect. 2 explains the various subsystem of UAV quadcopter design. Section 3 explains the functional block diagram of proposed UAV transreceiver. Simulation and experimental results are discussed in Sect. 4 followed with the conclusion.

2 System Components of UAV Quadcopter

2.1 *Quadcopter System Design*

Airframe of quadcopter uses H-shape structure and mounting is provided to fix Brush Less DC (BLDC) motor at the end. Figure 1 shows the construction of proposed H-shape airframe structure with 4 BLDC motor. Two motors M1and M3 will rotate in clockwise direction, while the other two motors M2 and M4 rotate in anticlockwise direction. Two motors adjacent to each other are always in the opposite direction of rotation. Thrust produced by motors should be twice that of the total weight of the quadcopter. If the thrust generated by the motors is too little,

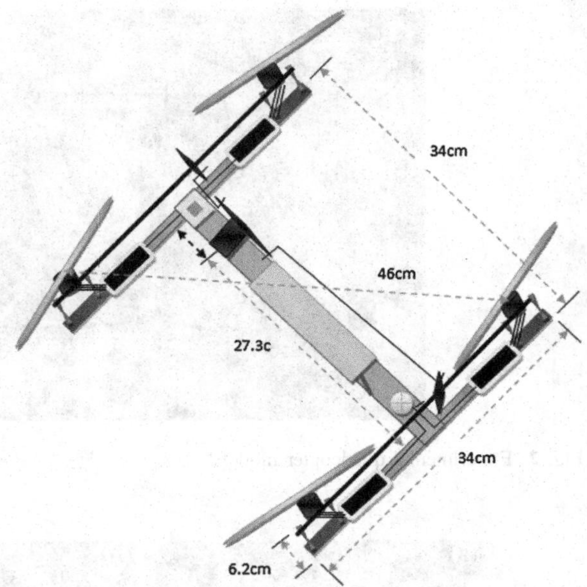

Fig. 1 H-airframe structure design (all dimensions are in cm)

the quadcopter does not take OFF. However, if the thrust is more than the design level, the quadcoptor might become too nimble and hard to control.

2.2 *Propulsion System*

BLDC electric motor used to produce thrust along with the propeller is shown in Fig. 2. DJI 2212 low weight BLDC motor produces 920 rpm/V; a peak power of 370 W is used in the modeling [8]. A propeller is a type of fan that transmits power by converting rotational motion into thrust. Four propellers of span 10 in., with fixed pitch angle of 4.5° are used in modeling [9]. Diagonal propeller pairs spin in the same direction and also have opposite tilting combination producing lifting thrust.

2.3 *Electrical System*

Power Bank LiPo battery is used as a power source for quadcopter due to high power to weight ratio [10]. NiMH is more cost-effective, but heavier than LiPo, hence it is not preferred. LiPo battery can be found in a single cell (3.7 V) in a pack of over 10 cells connected in series produces 37 V. A popular choice of battery for a Quadcopter is the 3SP1 batteries with terminal voltage of 11.1 V. Battery has peak

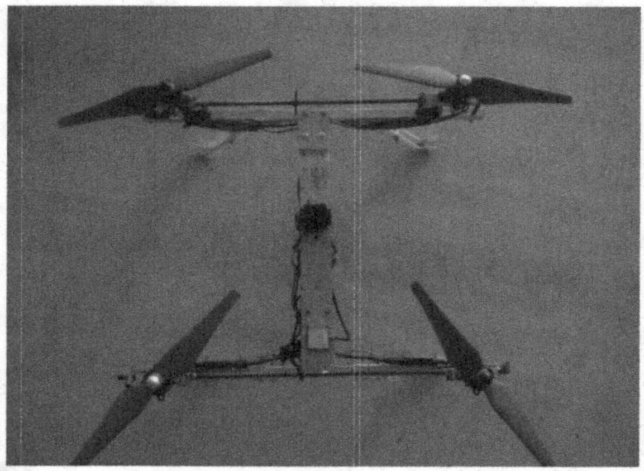

Fig. 2 Experimental quadcopter model

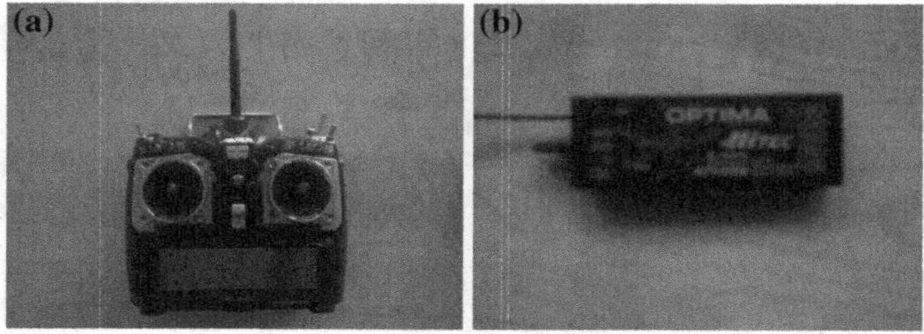

Fig. 3 Radio communication module. **a** Transmitter. **b** Receiver [12]

discharge rate of 25 times the rated capacity and a capacity of 5000 mAh will deliver maximum current of 25 C × 5000 mAh = 125 A.

Engine Speed Controller (ESC) ESC is used to control the speed of the motor by pulse-width modulation (PWM). PWM output to the ESC varies the output voltage by increasing or decreasing PWM signal duty cycle at the input to the ESC. The speed of a BLDC motor is controlled by changing the input DC supply voltage using DJI E300 ESC [11].

Radio Communication System Hitec Aurura 9 has line channel frequency 2.4 GHz transmitters and receiver pair [12]; that used in this work is shown in Fig. 3. It uses Adaptive Frequency Hopping Spread Spectrum (AFHSS). When the transmitter is in normal mode power of transmitter and receiver every time, it always uses the same distinct channels in the 2.4 GHz band spectrum. When flying the quadcopter, if the system does not respond, adjust the setting. Jerky movements are observed if there is co-channel interference. Hence, put Hitec system in scan mode to adjust the channels to avoid interference. The system will continue to adjust its channels used in the 2.4 GHz band every time it is powered in scan mode, and hence AFHSS is used.

2.4 IMU Controller

MPU-6050 IMU controller, which has 6 degrees of freedom (DOF), is used in this work. It has accelerometer to measure acceleration and gyroscope to measure orientation or angular velocity, to calculate the angular position (angle) by integrating the angular velocity.

2.5 PID Control System

Proportional, integral, and derivative (PID) controller fixed on UAV quadcopter is shown in Fig. 4. The variation of each of the parameters alters the effectiveness of the stabilization of the quadcopter. Generally, the 3 PID loops with their own PID coefficients were present, one per axis. Hence P, I, and D values for each axis of roll, pitch, and yaw have to be set independently. These parameters can cause the following behavior.

Proportional Gain coefficient (P) Stability of the quadcopter depends on this coefficient, which determines the sensitivity of the actuators toward each error in the three axes. If the coefficient is kept higher, the quadcopter is more sensitive and more reactive to angular change and vice versa.

Integral Gain coefficient (I) This coefficient is used to maintain the precision of the angular position. When disturbance is added to the quadcopter, its angle changes. In theory, it is in how much time the angle has changed and time to return

Fig. 4 Flight controller [13]

to 0° deviation. This term is useful with irregular wind and turbulence from the motors. However, when the I value is too high, the quadcopter might begin to have a slow reaction and a decreasing effect of the proportional gain. It oscillates, if P gain is high, however, with lower frequency.

Derivative Gain coefficient (D) This coefficient makes the quadcopter reach the desired attitude fast. Hence, it is known as the accelerator parameter as it amplifies the user input. It also decreases the control action faster, when the error is decreasing faster. In practice, it increases the reaction speed for certain cases and it increases the effect of the P gain.

Steady-state equations are obtained from [14] for the moment of inertia as given in Eq. 1.

$$I = \begin{bmatrix} I_{xx} & I_{xy} & I_{xz} \\ I_{yx} & I_{yy} & I_{yz} \\ I_{zx} & I_{zy} & I_{zz} \end{bmatrix} \tag{1}$$

where, I_{xx}, I_{yy}, and I_{zz} are the moments of inertia and the moment of inertia (I) is mass times the square of perpendicular distance to the rotation axis, $I = mr^2$. Therefore, I_{xy}, I_{xz}, I_{yz}, I_{yx}, I_{zx}, and I_{zy} are the products of inertia decentralized equation of motion.

Assuming perfect symmetry of quadcopter about the three axes, the products of inertia becomes zero [15]. According to Lagrange mechanics, the Lagrangian function L is given in Eq. 2.

$$L = T - V \tag{2}$$

where T is the total kinetic energy and V is the total potential energy of the system. But the potential energy is zero initially and rotation is analyzed, T the rotational kinetic energy of the quadcopter is finally given by Lagrange equation given in Eq. 3.

$$\frac{\partial}{\partial t}\left(\frac{\partial T}{\partial q}\right) - \frac{\partial T}{\partial q} = 0 \tag{3}$$

where q represents the generalized coordinates and its derivation is generalized velocity [16].

$$\begin{aligned} I_{xx} &= \Sigma_i m_i (y_i^2 + z_i^2) \\ I_{yy} &= \Sigma_i m_i (x_i^2 + z_i^2) \\ I_{zz} &= \Sigma_i m_i (x_i^2 + y_i^2) \end{aligned} \tag{4}$$

where m_i is the mass of a quadcopter (e.g., motor, ESC, battery, etc.), and x_i, y_i, and z_i are the perpendicular distances between the center of mass and the specified axis. Six generalized coordinates, corresponding to x, y, z are φ (roll), θ (pitch), and ψ (yaw). These generalized coordinates are reduced to just the Euler angles (φ, θ, ψ) when strictly rotation is analyzed. Yaw rotation without restricting the frame physically by spinning M1 and M3 in the clockwise direction and M2 and M4 in the anticlockwise direction, will bring the net moment about the z-axis to zero. Under the stabilization process it was assumed that the quadcopter was not translating in the z direction. Hence, the only generalized coordinates of concern were φ and θ. As rotational kinetic energy is used, instead of mass, it is moment of inertia and velocity as it is angular velocity.

Therefore, the rotational kinetic energy, T_{rot} is given as

$$\begin{aligned} T_{\mathrm{rot}} &= \frac{1}{2} I_{xx}(\dot{\varphi} - \dot{\psi}\sin\theta)^2 + \frac{1}{2} I_{yy}(\dot{\theta}\cos\varphi + \dot{\psi}\cos\theta\sin\varphi)^2 \\ &+ \frac{1}{2} I_{zz}(\dot{\psi}\cos\theta\cos\varphi + \dot{\theta}\sin\varphi)^2 \end{aligned} \tag{5}$$

Assume that ψ (and therefore dψ/dt) are zero, and the following equations of motion are obtained:

$$\mathrm{roll}(\varphi) : \ddot{\varphi} = \frac{\mathrm{M}_{\varphi} - I_{yy}\theta^2\cos\theta\sin\varphi}{I_{xx}} \tag{6}$$

$$\mathrm{pitch}(\theta) : \ddot{\varphi} = \frac{\mathrm{M}_{\theta} + (I_{yy} - I_{zz})2\dot{\theta}\cos\varphi\sin\varphi}{I_{yy}\cos^2\varphi + I_{zz}\sin^2\varphi} \tag{7}$$

where M is the applied moment.

3 Block Diagram of UAV Quadcopter

Functional block diagram of UAV quadcopter transreceiver is shown in Fig. 5. This receiver which uses NAZA flight controller controls 4 BLDC through ESC. It also receives signal from IMU to measure the attitude of quadcopter. It also receives control signal from RF radio transmitter.

Functional block diagram of onboard flight control interface is shown in Fig. 6. Flight controller has 9 pins, out of which 6 pins are connected to that of the radio receiver and 4 to the ESC. Each ESC is connected to the flight controller and also a

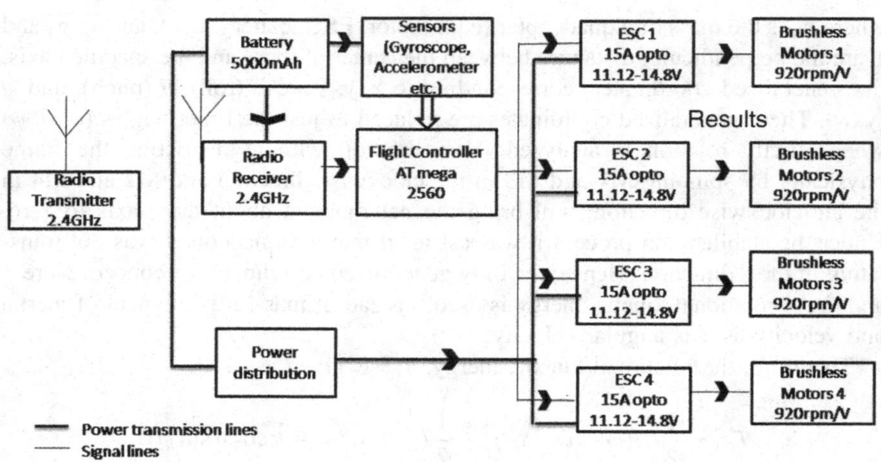

Fig. 5 Functional block diagram of UAV quadcopter

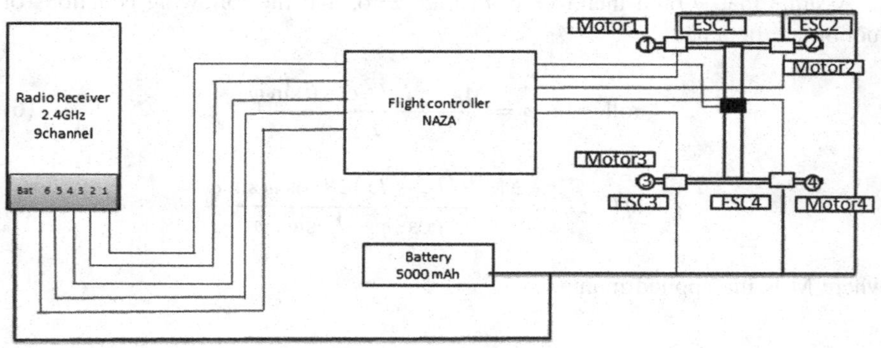

Fig. 6 Onboard controller interface

BLDC motor. Battery power is connected to all the components either by power distribution board or by direct connection. Inputs from IMU are combined and synchronized with signal clock. Single output is received due to Kalman filter.

4 Results and Discussion

Modeled UAV quadcopter under experimental testing is shown in Fig. 7.

Fig. 7 Flying quadcopter

4.1 *Simulation Result*

Case 1: 0° roll along *x*-axis For 0° roll angle along *x*-axis, the response time to stability is 500 ms and no overshoot. It is observed that if the initial angle is zero the set point is zero and if the controller is ON, under quadcopter flying condition. It produces angle decrease by −2.3° and then goes back to zero as shown in Fig. 8a. The reason is due to air turbulence, which induces additional damping.

Case 2: 5° roll along *x*-axis When set point is 5°, fast response of 765 ms is as in shown Fig. 8b. Steady state is reached within 1 s.

Case 3: 15° roll along *x*-axis For the set point of 15°, fast response for 80 ms as shown by the steep slope is shown in Fig. 8c. The slope changes at 0.86 s and reaches the steady state.

Case 4: 30° roll along *x*-axis For the set point of 30°, fast response of 235 ms and overshoot of 3.04438 % is observed as shown in Fig. 8d. The slope changes at 300 ms and reaches steady state.

Case 5: −5° roll along *x*-axis For the set point of −5°, response time is about 2.075 s and overshoot of 5.4229 % is observed is as shown in Fig. 8e.

Case 6: −30° roll along *x*-axis For the set point of −30°, response time is about 2.075 s and overshoot of 5.5 % is observed as shown in Fig. 8f.

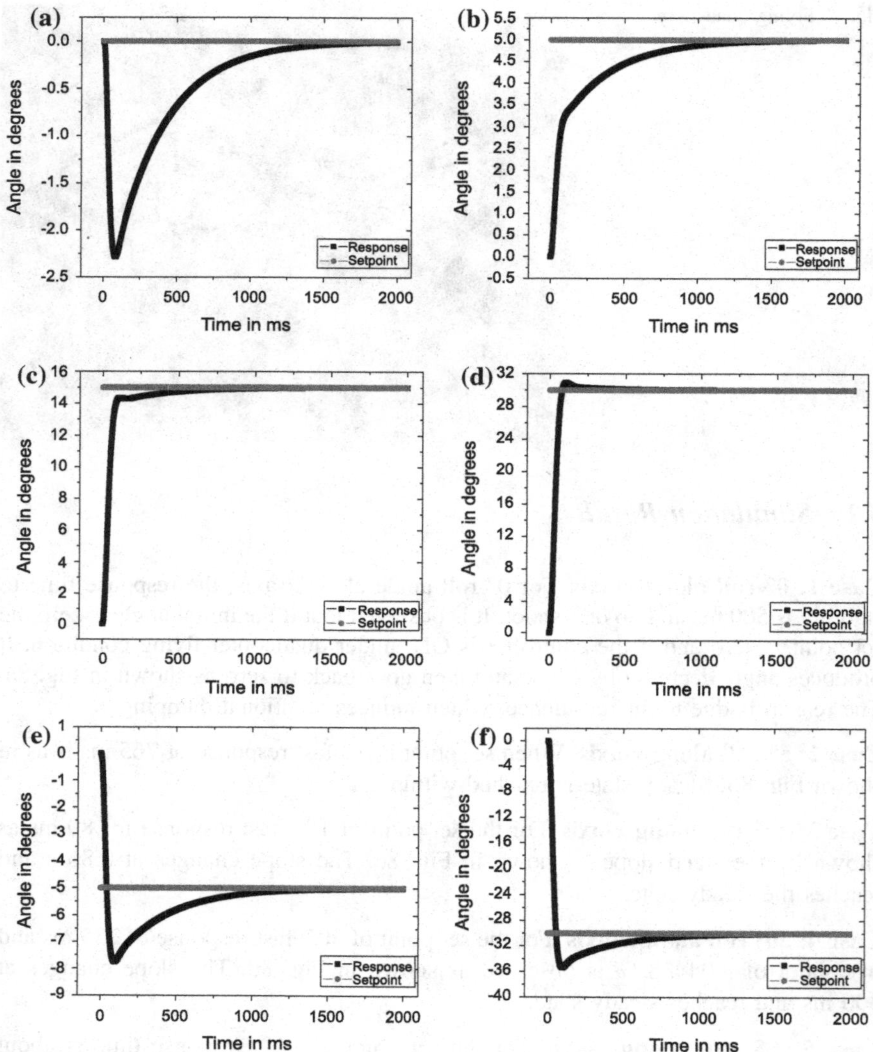

Fig. 8 Simulation results of PID controller. **a** 0° roll, **b** 5° roll, **c** 15° roll, **d** 30° roll, **e** −5° roll, **f** −30° roll

5 Conclusion

Low cost stabilization PID controller for UAV quadcopter is designed, implemented, and experimented and its performance is analyzed. From the simulation result, it is observed the designed flight controller quickly responses to the change in roll angle with very less overshoot. The practical quadcopter output confirms the simulation result. Thus, a low cost quadcopter control system includes stability and withstands small external disturbances. Only the roll manoeuvre along *x*-axis was tested. The result is assumed to be similar for pitch and yaw angles. Further work may include testing all the three angles along *x*, *y*, and *z* directions.

Acknowledgments The authors would like to thank the Department of Electronics and Communication Engineering for providing the facility and support.

References

1. Nemati A, Kumar M (2014) Modeling and control of a single axis tilting quadcopter. In: American control conference (ACC), 2014, pp 3077–3082
2. Tanveer MH, Hazry D, Ahmed SF, Joyo MK (2014) NMPC-PID based control structure design for avoiding uncertainties in attitude and altitude tracking control of quad-rotor (UAV). In: IEEE 10th international on signal processing & its applications (CSPA), Colloquium, 2014, pp 117–122
3. Sangyam T, Thailand Laohapiengsak P, Chongcharoen W, Nilkhamhang I (2010) Autonomous path tracking and disturbance force rejection of UAV using fuzzy based auto-tuning PID controller. In: International conference in electrical engineering/electronics computer telecommunications and information technology (ECTI-CON), (2010), pp 1–4
4. Salih AL, Moghavvemi M, Mohamed HAF, Gaeid KS (2010) Modelling and PID controller design for a quadrotor unmanned air vehicle. In: IEEE international conference in automation quality and testing robotics (AQTR), (2010), pp 1–5
5. Joyo MK, Hazry D (2013) Altitude and horizontal motion control of quadrotor UAV in the presence of air turbulence. In: IEEE conference in ICSPC, 2013, pp 16–20
6. Goodarzi F, Daewon L, Taeyoung L (2013) Geometric nonlinear PID control of a quadrotor UAV on SE(3). In: European control conference (ECC), 2013, pp 3845–3850
7. Sufendi Trilaksono BR, Nasution SH, Purwanto EB (2013) Design and implementation of hardware-in-the-loop-simulation for UAV using PID control method. In: 3rd international conference in instrumentation, communications, information technology, and biomedical engineering (ICICI-BME), 2013, pp 124–130
8. http://multirotorforums.com/threads/specifications-dji-2212-920kv.3060/
9. http://rctimer.com/product-946.html
10. http://www.gensace.de/gens-ace-5000mah-7-4v-50c-2s1p-hardcase-lipo-battery-10.html
11. http://www.buildyourowndrone.co.uk/dji-e300-15a-opto-esc.html
12. http://hitecrcd.com/products/aircraft-radios-receivers-and-accessories/2.4ghz-aircraft-receivers-modules/optima-9-9-channel-2.4ghz-receiver/product
13. http://www.dji.com/product/naza-m-v2
14. https://sites.google.com/site/quadcopterdesignfabrication/system-analysis-and-simulation
15. http://encyclopedia2.thefreedictionary.com/Product+of+Inertia
16. http://en.wikipedia.org/wiki/Generalized_coordinates

A Novel Encryption Method with Super Resolution Imaging Using DWT

K. Kishore Kumar, S. Jayanthi, M. Saranya and V. Elamaran

Abstract Secure transmission of images have become an important in case of its application and research. This study proposes a discrete wavelet-based resolution synthesis using interpolation in discrete wavelet transform (DWT) with a novel encryption method. Highly resoluted image has its importance in various applications like radar, electron microscopy, digital X-ray mammography, optical coherence tomography, astronomy, etc. High-resolution image is recovered from a low-resolution image where the image is transformed using Cohen-Daubechies-Feauveau 9/7 wavelet to obtain significant effectiveness in image resolution. By using the DWT, low-quality image is divided into various subbands of different frequency levels. These low-resoluted images are then interpolated using Cohen-Daubechies wavelet 9/7 coefficients and this interpolation provides nonlinearity over the image in order to preserve the visual smoothness edges. Preserved subband images are projected using fanbeam projection. These preserved various level images are used for reconstruction. High-resoluted image is then obtained using inverse discrete wavelet transform. These images are encrypted using the Pascal series. A significant improvement is obtained over other 2-D wavelet decomposition techniques such as bilinear, bicubic, nearest neighbor, and box wavelets. This study also implements this technique with the Digital Imaging and Communications in Medicine (DICOM) images for the betterment of performance metrics like peak signal-to-noise ratio (PSNR) and structural similarity index (SSIM). These results are produced using MATLAB simulation software.

Keywords Daubechies wavelet · DICOM · DWT · Image resolution · PSNR · SSIM

K.K. Kumar (✉) · S. Jayanthi · M. Saranya · V. Elamaran
Department of ECE, SEEE, SASTRA University, Thanjavur, Tamil Nadu, India
e-mail: kumar.kishore077@gmail.com

L.P. Suresh and B.K. Panigrahi (eds.), *Proceedings of the International Conference on Soft Computing Systems*, Advances in Intelligent Systems and Computing 397, DOI 10.1007/978-81-322-2671-0_79

1 Introduction

Transmission of multimedia technologies over an insecure channel is highly vulnerable to attacks. To achieve confidentiality and security over an insecure channel, a number of encryption algorithms have been proposed [1]. Cryptography is nothing but scrambling of the data to introduce security in transmission images. In this paper, a new encryption method based on using Pascal series is proposed [2].

Wavelet-based image resolution synthesis had its importance in various fields evidently. Interpolation method evolved as a major advantage in the fields of both signal processing and image processing. It performs the process of obtaining enlarged images from the original image with quality to any factor of magnification. The image resolution synthesis has been used in many practical applications where multiple frames of the same scene can be obtained, such as video applications, medical imaging, and satellite imaging.

Discrete wavelet transform decomposes an image into low and high-frequency subband images. These low-frequency components are dissected into low–low (LL) and low–high (LH) subband images. Similarly, the high-frequency components are dissected into high–low(HL), high–high(HH) subband images. Low-frequency components are named as approximation coefficients and high-frequency components are named as detailed coefficients. More information content is available in low-frequency coefficients, while detailed information (i.e., edges) is available in high-frequency components [3, 4].

Interpolation is a method which is used to increase the number of pixels in an image. Interpolation has been widely used in many applications like multiple description coding, facial reconstruction, etc. Some important methods are nearest neighbor interpolation, bilinear interpolation, bicubic interpolation, etc. Nearest neighbor interpolation can be performed by assigning gray value to the nearest integral value. Bilinear interpolation can be performed by obtaining a weighted average of two intermediate pixels. Similarly, bicubic interpolation can be performed by its weighted average of 16 surrounded pixels. Among these methods, bicubic interpolation provides better results than bilinear [5, 6].

Various wavelet filters are available such as Daubechies, Coiflets, Symlets, Discrete Meyer, Biorthogonal. To overcome the issues in the existing methods and also to improve the image quality, Cohen-Daubechies-Feauveau 9/7 wavelet is used. It is a biorthogonal wavelet obtained by lifting a 5/3 filter to a 7/5 filter, and then to a 9/7 filter. Some important features of this wavelet are symmetrical and obtaining the maximum number of smoothness in an image over its length [7, 8].

The fanbeam projection technique is used to project an image at different distances and angles in order to obtain sinogram information from the image. Fanbeam works on the basis of central pixel that acts as the center of the image.

Image vertices are calculated at a specific distance from the image. Angles vary from 0 to 360°. Complete fanbeam projections provide necessary information for image reconstruction [9].

2 Resolution Synthesis Using DWT

2.1 *Single Level 2-D Decomposition Using DWT*

In existing methods, DWT is used to decompose an image into various subband images. LH, HL, and HH subband images comprise the high-frequency components. Interpolation is applied to high-frequency components. In parallel, LL is obtained by low-pass filtering of the original image. The factor a/2 is used to interpolate the input image to obtain low subband images which act as low–low (LL) subband image [5].

Inverse discrete wavelet transform is used to obtain high-resolution image as illustrated in Fig. 1 [5]. Resultant images obtained from the conventional technique outperforms over other state-of-art techniques [6].

The original image of size (256 × 256) is downsampled to generate a low resolution image of size (128 × 128). DWT decomposes the image into various frequency subband images of size (64 × 64). High-frequency components are interpolated using the factor a to increase the image size to (256 × 256). Using these mathematical calculations several values are obtained in conventional techniques.

2.2 *DWT and Fanbeam Projection Using Fusion*

Initial loss of image quality after applying interpolation is on its high-frequency components. In order to increase the image quality to a subsequent extent,

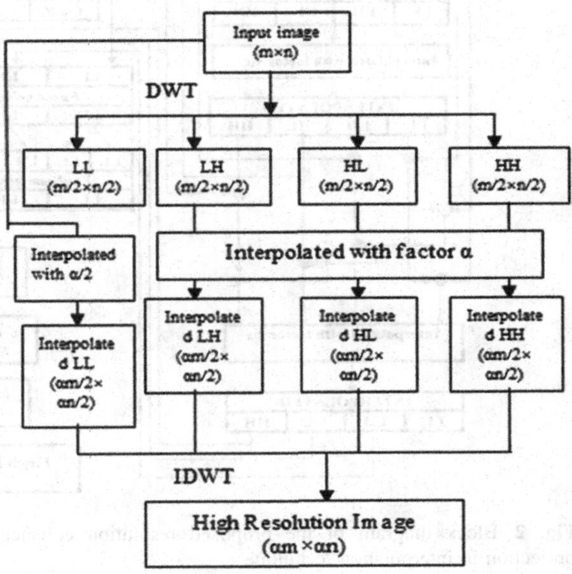

Fig. 1 Block diagram of resolution enhancement using single-level DWT [5]

preserving the edges is required. DWT is an efficient tool for processing an image. It is employed to preserve the edges in an image.

The proposed technique uses DWT to decompose the low-resolution image into different subband images. In wavelet domain, LL subband image is the low-resolution or low-pass filtering of the original image. The low-frequency components contain more information content. For resolution synthesis in an image, the original image should be taken as low-low subband image. Other high-frequency components obtained using DWT are interpolated using the factor a. IDWT is applied over these subbands to obtain the resolution synthesized image [10].

The input low-resolution image is obtained by downsampling the original image by factor a is shown in Fig. 2. By applying DWT, single-level decomposition is performed over the image to dissect the image into two subband images. Further, low and high subband images are again subdivided into four subband images. Bicubic interpolation is applied over these subbands to obtain interpolated image with factor a [5, 6].

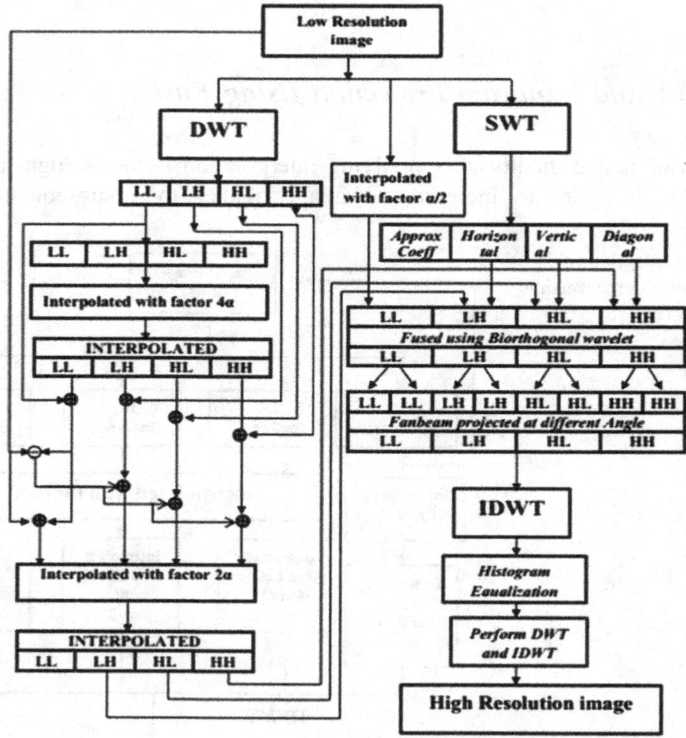

Fig. 2 Block diagram of the proposed resolution enhancement using DWT and fanbeam projection in interpolation technique

Interpolated high-frequency images act as subband images for IDWT. The original image is interpolated with half of the factor a to generate low–low subband images. Interpolated high-frequency components are used as the input image for fanbeam projection [11].

Fanbeam projection performs projection at different angles and distances to calculate its sinogram information. Variable distance should be localized in order to obtain its vertex position of the image. The obtained differently projected image are fused together to produce subband images. Biorthogonal uses invertible wavelet transform for fusion of image [12]. The data acquisition geometry is used to measure the vertex path and the vertex point. The vertex path refers to the trajectory along the beam projection and the vertex point is the converging point in line integral.

The input low resolution image of size ($m \times n$) is decomposed using DWT to obtain various frequency subband images of size ($m/2 \times n/2$). These subband images are interpolated using bicubic interpolation. The obtained interpolated information is projected using fanbeam projection. In fanbeam projection, each line that passes through the region-of-interest (ROI) intersects the path in non-tangential way provides the respective frequency values. Projected image at different angles are fused together to form a high-frequency subband image. High-frequency components obtained from the fusion technique are treated as the input subband images for IDWT. The interpolated original image turns out be LL band for IDWT. The final high resolution output image is created by using the IDWT of the interpolated subband images and the input image as illustrated in Fig. 2.

Nearest neighbor interpolation method can assign the gray value from the integral of the nearest pixel, bilinear uses surrounding 4 values. While bicubic uses nearest 16 pixel values to perform interpolation [5, 10, 13, 14] and its values are shown in Tables 1 and 2.

Table 1 Comparison of various techniques using Haar wavelet-PSNR values

Test images	Nearest neighbor	Bilinear	Bicubic	Lanczos3	Box
Cameraman	52.5222	52.5375	52.5411	52.5222	52.5222
Pout	55.2355	55.2318	55.2362	55.2355	52.2375
Cell	54.6393	54.6414	54.6460	54.6389	54.6393
Eight	49.3944	49.3977	49.4002	49.3963	49.3944

Table 2 Comparison of various techniques using biorthogonal wavelet-PSNR values

Test images	Nearest neighbor	Bilinear	Bicubic	Lanczos3	Box
Cameraman	53.3601	53.3603	53.3691	53.3683	53.3684
Pout	55.2284	55.2270	55.2291	55.2264	55.2284
Cell	54.6469	54.6509	54.6547	54.6489	54.6469
Eight	49.3848	49.3874	49.3888	49.3858	49.3848

The visual and PSNR values in the upcoming section show that the proposed technique with Haar and biorthogonal wavelet functions out performs other existing techniques.

3 High-Resolution Encryption Using Pascal's Series

By using Pascal's method refer to (1) and (2), array of binomial coefficients are generated. In terms of triangle sequence [2, 15] 1, 3, 6, 10, 15, 21, 28, 36, 45….

Similarly in tetrahedral number sequence 1, 4, 10, 20, 35, 56….

$$\binom{n}{k} \equiv \frac{n!}{(n-r)!r!} \tag{1}$$

$$= \binom{n-1}{r} + \binom{n-1}{r-1} \tag{2}$$

where triangular sequence for variable size n.

Algorithm

Step 1: Original image decimal value is obtained.
Step 2: Triangle and tetrahedral sequence are generated and bitwise XOR to produce new sequence.
Step 3: Bitwise manipulation is done between original image decimal value and sequence.
Step 4: Mapping based on the generated sequence to obtain the encrypted image.
Step 5: The inverse manipulation is performed to obtain the original image.

4 Results and Discussion

The proposed method has been tested on number of test images (open source). The original cameraman image of size (256 × 256) [16, 17] is the ground truth to calculate the performance evaluation which is shown in Fig. 3a. Original image is downsampled using the low-pass filtering technique to obtain low-resolution image which is shown in Fig. 3b. The input low resolution image is dissected into subband images and interpolation is processed over these subband images. Detailed information is available in the high-frequency subbands images. Fanbeam projection preserves these high-frequency components. Fusion technique performed over these

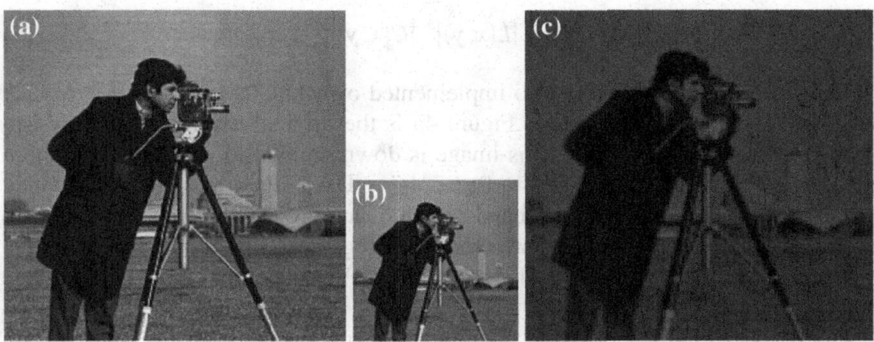

Fig. 3 **a** Original image (*open source*). **b** Low-resolution images. **c** Resolution-synthesized image using the proposed method

projected images are used as the input subband images for IDWT. The proposed technique preserves the edges information more than the conventional interpolation techniques, where Fig. 3c reflects this by including more edges' detail information [1].

4.1 Mean Square Error (MSE) and PSNR

The most common measures of image quality are the mean square (MSE) and the peak signal-to-noise ratio [13, 14, 18] refer (3) and (4). Let $f(x, y)$ be the original image with size $(m \times n)$ and $f'(x, y)$ be the modified image with the same size. MSE is expressed as in

$$\mathrm{MSE} = \frac{1}{\mathrm{MN}} \sum_{xy} (f'(x,y) - f(x,y))^2 \qquad (3)$$

and the PSNR in decibel is expressed as in

$$\mathrm{PSNR} = 10\log_{10}\left(\frac{255^2}{\mathrm{MSE}}\right) \qquad (4)$$

4.2 Structural Similarity Index (SSIM)

SSIM is another performance metric used for image analysis using the computation of the terms like luminance, contrast, and structural. The SSIM can be expressed as in (5)

$$\mathrm{SSIM}(x,y) = [L(x,y)]^{\alpha}\ [C(x,y)]^{\mu}\ [S(x,y)]^{\beta} \tag{5}$$

This proposed technique is also implemented over DICOM images [19] to take out resolution-synthesized image. Figure 4a is the original image to be processed using the proposed technique. This image is downsampled to get a low resolution image is which is shown in Fig. 4b. By applying DWT, the input image is divided into LL, LH, HL, and HH subbands, which are projected using the fanbeam technique. The projected images are then fused together using the fusion technique to obtain the subband images for the process of IDWT. The final synthesized image is shown in Fig. 4c.

Proposed method also provides significant results in satellite view images [16]. Original image of satellite view is shown in Fig. 5a is downsampled to get a low resolution image which is shown in Fig. 5b. The proposed algorithm performs the image resolution synthesis and the resulting image is shown in Fig. 5c [20].

The SSIM values for the test images [1] are shown in Table 3.

Using the proposed encryption algorithm, cameraman as original image as shown in Fig. 6a is encrypted and decrypted as shown in Fig. 6b, c respectively. Whereas Fig. 6d shows the information decrypted with wrong key.

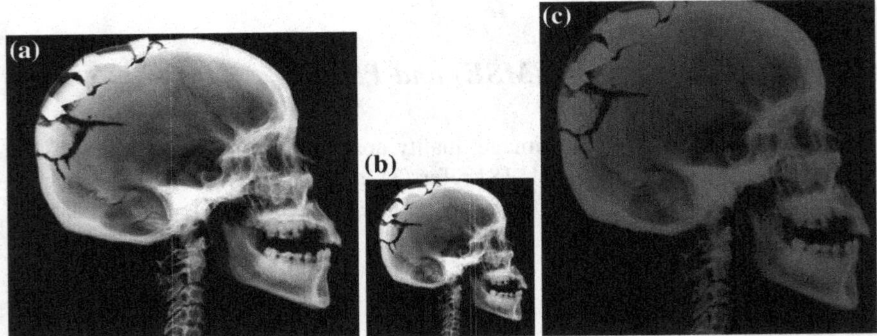

Fig. 4 a Original DICOM image (*open source*). **b** Low-resolution DICOM image. **c** Resolution-synthesized image using the proposed method

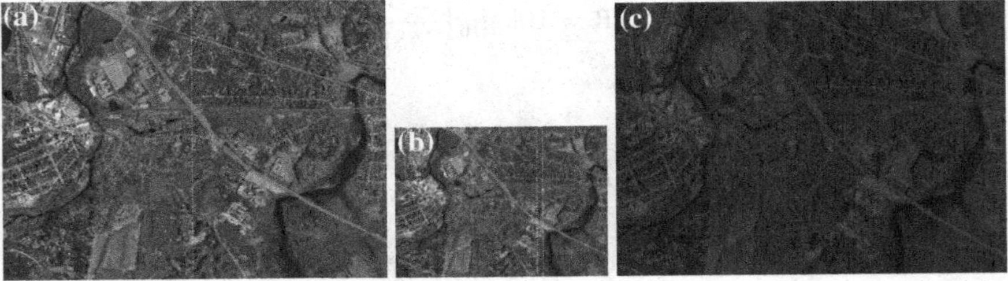

Fig. 5 a Original image (*open source*). **b** Low-resolution image. **c** Resolution-synthesized image using the proposed method

Table 3 Comparison of various techniques using biorthogonal wavelet-SSIM values

Test images	Nearest neighbor	Bilinear	Bicubic	Lanczos3	Box
Cameraman	0.4757	0.4778	0.4819	0.4749	0.4757
Pout	0.5864	0.5880	0.5912	0.5868	0.5864
Cell	0.6066	0.6092	0.6129	0.6071	0.6066
Eight	0.5461	0.5479	0.5489	0.5456	0.5461

Fig. 6 **a** Original image (*open source*). **b** Encrypted image. **c** Decrypted image. **d** Decrypted image with wrong key

4.3 Correlation

In digital images, correlation value between adjacent pixels seems to be high [21]. Using these equations refer (6), correlation value with adjacent pixels are calculated in horizontal, vertical and diagonal orientations.

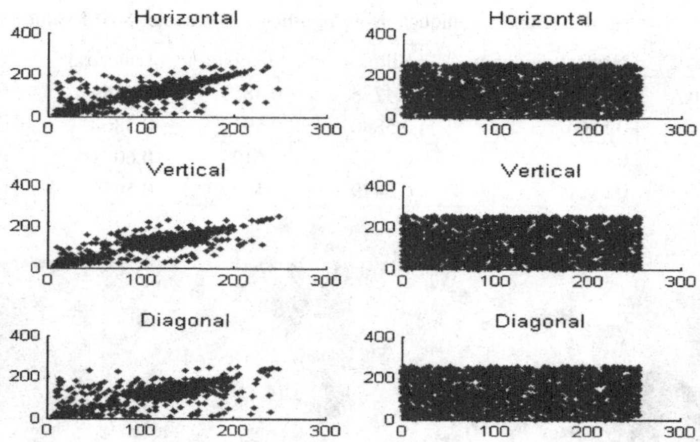

Fig. 7 Correlation value of the original and encrypted image

$$r_{xy} = \frac{\text{cov}(x,y)}{\sqrt{D(x)}\sqrt{D(y)}} \tag{6}$$

x and y are values of two adjacent pixels in an image.

Correlation value of the original image and encrypted image with different orientation is shown in Fig. 7.

It is observed that the pixels are evenly spread in the encrypted image. For original image, the pixel values are spread along diagonally to represent the mutual correlation between the two pixels are explained.

5 Conclusion

Using DWT and fanbeam projection technique, the visual quality such as contrast invariance, edges preservation, noise, sharpness, etc. of the interpolated image and the information content are better than other formulaic techniques. The PSNR values are compared with other conventional techniques such as nearest, bilinear, and bicubic interpolation methods. The resultant image is used in various applications like medical diagnostic imaging, biological imaging, human–machine interface, satellite imaging, cinematography, remote sensing, document processing, automation and robotics, etc.

Acknowledgments The authors would like to thank the open source community for their valuable contributions by providing their test images. Any opinions, findings, and conclusions expressed in this paper are those of the authors and do not necessarily reflect those of open sources.

References

1. Wang Y, Wong KW, Liao XF, Xiang T, Chen GR (2009) A chaos-based image encryption algorithm with variable control parameters. Chaos, Solitons Fractals 41:1773–1783
2. Cooper RH, Hunter-Duvar R, Patterson W (1989) A more efficient public-key cryptosystem using the pascal triangle. In: ICC/89: IEEE international conference on communications 1989. Boston, 11–14 June, pp 1165–1169
3. Piao Y, Shin I, Park HW (2007) Image resolution enhancement using inter-subband correlation in wavelet domain. In: Proceedings of International Conference on Image Processing, vol. 1, pp 445–448
4. Zeinali M, Ghassemian H, Moghaddasi MN (2014) A new magnification method for RGB color images based on subpixels decomposition. Signal Process Lett IEEE 21(5):577–580
5. Anbarjafari G, Demirel H (2010) Image super resolution based on interpolation of wavelet domain high frequency subbands and the spatial domain input image. ETRI J 32(3):390–394
6. Demirel H, Anbarjafari G (2011) Image resolution enhancement by using discrete and stationary wavelet decomposition. IEEE Trans Image Process 20(5):1458–1460
7. Khattak NS, Sarwar T, Arif F (2013) Single image magnification with edge enhancement. In: IEEE international symposium on signal processing and information technology (ISSPIT), 2013, pp 321–326
8. Temizel A, Vlachos T (2005) Wavelet domain image resolution enhancement using cycle-spinning. Electron Lett 41(3):119–121
9. Li H, Lam KM (2013) Guided iterative back-projection scheme for single-image super-resolution. In: 2013 IEEE global high tech congress on electronics (GHTCE), pp 175–180
10. Elamaran V, Praveen A (2012) Comparison of DCT and wavelets in image coding. In: Proceedings of the international conference on computer communication and informatics (ICCCI-2012), 2012, pp 1–4
11. Rener Y, Wei J, Ken C (2008) Downsample-based multiple description coding and post-processing of decoding. In: 27th Chinese control conference, July 2008, pp 253–256
12. Wang Z, Ziou D, Armenakis C, Li D, Li Q (2005) A comparative analysis of image fusion methods. IEEE Trans Geosci Remote Sens 43(6):1391–1402
13. Tan L (2008) Digital signal processing—fundamentals and applications. Elsevier
14. Marques O (2011) Practical image and video processing using MATLAB. Wiley-IEEE Press
15. Aburdene MF, Goodman TJ (2005) The discrete Pascal transform and its applications. IEEE Signal Process Lett 12(7):493–495
16. MATLAB and statistics toolbox. The MathWorks, Inc., Natick, Massachusetts, United States. http://www.mathworks.com
17. DEIMOS. http://wc.multimediatech.cz/tag/deblurring
18. Russ JC, Russ JC (2007) Introduction to image processing and analysis. CRC Press
19. Dicom library. http://www.dicomlibrary.com
20. Satellite imaging corporation. http://www.satimagingcorp.com
21. Rivest RL, Shamir A, Adelman L (1978) A method for obtaining digital signatures and public-key cryptosystems. Commun ACM 21:120–126

Engender Product Ranking and Recommendation Using Customer Feedback

V. Gangothri, S. Saranya and D. Venkataraman

Abstract In our day-to-day life we tend to buy products on the Internet. There are plenty of consumer reviews on the Internet. If a customer wants to know about a product, he sees the review and rating of the product given by the product users. In this case we come to know about the importance of rating and review of the product which impacts the product's market value. This article proposes a framework for calculating an accurate rating using customer feedback. In particular, we first take the consumer review as an input then remove all common words by using the information retrieval concepts like stop word removal and stemming. The next step is parts of speech tagging and finding the opinion word extraction to the rest of the phrases. Then we have to match the keywords with the ontology and finally we develop a probabilistic aspect ranking algorithm to rank the product. We see elaborately about our concept in this article.

Keywords Ontology · Recommender system · Opinion words · Customer feedbacks

1 Introduction

The impact of technological growth is people engaging more with gadgets. In this decade we can see tremendous changes in the ecommerce and electronics field. People start encouraging these fields because today's world mostly depends on

V. Gangothri (✉) · S. Saranya · D. Venkataraman
Department of Computer Science and Engineering, Amrita School of Engineering, Amrita Vishwa Vidyapeetham (University), Coimbatore, India
e-mail: gangothribsc@gmail.com

S. Saranya
e-mail: mayukamila@gmail.com

D. Venkataraman
e-mail: d_venkat@cb.amrita.edu

L.P. Suresh and B.K. Panigrahi (eds.), *Proceedings of the International Conference on Soft Computing Systems*, Advances in Intelligent Systems and Computing 397, DOI 10.1007/978-81-322-2671-0_80

these two grounds. In India we use ecommerce sites like FLIPKART, SNAPDEAL, JUNGLEE, AMAZON, etc. These are the most famous and the most used sites in India for shopping. Every year BBC, TIMES OF INDIA, and a few more news media survey these ecommerce sites to know the current status of Indian people and their mindsets. The eBay shopping mart was started in 1995 according to a BBC survey in 2013; it says that 606 million dollars have been invested for this site. There are 33,500 employees in eBay and the revenue is 16.05 billion dollars. Another ecommerce site is FLIPKART in India; it started in 2007, the investment was 210 million dollars till 2013 and the survey says that most of the investors are foreigners (BBC). After the US succeeded in ecommerce sites, many people started investing in India in E-Business. There are 15,000 employees working for FLIPKART and the revenue is 1billion dollars. The business today says that FLIPKART has 5 million products and it sells more than 600 crore products in a day.

From these surveys we can see that most of us wish to get products online rather than direct shopping, because it reduces time and it is not a complex task to purchase compared to the older method. These sites becomes an important factor in our day-to-day life, so it is necessary to take care of things like organizing the products in a correct domain, protecting a customer's details like credit and debit card pin numbers, interactive and attractive web pages, etc. There are many aspects to concentrate on the ecommerce site to give better enrichment in the process of buying a product. In this paper we have concentrated on reviews and ratings of a product which is a vital factor for sales. All the ecommerce sites encourage customers to write feedback about the product to help the customers to know about the product's positives and negatives.

For example, a customer who wants to buy a Sony laptop wants to know the feedback from the laptop users to know which laptop is better in the market. In this case the reviews and ratings given by the customers will be useful for him to get the best one from the pool. Hence this example reveals the importance of feedback and rating method in the ecommerce field. Nowadays all ecommerce sites expect comments and ratings from the customer to correct their mistakes in the future version of that product. Some of the websites get only ratings from the customers, some get the A consumer who mostly gives the genuine and quality reviews and ratings from persons who own it. But we cannot say that all of us give a very genuine rating in blogs which leads to consumers to get confused or negative about the product. So in order to make accurate ratings to facilitate customers we generate ratings from the customer feedback.

We could think a product from many aspects, i.e., "price," "Quality," "Quantity," and much more. So we generate a system to analyze ratings from all the maximum aspects which we could think as much as possible. For example, if you want to buy a laptop you come to a decision after thinking from many aspects like the price of a laptop, RAM, hard disk space, if display and battery package is good, warranty and guaranty of a product, and much more. Only after we get good feedback of all these things we tend to buy a laptop. But in a real case customer we find difficulties in finding feedback about a particular product in all aspects. So we design a system that gets a feedback from the customer and generates automatic result of accurate rating by considering a product in all its aspects.

Most ecommerce sites expect both feedback and rating from the customer without knowing the customer's mentality. Customer feels flexible in giving feedback rather than ranking a product because when he gives feedback he will explain a product in various aspects but in case of raking he cannot apply this method. A customer always gives genuine feedback than rating. So we propose a framework where we can get genuine rating from the customer feedback. The existing system also has the same features of proposed system but we tried to implement some concepts to overcome from a number of problems which the existing system has.

2 Related Works

There are many works done on this framework to make ratings. They use mainly two methods called the document level sentiment classification and extractive review summarization. The document level sentiment classification help us to conclude a review level documents, which is expressing a document's overall opinion whether it is positive or negative. The extractive review summarization method helps us to generate rating by extracting useful information from the customer feedback [1]. They implement information retrieval concepts to remove some words that are not needed to calculate the rating. The project has been divided into four stages according to the concepts that have to be implemented [2]. The first stage consists of stemming and stop word removal, the second stage has opinion word extraction and extracting common words, the third stage is clustering sentiment analysis with classification of polarity, and finally the product aspect ranking, document level classification, and extractive review summarization is the fourth stage [3].

From the title of the paper we understand what the paper is about (toward the next generation of recommendation systems: a survey of the state of the art and possible extensions) [4, 5]. This paper speaks elaborately about the present generation of recommendation system and its limitations. It gives an idea to overcome from all the limitations, how we could extend the recommendation systems and make it available even for a broader range of applications. These extension ideas make us understand about the users and their point of view toward the item they buy. They discuss the three-recommendation system which plays a vital role in today's world of ecommerce to do product recommendation. The important note in this discussion is to make us to realize how the rating part varies in each recommendation [6].

Recommendation technique has been widely discussed in the research communities of information retrieval, data mining, and machine learning. Due to all these values the recommendation system is commercially successful to do product, movie, and other recommendations. In this paper they has been compared three categories of getting recommendation of a product [7, 8]. They are real world social recommendation, social network, and the user item rating matrix. The real-world social recommendation is all about the direct recommendation, i.e., if a customer wants to buy a laptop, he gets recommendation from laptop users whom he knows. Then the social network recommendation gets recommendation of a product from the social media like Facebook, Twitter, etc. The last one is the user item rating matrix; it is in the form of matrix by considering the items and user's interest toward an item. Importance is also given to the low rank matrix factorization [9].

From these three papers we understand the importance of reviews given by the user. Each and every review and rating is an important aspect for the inflation of a product value in the market. We cannot generate a system to calculate a very accurate rating because people's mind varies. But we can try to produce a genuine rating from our system with the help of customer reviews, hence in our proposed system we develop a framework where the customer can post their feedbacks and get the ratings.

3 Proposed System

In our proposed system we develop a framework: (i) To automatically identify the aspects of a product with the help of online consumer reviews; (ii) Generate product ranking using ontology; (iii) It make us to understand the importance of product ranking in this ecommerce field; (iv) We demonstrate this framework with the real-time application. The main objective of this article is to facilitate better rating architecture for the consumers. To assist all the above features in the proposed system we need to take care of (a) Preprocessing of Review, (b) Opinion Word Extraction, (c) Opinion Word Feature Mining in Ontology, and (d) Product Aspect Ranking.

3.1 Proposed Architecture

In our architectural diagram (Fig. 1) we explain the mechanism of our proposed system. Three steps are important in the case of ranking a product and for recommendation. The first is preprocessing the review; discover the aspects in the review to find the opinion of the online customer about the product and mapping the keywords with the ontology. These three components play a major role in giving ratings for a product.

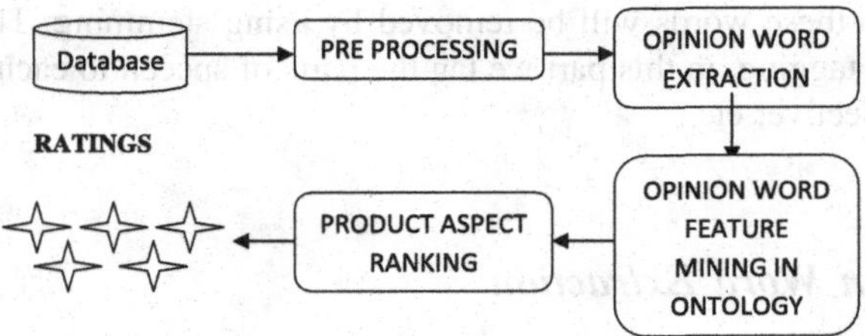

Fig. 1 The proposed system architecture

4 Module Description

In this section we clearly explain each module of architectural diagram step by step.

4.1 Preprocessing of Review

Consumer reviews will differ from each online shopping mart. Popular ecommerce websites like Flipkart, Snapdeal make the consumer to give star rating of a product and also expect detailed opinion about the product. But some other websites only ask the opinion rather than ratings. So the reviews and ratings differ from each websites. When we talk about the paragraph review giving by the customer they represent positives and negatives of a product. In this module we remove all common words which appear in the customer review. There are three submodules in this first component (Fig. 2).

(i) Stop word removal
(ii) Stemming and
(iii) Parts of Speech Tagging.

The first step is the reviews taken from the database and sent for Stop word removal. It is the one which removes all meaningless words in the consumer review. The meaningless words are "the," "is," "it," "that," "to," "be," etc. Then the reviews will be passed for the next process, i.e., stemming (NLP Method). This term is used in the information retrieval domain to express the method of reducing the root words. For example, we take a word "is," the root words of "is" is "are"

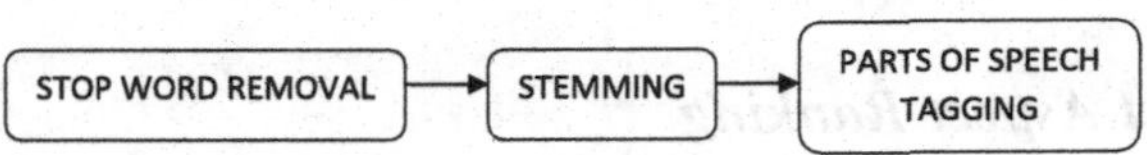

Fig. 2 Architecture of preprocessig of review

and "were". So these words will be removed by using stemming. The final one is parts of speech tagging, in this part we tag the parts of speech to each word such as noun, verb, adjective, etc.

4.2 Opinion Word Extraction

In this module we identify the aspect of opinion given by the online customer in the review about a product. After the first module, the review will be passed to this component. From the title itself we understand that this module is to extract the opinion words from the review [10]. The advantage is to reduce the redundant words from the box. For example if a customer reviews a mobile he says that "this mobile display is good and i like this mobile lot because of good display." We can see the same meaning repeated in this review. These things should be removed to get the accurate rating and also it occupies space in the database. So this module is helpful to remove the redundant datasets from the review [11]. The first two modules are the one to be concentrated only to remove the unwanted words and to identify the aspects of a consumer review about the product. Therefore, the next two modules are working toward finding the accurate ratings from the customer review.

4.3 Opinion Word Feature Mining in Ontology

In this decade people always wish to do online shopping rather than the older method. So business people are much interested to get the opinion from the public about the product they sell and also about their services from ecommerce websites. Moreover, this is the third module in our proposed system which has been predefined with the possible keywords in an ontology graph [1]. It is helpful to identify the semantic orientations, i.e., whether it is positive, negative, or neutral comment given by the customer using an ontology. To build an ontology we have used the FLIPKART's Mobile Review Datasets in our framework for experiment. We extract the opinion of a customer with the help of the previous two modules [12]. The extracted words are passed to this opinion word feature mining module to match the keywords with the predefined graph for identifying the polarity values. This module helps us to identify whether the user comment is positive, neutral, or negative. These values are passed to the next component called Aspect Ranking.

4.4 Product Aspect Ranking

This is the final module in our framework for getting accurate ratings using customer feedback. The input of our system is the customer review and the final output will be the rating. So this module will get the input as polarity values from the opinion word feature mining to calculate the overall ratings for the product which customer gave feedback on. In this part the important aspect will be found with the help of the following characteristics: (a) Maximum number of comments a product received (b) Each consumer's opinion about a product will influence the product when it comes for the overall ratings. The product's overall rating depends on each aspect of the consumer review. This is the mechanism of our framework to get the resultant accurate rating. For your further understanding we did experiment analysis with the dataset of FLIPKART mobile datasets and it is explained below in detail with an appropriate example.

5 Evaluation

In this section we did a broad range of experiments to check how our proposed system effectively works in the real-time application. It includes all the main process like preprocessing of review, opinion word extraction, opinion word feature mining in ontology, and product aspect ranking.

5.1 Datasets

To verify and for further clarification about the proposed system we have taken a FLIPKART's mobile review datasets.

5.2 Experimental Results and Discussions

The figure shows the result of the first and second components (Preprocessing of Review and Opinion Word Extraction). As explained above, this component does stop word removal, stemming, parts of speech tagging, and opinion word extraction. As soon as these two modules are done with their work they pass this data to the next component, that is, ontology. The noun and adjective will be taken as input for the ontology graph. And according to the dataset we used we create a class using the maximum possible keywords which the customer can use in their comments..

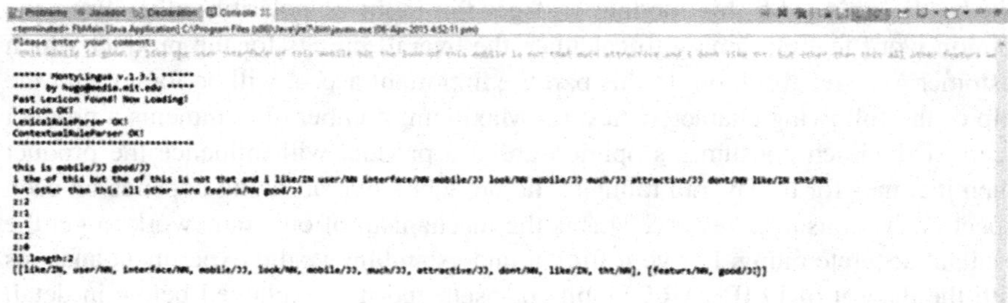

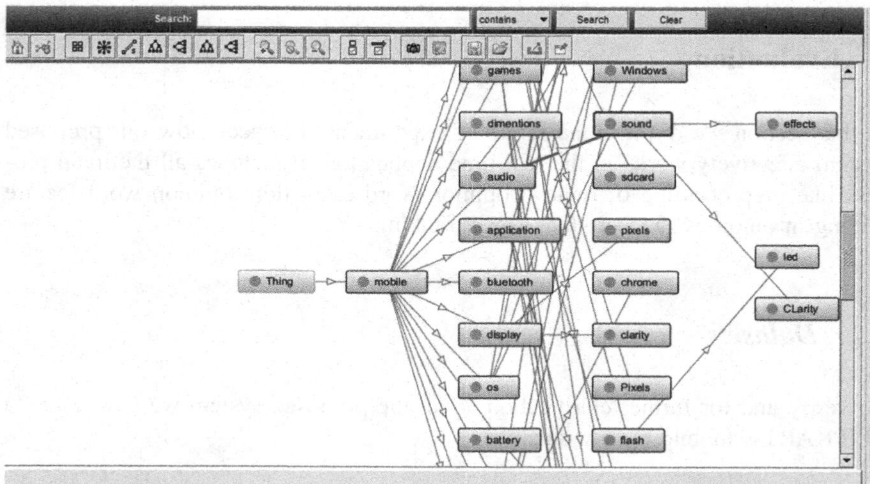

Fig. 3 The ontology graph

Figure 3 shows the ontology graph for our framework.

6 Conclusion and Future Work

In this paper, we proposed a framework for efficient ranking for identifying the various aspects of consumer's opinion toward a product. The evaluation section demonstrates the usefulness of a proposed system in the real-world application. We have done the preprocessing of review, opinion word extraction, opinion word feature mining in ontology, and product aspect ranking. For finding the accurate details we need to carry all the steps listed above.

In the future, we will develop an efficient algorithm for calculating the rating through the graph traversal in ontology.

References

1. Ricci F, Rokach L, Shapira B (2011) Introduction to recommender systems handbook. Springer, US
2. Resnick P, Varian HR (1997) Recommender systems. Commun ACM 40(3):56–58
3. Burke R (2002) Hybrid recommender systems: Survey and experiments. User Model User-Adap Inter 12(4):331–370
4. Ekstrand MD, Riedl JT, Konstan JA (2011) Collaborative filtering recommender systems. Found Trends Hum Comput Inter 4(2):81–173
5. Noy NF, McGuinness DL (2001) Ontology development 101: a guide to creating your first ontology
6. Guarino N, Welty C (2000) A formal ontology of properties. Knowledge engineering and knowledge management methods, models, and tools. Springer Berlin Heidelberg, pp 97–112
7. Schafer JB, Konstan J, Riedl J (1999) Recommender systems in e-commerce. In: Proceedings of the 1st ACM conference on electronic commerce. ACM, 1999
8. Sarwar BM et al (2002) Recommender systems for large-scale e-commerce: Scalable neighbourhood formation using clustering. In: Proceedings of the fifth international conference on computer and information technology, 2002, vol 1
9. Prasad RVVSV, Valli Kumari V (2012) A categorical review of recommender systems. System 1. U2 2012: U3
10. Wang XH et al (2004) Ontology based context modeling and reasoning using OWL. In: Proceedings of the second IEEE annual conference on pervasive computing and communications workshops, 2004. IEEE, 2004
11. Noy NF (2004) Semantic integration: a survey of ontology-based approaches.ACM Sigmod Rec 33(4): 65–70
12. Guarino N (1995) Formal ontology, conceptual analysis and knowledge representation. Int J Hum Comput Stud 43(5):625–640
13. Hiralall M (2011) Recommender systems for e-shops. Vrije Universiteit, Amsterdam, Business Mathematics and Informatics paper
14. Lops P, De Gemmis M, Semeraro G (2011) Content-based recommender systems: State of the art and trends. Recommender systems handbook. Springer US, pp 73–105
15. Burke R (2000) Knowledge-based recommender systems. Encyclopaedia Libr Inf Syst 69 (Supplement 32):75–186

Optical Flow Reckoning for Flame Disclosure in Dynamic Event Using Hybrid Technique

Ankit Kumar, P. Sujith and A. Veeramuthu

Abstract Computational vision-based discovery has drawn critical consideration and the cam reconnaissance frameworks have gotten to be to show up all over. Along these lines, we propose a PC vision approach for flame fire identification to be utilized as an early-cautioning blaze checking framework. Vision-based discovery is made out of three stages. First, preprocessing is important to make up for known wellsprings of variability, e.g., cam equipment and enlightenment. Besides, emphasis extraction is intended for the recognition of a particular focus on; a processing maps the crude information to a standard set of parameters to describe the target. At long last, characterization calculations utilize the processed peculiarities as information and settle on choice yields with respect to the target's vicinity. The regulated machine-learning-based characterization calculation, e.g., neural systems (NN) is deliberately prepared on an information set of gimmicks and the ground truth. While parts of preprocessing and grouping are viewed as, this paper concentrates on gimmick extraction: Spurred by the physical properties of flame and a set of novel optical stream characteristics which are intended for vision-based blaze discovery.

Keywords Preprocessing · Neural network · Blaze · Vision-based detection · Classification · Feature extraction

---

A. Kumar (✉) · P. Sujith · A. Veeramuthu
Sathyabama University, Rajiv Gandhi Road, Jeppiaar Nagar, Sholinganallur, Chennai 600119, Tamil Nadu, India
e-mail: ankitguptatamil@gmail.com

P. Sujith
e-mail: suzzu93@gmail.com

A. Veeramuthu
e-mail: aveeramuthu@gmail.com

L.P. Suresh and B.K. Panigrahi (eds.), *Proceedings of the International Conference on Soft Computing Systems*, Advances in Intelligent Systems and Computing 397, DOI 10.1007/978-81-322-2671-0_81

1 Introduction

Distinguishing the breakout of a flame quickly is a key for anticipation of material harm and human setbacks; this is an especially genuine issue in circumstances of congested vehicles activity, maritime vessels, and overwhelming industry. Customary point-sensors identify high temperature or smoke particles and are truly effective for indoor blaze recognition. On the other hand, they cannot be connected in expansive open spaces, e.g., shelters, ships, or in backwoods. This paper introduces a feature identification methodology intended for these situations where point-sensors may fall flat. Notwithstanding covering a wide review range, camcorders catch information from which extra data can be removed; for instance, the exact area, degree, and rate of development. Reconnaissance cams have as of late ended up pervasive, introduced by governments and organizations for applications like tag distinguishment and theft prevention. Solid vision-based flame location can plausibly exploit the current base and altogether add to open well-being with minimal extra cost.

It uses shading and movement data processed from feature successions to spot fire. This is carried out by first utilizing a methodology that is based after making a Gaussian-smoothed shading histogram to focus the flame-shaded pixels, and afterward utilizing the transient variety of pixels to figure out which of these pixels are really fires.

It utilizes fluffy rationale upgraded nonspecific shading model for flame pixel order The model uses YCbCr color space to partition the luminance from the chrominance more viably than shading spaces, e.g., RGB. Ideas from fluffy rationale are utilized to supplant existing heuristic guidelines and make the characterization more powerful in viably segregating fire and flame-like hued items. Further segregation in the middle of flame and no blaze pixels are attained to by a factually determined chrominance model which is communicated as a district in the chrominance plane. The execution of the model is tried on two huge sets of pictures; one set contains blaze while the other set contains no flame yet has locales like flame shading. The model accomplishes up to 99.00 % right fire location rate with a 9.50 % false caution rate.

It utilizes computational vision-based fire and fire recognition by utilizing a compound calculation and a choice combination system with naive Bayes classifier as order device. This methodology is to enhance the exactness of flame and fire recognition in features and to decrease the false caution rate as it were.

The partition of smoke and flame pixels utilizing shading data is performed. In parallel, a pixel choice in light of the motion of the zone is done keeping in mind the end goal to lessen false discovery. The yields of the three parallel calculations are inevitably combined by method for a MLP (multilayer perceptron).

2 Related Works

Recognition in feature successions, there are numerous past techniques to identify fire, be that as it may, all aside from two utilization spectroscopy or molecule sensors. The two that utilizes visual data experience the ill effects of the powerlessness to adapt to a moving camera or a moving scene. One of these is remarkable to chip away at general information, e.g., motion picture groupings. The other is excessively short-sighted and unrestrictive in figuring out what is considered flame, with the goal that it can be utilized dependably just as a part of flying machine dry straits. Our framework uses shading and movement data figured from feature successions to find fire [1]. This is carried out by first utilizing a methodology that is based on making a Gaussian-smoothed shading histogram to focus the flame hued pixels, and afterward utilizing the worldly variety of pixels to figure out which of these pixels is really fire. Dissimilar to the two past vision-based techniques for flame discovery, our strategy is relevant to more regions in view of its lack of care to cam movement [2]. In two particular applications impractical with past calculations are the distinguishment offered in the vicinity of worldwide cam movement or scene movement and the distinguishment of flame in motion pictures for conceivable use in a programmed rating framework. It demonstrates that the strategy lives up to expectations in an assortment of conditions [3], and that it can naturally focus when it has deficient data.

In this paper [4], fluffy rationale improved bland shading model for flame pixel grouping is proposed. The model uses YCbCr shading space to particular the luminance from the chrominance more viably than shading spaces, for example, RGB. Ideas from fluffy rationale are utilized to supplant existing heuristic principles and make the arrangement more vigorous in adequately separating fire and flame-like hued items. Further separation in the middle of flame and non-fire pixels are accomplished by a factually determined chrominance model which is communicated as an area in the chrominance plane. The execution of the model is tried on two huge sets of pictures; one set contains blaze while the other set contains no flame however has locales like flame shading. The model accomplishes up to 99.00 % right fire location rate with a 9.50 % false alert rate.

Computational vision-based fire and fire location [5, 6] has attracted noteworthy consideration the previous decade with cam reconnaissance frameworks getting to be omnipresent. Flag and picture handling routines are produced for the recognition of flame, blazes and smoke in open and extensive spaces with a scope of up to 30 m to the cam in obvious extent (IR) feature. This paper proposes another way to computational vision-based fire and fire recognition by utilizing a compound calculation and a choice combination system with naive Bayes classifier as grouping apparatus. The compound calculation comprises of a few sub-calculations, the combination system is to break the outcomes got by each of these sub-calculations and naive Bayes classifier is helpful for the last arrangement. This methodology is to enhance the precision of flame and fire discovery in features and to decrease the false caution rate all things considered. Since flame is a complex however strange visual wonder, not at all like typical items, it has dynamic composition [7]. Because of its incessant shape and size modifications, computational vision-based fire and fire location calculations are upon multi-peculiarity based methodologies. The trust and the objective of such calculations are to discover a mix of gimmicks whose shared event leaves fire as their just joined conceivable reason [1]. Shading, movement, shape, development, glimmering, and smoke conduct and so forth are a percentage of the low level unique peculiarities of flame areas. Alongside these unique gimmicks, unearthly, spatial and transient peculiarities are likewise utilized for recognizing blaze areas misleadingly.

In this paper [8], a feature observation system, Identification of flame and smoke pixels is right away accomplished by method for a movement recognition calculation furthermore, partition of smoke and blaze pixels utilizing shading information (within proper spaces, particularly picked with a specific end goal to upgrade particular chromatic gimmicks) is performed. In parallel, a pixel determination in light of the progress of the region is completed with a specific end goal to diminish false discovery. The yield of the three parallel calculations is in the end intertwined by method for a MLP. Best in class frameworks make utilization of sensors whose sort relies on upon the thickness of the sensors themselves: the more the thickness, the more the accuracy. For example, smoke locators work exceptionally well in generally little rooms, where every piece of the room can be viably observed. This is not possible in open air situations, since their vast augmentation would bring about an immense number of sensors to have a satisfactory thickness. A sample of a business framework is proposed in by [3]. They utilize ease CCDs to recognize hotbeds in baggage stockpiling holds in airplane. Their system uses dim-scale factual gimmicks of the picture and outside peculiarities, e.g., stickiness and temperature to diminish bogus alert of smoke identifiers. A surplus quality is given by the visual criticism given to the team, who can screen the hold with no compelling reason to enter it.

3 Proposed Works

3.1 Problem Description

To create a warning mechanism when any source of fire appears in a frame of a simple household camera and alert authorities via SMS, we propose a set of motion features based on motion estimators. The key idea consists of exploiting the difference between the turbulent, fast, fire motion, and the structured, rigid motion of other objects.

3.2 System Architecture

The detailed system architecture is shown in Fig. 1, which consists of the following components like, preprocessing, feature extraction, classification, fire detection, and video analysis.

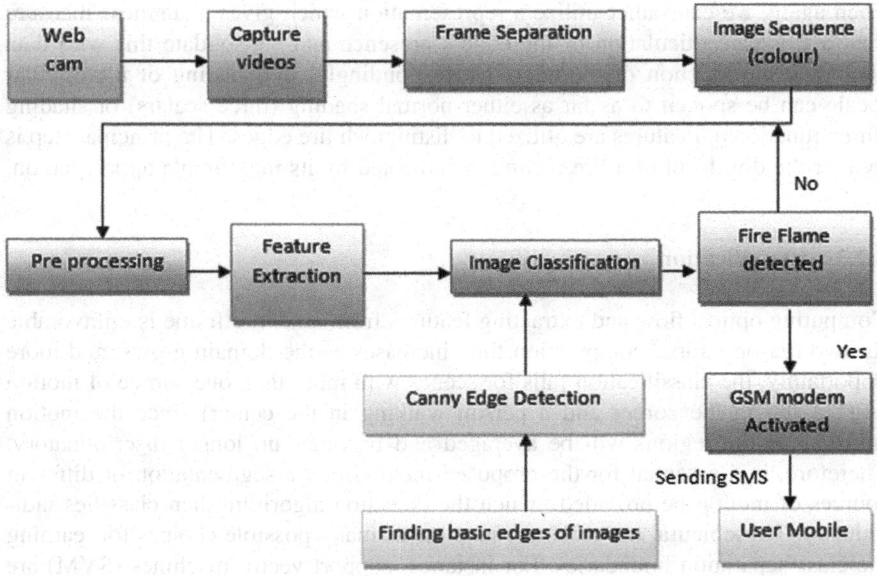

Fig. 1 System architecture

3.2.1 Preprocessing

In this method we use color transformation that means to convert RGB frames to scalar image. In the method, we implement color histogram techniques. As this technique is only used to convert to scalar image that only we find the image to be fire image or not. One major difference, we compute a spatial structure preserving feature based on a unique property of the optimal mass transport (OMT) solution. Regarding classification results, it is not comparable to our work, since that only classifies between 26 particular instances of dynamic textures (each one representing one class), and not between the presence/absence of an event (e.g.,"fire") among many videos.

3.2.2 Feature Extraction

In this technique we utilize two systems. First and foremost is to actualize optical stream estimation and second is non-smooth information. To begin with, we discover the force that happens in the copying process because of quick weight and high temperature dynamic. Second, smoothness regularization may be counter-gainful to the estimation of flame movement, which is required to have a turbulent, i.e., non-smooth movement field. A particular picture peculiarity, characterized by a particular structure in the picture information, can regularly be spoken to in distinctive ways. For instance, an edge can be spoken to as a Boolean variable in every picture point that depicts whether an edge is available by then.

Then again, we can rather utilize a representation which gives a sureness measure than a Boolean articulation of the edge's presence and consolidate this with data about the introduction of the edge. Correspondingly, the shading of a particular locale can be spoken to as far as either normal shading (three scalars) or shading (three functions). Features are utilized to distinguish fire edges. The principal step is to alter the dim level of a fire picture as indicated by its measurable appropriation.

3.2.3 Classification

Computing optical flow and extracting features from an entire frame is unfavorable for two reasons. First, computation time increases as the domain grows, and more importantly, the classification fails for scenes with more than one source of motion (e.g., a fire in the corner and a person walking in the center) since the motion statistics of the regions will be averaged and becomes no longer discriminatory. Therefore, it is essential for the proposed method that a segmentation of different sources of motion be provided, which the detection algorithm then classifies individually. Those neural networks are just one of many possible choices for learning the class separation boundaries. For instance, support vector machines (SVM) are expected to produce similar results. Selecting discriminatory features, not choosing the classifier, is most important for correct classification.

3.2.4 Fire Detection

The edge recognition techniques are gathered into two classes as per the calculation of image angles as first-request or second-request subsidiaries. In the first class, edges are identified through registering measure of edge quality with a first-request subordinate outflow. All in all, a fire locale has a stronger luminance in correlation to its encompassing foundation, and the limit between the fire district and its experience is basically nonstop. Besides, as a rule, there is just a primary fire in the picture; overall, the picture can be divided with the goal that every sectioned range contains standout principle fire. If need be, a processing calculation is proposed where these peculiarities are utilized to recognize fire edges. The essential method is to identify the coarse and pointless edges in a fire picture then recognize the fire's important edges and evacuate superfluous ones.

3.2.5 Video Analysis

The database features various scenarios including indoor/outdoor, far/close distance, different types of flames (wood, gas, etc.), changing lighting conditions, partial occlusions, etc. The videos have a frame rate of 30 frames/s and spatial dimension of 300 by 300 pixels. From each of the 263 scenarios (containing 169 fire and 94 non-fire sequences), 10 consecutive frames are labeled as ground truth

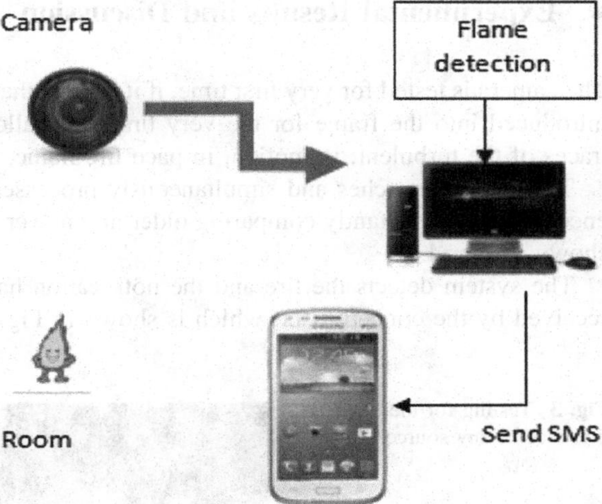

Fig. 2 Overall system process

providing a test database of 2740 frames. Note that the non-fire scenarios are chosen to be probable false positives, namely moving and/or fire-colored objects such as cars, people, red leaves, lights and general background clutter. Occasional false negative detections were observed in four types of scenarios. First, horizontal lines resulting from structured noise. Second, the flame color is oddly distorted to a blue tinge which causes the color transformation. Third, insufficient spatial resolution leads to very few pixels belonging to fire and motion structure cannot be detected. Lastly, partial occlusions may make the detector fail.

The overall process of this work is shown in Fig. 2 and hybrid technique is given in Algorithm 1.

3.3 Algorithm

Input: Sequence of frames ($I_{f=1,2,....nf}$)
Output: Fire detected, notification through SMS
Begin
  To preprocess the image like, noise removal and convert from RGB to gray
  The features are extracted based on Gaussian method
  For (all frames) do
    Apply canny edge detection method
   Compute the distance between any two frames, Martin distance method like,

$$M_{fa} = (F_{aa}, F_{ca}) for\ first\ dynamic\ feature$$

$$M_{fb} = (F_{bb}, F_{cb}) for\ second\ dynamic\ feature$$

    If (threshold >= MAX)
      Fire detected and SMS generated
    End if
  End for
End

Algorithm 1. Hybrid Disclosure

4 Experimental Results and Discussion

The camera is tested for very first time, if it detects the fire or not. A source of fire is introduced into the frame for the very first time, allowing the camera to pick up traces of the turbulent, in motion, in pace fire flame, which is shown in Fig. 3.

The camera searches and simultaneously processes towards detecting the presence of fire by constantly comparing older and newer frames using edges, which is shown in Fig. 4.

The system detects the fire and the notification has been sent, which has been received by the concern user, which is shown in Fig. 5.

Fig. 3 Testing for the presence of any source of fire

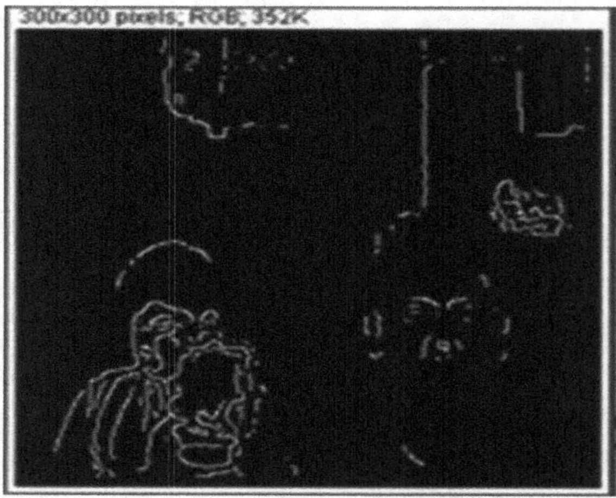

Fig. 4 Flame edge detected

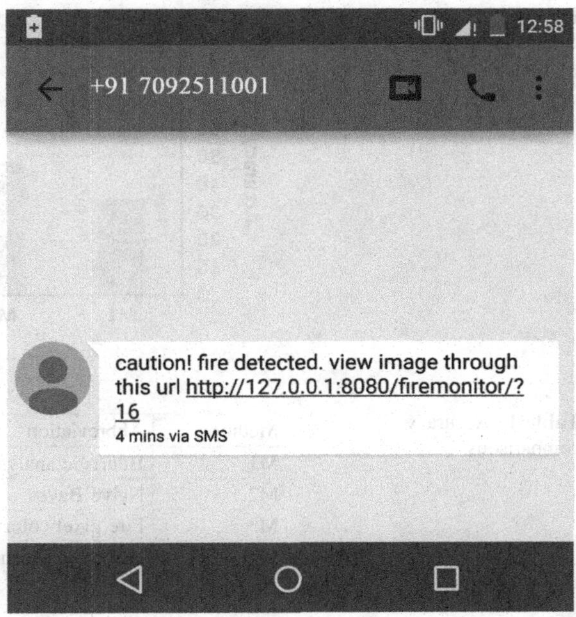

Fig. 5 Fire detected and SMS received

5 Performance Analysis

The outcomes area assesses the proposed characteristics' every performance in three ways. Initially, the calculation is tried on an extensive feature database. Since unquestionably the estimations of false discovery rates are exceptionally subject to the testing and preparing information, a subjective evaluation of the outcomes takes after. At last, a novel system for examining fire locators is presented by utilizing engineered fire recreations; they take into consideration a quantitative examination concerning how flame immersion, spatial determination, casing rate, and commotion influence recognition.

We make an arrangement of images where fire is mimicked (fire succession). We then make an alternate grouping of images made up of indistinguishable fire casings moved on a level plane left to right (inflexible arrangement). The measure of the movement is picked such that the vitality characteristics *f*1 and *f*2 are like the ones in the fire grouping. Outlines from the unbending succession are demonstrated. The same methodology is rehashed for a greater number of foundations other than dark, as indicated. Note that the foundation influences the rendering of the fire flame; along these lines, diverse foundations present some variability in the presence of the flame. By and large, there are five fire and five inflexible groupings. The objective is to recognize a fire and a moving image of a fire in a feature taking into account the directional gimmicks *f*3 and *f*4. We watch that pretty much any static peculiarity utilized as a part of the writing will regularly come up short this analysis, as will numerous transient peculiarities if they measure the magnitude using Eq. (1).

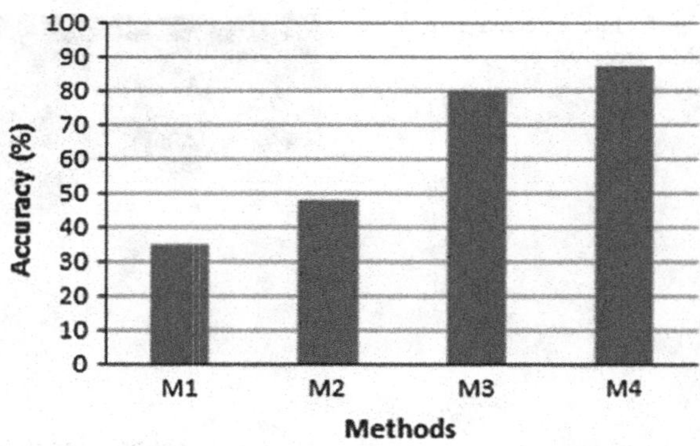

Fig. 6 Accuracy comparisons with hybrid technique

Table 1 Accuracy comparisons

Methods	Abbreviation	Accuracy (%)
M1	Heuristic analysis	35
M2	Naive Bayes	48
M3	Fire pixel color modeling	80
M4	Hybrid disclosure	87

$$G = \sum_{j=\text{frames}} Q(j)\left(\text{dist}(L, f_j)\right)^2 \quad (1)$$

Is minimized, here $f_j = (f_{j3}, f_{j4})$ is the feature point for frame, j in the $f3 - f4$ space. The weighting function $w(j)$ is defined as,

$$w(j) = \begin{cases} 1, \text{ if } j \text{ is fire (rigid) and below (above) L} \\ 100, \text{ if } j \text{ is fire (rigid) and above (below) L} \end{cases}$$

The performance of the existing methods and proposed hybrid technique are compared, which is shown in Fig. 6 and the methods and accuracy values are listed in Table 1.

6 Conclusion

After the fire attributes are dissected, another fire edge recognition technique has been created and assessed in correlation with customary routines. Exploratory results have showed that the calculation created is viable in distinguishing the edges of sporadic flares. The point of interest of this system is that the fire and blaze edges identified are clear and constant. Two novel optical stream estimators, OMT and NSD, have been introduced that overcome deficiencies of established optical stream models when connected to flame content. They got movement fields give valuable space on which to characterize movement characteristics. These peculiarities dependably recognize fire and reject non-fire movement, as exhibited on a vast dataset of genuine features. Little false identification is seen in the vicinity of critical clamor, halfway impediments, and fast edge change. In an analysis utilizing blaze reproductions, the prejudicial force of the chose gimmicks is exhibited to particular flame movement from inflexible movement. The work displayed was gone for the preparing of fire and fire pictures caught in research facilities. Further work is obliged to assess the execution of the calculation all things considered fire location situations.

References

1. Celik T, Demirel H (2009) Fire detection in video sequences using a generic color model. Fire Safety J 44(2):147–158
2. Borges P, Izquierdo E (2010) A probabilistic approach for vision-based fire detection in videos. IEEE Trans Circuits Syst Video Technol 20(5):721–731
3. Marbach G, Loepfe M, Brupbacher T (2006) An image processing technique for fire detection in video images. Fire Safety J 41(4):285–289
4. Liu C, Ahuja N (2004) Vision based fire detection. Proc Int Conf Pattern Recognit 4:134–137
5. Zhao J, Zhang Z, Han S, Qu C, Yuan Z, Zhang D (2011) SVM based forest fire detection using static and dynamic features. Comput Sci Inf Syst 8(3):821–841
6. Ko B, Cheong K, Nam J (2009) Fire detection based on vision sensor and support vector machines. Fire Safety J 44(3):322–329
7. Toreyin B, Dedeoglu Y, Gudukbay U, Cetin A (2006) Computer vision based method for real-time fire and flame detection. Pattern Recognit Lett 27(1):49–58
8. Ho C-C (2009) Machine vision-based real-time early flame and smoke detection. Meas Sci Technol 20(4)

A Novel Approach for Balancing the Loads on Virtual Machines for Scheduling Parallel Jobs Based on Priority-Based Consolidation Method

P. Mohamed Shameem, R.S. Shaji and Jyothi Vijayan

Abstract Cloud computing is an emerging computing technology that uses the internet and central server to maintain data and application. In computing, load balancing distributes workloads across multiple computing resources. Scheduling of parallel jobs is an important issue in the cloud data centers. Node number requirement of each parallel job is different in case of complex applications. A load balancing algorithm attempts to improve the response time of user-submitted applications by ensuring maximal utilization of available resources. The main contribution of the work is to analyze different scheduling algorithms and to reduce the migration cost, waiting time, makespan, and response time. So the proposed work introduced a new algorithm and compared makespan, waiting time, response time, number of migrations, and idle time of different algorithms such as FCFS, backfill, CMCBF, AMCBF. The number of migrations required in the proposed algorithm is less compared to that of other algorithms.

Keywords Load balancing · Cloud computing · Bee backfill with migration · Priority-based consolidation method

P. Mohamed Shameem (✉)
Faculty of Computer Science and Engineering/Computer Applications,
Noorul Islam University, Kumaracoil, India
e-mail: pms.tkmit@yahoo.in

R.S. Shaji
Faculty of Information Technology, Noorul Islam University, Kumaracoil, India
e-mail: shajiswaram@yahoo.com

J. Vijayan
Department of Computer Science and Engineering, TKM Institute of Technology,
Kollam, India
e-mail: jyothychinchu@gmail.com

L.P. Suresh and B.K. Panigrahi (eds.), *Proceedings of the International Conference on Soft Computing Systems*, Advances in Intelligent Systems and Computing 397, DOI 10.1007/978-81-322-2671-0_82

1 Introduction

Cloud Technology standardizes and pools IT resources and automates many of the maintenance tasks done manually today. Its architecture model facilitates self-service, elastic consumption and pay-as-you-go pricing. One of the advanced concepts of cloud computing is virtualization. With virtualization it is possible to run multiple operating systems and multiple applications on the same server at the same time, by improving the utilization and flexibility of hardware.

Load balancing is an issue in cloud datacenters. Load balancing is the process of reassigning the total load to the individual nodes of the collective system. It makes resource utilization effective and improves the response time of the job. The aim of load balancing algorithm is dynamic in nature which does not consider the previous state. Parallel computing is the simultaneous use of multiple compute resources to solve a computational problem. Jobs are the sets of smaller components where the applications and services can be decomposed. Virtual machines are the emulation of a particular computer system and their operations are based on the computer architecture and functions of a real or hypothetical computer. And these virtual machines are treated as the processing units of cloud computing environments. Scheduling of jobs is of great consequence in cloud. Schedulers for cloud computing is accountable to schedule jobs efficiently by fully utilizing the available resources. The categories of scheduling include static scheduling and dynamic scheduling. In static method the scheduling decision should be computed before executing the job. Dynamic method does not allow the prior knowledge about the time of termination of jobs. The general job scheduling problem mainly includes (i) selection of processing resource for every job (ii) selection of job processing order/time for every resources, and they are driven by different constraints such as QoS requirement of jobs, date/time dependencies between jobs and the processing limitation of resources.

Algorithms having desirable qualities should follow the scheduling criteria maximum CPU utilization, maximum throughput, turnaround time, minimum waiting time, and minimum response time.

2 Related Works

By far the simplest scheduling algorithm is the first come first served (FCFS) [1]. With this scheme, the process that requests the CPU first is first allocated with the CPU. Each job should mention particularly about their node number requirement and it is the responsibility of the scheduler to process these jobs based on their arrival order. If the number of nodes is adequate to the needs of job then the scheduler allocates the job to run on these nodes. Otherwise it has to wait until the currently running job stops its execution. The average waiting time under FCFS policy is often quite long. Fragmentation is the major drawback of this FCFS algorithm. If the first job cannot run then the processing power should be wasted. No starvation occurs but the utilization is poor.

Backfilling [2] permits small jobs from the back of the queue to execute before larger jobs that arrived earlier. And it requires job runtimes to be known in advance and often specified as runtime upper-bound. Backfilling can be of two types, conservative and aggressive. Gang scheduling [3] executes related processes together on a machine. It allows sharing of resources among multiple parallel jobs in which the computing capacity of a node is divided into time slices. Time slices are created and within a time slice processors are allocated to jobs. Jobs are context switched between time slices. It leads to increased utilization.

Paired Gang Scheduling [4] tries to overcome the drawbacks of gang scheduling in which it utilizes the system resources well without causing interference between the processes of competing jobs. The processes will not have to wait longer because a process which occupies the CPU most of the time will be matched with a process that occupies an I/O device most of the time, so they will not interfere with each other's work. On the other side, the CPU and the I/O devices will not be idle while there are jobs which can be executed.

By using gang scheduling along with backfilling and migration [5] is an effective scheduling strategy to improve time of response, throughput and resource utilization in cloud. Gang scheduling and backfilling are two optimization techniques that operates on orthogonal axes, space for backfilling and time for gang scheduling. And the proposed technique is made by treating each of the virtual machines created by gang scheduling as a target for backfilling. The difficulty arises in estimating the execution time for parallel jobs so migration is taken into account which improves the performance of gang scheduling without the need for job execution time estimates.

A Double level priority-based task scheduling [6] in which three different waiting queues are considered. They are low-priority queue, medium-priority queue, and high-priority queue, and the local schedulers maintain these queues. The scheduler needs to effectively schedule tasks in terms of both performance and energy consumption. For this, power threshold of processor is monitored. When processor reaches its power threshold, the task is assigned into another processor.

A QoS aware honey bee scheduling algorithm is proposed for cloud infrastructure as a service (IaaS) [7] and the fault rate of datacenters or the nodes are taken into consideration. Based on the lines of food search behavior of honey bees, it optimizes the search process by selectively going to more promising honey sources and scan through a sizeable area [8]. Loads on virtual machine can be found using the concept of honey bee foraging the food source [9]. Based on the loads virtual machine can be categorized into three [10]. For scheduling parallel jobs, two-tier virtual machine is introduced to improve the responsiveness of the jobs [11].

3 Proposed Method

In order to overcome the challenges and difficulties of existing algorithms a new algorithm "Bee Backfill with Migration" is introduced. Cloud environment is created by implementing the data centers, cloud broker/scheduler and virtual machines.

Fig. 1 System architecture

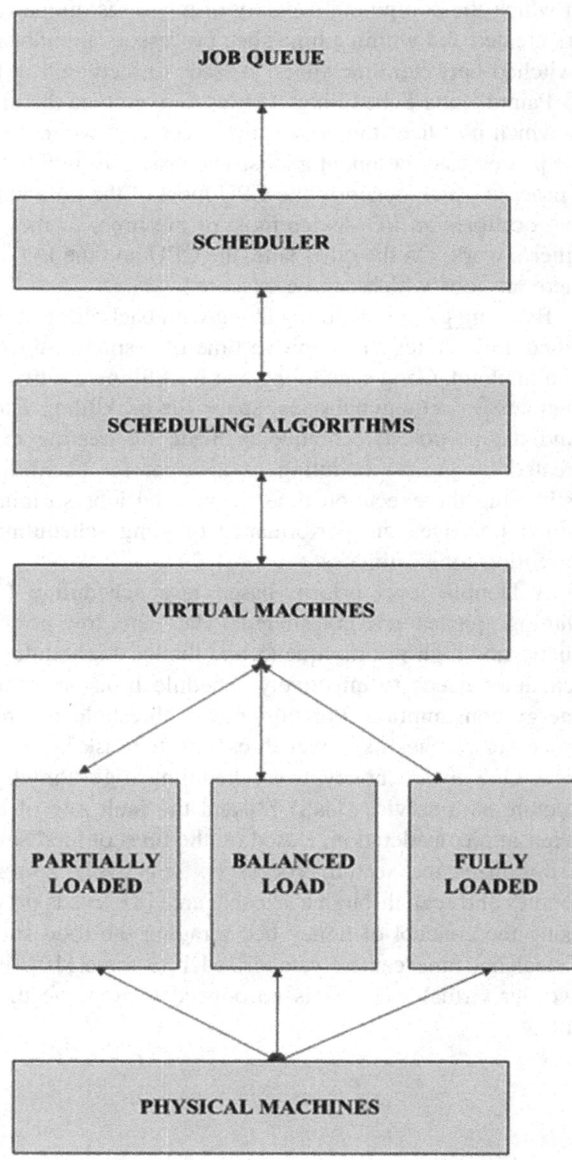

3.1 System Architecture

Cloud broker is the mediator to perform the job in the resources. Upon the submission of job to the job queue in Fig. 1 [1], the scheduler arranges the jobs according to the order of their arrival time and will specify their node requirement. "Bee Backfill with Migration" schedules the jobs according to the order of their arrival time when there is enough number of nodes. When the number of inactive (idle) nodes is inadequate for a job, another job with a later arrival time but smaller node number requirement maybe scheduled and the remaining jobs are backfilled to the queue. The backfilled jobs can check the loads on the virtual machines while deciding to migrate to some other nodes. Based on the loads the virtual machines can be categorized into three: Partially loaded, Fully loaded, and Balanced virtual machines.

3.2 Pseudo-code for Bee Backfill with Migration

Input : The queue of incoming jobs
Allocation matrix : Rows containing nodes and columns with jobs.
Output : The updated allocation map

i) Begin
ii) job←Get the first job from job_queue
iii) While queue is not null do
  (1) check the number of nodes(num_{nodej}) required by job
  (2) check the number of idle nodes(num_{idlej})
  (3) If $(num_{nodej}) \leq (num_{idlej})$
    (a) Remove job from job queue(Q) and dispatch it to any (num_{idlej}) idle nodes
    (b) Update the allocation matrix
    (c) If job is not at the head of the job_queue(Q) then insert the job into $backfill_{queue}$
  (4) Else
    (a) If the $(num_{nodej}) \leq (num_{backfill} + num_{idlej})$ is less than or equal to the summation of the number of nodes running jobs arriving later than the job and number of idle nodes then
      (i) Suspend the jobs in backfill queue that arrive later than job and move them back to queue according to descending order of their arrival time until the number of idle nodes is greater than the number of nodes required by job.
        a) Find the capacity of remaining virtual machines
        $capacity = \frac{sum\ of\ job\ size}{MIPS}$, MIPS is millions of instruction per second.
        b) Find the loads on virtual machines
        $$load = \frac{number\ of\ tasks}{service\ rate\ of\ virtual\ machines}$$
        c) Find the time required to process a job in the virtual machine
        $$time\ to\ process\ a\ job = \frac{load}{capacity}$$
        d) Based on the loads $(Partially\ loaded, Fully\ loaded\ or\ Balanced)$ of the virtual machines select the best suited Virtual machine for migration.
      (ii) Remove job from queue(Q) and dispatch it to num_{idlej} idle nodes and call check load function
      (iii) Update the allocation matrix

iv) Get the next job from job_queue

4 Performance Evaluation

4.1 Performance Metrics

- Makespan is the total completion time of task.
- Response time is the time a system or functional units takes to react to a given input.
- Waiting time is the period of time between an action requested or mandated when it occurs.
- Number of Migration is the process of transferring jobs between different virtual machines.

The performance of "Bee Backfill with Migration" is compared with FCFS, Backfill, Backfill with Migration, CMCBF, and AMCBF algorithms. And from this evaluation, it is found that number of migrations (Fig. 2), response time (Fig. 3), waiting time (Fig. 4), and idle time (Fig. 5) of "Bee Backfill with Migration" is less compared to that of other algorithms.

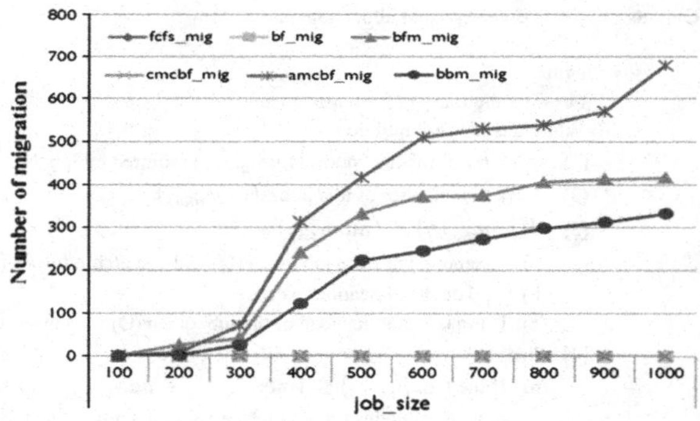

Fig. 2 job_size versus number of migrations

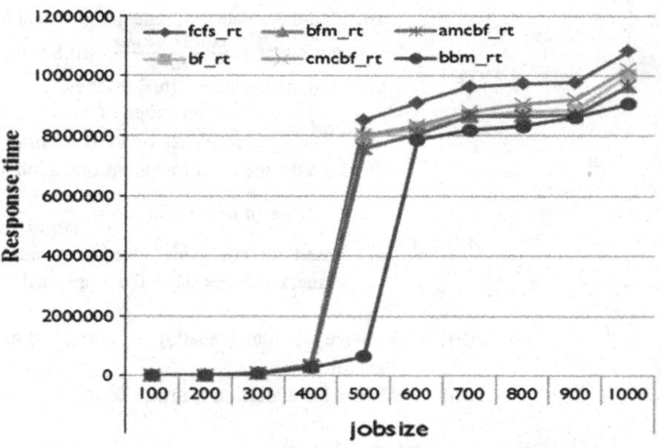

Fig. 3 job_size versus response time

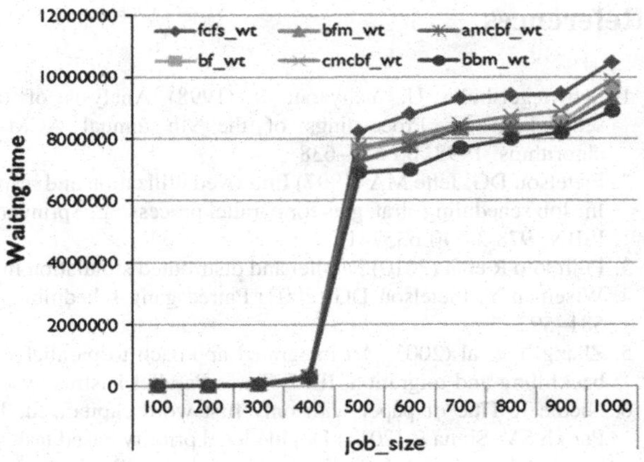

Fig. 4 job_size versus waiting time

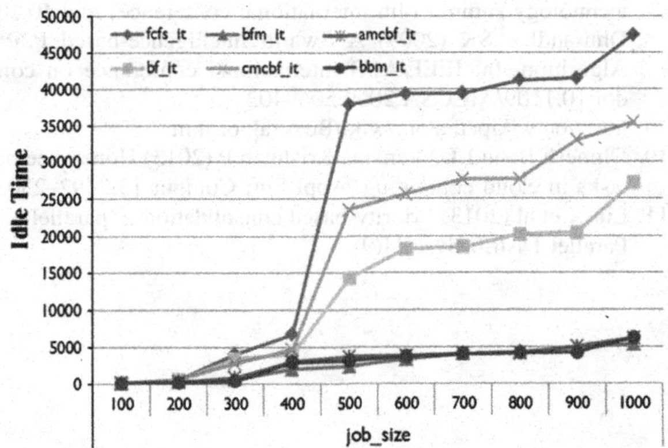

Fig. 5 job_size versus idle time

5 Conclusion

Scheduling of parallel jobs is an important issue in the cloud data centers. Node number requirement of each parallel job is different in case of complex applications. So it is necessary to improve the throughput, response time, and utilization of the nodes. Existing scheduling mechanisms normally take responsiveness as high priority and need nontrivial effort to make them work for datacenters. This work proposes a new algorithm that considers priority of task as the main QoS factor. From the evaluation of performance it is clear that the "Bee Backfill with Migration" algorithm shows better performance than the other algorithms.

References

1. Schwiegelshohn U, Yahyaour R (1998) Analysis of first-come-first-serve parallel job scheduling. In: Proceedings of the 9th annual ACM-SIAM symposium on discrete algorithms, 1998, pp 629–638
2. Fietelson DG, Jette MA (1997) Improved utilization and responsiveness with gang scheduling. In: Job scheduling strategies for parallel processing. Springer, Berlin, Heidelberg, pp 238–261. ISBN: 978-3-540 635741
3. Fujimoto R et al (2010) Parallel and distributed simulation in the cloud. SCS M&S Magazine 3
4. Wiseman Y, Fietelson DG (2003) Paired gang scheduling. IEEE Trans Parallel Distrib Syst 581–592
5. Zhang Y et al (2003) An integrated approach to parallel scheduling using gang scheduling, backfilling and migration. IEEE Trans Parallel Distrib Syst 14:236–247
6. Nicole R Title of paper with only first word capitalized. J Name Stand Abbrev (in press); Parikh SV, Sinha R (2011) Double level priority based task scheduling with energy awareness in cloud computing. Int J Eng Technol 142–147
7. Kumar P, Dept. of Comput. Sci. Eng., JIIT (2013) Fault aware Honey Bee scheduling algorithm for cloud Infrastructure. In: Confluence 2013: the next generation information technology summit (4th international conference). doi:10.1049/cp.2013.2306
8. Dhurandher SK (2009) A Swarm Intelligence-based P2P file sharing protocol using Bee Algorithm. In: IEEE/ACS international conference on computer systems and applications. doi:10.1109/AICCSA.2009.5069402
9. http://en.wikipedia.org/wiki/Bees_algorithm
10. Dhinesh Babu LD, Venkata Krishnab P (2013) Honey bee behavior inspired load balancing of tasks in cloud computing. Appl Soft Comput 13:2292–2303
11. Liu X et al (2013) Priority-based consolidation of parallel workloads in the cloud. IEEE Trans Parallel Distrib Syst 24(9)

Prevention of Damage of Motor Using Water Flow Sensor

Bhatter Abhinay, S. Roji Marjorie, M. Aravind and K. Jayanth

Abstract Water scarcity is one of the major problems being faced by major cities of the world. Many works have been done to control the switching on and off of the motor by monitoring the water level. This work can be claimed to be novel since it controls the motor operation by monitoring the flow of water which connects the pipe to the tank in addition to monitoring the water level. One has to keep on observing the water level of the tank to switch off the motor when its level goes below the predetermined level. The motor coil might also be burnt in the absence of water in the sump. An analogous water flow sensor is connected to the pipe which connects it to the tank. If the rate of flow of water is lower than the optimum level and when water level is high in the tank the motor is turned off automatically. Thus it saves the life of the water motor.

Keywords Water flow sensor · AT89S51 · ADC0808 · L293D

1 Introduction

Maximum portion of the earth's surface is covered with water and out of that only a small portion of the planet's water is fresh. This issue is identified with poor water designation, improper utilization, and absence of satisfactory and incorporated water administration. Numerous works have been completed on water level control and automization. In a recent work [1], a smart water tank administration framework was used along with Atmega 128A microcontroller. In this work manual mediation is not needed for nonstop water supply. Tanks have water level markers appended to them and contact sensors are utilized for water level detection. When the tank is void the water pump will draw water from the supply until the tank is

B. Abhinay · S. Roji Marjorie (✉) · M. Aravind · K. Jayanth
Saveetha School of Engineering, Electronics and Communication Department, Saveetha University, Chennai, India
e-mail: roji_marjorie@yahoo.co.in

L.P. Suresh and B.K. Panigrahi (eds.), *Proceedings of the International Conference on Soft Computing Systems*, Advances in Intelligent Systems and Computing 397, DOI 10.1007/978-81-322-2671-0_83

full. In another work [2], programmed water level detecting and controlling is done utilizing remote correspondence between controllers set at the tank and the sump. The framework fundamentally works with two controllers and RF handset modules. In another work [3], a programmed pump controller fabricated utilizing rationale entryways minimizes the need for any manual exchanging. In a recent work [4], a PIC 16F84A microcontroller is utilized to recognize the presence of water in the store tank. A detecting unit monitors the level of water in the tank and switches off the motor when the level is exceeded. In another work [5], the framework utilizes a microcontroller to mechanize the procedure of water pumping and can recognize the level of water in a tank by switching on/off the pump likewise and it shows the status on a LCD screen. In this work the water is allowed to flow through the funnel which joins the channel to the water tank. All these methods are aimed at enhancing the life of the water motor.

2 Block Diagram

The proposed system consists of a water flow sensor and a water level sensor along with a microcontroller. There is a high possibility of a motor getting damaged by reduced water pressure (Fig. 1).

Water starts flowing through the motor's outlet to the tank when it is switched on. The release of water is measured by utilizing the water flow sensor at the outlet funnel of the motor and the motor is halted when the released quantity measured is low. The water flow sensor is connected to an analog-to-digital converter ADC0808. This ADC is associated with microcontroller AT89S51. An embedded C is encrypted into this AT89S51 microcontroller for controlling the motor [6]. The code is created considering the threshold levels such as level control and flow control of the water. These controls are determined by considering the pumping limit of engine and limit of tank separately.

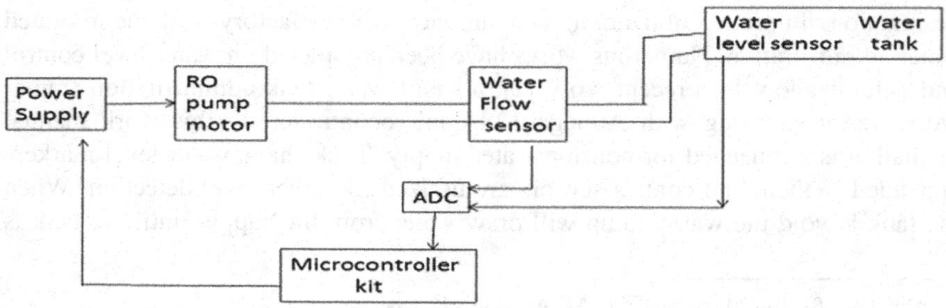

Fig. 1 Block diagram of proposed system

3 Construction

The RO water motor pump is driven using a motor-driving circuit L293D, since a microcontroller cannot drive a motor pump straightforwardly. AT89C51 is utilized which contains four ports P1, P2, P3, and P0. Each port is 8-bit bi-directional ports, i.e., they can be used as both data and yield ports. ATMEL 89C51 has 4 kB of flash programmable and erasable read-only memory (PEROM) and 128 bytes of RAM. It can be modified to work at differing bit rates. This microcontroller is associated with supply to control the motor (Fig. 2).

L293D is an H-bridge motor driver integrated circuit (IC). Motor drivers go about as flow intensifiers and flow enhancers. The motor driver circuit information pins 2 and 7 are associated with two pins of AT89S51 port 1 (P1). Motor is joined with the output pin 3 and 6 of L293D. The pins (4, 5, 12, 13) are GND and the pin (8, 16) are given to +VCC1 and +VCC2 separately. The motor operations can be directed by data rationale at pin 2 and 7, if the information rationales 00 or 11 will stop the motor. Enable pin 1 must be high for motors to begin working. When an enable input is high, the driver is enabled and the output becomes active, and works in phase with their inputs. Similarly, when the input is low, the motor driver is disabled, and their output becomes inactive, and works out of phase with their inputs [7].

ADC0808 is an 8-bit analog-to-computerized digital with eight information analog channels. The voltage reference can be set utilizing the V_{ref+} and V_{ref-} pins.

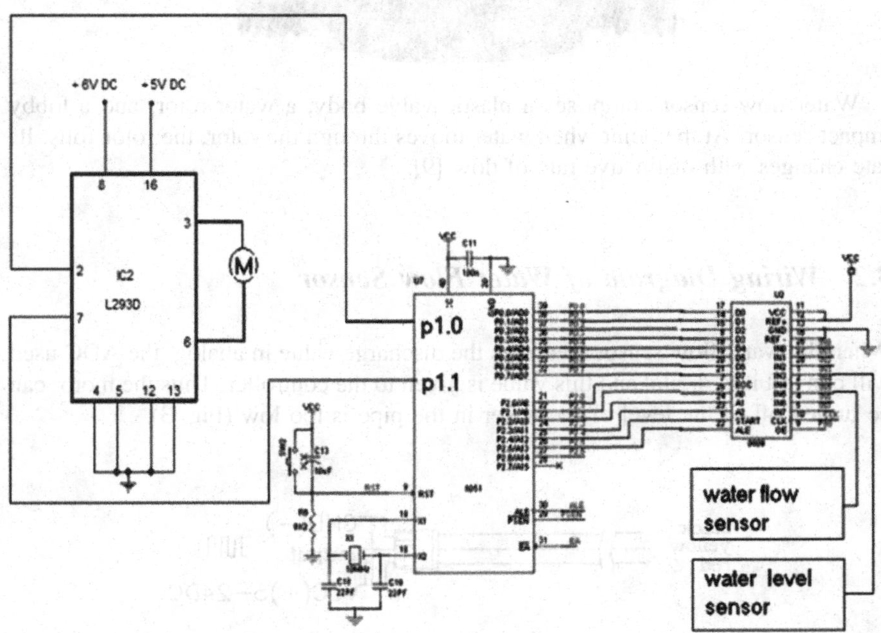

Fig. 2 Circuit diagram of proposed system

The step size is chosen depending on the reference input. The default step size is 19.53 mV comparing to 5 V reference voltage. At the point when the change is finished, the EOC pins goes low to demonstrate the end of conversion [8].

This ADC0808 IC is connected to the information port P0 of the AT89s51. The port P0 is joined with the advanced output of the AT89s51 and the simple data port of ADC0808 where it contains eight pins and is associated with the water flow sensor. The water flow sensor measures the analog input and is associated with the data pin (4) of the ADC0808.

3.1 *Pictorial Representation of Water Flow Sensor*

Water flow sensor comprises a plastic valve body, a water rotor, and a lobby impact sensor. At the point when water moves through the rotor, the rotor rolls. Its rate changes with distinctive rate of flow [9].

3.2 *Wiring Diagram of Water Flow Sensor*

When the water flow sensor measures the discharge value in analog, the ADC used will convert it to digital and this value is given to the controller. Thus the motor can be turned off, if the level of the water in the pipe is too low (Fig. 3).

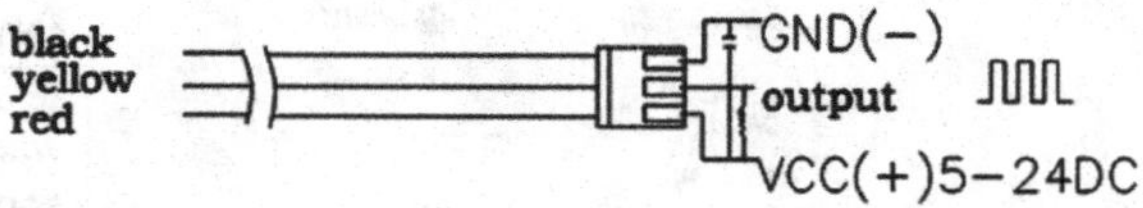

Fig. 3 Wiring diagram of water flow sensor

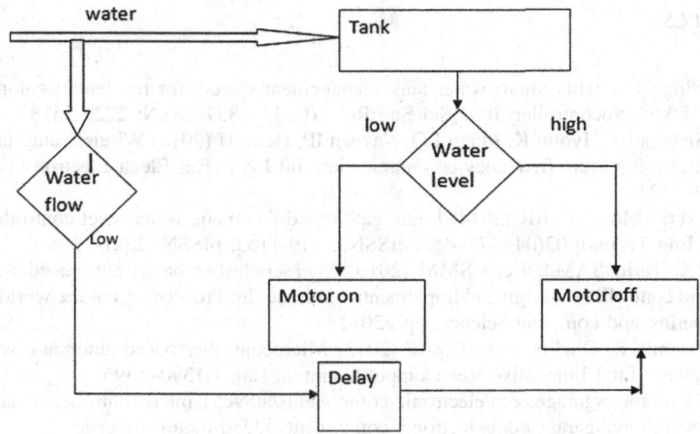

Fig. 4 Flow diagram of proposed system

4 Work Flow

The water flow and control of motor individually are shown in Fig. 4. The water level in the tank is measured and the motor is turned on when the level is low and motor stops running if the water level is high. This level is measured utilizing the water level sensor. The water flow is measured using the water flow sensor and if the flow measured is low the motor stops running after a certain delay. Accordingly the motor is controlled by measuring both the water level and the flow rate and is safeguarded from damage (Fig. 4).

5 Conclusion

Our proposed gadget is by regional standards planned and created novel system for preventing damage to the motor. This controller controls the working of the motor based on the flow of water through the pipes in addition to acting as a water level controller. The system utilizes a microcontroller to computerize the methodology of measuring the release of water to a tank and can recognize the release of water from the outlet channel to the tank and can switch on/off the water motor.

References

1. Patil Y, Singh R (2014) Smart water tank management system for residential colonies using Atmega128A microcontroller. Int J Sci Eng Res 5(6):335–337. ISSN: 2229-5518
2. Muktha Shankari K, Jyothi K, Manu EO, Naveen IP, Herle H (2013) Wireless automatic water level control using radio frequency communication. Int J Adv Res Electr Electron Instrum Eng 2(4):1320–1324
3. Moyeed Abrar Md, Patil RR (2014) Logic gate based automatic water level controller IJRET: Int J Res Eng Technol 03(04):477–482. eISSN: 2319-1163. pISSN: 2321-7308
4. Reza SMK, Tariq SAMd, Reza SMM (2010) Microcontroller based automated water level sensing and controlling: design and implementation issue. In: Proceedings of the world congress on engineering and computer science, pp 220–224
5. Ejiofor Virginia E, Oladipo Onaolapo F (2013) Microcontroller based automatic water level control system. Int J Innovative Res Comput Commun Eng 1:1390–1396
6. http://www.motorersgarage.com/electronic-components/at89c51-microcontroller-datasheet
7. http://www.motorersgarage.com/electronic-components/l293d-motor-driver-ic
8. http://www.motorersgarage.com/electronic-components/adc0808-datasheet
9. http://www.seeedstudio.com/depot/datasheet/water%20flow%20sensor%20datasheet.pdf

Nano Silver-Based Monopole Paper Antenna for High-Frequency Application

J. Arul Linsely and A. Shajin Nargunam

Abstract The rapid growth of wireless technologies requires compact, easily designed, and fabricated high-frequency antennas with a new technique. The antenna is designed using nano silver material which is coated on flexible cellulosic paper and then sintered to make a low-profile monopole paper antenna for UHF and VHF applications. Previously, metal antennas were used for receiving signals which have been replaced by using nano particles. The monopole antennas thus developed will be flexible, cost effective, biodegradable, and easy to be fabricated. Nowadays, due to rapid development in communication, flexible antennas are required for RF applications. This can be eco-friendly and also biodegradable. This monopole antenna is tested by using Antenna Training System (ATS-2002) developed by FALCON. These results are discussed here in this research work.

Keywords Antenna Training System (ATS-2002) · FALCON · Nano silver material · Artificial Transmission Line (ATL)

1 Introduction

The electronics and communication industries largely depend on PCB fabrication methods such as photolithography and screen printing technologies. Most of the circuits fabricated in PCB use FR4 which is nonbiodegradable and it also takes decades to breakdown in landfills and this largely contributes to the bulk of the electronic waste that is generated annually.

J. Arul Linsely (✉)
Department of EEE, Noorul Islam University Kumaracoil, Kanyakumari
Tamil Nadu, India
e-mail: arullinsely@gmail.com

A. Shajin Nargunam
Department of Computer Science and Engineering,
Noorul Islam University Kumaracoil, Kanyakumari, Tamil Nadu, India

L.P. Suresh and B.K. Panigrahi (eds.), *Proceedings of the International Conference on Soft Computing Systems*, Advances in Intelligent Systems and Computing 397, DOI 10.1007/978-81-322-2671-0_84

Paper is an organic-based substrate which is universally available; the high demand and mass production of paper has made it as one of the cheapest materials that is available. In addition, from a manufacturing point of view, paper is well suited for reel-to-reel processing and hence, mass fabrication on paper becomes more feasible. Paper is highly biodegradable and it requires just a few months to breakdown completely in landfills. The design of compact antennas to support wireless devices is an interesting topic of research. As electronic technologies have progressed, the market demands devices to become smaller, lighter, and low profile than ever before and it is a highly promising research area. This has led to the development of a technology to fabricate the monopole antennas on a low-cost substrate such as cellulosic material.

The low-profile antennas are meant to have a height of $\lambda/10$ or less, where λ is the free space wave length at low end frequency. Thus, the occupation surface is the vital point more than the height to get a smaller device. The characteristics of the antenna are analyzed and the experimental results of the proposed antenna are discussed here.

2 Literature Review

Organic substrates offer a unique advantage of being much cheaper, lighter, and more eco-friendly in processing and also in disposal of antennas than current substrates like FR4. They have the additional advantage of providing flexibility as well.

Flexible devices can be fabricated quickly. Eco-friendly and disposable antennas can be fabricated by this proposed technology also. The first paper published on printed monopole antennas started to appear in literatures around 2007 for the RFID integration on low-cost flexible substrates to target the inventory control and supply chain management markets [1].

The cellulosic paper used for monopole antennas can potentially revolutionize the market for electronics. This could eventually take the first step in creating eco-friendly RF electronics and modules. Additionally, cellulosic paper is one of the most low-cost materials that have been produced [2].

The antennas thus developed have the advantages of small size, low profile, and also simple configurations. The low profile of the antennas makes it a promising device for compact and slim wireless devices of the future [3].

Sangkil Kim et al. designed a micro strip antenna for wearable applications which has better antenna gain and also maintains a low profile. It has a bandwidth of −10 dB which covers the entire frequency range of IEEE 802.15 standard which is a good application for wearable electronic devices [4].

The feasibility of realizing ultra-wideband antennas through ink jetting of conductive inks on commercially available paper sheets are reported for frequencies up to 10 GHz and above that also [5].

Development of compact five band printed antenna for fixed or reconfigurable communication systems was designed by using four varactors. This design enables the four bands to be tuned independently over wide frequency ranges also [6].

Ma and Vandenbosch designed an antenna of operational frequency band targeted which neglected the oxidation effects. In this, metallic silver was the best choice for communication signals having a wavelength of above 500 nm and aluminum was the best choice for wavelengths of below 500 nm. Due to the oxidation effects of this metal, the threshold shifts to a wavelength of around 700 nm [7].

Effects of the dispersive properties of some important metals were studied and the input impedance of Ag > Al was observed [8].

Excellent conductive values of metallic silver was attained by using very small amounts of organic additives without any strong adsorbing groups such as amides and amines. Even at low temperatures, conductivity of the printed antennas by using nano materials was highly dependent on the organic additives used. The low sintering temperature enhances the conductivity of the flexible materials [9].

Advantages of designing a low-profile antenna on a high-impedance surface offer a viable solution for naval, vehicle, and aerospace platforms [10].

Planar modified dipole antennas have an enhanced radiation performance in terms of maximum gain and stability of gain response. This type of antenna has much wider impedance bandwidth characteristics as well [11].

In modern multipurpose handheld gadgets, different communication technologies are embedded in the same device which shares the same antenna. Nowadays, there is a huge demand for wide band antennas that are small in size having low profile, light weight, low cost, and are easy to fabricate and install [12].

These paper-based RFID tag antennas are fabricated by using commercially available ink-jet printers by using nano silver ink. The performance is improved by modifying the ink's properties as well [13].

Planar dipole antenna with a flat reflector and a nonplanar antenna have a good balance between the impedance bandwidth and pattern bandwidth. Thus, the operating bandwidth was improved significantly. The combination of nonplanar dipoles and nonplanar reflectors is very attractive for a larger operating bandwidth also [14].

The compact omnidirectional printed quasi-Yagi antenna decreases the antenna's size and its radiating elements operate at 915 MHz. The Artificial Transmission

Line (ATL) structure has compact arms of a quasi-Yagi antenna for RFID applications. Presence of this ATL structure results in reduction of the quasi-Yagi antenna arm's length from 0.25 λ to 0.1 λ [15].

The broadband antenna inherits a wide bandwidth with its height that is considerably reduced from λ to 0.03 λ of its lowest operating frequency. This design consists of a top loaded monopole antenna with several shorting pins and inductive loading [16].

A high-gain low-profile miniaturized antenna with omnidirectional and vertically polarized radiation, similar to a short dipole is fabricated and the gain and polarization improvements are achieved by isolating the feed structure from the miniaturized resonant radiating structure by using inductive, top loaded in-plane capacitive couplings.

Printing additional layers only on critical areas were shown to be highly effective in terms of attenuation loss of transmission lines and total efficiency of antennas. The result of the two printed layers was achieved by printing the second layer only on high-current-density areas [17].

For printing, the process was adapted optimally for multiple layers that have minimum ink in the first layer. Decrease in attenuation was seen to be limited, showing a promising option for decreasing loss without highly increasing ink usage [18].

This printed monopole antenna for ultra-wideband applications with a frequency band-notch characteristic is presented here. The proposed antenna consists of a stepped rectangular radiating patch and a modified ground plane which provides a wide usable fractional bandwidth of more than 120 % [19].

The circularly polarized printed antenna, capable of operating over an octave bandwidth is designed and fabricated. This design is evolved from classical printed monopole. The antenna employs a micro strip-line fed rectangular radiator, printed on the top of a substrate [20].

The demonstrated UWB monopole antenna with fractal matching network is the smallest reported antenna with operation over the entire UWB band, which is important in the realization of low-cost high data rate wireless sensor networks and wearable wireless devices [21].

The emerging wireless technologies increase the needs for small size, lightweight, and easily fabricated antennas. The quarter wave monopole antenna is the most ubiquitous antenna which is being used for many applications such as unattended ground sensors and ground-based communication systems at various frequency bands. However, the size of such an antenna is prohibitively large for portable devices operating at low frequencies [22].

The wideband antenna design standard had been proposed from investigation of the volcano smoke antenna. The dimension of the main radiation body of a wideband antenna should be about a quarter of wavelength at the lowest operating frequency [23].

3 Methodology

This paper aims at developing a compact low-profile monopole paper antenna by coating the nano silver mixed in organic additives over cellulosic paper and sintering it to increase the conductivity of the antenna designed for 800 MHz, which is cost effective, eco–friendly, and easy to fabricate.

3.1 Antenna Design and Fabrication

The cellulosic paper used for fabricating the antenna is cut as a square measuring 8–8 cm. Two cellulosic papers measuring 8–8 cm size is pasted using glue to make a thick base and to give mechanical strength as well. The cellulosic paper is then preheated by using a hot air oven at a temperature of about 60 °C for 45 min.

The nano silver powder is made as a paste by mixing it with organic additives having nonpolar solvents which have long alkyl chains as it is used to stabilize the nano particles. This paste is then applied on the cellulosic paper to a length of 9.375 cm to form a monopole antenna using a very thin brush.

This is then sintered by using a hot air oven for 1 h at 80 °C; by this process, the conductivity increases gradually as the contact between the particles become better. By selecting the quarter wavelength, the total length of the monopole antenna that is pasted is 9.375 cm.

3.2 Materials

The nano silver mixed with an organic additive is coated on cellulosic paper to develop an environmentally friendly monopole antenna in the frequency range above 2 MHz to 10 GHz.

Cellulosic paper has a number of distinct advantages and is possible for low-cost green electronics. Moreover, cellulose is abundantly available which can be considered as a renewable resource; in addition it can be easily processed in a reel-to-reel fashion enabling low-cost manufacturing solutions.

3.3 Conductivity

Attaining high nanoparticles conductivity after deposition is one among the most challenging processes. To form a continuous conductive structure, the polymer coating must be removed or burnt off and the particles sintered, or melted, together.

The metallic nanoparticles are so small; they display a unique quality of having melting points much lower than that of bulk metals which allows for deposition and sintering on substrates that cannot handle high processing temperature as paper.

Following the sintering techniques, packing density of nano particles upon deposition is an important factor in the final conductivity of this sintered structure. Tighter packing of nanoparticles means that the final sintered structures will have fewer gaps and breaks caused by dispersed nanoparticles melting together that leads to higher final conductivities.

4 Results and Discussions

4.1 Nano Silver Antenna Characterization

After fabrication, the processed cellulosic paper antenna is tested for its properties. The characteristics of the antenna showed the properties of a monopole antenna with a frequency range of 800 MHz. The monopole antenna has a very good radiation pattern. The characterization was done by using ATS-2002 developed by FALCON.

4.2 Equation

$\lambda = C/\vartheta$, here C is approximately 3 $*$ 108 m/s, ϑ is selected as 800 MHz, and the value of λ comes to around 37.5 cm. Since the quarter wavelength is considered for the monopole antenna, the value of 37.5 cm is divided by four and the value thus obtained is 9.375 cm. Velocity is equal to distance traveled divided by time. Here, velocity means velocity of electromagnetic waves in free space, distance is the wavelength, and frequency is the inverse of the time period.

4.3 Figures and Tables

Figures 1 and 2 show the output of this prototype monopole antenna which is placed in the YZ plane (E-plane) which determines the polarization or orientation of the radio wave and the XZ plane (H-Plane) which is at right angles to the E-plane, respectively.

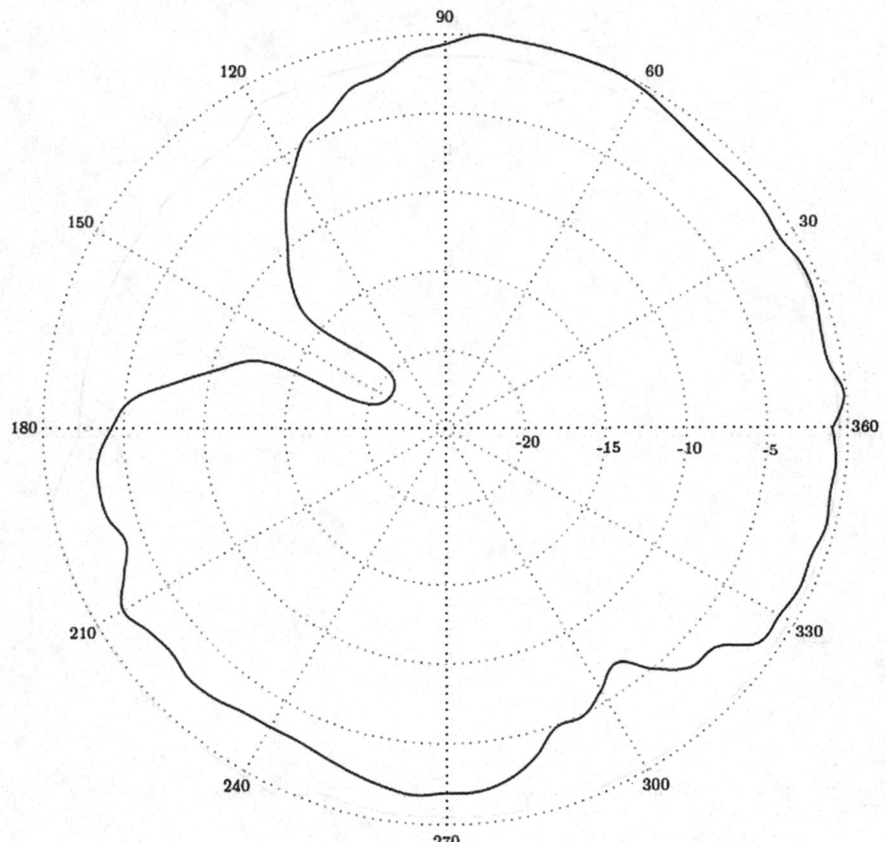

Fig. 1 YZ plane or E-plane

4.4 Discussion

When compared to a low profile, miniaturized, inductively coupled, and capacitively loaded monopole antenna discussed by Jungsuek Oh and Kamal Sarabandi, this monopole antenna has a better radiation pattern. An antenna with a higher gain is more effective in its radiation pattern.

The Azimuth pattern

Transmitter and Receiver – Monopole antenna

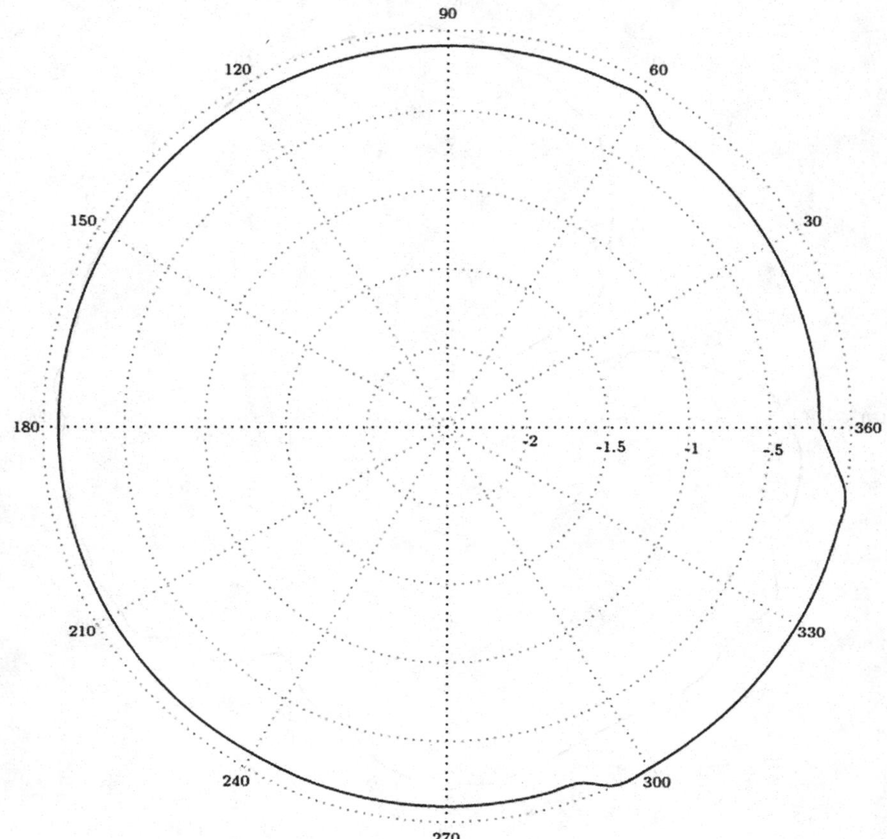

Fig. 2 XZ plane or H-plane

5 Conclusion

Proposed in this paper is a folded dipole antenna designed on the flexible cellulosic base with nano silver particles, which is found to be cost effective and easy to fabricate for a low number of antennas with minimum time by using simple techniques as discussed.

With this simple technique, a normal angle of 80° is obtained while comparing it with other nano particle antennas. Feasibility of this antenna is from 2 MHz to 10 GHz. This work will pave the way to fabricate the next generation low cost, biodegradable, and eco-friendly folded dipole antenna receivers in RFID applications.

References

1. Cook BS, Shamim A (2012) Inkjet printing of novel wideband and high gain antennas on low-cost paper substrate. IEEE Trans Antenn Propag 60(9)
2. Rida A, Yang L, Vyas R, Tentzeris MM (2009) Conductive inkjet-printed antennas on flexible low—cost paper-based substrates for RFID and WSN applications. IEEE Antennas Propag Magaz 51(3)
3. Abutarboush HF, Shamim A (2012) Paper-based inkjet-printed tri-band u–slot monopole antenna for wireless applications. IEEE Antennas Wirel Propag Lett 11:1234
4. Kim S, Ren Y-J, Lee H, Rida A Monopole antenna with inkjet-printed EBG array on paper substrate for wearable applications. IEEE Antennas Wirel Propag Lett 11:663
5. Shaker G, Safavi-Naeini S, Sangary N, Tentzeris MM (2011) Inkjet printing of ultrawideband (UWB) antennas on paper-based substrates. IEEE Antennas Wirel Propag Lett 10:111
6. Abutarboush HF, Nilavalan R, Cheung SW, Nasr KM (2012) Compact printed multiband antenna with independent setting suitable for fixed and reconfigurable wireless communication systems. IEEE Trans Antennas Propag 60:3867
7. Ma Z, Vandenbosch GAE (2013) Fellow. In: IEEE, systematic full-wave characterization of real-metal nano dipole antennas. IEEE Trans Antennas Propag 61:4990
8. Pelayo García De Arquer F, Volski V, Verellen N, Vandenbosch GAE, Moshchalkov VV (2008) Engineering the input impedance of optical nano dipole antennas: materials, geometry and excitation effect. IEEE Trans Antennas Propag 59:3209–3215
9. Perelaer J, Antonius A, De Laat WM, Chris B, Hendriksa E, Schubert USS (2008) Inkjet-printed silver tracks: low temperature curing and thermal stability investigation. The Royal Society of Chemistry 2008. J Mater Chem 18:3144
10. Vallecchi A, De Luis JR, Capolino F, De Flaviis F (2012) Low profile fully planar folded dipole antenna on a high impedance surface. IEEE Trans Antennas Propag 60:51
11. Gao F, Zhang F, Lu L, Ni T, Jiao Y (2013) Low-profile dipole antenna with enhanced impedance and gain performance for wideband wireless applications. IEEE Antennas Wirel Propag Lett 12:372
12. Behera AR, Harish AR (2012) A novel printed wideband dipole antenna. IEEE Trans Antennas Propag 60:4418
13. Pranonsatit S, Worasawate D, Sritanavut P (2012) Affordable ink-jet printed antennas for RFID applications. IEEE Trans Compon Packag Manufact Technol 2:878
14. Wu Q, Wang S, Sun X (2012) Nonplanar dipole antennas for low-profile ultra wideband applications: design. Modell Implement IEEE Antennas Wirel Propag Lett 11:897
15. Hajizadeh P, Hassani HR, Hassan Sedighy S (2013) Planar artificial transmission lines loading for miniturization of RFID printed quasi-yagi antenna. IEEE Antennas Wirel Propag Lett 12:464
16. Akhoondzadeh-Asl L, Hill J, Laurin J-J, Riel M (2013) Novel low profile wideband monopole antenna for avionics applications. IEEE Trans Antennas Propag 61(11)
17. Oh JE, Saraband K (2012) Low profile, miniaturized, inductively coupled capacitively loaded monopole antenna. IEEE Trans Antennas Propag 60(3):1206
18. Pynttäri V, Halonen E, Sillanpää H, Mäntysalo M, Mäkine M (2012) RF design for inkjet technology: antenna geometries and layer thickness optimization. IEEE Antennas Wirel Propag Lett 11:188
19. Movahedinia R, Ojaroudi M, Madani SS (2011) Small modified monopole antenna for ultra-wideband application with desired frequency band-notch function. IET Microw Antennas Propag 5(11):1380–1385
20. George Thomas K, Praveen G (2012) A novel wideband circularly polarized printed antenna. IEEE Trans Antennas Propag 60(12):5564

21. Maza AR, Cook B, Jabbour G, Shamim A (2012) Paper-based inkjet-printed ultra-wideband fractal antennas. IET Microw Antennas Propag 6(12):1366–1373
22. Oh J., Choi J, Dagefu FT, Sarabandi K (2013) Extremely small two-element monopole antenna for HF band applications. IEEE Trans Antennas Propag 61(6)
23. Liu J, Zhong S, Esselle KP (2011) A printed elliptical monopole antenna with modified feeding structure for bandwidth enhancement. IEEE Trans Antennas Propag 59(2)

Balanced Permutation Graphs

T.M. Selvarajan and M.K. Anitha

Abstract Given any permutation π of the set $\{1, 2, 3, \ldots, n\}$, the permutation graph G_π is classical. (If $V = \{1, 2, 3, \ldots n\}$, then $G_\pi = (V, E)$ with V as vertex set and $\pi(i)\ \pi(j) \in E$ if and only if $i < j \Rightarrow \pi(i) > \pi(j)$). We define another graph called balanced permutation graph, with the same vertex set $\{1, 2, 3, \ldots, n\}$ as G_π, but with adjacency condition $ij \in E$ if and only if $i + j = \pi(i) + \pi(j)$. Some properties of balanced permutation graph are discussed in this paper.

Keywords Permutation graph · Balanced permutation graph and connected components

1 Introduction

Permutation graphs are well studied in the past decade or so [1]. This study perhaps originated from the study of 'perfect graphs'(chromatic numbers and clique numbers coincide for the graph and all its induced subgraphs). We have invented a new graph based on similar construction of permutation graphs, which can be termed a "Balanced Permutation Graph". Some interesting fundamental properties of the balanced permutation graphs are studied in this paper.

2 Permutation Graphs

First, we describe permutation graphs, highlighting some results (without proofs, but with references) and then pave the platform for balanced permutation graphs in the next section.

T.M. Selvarajan (✉) · M.K. Anitha
Department of Mathematics, Noorul Islam Centre for Higher Education, Kanyakumari 629175, Tamilnadu, India
e-mail: selvarajan1967@gmail.com

L.P. Suresh and B.K. Panigrahi (eds.), *Proceedings of the International Conference on Soft Computing Systems*, Advances in Intelligent Systems and Computing 397, DOI 10.1007/978-81-322-2671-0_85

2.1 Definition

For any positive integer n, we denote by $[n]$ the set $\{1, 2, 3,\ldots, n\}$. Let π be a permutation of the set $[n]$, which is thus an element of the symmetric group S_n which contains $n!$ elements. We sometimes write $\pi = \pi(1)\ \pi(2)\ldots\ \pi(n)$ (the image integers are put in an ordered sequence even though this is not the best notation) or in terms of the (unique) cycle decomposition of π (which is standard).

The permutation graph $G_\pi = (V, E)$ corresponding to π is defined as follows: $V = [n]$ and $E = \{\pi(i)\ \pi(j)/\text{if and only if } i \text{ and } j \text{ satisfy the conditions } i < j \Rightarrow \pi(i) > \pi(j)\}$. This is equivalent to: $ij \in E \Leftrightarrow (i - j)\ (\pi(i)^{-1} - \pi(j)^{-1}) < 0$. (Notice that $(\pi^{-1})(i)$, denote here as $\pi(i)^{-1}$). But we prefer to follow the first formulation. Clearly G_π is a finite, simple, undirected graph. Koh and Ree in [2] found some interesting properties of the graph G_π. We shall recall some of them, without proof, in order to compare similar situations in our new graph.

2.2 Proposition

(i) G_π is the null graph if and only if π = identity.
(ii) G_π is the complete graph K_n if and only if $\pi = n(n-1)\ldots 2.1$.

2.3 Proposition

Let $\pi \in S_n$ such that n comes ahead of 1 in the arrangement $\pi(1)\pi(2)\ \ldots\ \pi(n)$. Then G_π is connected.

2.4 Proposition

Let $\pi = \pi(1)\pi(2)\ldots\pi(n) \in S_n$. Then G_π is disconnected if and only if there exists $i < n$ such that $\{\pi(1), \pi(2),\ldots, \pi(i)\} = \{1, 2,\ldots, i\}$ and $\{\pi(i+1), \pi(i+2),\ldots,\pi(n)\} = \{i+1, i+2,\ldots,n\}$.

We now construct a new graph.

3 Balanced Permutation Graphs

3.1 Definition

The balanced permutation graph $B_\pi = (V, E)$ for any permutation π has $V = \{1, 2, 3, \ldots, n\}$ and $ij \in E$ if and only if $i + j = \pi(i) + \pi(j)$. Since the definition of adjacency depends on a 'sort of balancing' the sums $i + j$ and $\pi(i) + \pi(j)$, we call this a balanced permutation graph, which does not in any way depend on the classical permutation graph G_π, as may be seen below.

3.2 Remark

The two graphs G_π and B_π are distinct in general.

Example 1 Take $n = 4$ and $\pi = (2413)$. Then G_π and B_π are represented as

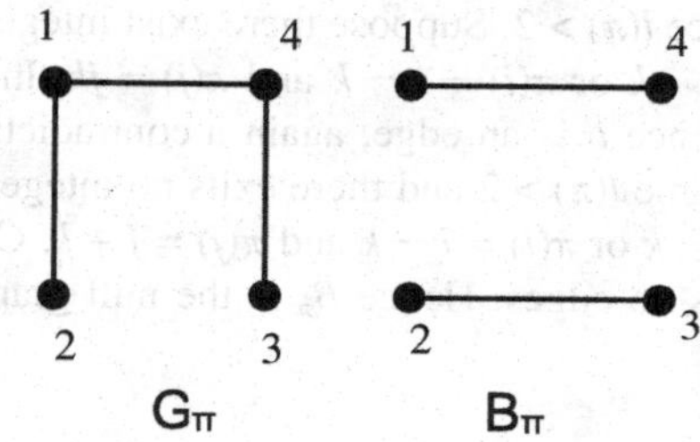

3.3 Proposition

If $n > 2$, B_π is complete if and only if π = identity.

Proof First assume that π = identity. Then for any pair of vertices i and j, $i + j = \pi(i) + \pi(j)$. Hence ij is an edge for all $i \neq j$ proving B_π is complete.

For the converse, assume B_π is complete and let that $\pi \neq$ identity.

Case 1 In the unique cycle decomposition of π, (we change the notation here) assume that there is at least one cycle $(i_1 i_2 \ldots i_r)$, $r \geq 3$. Since B_π is complete, $i_1 i_2 \in E$. Hence $i_1 + i_2 = \pi(i_1) + \pi(i_2)$. But Since $\pi(i_1) = i_2$ and $\pi(i_2) = i_3$ the above gives $i_1 + i_2 = i_2 + i_3$ forcing $i_1 = i_3$ not correct since $r > 2$. Hence π is identity.

Case 2 π is a disjoint product of transpositions. Since disjoint transpositions commute, we can write $\pi = (1j)\,(ik)\ldots\,(rs)$. Clearly $1 < j$ and we can take $i < k$. Since $n > 2$, i, j, k are distinct.

Now $1 + i < j + i < j + k = \pi(1) + \pi(i)$. This clearly shows that $1i$ is not an edge, contrary to the assumption that B_π is complete. Therefore π must be the identity. Let $l(\pi)$ denote the length of the smallest cycle in the (unique) cycle decomposition of π.

3.4 Proposition

Let $n > 2$. Then B_π is the null graph if and only if $l(\pi) > 2$ and there exits no integers i, j, and k such that $\pi(i) = i + k$ and $\pi(j) = j - k$ or $\pi(i) = i - k$ and $\pi(j) = j + k$.

Proof Assume that B_π has no edges. Suppose $l(\pi) = 1$. If there are two or more vertices i and j such that $\pi(i) = i$ and $\pi(j) = j$; hence $i + j = \pi(i) + \pi(j)$. This shows that ij is an edge, contrary to the assumption. Next suppose there is a unique vertex i such that $\pi(i) = i$. Let (jk) be a transposition in π. Then $j + k = k + j = \pi(j) + \pi(k)$, proving that jk is an edge, again a contradiction. Thus $l(\pi) \geq 2$. If $l(\pi) = 2$, then π is a distinct product of transpositions, and as before, any transposition (jk) will give an edge, contradiction. Hence $l(\pi) > 2$. Suppose there exist integers i, j, and k such that $\pi(i) = i + k$ and $\pi(j) = j - k$ or $\pi(i) = i - k$ and $\pi(j) = jk$. In both cases, we have $i + j = \pi(i) + \pi(j)$ and hence ij is an edge, again a contradiction.

For the converse, assume $l(\pi) > 2$ and there exits no integers i, j, and k such that $\pi(i) = i + k$ and $\pi(\mathrm{i}) = j - k$ or $\pi(i) = i - k$ and $\pi(j) = j + k$. Construct B_π using the above conditions, B_π has no edges. Hence B_π is the null graph.

3.5 Remark

$l(\pi) > 2$ (only) need not imply that B_π is null has been seen already in our Example 1 given earlier: $\pi = (2413)$: $1 + 4 = \pi(1) + \pi(4)$.

3.6 Theorem

For $n > 2$, B_π includes no connected tree of $(n - 1)$ edges.

Proof Obviously, we need the assumption that $n > 2$. For $n = 2$, $\pi = (12)$, the edge 12 is a tree. We proceed to prove the theorem using induction on n. First we note that $X = \{1, 2, 3\}$. For any π on X, B_π cannot be a tree of length 2. For any integer k such that $3 < k < n$, that is for any permutation π on $\{1, 2, 3, \ldots, k\}$, B_π is not a connected tree with k vertices. Now $X = \{1, 2, 3, \ldots, k, k + 1\}$. Put $Y = X - \{k + 1\}$, clearly Y contains $k > 2$ elements. Let π' be a permutation on Y, then by induction assumption $B_{\pi'}$ is not a connected tree with k vertices. That is $B_{\pi'}$ has a circuit. Using balanced permutation definition, the new edges formed between the vertex

$k + 1$ and the graph $B_{\pi'}$ will not affect the circuit in graph $B_{\pi'}$. Hence B_π has a circuit, where π is a permutation on $X = \{1, 2, 3, \ldots, k, k + 1\}$. Hence by induction, B_π includes no connected tree of $(n - 1)$ edges.

3.7 Theorem

If $n > 2$, connected B_π has no pendant vertices.

4 Connectedness Properties of B_π

4.1 Proposition

Suppose B_π is connected and $\pi(i) = i$ for some i, then B_π must be complete.

Proof Let B_π is connected. First, let $\pi(i) = i$ for exactly one i. Now ij is an edge for some $j(\neq i)$, due to connectivity. Then $i + j = \pi(i) + \pi(j) = i + \pi(j)$ gives $\pi(j) = j$, contrary to the assumption. Next suppose exactly two vertices i, j are fixed by π. Then for $k \neq i, j$, $i + k = \pi(i) + \pi(k) = i + \pi(k)$, forcing $\pi(k) = k$, contrary to the assumption. Next if exactly three vertices, say, i, j, k are fixed by π, then we get a triangle with vertices i, j and k in B_π, which is clearly a connected component, destroying connectivity of B_π. We continue this procedure until we arrive at $\pi(i) = i$ for every i, which means that π is the identity. By proposition (3.3), B_π is complete.

4.2 Theorem

Suppose π is a product of n transpositions where $n \geq 1$ and $(a_i - b_i) = k$ for $i = 1, 2, 3, \ldots, n$ then B_π is perfect.

Proof Let $\pi = = (a_1, b_1)(a_2, b_2)\ldots(a_n, b_n)$. $\pi(a_i) = b_i$, $\pi(b_i) = a_i$, and given $(a_i - b_i) = k$ for $i = 1, 2, 3, \ldots, n$. From the first transposition, $\pi(a_1) = b_1$, $\pi(b_1) = a_1$, $a_1 + b_1 = \pi(b_1) + \pi(a_1)$, a_1 is adjacent to b_1. Therefore each a_i is adjacent to b_i, for all $i = 1, 2, 3, \ldots n$. Also $(a_1 - b_1) = k$ and $(a_2 - b_2) = k$, $(a_1 - b_1) = (a_2 - b_2) = k$, $a_1 + b_2 = a_2 + b_1$, $a_1 + b_2 = \pi(a_1) + \pi(b_2)$. a_1 is adjacent to $b_1 (a_1 - b_1) = k$ and $(a_3 - b_3) = k$, $(a_1 - b_1) = (a_3 - b_1) = k$, $a_1 + b_3 = a_3 + b_1$, $a_1 + b_3 = \pi(a_1) + \pi(b_3)$. a_1 is adjacent to b_3. In the similar manner, the first element in the first transposition is adjacent to second element of every other transposition. Therefore first element of every transposition is adjacent to second element of every other transposition. Hence B_π is a connected graph, the graph π looks like

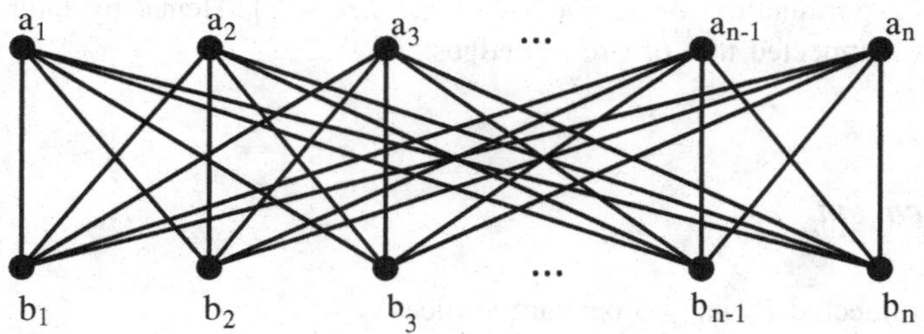

Suppose a_1 is adjacent to a_2, then, $a_1 + a_2 = \pi(a_1) + \pi(a_2)$, $a_1 + a_2 = b_1 + b_2$, $a_1 - b_1 = b_2 - a_2$, $a_1 - b_1 = -(a_2 - b_2) = -k$, $a_1 - b_1 = -k$ gives a contradiction since $a_1 - b_1 = k$. Therefore $a1$ is not adjacent to a_2, Hence a_i's are not adjacent to each other, $i = 1, 2, 3,\ldots, n$. Also if b_1 is adjacent to b_2 then $a_1 - b_1 = -k$. Therefore b_i's are not adjacent to each others, $i = 1, 2, 3,\ldots, n$. B_π contain $2n$ vertices, each vertex of degree $(n/2)$. The vertex set of B_π can be partition into two disjoint sets

$X = \{a_1, a_2, \ldots, a_n\}$ and $Y = \{b_1, b_2, b_3, \ldots, b_n\}$, B_π is a regular graph of degree $(n/2)$. Therefore B_π is a complete bipartite graph. Hence B_π is perfect.

4.3 Theorem

Suppose π is a product of n transpositions where $n \geq 1$ and $(a_i - b_i) = k$ for $i = 1, 2, 3,\ldots, n$ and B_π contains odd number of vertices then B_π is disconnected.

4.4 Theorem

Suppose π is a product of m transpositions where $m > 2$. Then, if $(1m)$ is a part of π, then B_π is disconnected.

Proof Let B_π be connected. Write $n = 2m$ and take $(1m)$ as the first transposition occurring in π. we can write $\pi = (1\ a_1)(2\ a_2)(3\ a_3)\ldots(m+1 a_m)$ (taking $a_1 = m$). Then we get the edges between the vertices (1 and m), (2 and a_2),…,($m+1$ and a_m). Clearly $a_i \geq m+2$ for all $i = 2, 3, 4,\ldots m$. Without loss of generality, we can assume that 1 is adjacent to some vertex $v_j \neq m$ (due to connectivity) we prove that v_j cannot be any vertex in the top row. In fact, suppose $v_j = j$ for some j such that $2 \leq v_j \leq m+1 (v_j \neq m)$. Then $1 + v_j = \pi(1) + \pi(v_j) = m + w_j$ with w_j = one of the a_j's and is hence $\geq m+2$. This is clearly not possible (as $1 < m$ and $v_j < w_j$). Hence 1 is adjacent to some a_j in the bottom row. Then by adjacency condition, $1 + a_j = m + j$. Also $m+1$ must be adjacent to some other a_t, due to connectivity. Hence $m + 1 + a_t = a_m + t$. Already we have the inequalities: $a_i > m + 1$ for all i and $a_m > a_t$

Case 1 The bottom row numbers strictly increase from m to a_m.

Let $m = a_1 < a_2 < \ldots < a_j < \ldots < a_t < \ldots < a_m$. This means that the last entry in the top row is $m + 1$ and the last entry am in the bottom row is $2m$.

Hence $1 + 2m = m + m + 1 = \pi(1) + \pi\,(2m)$, meaning that 1 and $2m$ are adjacent. Also by the same argument, m is adjacent to $m + 1$. Hence we get a circuit of length four joining the vertices 1, $2m$, m and $(m + 1)$ as subgraph in B_π. To ensure connectivity, 1 (and hence m) must be adjacent to some other vertices. (since $m > 2$):1 to a bottom row vertex $m + y$ and m to the corresponding top row vertex y. But since the bottom row vertices steadily increase, $1 + m + y = \pi(1) + \pi(m + y) = m + y$ which is absurd. Hence in this case, B_π must be disconnected. The final graph B_π looks like

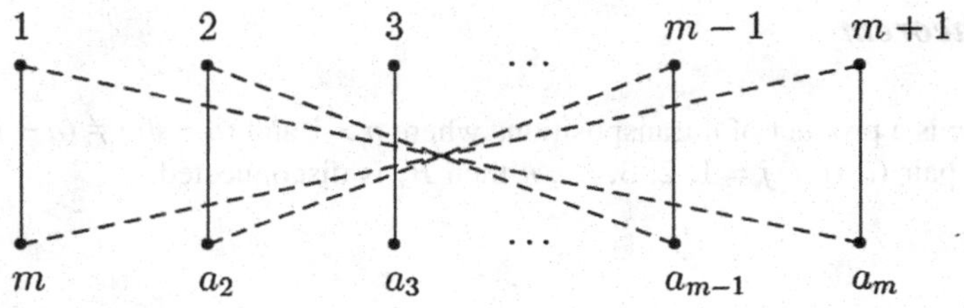

Redrawing B_π, we get (n/4) disconnected (components) circuits of length four when $n \equiv 0 \pmod 4$ as in the following figure.

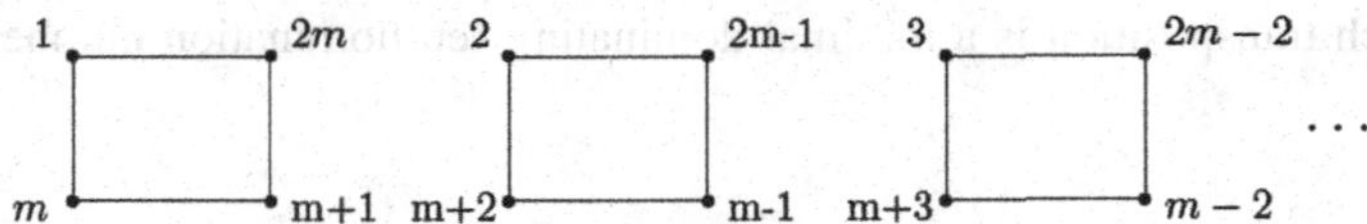

If n is not a multiple of 4, B_π has disconnected circuit of length four together with disconnected component K_2, complete graph of two vertices. Clearly B_π is disconnected with minimum (n/4) components, contrary to our original assumption that B_π is connected.

Case 2 There exist entries a_i, a_j with bottom row and i, j in the top row such that $a_i > a_j$, $i < j$. In this case, we get a circuit like component of length four joining the vertices i, a_i, j, a_j and i in B_π.

In the case, $j + a_i = a_j + i$ or $i + a_j = a_i + j$ (in which case there cannot be another k different from i, j such that $\pi(k) = a_k = a_i$ or a_j), otherwise we will simply have disconnected edges $i\ a_i$ and $j\ a_j$ only. After renumbering (if necessary) B_π, we get disconnected circuits of length four or disconnected circuits of length four together with disconnected components k_2 as in the following figure.

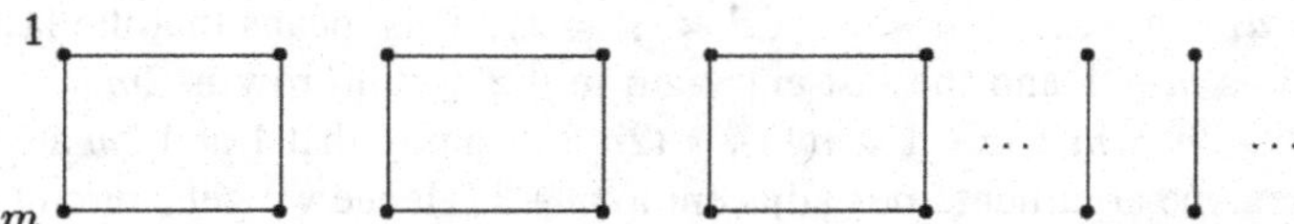

Which is clearly B_π is disconnected contradicting the original assumption. Thus we have proved finally that B_π must be disconnected.

4.5 Theorem

Suppose π is a product of n transpositions where $n > 1$ and $(a_i - b_i) \neq (a - b_j)$ for at least one pair (i, j), $i, j = 1, 2, 3, \ldots, n$ then B_π is disconnected.

4.6 Theorem

π is a product of n transpositions then domination number of connected B_π is two.

Proof Each transposition is a minimal dominating set, domination number is two.

5 Number of Connected Components of B_π

We shall give an algorithm to work out the number of connected components of B_π, through partitions.

5.1 Definition

If we write $n = n_1 + n_2 + \ldots + n_r$ with $n_1 \geq n_2 \geq \ldots \geq n_r > 0$, then we say $\lambda = (n_1, n_2, \ldots, n_r)$ is a partition of n and we write this as $\lambda \to n$.

If $\mu = (a_1, a_2, \ldots, a_{n1})(a_{n1+1}, \ldots, a_{n2})(a_{r-1+1}, \ldots, a_{nr})$ is the unique disjoint cycle decompositions of π, then the partition $n_1 + n_2 +, \ldots, + n_r = n$ corresponds to μ. The number of connected components of the graph B_π is given by the following algorithm. Let π denote the permutation corresponding to the standard partition $\lambda = (1, 2, \ldots, n_1)(n_1 + 1, \ldots, n_2)(n_2 + 1, \ldots, n_3) \ldots (n_{r-1} + 1, \ldots, n_r)$, we simply denote this as $\lambda = (n_1, n_2, \ldots, n_r)$, $\Sigma\, n_i = n$.

5.2 Theorem

The number of connected components of B_π is exactly equal to the number $Q + 1$, where Q is the number of components arising from the adjacencies of vertices in the top rows of λ in the Young diagram of λ such that $l(\lambda) \geq 2$.

Proof Let Q denote the number of components arising out of the parts of λ such that $l(\lambda) \geq 2$.

Let the Young diagram corresponding to λ be given in the following figure.

1	2	3				n_1
n_1+1	...	.	...	n_2		
...						
...		...				
	...	n_k				
n_{k+1}						
⋮						
n_r						

Suppose k of the top rows has length ≥ 2. These k rows will account for $n_1 + n_2 + \ldots + n_k$ vertices. We can draw the edges among these vertices according to our rule: ij is an edge if and only if $i + j = \pi(i) + \pi(j)$. This gives rise to Q components. The rows $k + 1, \ldots\ r$ have just one vertex in each row, which gives rise to just one component. Hence the total number of connected components equals $Q + 1$, which proves our theorem.

5.3 Remark

The above arguments open the gateway to character theory of the symmetric group S_n. It is well known that the number of (complex) irreducible characters = number of conjugacy classes = number of partitions of n. The above algorithm therefore would have established a nice connection between the class of irreducible characters of S_n and the class of balanced permutation graphs corresponding to the standard partition n. But such a neat connection is not so easy to obtain as yet, for the simple reason that B_π is not invariant under conjugacy in general. In other words, if $\pi, \sigma \in S_n$ then B_π and $B_{\sigma\pi\sigma^{-1}}$ need not be the same. This can be easily seen by the following example.

Example 2 $\pi = (14)(23)$, $\sigma = (1234)$, $\sigma\pi\sigma^{-1} = (12)(34)$. The graph B_π and $B_{\sigma\pi\sigma^{-1}}$ are represented as

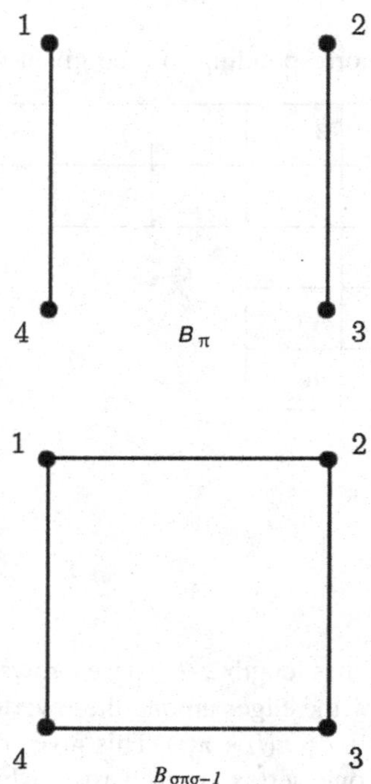

6 Conclusion

(i) The above difficulty may be circumvented by a 'trick' which we found was close to the one used by R.C. Grellana [3] in her recent work on 'Centralizer Algebras and Kronecker Products' (Conference on Non-commutative rings and Representation Theory, Pondicherry, India, 2010). Work is in progress on the exact result (or algorithm) involving the number of component of B_π.

(ii) Theorem 4.2 and 4.4 especially, the zigzag case, may be viewed as an opening for another major area of research now going on, namely, Brauer algebras and their generalizations.

References

1. Golumbic MC (1980) Algorithmic graph theory and perfect graphs. Academic Press, Inc
2. Koh Y, Ree S (2007) Connected permutation graphs. Discrete Math 307:2628–2635
3. Grellana RC, Kronecker product of representations. In: International conference on non-commutative rings and representations, Pondicherry, India

Pi Doctor: A Low Cost Aquaponics Plant Health Monitoring System Using Infragram Technology and Raspberry Pi

V.V. Sajith Variyar, Nikhila Haridas, C. Aswathy and K.P. Soman

Abstract The technological and scientific advancement in the field of agriculture has opened a new era for design and development of modern devices for plant health monitoring. The analysis of various parameters, which affects the plant health such as soil temperature, moisture level and pH are easier with the use of advanced devices like Raspberry Pi and Arduino integrated with different types of sensors. The development of infragram technology has created new possibilities to capture infragram images, where both infrared and visible reflectance are obtained in a single image. The rationale of this paper is to monitor the health of a small scale aquaponics vegetation using Infragram technology and Raspberry Pi. The proposed experimental setup captures infragram images using a low cost modified web camera containing infra-blue filter. These images are post-processed to calculate normalized difference vegetation index (NDVI), which is a good indicator of photosynthetic activity in plants. The study also assesses and monitors the influence of various parameters in the aquaponics system such as nitrogen usage by plants and pH change in the system under different illumination conditions. The study shows that the change in pH and health condition of the plant due to the variation in photosynthesis are the major factors that affects the balance of nitrogen cycle in the aquaponics system.

Keywords Infragram technology · Normalized difference vegetation index · Plant health monitoring · Raspberry Pi

V.V. Sajith Variyar (✉) · N. Haridas · C. Aswathy · K.P. Soman
Centre for Excellence in Computational Engineering and Networking,
Amrita Vishwa Vidyapeetham, Coimbatore 641112, India
e-mail: vv_sajithvariyar@cb.amrita.edu

N. Haridas
e-mail: nikhila.haridas92@gmail.com

C. Aswathy
e-mail: aswathy0257@gmail.com

K.P. Soman
e-mail: kp_soman@amrita.edu

L.P. Suresh and B.K. Panigrahi (eds.), *Proceedings of the International Conference on Soft Computing Systems*, Advances in Intelligent Systems and Computing 397, DOI 10.1007/978-81-322-2671-0_86

1 Introduction

Agriculture forms the basis of food supply. However, the increasing pressure on soils, unexpected changes in weather and climate, availability of water and unsustainable farming practices are some of the major challenges faced in this field. In this modern world, we have new generation food production system that consumes less power, space and time which overcomes all the above-mentioned challenges. Aquaponics [1] is such a system where it combines conventional aquaculture with hydroponics in a symbiotic environment.

There is a long history of doing things from space, where LANDSAT vehicles were looking at the Earth across a broad spectrum, for sophisticated agricultural and ecological assessment. The development of such sophisticated system is difficult in the case of a researcher or in an educational sector. The use of satellite images for monitoring the vegetation belongs to a global scale of enquiry, but in case of an aquaponics system, we need a local scale of enquiry that can determine the overall performance of the system. As the cameras are ground based and the enquiry is restricted to a few numbers of plants or single one, critical information about the health of vegetation could be easily revealed. The vegetation information from infragram technology relies on indices which compare the vegetation reflectance in multiple spectral regions.

In literature, several approaches have been proposed to determine the effect of various parameters for the growth and survival of plant and fish in an aquaponics system. The authors in [2, 3] determined the impact of pH in nitrification process of aquaponics system. Various sources of micronutrients and their potential suitability for the aquaponics setup are discussed in [4]. Fox et al. [5] conducted a study in commercial aquaponics farm to find out the relation between microbial water quality to food safety in recirculating aquaponic vegetable production system and fish. Jose et al. [1] calculated NDVI of the satellite image captured using NOAA/AVHRR to quantify the health and status of vegetation in a particular zone.

In this paper, we propose a low cost aquaponics plant health monitoring system utilizing a modified web camera that captures both IR and visible information in a single image. The resulting NGB image (near-infrared, green, blue) is post-processed to calculate the NDVI for quantifying the health and status of vegetation in the system. The analysis of other parameters in aquaponics system such as pH and nitrogen usage by the plants is also done by exposing the system to different conditions such as moderate, sunny and dark. The study also finds optimum pH condition for the water in fish tank and soil in the grow bed for the aquaponics system.

2 Supporting Concepts and Technologies

This section describes about the various technological concepts and devices involved in the overall system. First, an overview of the aquaponics system is given followed by Raspberry PI, Infragram technology and NDVI.

2.1 *Aquaponics*

Aquaponics is a fastest growing food producing sector which involves combination aquaculture (farming of aquatic organisms) and hydroponics (growing plants using nutrient-rich solutions without soil). Aquaponics system has many benefits over the traditional aquaculture and hydroponics done separately. In a normal aquaculture system, excretions from aquatic animals build up the waste which increases the toxicity of the water and in a hydroponics system nutrient-rich water is required for the cultivation of plants. Aquaponics is an integrated system where both fish and plants are grown by creating a symbiotic relationship between the two. The system maintains a proper balance in the nitrogen cycle by circulating the nutrient-rich waste water from the fish tank to the grow bed where the plants are grown.

Nitrogen Cycle The most important process in an aquaponics system is nitrogen cycle which is responsible for the conversion of fish waste into nutrients required for the plants. The overall process can be described as shown in Fig. 1. The fish takes the protein-rich feed and excretes waste rich in ammonia. The ammonia is

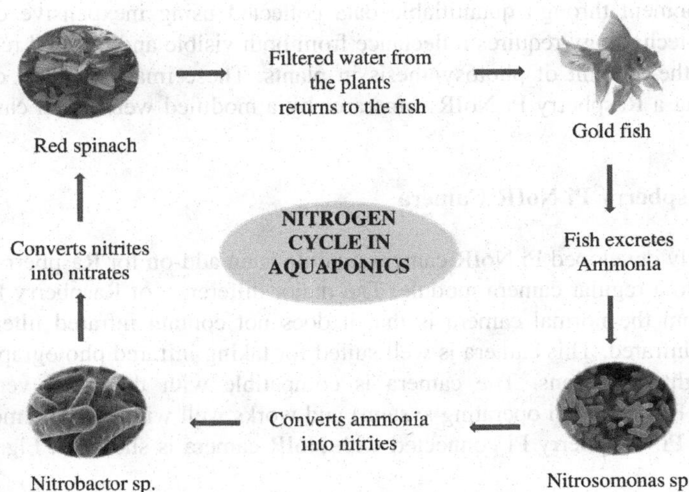

Fig. 1 Aquaponics life cycle

converted into nitrites by *Nitrosomanas* sp. bacteria. The *Nitrobacter* sp. converts the nitrites to nitrates which are absorbed by the plants in the grow bed. This process is called nitrification. If this process is not balanced properly, the water quality in the aquaponics system deteriorates rapidly and becomes toxic for both fish and plants.

2.2 Raspberry Pi

Raspberry Pi is a cheap credit card-sized single board computer developed by Raspberry Pi foundations [6]. This low cost embedded module is one of the widely used systems in academics and hobby projects. It mainly consists of a processor, RAM, graphics chips, different interfaces and connectors. Pi with a monitor and keyboard is similar to a standard desktop PC. Among the two versions of Pi, Raspberry Pi 1 is available in different models. Models A, A+, B Rev 1 are with 256 MB RAM, whereas Models B Rev 2, B+ and CM possess 512 MB RAM. The latest version of Pi (Raspberry Pi 2) is currently available only in model B configuration with 1 GB RAM.

2.3 Infragram

Near-infrared photography is one of the latest techniques used in large farms for plant health analysis using expensive sensors mounted on satellites and airplanes [7]. In small scale farming, individuals can utilize the same technology to monitor the environment through quantifiable data collected using inexpensive cameras. Infragram technology requires reflectance from both visible and infrared region for assessing the amount of photosynthesis in plants. These images can be captured either using a Raspberry Pi NoIR camera or by a modified web/digital camera.

2.3.1 Raspberry Pi NoIR Camera

The recently developed Pi NoIR camera module is an add-on for Raspberry Pi that is similar to a regular camera module. The major difference of Raspberry Pi NoIR camera from the normal camera is that it does not contain infrared filter. NoIR means no infrared. This camera is well suited for taking infrared photographs even in low light conditions. The camera is compatible with the latest version of Raspbian (Raspberry Pi operating system) and works well with both the models of Raspberry Pi. Raspberry Pi connected with NoIR camera is shown in Fig. 2.

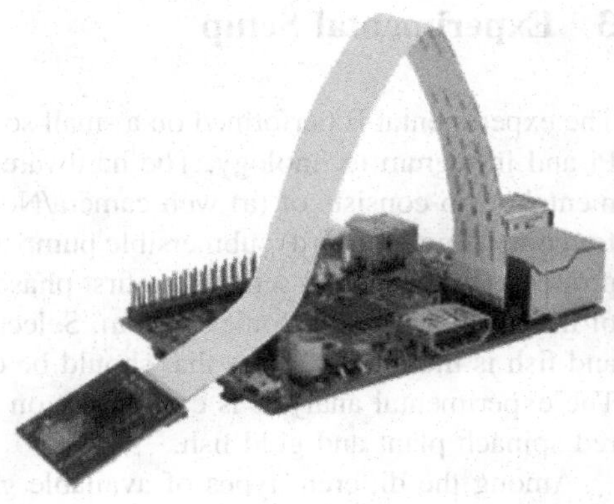

Fig. 2 Raspberry Pi NoIR camera

2.3.2 Digital/Web Camera

A conventional digital camera or web camera can be converted to an infragram camera by removing its infrared blocking filter and replacing the same with an infra-blue filter. Thus red light is filtered out and infrared light is captured in the place of red region to produce NGB images. Since the Raspberry Pi has USB ports in it, the web camera can be easily mounted on it to capture the infragram images. With the help of a LAN connection, these images can be remotely captured.

2.4 Normalized Difference Vegetation Index (NDVI)

The productivity of vegetation can be estimated by analysing the amount of light absorbed and reflected by the plants. The plants appear green because they reflect green light and absorb red and blue wavelengths. The pigments present in the leaves absorb light to power the photosynthesis for the conversion of CO_2, water and nutrients into sugar (food). Comparison of the amount of red light reflected to the amount of infrared light reflected helps to analyse the proportion of sunlight used by the plants. Normalized difference vegetation index (NDVI) is a good indicator of plant productivity. The NDVI index is calculated by taking the difference of red and near-infrared light reflected and normalizing the difference by dividing it with the total amount of infrared and red light.

$$\text{NDVI} = \frac{\text{NIR} - \text{VISIBLE}}{\text{NIR} + \text{VISIBLE}} \tag{1}$$

NIR: Near-InfraRed

3 Experimental Setup

The experimental is performed on a small scale aquaponics system using Raspberry Pi and infragram technology. The hardware configuration of the proposed experimental setup consists of (a) web camera/NoIR camera (b) Raspberry Pi (c) circuit for controlling motor (d) submersible pump and (e) a pH sensor. Figure 3 shows the proposed experimental setup. The first phase of the experiment involves setting up of the small scale aquaponics system. Selection of the appropriate grow bed, plant and fish is the major factors that should be considered before building the system. The experimental analysis is carried out on a system with gravel-based grow bed, red spinach plant and gold fish.

Among the different types of available growing medium, the proposed experimental setup uses gravel-based grow bed which is cheap and easily available. Best results are obtained with leafy vegetables like spinach which are fast growing when compared with other aquaponics plants. Gold fish is used for the analysis because of its ability to adjust with the change in temperature and pH. The overall system is controlled by Raspberry Pi. Raspberry Pi has been chosen as the processing unit because of its low cost and user friendly features. The experiment uses Raspberry Pi model B with 512 MB RAM. This module controls the pumping of nutrient-rich water from the fish tank to the grow bed every 2 h. The analysis of pH and NDVI is

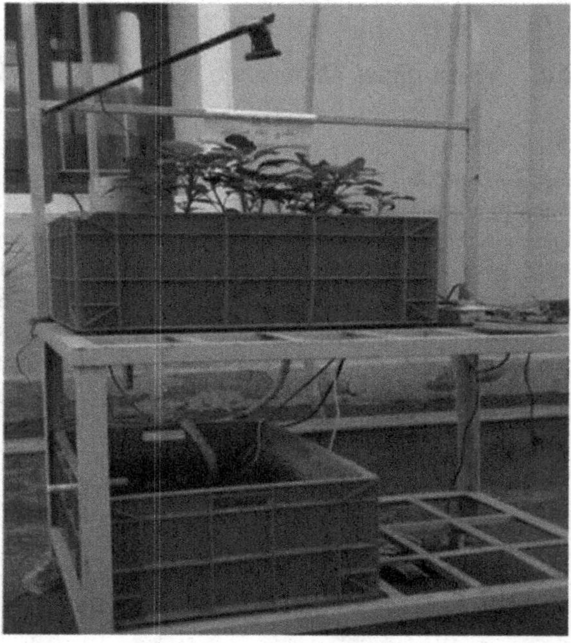

Fig. 3 Experimental setup

carried out thrice a day. NDVI index is calculated from the images captured using infragram technology. As the aim of this project is to develop a low cost plant health monitoring system, we used a modified web camera for the experiment.

4 Results and Discussion

The experiment analyses the effect of various parameters in the aquaponics system such as normalized difference vegetation index (NDVI), nitrogen usage by the plants and pH. The health condition of the red spinach plant in the aquaponics system is monitored by calculating the NDVI index during different illumination conditions. The presence of anthocyanin pigment gives red colour to spinach. Hence, it reflects red colour instead of green. A low cost modified web camera containing infra-blue filter is used to capture reflectance of infrared and visible region in a single image. The reason behind the use of infra-blue filter is that it filters out the red region and infrared light is captured in its place producing NGB (N-Near-infrared, G-Green, B-Blue) images. The obtained NGB images are post-processed to obtain NDVI image which uses a false colour map where dark red region shows dense vegetation and blue represents no vegetation. Figure 4a–c shows the images of aquaponics vegetation captured at morning, noon and evening on a sunny day using modified web camera, and Fig. 4d–f shows the corresponding post-processed NDVI images. It is observed that the highest NDVI (0.847288) value is obtained in the afternoon due to the increase in photosynthesis.

Fig. 4 Images captured using infragram technology along with post-processed NDVI images. **a** Morning. **b** Afternoon. **c** Evening. **d** NDVI = 0.733302. **e** NDVI = 0.847288. **f** NDVI = 0.729606

Table 1 Range of calculated NDVI values for the three conditions

Condition	pH range	NDVI value
Dark	4.3–5.6	0.4–0.58
Sunny	7.3–8.2	0.7–0.85
Moderate	6.6–7.9	0.6–0.73

Nitrogen cycle is the most significant process happening in an aquaponics system. Drastic change in any of the parameters such as pH, nitrogen usage and NDVI may affect the balance of the aquaponics nitrogen cycle. Fish takes the feed and excrete waste rich in ammonia. This nutrient-rich water containing ammonia is circulated to the grow bed every 2 h. Without this process, the quality of fish tank water decreases rapidly and becomes toxic for both the plant and fish in the system. The conversion of ammonia to nitrates and nitrites is carried out by the bacteria colony in the grow bed. In the nitrogen cycle, the nitrosomonas bacteria converts ammonia into nitrites which are further converted to nitrates by nitrobacter bacteria. The nitrogen usage of the plants in the system depends on the variation in photosynthetic activity. During sunny days, plants utilize much of the sunlight for photosynthesis and thereby increase the absorption of nitrates presence in the grow bed whereas in dark, the consumption of nitrates is comparatively very less. This results in accumulation of nitrates in the grow bed which disturbs the aquaponics nitrogen cycle. Table 1 shows the range of obtained results for pH and NDVI for the three different illumination conditions.

pH management in aquaponics system is very important as its variation can affect the health of the plant, fish and bacteria. Therefore, pH should be maintained at a particular range suitable for the three living constituents. From the study, it is analysed that lowering pH in aquaponics system is very dangerous. As the pH lowers, the water becomes more acidic which is harmful for the health of fish. Increasing pH is usually not a serious issue in aquaponics as the nitrification process will cause the system to become more acidic over time. From the experimental analysis, it is observed that the suitable survival pH range for gold fish is 7.2–8.0 and for spinach is 5.8–7.0. Table 1 shows the range of pH and NDVI values obtained for different illumination conditions.

5 Conclusion

This paper presents a low cost aquaponics plant health monitoring system using Infragram technology and Raspberry Pi. The experiment is performed by continuously monitoring the pH change and vegetation index (NDVI) under three different conditions (moderate, sunny and dark). The NDVI index is calculated from the NGB images captured using infragram technology. From the analysis it is observed that, during sunny condition the high NDVI value is obtained due to the increase in photosynthetic activity. The effect of variation in pH on the balance of nitrogen cycle is also discussed in this experiment. In future, the system can be expanded to analyse the effect of various other parameters such as soil tillage, fish feed, type of fish, and type of plant in increasing the productivity of aquaponics system.

Acknowledgments The authors would like to thank Ms. V. Sowmya, Assistant Professor, Center for Excellence in Computational Engineering and Networking, Amrita Vishwa Vidyapeetham for her valuable comments and suggestions.

References

1. Rakocy JE, Masser MP, Losordo TM (2006) Recirculating aquaculture tank production systems: aquaponics integrating fish and plant culture. SRAC publication 454:1–16
2. Tyson RV, Simonne EH, White JM, Lamb EM (2004). Reconciling water quality parameters impacting nitrification in aquaponics: the pH levels. Proc Florida State Hort Soc 117:79–83
3. Foster I, Kesselman C, Nick J, Tuecke S (2002) The physiology of the grid: an open grid services architecture for distributed systems integration. Technical report, Global Grid Forum
4. Treadwell D, Taber S, Tyson R, Simonne E (2010) Foliar-applied micronutrients in aquaponics: a guide to use and sourcing. Horticultural Sciences Department, Florida Cooperative Extension Service, Institute of Food and Agricultural Sciences, University of Florida, pp 1–8
5. Fox BK, Tamaru CS, Hollyer J, Castro LF, Fonseca JM, Jay-Russell M, Low T (2012) A preliminary study of microbial water quality related to food safety in recirculating aquaponic fish and vegetable production systems. University of Hawaii at Manoa. Food Safety and Technology, College of Tropical Agriculture and Human Resources
6. Raspberry Pi. www.raspberrypi.org
7. Near-infrared camera, Public Lab. http://publiclab.org/wiki/near-infrared-camera
8. Jose B, Nicolas M, Danilo C, Eduardo A (2014) Multispectral NDVI aerial image system for vegetation analysis by using a consumer camera. In: 2014 IEEE international autumn meeting on power, electronics and computing (ROPEC), pp 1–6. IEEE
9. Pi NoIR and Catch Santa Challenge. http://in.element14/community/roadTests/1220.com
10. Infragram. http://publiclab.org/wiki/infragram

Reduction of Optical Background Noise Impact in Light-Emitting Diode (LED)-Based Optical Wireless Communication Systems by Hadamard Codes

Sapna Gour and Saket Kumar

Abstract Light-emitting diode (LED)-based Optical wireless communications is capable to provide ultracheap wireless communication for many applications, especially for the indoor applications, it can be used to provide flexible wireless communication system for data communication (to connect personal computers/laptops with printers, and digital imaging devices) or creating access points in the floors the LED-based optical wireless communication has many advantages over RF communications like lower system complexity, cheaper components, simple interfacing, and security. Besides these advantages, the LED optical wireless communications face many challenges for its implementation in indoor applications because of high interference from other light sources like florescent lamps, CFL's, which not only emits the light at same spectrum but may also use the switching frequencies (including harmonics) similar modulation frequencies. However, many solutions have been already presented to overcome these challenges. This paper provides the simple and efficient spectrum spreading approach which not only reduces the effect of interference but also enables multiuser communication without using modulating carrier. The simulation results shows that the proposed technique provides much better results than the Manchester coded systems.

Keywords Free-space optical communication · Spreading codes · Manchester encoding

S. Gour (✉) · S. Kumar
NRI Institute of Information Science and Technology, Bhopal, India
e-mail: sapnagour.2222@gmail.com

S. Kumar
e-mail: saket.subodh@gmail.com

L.P. Suresh and B.K. Panigrahi (eds.), *Proceedings of the International Conference on Soft Computing Systems*, Advances in Intelligent Systems and Computing 397, DOI 10.1007/978-81-322-2671-0_87

1 Introduction

The LED optical wireless communications has emerges as an efficient alternative to radio frequency (RF) communications for indoor applications, because of its lower cost, simple implementation, and security. Although the LED optical communication system does not provides the comparable bandwidth with RF systems for broadcasting applications but for indoor uses the use of radio frequency communications bandwidth are limited because in the unlicensed band, the RF has limited usable bandwidth. Besides of complexity and cost the RF system also suffers from interference by other RF systems. The indoor optical wireless systems use the light sources and photodetectors which avoid the interference with the signals other parts of the building or environment because the light does not passes through walls, this makes it a much preferable for cell-based secure networks which can reuse the same spectrum in different rooms of a building.

Thus, there are several challenges for the efficient practical implementation of the LED-based optical wireless communication. In scope of this paper, we are considering only the most important two. The first one is the direct modulation which is limited by the characteristics of the diode itself [1] and the second is optical noise generated by other optical sources like AC-LEDs or fluorescent lamps [2]. The most of the solution approach for the first problem is use of equalization techniques or advanced modulation techniques (like discrete multitone (DMT)) [2, 3]. In [4], adaptive filtering is presented for optical noise reduction. However, the adaptive filtration increases the system complexity and requires continuous monitoring [2].

In this paper, we extended the work of [2] by introducing the applicability of Hadamard spreading codes for the LED optical communication to mitigate the optical noise. The advantage of system is that like [2] it also does not requires adaptive monitoring, feedback, or optical filtering while adding the facility for multiuser communication with extra security.

Further simulations are carried out to validate the performance of proposed optical wireless communication system under different operating conditions, and the simulation results show that proposed technique performs better than the conventional non-return-to-zero (NRZ) and Manchester coding [2] for the under 500 kHz optical noise frequencies.

The rest of the paper is arranged as follows. The second section presents a brief review of related work. The third section explains the structure and properties of Hadamard codes. The fourth section presents the characteristics of optical background noise. He proposed system structure is presented in fifth section followed by simulation results and conclusion in sixth and seventh sections.

2 Literature Review

A number of literatures are available on LED optical wireless communication system which covers the different aspects of systems like impact of noise [2, 5, 6], impact of modulation techniques [7, 8], and new developments in the architecture [9–12]. On the impact of noise the authors of [2] Chow et al. presented an experimental characterization of optical background noise impact on the performance of LED optical wireless communication system. They also proposed a Manchester coding scheme for such systems to mitigate the optical noise. The advantage of the proposed algorithm in [2] is that no adaptive monitoring, feedback, or optical filtering is needed and is founded useful for the noise of frequency under 500 kHz. Navidpour et al. [13] presented the applicability of spatial diversity for FSO (free-space optical) communication to improve its performance under log-normal atmospheric turbulence fading channels the technique involves the deployment of multiple transmitters/receivers. Qazi et al. [5] discusses the challenges for outdoor systems like weather conditions, line of sight alignment, scintillation and for indoor systems like ambient light, (a combination of incandescent, fluorescent, and sunlight) of FSO communication. Noshad et al. [14] discusses the feasibility of visible light communications (VLC) to provide gigabits communication link and presented different modulation techniques that have the capabilities to boost the performance of VLC to multi-Gb/s indoor wireless communication system buy overcoming the many of the challenges which limits the capabilities of VLC. A real-time visible light communication (VLC) with 37 Mbit/s total throughput under a 1.5 m free-space transmission length, using white light phosphor-LED with compact size is presented in [6] their system utilizes their proposed pre-equalization technique, which extends the 1 MHz bandwidth of phosphor LED to ~ 12 MHz without using blue filter. Yan et al. [7] presented their research on performance of modulation techniques (OOK, PPM, DPPM, DPIM, and DH-PIM) for optical wireless communication system. The article analyzes the bandwidth efficiency, transmission capacity, power efficiency, and slot error rate for typical modulation schemes as OOK, PPM, DPPM, DPIM, and DH-PIM for combination of the characteristic of the atmospheric optical wireless channel.

3 Walsh-Hadamard Matrix

Walsh codes are used in direct sequence spread spectrum (DSSS) systems and the Hadamard matrix is used to generate the Walsh codes. Since Walsh codes are mutually orthogonal, hence the correlation of two similar and different codes results positive constant and zero, respectively (Eq. (1)), this particular property of Walsh code is used for the code division multiplexing where the receiver with similar codes can recover the signal.

Orthogonal codes have the following characteristics [15]:

$$\sum_{k=0}^{N-1} W_i(k\tau)W_j(k\tau - n\tau) = 0, i \neq j, n = 0, 1, \ldots, N-1, \tag{1}$$

Walsh codes are generated using the Hadamard matrix with $H_1 = [0]$, where H_1 is a 1×1 matrix and is an order 1. The Hadamard matrix is built by [15].

$$H_{2n} = \begin{bmatrix} H_n & H_n \\ H_n & \overline{H_n} \end{bmatrix}, \tag{2}$$

For example, the Hadamard matrix of order 2 and 4 will be:

$$H_2 = \begin{bmatrix} 0 & 0 \\ 0 & 1 \end{bmatrix}, H_4 = \begin{bmatrix} 0 & 0 & 0 & 0 \\ 0 & 1 & 0 & 1 \\ 0 & 0 & 1 & 1 \\ 0 & 1 & 1 & 0 \end{bmatrix}$$

In the bipolar form

$$H_2 = \begin{bmatrix} -1 & -1 \\ -1 & 1 \end{bmatrix}, H_4 = \begin{bmatrix} -1 & -1 & -1 & -1 \\ -1 & 1 & -1 & 1 \\ -1 & -1 & 1 & 1 \\ -1 & 1 & 1 & -1 \end{bmatrix}$$

To spread the signal, it is multiplied by the any of the row or column of Hadamard matrix. This operation causes the large switching activities in the output signal stream which ultimately results in spreading of spectrum over large frequency band as shown in Fig. 1.

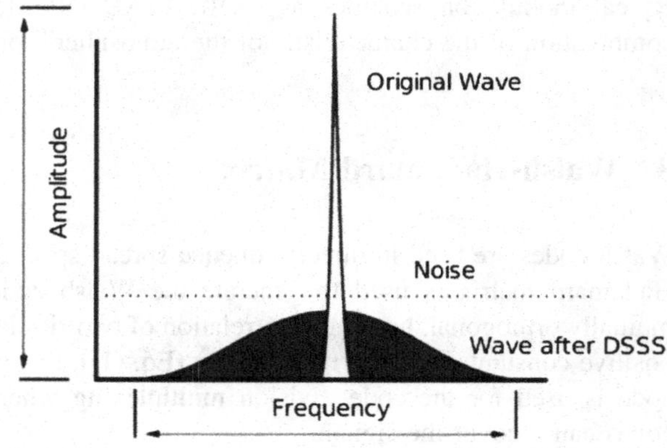

Fig. 1 Spreading of spectrum by Hadamard codes [16]

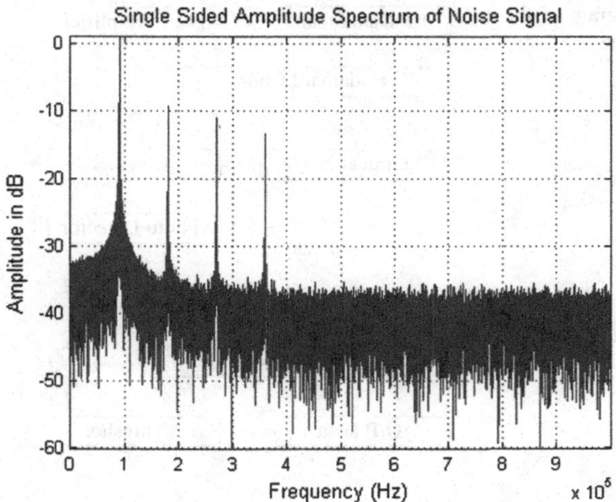

Fig. 2 The frequency spectrum of the optical noise (combining thermal noise and CFL's noise operating at 90 kHz with three harmonic components at 180, 270, and 360 kHz). The relative amplitude (power) of harmonic components with respect to fundamental components are [−10 dB (−20 dB), −12.5 dB (−25 dB), −14.5 dB (−29 dB)]

4 Optical Noise Characteristics

The optical noise in generated by the florescent bulbs and CFL's because of the high frequency converters used to convert the voltage levels. These converters operate at different switching frequencies which cause the flickering and hence the noise generated by these devices depends upon the operating frequency the device. However, presently the most of the converters used in CFL's are work at the frequency of 90 kHz [b]. Hence in this paper, it is considered as fundamental frequency of flickering noise. The other form of noise is known as thermal noise which is cause by random motion of charge carriers induced by thermal energy. The characteristics thermal noise is considered as white noise (Fig. 2).

5 Proposed Technique

The proposed technique can be explained by the following block diagram:

As the diagram in Fig. 3 shows: an arbitrary waveform generator (AWG) generated a normalized binary sequence of length 10000 after that spreading of these bits are performed by multiplying it with Hadamard codes these signal are then amplified to sufficient level to derive the optical source (LED) the data transmission rate is maintained to 1.25 Mbps with the sampling rate of 20 Mega sample per seconds. These signals were then pass through the channel which mixes the optical noise from

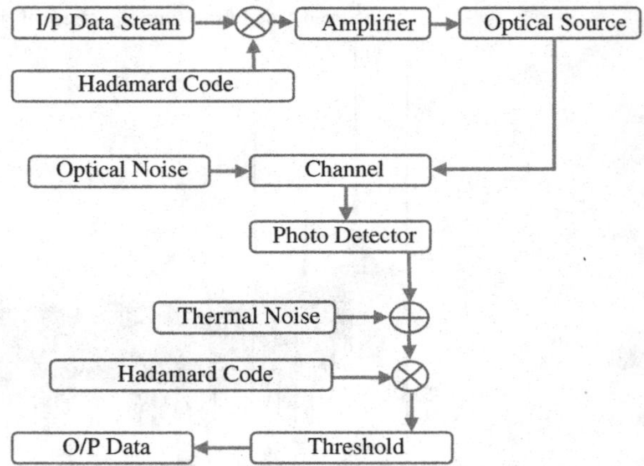

Fig. 3 Block diagram of the proposed technique

CFL's or the optical signals are then converted to electrical signals using photo detector now the thermal noise of the system is added as electrical signal. After mixing the thermal noise, the signal dispreading is performed by again multiplying with same Hadamard code; finally these signals are passed through threshold detector to convert it into binary sequences.

6 Simulation Results

The simulation was carried out to evaluate the performance of Hadamard-coded optical wireless communication system. Here, we used LED electrically driven by 1.25 Mbps data rate (frequencies), and the background noises generated by different

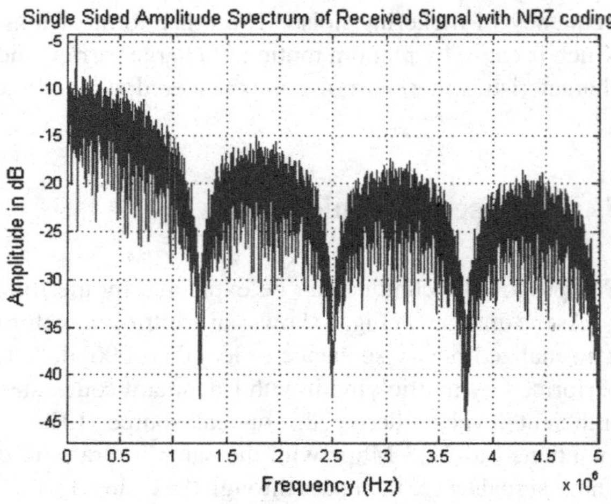

Fig. 4 The amplitude frequency spectrum of the NRZ coded signal at 1.25 Mbps

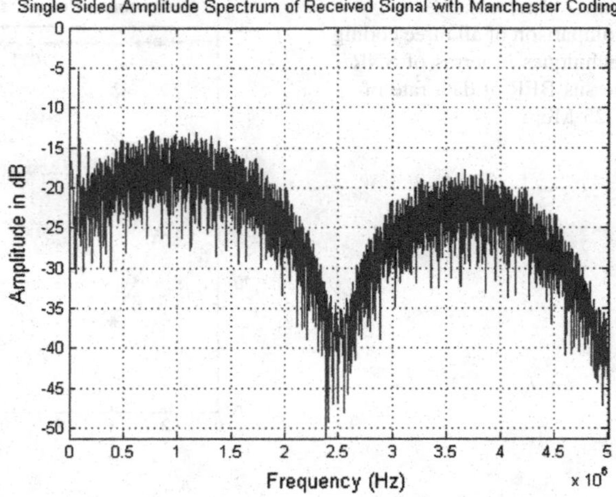

Fig. 5 The amplitude frequency spectrum of the Manchester coded signal at 1.25 Mbps

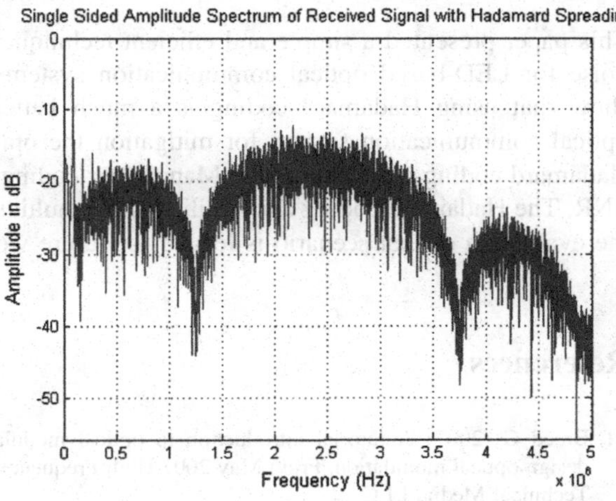

Fig. 6 The amplitude frequency spectrum of the Hadamard coded signal at 1.25 Mbps

source are considered at 90 kHz as fundamental frequency. The simulation setup was similar to that shown in Fig. 3. In this evaluation, the noise optical power was equal to the signal optical power. All three coding techniques Hadamard, Manchester and NRZ were compared for same operational conditions. The results show that Hadamard coding performs much better than Manchester and NRZ (Figs. 4, 5, 6 and 7).

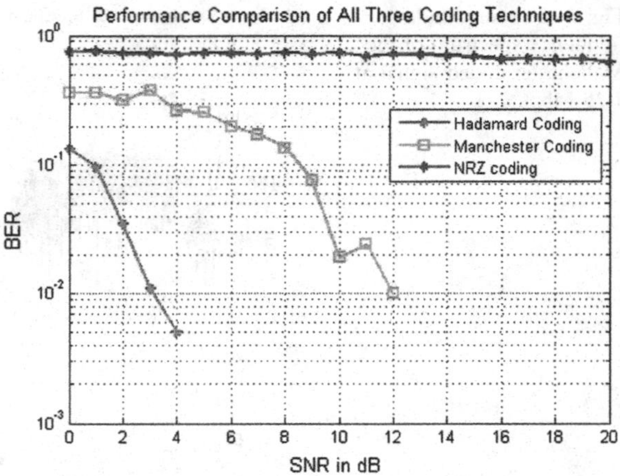

Fig. 7 Performance comparison of all three coding techniques in terms of SNR versus BER at data rate of 1.25 Mbps

7 Conclusion

This paper presented a simple and efficient technique to mitigate the back ground noise for LED-based optical communication systems, and the simulation results show that using Hadamard coding is a much better option for the LED-based optical communication system for mitigation the optical background noises. The Hadamard coding outperforms the Manchester coding by the ration of 100 at 5 dB SNR. The Hadamard coding also facilitates the multiuser communication; however, the evaluation of that scenario is leaved for future work.

References

1. Breed G (2007) A tutorial introduction to optical modulation techniques. High frequency design optical modulation, From May 2007 High Frequency Electronics Copyright © Summit Technical Media, LLC
2. Chow CW, Yeh CH, Liu YF, Huang PY (2013) Mitigation of optical background noise in light-emitting diode (LED) optical wireless communication systems. IEEE Photon J 5(1)
3. Lee D, Choi K, Kim K-D, Park Y (2012) Visible light wireless communications based on predistorted OFDM. Opt Commun 285(7):1767–1770
4. Ya´nez VG, Torres JR, Alonso JB, Borges JAR, Sa´nchez CQ, Gonza´lez CT, Jime´nez RP, Rajo FD (2009) Illumination interference reduction system for VLC Communications. In: Proceedings WSEAS international conference mathematical methods, computer technical intelligent systems 2009, pp 252–257
5. Qazi S Challenges in outdoor and indoor optical wireless communications. School of Information Systems and Engineering Technology State University of New York Institute of Technology P.O. Box 3050, Utica, NY 13504, USA
6. Yeh C-H, Liu Y-L, Chow C-W (2013) Real-time white-light phosphor-LED visible light communication (VLC) with compact size. Optics Express 21(22)

7. Gao Y, Wu M, Du W (2011) Performance research of modulation for optical wireless communication system. J Netw 6(8)
8. Elgala H, Mesleh R, Haas H (2011) Indoor optical wireless communication: potential and state-of-the-art. IEEE Commun Magaz
9. Juan-de-Dios S-L, Arturo A Trends of the optical wireless communications. Adv Trends Wireless Commun
10. Green RJ, Joshi H, Higgins MD, Leeson MS Recent developments in indoor optical wireless systems. University of Warwick Institutional Repository. IET 2(1):3–10. http://go.warwick.ac.uk/wrap
11. Tang X, Ghassemlooy Z, Rajbhandari S, Popoola WO, Lee CG (2011) Coherent polarization shift keying modulated free space optical links over a gamma-gamma turbulence channel. American J Eng Appl Sci 4(4):520–530
12. Haddad S, L´ev^eque O (2014) Diversity analysis of free-space optical networks with multihop transmissions. In: 2014 IEEE international conference on communications (ICC), 10–14 June 2014
13. Mohammad Navidpour S, Uysal M, Kavehrad M (2007) BER performance of free-space optical transmission with spatial diversity. IEEE Trans Wirel Commun 6(8)
14. Noshad M, Brandt-Pearce M (2014) Can visible light communications provide Gb/s service? Opt Lett
15. Garg V (2007) Appendix D: spreading codes used in CDMA. Wirel Commun Netw
16. Dixon RC (1984) Spread spectrum systems, 2nd edn. Wiley, New York

Analysis of ST/QT Dynamics Using Independent Component Analysis

S. Thulasi Prasad and S. Varadarajan

Abstract Heart diseases such as arrhythmia and myocardial infarction are the leading causes of death. If these cardiac-related deceases are detected at early stage then the people can be saved from death. With our proposed method, it is possible to estimate the occurrence of these diseases easily at early stage. Generally the variations in QT-interval are differentiated by the occurrence of premature activation (PA) beats in electrocardiogram (ECG). Similarly the elevation or depression of ST-segments is the indication of ischemia. We used the ratio of ST-segment to QT-interval as an index to specify the probability of ischemia and cardiac injury. The ΔST/QT can be used as the index of the severity of the disease. The aim of this paper is to investigate the instability in the ventricular repolarization process by detecting ST/QT. In our proposed method, four 1-min-long ECG signals of same subject were taken and applied independent component analysis (ICA) to extract the cleaned ECG from various artifacts and non-Gaussian noise background. The performance of proposed method can also be measured using SNR. The ST/QT slope signifies the peak rate of ST-segment change with heart rate, whereas The ΔST/QT index represents the average change of ST-segment depression with QT-interval. In our method, we used a well-known BSS algorithm, JADE algorithm, to obtain ICs. The JADE is able to separate ECM artifact from ECG activity in component domain.

Keywords ECG · Arrhythmia · QRS complex · QT-interval · ST-segment · Baseline wander · ICA · Kurtosis · JADE · MIT-BIH database · MATLAB

S. Thulasi Prasad (✉)
Department of Electronics and Communication Engineering,
Sree Vidyanikethan Engineering College, Tirupati, India
e-mail: stprasad123@yahoo.co.in

S. Varadarajan
Department of Electronics and Communication Engineering,
Sri Venkateswara University College of Engineering, Tirupati, India
e-mail: varadasouri@gmail.com

L.P. Suresh and B.K. Panigrahi (eds.), *Proceedings of the International Conference on Soft Computing Systems*, Advances in Intelligent Systems and Computing 397, DOI 10.1007/978-81-322-2671-0_88

1 Introduction

ECG is a waveform representation of the electrical activity of human heart. The bioelectrical action potentials are communicated through the human heart and resulted in contractions and relaxations of cardiac muscles in a rhythmic way. These electrical potentials from the heart are sensed at the surface of chest and recorded as an ECG signal [1, 2]. It is used as a basic investigative and major diagnostic tool by the cardiologists. The ECG of one cycle is shown in Fig. 1. The ECG wave consists of the P-wave, QRS complex, T-wave, PR-segment, QT-interval; ST interval; PR-segment; and ST-segment. Doctors can use the ECG to analyze the conditions leading to the cardiac ventricular fibrillation (CVF) thereby to make correct clinical diagnoses and to give right treatment at right time. Total time of ventricular depolarization and repolarization of a cardiac cycle in ECG is represented by the duration of the QT-interval. The separation between J point and K is defined as ST-segment. The elevated beat-to-beat QTV has also been identified as a marker of increased risk for sudden cardiac death [3–5].

The accuracy of existing measurement techniques for beat-to-beat QT-interval and ST-segment is limited due to different reasons. Usually, the accurate R-peak detection followed by the detection of Q-wave onset and T-wave offset is required for accurate measuring of QT-interval. As the baseline wander influences, the

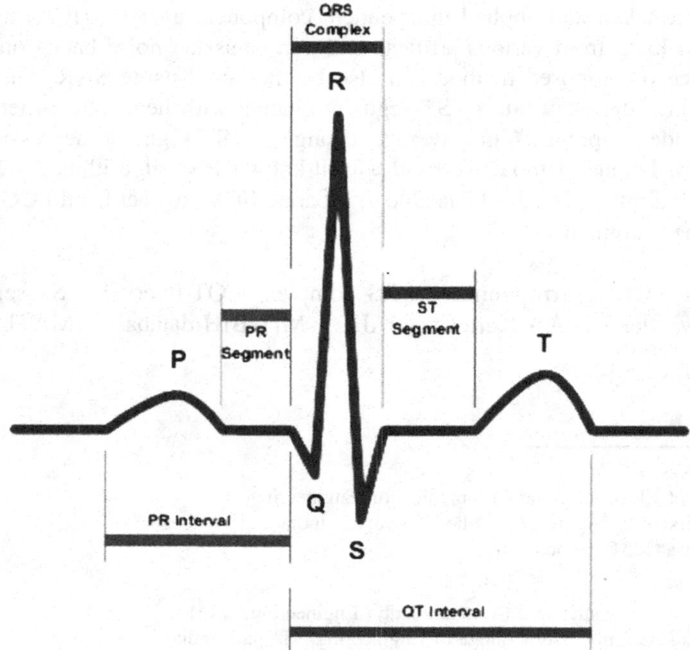

Fig. 1 ECG wave representing heart beat [12]

T-wave offset it will affect the measurement of QT-interval and ST-segment. This may cause the T-peak higher than the R-peak and detected as an R-peak instead. Normally, the changes in the heart period will affect the response of the ST-segment and QT-interval during pacing-induced sudden changes [6]. It was observed that 90 % of QT-interval adaptation to abrupt changes in heart rate takes place approximately within 2–3 min.

QRS complexes are detected normally using linear filtering and Pan and Tompkins [7]. However, it was observed from a number of studies that when the QRS complexes in ECG are complicated by cardiac disorders, this method of QRS detection is giving more difficulties. In particular, most of the studies reported miss and false detection of true R-peak with less effective baseline removal in ECG signal. Therefore, we incorporated an alternative R-peak detection technique and baseline removal approach.

2 Methodology

The accurate analysis, extraction and measurement of features from ECG require the accurate preprocessing of ECG signal. The block diagram of our proposed method is shown in Fig. 2. First the ECG signal is preprocessed using digital filters and then post-processed using independent component analysis (ICA), and finally statistical analysis is performed on cleaned ECG signal to find ST-segment and QT-interval, ST/QT, ΔST/QT, etc.

2.1 Preprocessing

The noise which haphazardly extends over the entire ECG spectrum is removed through preprocessing. The preprocessing improves the signal-to-noise ratio (SNR) to such an extent that the measurements made on original signals become more reliable. The preprocessing involves three steps: 1. High-frequency noise removal; 2. Baseline wander removal; 3. QRS complex detection. The high-frequency noise is removed from ECG signal using a low-pass filter with a cutoff frequency of 45 Hz. This filter can be implemented with a Butterworth low-pass digital filter having 4-poles or 6-pole [8]. The baseline wander can be removed from ECG by using a high-pass filter with a cutoff frequency of 0.8 Hz.

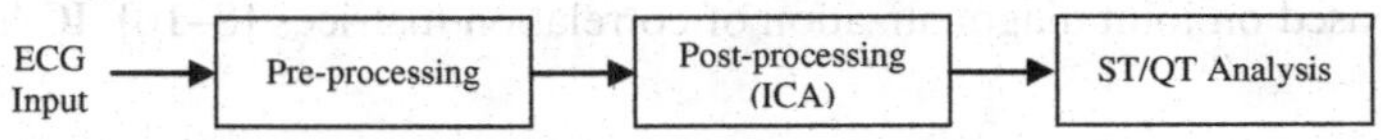

Fig. 2 Block diagram of the proposed method

2.2 Post-processing

Independent component analysis is one of the best solutions to the blind source separation problems (BSS) [8]. In ICA, a set of signals are extracted based on their mixtures. A set of observations or signals are transformed into another set of observations or signals to maximize the separation. This separation also performed based on independence rather than its frequency content as in case of Fourier analysis. The block diagram of our proposed method is shown in Fig. 2. In our proposed first, the ECG is preprocessed for removing baseline wander. Second, ICA is used to remove the non-Gaussian noise and artifacts from ECG, and then the desired ECG is reconstructed by eliminating ICs which are embedded with the noise and artifact by setting to zero in (2) [9, 10]. This may be achieved with the help of kurtosis and variance of ICs. ICA is a quite powerful technique and is able (in principle) to separate independent sources linearly mixed in several sensors. Third, all the fiducial points P-peak, QRS complex, and T-peak are identified. Finally QT episodes are analyzed using suitable algorithms and MATLAB.

The ECG is considered as a mixture of signals from nodes presented in the heart and various artifacts and noise. Basic ICA model assumes linear combination of source signals (called components)

$$X = AS. \tag{1}$$

where X and S are the two vectors representing the observed signals and source signals, respectively, and A is an unknown matrix called the mixing matrix. The matrix A is then of size $n \times n$ (in general A does not need to be square, but many algorithms assume this 'property'), X and S get the size $n \times m$, where n is number of sources and m is length of record in samples. Incidentally, the justification for the description of this signal processing technique as blind is that we have no information on the mixing matrix, or even on the sources themselves. The objective is to recover the original signals, S, from only the observed vector X. Denoting the output vector by V, the aim of ICA algorithms is to find a matrix U to undo the mixing effect. That is, the output will be given by

$$V = UX. \tag{2}$$

where V is an estimate of the sources. The sources can be exactly recovered if U is the inverse of A up to a permutation and scale change. The BSS/ICA methods try to estimate components that would be as independent as possible and their linear combination is original data. Estimation of components is done by iterative algorithm, which maximizes function of independence, or by a noniterative algorithm, which is based on joint diagonalization of correlation matrices [8–10]. ICA has one large restriction that the original sources must be statistically independent. This is the only assumption we need to take into account in general. The reconstructed ECG can be derived using the following equation

$$X = U^{-1}V. \tag{3}$$

where V is the matrix of derived independent components with the row representing the noise or artifacts set to zero.

After getting ICs, it is necessary to determine the order of the independent components in order to identify normal ECG, noise, and abrupt alterations. As the ICs corresponding to noise and abrupt alterations have more distinctive properties than that of original signal both in time and frequency domains, we may employ the statistical properties of these waveforms to recognize the original ECG automatically instead of identifying visually. The noise is identified using kurtosis and abrupt changes using variance. The kurtosis is the fourth-order cumulant [10, 11]. For a signal $x(n)$, it is classically defined as in (4) by dropping n for convenience

$$\mathrm{Kurt}(x) = E(x^4) - 3[E(x^2)]^2. \tag{4}$$

Here the kurtosis is zero for Gaussian densities. The normal ECG will have large Kurtosis value than continuous noise. In our approach, a threshold is chosen from analysis of sample waveforms, and a component whose modulus of kurtosis is below this threshold will be considered as continuous noise. There are several ways to detect abrupt changes which are usually short transients. The variance or energy is more or less similar and negligibly small for all IC components except for those ICs containing abrupt changes. Thus the IC waves whose variance is large can be identified as abrupt variations or noise. The variance of signal $x(n)$ is given by

$$x_{\mathrm{var}} = \sum_{n=1}^{N-1} [x(n) - \overline{x(n)}]^2 \tag{5}$$

Here $x(n)$ is the mean value of $x(n)$. In our approach, we calculated the modulus of Kurtosis value of each ICA component and compared with the threshold. If the modulus of Kurtosis exceeds the threshold that IC is marked as continuous noise component. Then, the remaining ICA components are divided into 10 nonoverlapping blocks, each of one-second duration. The variances of the 10 segments for each component are calculated as shown in (5), and then the variance of these 10 variance values is obtained as the parameter x_{var}. The component whose x_{var} value is above a predetermined threshold is marked as an abrupt change component. Finally, the required ECG can be obtained using (2) and (3).

2.3 ST/QT Analysis

The QRS detection involves detecting R-peaks initially. This involves differentiation of ECG to convert ECG signal into slope signal and then filtered with moving average filter to enhance the peaks of the signal. A threshold is set based on mean and maximum values of the ECG. The peak point above this threshold is identified as R-peak. Based on physiological considerations, it is assumed that the Q-peaks are about 60 ms before the R-peak and S-peaks are about 30 ms after R-peak. Therefore Q and S peaks are identified by searching backward and forward from R-peak. As the slope of the signal is either zero or changes its sign at its Q and S peaks, they can be easily identified as Q-peak and S-peaks.

After finding QRS complex, the isoelectric segment (PQ-segment) is identified just before the Q-peak as the separation between onset of Q-wave and offset of P-wave by searching backward up to 80 ms. The point where the slope is either zero or changes its sign within the 20 ms from Q-peak can be identified as onset of Q-wave and the point where the slope is 'flattest' can be considered as onset of PQ-segment, when searching backward from Q-peak. The flattest slope is one that is calculated as the minimum of mean of slopes of each segment. The middle sample of this flattest waveform segment in ECG wave defines the position of the isoelectric reference point. The mean amplitude of this flattest segment is taken as an estimate of the isoelectric level. Similarly, the J point is the point within 100 ms from S-peak where the slope is flattest and can be identified by searching forward from S-peak. Once the J peak is identified, the ST-segment is defined as the separation between the J point and the K point. The location of K point depends upon the heart beat. If the heart beat is 60 beats/s, then the K point is at 60 ms away from J point. If the heart beat is less than 60 beats/s (Bradycardia), then the K point is located at 80 ms away from J point. If heart beat is more than 60 beats/s (Tachycardia), then the K point is located at 40 ms away from J point. Finally, the ST-segment and the QT_{peak} interval is determined using similar procedure as separation between Q-peak to T-peak. The signal-to-noise ratio (SNR) ST of processed signal, ST/QT, and ΔST/QT index for each beat are measured and plotted against every beat.

3 Results

We tested our proposed algorithm on abnormal ECG records 's20011 m' and 's20041 m.' These records are downloaded from MIT-BIH (Massachusetts Institute of Technology—Beth Israel Hospital) Arrhythmia database. Independent Components (ICs) are obtained using JADE algorithm as shown Fig. 3a. Later the cleaned ECG is obtained by choosing the relevant independent components based on kurtosis and variance of ICs. The cleaned ECG is shown plotted in Fig. 3b.

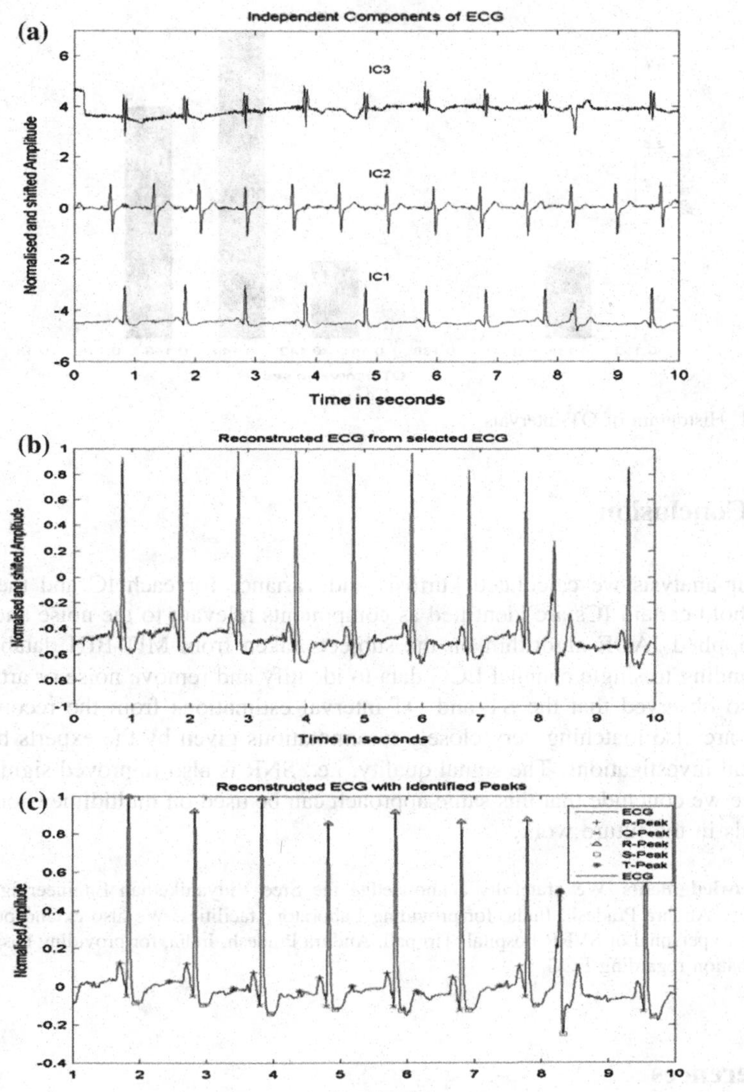

Fig. 3 a Extracted ICs from mixed ECG. **b** Reconstructed ECG from selected ICs. **c** Reconstructed ECG from selected ICs with peaks

Later we computed the R-peaks, R-R intervals, QRS amplitudes, ST-segments, and QT-intervals from cleaned ECG signal by applying simple algorithms. These peaks are shown with markings in Fig. 3c. We observed a significant improvement in SNR with ICA than without ICA. The histogram of QT-intervals is plotted in Fig. 4.

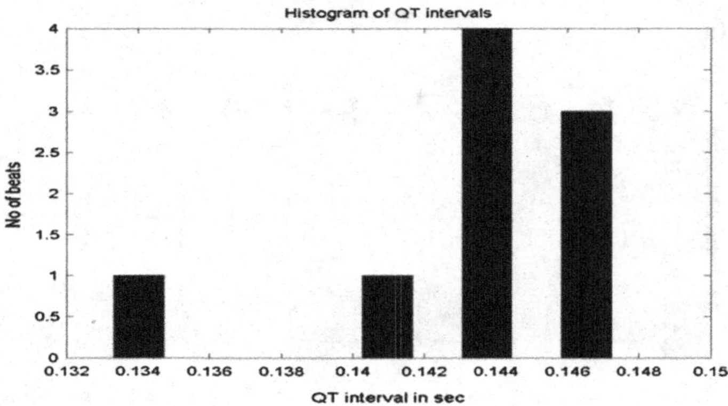

Fig. 4 Histogram of QT-intervals

4 Conclusion

In our analysis we calculated kurtosis and variance for each IC and then using threshold certain ICs are identified as components relevant to the noise and signal. We applied JADE algorithm on the subjects taken from MIT-BIH database corresponding to single channel ECG data to identify and remove noise or artifacts. It is also observed that the ST and QT-interval estimations from the reconstructed ECG are also matching very closely to annotations given by the experts based on clinical investigations. The signal quality, i.e., SNR is also improved significantly. Hence we conclude that this same approach can be used on multidimensional ECG signals in the future work.

Acknowledgments We gratefully acknowledge the Sree Vidyanikethan Engineering College, Tirupati, Andhra Pradesh, India, for providing Laboratory facilities. We also extend our sincere thanks to personal of SVRR hospital, Tirupati, Andhra Pradesh, India, for providing basic clinical information regarding ECG.

References

1. Goldschlager N. Principles of clinical electrocardiography, 13th edn. Connecticut. ISBN:978-083-8579-510
2. Hampton JR (2013) The ECG made easy. International edition (English), 8th edn. Elsevier Health Sciences
3. Phlypo R, Zarzoso V, Lemahieu I (2008) Exploiting independence measures in dual spaces with application to atrial f-wave extraction in the ECG. In: Proceedings of the MEDSIP-2008, 4th international conference on advances in medical, signal and information processing, Santa Margherita Ligure, Italy
4. Benitez D, Gaydecki P, Zaidi A, Fitzpatrick A (2000) A new QRS detection algorithm based on the Hilbert transform. Comput Cardiol 2000:379–382

5. Hamilton PS, Tompkins WJ (2007) Quantitative investigation of QRS detection rules using MIT/BIH Arrhythmia database. IEEE Trans Biomed Eng 31(3):1157–1165 ISSN 0018-9294
6. Hyvarinen A, Karhunen J, Oja E (2001) Independent component analysis. Wiley, New York
7. Pan J, Tompkins WJ (1985) A real-time QRS detection algorithm. IEEE Trans Biomed Eng 32 (3):230–2
8. AbedMeraim K, Amin MG, Zoubir AM (2001) Joint anti-diagonalization for blind source separation. In: Proceedings of the ICASSP-01
9. Hyvarinen A, Oja E (2000) Independent component analysis: algorithms and applications. Neural Netw 13:411–430
10. Cardoso JF (1999) High-order contrasts for independent component analysis. Neural Comput 11:157–192
11. Chawla MPS (2009) Detection of indeterminacies in corrected ECG signals using parameterized multidimensional independent component analysis. Comput Math Methods Med 10(2):85–115
12. Thulasi Prasad S, Varadarajan S (2013) PC based digital signal processing of ECG signals. Int J Adv Res Comput Commun Eng 2(12)

Modeling of Solar Wind Hybrid Renewable Energy Sources in Simulink

Subash Chandra Sahoo, Bhagabat Panda, Ritesh Dash, Babita Panda and Sasmita Kar

Abstract The dependence of energy sector is gradually showing a substantial growth toward renewable sources of energy. This dependence is the result of factors like unsure future of nonrenewable energy sources along with the concern for severe climate change which primarily takes place owing to the high amount of concentrated carbon dioxide in the atmosphere. In this scenario, the efficient and least-polluting renewable energy hybrid systems are considered owing to their manifold advantages. The modeling of solar and wind energy using Simulink has been done in this paper, and the results are shown for this hybrid system.

Keywords Hybrid renewable energy systems (HRES) · Solar · Wind · Simulink

1 Introduction

Although the Sun is the ultimate source of energy, nevertheless, the Sun leaves us with alternatives generated from it which can be sustainable and simultaneously convenient to be used by us also it is available in free of cost and pollution free. However, every system is attached with its pros and cons. The main factor in the solar energy for which it lags behind other energy sources is the time factor, because it can be obtained in the day time only. So, in this article, two renewable

S.C. Sahoo · B. Panda · R. Dash (✉) · B. Panda
School of Electrical Engineering, KIIT University, Bhubaneswar, India
e-mail: rdasheee@gmail.com

S. Kar
Maharaja Institute of Technology, Bhubaneswar, India

L.P. Suresh and B.K. Panigrahi (eds.), *Proceedings of the International Conference on Soft Computing Systems*, Advances in Intelligent Systems and Computing 397, DOI 10.1007/978-81-322-2671-0_89

energy systems have been combined so as to obtain maximum efficiency. This system is known as a Hybrid system which primarily helps in enhancing the reliability of the whole energy system making it more cost-effective [1–3]. The rapid advances in renewable energy technologies and rise in prices of petroleum products have resulted in the increasing popularity of hybrid renewable energy systems (HRES) [4]. A solar and a wind system have been considered in this article. The block diagram of a solar wind hybrid energy system has been shown in Fig. 1.

2 System Modeling

The current system is built keeping an eye on the realisting existing system configurations. It contains a wind park and PV generating station with the demand of 200 households.

3 Simulation

The simulation for the proposed system has been developed with Matlab/Simulink model. The system consists of a radial bus system having four number of alternator each of 250 MVA. The grid is supposed to supply a load of 20 kW located at a remote location of 52 km. Figure 1 shows the daily load profile of the remote location. The remote location basically includes the house hold machines like fan, induction generator, washing machine, and other daily use systems.

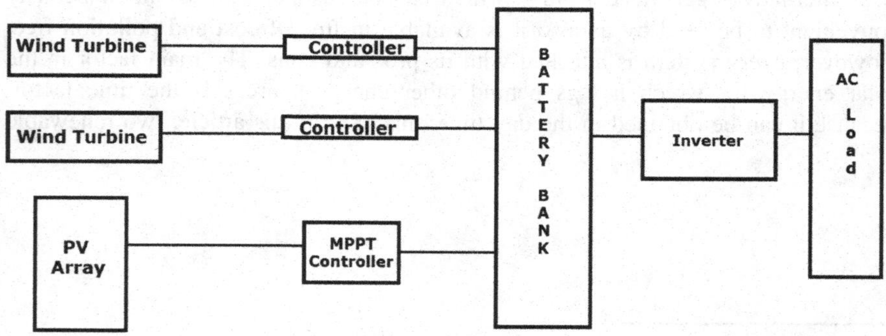

Fig. 1 Block diagram of a solar wind hybrid energy system

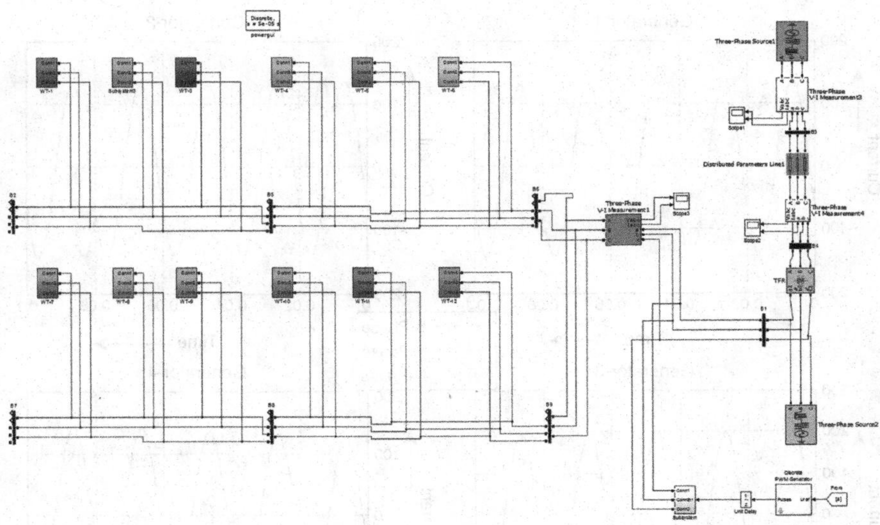

The minimum load on the grid is only 600 W. So a voltage profile of 500 V DC is considered as the reference for the system. To avoid the un-necessary discharge of battery and frequent turn on and off the biogas generator system, a time limit of 6 h is provided in the system.

Figures 2 and 3 show the current and voltage wave of the alternator at the different location. From figure, the total current shared by first and second alternator

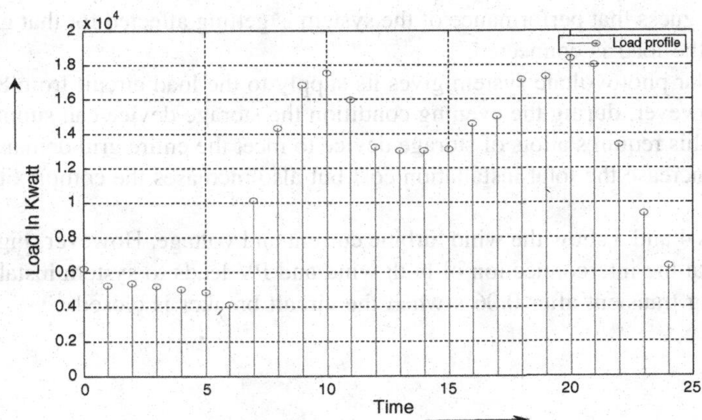

Fig. 2 Daily load profile for 24 h

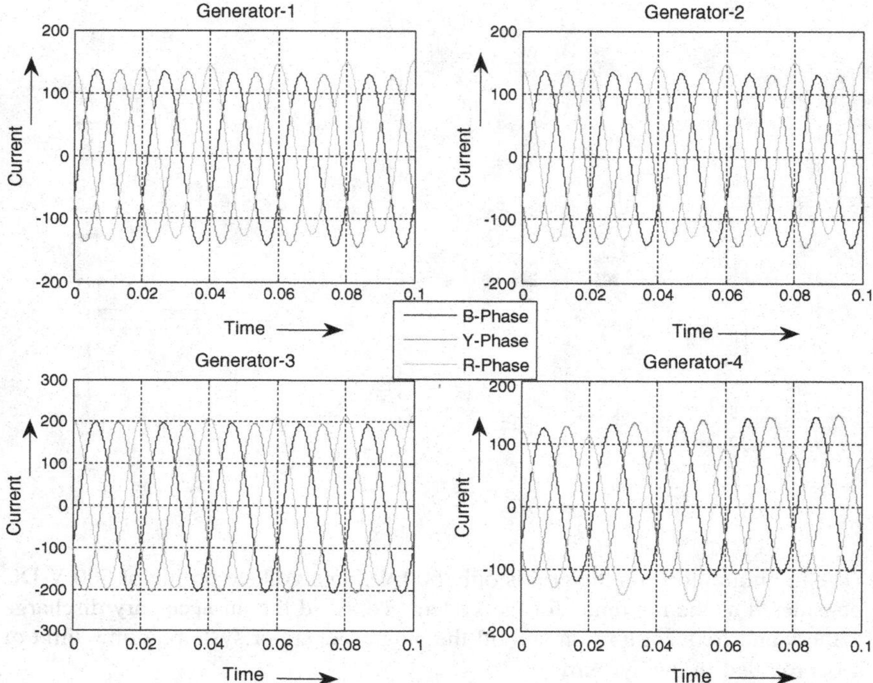

Fig. 3 Current wave form at different generator

is 100 A where as that of the 200 and 110 A shared by 3 and 4, respectively. So one can easily guess that performance of the system is getting affected for that generator to which the load is connected.

The solar photovoltaic system gives its supply to the load circuit from 8 a.m. to 6 p.m. However, during the evening condition the storage device can supply to the grid. But this requires a lots of storage device to meet the entire grid demand which not only increase the total installation cost but also increases the complexity of the system.

Figures 4 and 5 show the wind turbine current and voltage. However, Figs. 6 and 7 show that the interconnection of both wind and PV leads to system instability by introducing transient after 0.06 s when the circuit breaker is closed.

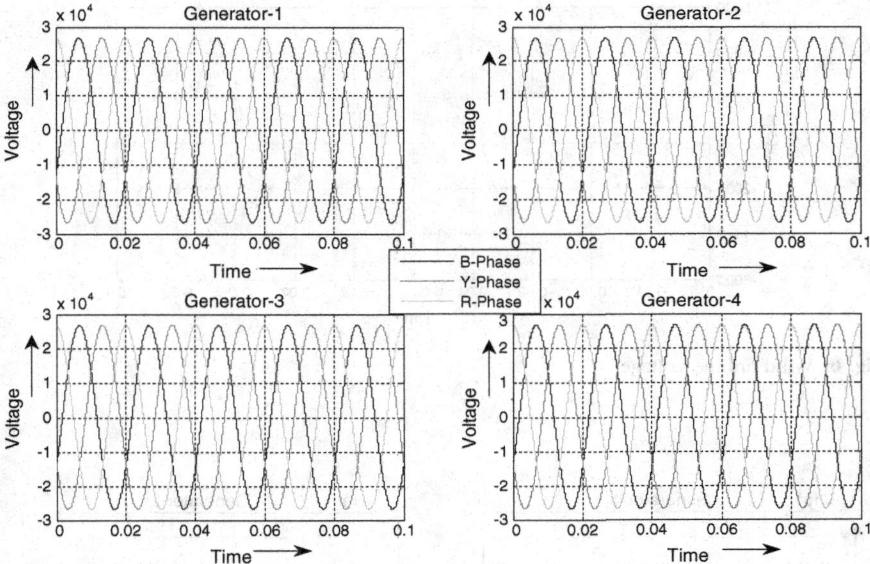

Fig. 4 Voltage wave form

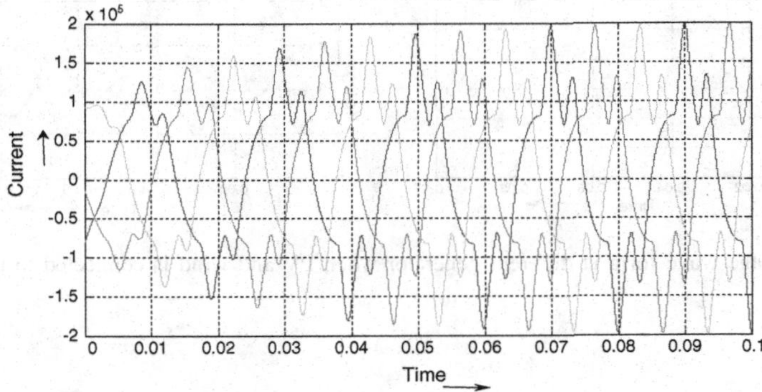

Fig. 5 Wind turbine current

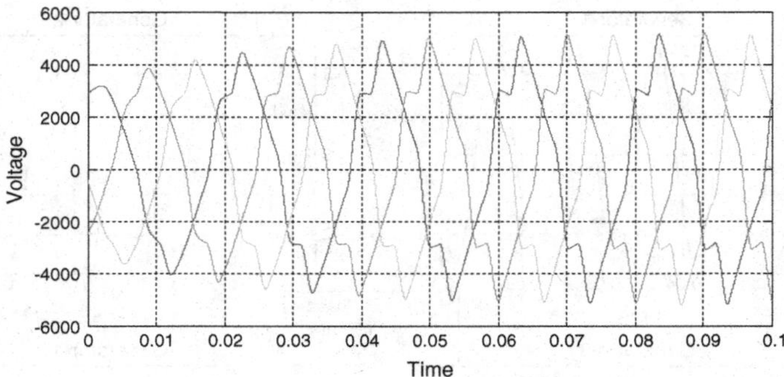

Fig. 6 Wind turbine voltage

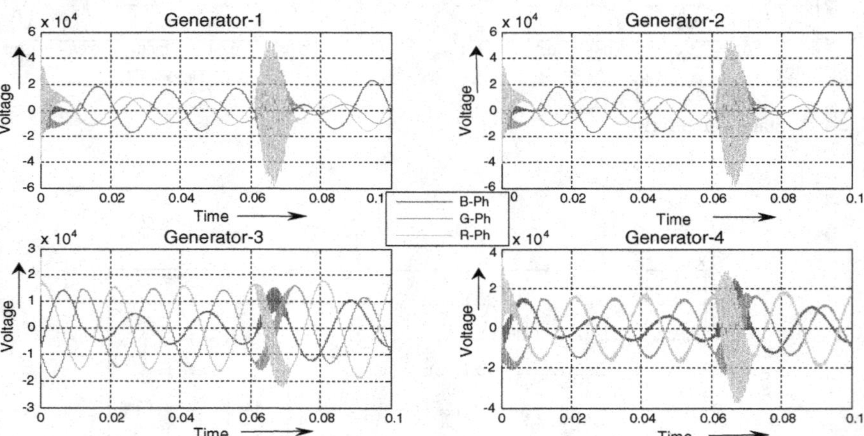

Fig. 7 Voltage wave form at different generator when PV and wind is connected to the grid at load no-2

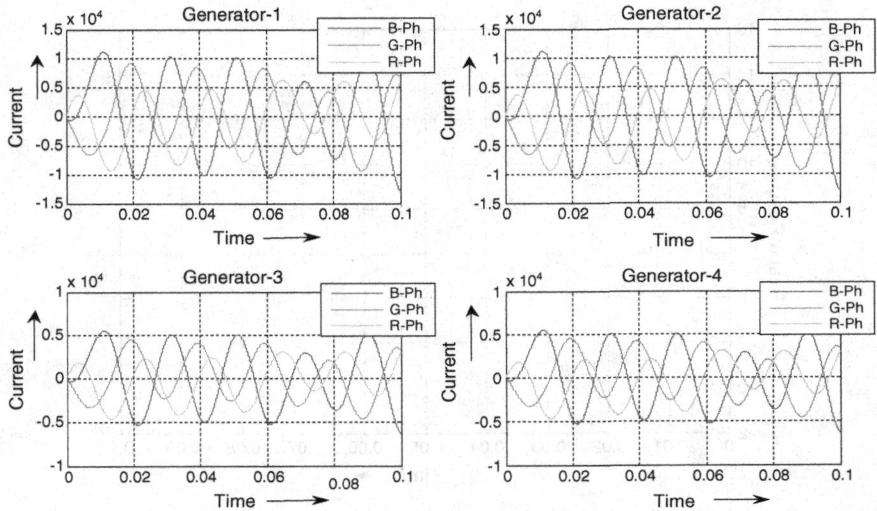

Fig. 8 Current wave form at different generator when PV and wind is connected to the grid at load no-2

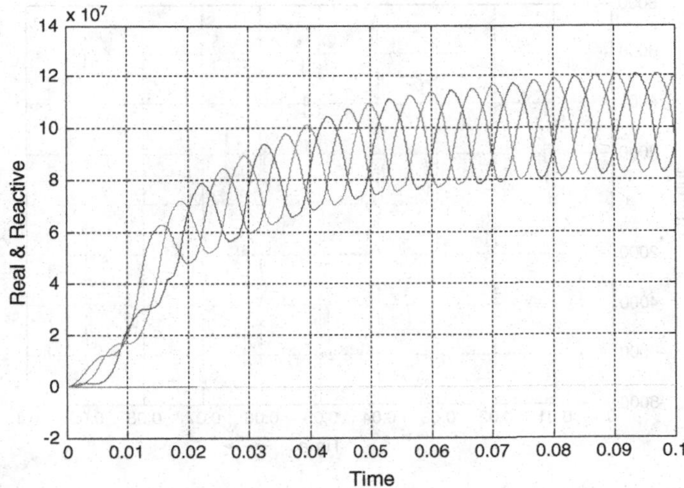

Fig. 9 Real and reactive power supplied by the load-2

Again from Figs. 8 and 9 one can easily noticed that with the help of hybrid system the real and reactive power drawn by the load is much better than the previous one. However, the interconnection of renewable energy affects the frequency and stability of the system (Figs. 10, 11 and 12).

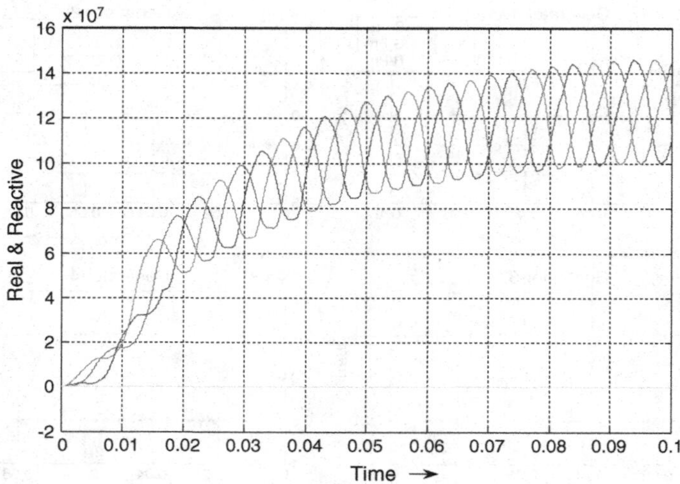

Fig. 10 Current wave form at different generator when PV is connected to the grid at load no-1

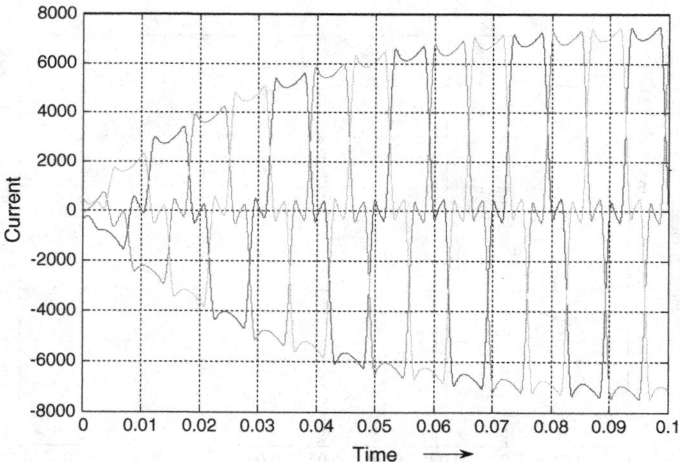

Fig. 11 Current supplied by bio gas

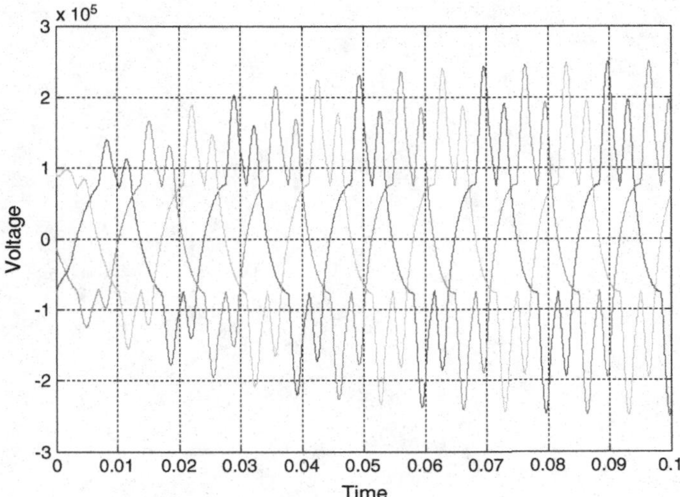

Fig. 12 Voltage supplied by bio gas

4 Conclusion

Paper presents a detailed solar photovoltaic and wind system model with an aim to calculate the maximum penetration level of a grid connected system. The paper has also focused on the implementation of battery storage system with the grid. The simulated result also shows the real and reactive power injection level of the system under balanced and unbalanced condition. Further in future, the system can be analyzed with one or more optimization technique to arrive at a particular point.

References

1. Pachori A, Suhane P (2014) Modeling and simulation of photovoltaic/wind/diesel/battery hybrid power generation system. Int J Electr Electron Comput Eng 3(1):122–125. ISSN: 2277-2626
2. Singh M et al (2014) Simulation for wind turbine generators—with FAST and MATLAB-Simulink modules. Technical report, NREL
3. Angelis-Dimakis A et al (2011) Methods and tools to evaluate the availability of renewable energy sources. Renew Sustain Energy Rev 15:1182–1200, Elsevier
4. Notton G, Diaf S, Stoyanov L (2011) Hybrid photovoltaic/wind energy systems for remote locations. Energy Procedia 6:666–677. ISSN: 1876-6102. http://dx.doi.org/10.1016/j.egypro.2011.05.076

Dynamic Resource Allocation in OFDM-Based Cognitive Radio System Based on Estimated Channel Information

Kishore V. Krishnan, A. Bagubali and Sibaram Khara

Abstract In this paper, resource allocation scheme for an OFDMA-based multiuser cognitive radio (CR) systems is considered based on the channel state Information (CSI). The secondary user (SU) network capacity can be enhanced by a power allocation in cognitive radio systems. The primary users (PU) interference power limit introduces subchannel transmit power constraint for the SU. The resource allocation among the subcarriers of the SU should be constrained to both the total transmitted power and the transmitted power of each subchannel. Maximization of SU system capacity based on an error free estimated channel state information at the SU source is the main motivation of this paper. Different form the previous works, we considered an optimal number of pilots and the optimal power allocated to each pilot and the data subcarriers. Simulation results show that, the proposed scheme maximizes the sum capacity while considering the CSI.

Keywords Cognitive radio · OFDM · Channel state information · Spectrum allocation

1 Introduction

Access to the wireless spectrum is regulated by the government by the static spectrum allocation schemes. The requirement of advanced radio systems which are capable of providing higher data rate and which can accommodate large number of user are increasing. The static spectrum allocation schemes are ineffective. Also recent studies exposed that some of the spectrum which are allocated to some radio network by the static spectrum allocation schemes are underutilized [1]. Cognitive

K.V. Krishnan (✉) · A. Bagubali
VIT University, Vellore, Tamilnadu, India
e-mail: kishorekrishnan@vit.ac.in

S. Khara
Galgotias University, Noida, India

L.P. Suresh and B.K. Panigrahi (eds.), *Proceedings of the International Conference on Soft Computing Systems*, Advances in Intelligent Systems and Computing 397, DOI 10.1007/978-81-322-2671-0_90

radio is key candidate to recover from this underutilization of frequency spectrum. In a CR system the SU, also called as unlicensed user, intelligently detects the spectrum white spaces (unused spectrums) and starts transmitting in the same spectrum thereby improving the spectrum efficiency.

OFDM is the key technology for cognitive radio [1]. In OFDM, multiple SU signals are transmitted over different subcarriers simultaneously in each OFDM symbol. OFDM also haves the advantage of finer granularity, better link budget simpler receiver design and better multiuser diversity [2]. As OFDM divides the spectrum into several orthogonal subcarriers and individually controls every subcarriers it is very adaptive and flexible. This flexibility and adaptiveness made to consider various resource allocation problems in recent years [3]. Best quality of service can be achieved at low cost if an optimized resource allocation (subcarrier and power in case of OFDM) is performed. Different users have different channel state information (CSI) for different subcarriers; a channel estimation along with resource allocation will improve the overall performance of the SU systems (Fig. 1).

Channel state information is required at the receiver for coherent reception. Also, the computation of the channel capacity of a cognitive radio link and the power control algorithm in the transmitter necessitate information of channel state information [4]. Channel identification algorithms can be classified into three categories: data-aided, non-data-aided and decision-directed methods. In Data-aided channel estimation methods the transmitted data is known and channel estimation is performed using these information. Non-data-aided channel estimation methods assume unknown transmitted data and remove the data by averaging. Decision-directed methods approximate the data-aided methods by detecting the data and using this data as a reference signal to the estimator. An example of a possible signal model is illustrated in Fig. 1.

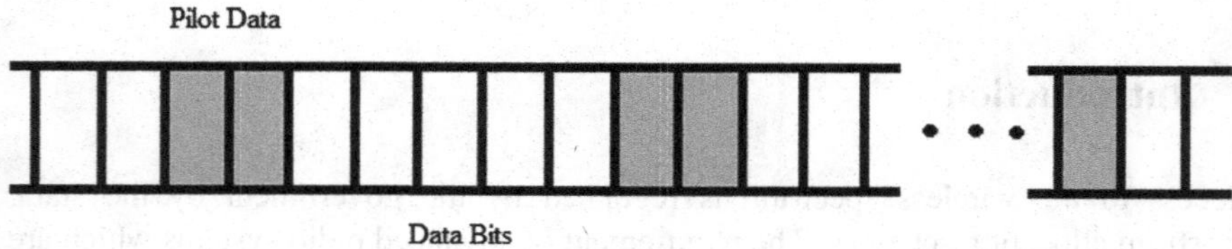

Fig. 1 An example of signal model

Fig. 2 System model

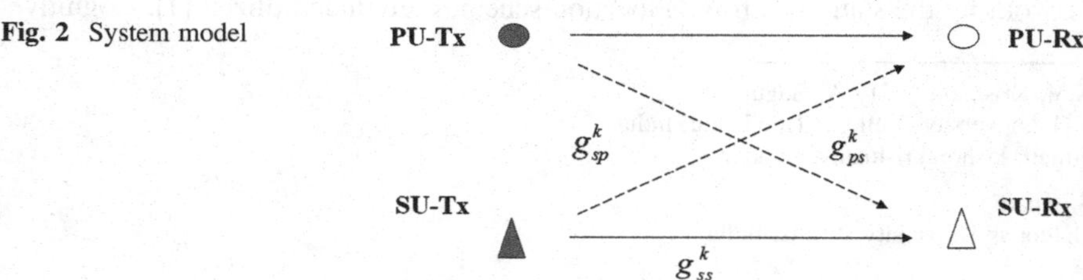

In [2] a joint spectrum and power allocation framework is proposed that addresses the issues regarding negotiation of spectrum based on licensed user opportunistically. A suboptimal power allocation algorithm with less complexity is proposed in [5]. Beamforming is considered as an efficient technique to overcome interference. In [6] Zhang et al. has proposed a method in which beamforming based on instantaneous CSI from the SU transmitter to the PU receiver, the SU suppresses the interference to the PU. In [6], interference effect of subcarriers side lobes is mathematical provided in details. In [8] a MIMO-OFDM cognitive radio network is considered where wireless users with multiple antennas communicate over several noninterfering frequency bands, the SU user transmit profiles which are adapted on the fly according to the primary user activity thereby maximizing their data rate.

In all the aforementioned schemes, a perfect knowledge of instantaneous gains of the subchannel from SU transmitter to SU receiver and the subchannel form SU transmitter to PU receivers is considered. But in practice it is possible to estimate the channel theoretically and track the short-term changes of the channel form the SU transmitter to the PU receiver. In this paper power distribution for the different subchannels is find out, and the optimum number of pilots and the amount of power allocated between pilot and data subcarriers is obtained.

2 System Model and Problem Definition

In this paper, an OFDM-based cognitive radio network with one SU sharing the resource of one PU is modeled. The total frequency band is divided into N subcarriers. The system model is shown in Fig. 2. The SU-SU and the SU-PU channel are point-to-point and flat fading channel with channel parameters h_{ss}^i, h_{sp}^i and the AWGN noise are n_s^i, n_p^i. Then, the output of the channel can be expressed as

$$\begin{bmatrix} y_s^i \\ y_p^i \end{bmatrix} = \begin{bmatrix} h_{ss}^i & h_{sp}^i \\ h_{ps}^i & h_{pp}^i \end{bmatrix} \begin{bmatrix} x_s^i \\ s_p^i \end{bmatrix} + \begin{bmatrix} n_s^k \\ n_p^i \end{bmatrix} \tag{1}$$

An OFDMA-based cognitive radio system with comb-type pilot is assumed. The signal-to-noise ratio (SNR) in each subcarrier considering the minimum mean square estimation (MMSE) in each block can be expressed as [9]

$$\gamma_i = \frac{P_i |\overline{H}_i|^2}{\sigma^2 + P_i \sigma_{\epsilon,i}^2} \tag{2}$$

where $\overline{H}_i$ is the ith channels estimated channel response in frequency domain, σ^2 is the AWGN noise and is given by $\sigma^2 = \frac{N_0 B}{N}$, P_i is ith subchannel transmit power and $\sigma_{\epsilon,i}^2$ is the noise created due to channel estimation based on MMSE. As equispaced pilots insertion is optimum in BER performance in the system, we consider N_k equispaced pilots in each block and we allot equal power P_k. It is found that the noise created by the pilot $\sigma^2_{\epsilon,i}$ is same for all the subchannels and is given by

$$\sigma^2_{\varepsilon,i} = \sigma^2_{\varepsilon} = \sum_{j=1}^{L} \left(\frac{1}{\sigma^2_{h,j}} + N_k \frac{P_k}{\sigma^2} \right) \tag{3}$$

where k is the number of channel taps. $\sigma^2_{h,j}$ is the channel delay profile of the jth tap.

3 Computation of Traffic Parameters

This section deals with traffic parameters of the primary and secondary networks. These are calculated using statistical methods [10]. In the secondary network, the channel allocated is adaptive to the traffic parameter variations. In this system, we are assuming the arrival rates of PUs and SUs to be following the Poisson distribution. Let us assume the average arrival rates of PUs and SUs to be λ_p, λ_s, respectively. The average transmission duration is represented as λ_p^T, λ_s^T. The presence probability of the primary user can be obtained by using exponential distribution type [11],

$$P^p(k) = \frac{e^{-\lambda_p \lambda_p^T} \left(\lambda_p \lambda_p^T \right)^{-1}}{1!} \tag{4}$$

In the secondary user, the kth subcarrier is selected with a probability of $P_s(k)$. The calculation of this probability includes two parameters: interference power constraint and outage probability [7]. Thus, it is noted in mathematical form as

$$P^s(k) = P(\rho < \rho_{\text{th}}) P(\tau_s < \tau_{\text{th}}) \tag{5}$$

The outage probability of SU will increase with decrease in SNR. The assigned power is setting SNR to zero. But it does not contain enough power to do so. Thus, the SNR $\tau_s^k = \frac{e_s^k h_{ss}^k}{n_s^k}$ is proportional to $e_s^k h_{ss}^k$.

$$P\left(e_s^k h_{ss}^k < \tau_{\text{th}}\right) = P\left(h_{ss}^k < \frac{\tau}{e_s^k}\right) = 1 - \exp\left(-\frac{\tau_{\text{th}}}{e_s^k \lambda_s}\right) \tag{6}$$

Table 1 Simulation Parameters

Channel type	Rayleigh flat fading channel
Physical layer	OFDM
Noise power mean	10^{-8} W
Number of subchannels N	16
Number of users N_s	10
Total power Pt	3.1623e-005 W
Average signal to noise ratio (ASNR) $\frac{Pt}{\sigma}$	15 dB
Pilot signal to noise ratio SNRP	15 dB
Pilot space interval	8
Channel length	16
Total bandwidth B	1 MHz

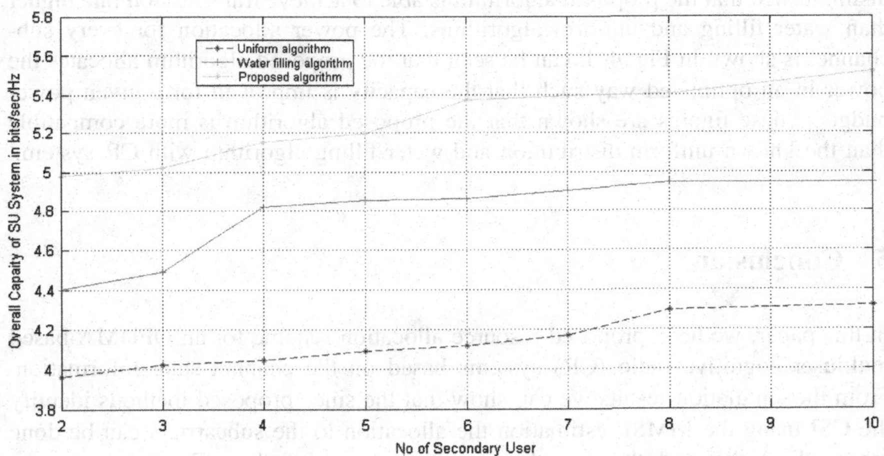

Fig. 3 Capacity versus number of secondary user

4 Simulation Results

The simulation results are presented here to evaluate the effects of channel estimation on the performance of the proposed cognitive radio system. An OFDM-based cognitive radio system with subcarriers, $N = 16$ subcarriers and one PU is considered for simulation. Table 1, shows the simulation parameters adapted for CR system. The Fig. 3 shows the performance of the proposed algorithm.

The maximum achieved capacity for the CR user with the total power is plotted. In this, we have compared our proposed method with uniform algorithm and water filling algorithms. It is found that for a given total power the proposed method

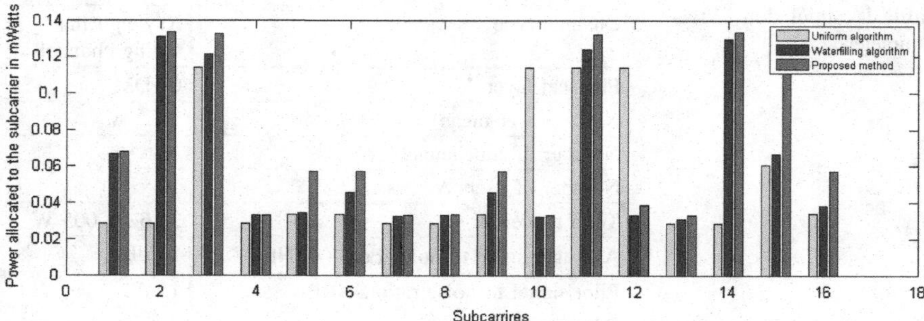

Fig. 4 Power allocated to the subcarriers

outperformed those others. The achievable transmission data rate for the CR user versus the total power budget for various algorithms is plotted. From this figure, it is distinguished that the proposed algorithm is able to achieve transmission rate higher than water filling and uniform algorithms. The power allocation for every sub-channel is shown in Fig. 4. It can be seen that the proposed algorithm allocates the power in an optimized way such that the capacity is improved for a given power budget. These figures are shown that the proposed algorithm is more compatible than the known uniform distribution and water-filling algorithm with CR system.

5 Conclusion

In this paper, we have proposed resource allocation scheme for an OFDMA-based multiuser cognitive radio (CR) systems based on the channel state Information. From the simulation results we can show that the since proposed methods identify the CSI using the MMSE estimation the allocation to the subcarriers can be done more efficiently, and thereby, the overall capacity of the CR system can be improved. In this, we have found that the CSI helps to reduce the interference introduced to the primary user.

References

1. Federal communication commission spectrum policy task force. FCC Report of Spectrum Efficiency Working Group. November 2002
2. Hossain E, Bhargava V (2007) Cognitive wireless communication network. Springer Science Publication, Berlin
3. Zhao Q, Sadler BM (2007) A survey of dynamic spectrum access: signal processing, networking and regulatory policy. IEEE Signal Process Magaz 24(3):78–89
4. Zheng L, Wei Tan C (2014) Maximizing sum rates in cognitive radio networks: convex relaxation and global optimization algorithms. IEEE J Select Areas Commun 32(3)

5. Dingham FF (2008) Joint power and channel allocation for cognitive radios. In: Procceedings IEEE Wireless Communication Networking Conference, pp 882–887
6. Zhang R, Liang Y-C (2008) Exploiting multi-antennas for opportunistic spectrum sharing in cognitive radio networks. IEEE J Select Topics Signal Process 2(1):88–102
7. Elham H, Falahati AF (2012) Improving water filling algorithm to power control cognitive radio system based upon traffic parameters and QoS. Springer, pp 1747–1759
8. Tse D, Viswanath P (2005) Fundamentals of wireless communication. Cambridge University Press
9. Wang P, Zhao M, Xiao L, Zhou S, Wang J (2007) Power allocation in OFDM based cognitive radio system. In: Proceedings IEEE global telecommunication conference, pp 4061–4065
10. Xiao Y, Bi G, Niyato D (2011) A simple distributed power control algorithm for cognitive radio networks. IEEE Trans Wireless Commun 10(11):3594–3600
11. Chen C-H, Wang C-L, Chen C-T (2011) A resource allocation scheme for cooperative multiuser OFDM based cognitive radio system. IEEE Trans Commun 59(11):3024–3215

Mobile Governance Framework for Emergency Management

K. Sabarish and R.S. Shaji

Abstract Mobile Technologies (MT) have dramatically transformed the nature of communication across the globe and have profoundly impacted every field of human activity. They have opened up new vistas allowing instant communication and collaboration on anywhere, anytime basis, allowing rapid transfer of information and services across the globe sans geographic boundaries. Many governments across the globe have started harnessing the potential of MT's. The adoption of MT's have not only benefited the end-users but also has enabled governments to reinvent themselves by replacing obsolete and less efficient processes with improved efficient systems and practices. In order to provide a bouquet of advanced mobile services to its customers, viz., citizens, intergovernment departments, employees, businesses, and tourists, governments have started establishing service delivery platforms by deploying next generation networks and cloud-enabled data centres resulting in convergence of many technologies. This paper describes how a cloud-based interoperable framework based on Service-Oriented Architecture (SOA) and web service-based technologies were developed for delivering multifarious public services on mobile devices. It discusses the various technological components required for developing a mobile service delivery platform (MSDP) and its role in providing greater agility in synchronizing two or more disparate e-governance applications, allowing a group of departments/organizations to provision and consume services as a shared infrastructure. The paper also illustrates a case study on how MSDP was leveraged for provisioning emergency management services.

K. Sabarish (✉)
Information Systems Division, KSCSTE, Pattom, Trivandrum, Kerala, India
e-mail: sabarishtvm@gmail.com

R.S. Shaji
Department of Information Technology, Noorul Islam University, Kumaracoil, India
e-mail: shajiswaram@yahoo.com

L.P. Suresh and B.K. Panigrahi (eds.), *Proceedings of the International Conference on Soft Computing Systems*, Advances in Intelligent Systems and Computing 397, DOI 10.1007/978-81-322-2671-0_91

1 Introduction

E-governance (e-gov) is making strides, but it remains restricted primarily to the use of computer-based internet access to deliver services. Low internet penetration, lack of last mile connectivity, low PC ownership, and net connection vis-à-vis poor technical skills to access internet are the major impediments in expanding the reach of e-gov services [1]. Though e-gov was seen as a welcome change in achieving good governance, it could not permeate down to the grassroots and failed to bring citizens closer to government [2]. On the other hand, the mobile revolution that was making great strides, evidenced by the mind boggling number of mobile phone subscribers and its permeability down to the grassroots, has made it imperative for the governments across the globe to deliver services over mobile phones to ensure that government services reach out every citizen at their doorstep, ensuring equitable and enhanced government–citizen interaction [3].

As the global mobile cellular subscriptions soars and crossed the 7.2 billion mark in 2014, with a penetration level of 99 %, it is projected to reach 9.2 billion at the close of 2015 ITU [4]. With ever increasing adoption of mobile phones, especially smart phones, governments across the globe are under sheer pressure to reinvent and to provide government services to its citizens through mobile devices. Mobile governance (m-gov) can be seen as the new channel for the delivery of government services and information to the citizens and for improving government–citizen interaction. It will complement e-governance and will not replace e-government because of the inherent limitations of mobile devices [5]. Antovski and others provide a rather loose definition of m-gov. They define m-gov as an inter-operable overlay over the existing e-governance infrastructure enabling the e-services to be ported to citizen's mobile devices [6]. With the adoption of smartphones equipped with high-resolution cameras, Global Position System, various sensors, social networking, mobile web, and video conferencing in an unprecedented scale, coupled with innovative features offered by 3G, 4G, more and more services are routed the mobile way, and m-gov has become inseparably intertwined with the social fabric [7]. Apart from the utility of mobile phone as voice-based communication tool to share information from the field where the crisis has occurred and to provide updates about its evolution, m-gov services can also be provisioned in the area of emergency management, like early warning systems providing storm alerts and rough sea information to fishermen and people inhabiting the coastal regions, flood warnings to people who reside on the banks of rivers and water bodies, landslide warnings to people who lives in the hilly terrains, and information relating to road blocks to traffic commuter as SMS alert messages.

This paper describes how a cloud based interoperable framework based on Service Oriented Architecture and Web Service (WS) based technologies were developed for integrated delivery of multifarious public services on mobile devices. As Tarek [8] points out, lack of interoperability is one of the major impediments that hamper the successful rollout of m-gov services and projects. The framework developed will address this lacunae and enables seamless rollout of m-gov services. It discusses the various technological components required for developing a mobile service delivery platform (MSDP) and its role in providing greater agility in synchronizing two or more disparate e-governance applications, allowing a group of departments/organizations to provision and consume services as a shared infrastructure. The paper also illustrates a case study on how MSDP and Quick Response code technologies were leveraged for provisioning emergency management services using crowdsourcing.

2 Related Works

A variety of work has been carried out in the realm of mobile governance. The following list is not intended to be a complete list of all related solutions, but a few samples drawn in from the m-gov services deployed, mobile technologies in emergency management/disaster management and early warning system and use of quick response code (QRC) in e-gov.

2.1 Mobile Government Services

Mobile government, also labelled as mobile e-Government or "m-Government" or m-gov, emerges as a significant area of intervention for administrative and government-linked actions leading to a positive transformation by enhancing the relationship between public services and their users; with the aid of mobile communication technologies [9]. The introduction of mobile communication technologies has brought in hither to unimagined possibilities for governments to extend services to most of the population. This has also enabled information access and related services of public and private organisations beyond time and space OECD [10]. European Union Regional Development Fund, in a white paper published, points out that m-gov helps governments to create an integrated digital nervous system EURDF [11]. The benefits afforded by m-governance such as immediacy and convenience reduce the previous barriers to public service operations, encouraging citizens or service providers to make use of this technology. Citizens are motivated to utilize public services more easily and can contribute to the government's decision process. Encouraged by the benefits, mobile phones that were primarily used for voice communication have come to be effectively employed in multiple sectors such as healthcare, agriculture, emergency management, law enforcement, financial inclusion, and entrepreneurial development, to name a few. These changes have opened up wide vistas for development and empowerment in developing countries. As interest in m-services increases and more services are routed the mobile way, it becomes essential that all citizens can enjoy the benefits of these services equally.

2.2 *Delivery Models of M-Governance*

OECD/ITU [12] identifies four delivery models of M-Government as described below.

(a) Government to Citizens (G2C): This will be by far the most intense and far-reaching of all delivery models. The Mobile apps of G2C will enable the government to push real-time information and alerts and provide interactive services to complete transactions. An example is The mCity project, city of Stockholm [providing health care, education, tourism, and business].

(b) Government to Employee (G2E): The government being the biggest employer, its employees are its asset and their well-being, cooperation, rigorous engagement, and continuous communication are essential for practicing good and responsible governance. The mobile apps of G2E will facilitate the mobile workers to carry a smartphone or a tablet to carry out their tasks with much more efficiency in the field. This will greatly facilitate staff works in remote locations and in difficult circumstances. Example e-challan—a G2E application developed by Motor Vehicles Department, Government of Kerala which equip patrol officers with tablet PCs to quickly conduct queries regarding offending drivers' license and vehicle information which increases the efficiency of the mobile traffic units. The location of each mobile traffic unit can also be found and dispatched to a particular location such as a traffic incident instantly.

(c) Government to Business (G2B): Private sector is the revenue earner and execution arm of government for carrying out almost all the projects. G2B applications will help in monitoring the projects and to promote a business-friendly environment, which would bring more investment and help in the overall development.

(d) Government to Government (G2G): This will promote intra- and intergovernmental collaboration and data sharing by eliminating duplication of efforts. The major beneficiaries of G2G services are security and law enforcement agencies, coordinated inspection and emergency response teams, where real-time data is essential from staff working in multiple agencies so as to get a common operative picture. Example, Mobile Crime and Accident Reporting Platform (MCARP) is G2G solution developed for Kerala police force for the efficient tackling of crime, accidents and traffic issues. MCARP helps the police to not only control incidents such as riots but also record traffic violations and untoward incidents, for maintaining apposite visual evidence. Images are captured using mobile phone cameras by policemen and uploaded to the central server instantly via MMS/GPRS which can be shared with Fire and Rescue services, Ambulance services, etc. [3].

2.3 Mobile Application in Emergency Management

Due to technological advancements made in Mobile Technologies (MT) [13] and the ability of systems to converge multiple agent systems has enabled the use of MT in Emergency Management (EM) [14]. The tracking capability provided by GPS enabled mobiles, cell tower triangulation, etc., has enabled to track population in disaster zones and to provide them with timely predisaster warnings and updates about the disaster. Similarly, mobile enables a two way interaction; users marooned at the disaster sites can air videos, pictures, messages and location and progress of the disaster to emergency management agencies [15]. The feature rich web 2.0 applications aided by geo-tagging and geo-coding capabilities of smartphones have enabled users to transmit videos, pictures and audio almost in real-time with geolocation codes embedded which makes it easy to represent the same in 3D by plotting the contents in a map [16, 17]. The delivery of right information at the right time is a critical aspect in EM. Mobiles serve as a two way communication device in disseminating information from and to the site of a disaster. Zheng et al. [18] Early warnings can be generated for predictable disasters like flooding, cloud burst, cyclones, etc. Since the reach of mobiles are good, it can be an ideal device to deliver early warning messages and enables to spread the message across to their neighbours, relatives and friends easily and save precious lives [19]. Without timely warnings, the consequences of an impending disaster will be tragic and costly.

2.4 Q-R Codes in e-/m-Governance

A quick response code (QRC) is a type of matrix barcode (or two-dimensional code) patterns of small black squares arranged on a white background that encodes text strings. They are able to encode information in both vertical and horizontal direction, so can pack several times more information than the one dimensional barcodes. QRC can store information in the form of a two-dimensional black-and-white square barcode and can encode URL addresses, contact information, coupon no, maps, call a phone number, and/or send SMS messages and emails, etc. In order to access the encoded data in a QR code, a built-in smartphone camera is used to capture an image of the QR code and then decode it using QR code reader software. Forty different versions of QR codes with different data capacities are in existence. QRC can be small (on a business card) or huge (hoardings on the side of a building). The technical specifications of QRC are laid down in the ISO-18004 standard so it's an open standard and the same all over the world. Largest standard QRC is version 40, 177 × 177 modules in size and can hold up 4296 characters of alphanumeric data. Though there exist many 2-D barcode formats, Alapetite [20] the QRC is the most popular and has found applications as diverse as product identification, marketing, entertainment, tickets/coupons, etc. Rukzio et al. [21]. This was attributed mainly to the ease of use—QRCs can be read with any smart phone with a camera and one of the free QRC reader apps installed. MobileBarcodes.com [22]. Many open source libraries and applications are readily available for free download which helps in generating QRC from a variety of data source on condition that the data is not crossing prescribed limit of characters which can be represented using a QRC.

The usage of mobile barcodes is increasing in the government sector [23]. QRC have helped both government and private organisations to effectively distribute valuable information to the public, market the services, programs offered and also for receiving real time data from the field, etc. QRCs are used to increase citizen participation [24] and to navigate users through park trails and museums [25]. Furthermore, they are used as supplementary material for education and within games [26]. Huang et al. [27] used the example of a museum guiding system which allows tourists through handheld devices to browse exhibition contents including text, pictures, audio, and video to study multi-platform information transfer. QRCs are also used to share information between people who participate in the same social event to share information in order to support the learning process [28]. QRC was used as the calibration mechanism for an augmented reality system by linking QRC with geocodes [29]. In short, QRC is an excellent tool which offers the general public as well as government officials a shortcut to access web links, add contacts, navigate maps, and much more without the hassle of remembering the codes or keying in codes and URLs.

3 Materials and Methods

As Antovski et al. pointed out, "mGovernment is largely a matter of getting different and disparate information systems owned and operated by different departments and organisations geared to interoperability with citizen's mobile devices" the same is partially true. The real landscape is that many departments which have automated their workflows and provisioned e-services independently has resulted in the creation of "information silos" which does not talk to each other resulting in failure to achieve the highest level of e-gov maturity "Connected governance model" put forward by United Nations [30]. It is characterized by achieving horizontal connections (among government agencies), vertical connections (central and local government agencies), infrastructure connections (interoperability issues), connections between governments and citizens, connections among stakeholders (government, private sector, academic institutions, non-government organizations and civil society), and suggests that smart governments should provide a standardized and integrated unified access point—"one shop stop" platform to the citizens through interoperability of legacy systems and provisioning multiple channel service delivery including mobile devices [31].

If this is the kind of disparity exists in the e-world, the problem of achieving interoperability in the mobile space will be still complex due to the fact that internet is not owned by anyone; but mobile networks are owned by different Mobile

Network Operators (MNO) and the kind of device used in mobile networks also vary greatly in their processing power and display capabilities. It is in this context to simplify the access it was decided to develop a Mobile Service Delivery Platform in the Cloud (MSDC), with the following objectives: to provide device and MNO agnostic services to the citizens, uniformity in services across government departments for citizens, reduced timelines for launching new services, avoid wastage of resources by duplication of efforts, overall expense reduction with economies of scale, provide open API's for interoperability and legacy and third party system integration [3].

3.1 Mobile Technologies and MSDC

MSDC combines the three channels of mobile communication, viz., voice, signalling, and data and a wide range of applications (APPS). The voice channel providing IVR and OBD, signalling channel providing SMS, MMS, USSD, and cell broadcast capabilities and the data channel providing WAP/GPRS, 3G and 4G support, as well as Government App Store for downloading mobile applications has to be developed for specific mobile platforms. MSDC provides a unified view to the citizen and the government departments. These channels helps in either pushing the information/data to the citizens by the departments or allow the citizens to seek (pull) information/data from various departments. In short, MSDC serves as an Infrastructure as a Service (IaaS) platform for the departments and citizens to consume.

(i) Voice channel: The voice remains an important function for telecommunications, be it mobile or land lines, and has the unique distinction that it works on all telephony networks and all phones; it has greater capacity for information exchange; and can be consumed by people with lower literacy; voice is a familiar and trusted communication channel; Out Bound Diallers (OBD); and Interactive Voice Response (IVR) systems can be developed easily in multiple or local languages using open source tools like Asterisk.

  (a) Out Bound Diallers (OBD): It helps in initiating simultaneous calls and can be scaled up to dial a very large number of land-line as well as mobile subscribers simultaneously and relay prerecorded voice messages and alerts.
  (b) Interactive Voice Response (IVR) is a technology that allows a computer to interact with humans through the use of voice and dual tone multi frequency tones input via keypad after which they can service their own inquiries by following the voice prompts. IVR systems can respond with prerecorded or dynamically generated audio to further direct users on how to proceed. To setup an IVR one needs a computer and a special hardware called telephony card to understand the DTMF signals produced by a phone and an IVR software, to play the prerecorded messages and menu options. In the context of m-gov, the IVR is intended to serve the C2G and G2C services through

which status enquiries for a large number of services can be automated and the requisite information can be provided to the service seekers and also for getting feedbacks/survey information from public on a 24 × 7 basis. IVR systems deployed will have surges in consumption pattern and have to be scaled up on demand to handle large call volumes expected from citizens and a candidate for cloud deployment.

(c) Voice XML (VXML) Voice applications can be developed and deployed using W3C's standard format for interactive voice dialogues between a human and a computer. VXML-based applications can process millions of telephone calls daily to provide audio-guides, driving directions, information about artefacts in a museum, etc. Additional modules, such as text-to-speech and speech recognition will enhance the service delivery greatly especially for the differently abled citizens.

(ii) Signalling channel: The service offerings in this segment include the most popular mobile applications, viz., Short Messaging Service (SMS), Multimedia Messaging Service (MMS), Unstructured Supplementary Service Data (USSD), and Cell Broadcast.

(a) Short Message Service (SMS) represents a communication protocol which permits the exchange of short text messages up to 160 characters between mobile devices. SMS is a globally accepted service and can be categorized into the following types Push SMS (Mobile Terminated) and Pull SMS (Mobile Originated).

(b) Multimedia Message Service (MMS) is an extension of messaging services similar to SMS for data transfer, which enables sending rich content, such as images, video or audio files back and forth between governments and citizens.

(c) Unstructured Supplementary Service Data (USSD) messages are transferred directly over network signalling channels. It is a session-based service unlike SMS, which is a store and forward service. USSD can be used by the user to send command to an application in text format. USSD acts as a trigger for the application. Currently, this service is mainly being used for checking balance in financial accounts and mobile prepaid recharge. USSD is more interactive as compared to SMS but nothing is stored on the phone. This can be very useful for submitting requests for a service through an interactive menu and for tracking their status and also has great potential for mobile banking, accessing news services, submission services, feedback, voting, and directories. With interactive navigation, USSD is fast and allows for mass usage. However, messages cannot be saved or forwarded, the codes may be difficult to remember, and usage is not always reliable due to session-based timeouts, [32].

(d) Cell broadcast is an under-utilised signalling channel on the air interface between the handset and the transceiver and in basic terms allows the GSM network to function as a traditional radio, so every handset in the area can pick up that broadcast message. Furthermore, cell broadcast has the unique feature of not being subjected to or contributing to network congestion. Since it uses a signalling channel and is not affected by traffic channel availability, it does not carry voice traffic.

Cell Broadcast has several other key features that make it readily applicable as a public alerting tool. These include

- Message Display—The message is displayed on the handset with no user interaction and a distinct warning tone is sounded.
- Message Delivery—Cell broadcast works on a broadcast, i.e., one-to-many basis; one message can be sent to millions of devices quickly and the message is broadcast to all connected handsets within a designated target area. The area can be as large as an entire network or as small as a single cell.
- Anonymity—Another key advantage of Cell Broadcast is that the recipients remain anonymous since it does not require registration of numbers or maintenance of a number database and messages are sent to all users within a geographic area.

(iii) Data channel–Technologies including WAP, GPRS, APPS, etc., enable data transmission using mobile networks. With the smart phone shipment over-taking ordinary handsets, a large number of applications designed for smart phone platforms (Android, Blackberry-OS, iOS, Windows Phone, etc.) utilises data services to transact.

In addition to the above mobile technologies, open API-based mapping and other such open standards-based technologies can be plugged into the MSDC.

3.2 Architecture of MSDC

A layered approach has been followed in developing MSDC. It has been developed with the goal of providing inclusive access, keeping in mind the diversity of access devices used by citizens ranging from basic phones, feature phones, and smart phones. So, the MSDC should support all such devices and provide the departments with an easy to use interface, that requires less technological skills and ensure interoperability and data sharing capabilities between technologies. Free and Open-Source Software have been leveraged to the maximum extent possible and also Open Standards. Cloud offers the government department's on-demand ICT infrastructure provided as a service (IaaS) to host departmental application with ease and provision user services, while for the user it provides virtualized infrastructure and a unified access platform to consume mobile services heterogeneous mobile devices whenever and wherever. MSDP has been designed using open standards to ensure integration capabilities are compatible and seamlessly integrated with any existing legacy or third party system when on demand scaling up is initiated, without disturbing the existing services and thus capable of accommodating future expansions with ease, enabling to derive optimal output from the infrastructure investment.

The uppermost layer is the interface engine, which utilizes XML-based device and content profiles to render the contents, either in (HTML, WML, HDML, etc.) formats, properly in consonance with the access device capabilities.

The second layer acts like a proxy, which is basically a short-code or long code number where all the signalling channel services converge, e.g., the short code 537252(corresponds to spelling of Kerala in non-qwerty mobile keypad) for Kerala specific m-services, 166-for national m-gov services World Bank, IC4D (2012). This layer intercepts the requests from various wireless devices subscribed to services from different MNOs. A WS will then push the data received in different formats at the short-code number of a particular MNO to this layer, the intelligent application middleware parses the forwarded data and extracts the metadata keywords and routes the requests received in different formats specific to the channel (SMS/USSD/IVRS) used, to specific gateway (SMS Gateway/USSD Gateway/Voice Gateway). These gateways, after processing the data, forwards the same to the MSDC which applies the various routing logic and routes the data to the appropriate government department directly or through standards-based messaging middleware. The reply from the department is sent back to the citizen through the same path, and the pull request from the citizen is fulfilled by way of receiving information in appropriate formats. This layer also accommodates the caching module, which acts as a repository for temporarily storing the contents optimized for a particular device, based on the XML profile. This helps in dynamically serving the contents as and when the same contents are sought later. The metadata storage facility, in the previous layer where in the XML-based device profiles and content profiles are stored is consulted before rendering the content in a proper format in accordance with the device capabilities. This layer is responsible for rendering the contents derived by invoking a single WS or through service orchestration–combining two or more WS from the next layer-the messaging middleware (State and National Service Delivery Gateways).

The MSDC developed which is embodied on the principles of Service-Oriented Architecture, will make the heterogeneous ICT infrastructure already deployed as part of independent departmental information systems interoperable and thus allowing seamless intra-departmental flow of information possible, and aids in presenting the user with contents rendered based on single (WS) or through service orchestration–combining two or more WS. For example a G2E service ‘e-challan’ deployed in the mobile devices of Police Patrol officers enables to verify the authenticity of the vehicle’s identity, tax paid as well as the validity of driver license, and any historical offences/traffic rule violation committed by the driver-the patrol vehicle can fine the driver on the field and show evidence of his violation (videos and pictures). The application combines WS offered by the Motor Vehicle Department (MVD) who are the custodians of vehicle and licence data and combining with WS from police CCTV-based violations database of the Police Department, before the fine is collected another WS will check the MVD as well as Police databases as to there is any un-paid offences and the pending fines will also be added to the current fine, which is presented to the patrol officer’s PDA. Once the fine is paid another WS will update the offence database in both police as well as MVD databases. As you can see from the example the same infrastructure is being leveraged for pushing and pulling information using API calls. Now the author focuses on how MSDC was leveraged for emergency management. Weather alerts, natural disaster alerts, etc., are aired to the concerned people using the SMS channel, and if the magnitude of the disaster is high and the response time is too low, cell broadcast channel will be used.

3.3 Geo-referenced Emergency Management System

The services offered by government departments in the 'e'lectronic, 'm'obile, and 'g'eographical domains exist as islands and there is no such integrated system in existence. This paved the way for leveraging MSDC by the Emergency Management Agency (EMA) and to bring all the departments which deals with emergencies in an equal plane to: Provide guidance, safety and overall security to the citizens. Provide security to the weaker and vulnerable strata of the society especially elderly staying alone at home, women and children at bus stops and other crowded public places by utilizing innovative QRC technologies; provide a dashboard regarding various crimes and accidents in a spatial perspective and graphically to reveal the hot spots. Use the data generated to intelligently reroute traffic, reduce accidents, identify crime hot spots and target resources for crime reduction, and act as a channel for effectively communicating with the citizens. During an emergency of higher grades which involves large number of casualty mutipronged action can be initiated in tandem, where in a single alert that reaches the Control Room, can be passed onto fire and rescue services, ambulance services, hospital casualties, and the Bureaucracy in real time.

3.4 Methodology

Step 1 The master geo-tagged mapping table creation was the first step, Global Positioning System (GPS) coordinates of vulnerable locations and popular locations like tourist/pilgrim locations were mapped using GPS navigator equipment (GARMIN etrex VISTA) and table was populated. Synergy was also drawn with other departments like tourism, forests, health, telecom operators, etc., by porting the geo-referenced location details shared by them to our database. The existing Short-code number "537252"–easy to remember, spelling of Kerala in non-qwerty mobile keypad, opened by Government of Kerala was used. The short-code number and formats for making a call or sending the SMS's were popularized by way of targeted campaigns through FM Radios and other Channels like TV, print, etc.

Step 2 The second step involves generating four sets of QRCs encoded with [E] mergencies, [A]ccident, [W]omen attack and [Fire] corresponding to each and every location code in the database. The Barcodes were generated in a batches using TEC-IT QR-Code Studio–a freeware from, TEC-IT, Datenverarbeitung GmbH, Austria and the sample QRCs generated are as given in Fig. 1, when scanned it triggers sending SMS to "537252" configured in the MSDC.

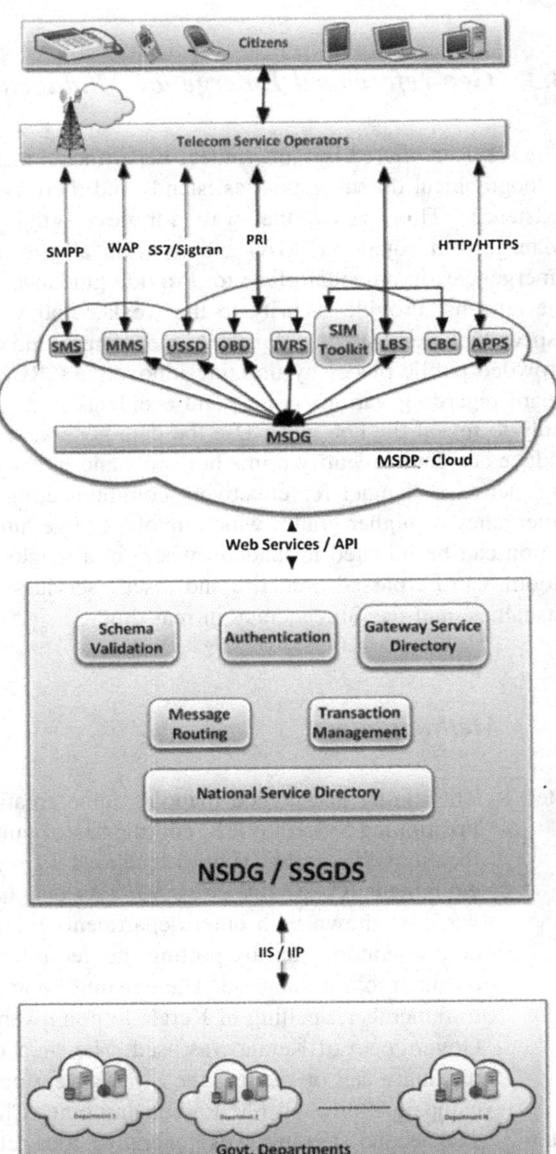

Fig. 1 Technical architecture of MSDC

Fig. 2 QRCs generated using QR-Code studio with location and emergency codes

The QRCs thus generated along with graphics which conveys the type of event as depicted above was pasted in vulnerable locations or displayed as part of hoardings near the spots with the number SMS has to be triggered, unique location code assigned to the location (say 1001 as in Fig. 2) and the incident type code (E, A, W F).

Step 3 **(User initiated)** If a person meets with an emergency at a specific location one has to scan the QRC with his smartphone and it in turn automatically triggers the sending of SMS to "537252", with the details encoded in the QRC getting translated into a predesignated format. As soon as the SMS hits the short-code, it pushes the same to the XML messaging-based framework MSDC, which has inbuilt logic to parse the SMS message received and make out the location code, nature of the event, as well as the sender mobile number and stores the details into the database. Based on the location code received the application references the mapping table and extracts the geo-coordinates of the corresponding location. Using simple "GMap API" the actual location is plotted in the map in juxtaposition with the location of GPS enabled Control Room Vehicles (CRV) patrolling the streets, and based on the nature of the event (fire, accident, etc.) an audio alert will be played which will alert the Control Room Operator (CRO). This will help the CRO to identify one or two CRV's nearest to the event location and command can be passed to take action immediately.

MSDC not just caters to smartphone users but also addresses the ordinary mobiles and even land line users, thus ensuring open and equitable access. If the user does not carry a smartphone one can dial the short number (say 10801 for emergency 10802 for Accident, 10803 for women attack 10804 for fire 1080 for any other) and an Interactive Voice Response (IVR) system built into the framework based on open source Asterisk software, and digium Voip-card gets activated and prompts the dialer to enter the location code and the code corresponding to the type of the incident, which is also printed below the QR-code in a human readable format, as depicted in Fig. 2.

4 Results

4.1 Pre-implementation Situation—"As-Is"

Before the implementation of this innovative multimodal communication system, in case of an emergency citizens need to contact the Police Control Room over phone for necessary assistance. The CRO asks over the phone location details, "what and when" type of questions and the operator in-turn dials the mobile phone numbers of highway patrols or CRVs, these vehicles after getting the leads from the operator has to travel to the location and will take more time to identify the spot of the emergency. This is a pain area for the CRVs' and may lead to loss of precious time, and in case of accidents and criminal assaults this delay may lead to the death of the victim. Approximately 30 minutes is the response time in the existing system (Fig. 3).

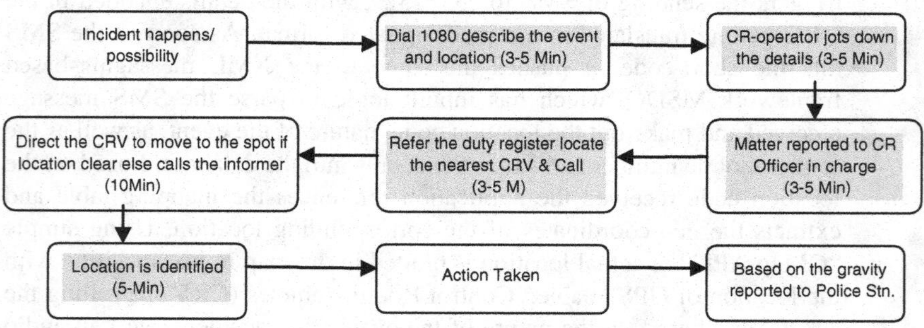

Fig. 3 Process flow and the time taken in current system

4.2 *Post Implementation Process After New System Implemented*

It is necessary that a system should be introduced in such a way that the Control room vehicles should reach the spot within no time. If a Quick Response Code which contains details of the location and a predefined number for sending SMS is pasted on vulnerable locations and is scanned by a citizen, the mobile phone automatically triggers an SMS in the predesignated format as deciphered from QRC to the Short Code Number "537252". Once the SMS hits the operator gateway it pushes the same to the XML messaging-based framework developed, the framework has inbuilt logic to parse the SMS message received and decodes it to make out the nature of the event, event location code (which is already mapped and stored in the backend) and sender's mobile number. Based on the location code received the application references the mapping table and decodes the location number to extract the geo-coordinates of the location and using simple "GMap API" the actual location is plotted in the map in juxtaposition with the location of GPS enabled CRVs who are patrolling the streets, and based on the nature of the event (fire, accident, etc.) an audio alert will be played which will alert the Control Room Operator. This will help the control room party to identify one or two CRV's nearest to the location and command can be passed onto them to take action immediately.

The framework developed not just caters to users with smartphone but also addresses the user base with ordinary mobiles or even land lines, thus ensuring open and equitable access. If the user does not carry a smartphone one can dial the short number (say 1080) and an Interactive Voice Response (IVR) system built into the framework based on open source Asterisk software, and Digium® voice over IP card prompts the user to enter the location code followed by incident type code, which is also printed below the QR-code displayed in public in a human readable format.

It may be noted that by implementing the new innovative system considerable time and effort are saved by eliminating those non-productive processes coloured yellow in the process flow depicted above by applying Business Process Re-engineering (Fig. 4).

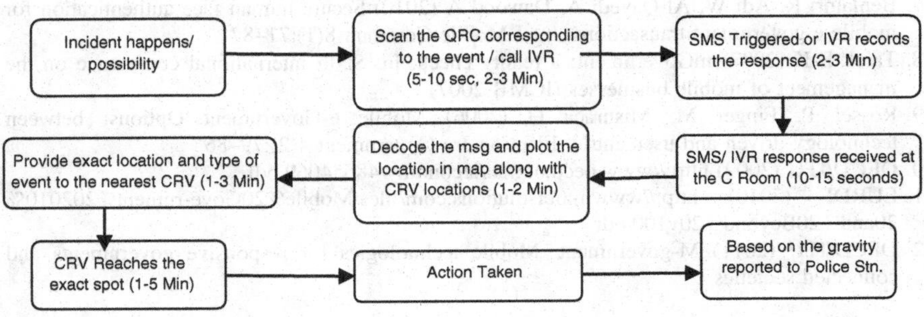

Fig. 4 Process flow and the time taken by implementing new system

5 Discussion

This paper is an attempt to use MT, GIS tools like Google Maps and QRC in an integrated fashion and in a collaborative way during emergency situations. It also proves that MT allows governance to shift from one-way service delivery to a more collaborative, codesigned, and co-created model and through crowdsourcing allowing the citizens to play a more active role as a participant rather than being passive recipients. The framework developed also allows citizens to add newer geo-coordinates of potential 'hot spot', which after validation will be accepted into the database and a new location code is granted and thereafter QRC can be printed and pasted at the site. The authors will like to end this paper with a note that technological obsolescence is happening at a rampant pace and any technology-based solution that is suggested today may be obsolete within a couple of years, so it is suggested to innovate and re-invent so that it brings efficiency, economy and benefit all the stakeholders in the emergency management ecosystem. As the study was restricted only to the use of MT the same may be expanded by the use of other wireless technologies, sensor network integration and to mine the fast growing social media feeds using business intelligence tools.

References

1. Al-Khamayse S, Lawrence E, Zmijewska A (2007) Towards understanding success factors in interactive mobile government. http://www.mgovernment.org
2. Veljkovic N et al (2010) E-Local self-government in Serbia. In: EGOV 2010 Conference Proceedings, August 29–September 2 2010
3. Sabarish K (2011) Mobile governance: the kerala experience and insights for a comprehensive strategy. Eur J ePractice www.epracticejournal.eu No 12 March/April 2011. ISSN:1988-625X
4. ITU (2015) World Telecommunication/ICT Indicators database. http://www.itu.int/en/ITU-D/Statistics/Pages/publications/wtid.aspx. Accessed on 01 April 2015
5. Kushchu I, Kuscu H (2003) From e-government to m-government: facing the inevitable. Paper Presented at the European conference on e-government (ECEG 2003). July 3–4, 2003. Trinity College, Dublin
6. Antovski L, Marjan G (2005) M-government framework. In: Kushchu I, Huschu M (eds) Proceedings EURO mGov 2005. Sussex, UK, pp 36–44
7. Benlamri R, Adi W, Al-Qayedi A, Dawood A (2010) Secure human face authentication for mobile e-government transactions. Int J Mobile Commun 8(1):71–87
8. Tarek E-K (2007) mGovernment: a reality check. In: Sixth international conference on the management of mobile businesses (ICMB 2007)
9. Rossel P, Finger M, Misuraca G (2006) Mobile e-Government Options: between technology-driven and usercentric. Electron J e-Government 4(2):79–86
10. OECD/ITU (2009) http://www.oecd.org/sti/ieconomy/48654069.pdf
11. EURDF (2010) http://www.mobisolutions.com/files/Mobile%20Gove-rnment%202010%20and%20Beyond%20v100.pdf
12. OECD/ITU (2011) M-government: Mobile technologies for responsive governments and connected societies

13. Araujo F, Borges M (2012) Support for systems development in mobile devices used in emergency management. In; Proceedings of the 2012 IEEE 16th international conference on computer supported cooperative working design (CSCWD), pp 200–206. IEEE
14. Singh VK, Modanwal N, Basak S (2011) MAS coordination strategies and their application in disaster management domain. In: 2nd International conference on intelligent agent and multi-agent systems (IAMA), pp 14–19. IEEE
15. Ling R, Donner J (2009) Mobile phones and mobile communication. Wiley
16. Liu SB, Palen L, Sutton J, Hughes AL, Vieweg S (2008) In search of the bigger picture: the emergent role of online photo sharing in times of disaster. In: Proceedings of the information systems for crisis response and management conference (ISCRAM)
17. Roche S, Propeck-Zimmermann E, Mericskay B (2013) GeoJ 78:21–40
18. Zheng L, Shen C, Tang L, Li T, Luis S, Chen S-C (2011) Applying data mining techniques to address disaster information management challenges on mobile devices. In: Proceedings of the 17th ACM SIGKDD international conference on knowledge discovery and data mining. San Diego, California, USA, pp 283–291. ACM
19. Min GY, Shim HS, Jeong DH (2013) Ubiquitous information technologies and applications. Springer, pp 689–697
20. Alapetite A (2010) Dynamic 2-D barcodes for multi-device web session migration including mobile phones. Pers Ubiquit Comput 14(1):45–52
21. Rukzio E, Broll G, Leichtenstern K, Schmidt A (2007) Mobile interaction with the real world: an evaluation and comparison of physical mobile interaction techniques. Ambient Intelligence (pp 1–18). Springer, Berlin Heidelberg
22. Mobile Barcodes.com (2012) QR-code readers. http://www.mobile-barcodes.com/qr-code-software/
23. Using QR-codes in government (2014). http://www.708media.com/qrcode/using-qr-codes-ingovernment/
24. Lorenzi D, Shaq B, Vaidya J, Nabi G, Chun S, Atluri V (2012) Using QR codes for enhancing the scope of digital government services. In: Proceedings of 13th annual international conference on digital government research, pp 21–29
25. Rouillard J, Laroussi M (2008) Perzoovasive: contextual pervasive qr codes as tool to provide an adaptive learning support. In: Proceedings of the 5th international conference on soft computing as transdisciplinary science and technology, pp 542–548. ACM
26. Ceipidor UB, Medaglia CM, Perrone A, De Marsico M, Di Romano G (2009) A museum mobile game for children using QR-codes. In: Proceedings of the 8th international conference on interaction design and children, IDC 09, pp 282–283
27. Huang Y-P, Chang Y-T, Sandnes FE (2010) Ubiquitous information transfer across different platforms by QR codes. J Mobile Multimedia 6:3–13
28. Pirrone D, Andolina S, Santangelo A, Gentile A, Takizava M (2012) Platforms for human-human interaction in large social events. In: Seventh international conference on broadband, wireless computing, communication and applications, pp 545–551
29. Nikolaos T, Kiyoshi T (2010) QR-code calibration for mobile augmented reality applications: linking a unique physical location to the digital world. ACM SIGGRAPH Posters, SIGGRAPH'10. NY, USA, ACM 144:1
30. United Nations (2008) UN e-government survey. From e-Government to connected governance ST/ESA/PAD/SER.E/112. UNDESA Division for Public Administration and Development Management, New York
31. Scholl HJ, Klishewski R (2007) E-government integration and interoperability: framing the research agenda. Int J Public Admin July 2007, 30(8 & 9):889–920
32. Wang H, Song Y, Hamilton A, Curwell S (2007) Urban information integration for advanced e-planning in Europe. Govern Inf Quar 24(4):736–754

Fuzzy Logic Control of Shunt Active Power Filter for Power Quality Improvement

S. Anjana and P. Maya

Abstract With the advent of semiconductor technology, the usage of nonlinear loads in the power system has increased tremendously thereby creating different power quality issues. Shunt active power filters are used to eliminate current harmonics and also for reactive power compensation, which are major power quality issues. In this paper, simulation study of fuzzy logic for shunt active filter is presented. Synchronous reference frame theory is used for current control and fuzzy logic is used for DC link voltage control. The output from current control and voltage control is used for generating switching pulses for shunt active filter using a hysteresis controller. The proposed scheme is simulated using MATLAB/SIMULINK. Results are compared with a conventional control scheme based on PI controller.

Keywords Shunt active filter · Power quality · Harmonics · Fuzzy logic controller · Hysteresis controller · Harmonic compensation

1 Introduction

The developments in the semiconductor technology have led to the increased usage of power electronic devices. Residential appliances, industrial equipments, adjustable speed drives, rectifiers, and inverters make use of switching devices having non linear characteristics thereby distorting the waveforms that were actually sinusoidal at the generation side [1], thus creating power quality issues in the distribution system. Some of the problems are harmonic generation, poor power

S. Anjana (✉) · P. Maya
Department of Electrical and Electronics, Amrita Vishwa Vidyapeetham, Coimbatore, Tamil Nadu, India
e-mail: anjana.s191@gmail.com

L.P. Suresh and B.K. Panigrahi (eds.), *Proceedings of the International Conference on Soft Computing Systems*, Advances in Intelligent Systems and Computing 397, DOI 10.1007/978-81-322-2671-0_92

factor, low system efficiency, heating of devices, etc. Harmonics causes many problems in power system and consumer products such as heating of the electrical equipment, tripping of circuit breaker, eddy current loss, communication interference, damage of sensitive electronic equipments, etc [2]. Hence, it is necessary to reduce the mains harmonics below the limit specified in IEEE 519 harmonic standards.

To mitigate harmonics, passive filters, active filters, or a combination of both may be used. Passive filters make use of reactive components to reduce the flow of harmonic currents in the distribution system. However their performance is limited due to large size, fixed compensation characteristics and causing resonance in the power system. Active filter which is a better choice consists of an inverter circuit, DC link capacitor and control scheme [3, 4]. This control scheme is responsible for estimating the current harmonics, maintaining DC link voltage and generating switching pulses. Figure 1 shows the basic principle of compensation using shunt active power filter (SAPF). According to the signals from the control scheme, active filter injects equal and opposite harmonics to the point of common coupling (PCC) thereby maintaining source current purely sinusoidal [5–7].

In an uncompensated system with nonlinear loads the current and voltage waveforms will be composite in nature. There are different mathematical transformations for separating fundamental and higher frequency components from these composite waveforms. One such transformation technique is synchronous reference frame theory [8]. Maintaining DC link capacitor voltage is done by PI controller in conventional control schemes. This methodology requires precise linear mathematical model. Its performance is not satisfactory under parameter variations. By using fuzzy logic to control DC link voltage the inadequacies of PI control can be eliminated. Fuzzy logic control can work with imprecise inputs and can handle nonlinearities [9, 10]. Among different current controllers hysteresis current control technique is most suitable for generating switching pulses for the switching devices of VSI-based active filter [11–14].

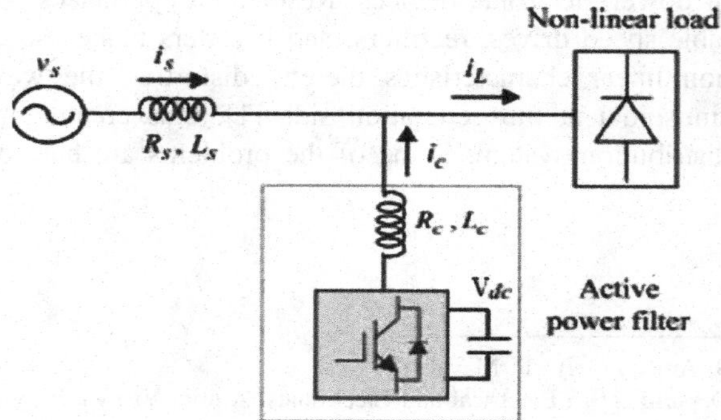

Fig. 1 Working of shunt active power filter [1]

2 Proposed Control Scheme

Figure 2 shows the proposed shunt active power filter with fuzzy and hysteresis controller for a sample power system in which a balanced three-phase supply is connected to a diode bridge rectifier with RL load. The subsequent sections describe working of different blocks in the control scheme.

2.1 Synchronous Reference Frame Theory (SRF)

The phase voltages V_a, V_b, and V_c and load currents I_{la}, I_{lb}, and I_{lc} are sensed and converted to $dq0$ components according to the equations given below.

$$\begin{aligned} i_d &= \frac{2}{3}\left(i_{1a} * \sin\ \omega t + i_{1b} * \left(\sin \omega t - \frac{2\pi}{3}\right)\right) + i_{1c} * \sin\left(\omega t + \frac{2\pi}{3}\right) \\ i_c &= \frac{2}{3}\left(i_{1a} * \cos\ \omega t + i_{1b} * \cos\left(\omega t - \frac{2\pi}{3}\right)\right) + i_{1c} * \cos\left(\omega t + \frac{2\pi}{3}\right) \\ i_0 &= \frac{1}{3}\left(i_{1a} + i_{1b} + i_c\right) \end{aligned} \tag{1}$$

The DC component present in the d axis current represents of active part of fundamental current. It is obtained by low-pass filtering. Similarly the other two components are passed through low-pass filters as shown in Fig. 3. The three phase fundamental currents are obtained by transforming $dq0$ components to abc components using the following equations.

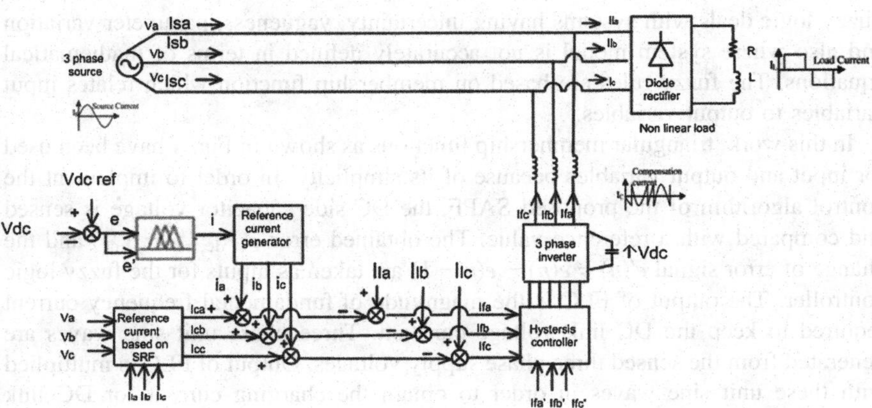

Fig. 2 Proposed control scheme

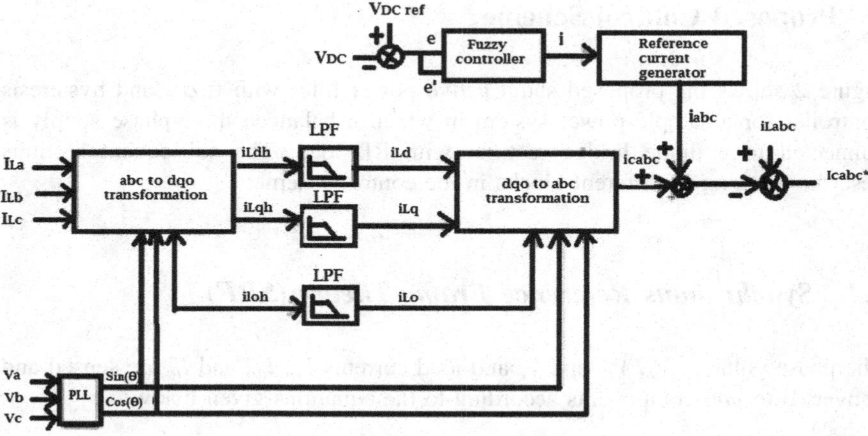

Fig. 3 Diagram for reference compensating current generation

$$
\begin{aligned}
i_a &= i_d * \sin\,\omega t + i_c * \cos(\omega t) + i_0 \\
i_b &= i_d * \sin\left(\omega t - \frac{2\pi}{3}\right) + i_c * \cos\left(\omega t - \frac{2\pi}{3}\right) + i_0 \\
i_c &= i_d * \sin\left(\omega t + \frac{2\pi}{3}\right) + i_c * \cos\left(\omega t + \frac{2\pi}{3}\right) + i_0
\end{aligned}
\tag{2}
$$

2.2 Fuzzy Logic Controller (FLC)

Fuzzy logic deals with systems having uncertainty, vagueness, parameter variation and also where system model is not accurately defined in terms of mathematical equations. The fuzzy rules are based on membership function which relates input variables to output variables.

In this work, triangular membership functions as shown in Fig. 4 have been used for input and output variables because of its simplicity. In order to implement the control algorithm of the proposed SAPF, the DC-side capacitor voltage is sensed and compared with a reference value. The obtained error $e(V_{\text{dc, ref}} - V_{\text{dc}})$ and the change of error signal $e'(n) = e(n) - e(n-1)$ are taken as inputs for the fuzzy logic controller. The output of FLC is the magnitude of fundamental frequency current required to keep the DC link voltage constant. Three phase unit sine waves are generated from the sensed three phase supply voltages. Output of FLC is multiplied with these unit sine waves in order to obtain the charging current for DC link capacitor. These currents are added with fundamental component of load currents obtained as the output of SRF block. The resulting current waveforms represent source current pattern for the given power system after compensation.

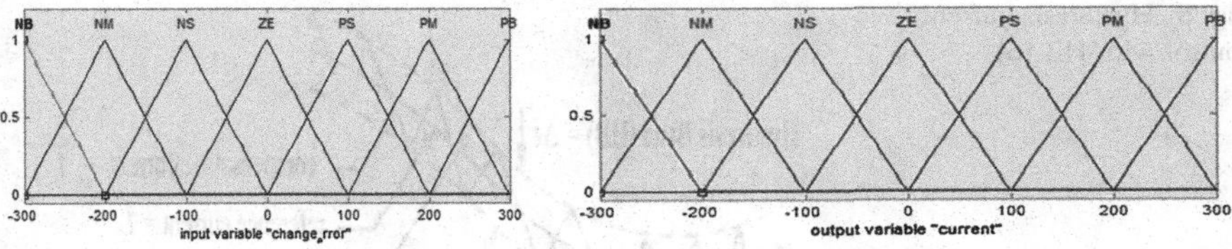

Fig. 4 Membership function for input variable and Membership function for output variable [1]

Load current is sensed after the point of common coupling. The source current profile obtained is subtracted from the load current waveforms. The resulting waveforms are the harmonics to be injected by active filter. These waveforms will be serving as the input for the hysteresis current controller as explained in the subsequent section.

The input and output variables are converted into linguistic variables. Each variable has seven fuzzy membership functions namely Negative Big (NB), Negative Medium (NM), Negative Small (NS), Zero (ZE), Positive Small (PS), Positive Medium (PM), and Positive Big (PB). A fuzzy rule base is formulated for the proposed system as shown in Table 1. Mamdani model is used for simulating for fuzzy logic controller.

2.3 Hysteresis Controller

Among different Pulse Width Modulation (PWM) techniques, hysteresis current control technique is used for generating switching pulses for the switching devices of VSI-based active filter because of its simplicity and robustness. Figure 5 shows the working of hysteresis current controller. The actual filter currents are monitored instantaneously and then compared to the reference filter currents generated by the proposed fuzzy logic-based control algorithm. A suitable value of hysteresis band (HB) is selected by trial and error. Here the positive group device and the negative group device in one phase leg of VSI are switched in complementary manner to avoid dead short circuit. The current controllers in the three phases are designed to operate independently. For phase *A*, if $I_{fa} * -I_{fa} >$ HB upper switch is OFF and

Table 1 Rule base table

Ce(*n*) / *e*(*n*)	NB	NM	NS	ZE	PS	PM	PB
NB	NB	NB	NB	NB	NM	NS	ZE
NM	NB	NB	NB	NM	NS	ZE	PS
NS	NB	NB	MN	NS	ZE	PS	PM
ZE	NB	NM	NS	ZE	PS	PM	PB
PS	NM	NS	ZE	PS	PM	PB	PB
PM	NS	ZE	PS	PM	PB	PB	PB
PB	ZE	PS	PM	PB	PB	PB	PB

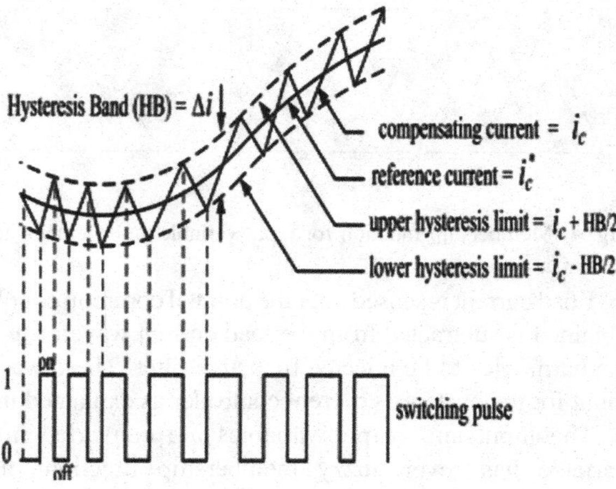

Fig. 5 Hysteresis current control with HB [6]

lower switch is ON and if $I_{fa}* - I_{fa} < HB$, upper switch is ON and lower switch is OFF. In the similar fashion, switching patterns for phase B and C are derived.

3 Simulation Results

3.1 Performance Under Steady State

A balanced three-phase source supplying power to three-phase diode bridge rectifier with RL load is the sample power system considered for simulation study of the proposed scheme. A shunt active power filter is connected to compensate for the current harmonics drawn by the load. The platform used for simulation study is MATLAB/SIMULINK. Specifications of all the components used are given in Appendix. A conventional control scheme using PI controller is also simulated for SAPF connected to the sample power system for a comparative study of the proposed scheme. It is found that the proposed scheme effectively eliminates the harmonics in the source current under steady state with constant load and balanced supply. Total Harmonic Distortion (THD) of source current is found to be reduced compared to an uncompensated system. Reactive power compensation is also

Table 2 Performance comparison: steady state

Control scheme	THD source current (%)	Power factor
Uncompensated system	30.9	0.9257
Conventional control	2.18	0.9983
Fuzzy control	3.22	0.998

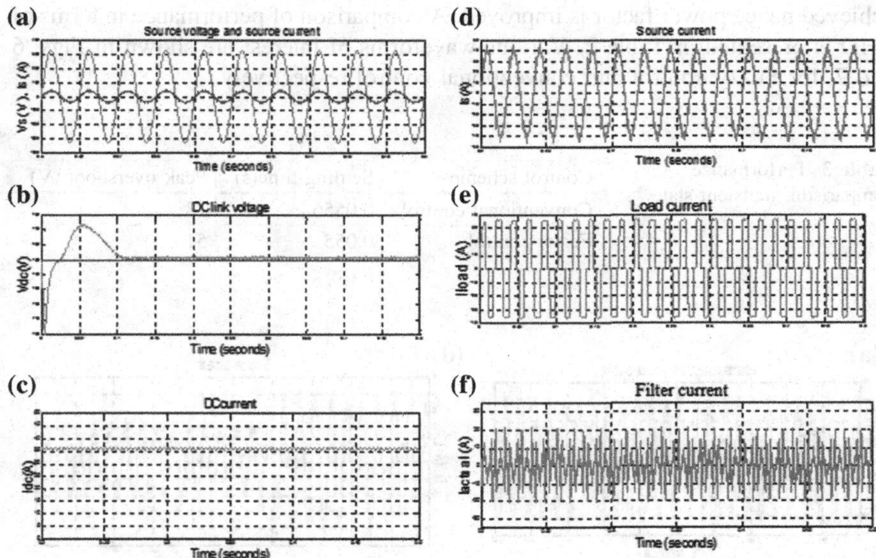

Fig. 6 **a** Source voltage and source current. **b** DC link voltage. **c** DC side current. **d** Source current. **e** Load current. **f** Filter current

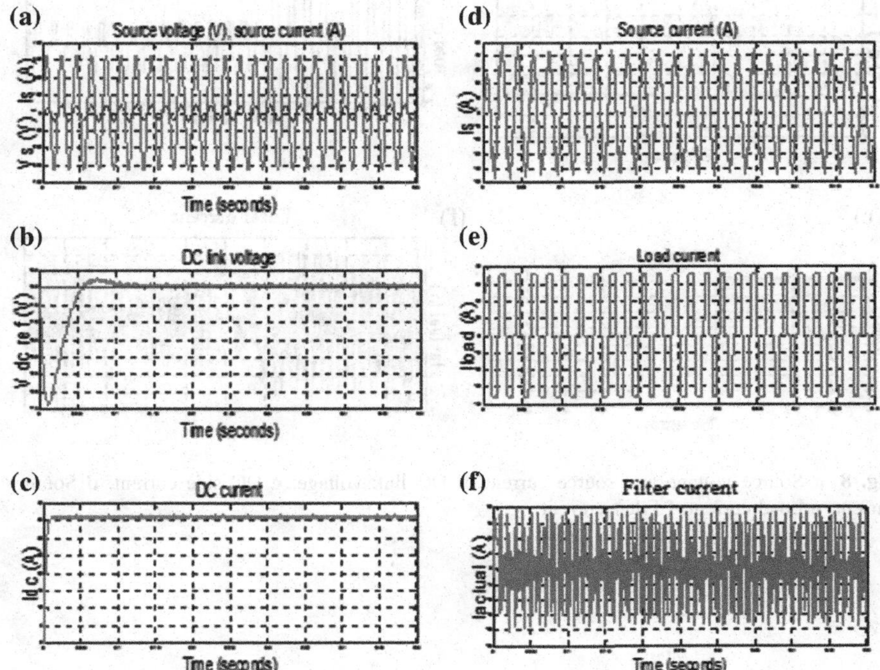

Fig. 7 **a** Source voltage and source current. **b** DC link voltage. **c** DC side current. **d** Source current. **e** load current. **f** Filter current

achieved hence power factor is improved. A comparison of performance in terms of THD is presented in Table 2. Various waveforms of interest are shown in Figs. 6 and 7, for Fuzzy control and conventional control respectively.

Table 3 Performance comparison: transient state

Control scheme	Settling time(s)	Peak overshoot (V)
Conventional control	0.0556	18
Fuzzy control	0.055	51

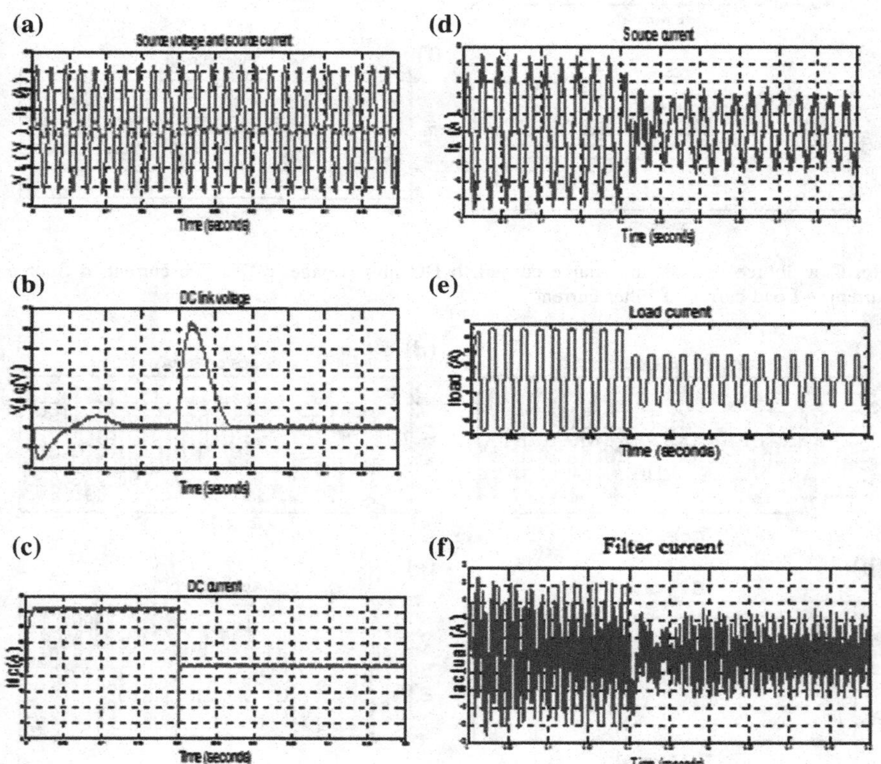

Fig. 8 **a** Source voltage and source current. **b** DC link voltage. **c** DC side current. **d** Source current. **e** Load current. **f** Filter current

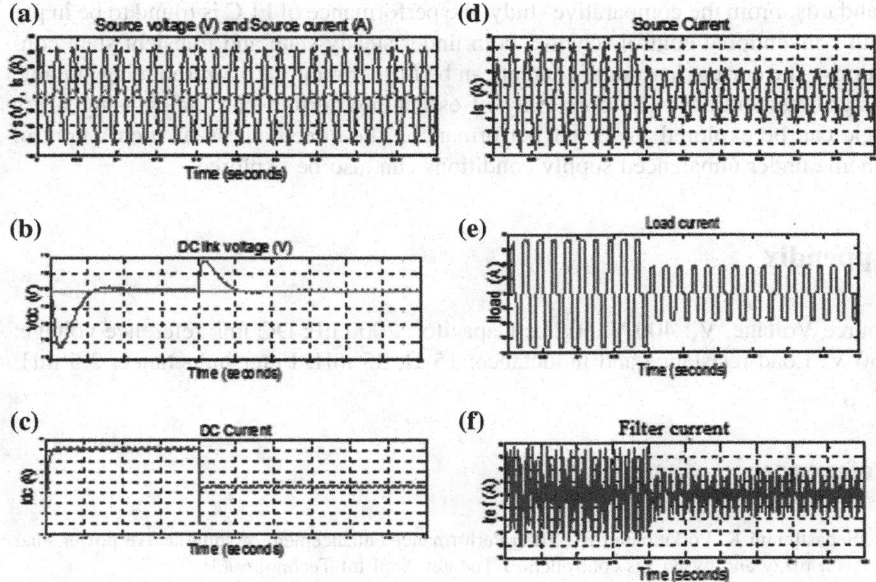

Fig. 9 **a** Source voltage and source current. **b** DC link voltage. **c** DC side current. **d** Source current. **e** Load current. **f** Filter current

3.2 *Performance Under Transient State*

Dynamic behaviour of the proposed scheme is studied by creating a load change in the sample power system with supply remaining balanced. DC link capacitor voltage *V*dc is disturbed during transient condition. It is observed that proposed scheme is capable of controlling the value of DC link voltage. Comparison of dynamic performance in terms of settling time and peak overshoot in DC link voltage waveform is presented in Table 3. Various waveforms of interest are shown in Figs. 8 and 9, for Fuzzy control and conventional control, respectively.

4 Conclusion

A control scheme based on fuzzy logic for shunt active power filter is simulated and results are presented. Under steady-state conditions fuzzy logic controller is capable of mitigating harmonics in the source current and improving the power factor. DC link voltage is also maintained constant by the fuzzy logic controller. Under transient conditions, fuzzy logic controller is capable of maintaining DC link voltage thereby achieving harmonic mitigation and reactive power compensation. It is seen that the THD of the source current is below the harmonic limits as specified by IEEE 519 standards. From the comparative study, the performance of FLC is found to be in par with conventional control scheme, both under steady state and transient state conditions. The current harmonic content can be further reduced by introducing suitable design interventions. The possibility of estimating current harmonics using fuzzy logic can be explored for better performance. The effectiveness of such a control scheme under unbalanced supply conditions can also be explored.

Appendix

Source Voltage, V_s: 400 V, 50 Hz; Capacitor: 2000 μF; DC link reference voltage: 750 V; Load resistance and inductance: 15 Ω, 35 mH; Filter inductance: 2.5 mH

References

1. Sebasthirani K, Porkumaran K (2014) Performance enhancement of shunt active power filter with fuzzy and hysteresis controllers. J Theoret Appl Inf Technol 60(2)
2. Gupta N, Singh S P, Dubey S P (2011) Fuzzy logic controlled shunt active power filter for reactive power compensation and harmonic elimination. In: International conference on computer and communication technology (ICCCT) 2011, pp 82–87
3. Belaidia R, Haddouchea A, Guendouza H (2012) Fuzzy logic controller based three-phase shunt active power filter for compensating harmonics and reactive power under unbalanced mains voltages. Energy Proc 18:560–570
4. El-Habrouk M, Darwish M K, Mehta P (2000) Active power filters: a review. Proc IEEE Electr Power Appl 147(5):403-412
5. Salmeron P, Litran S P (2010) Improvement of the electric power quality using series active and shunt passive filters. IEEE Trans Power Delivery 25(2):1058–1067
6. Shah A, Vaghela N (2014) Shunt active power filter for power quality improvement in distribution systems. Int J Eng Dev Res 1(2):23–27
7. Kumar A, Singh J, International (2013) Harmonic mitigation and power quality improvement using shunt active power filter. J Electric, Electron Mech Controls 2
8. Parithimar Kalaignan T, Sree Renga RajaTX (2011) Harmonic elimination by shunt active filter using pi controller. In: IEEE conference publications, international conference on computational intelligence and computing research (ICCIC), pp 1–5
9. Suresh Kumar B, Ramesh Reddy K, Lalitha V (2011) PI, fuzzy logic controlled shunt active power filter for three-phase four-wire systems with balanced, unbalanced and variable loads. J Theoret Appl Inf Technol 23(2)
10. Rahmani S, Mendalek N, Al-Haddad K (2010) Experimental design of a nonlinear control technique for three-phase shunt active power filter. IEEE Trans Indus Electron 57(10):3364–3375
11. Herrera R S, Salmeron P, Kim H (2008) Instantaneous reactive power theory applied to active power filter compensation: Different approaches, assessment, and experimental results. IEEE Trans Ind Electron Jan 55(1):184–196
12. Bhonsle D C, Kelkar R B (2011) Design and simulation of single phase shunt active power filter using MATLAB. In: IEEE international conference on recent advancements in electrical, electronics and control engineering, pp 237–241

13. Buso S, Malesani L, Mattavelli P (1998) Comparison of current control techniques for active power filter applications. IEEE Trans Indus Electron 45(5):722–729
14. Kale M, Ozdemir E (2005) An adaptive hysteresis band current controller for shunt active power filter. Electr Power Energy Syst 73(2):113–119

Fuzzy Logic and PI Controls in Speed Control of Induction Motor

E. Akhila, N. Praveen Kumar and T.B. Isha

Abstract Today, the most of the industries use variable frequency drives (VFDs) on a large scale. This paper presents a method for implementation of a rule-based fuzzy logic controller (FLC) for a constant speed operation of a three-phase induction motor (IM). The control strategy is based on keeping the voltage–frequency (V/f) ratio constant. A simulation study is done on a 10 HP (7.5 kW), 400 V, 50 Hz, and 1440 rpm three-phase squirrel cage IM. The system consists of a three-phase diode rectifier feeding a three-phase PWM inverter with a dc link capacitor. The inverter is switched at fundamental frequency as dictated by a fuzzy logic controller block. Actual speed of the motor and variation of speed error is the two inputs to the fuzzy logic controller block, and variation of frequency is available as the output of this block. This variation in frequency is added to the preceding sample frequency of the inverter to obtain the current fundamental frequency of the inverter. The V/f block determines the fundamental voltage magnitude of the motor corresponding to the current frequency. The same control algorithm is implemented using a PI controller. A complete simulation in closed loop of the model is carried out and the machine was found to be running at the set speed. The performance of fuzzy logic-based control is seen to be superior compared to that with a conventional PI controller.

Keywords Induction motor (IM) · Scalar control · V/f ratio · Membership functions · PI control · Fuzzy logic control (FLC)

E. Akhila (✉) · N. Praveen Kumar · T.B. Isha
Amrita School of Engineering, Coimbatore, Tamilnadu, India
e-mail: akhilaanu2234@gmail.com

L.P. Suresh and B.K. Panigrahi (eds.), *Proceedings of the International Conference on Soft Computing Systems*, Advances in Intelligent Systems and Computing 397, DOI 10.1007/978-81-322-2671-0_93

1 Introduction

Low cost, reliability, high robustness, self-starting capability, and high efficiency have made induction motor the most attractive in industry [1–3]. Induction motor is coupled and has complex multivariable structure, whereas a dc motor by virtue of its construction itself is of a decoupled structure [4]. So the implementation of induction motor needs a fast-acting processor [5]. To vary the speed and torque of an induction motor, the frequency and voltage can be varied. Either scalar control where only their magnitudes are controlled [6–8], or vector control where the phase and magnitude of the control variables are controlled [9]. Yet another control, viz., adaptive control is prevalent, where the controller parameters are continuously varied to adapt to the output variable variations [10]. Although scalar control is sluggish in nature, it is easy to implement and requires less number of resources. Scalar control technique can be either in open loop or closed loop. Closed-loop speed control has its own advantages and is required to study the steady state and the dynamic behavior of drives.

Induction motor control is highly challenging due to very fast motor dynamics and highly nonlinear motor model. Therefore, an adequate motor mode should be selected, depending on the application. Induction motor is spotted as the best candidate for variable speed applications. Even under variable load conditions, variable speed drives offer wide range of speed and fast torque response. This invokes the necessity of advanced methods for control. PI control was used for most of the applications. But certain short comings are pin pointed herein the form [11–13].

- Accuracy depends on the mathematical model of the systems
- Performance degradation with change in load, motor saturation, and thermal variations
- Sensitivity to parameter variations and operating conditions.

These disadvantages can be offset with soft computation [14] which is widely used in electric drives. They include:

- Fuzzy Logic Set
- Artificial Neural Network

Fuzzy logic embodies human-like thinking into a control system [15, 16]. Fuzzy linguistic descriptions are used for fuzzy control when applied to control of processes. In this work, a rule-based Mamdani type fuzzy logic controller is used to control induction motor in closed loop. The mathematical model of the motor is developed and parameters of the developed model decide the membership functions. The system is simulated in Simulink of MATLAB environment. The performance obtained is compared with that of a PI controller in the same environment.

2 Proposed Speed Control System

Figure 1 shows the block diagram of the proposed system. The voltage to frequency ratio is kept constant in order to avoid variation of flux. This method is cheap and easy to control the speed of induction motor. It is a combination of stator voltage and frequency control. When frequency is reduced, the air-gap flux increases and in order to reduce it, voltage is reduced correspondingly so that flux also reduces back to normal and hence preventing the core from saturation. The stator voltage and frequency are varied simultaneously to keep the ratio V/f a constant. By developing the mathematical model of the system in the form of an m file, an analytical simulation of the system is carried out and the torque-speed characteristic obtained is shown in Fig. 2. The machine details used for analytical simulation are given in Appendix.

In Fig. 2, it is seen that below the base speed, the motor works in constant torque mode with the voltage reaching the steady state value and above the base speed, since the voltage cannot be increased beyond its rated value, V/f ratio no longer remains constant and motor works in flux weakening mode and power remains constant in this mode.

In the system shown in Fig. 1, the actual speed (ω) of the induction motor and reference speed (ω_{ref}) are compared with each other; difference is given as an input to the FLC. Speed error (ω_{er}) and variation ($d\omega_{(er)}$) are the two inputs to the FLC. The memory requirements can be reduced, by quantifying both linguistic variables in a common discourse universe such that the same membership functions will be

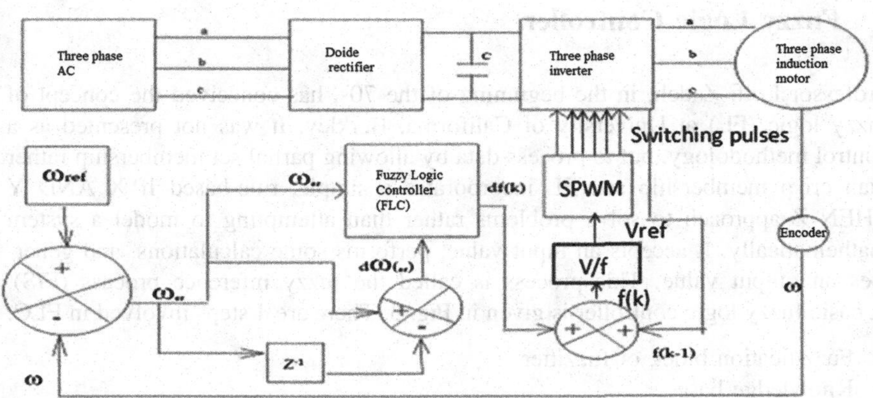

Fig. 1 Block diagram of the proposed system

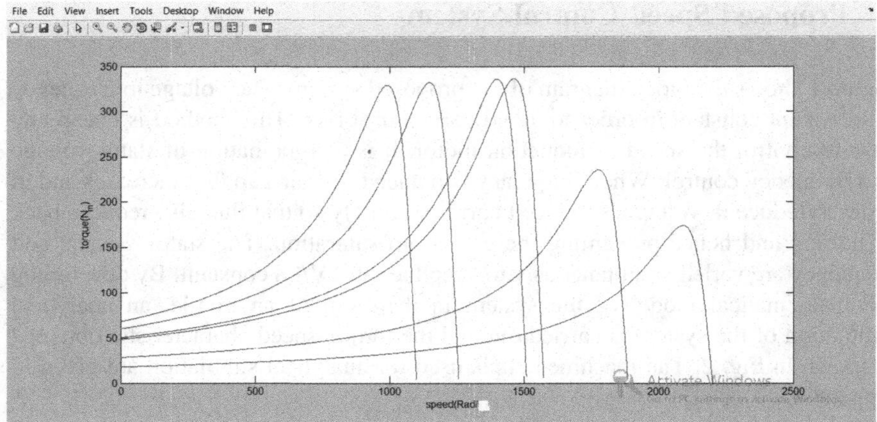

Fig. 2 Torque-speed characteristics of an adjustable speed drive

shared by them. The inputs are processed by the fuzzy system as per the rule data base and gives the variation of frequency ($df(_k)$) as output. Then, this is added to the previous frequency value ($f(_{k-1})$) of the inverter, to get the fundamental frequency (f_{ref}) of the inverter. The machine voltage is calculated from the V/f block, corresponding to fundamental frequency. The switching pulses of the inverter are generated in such a way that the inverter output is the required machine voltage at f_{ref}.

3 Fuzzy Logic Controller

Professor Lotfi Zadeh, in the beginning of the 70s, has conceived the concept of fuzzy logic (FL) at University of California, Berkley. It was not presented as a control methodology, but to process data by allowing partial set membership rather than crisp membership set. FL incorporates a simple, rule-based If X AND Y THEN Z approach to solve problems rather than attempting to model a system mathematically. It accepts an input value, performs some calculations, and generates an output value. This process is called the fuzzy inference process (FIS). A basic fuzzy logic controller is given in Fig. 3. There are 4 steps involved in FLC:

- Fuzzification block or fuzzifier
- Knowledge base
- Decision making block
- Defuzzification block or defuzzifier

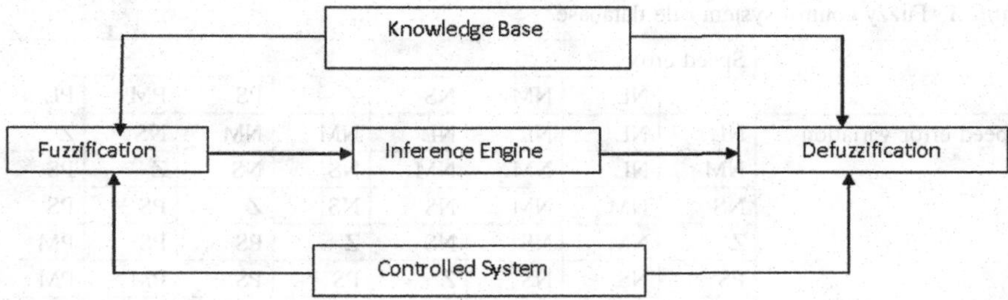

Fig. 3 Basic fuzzy logic controller

4 Fuzzy Logic Controller in the Proposed System

Fuzzy inference system having two inputs and frequency variation as output is tuned in order to obtain the best output. The range of fuzzy controller inputs, speed error, and speed error variation are characterized into seven membership functions (MFs) and seven MFs are defined for output. There are 49 rules based on which the fuzzy inference system infers the gains. Figure 4 shows the MFs of the fuzzy control system.

MFs for both inputs and output are:

- Negative Large (NL)
- Negative Medium (NM)
- Negative Small (NS)
- Zero Error (ZZ)
- Positive Small (PS)
- Positive Medium (PM)
- Positive Large (PL)

For positive and negative values of the variables, all the MFs are seen to be symmetrical. Table 1 shows the corresponding rule table for the speed controller.

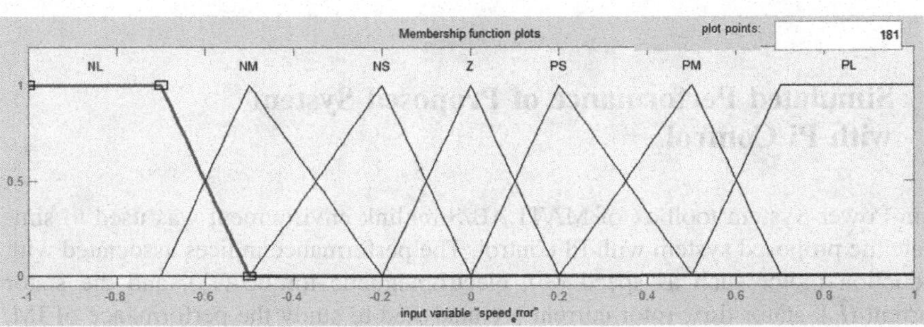

Fig. 4 MFs of the fuzzy control system

Table 1 Fuzzy control system rule database

	Speed error						
		NL	NM	NS	Z	PS	PM	PL
Speed error variation	NL	NL	NL	NL	NM	NM	NS	Z
	NM	NL	NM	NM	NS	NS	Z	PS
	NS	NM	NM	NS	NS	Z	PS	PS
	Z	NM	NS	NS	Z	PS	PS	PM
	PS	NS	NS	Z	PS	PS	PM	PM
	PM	NS	Z	PS	PS	PM	PM	PL
	PL	Z	PS	PM	PM	PL	PL	PL

The fuzzy sets of the variables and change in speed error are shown in the top row and left column of the matrix, respectively. MFs of the output variable, i.e., frequency variation are shown in the body of the matrix. Mamdani type defuzzification controller is chosen for this application. Table 1 gives knowledge database of all fuzzy rules. The control process, which is treated linguistically in a if–then structure, whose fuzzy rules are represented as follows:

If (speed error is NL) and (speed-error variation is NL) then (frequency variation is NL).

The interval values of the discourse universe were obtained experimentally, i.e., load test on three-phase IM. Table 2 shows the observations, which includes speed, speed error, torque, and rotor frequency. The range of discourse universe is fixed based on the actual load test conducted on the machine. All the membership functions were scaled into a common universe of discourse with values between −1 and 1. The universe of discourse of the "speed error" linguistic variable is designed for the [−100 100] interval. By dividing the speed error signal before the fuzzification process, it was normalized to [−1 1] r/min. Similarly for the "speed error variation," linguistic variable was adjusted to a [−50 50] r/min interval, which was divided by 50 before the fuzzification process. Similarly, the "frequency variation" linguistic variable with a [−1 1] Hz interval was multiplied by 3 to take it to the [−3 3] Hz interval. Fuzzy system should work under the universe of discourse, beyond the values, it may get saturated.

5 Simulated Performance of Proposed System with Pi Control

Sim-Power-System toolbox of MATLAB/Simulink environment was used to simulate the proposed system with PI control. The performance indices associated with induction motor such as speed (ω), electromagnetic torque (T_e), and the stator current (i_a), stator flux, rotor current are analyzed to study the performance of IM.

Table 2 Details of actual load test conducted on the machine

Sl. no	Input voltage (V)	Line current (A)	Input power (W)	S1 (Kg)	S2 (Kg)	S1 − S2 (Kg)	Speed (rpm)	Torque (Nm)	Output power (W)	Efficiency (%)	Power factor	Slip (pu)	Rotor frequency (hz)
1	415	2.65	210	1.5	0.6	0.9	1494	1.05	164.19	78.18	0.11	0.004	0.2
2	415	2.7	1020	5	0.6	4.4	1485	5.17	804.97	78.91	0.52	0.01	0.5
3	415	3	1220	6	0.8	5.2	1476	6.12	945.62	77.5	0.565	0.016	0.8
4	415	3.1	1420	7.5	1.3	6.2	1471	7.29	1123.63	79.129	0.637	0.0193	0.965
5	415	3.45	1620	11	2.8	8.2	1465	9.65	1480.15	91.36	0.653	0.023	1.1
6	415	3.8	2020	12.5	3.2	9.3	1458	10.94	1670.35	82.77	0.739	0.028	1.4
7	415	4.2	2460	15	4	11	1446	12.94	1959.80	79.66	0.814	0.036	1.8
8	415	4.45	2660	16	4.2	11.8	1432	13.89	2081.87	78.26	0.831	0.045	2.25

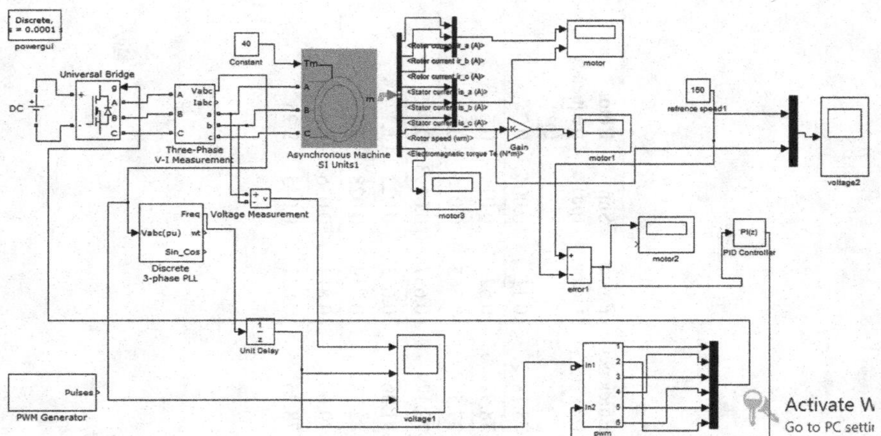

Fig. 5 Closed-loop V/f control of IM using PI control

Figure 5 shows the simulation block diagram of closed-loop V/f control of three-phase IM using PI control. This block diagram consists of DC source, three-phase IM which is fed from a PWM voltage source inverter built out of a universal bridge block. Here the reference speed is compared to the actual speed to obtain the error. The error is then amplified with the help of PI controller to obtain the frequency variation. The frequency variation is then added to the previous frequency of the PWM inverter to obtain the fundamental frequency for PWM inverter. This is given as input to the V/f controller. The V/f controller then generates the reference voltages. This reference voltage is then fed to the sinusoidal pulse width modulation block which generates the pulses to the voltage sourced inverter, such that the motor rotates at a speed equal to the reference speed. Figure 5 shows the closed-loop V/f control of IM for a reference speed 152 rad/s and 40 N-m load torque. Rotor speed is compared with measured speed and the error is processed through PI controller block to drive the frequency at which the inverter is to be switched. The voltage corresponding to this frequency is computed using the V/f block. This is the fundamental voltage output of the inverter. Results obtained from simulations using PI controller are given in Fig. 6. These results were obtained with V/f ratio 8 and for reference speed equal to the 152 rad/s and load torque 40 N-m. Fig. 6a shows the rotor speed, Fig. 6b shows the electromagnetic torque, and Fig. 6c shows the air-gap flux. Figure 6d, e shows the simulation results for stator current and rotor current, respectively.

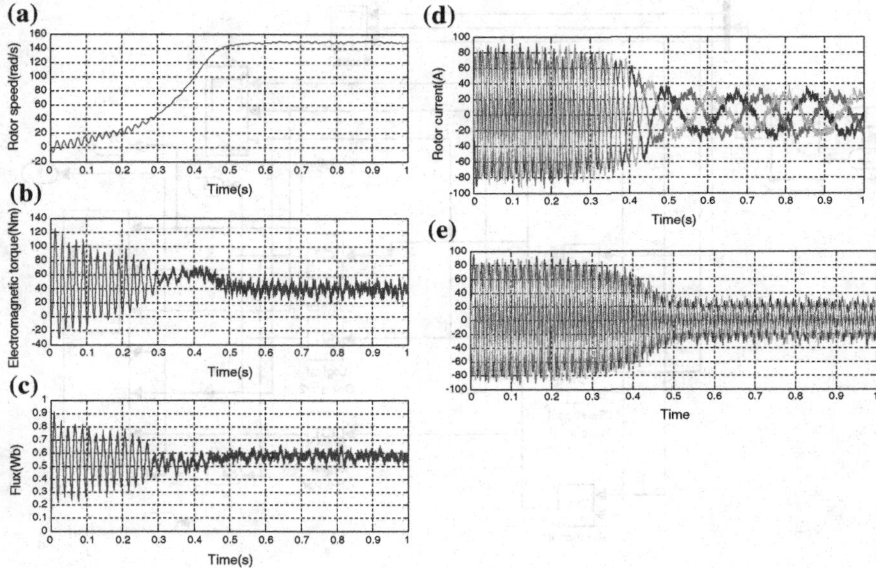

Fig. 6 Simulation results for PI controlled system; **a** speed, **b** electromagnetic torque, **c** stator flux, **d** stator current, **e** rotor current

6 Simulated Performance of Proposed System with FLC Control

The block diagram for closed-loop V/f control of IM with FLC is shown in Fig. 7. Here the reference speed is compared with the speed of the IM to obtain speed error. This is one of the inputs to the fuzzy controller to obtain frequency variation

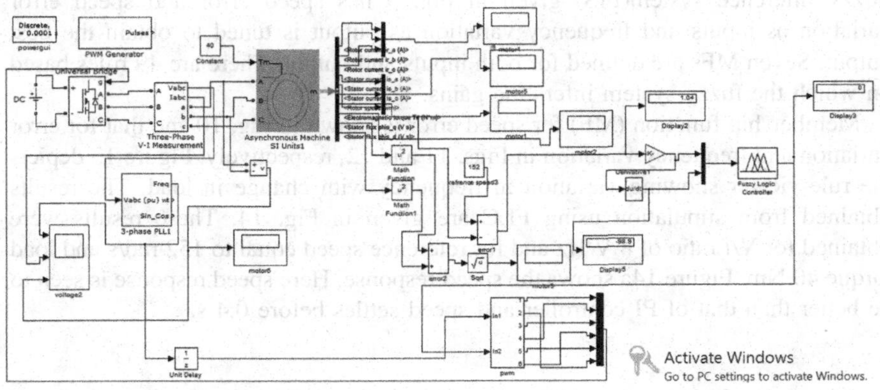

Fig. 7 Closed-loop V/f control of IM using FLC control

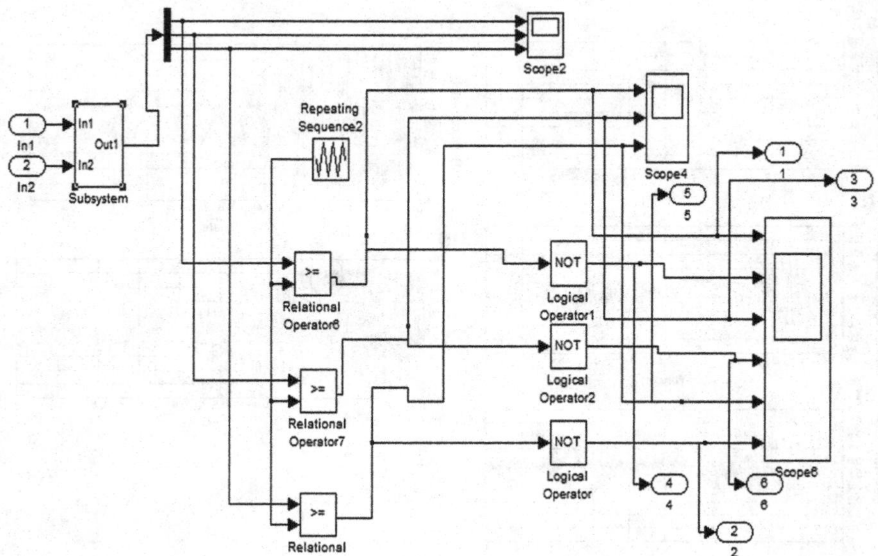

Fig. 8 Detailed diagram of the sinusoidal pulse width modulation block

as output. This signal is added to the previous frequency of voltage sourced inverter, to obtain the fundamental frequency. Rest of the control is the same as that of the PI controller. Figure 8 gives detailed diagram of the sinusoidal pulse width modulation block in simulink.

6.1 *Fuzzy Tuning*

Fuzzy inference system(FIS) given in Fig. 9 has speed error and speed error variation as inputs and frequency variation as output is tuned to obtain the best output. Seven MFs are defined for both inputs and output. There are 49 rules based on which the fuzzy system infers the gains.

Membership function (MF) for speed error is shown in Fig. 10 and that for error variation and frequency variation in Figs. 11 and 12, respectively. Figure 13 depicts the rule-viewer showing variation in frequency with change in load. The results obtained from simulation using FLC are given in Fig. 14. These results were obtained for V/f ratio of 8 V/Hz and for reference speed equal to 152 rad/s and load torque 40 Nm. Figure 14a shows the speed response. Here speed response is seen to be better than that of PI controller and speed settles before 0.4 s.

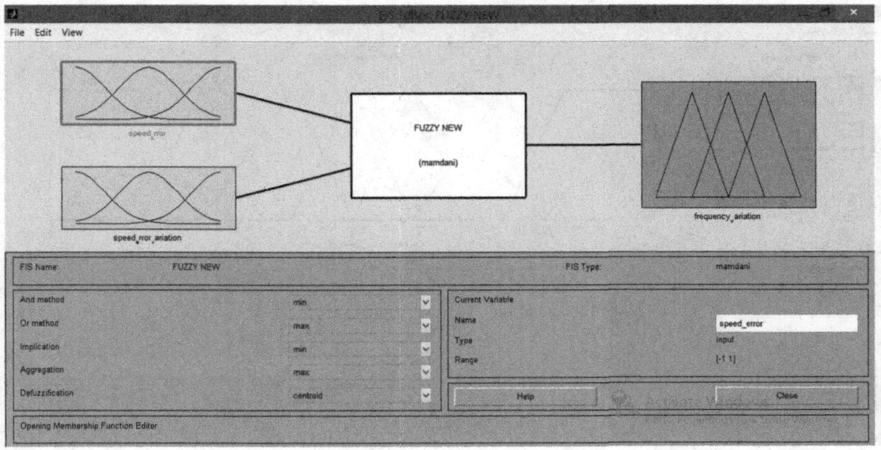

Fig. 9 FIS for fuzzy controller

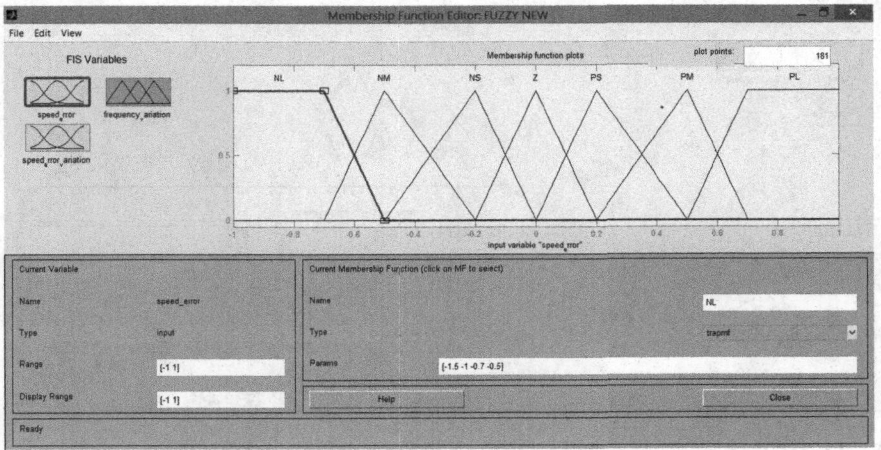

Fig. 10 MF for speed error

Figure 14b shows the electromagnetic torque. From simulation it is seen that the electromagnetic torque is equal to the load torque at steady state and the speed settles to reference speed. The air-gap flux is given in Fig. 14c. Figure 14d, e shows the simulation results for stator current and rotor current. It is seen that the speed

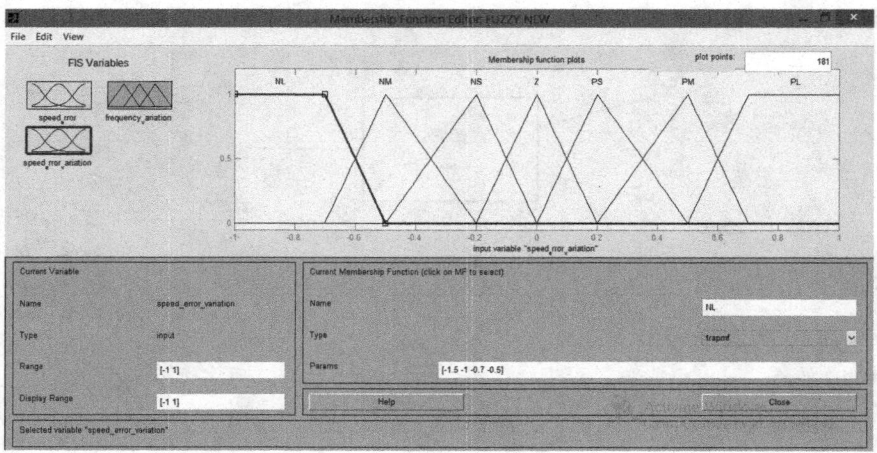

Fig. 11 MF for speed error variation

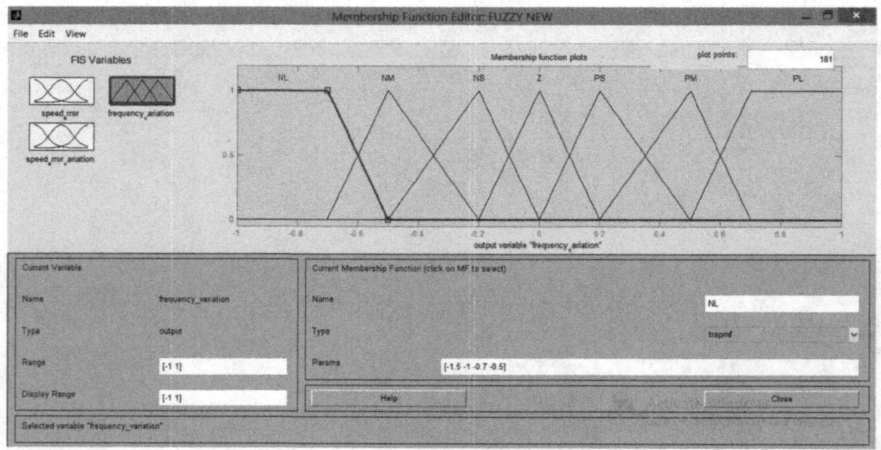

Fig. 12 MF for frequency variation

response is better and superior compared to that with PI controller. Hence it can be inferred that the control of the IM using FLC is superior when compared to traditional PI control. For nonlinear systems FLC offers a better dynamic response compared to a PI controller.

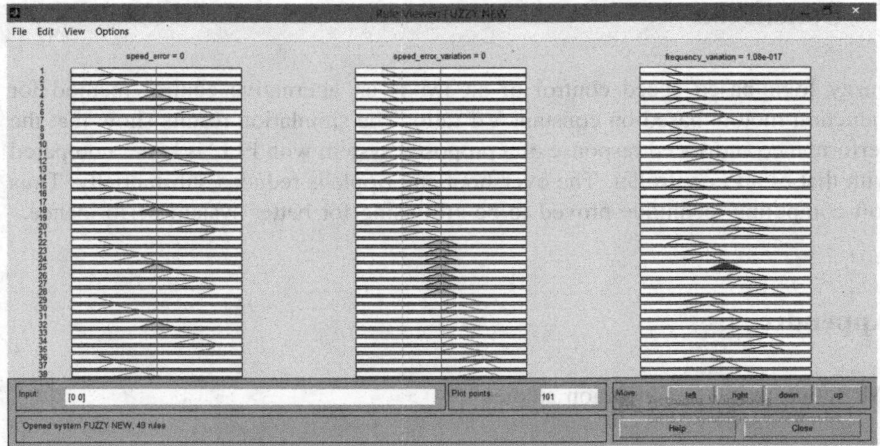

Fig. 13 Rule viewer showing variation in frequency with change in load

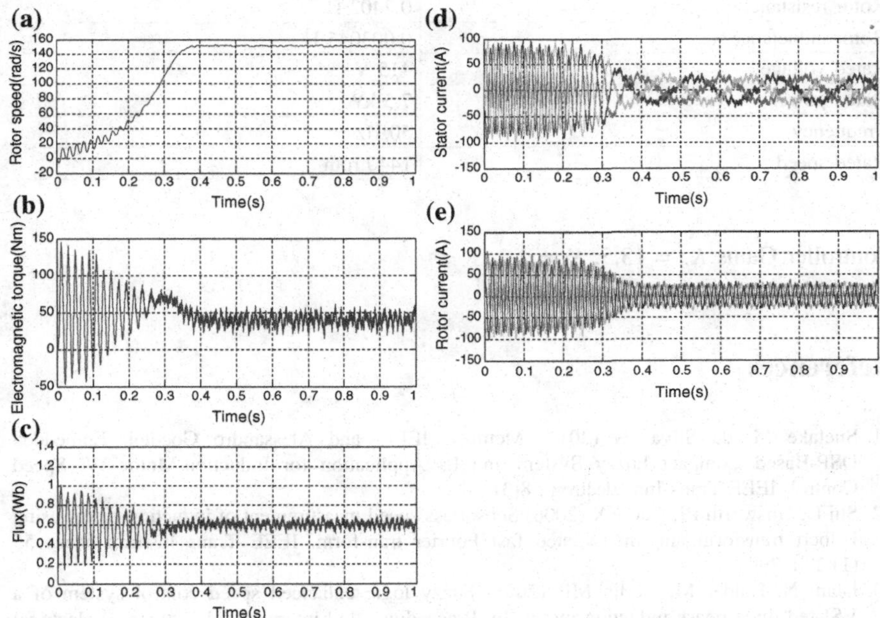

Fig. 14 Results obtained by simulation using FLC; **a** speed, **b** electromagnetic torque, **c** stator flux, **d** stator current, **e** rotor current

7 Conclusion

Fuzzy logic-based speed control of an IM is an alternative control method for induction motors based on constant V/f ratio. The simulation results show that the performance and speed response of a proposed system with FLC is better compared with that of a PI controller. The overshoot and ripple is reduced substantially. Thus soft-computing technique proved to be promising for better system performance.

Appendix

IM Rating used for simulation studies

Number of poles	4
Stator resistance	0.7384 Ω
Stator inductance	0.003045 H
Rotor resistance	0.7402 Ω
Rotor inductance	0.003045 H
Rated voltage	415 V
Rated power	7.5 kW
Frequency	50 Hz
Rated speed	1440 rpm

Controller Gains $K_p = 13.5$, $K_i = 1.1$.

References

1. Suetake M, da Silva IN (2011) Member, IEEE, and Alessandro Goedtel. Embedded DSP-Based Compact Fuzzy System and Its Application for Induction-Motor V/f Speed Control. IEEE Trans Ind Electron 58(3)
2. Shi D, Unsworth PJ, Gao RX (2006) Sensorless speed measurement of induction motor using Hilbert transform and interpolated fast Fourier transform. IEEE Trans Instrum Meas 55 (1):290–299
3. Islam N, Haider M, Uddin MB (2005) Fuzzy logic enhanced speed control system of a VSI-fed three phase induction motor. In: Proceedings 2nd international conference electrical electronics engineering 2005, pp 296–301
4. Aspalli MS, Asha R, Hunagund PV (2012) Three phase induction motor drive using igbts and constant v/fmethod. Int J Adv Res Electric Electron Instrum Eng 1(5)
5. Zeying, Rucheng Z, Long H, Zuo (2010) Closed-loop vector control scheme for induction motor based on DSP. In: IEEE International conference on computational aspects of social networks, pp 667–670

6. Tae-Chon A, Yang-Won K, Hyung-Soo H, Pedricz W (2001) Design of neuro-fuzzy controller on DSP for real-time control of induction motors. In: Proceedings IFSA World Congress 20th NAFIPS International Conference 2001, vol 5, pp 3038–3043
7. Islam N, Haider M, Uddin MB (2005) Fuzzy logic enhanced speed control system of a VSI-fed three phase induction motor. In: Proceedings 2nd international conference electrical electronics engineering, pp 296–301
8. El-Saady G, Sharaf AM, Makky A, Sherbiny MK, Mohamed G (1994) A high performance induction motor drive system using fuzzy logic controller. In: Proceedings 7th mediterranean electrotechnical conference 1994, vol 3, pp 1058–1061
9. Bim E (2001) Fuzzy optimization for rotor constant identification of an indirect FOC induction motor drive. IEEE Trans Ind Electron 48(6):1293–1295
10. Maiti S, Chakraborty C, Hori Y, Ta MC (2008) Model reference adaptive controller-based rotor resistance and speed estimation techniques for vector controlled induction motor drive utilizing reactive power. IEEE Trans Ind Electron 55(2):594–601
11. Krishnan R (2001) Electric motor drives—modeling, analysis, and control. Prentice-Hall, Upper Saddle River
12. Trzynadlowski AM (2001) Control of induction motors. Academic, New York
13. Xiang-Dong S, Kang-Hoon K, Byung-Gyu Y, Matsui M (2009) Fuzzylogic-based V/f control of an induction motor for a DC grid powerleveling system using flywheel energy storage equipment. IEEE Trans Ind Electron 56(8):3161–3168
14. Zidani F, Diallo D, Benbouzid MEH, Saïd RN (2008) A fuzzy-based approach for the diagnosis of fault modes in a voltage-fed PWM inverter inductionmotor drive. IEEE Trans Ind Electron 55(2):586–593
15. El-Saady G, Sharaf AM, Makky A, Sherbiny MK, Mohamed G (1994) A high performance induction motor drive system using fuzzy logic controller. In: Proceedings 7th mediterranean electrotechnical conference 1994, vol 3, pp 1058–1061
16. Ustun SV, Demirtas M (2008) Optimal tuning of PI coefficients by using fuzzy-genetic for V/f controlled induction motor. Expert Syst Appl 34(4):2714–2720

Classification of ECG Signals Using Hybrid Feature Extraction and Classifier with Hybrid ABC-GA Optimization

K. Muthuvel, L. Padma Suresh and T. Jerry Alexander

Abstract In this research, an efficient technique has been developed to classify the five abnormal beat (Afonso et al., IEEE Trans Biomed Eng 46:192–202 (1999) [1]; Kohler et al., IEEE Eng Med Biol Mag 21:42–57 (2002) [2] signals which includes Left Bundle Branch Block beat (LBBB), Right Bundle Branch Block beat (RBBB), Premature Ventricular Contraction (PVC), Atrial Premature Beat (APB), and Nodal (junction) Premature Beat (NPB) along with the normal beat. The proposed technique is composed into three stages, (1) preprocessing (2) Hybrid feature extraction (3) classifier. For efficient feature extraction, hybrid feature extractor is used. Hybrid feature extraction is done in two steps, (i) Morphological-based feature extraction (ii) Haar wavelet-based feature extraction. Once the features are extracted, a Feed Forward Neural Network (FFNN) classifier classifies the beat signal. Artificial Bee Colony (ABC) combined with genetic algorithm has been used for training the neural network. A best crossover rate has been chosen in order to achieve higher accuracy. The proposed technique gives an accuracy of 81 %, sensitivity of 75 %, and specificity of 79 % for the crossover rate of 0.8.

Keywords Hybrid feature extraction · Morphological · Haar wavelet · Neural network · ECG classification · ABC · GA · Crossover

K. Muthuvel · L. Padma Suresh
Noorul Islam University, Kanyakumari, Tamil Nadu, India
e-mail: er.muthuvel@gmail.com

L. Padma Suresh
e-mail: suresh_lps@gmail.com

T. Jerry Alexander (✉)
Department of E&C, Sathyabama University, Kanyakumari, Tamil Nadu, India
e-mail: tjerryalexander@gmail.com

L.P. Suresh and B.K. Panigrahi (eds.), *Proceedings of the International Conference on Soft Computing Systems*, Advances in Intelligent Systems and Computing 397, DOI 10.1007/978-81-322-2671-0_94

1 Introduction

The cardinal function of electrocardiograph is the sound management of electrical activity of the central organ of the blood circulatory system, i.e., the heart. An ECG beat signal contains important information that can aid medical diagnosis, thereby reflecting the cardiac activity of a patient, whether it is normal or a failing heart that has certain pathologies [3]. The ECG beat signal investigation is used to identify many heart ailments such as ischemia, arrhythmias, and myocarditis, or disorder of heart beat or rhythm. It is also used to modify the morphological model and monitoring drug effects or pacemaker action. Nowadays, electrocardiogram (ECG) is one of the most effective diagnostic tools to detect heart diseases. The typical ECG waveform shown in Fig. 1 records the electrical activity of the heart, where each heart beat is displayed as a series of electrical waves characterized by peaks and valleys. For normal case, the frequency range of an ECG signal is of 0.05–100 Hz and dynamic range of 1–10 mV [4]. Each cardiac cycle in an ECG signal consist of the P-QRS-T waves. Most of the clinically useful information in the ECG are found in the intervals and amplitudes as defined by its features [5]. The P wave symbolizes atria depolarization, Q, R, and S waves generally known by the name QRS complex signifies the ventricular depolarization and T wave stands for the repolarisation of the ventricle [6].

The aim of this work is to develop an effective algorithm for solving problems associated with arrhythmia recognition. In order to achieve the following aspects of ECG analysis are discussed.

1. The feature is extracted by two feature extraction process (1) morphological feature extraction (2) wavelet feature extraction-based feature extraction.
2. After feature extraction, the features are grouped by using hybrid classifier. The hybrid classifier uses two optimization techniques Artificial Bee Colony (ABC) and Genetic algorithm along with the neural network for classification. For better accuracy, best crossover rate has been calculated.

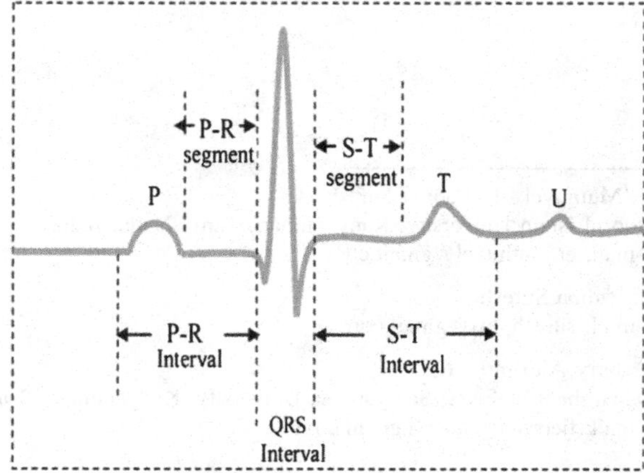

Fig. 1 ECG wave form

2 Feature Extraction

Hybrid feature techniques are used for effective feature extraction which involves (i) morphological feature extraction (ii) wavelet-based feature extraction.

2.1 *Morphological-Based Feature Extraction*

Morphological feature extraction is done in three steps. (i) Find the standard deviation of R-R interval, P-R interval, P-T interval, S-T interval, T-T interval, and Q-T interval. (ii) Second, the maximum values of P, Q, R, S, and T peaks (iii) number of R peaks count.

The processing of the information by the heart is reflected in dynamical changes of electrical activity in time, frequency and space. Mostly, features in time [7] and frequency [8] were extracted and combined with efficient classifiers.

2.2 *Haar Wavelet Feature Extraction*

Haar wavelet transform [9] is capable of detecting and characterizing specific phenomena in time and frequency planes.

In Haar wavelet, the ECG beat signals are disintegrated into coarse approximation and detailed information. For this disintegration, low-pass filter and high-pass filter are used. First, the ECG beat signals are passed to low pass filter. Low-pass filter screens the low frequency beat signals less than the cut off frequency. Second, the ECG beat signals are passed to high-pass filter. High-pass filter screens the high frequency beat signals beyond the cut off frequency. The resultant beat signal from the low-pass filter down sampled by two gives coarse coefficients and the resultant beat signal from the high-pass filter is down sampled to produce detail coefficients, which is shown in Fig. 2.

3 ECG Beat Classification

After extracting the features from ECG, beat signal classifier is used to classify the ECG beat signal. The classifier uses both ABC algorithm and genetic algorithm to train the beat signals in the neural network. Five abnormal beat signals along with the normal beat signals are used to train [10] the Neural Network. The five abnormal beat signals include left bundle branch block beat (LBBB), premature ventricular contraction (PVC), atrial premature beat (APB), right bundle branch

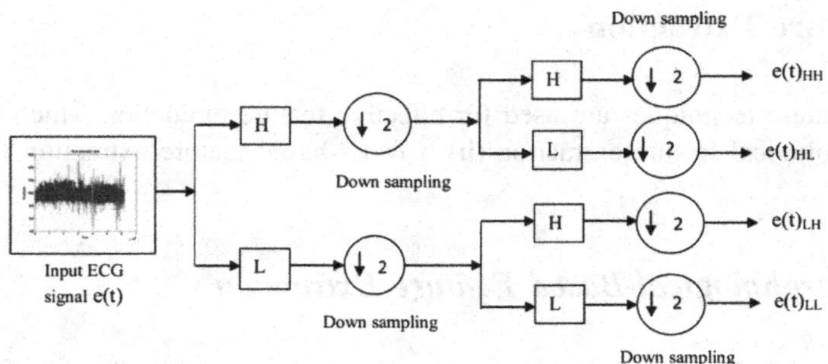

Fig. 2 Architecture of two-dimensional Haar wavelet decomposition

block beat (RBBB), and nodal (junctional) premature beat (NPB). The hybrid classifier involves the following steps.

3.1 Feed Forward Neural Network Layer Generation

In the artificial neural network [11], number of neurons required in the output layer depends on the target solutions in each sequence. Initially, generate an output layer [12, 13] model to optimize the weights.

3.2 Training Phase of FFNN

In this proposed method, five abnormal beat signals are trained along with normal beat signals by using both ABC and genetic algorithm. Most of the neural networks use back propagation algorithm for training. But it consumes more time to find the minimum error. To overcome the above problem, both ABC and genetic algorithm has been used. FFNN is first trained by ABC algorithm. Then use genetic algorithm to train the FFNN by initializing the optimized weight obtained from ABC algorithm. For training the neural network, 24 features extracted from the hybrid feature extraction is given as input layer. After optimizing the weights used in the hidden layer the six beats of ECG beat signals are get grouped in the output layer.

Step 1 Initialization of population
In neural network, data are mainly trained to optimize the weight and to detect the minimum error. To optimize the weights, ABC algorithm initially creates arbitrary population of solution.

Step 2 Fitness evaluation
After generating the initial population, the fitness of the solution is evaluated. The fitness of the solution is determined by calculating the error between the target and the output obtained.

Step 3 Modification of food source by employed bee
After initializing the solution, employed bees are allowed to search the neighboring food source. Employed bee examines the nectar quantity, i.e., fitness of the new food source. Based on the nectar quantity and the visual information employed bee updates the food source.

Step 4 Fitness selection by onlooker bee
After collecting the information of new food source all the employed bees return to the hive and reveal the data to the onlooker bees. The onlooker bees examines all the food source data provided by the employed bees and select a food source based upon the nectar quantity and distance. Then, onlooker bee stores the new food source in its memory and forgets the exiting data. Meanwhile, the employed bees check the selected food source and updates its memory.

Step 5 Generation of new food source by scout bee
After few iteration if there is no change in the food source location, the scout bees are allowed to find the new food source. If the nectar amount of the new food source is high, then the onlooker bee forgets the old one and stores the new food source in the memory. If the nectar quantity is not high, no modification is made. In this proposed work, crossover technique is used to generate random solution.

Step 6 Crossover
This step randomly generates solution based on the optimized weight obtained from ABC algorithm. To generate random solution, pair crossover technique is used in which the crossover rate is multiplied with the length of solution. In this proposed work chromosome length as 10 and best crossover rate is calculated. After generating random solutions the fitness value of the solution are examined and the best solution is given for testing.

4 Result and Discussion

In this section the results obtained from the proposed technique has been discussed.

In this proposed work, MIT-BIH Arrhythmia database have been chosen which contains almost 109,000 beats.

The evaluation of proposed ECG beat classification technique in MIT-BIH Arrhythmia Database are carried out using the following metrics.

Sensitivity: The sensitivity of the hybrid feature extraction and the feature classification is determined by taking the ratio of number of true positives to the sum of true positives and false negatives.

Specificity: The specificity of the hybrid feature extraction and the feature classification can be evaluated by taking the relation of number of true negatives to the combined true negatives and the false positives.

Accuracy: The accuracy of hybrid feature extraction and the feature classification can be calculated by taking the ratio of true values present in the population.

In this section, the experimental results obtained by using hybrid feature extraction and classification for grouping the beats in the ECG beat signals has been reviewed. The Fig. 3 shows the two sample input ECG beat signals. Figure 4 shows the marked P, Q, R, S, T for input beat signal and Fig. 5 shows for the accuracy, sensitivity, specificity of different crossover rates.

To prove the effectiveness of hybrid feature extraction and classification, the proposed technique is compared against individual feature extraction method for different crossover rates. The evaluation graphs for sensitivity, specificity and accuracy are shown in Fig. 5. From the Fig. 5a, it is observed that, when the crossover rate is 0.8 the proposed ECG beat classification technique achieves the overall accuracy value of 81 % which is high compared to that of the morphological feature (Morp+GABC) where the accuracy is 69 % and wavelet-based feature (Wavelet+GABC) gives an accuracy of 79 %. From the Fig. 5b, when the crossover

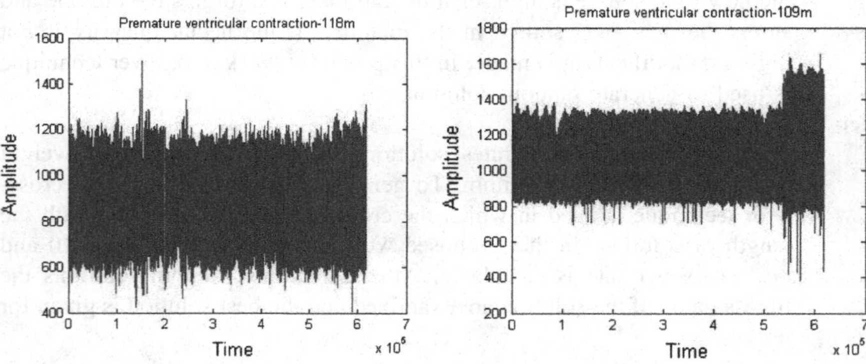

Fig. 3 Input ECG signal

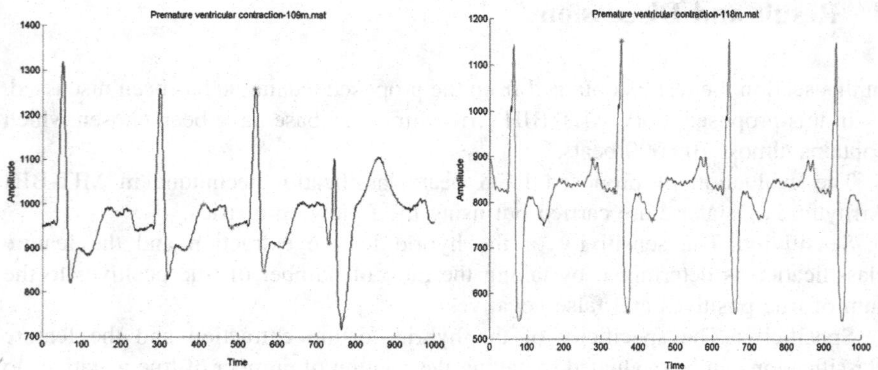

Fig. 4 Marked P, Q, R, S, and T for input beat signal

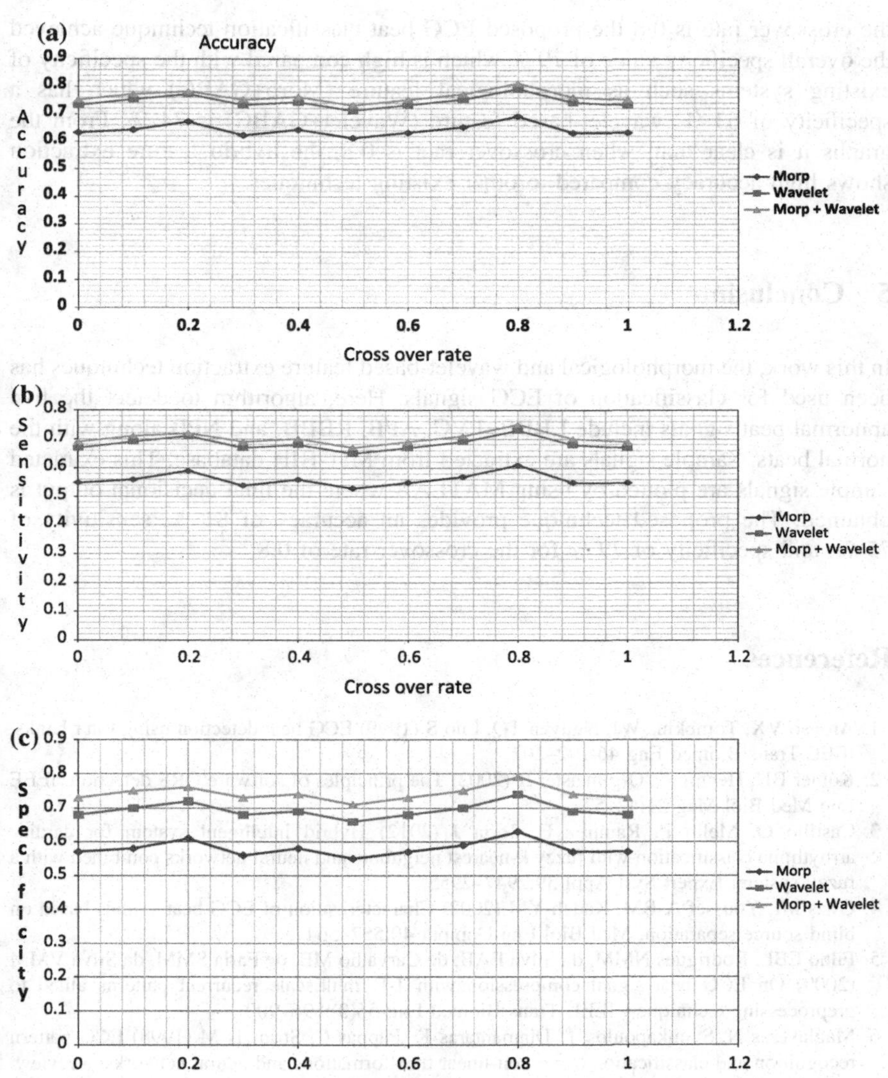

Fig. 5 **a** Accuracy plot for different crossover rate and different feature extraction. **b** Sensitivity plot for different crossover rate and different feature extraction. **c** Specificity plot for different crossover rate and different feature extraction

rate is 0.8, the proposed ECG beat classification technique achieves the overall sensitivity value of 75 % which is high compared with the sensitivity of existing systems such as morphological feature (Morp+GABC) is achieved 61 %, wavelet-based feature (Wavelet+GABC) is achieved 73 %. From the Fig. 5c, when the crossover rate is 0.8 the proposed ECG beat classification technique achieved the overall specificity value of 79 % which is high compared with the specificity of existing systems such as morphological feature (Morp+GABC) which has a specificity of 63 %, wavelet-based feature (Wavelet+GABC) is 74 %. From the graphs it is clear that, when crossover rate is 0.8, the hybrid feature extraction shows high accuracy compared to other existing techniques.

5 Conclusion

In this work, the morphological and wavelet-based feature extraction techniques has been used for classification of ECG signals. Here, algorithm to detect the five abnormal beat signals include LBBB, PVC, APB, RBBB, and NPB along with the normal beats. Sample signals are extracted from MIT-BIH database. This extracted sample signals are plotted by using MATLAB where the final decisional output is obtained. The proposed technique provides an accuracy of 81 %, sensitivity of 75 %, and specificity of 79 % for the crossover rate of 0.8.

References

1. Afonso VX, Tompkins WJ, Nguyen TQ, Luo S (1999) ECG beat detection using filter banks. IEEE Trans Biomed Eng 46:192–202
2. Kohler BU, Hennig C, Orglmeister R (2002) The principles of software QRS detection. IEEE Eng Med Biol Mag 21:42–57
3. Castillo O, Melin P, Ramírez E, Soria J (2012) Hybrid intelligent system for cardiac arrhythmia classification with fuzzy k-nearest neighbors and neural networks combined with a fuzzy system. Expert Syst Appl 39:2947–2955
4. Owis MI, Youssef A-BM, Kadah YM (2002) Characterization of ECG beat signals based on blind source separation. Med Biol Eng Comput 40:557–564
5. Filho EBL, Rodrigues NMM, da Silva EAB, de Carvalho MB, de Faria SMM, da Silva VMM (2009) On ECG beat signal compression with 1-D multiscale recurrent patterns allied to preprocessing techniques. IEEE Trans Biomed Eng 56(3):896–900
6. Maglaveras N, Stamkapoulos T, Diamantaras K, Pappas C, Strintzis M (1998) ECG pattern recognition and classification using non-linear transformations and neural networks: a review. Int J Med Inform 52:191–208
7. Jekova I, Bortolan G, Christov I (2008) Assessment and comparison of different methods for heartbeat classification. Med Eng Phys 248–257
8. Khadra L, Al-Fahoum A, Binajjaj S (2005) A quantitative analysis approach for cardiac arrhythmia classification using higher order spectral techniques. IEEE Trans Biomed Eng
9. Muthuvel K, Padma Suresh L (2014) ECG signal feature extraction and classification using harr wavelet transform and neural network. IEEE, pp 1396–1399
10. Ge D, Srinivasan N, Krishnan SM (2002) Cardiac arrhythmia classification using autoregressive modeling. Biomed Eng OnLine 1(5)

11. Afonso VX, Tompkins WJ, Nguyen TQ, Luo S (1999) ECG beat detection using filter banks. IEEE Trans Biomed Eng 46:192–202
12. Kohler BU, Hennig C, Orglmeister R (2002) The principles of software QRS detection. IEEE Eng Med Biol Mag 21:42–57
13. Muthuvel K, Dr Padma Suresh L (2013) Adaptive neuro-fuzzy inference system for classification of ECG signal. IEEE, pp 1162–1166

Author Index

L.P. Suresh and B.K. Panigrahi (eds.), *Proceedings of the International Conference on Soft Computing Systems*, Advances in Intelligent Systems and Computing 397, DOI 10.1007/978-81-322-2671-0

Printed in the United States
By Bookmasters

Printed in the United States
By Bookmasters